Predictability of Weather and Climate

The topic of predictability of weather and climate has advanced significantly over recent years, both through an increased understanding of the phenomena that affect predictability, and through development of techniques used to forecast state-dependent predictability.

This book brings together some of the world's leading experts on predictability of weather and climate. It addresses predictability from the theoretical to the practical, on timescales from days to decades. Topics such as the predictability of weather phenomena, coupled ocean–atmosphere systems and anthropogenic climate change are among those included. Ensemble systems for forecasting predictability are discussed extensively. Ed Lorenz, father of chaos theory, makes a contribution to theoretical analysis with a previously unpublished paper.

This well-balanced volume will be a valuable resource for many years. High-quality chapter authors and extensive subject coverage will make it appeal to people with an interest in weather and climate forecasting and environmental science, from graduate students to researchers.

TIM PALMER is Head of the Probability Forecasting and Diagnostics Division at the European Centre for Medium-Range Weather Forecasts (ECMWF). He has won awards including the American Meteorological Society Charney Award and the Royal Meteorological Society Buchan Award. He is a fellow of the Royal Society, and is co-chair of the Scientific Steering Group of the World Climate Research Programme's Climate Variability and Predictability (CLIVAR) project. He was a lead author of the Intergovernmental Panel on Climate Change Third Assessment Report, and coordinator of two European Union climate prediction projects: PROVOST and DEMETER.

RENATE HAGEDORN is the education officer for the ECMWF research department. She gained her Ph.D. at the Institute for Marine Sciences in Kiel, Germany, where she developed a coupled atmosphere ocean model for the Baltic Sea catchment area. Upon joining the ECMWF she was part of the DEMETER team. More recently she has been working on improving the ECMWF ensemble prediction systems. In conjunction with Tim Palmer and other colleagues, she was awarded the Norbert Gerbier-Mumm International Award 2006 by the World Meteorological Organization.

Predictability of Weather and Climate

Edited by

Tim Palmer and Renate Hagedorn

European Centre for Medium-Range Weather Forecasts, Reading, UK

CAMBRIDGE
UNIVERSITY PRESS

32 Avenue of the Americas, New York NY 10013-2473, USA

Cambridge University Press is part of the University of Cambridge.

It furthers the University's mission by disseminating knowledge in the pursuit of education, learning and research at the highest international levels of excellence.

www.cambridge.org
Information on this title: www.cambridge.org/9781107414853

First published 2006
Reprinted 2008
First paperback edition 2014

A catalogue record for this publication is available from the British Library

ISBN 978-0-521-84882-4 Hardback
ISBN 978-1-107-41485-3 Paperback

Contents

The colour plates are situated between pages 364 and 365.

Contributors

Myles Allen Department of Atmospheric Physics, University of Oxford, Parks Road, Oxford OX1 3PU, UK

David Anderson ECMWF, Shinfield Park, Reading RG2 9AX, UK

Roberto Buizza ECMWF, Shinfield Park, Reading RG2 9AX, UK

Arun Chakraborty Department of Meteorology, Florida State University, Tallahassee, FL 32306-4520, USA

H.-R. Chang School of Earth and Atmospheric Sciences, Atlanta, GA, USA

Susanna Corti Institute of Atmospheric Sciences and Climate, ISAC-SNR, Bologna, Italy

Francisco J. Doblas-Reyes ECMWF, Shinfield Park, Reading RG2 9AX, UK

Martin Ehrendorfer Inst. für Meteorologie und Geophysik, Universität Innsbruck, Innrain 52, A-6020 Innsbruck, Austria

Brian Farrell Harvard University, Division of Engineering and Applied Sciences, Pierce Hall 107d, Oxford Street Mail Area H0162, Cambridge, MA 02138, USA

David Frame Department of Physics, University of Oxford, UK

R. Grossman Colorado Research Associates, Boulder, CO, USA

Renate Hagedorn ECMWF, Shinfield Park, Reading RG2 9AX, UK

Thomas Hamill Physical Sciences Division, NOAA Earth System Research Laboratory, Boulder, CO, USA

T. Hopson Program in Atmospheric and Oceanic Sciences, University of Colorado, Boulder, CO, USA

Brian Hoskins Department of Meteorology, University of Reading, 2 Early Gate, Whiteknights, Reading RG6 6AH, UK

C. Hoyos School of Earth and Atmospheric Sciences, Atlanta, GA, USA

Brian Hunt Chaos Group, University of Maryland, College Park, MD, USA

Petros Ioannou Section of Astrophysics, Astronomy and Mechanics, Department of Physics, Panepistimiopolis, Zografos 15784, Athens, Greece

Fei-Fei Jin Department of Meteorology, Florida State University, Tallahassee, FL 32306-4520, USA

Eugenia Kalnay Department of Meteorology, University of Maryland, 3431 Computer and Space Sciences Building, College Park, MD 20742-2425, USA

Jamie Kettleborough Space Science and Technology Department, Rutherford Appleton Laboratory, Didcot, Oxon, UK

J. L. Kinter III Center for Ocean-Land-Atmosphere Studies, Calverton, MD, USA

Tiruvalam Krishnamurti Department of Meteorology, Florida State University, Tallahassee, FL 32306-4520, USA

T. S. V. Vijaya Kumar Department of Meteorology, Florida State University, Tallahassee, FL 32306-4520, USA

Fred Kucharski Abdus Salam International Centre for Theoretical Physics, Trieste

François Lalaurette Ecole Nationale de la Météorologie, Av. G. Coriolis, 31057 Toulouse, France

Mojib Latif IfM-Geomar, Düsternbrooker Weg 20, 24105 Kiel, Germany

Edward Lorenz Department of Earth, Atmospheric, and Planetary Sciences, Massachusetts Institute of Technology, Cambridge, MA 02139-4307, USA

Chiara Marsigli ARPA-SIM, Bologna, Italy

Franco Molteni Physics of Weather and Climate Dynamics, The Abdus Salam International Centre for Theoretical Physics, PO Box 586, I-34100 Trieste, Italy

Andrea Montani ARPA-SIM, Bologna, Italy

Ken Mylne Met Office, FitzRoy Road, Exeter EX1 3PB, UK

Fabrizio Nerozzi ARPA-SIM, Bologna, Italy

Edward Ott Chaos Group, University of Maryland, College Park, MD, USA

Tiziana Paccagnella ARPA-SIM, Bologna, Italy

Tim Palmer ECMWF, Shinfield Park, Reading RG2 9AX, UK

Wonsum Park Max-Planck-Institut für Météorologie, Hamburg, Germany

Goðrún Nína Petersen University of Reading, UK

Holger Pohlmann Max-Planck-Institut für Meteorologie, Hamburg, Germany

David Richardson ECMWF, Shinfield Park, Reading RG2 9AX, UK

Jagadish Shukla George Mason University, Institute of Global Environment and Society Inc, 4041 Powder Mill Road, Suite 302 Calverton, MD 20705-3106, USA

Adrian Simmons ECMWF, Shinfield Park, Reading RG2 9AX, UK

Leonard Smith OCIAM Mathematical Institute, 24–29 St Giles', Oxford OXl 3LB, UK

David Stainforth Department of Physics, University of Oxford, UK

Lydia Stefanova Department of Meteorology, Florida State University, Tallahassee, FL 32306-4520, USA

A. Subbiah Asian Disaster Preparedness Centre, Bangkok, Thailand

Istvan Szunyogh University of Maryland, College Park, MD, USA

Olivier Talagrand Laboratoire de Météorologie Dynamique, Paris, France

Alan J. Thorpe Department of Meteorology, University of Reading, 2 Early Gate, Whiteknights, Reading RG6 6AH, UK

Stefano Tibaldi Servizio Meteorologico Regionale, ARPA Emilia-Romagna, Viale Silvani, 640122 Bologna, Italy

Axel Timmermann Department of Oceanography, University of Hawaii at Manoa, 1000 Pope Road, Marine Sciences Building, Honolulu, HI 96822, USA

Zoltan Toth Environmental Modeling Center, NCEP, NWS/NOAA, Washington DC 20233, USA

Gerald van der Grijn ECMWF, Shinfield Park, Reading RG2 9AX, UK

Duane Waliser Jet Propulsion Laboratory, MS 183-505, California Institute of Technology, 4800 Oak Grove Drive, Pasadena, CA 91109, USA

Peter Webster School of Earth and Atmospheric Sciences, Georgia Institute of Technology, Atlanta, GA 30332-0340, USA

Won-Tae Yun Department of Meteorology, Florida State University, Tallahassee, FL 32306-4520, USA

Yuejian Zhu National Centers for Environmental Prediction, Washington DC, USA

Preface

In his biography of the great twentieth-century theoretical physicist Richard Feynman, Gleick (1993) writes: 'He (Feynman) believed in the primacy of doubt, not as a blemish on our ability to know, but as the essence of knowing'. Feynman's philosophy applies as much to weather and climate forecasting as to fundamental physics, as made explicit by Tennekes *et al.* (1987) when they wrote: 'no forecast is complete without a forecast of forecast skill'.

The estimation of uncertainty in weather and climate prediction is encapsulated in the word 'predictability'. If something is said to be predictable, then presumably it can be predicted! However, initial conditions are never perfect and neither are the models used to make these predictions. Hence, the predictability of the forecast is a measure of how these inevitable imperfections leave their imprint on the forecast. By virtue of the non-linearity of the climate, this imprint varies from day to day, just as the weather itself varies; predictability is as much a climatic variable as rainfall, temperature or wind.

Of course, it is one thing to talk about predictability as if it were just another climatic variable; it is another thing to estimate it quantitatively. The predictability of a system is determined by its instabilities and non-linearities, and by the structure of the imperfections. Estimating these instabilities, non-linearities and structures provides a set of tough problems, and real progress requires sophisticated mathematical analysis on both idealised and realistic models.

However, the big world out there demands forecasts of the weather and the climate: is it going to rain tomorrow, will the Arctic ice cap melt by the end of the century? The man in the street wanting to know whether to bring his umbrella to work, or the politician looking for advice on formulating her country's strategy on climate change, cannot wait for the analysis on existence or otherwise of heteroclinic state-space orbits to be finalised! The difference between the real world of prediction, and the more aesthetic world of predictability has been perfectly encapsulated by one of the pioneers of the subject, Kiku Miyakoda, who said: 'Predictability is to prediction as romance is to sex!'. Oscar Wilde, who wrote: 'The very essence of romance is uncertainty!', might well have approved.

However, as we enter the twenty-first century, is this still a fair characterisation? We would argue not! In the last decade, the romantic world of predictability has collided head-on with the practical world of prediction. No longer do operational centres make forecasts without also estimating forecast skill – whether for predictions one hour ahead or one century ahead. This change has come in the last few years through the development of ensemble forecast techniques made practical by mind-boggling developments in high-performance computer technology.

In late 2002, the European Centre for Medium-Range Weather Forecasts (ECMWF) held a week-long seminar on the topic of Predictability of Weather and Climate. A subtheme, borrowing from Kiku Miyakoda's aphorism, was to celebrate the 'reconciliation of romance and sex'! World leaders in the field of predictability of weather and climate gave pedagogical presentations covering the whole range of theoretical and practical aspects on weather and climate timescales, i.e. from a few hours to a century. It was decided, as this was a sufficiently landmark meeting and the presentations sufficiently comprehensive, that it was worth publishing the proceedings for the benefit of the larger scientific community. During 2004 and 2005 authors were asked to expand and update their presentations.

In fact there is one exception to this strategy. One of the greatest pioneers of the subject is Ed Lorenz – his prototypical model of chaos spawned a revolution, not only in meteorology, but in mathematics and physics in general. Ed was unable to come to the 2002 meeting, but a few years earlier had given a presentation at ECMWF on what has become known as the Lorenz-1996 model. This paper is widely cited, but has never been published externally. We decided it would be proper to acknowledge Ed's unrivalled contribution to the field of weather and climate predictability by publishing his 1996 paper in this volume.

Lorenz's contribution is one of the introductory chapters on predictability where both general and specific theoretical/mathematical aspects of predictability theory are discussed. These chapters are followed by contributions on data assimilation methods. The next chapters represent a journey through the predictability of different timescales and different phenomena. The link to real-world applications is made by discussing important developments in operational forecast systems, presenting methods to diagnose and improve forecast systems, and finally giving examples utilising predictability in decision-making processes.

We would like to acknowledge the help of Anabel Bowen, Rob Hine, Els Kooij-Connally, and Matt Lloyd during all stages of the production of this book. Last but not least, we would like to thank ECMWF for initiating and supporting the seminar on which the contributions of this book are based.

References

Gleick, J. (1992) *Genius: The Life and Science of Richard Feynman*. Pantheon.

Tennekes, H., A. P. M. Baede and J. D. Opsteegh (1987) Forecasting forecast skill. In *Proceedings, ECMWF Workshop on Predictability*. ECMWF, Reading, UK.

1

Predictability of weather and climate: from theory to practice

T. N. Palmer

European Centre for Medium-Range
Weather Forecasts, Reading

1.1 Introduction

A revolution in weather and climate forecasting is in progress, made possible by theoretical advances in our understanding of the predictability of weather and climate on the one hand, and by the extraordinary developments in supercomputer technology on the other. Specifically, through ensemble prediction, whose historical development has been documented by Lewis (2005), weather and climate forecasting is set to enter a new era, addressing quantitatively the prediction of weather and climate risk in a range of commercial and humanitarian applications. This chapter gives some background to this revolution, with specific examples drawn from a range of timescales.

1.2 Perspectives on predictability: theoretical and practical

Predictions of weather and climate are necessarily uncertain; our observations of weather and climate are uncertain and incomplete, the models into which we assimilate this data and predict the future are uncertain, and external effects such as volcanoes and anthropogenic greenhouse emissions are also uncertain. Fundamentally, therefore, we should think of weather and climate prediction in terms of equations whose basic prognostic variables are probability densities $\rho(X, t)$, where X denotes

Predictability of Weather and Climate, ed. Tim Palmer and Renate Hagedorn. Published by Cambridge University Press.

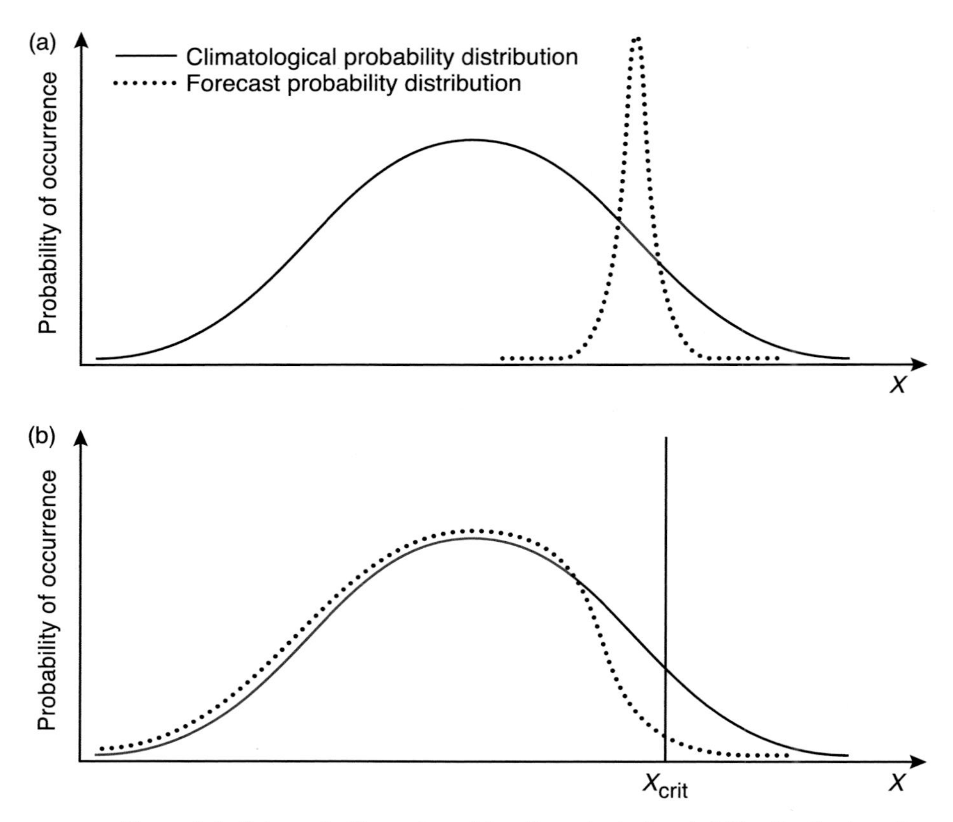

Figure 1.1 Schematic illustration of the climatological probability distribution of some climatic variable X (solid line) and a forecast probability distribution (dotted line) in two different situations. The forecast probability distribution in (a) is obviously predictable. In a theoretical approach to predictability, $\rho(X, t) - \rho_C(X)$ in (b) may not be significantly different from zero overall. However, considered more pragmatically, the forecast probability distribution in (b) can be considered predictable if the prediction that it is unlikely that X will exceed X_{crit} can influence decision-makers.

some climatic variable and t denotes time. In this way, $\rho(X, t)dV$ represents the probability that, at time t, the true value of X lies in some small volume dV of state space. Prognostic equations for ρ, the Liouville and Fokker–Planck equations, are described in Ehrendorfer (this volume). In practice these equations are solved by ensemble techniques, as described in Buizza (this volume).

The question of whether or not X is predictable depends on whether the forecast probability density $\rho(X, t)$ is sufficiently different from some prior estimate $\rho_C(X)$, usually taken as the climatological probability density of X. What do we mean by 'sufficiently different'? One could, for example, apply a statistical significance test to the difference $\rho(X, t) - \rho_C(X)$. On this basis, the hypothetical forecast probability distribution shown as the dotted curve in Figure 1.1(a) is certainly predictable;

but the forecast probability distribution shown in Figure 1.1(b) may well not be predictable.

However, this notion of predictability is a rather idealised one and takes no account of how $\rho(X, t)$ might be used in practice. In a more pragmatic approach to predictability, one would ask whether $\rho(X, t)$ is sufficiently different from $\rho_C(X)$ to influence decision-makers. For example, in Figure 1.1, an aid agency might be interested only in the right-hand tail of the distribution, because disease A only becomes prevalent when $X > X_{\text{crit}}$. On the basis of Figure 1.1(b), the agency may decide to target scarce resources elsewhere in the coming season, since the forecast probability that $X > X_{\text{crit}}$ is rather low.

These two perspectives on the problem of how to define predictability reflect the evolving nature of the study of predictability of weather and climate prediction; from a rather theoretical and idealised pursuit to one which recognises that quantification of predictability is an essential part of operational activities in a wide range of applications. The latter perspective reflects the fact that the full economic value of meteorological predictions will only be realised when quantitatively reliable flow-dependent predictions of weather and climate risk are achievable (Palmer, 2002).

The scientific basis for ensemble prediction is illustrated in Figure 1.2, based on the famous Lorenz (1963) model. Figure 1.2 shows that the evolution of some isopleth of $\rho(X, t)$ depends on starting conditions. This is a consequence of the fact that the underlying equations of motion

$$\dot{X} = F[X] \tag{1.1}$$

are non-linear, so that the Jacobian dF/dX in the linearised equation

$$\frac{d\,\delta X}{dt} = \frac{dF}{dX}\,\delta X \tag{1.2}$$

depends at least linearly on the state X about which Equation (1.1) is linearised. As such, the so-called tangent propagator

$$M(t, t_0) = \exp \int_{t_0}^{t} \frac{dF}{dX}\, dt' \tag{1.3}$$

depends on the non-linear trajectory $X(t)$ about which the linearisation is performed. Hence, the evolved perturbations

$$\delta X(t) = M(t, t_0)\,\delta X(t_0) \tag{1.4}$$

depend not only on $\delta X(t_0)$, but also on the region of phase space through which the underlying non-linear trajectory passes.

It is of interest to note that the formal solution of the Liouville equation, which describes the evolution of $\rho(X, t)$ arising from initial error only (Ehrendorfer, this volume, Eq. (4.49)), can be written using the tangent propagator (for all time in

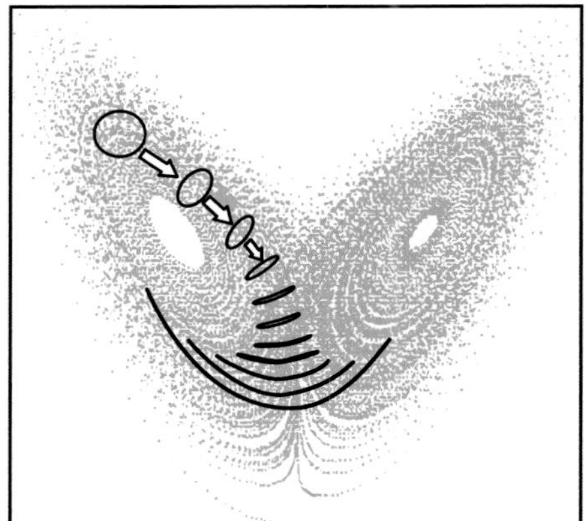

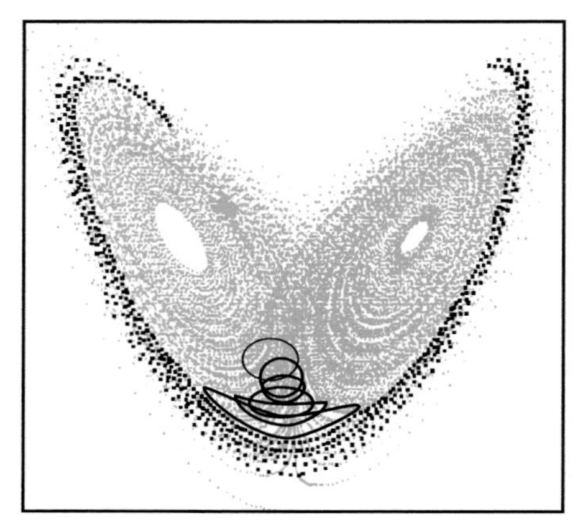

Figure 1.2 Finite time ensembles of the Lorenz (1963) system illustrating the fact that in a non-linear system, the evolution of the forecast probability density $\rho(X, t)$ is dependent on initial state.

the future, not just the time for which the tangent-linear approximation is valid). Specifically

$$\rho(X, t) = \rho(X', t_0)/|\det M(t, t_0)| \tag{1.5}$$

where X' corresponds to the initial state which, under the action of Eq. (1.1), evolves into the state X at time t. Figure 1.2 shows solutions to Eq. (1.5) using an ensemble-based approach.

To illustrate the more practical implications of the fact that $\rho(X, t)$ depends on initial state, I want to reinterpret Figure 1.2 by introducing you to Charlie, a builder by profession, and a golfing colleague of mine! Charlie, like many members of my golf club, takes great pleasure in telling me when (he thinks) the weather forecast has gone wrong. This is mostly done in good humour, but on one particular occasion Charlie was in a black mood. 'I have only four words to say to you,' he announced,

'How do I sue?' I looked puzzled. He continued: 'The forecast was for a night-time minimum temperature of five degrees. I laid three thousand square yards of concrete. There was a frost. It's all ruined. I repeat – how do I sue?'

If only Charlie was conversant with Lorenz (1963) I could have used Figure 1.2 to illustrate how in future he will be able to make much more informed decisions about when, and when not, to lay concrete! Suppose the Lorenz equations represent part of an imaginary world inhabited by builders, builders' customers, weather forecasters and lawyers. In this Lorenz world, the weather forecasters are sued if the forecasts are wrong! The weather in the Lorenz world is determined by the Lorenz (1963) equations where all states on the right-hand lobe of the attractor are 'frosty' states, and all states on the left-hand lobe of the attractor are 'frost-free' states. In this imaginary world, Charlie is planning to lay a large amount of concrete in a couple of days' time. Should he order the ready-mix concrete lorries to the site? He contacts the Lorenzian Meteorological Office for advice. On the basis of the ensemble forecasts in the top left of Figure 1.2 he clearly should not – all members of the ensemble predict frosty weather. On the basis of the ensemble forecasts in the bottom left of Figure 1.2 he also should not – in this case it is almost impossible to predict whether it will be frosty or not. Since the cost of buying and laying concrete is significant, it is not worth going ahead when the risk of frost is so large.

How about the situation shown in the top right of Figure 1.2? If we took the patronising but not uncommon view that Charlie, as a member of the general public, would only be confused by a probability forecast, then we might decide to collapse the ensemble into a consensus (i.e. ensemble-mean) prediction. The ensemble-mean forecast indicates that frost will not occur. Perhaps this is equivalent to the real-world situation that got Charlie so upset. Lorenzian forecasters, however, will be cautious about issuing a deterministic forecast based on the ensemble mean, because, in the Lorenz world, Charlie can sue!

Alternatively, the forecasters could tell Charlie not to lay concrete if there is even the slightest risk of frost. But Charlie will not thank them for that either. He cannot wait forever to lay concrete since he has fixed costs, and if he doesn't complete this job, he may miss out on other jobs. Maybe Charlie will never be able to sue, but neither will he bother obtaining the forecasts from the Lorenzian Meterorological Office.

Suppose Charlie's fixed costs are C, and that he loses L by laying concrete when a frost occurs. Then a logical decision strategy would be to lay concrete when the ensemble-based estimate of the probability of frost is less than C/L. The meteorologists don't know Charlie's C/L, so the best they can do is provide him with the full probability forecast, and allow him to decide whether or not to lay concrete.

Clearly the probability forecast will only be of value to Charlie if he saves money using these ensemble forecasts. This notion of 'potential economic value' (Murphy, 1977; Richardson, this volume) is conceptually quite different from the notion of skill (in the meteorological sense of the word), since value cannot be assessed by analysing meteorological variables alone; value depends also on the user's economic parameters.

The fact that potential economic value does not depend solely on meteorology means that we cannot use meteorological skill scores alone if we want to assess whether one forecast system is more valuable than another (e.g. to Charlie). This is relevant to the question of whether it would be better to utilise computer resources to increase ensemble size or increase model resolution. As discussed in Palmer (2002), the answer to this question depends on C/L. For users with small C/L, more value may accrue from an increase in ensemble size (since decisions depend on whether or not relatively small probability thresholds have been reached), whilst for larger C/L more value may accrue from the better representation of weather provided by a higher-resolution model.

In the Lorenz world, Charlie never sues the forecasters for 'wrong' forecasts. When the forecast is uncertain, the forecasters say so, and with precise and reliable estimates of uncertainty. Charlie makes his decisions based on these forecasts and if he makes the wrong decisions, only he, and lady luck, are to blame!

1.3 Why are forecasts uncertain?

Essentially, there are three reasons why forecasts are uncertain: uncertainty in the observations used to define the initial state, uncertainty in the model used to assimilate the observations and to make the forecasts, and uncertainty in 'external' parameters.

Let's consider the last of these uncertainties first. For example, the aerosol content of the atmosphere can be significantly influenced by volcanic eruptions, which are believed to be unpredictable more than a few days ahead. Also, uncertainty in the change in atmospheric CO_2 over the coming decades depends on which nations sign agreements such as the Kyoto protocol.

In principle, perhaps, 'stochastic volcanoes' could be added to an ensemble prediction system – though this seems a rather fanciful idea. Also, uncertainties in humankind's activities can, perhaps, be modelled by coupling our physical climate model to an econometric model. However, we will not deal further with such uncertainties of the 'third kind' but rather concentrate on the first two.

1.3.1 Initial uncertainty

At ECMWF, for example, the analysed state X_a of the atmosphere is found by minimising the cost function

$$J(X) = \tfrac{1}{2}(X - X_b)^T B^{-1}(X - X_b) + \tfrac{1}{2}(HX - Y)^T O^{-1}(HX - Y) \quad (1.6)$$

where X_b is the background state, B and O are covariance matrices for the probability density functions (pdf) of background error and observation error, respectively, H is

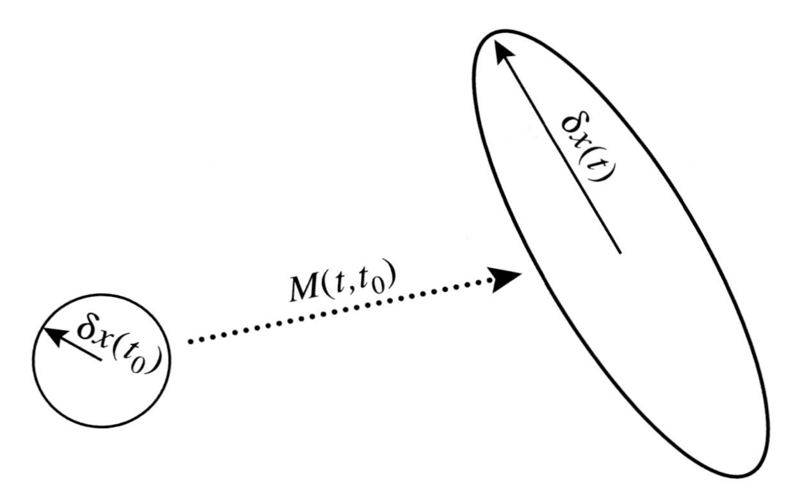

Figure 1.3 Isopleths of probability that the region enclosed by the isopleths contains truth at initial and forecast time. The associated dominant singular vector at initial and final time is also shown.

the so-called observation operator, and Y denotes the vector of available observations (e.g. Courtier *et al.*, 1998). The Hessian

$$\nabla\nabla J = B^{-1} + H^T O^{-1} H \equiv A^{-1} \tag{1.7}$$

of J defines the inverse analysis error covariance matrix.

Figure 1.3 shows, schematically, an isopleth of the analysis error covariance matrix, and its evolution under the action of the tangent propagator M (see Eqs. 1.3 and 1.4). The vector pointing along the major axis at forecast time corresponds to the leading eigenvector of the forecast error covariance matrix. Its pre-image at initial time corresponds to the leading singular vector of M, determined with respect to unit norm in the metric given by A. The singular vectors of M correspond to the eigenvectors of $M^T M$ in the generalised eigenvector equation

$$M^T M\,\delta x(t_0) = -\lambda A^{-1}\delta x(t_0). \tag{1.8}$$

Given pdfs of uncertainty based on Eq. (1.6), we can in principle perform a Monte Carlo sampling of the Hessian-based initial pdf and produce an ensemble forecast system based on this initial sampling.

There are three reasons for not adopting this strategy.

Firstly, there is the so-called 'curse of dimensionality'. The state space of a weather prediction model has about 10^7 dimensions. Many of these dimensions are not dynamically unstable (i.e. are not associated with positive singular values). In this sense, a random sampling of the initial probability density would not be a computationally efficient way of estimating the forecast probability density. This point was made explicitly in Lorenz's analysis of his 28-variable model (Lorenz, 1965):

> If more realistic models . . . also have the property that a few of the eigenvalues of MM^{T} are much larger than the remaining, a study based upon a small ensemble of initial errors should . . . give a reasonable estimate of the growth rate of random error. . . . It would appear, then, that the best use could be made of computational time by choosing only a small number of error fields for superposition upon a particular initial state.

Studies of realistic atmospheric models show that the singular values of the first 20–30 singular vectors are indeed much larger than the remainder (Molteni and Palmer, 1993; Buizza and Palmer, 1995; Reynolds and Palmer, 1998).

The second reason for not adopting a Monte Carlo strategy is that in practice Eq. (1.6) only provides an estimate of part of the actual initial uncertainty; there are other sources of initial uncertainty that are not well quantified – what might be called the 'unknown unknowns'. Consider the basic notion of data assimilation: to assimilate observations that are either made at a point or over a pixel size of kilometres into a model whose smallest resolvable scale is many hundreds of kilometres (bearing in mind the smallest resolvable scale will be many times the model grid). Now sometimes these point or pixel observations may be representative of circulation scales that are well resolved by the model (e.g. if the flow is fairly laminar at the time the observation is made); on other occasions the observations may be more representative of scales which the model cannot resolve (e.g. if the flow is highly turbulent at the time the observation is made, or if the observation is sensitive to small-scale components of the circulation, as would be the case for humidity or precipitation).

In the latter case, the practice of using simple polynomial interpolation in the observation operator H in Eq. (1.6) to take the model variable X to the site of the observation, is likely to be poor. However, this is not an easily quantified uncertainty– since, ultimately, the uncertainty relates to numerical truncation error in the forecast model (see the discussion below). Similarly, consider the problem of quality control. An observation might be rejected as untrustworthy by a quality-control procedure if the observation does not agree with its neighbours and is different from the background (first-guess) field. Alternatively, the observation might be providing the first signs of a small-scale circulation feature, poorly resolved by either the model or the observing network. For these types of reason, a Monte Carlo sampling of a pdf generated by Eq. (1.6) is likely to be an underestimate of the true uncertainty.

The third reason for not adopting a Monte Carlo strategy is not really independent of the first two, but highlights an issue of pragmatic concern. Let us return to Charlie, as discussed above. Charlie is clearly disgruntled by the occasional poor forecast of frost, especially if it costs him money. But just imagine how much more disgruntled he would be, having invested time and money to adapt his decision strategies to a new weather risk service based on the latest, say, Multi-Centre Ensemble Forecast System, if no member of the new ensemble predicts severe weather, and severe

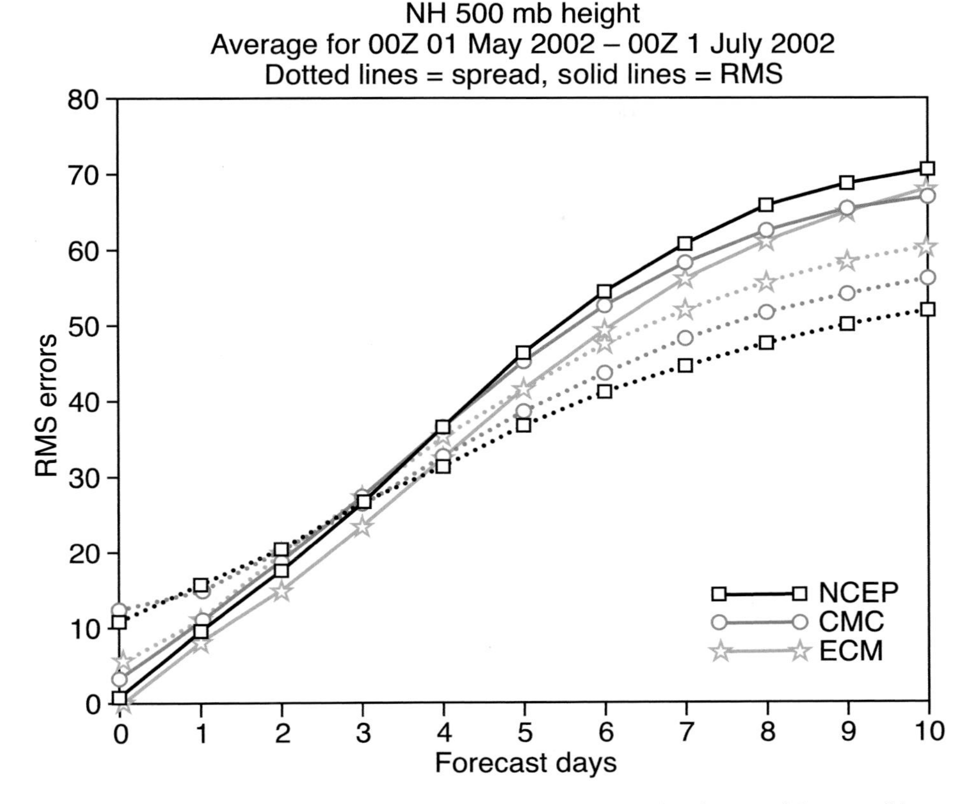

Figure 1.4 May–July 2002 average root-mean-square (rms) error of the ensemble-mean (solid lines) and ensemble standard deviation (dotted lines) of the ECMWF, NCEP and MSC ensemble forecast systems. Values refer to the 500 hPa geopotential height over the Northern Hemisphere latitudinal band 20–80 N. From Buizza *et al.* (2003, 2005).

weather occurs! Just one failure of this sort will compromise the credibility of the new system.

To take this into account, a more conservative approach to sampling initial perturbations is needed, conservative in the sense of tending towards sampling perturbations that are likely to have significant impact on the forecast.

For these three reasons (together with the fact that instabilities in the atmosphere are virtually never of the normal-mode type: Palmer, 1988; Molteni and Palmer, 1993; Farrell and Ioannou, this volume and Ioannou and Farrell, this volume), the initial perturbations of the ECMWF ensemble prediction system are based on the leading singular vectors of M (Buizza, this volume).

The relative performance of the singular-vector perturbations can be judged from Figure 1.4 (Buizza *et al.*, 2003), based on a comparison of ensemble prediction systems at ECMWF (Palmer *et al.*, 1993; Molteni *et al.*, 1996), NCEP (US National

Centers for Environmental Prediction; Toth and Kalnay, 1993) and MSC (Meteorological Service of Canada; Houtekamer *et al.*, 1996); the latter systems based on bred vectors and ensemble data assimilation respectively. The solid lines show the ensemble-mean root-mean-square error of each of the three forecast systems, the dashed lines show the spread of the ensembles about the ensemble mean. At initial time, both NCEP and MSC perturbations are inflated in order that the spread and skill are well calibrated in the medium range. The growth of perturbations in the ECMWF system, by contrast, appears to be more realistic, and overall the system appears better calibrated to the mean error.

1.3.2 Model uncertainty

From the discussion in the last section, part of the reason initial conditions are uncertain is that (e.g. in variational data assimilation) there is no rigorous operational procedure for comparing a model state X with an observation Y. The reason that there is no rigorous procedure is directly related to the fact that the model cannot be guaranteed to resolve well the circulation or weather features that influence the observation. In this respect model error is itself a component of initial error. Of course, model error plays an additional role as one integrates, forward in time, the model equations from the given initial state.

Unfortunately, there is no underlying theory which allows us to estimate the statistical uncertainty in the numerical approximations we make when attempting to integrate the equations of climate on a computer. Moreover, an assessment of uncertainty has not, so far, been a requirement in the development of subgrid parametrisations.

Parametrisation is a procedure to approximate the effects of unresolved processes on the resolved scales. The basis of parametrisation, at least in its conventional form, requires us to imagine that within a grid box there exists an ensemble of incoherent subgrid processes in secular equilibrium with the resolved flow, and whose effect on the resolved flow is given by a deterministic formula representing the mean (or bulk) impact of this ensemble. Hence a parametrisation of convection is based on the notion of the bulk effect of an incoherent ensemble of convective plumes within the grid box, adjusting the resolved scales back towards convective neutrality; a parametrisation of orographic gravity-wave drag is based on the notion of the bulk effect of an incoherent ensemble of breaking orographic gravity waves applying a retarding force to the resolved scale flow.

A schematic representation of parametrisation in a conventional weather or climate prediction model is shown in the top half of Figure 1.5. Within this framework, uncertainties in model formulation can be represented in the following hierarchical form:

- the multimodel ensemble whose elements comprise different weather or climate prediction models;

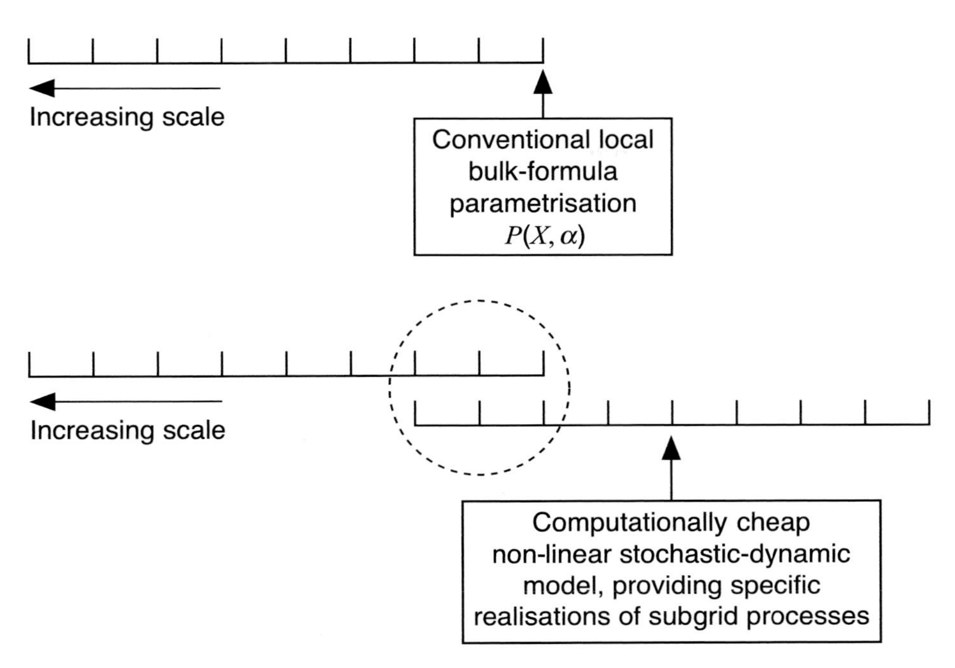

Figure 1.5 (Top) Schematic for conventional weather and climate prediction models (the 'Reynolds/Richardson Paradigm'). (Bottom) Schematic for weather and climate prediction using simplified stochastic-dynamic model representations of unresolved processes.

- the multiparametrisation ensemble whose elements comprise different parametrisation schemes $P(X, \alpha)$ within the same dynamical core;
- the multiparameter ensemble whose elements are all based on the same weather or climate prediction model, but with perturbations to the parameters α of the parametrisation schemes.

The DEMETER system (Palmer *et al.*, 2004; Hagedorn *et al.*, this volume) is an example of the multimodel ensemble; the ensemble prediction system of the Meteorological Service of Canada (Houtekamer *et al.*, 1996) is an example of a multiparametrisation scheme; the Met Office QUMP system (Murphy *et al.*, 2004) and the climateprediction.net ensemble system (Stainforth *et al.*, 2005) are examples of multiparameter ensemble systems.

The hierarchical representation of model error as discussed above should be considered a pragmatic approach to the problem – it certainly should not be considered a complete solution to the problem. The fundamental reason why parametrisations are uncertain is that in reality there is no scale separation between resolved and unresolved motions: according to Nastrom and Gage (1985), the observed spectrum of atmospheric motions shallows from a -3 slope to a $-5/3$ slope as the truncation limit of weather and climate models is approached. That is to say, the spectrum of

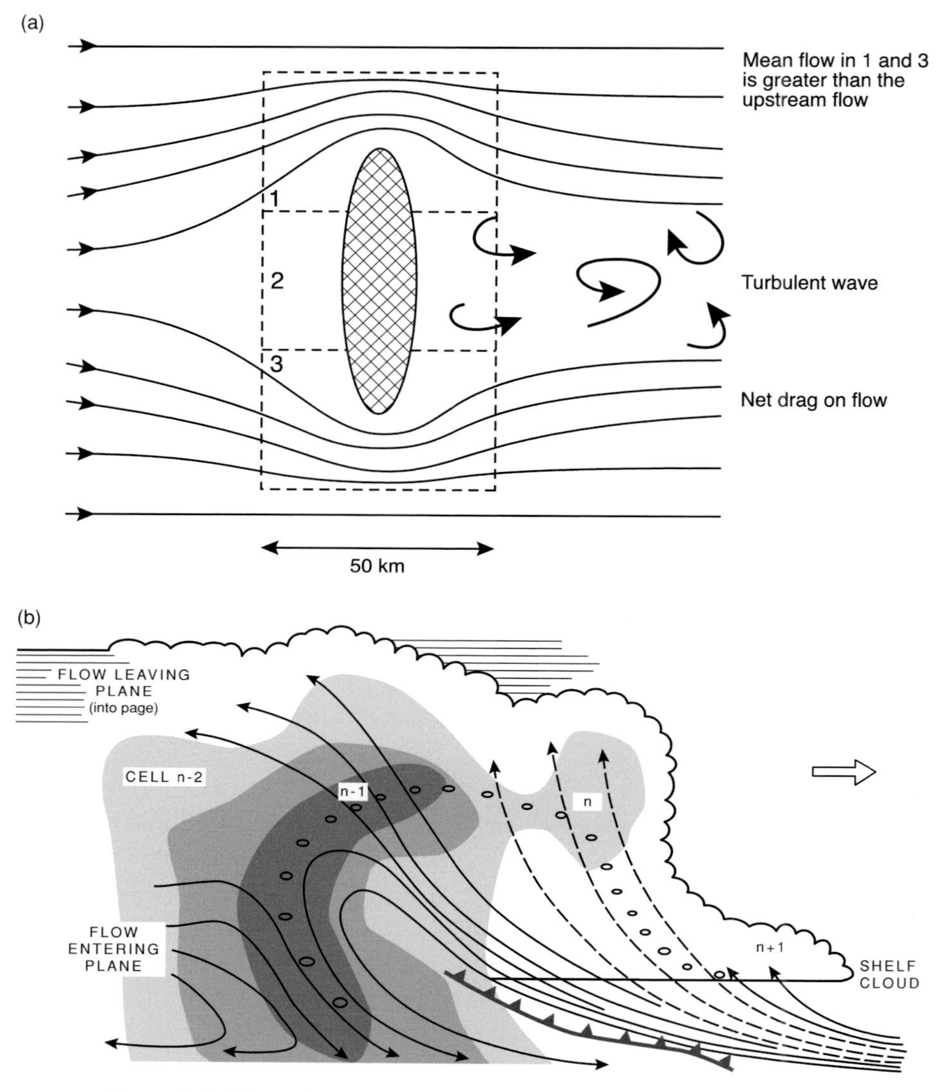

Figure 1.6 Schematic examples of the failure of conventional parametrisation to account for tendencies associated with subgrid processes: (a) when the subgrid topographic forcing is coherent across grid boxes; (b) when the convective motions have mesoscale organisation.

unresolved motions is dominated by a range of near grid-scale motions. Figure 1.6 gives schematic examples of near-grid-scale motions (a) for orographic flow and (b) for organised deep convection. In the case of orography, the flow is forced around the orographic obstacle. In the grid box containing the tip of the orography, a parametrisation will detect unresolved orography and apply a drag force, the very opposite of what may actually be required. In the case of convection, the grid box containing the bulk of the updraught may not be warming (through environmental subsidence); moreover, there is no requirement for the vertical momentum transfer to be upgradient, or for the implied kinetic energy generated by convective available potential energy within a grid box to be dissipated within that grid box.

Hence, in neither case can the impact of the unresolved orographic or convective processes be represented by conventional parametrisations, no matter what formulae are used or what values the underlying parameters take. In other words, part of the uncertainty in the representation of unresolved scales is structural uncertainty rather than parametric uncertainty.

In order to represent such structural uncertainty in ensemble prediction systems, we need to broaden the standard paradigmatic representation of subgrid scales. In considering this generalisation, it can be noted that in the conventional approach to weather and climate modelling, there is in fact a double counting of subgrid processes within an ensemble prediction system. By averaging across an ensemble prediction system with identical resolved-scale flow but different subgrid circulations, the ensemble prediction system effectively provides us with a mean of subgrid processes. But this averaging process has already been done by the parametrisation scheme, which itself is defined to be a mean of putative subgrid processes. Apart from possible inefficiency, what danger is there in such double counting?

The danger is that we miss a key element of the interaction between the resolved flow and the unresolved flow, leading to a component of systematic error in the climate models. Let us represent the grid-box tendency associated with unresolved scales by a probability density function $\rho_m(X)$ where X is some resolved-scale variable. Consider an ensemble prediction system where the grid-box mean variable is equal to X_0 across all members of the ensemble. Suppose now that instead of using the deterministic subgrid parametrisation $P(X_0; \alpha)$ across all ensemble members, we force the ensemble prediction system by randomly sampling $\rho_m(X_0)$. Would the ensemble-mean evolve differently? Yes, because of non-linearity!

It is therefore being suggested here that we change our paradigm of parametrisation as a deterministic bulk formula representing an ensemble of some putative 'soup' of incoherent subgrid processes, to one where each ensemble member is equipped with a possible realisation of a subgrid process. Hence, Figure 1.6(b) illustrates a possible generalisation in which the subgrid scales are represented by a simplified computationally efficient stochastic-dynamic system, potentially coupled to the resolved scales over a range of scales.

There are three reasons why the representation of model uncertainty through stochastic-dynamic parametrisation may be superior to the multimodel and related representations. Firstly, model uncertainty may be more accurately represented, and the corresponding ensembles may be more reliable. Secondly, as discussed above, noise-induced drift may lead to a reduction in model systematic error in a way impossible in a multimodel ensemble. Thirdly, estimates of 'natural climate variability' may be more accurate in a model with stochastic-dynamic representation of unresolved scales. This is important for the problem of detecting anthropogenic climate change.

A simple example of stochastic parametrisation has been discussed in Buizza *et al.* (1999) and Palmer (2001). Let us write, schematically, the equations of motion of our climate or weather prediction model as

$$\dot{X} = F[X] + P + e \qquad (1.9)$$
$$e = \varepsilon P$$

where P denotes the conventional parametrisation term and ε is a non-dimensional stochastic parameter with mean zero. The physical basis for such a multiplicative-noise form of stochastic parametrisation is that stochastic model perturbations are likely to be largest when the parametrisation tendencies themselves are largest, e.g. associated with intense convective activity, when the individual convective cells have some organised mesoscale structure, and therefore where the parametrisation concept breaks down. This notion of multiplicative noise has been validated from a coarse-grained budget analysis in a cloud-resolving model (Shutts and Palmer, 2004, 2006; Palmer *et al.*, 2005). Buizza *et al.* (1999) showed that probabilistic skill scores for the medium-range ensemble prediction systems (EPS) were improved using this stochastic parametrisation scheme.

From a mathematical point of view the addition of stochastic noise in Eq. (1.9) is straightforward. However, adding the noise term e makes a crucial conceptual difference to Eq. (1.9). Specifically, without stochastic noise, the subgrid parametrisation represents an averaged tendency associated with a supposed ensemble of subgrid processes occurring inside the grid box. With stochastic noise, the subgrid parametrisation represents a possible realisation of the subgrid tendency.

We can go further than this and ask whether some dynamical meteorology could be built into this stochastic realisation of the subgrid world. A possible stochastic-dynamic model could be associated with the cellular automaton, a computationally simple non-linear dynamical system introduced by the mathematician John von Neumann (Wolfram, 2002). Figure 1.7 (from Palmer, 2001) is a snapshot from a cellular automaton model where cells are either convectively active ('on') or convectively inhibited ('off'). The probability of a cell being 'on' is dependent on the convective available potential energy in the grid box and on the number of adjacent 'on' cells. Agglomerations of 'on' cells have the potential to feed vorticity from the parametrisation to the resolved scales. 'On' cells can be made to advect with the grid-mean wind. In a further development of this scheme (J. Berner, personal communication) a two-level multiscale cellular automaton has been developed. The smallest level represents individual convective plumes, whilst the intermediate level represents convectively coupled wave motions which can force the Madden–Julian Oscillation. It is planned to try to include characteristics of the dispersion equation of convectively coupled Kelvin waves in the intermediate cellular automaton.

Independently, cellular automata based on the Ising model have been developed as a stochastic-dynamic parametrisation of deep convection (Khouider *et al.*, 2003). Recently Shutts (2005) has built a hybrid stochastic-dynamic parametrisation scheme which combines the cellular automaton with the notion of stochastic

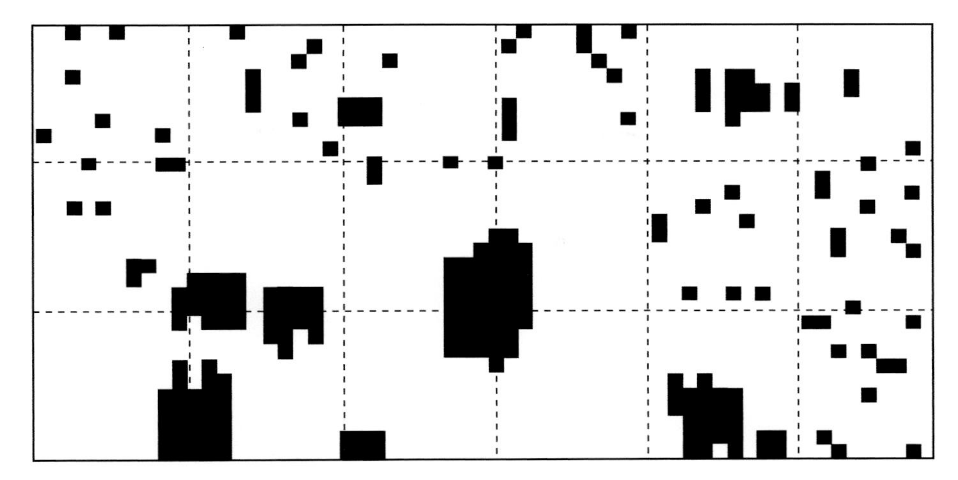

Figure 1.7 A snapshot in time from a cellular automaton model for convection. Black squares correspond to convectively active sites, white squares to convectively inhibited sites. From Palmer (2001).

backscatter (e.g. Leith, 1990; Frederiksen and Davies, 1997). Specifically, estimates of dissipation are made based on four components of the ECMWF model: convection, orographic gravity-wave drag, numerical diffusion, and the implicit dissipation associated with the semi-lagrangian scheme. The basic assumption is that a fraction (e.g. 10%) of the implied energy dissipation is actually fed back on to the model grid, through the cellular automaton.

Preliminary results with this scheme are very promising: there is quantifiable reduction in midlatitude systematic error which can be interpreted in terms of an increased frequency of occurrence of basic circulation regimes (Molteni *et al.*, this volume; Jung *et al.*, 2005) which are underestimated in the version of the model with conventional parametrisation. In this way, one can say that the stochastic parametrisation has reduced systematic error through a non-linear noise-induced drift effect.

1.4 Ensemble forecasts: some examples

In this section, some examples of ensemble forecasts from different forecast timescales will be shown, illustrating some of the ideas discussed above. Logically, one should perhaps order these examples by timescale, e.g. starting with the short range, finishing with centennial climate change. However, in this section the examples will be presented more or less in the historical order in which they were developed.

1.4.1 Extended range

Much of the early work on ensemble forecasting with comprehensive weather prediction models arose in trying to develop dynamical (as opposed to statistical-empirical)

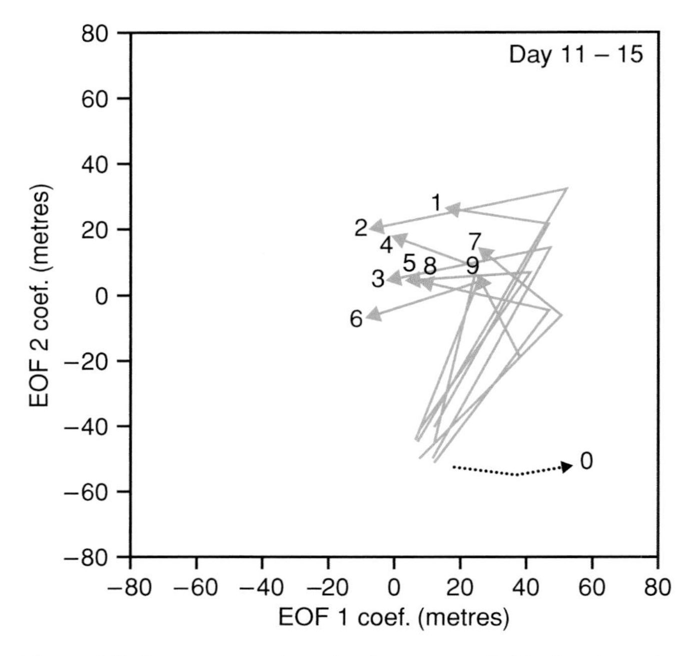

Figure 1.8 State-space trajectories from an unreliable time-lagged ensemble (solid), and verifying analysis (dashed). See Palmer *et al.* (1990).

techniques for monthly forecasting. One of the basic motivations for such work was that whilst the monthly timescale was clearly beyond the mean limit [P_W] of predictability of daily weather, as shown by Miyakoda *et al.* (1986), from time to time the actual predictability of the atmospheric circulation would far exceed [P_W]. Ensemble forecasts were seen as a necessary means of determining ahead of time whether such enhanced predictability existed. The first ever operational probabilistic ensemble forecast was made for the 30-day timescale (Murphy and Palmer, 1986).

In these early days, methodologies to produce initial ensemble perturbations were rather simple, e.g. based on adding random noise to the initial conditions, or using the time-lagged technique, where ensemble members were taken from consecutive analyses (Hoffman and Kalnay, 1983).

Unfortunately, early results did not always live up to hopes. Figure 1.8 shows one of the pitfalls of ensemble forecasting. The figure shows the evolution in phase space (spanned by leading empirical orthogonal functions of the forecast ensemble) of an ensemble of five-day mean forecasts made using the time-lagged technique, based on an early version of the ECMWF model (Palmer *et al.*, 1990). Unfortunately the evolution of the verifying analysis is in a class of its own! A probability forecast based on this ensemble would clearly be unreliable. Charlie would not be impressed!

During the 1990s, work on 30-day forecasting went into a period of decline. However, in 2004, an operational 30-day forecast system was finally implemented at ECMWF, using singular-vector ensemble perturbations in a coupled ocean–atmosphere model (Vitart, 2004). Unlike the early example shown above, probability forecasts have been shown to be reliable in the extended range. This improvement results from developments in data assimilation, in deterministic forecasting, and in medium-range ensemble forecasting as discussed in this chapter and in Buizza (this volume).

1.4.2 Medium range

A general assessment of predictability in the medium range, based on the ECMWF system, is given in Simmons (this volume). Even though early results on 30-day forecasting were disappointing, it was nevertheless clear that the idea of using ensemble forecasts to determine periods where the atmospheric circulation was either especially predictable, or especially unpredictable, was also relevant to the medium range. Based on the experience outlined above, there was clearly a need to ensure that the resulting ensembles were not underdispersive. The initial work in this area was done at ECMWF and NCEP, using different methods for obtaining initial perturbations (see section above). The ECMWF ensemble prediction system comprises 51 forecasts using both singular vector initial perturbations and stochastic physics (more details are given in Buizza, this volume).

In late December 1999, two intense storms, subsequently named Lothar and Martin, ran across continental Europe leaving behind a trail of destruction and misery, with over 100 fatalities, over 400 million trees blown down, over 3 million homes without electricity and water. Figure 1.9 shows the ensemble 'stamp maps' (based on a TL255 version of the ECMWF model) for Lothar, at initialisation time on 24 December and for forecast time 6 UTC on 26 December. This storm was exceptionally unpredictable, and even at 42 hours lead time there is considerable spread in the ensemble. The best-guidance deterministic forecast only predicts a weak trough in surface pressure. A number of members of the ensemble support this forecast; however, a minority of ensemble members also show an intense vortex over France. In this sense, the ensemble was able to predict the risk of a severe event, even though it was impossible to give a precise deterministic forecast. More recent deterministic reforecasts with a T799 version of the ECMWF model also fail to predict this storm (Martin Miller, personal communication) – this is clearly a case which demonstrates the value of ensemble forecasts even at intermediate resolution (Palmer, 2002).

1.4.3 Seasonal and decadal prediction

The scientific basis for seasonal prediction lies in the interaction of the atmosphere with slowly varying components of the climate system: notably the oceans and land surface (Timmermann, this volume; Shukla and Kinter, this volume). Early work showed firstly that El Niño events are predictable seasons ahead of time using intermediate-complexity coupled ocean–atmosphere models of the tropical Pacific

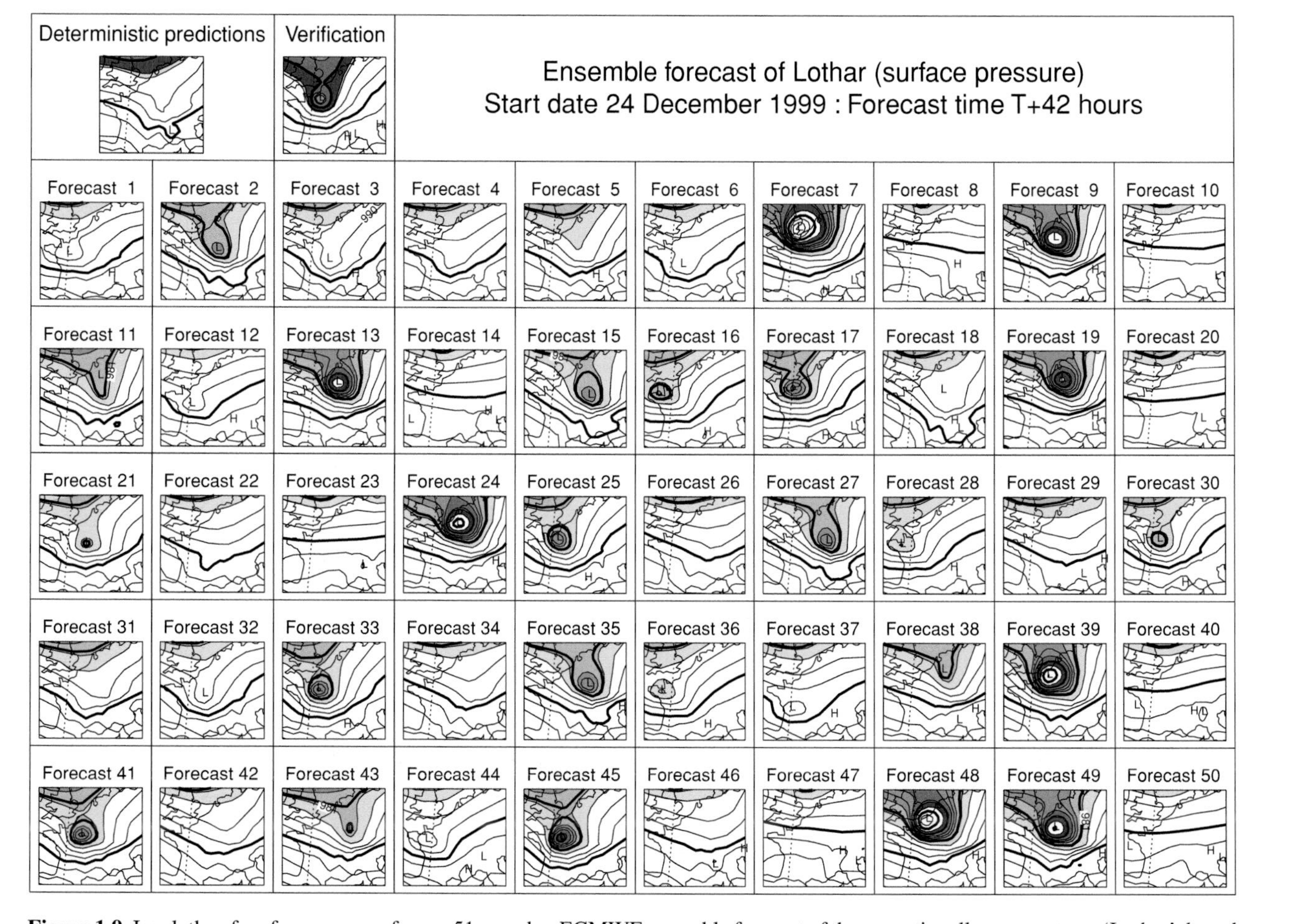

Figure 1.9 Isopleths of surface pressure from a 51-member ECMWF ensemble forecast of the exceptionally severe storm 'Lothar', based on initial conditions 42 hours before the storm crossed northern France in December 1999. The top left shows the forecast made from the best estimate of these initial conditions. This did not indicate the presence of a severe storm. The 50 forecasts which comprise the ensemble indicated exceptional unpredictability and a significant risk of an intense vortex.

(Zebiak and Cane, 1987), and secondly that sea surface temperature (SST) anomalies in the tropical Pacific Ocean have a global impact on atmospheric circulation (e.g. Shukla and Wallace, 1983). Putting these factors together has led to global seasonal prediction systems based on comprehensive global coupled ocean–atmosphere models (Stockdale *et al.*, 1998). Inevitably such predictions have been based on ensemble forecast techniques, where initial perturbations represent uncertainties in both atmosphere and ocean analyses.

In addition to initial uncertainty, representing forecast model uncertainty is a key element in reliably predicting climate risk on seasonal and longer timescales. The ability of multimodel ensembles to produce more reliable forecasts of seasonal climate risk over single-model ensembles was first studied by the PROVOST (Prediction of Climate Variations on Seasonal to Interannual Timescales) project funded by the European Union IVth Framework Environment Programme, and a similar 'sister' project DSP (Dynamical Seasonal Prediction) undertaken in the United States (Palmer and Shukla, 2000).

As part of the PROVOST project, three different atmospheric general circulation models (including one model at two different resolutions) were integrated over four-month timescales with prescribed observed SSTs. Each model was run in ensemble mode, based on nine different initial conditions from each start date; results were stored in a common archive. One of the key results from PROVOST and DSP was that, despite identical SSTs, ensembles showed considerable model-to-model variability in estimates both of the SST-forced seasonal-mean signal, and the seasonal-mean 'noise' generated by internal dynamics (Straus and Shukla, 2000). Consistent with this, probability scores based on the full multimodel ensemble scored better overall than any of the individual model ensembles (e.g. Doblas-Reyes *et al.*, 2000; Palmer *et al.*, 2000).

Based on such results, the DEMETER project (Development of a European Multimodel Ensemble System for Seasonal to Interannual Prediction; Palmer *et al.*, 2004, Hagedorn *et al.*, this volume) was conceived, and successfully funded under the European Union Vth Framework Environment Programme. The principal aim of DEMETER was to advance the concept of multimodel ensemble prediction by installing a number of state-of-the-art global coupled ocean–atmosphere models on a single supercomputer, and to produce a series of six-month ensemble reforecasts with common archiving and common diagnostic software.

Figure 1.10 shows an example of results from DEMETER. Forecasts of El Niño are seen to be more reliable in the multimodel ensemble than in the ECMWF single-model ensemble; more specifically, the observed SSTs do not always lie in the range of the ECMWF-model ensembles, but do lie in the range of the DEMETER multimodel ensembles. Other results supporting the notion that multimodel ensembles are more reliable than single-model ensembles are given in Hagedorn *et al.* (this volume). However, it is not necessarily the case that multimodel ensembles are reliable for all variables. As discussed in Palmer *et al.* (2005), seasonal forecasts of

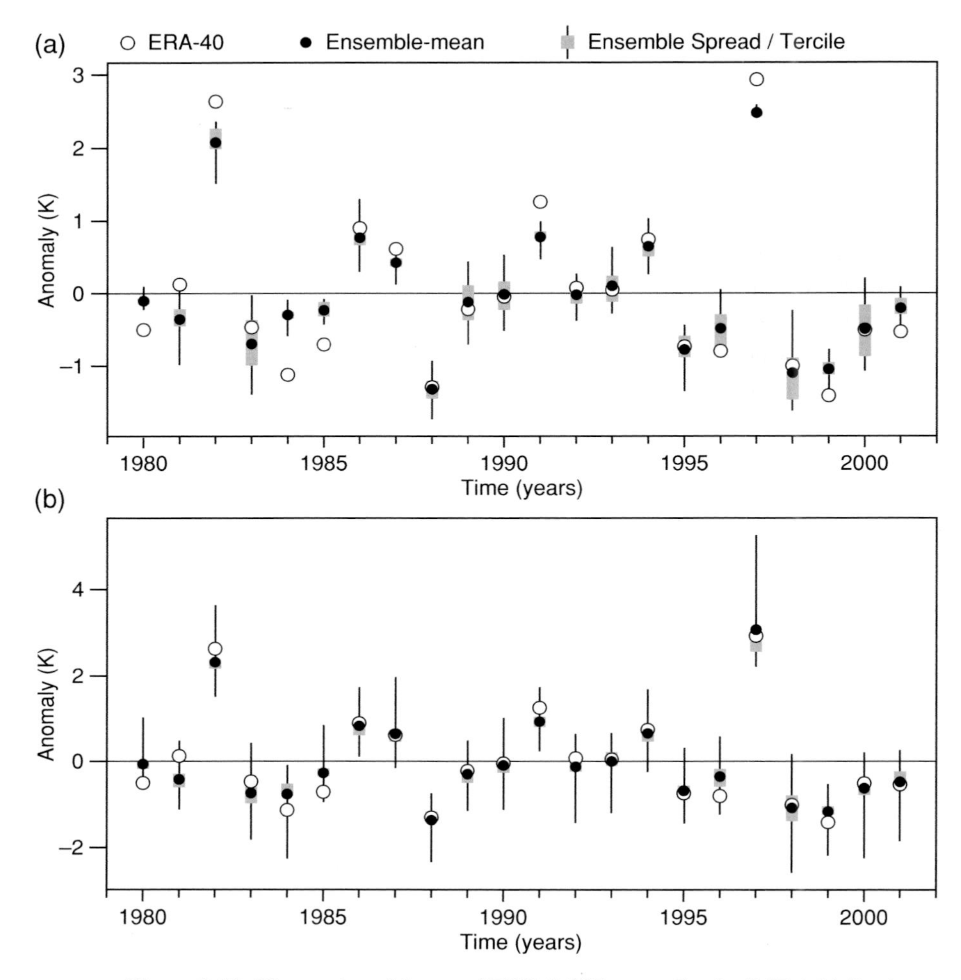

Figure 1.10 Time series of forecast NINO-3 SST anomalies for DJF, initialised 1 November, based on (a) ECMWF ensemble, (b) DEMETER multimodel ensemble. Bars and whiskers show terciles, the ensemble-mean values are shown as solid circles, and the actual SST anomalies are shown as open circles.

upper-tercile precipitation over Europe are neither reliable in single-model nor DEMETER multimodel ensemble systems. Consistent with the discussion above, this latter result suggests that multimodel systems may not represent model uncertainty completely. In the European Union FP6 project ENSEMBLES, it is proposed to compare the reliability of seasonal ensemble forecasts, made using the multimodel technique, with stochastic parametrisation.

At the beginning of this chapter, the existence or otherwise of predictability was discussed from the perspective of decision-making: are the forecast probability densities sufficiently different from climatological densities to influence decision-making? In DEMETER quantitative crop and malaria prediction models were linked to individual ensemble members; based on this, the probability of crop failure or malaria epidemic could be estimated. See Hagedorn *et al.* (this volume) for details.

Many observational and modelling studies document pronounced decadal and multidecadal variability in the Atlantic, Pacific and Southern Oceans. For example, decadal variability in Atlantic sea surface temperatures is in part associated with fluctuations in the thermohaline circulation (e.g. Broecker, 1995); decadal variability in the Pacific is associated in part with fluctuations in the Pacific Decadal Oscillation (e.g. Barnett *et al.*, 1999). Variations in the Atlantic thermohaline circulation appear to be predictable one or two decades ahead, as shown by a number of perfect model predictability studies, e.g. Griffies and Bryan (1997), Latif *et al.* (this volume). The SST anomalies (both tropical and extratropical) associated with decadal variations in the thermohaline circulation appear to impact the North Atlantic Oscillation in the extratropics. A weakening of the thermohaline circulation is likely to lead to a cooling of northern European surface temperatures. There is also evidence (Landerer *et al.*, 2006) that decadal variability in the thermohaline circulation could lead to significant decadal sea level fluctuations in Europe.

Variations in the thermohaline circulation may also be associated with climate fluctuations in the tropics. For example, there is some evidence that long-lasting drought over the African Sahel is associated with decadal-timescale variability in the so-called sea surface temperature dipole in the tropical Atlantic (Folland *et al.*, 1986).

In order for decadal prediction to evolve into a possible operational activity, suitable observations from which the thermohaline circulation can be initialised must exist. Programmes such as ARGO (Wilson, 2000) may help provide such observations. These need to be properly assimilated in the ocean component of a coupled forecast model. It is clearly essential that the model itself has a realistic representation of the thermohaline circulation.

On the timescale of a decade, anticipated changes in greenhouse gas concentrations will also influence the predictions of future climate (Smith *et al.*, 2006). In this sense, decadal timescale prediction combines the pure initial value problem with the forced climate problem, discussed in the next subsection.

1.4.4 Climate change

Climate change has been described by the UK Government's Chief Scientific Advisor as one of the most serious threats facing humanity – more serious even than the terrorist threat. Nevertheless, there is uncertainty in the magnitude of climate change; this uncertainty can be quantified using ensemble techniques (Allen *et al.*, this volume). For example, Palmer and Räisänen (2002) used the multimodel ensemble technique to assess the impact of increasing levels of CO_2 on the changing risk of extreme seasonal rainfall over Europe in winter (Figure 1.11; colour plate), and also for the Asian summer monsoon, based on the CMIP multimodel ensemble

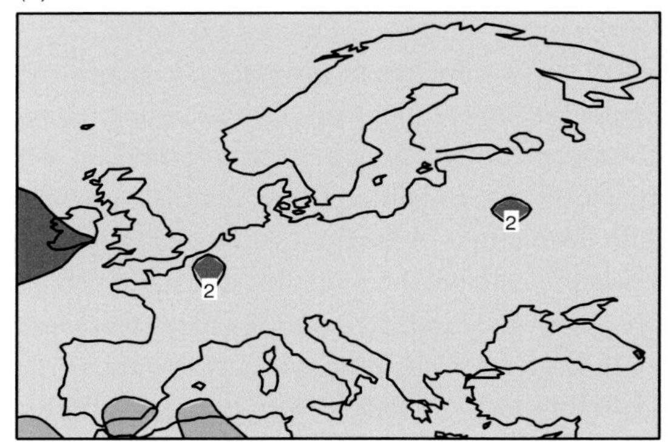

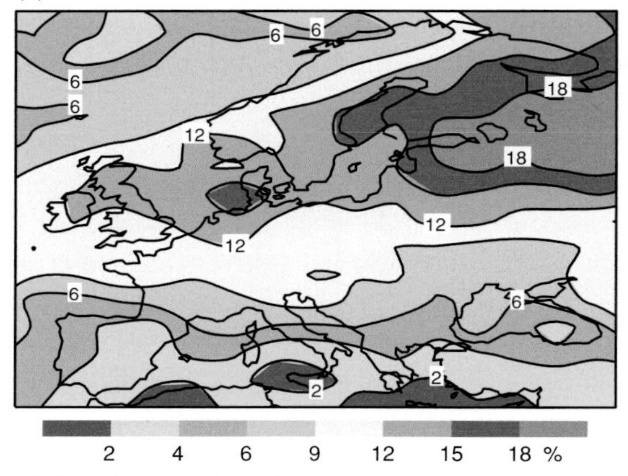

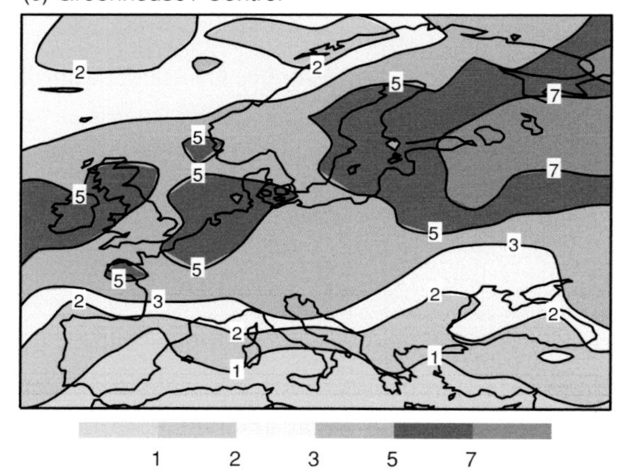

Figure 1.11 (See also colour plate section.) The changing probability of extreme seasonal precipitation for Europe in boreal winter. (a) The probability (in %) of a 'very wet' winter defined from the control CMIP2 multimodel ensemble with twentieth-century levels of CO_2 and based on the event E: total boreal winter precipitation greater than the mean plus two standard deviations. (b) The probability of E but using data from the CMIP multimodel ensemble with transient increase in CO_2 and calculated around the time of CO_2 doubling (years 61–80 from present). (c) The ratio of values in (b) to those in (a), giving the change in the risk of a 'very wet' winter arising from human impact on climate. From Palmer and Räisänen (2002).

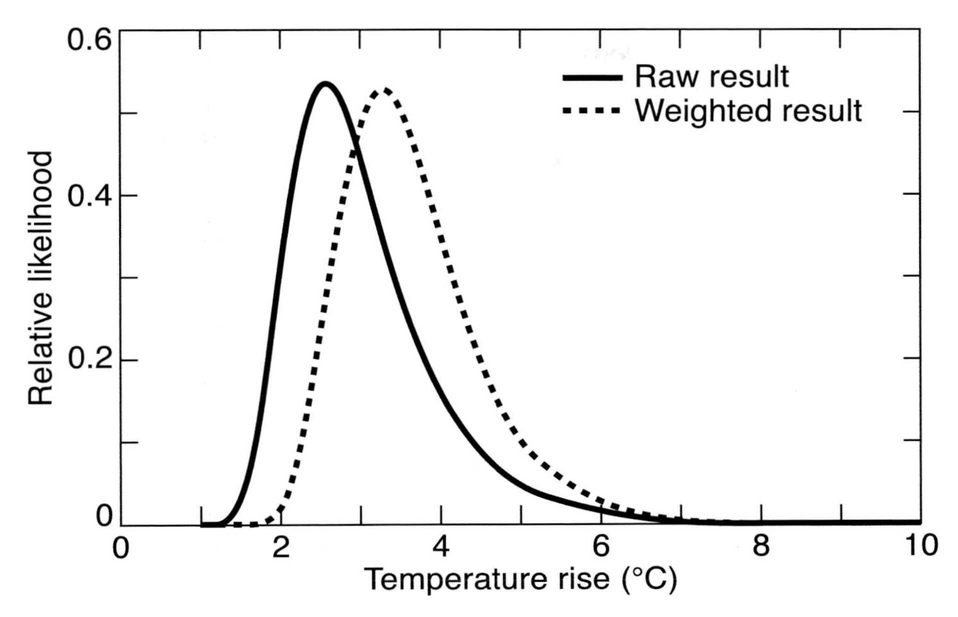

Figure 1.12 Probability distributions of global average annual warming associated with a 53-member ensemble for a doubling of carbon dioxide concentration. Ensemble members differ by values of key parameters in the bulk formulae used to represent unresolved processes, in a version of the Hadley Centre climate model. Solid curve: based on 'raw model output'. Dashed curve: the probability distribution weighted according to the ability of different model versions to simulate observed present day climate. From Murphy *et al.* (2004).

(Meehl *et al.*, 2000). More recently Murphy *et al.* (2004) and Stainforth *et al.* (2005) have quantified uncertainty in climate sensitivity (the global warming associated with a doubling of CO_2) based on multiparameter ensembles (see also Allen *et al.*, this volume).

Figure 1.12 shows a probability distribution of climate sensitivity based on the multiparameter ensemble of Murphy *et al.* (2004). The solid line shows the raw output; the dashed line shows results when the individual ensemble members are weighted according to the fit of control integrations to observations. It is interesting to note that this fit to data does not change the range of uncertainty – rather the forecast probability distribution is shifted to larger values of climate sensitivity.

As yet, probability distributions in global warming have not been estimated using the stochastic physics approach. It will be interesting to see how estimates of uncertainty in climate sensitivity are influenced by the methodology used to represent model uncertainty.

How can the uncertainty in global warming be reduced? It can be noted that much of the ensemble spread in Figure 1.12 is associated with uncertainty in parameters from parametrisations of clouds and boundary layer processes. These are fast-timescale processes. Hence it may be possible to reduce the spread in Figure 1.12 by assessing how well the contributing models perform as short-range weather prediction models. More specifically, the type of budget residual technique pioneered by Klinker and Sardeshmukh (1992) could be applied to the multiparameter ensemble. On this basis, models with large residuals, obtained by integrating over just a few time steps but from many different initial conditions, could be rejected from the ensemble (Rodwell and Palmer, 2006).

Ultimately, the uncertainties in climate prediction arise because we are not solving the full partial differential equations of climate – for example, the important cloud processes mentioned above are parametrised, and, as noted, parametrisation theory is rarely a justifiable procedure. On this basis, further reduction of the uncertainty in global warming may require significantly larger computers so that at least major convective cloud systems can be resolved. Of course, this will not eliminate uncertainty, as cloud microphysics will still have to be parametrised.

In truth, reducing uncertainty in forecasts of climate change will require a combination of significantly greater computer resources and the use of sophisticated validation techniques as used in numerical weather prediction studies.

1.4.5 Short-range forecasting

For many years, it was generally assumed that while ensemble techniques may well be important for medium and longer range predictions, the short-range weather prediction problem, up to day 2, let's say, should essentially be considered deterministic. Such a view is no longer held today – predictions of flash floods and other types of mesoscale variability are not likely to be strongly predictable on timescales of a day. To quantify uncertainty in such forecasts, ensemble prediction systems based on multiple integrations of fine-scale limited area models are now being actively developed around the world. To create such ensembles, ensemble boundary conditions are taken from a global ensemble prediction system, and these are combined with initial perturbations within the limited-area model domain. Insofar as some of the principal forecast variables are related to processes close to the model resolution, a representation of model uncertainty is also necessary.

An example of an ensemble of limited-area model integrations is shown in Figure 1.13, based on the COSMO-LEPS system (Tibaldi *et al.*, this volume; Waliser *et al.*, 2006). The boundary conditions for this limited-area model ensemble have been taken from ECMWF ensemble integrations (in this case using moist singular vectors; M. Leutbecher, personal communication). The example shown here is for the storm Lothar (see Figure 1.9). It can be seen that the limited-area model ensemble is predicting a significant risk of damaging wind gusts – in a situation where the deterministic forecast from the most likely initial state had no warning of severe gusts at all.

Development of short-range ensemble prediction systems using limited-area models is now a significant growth area.

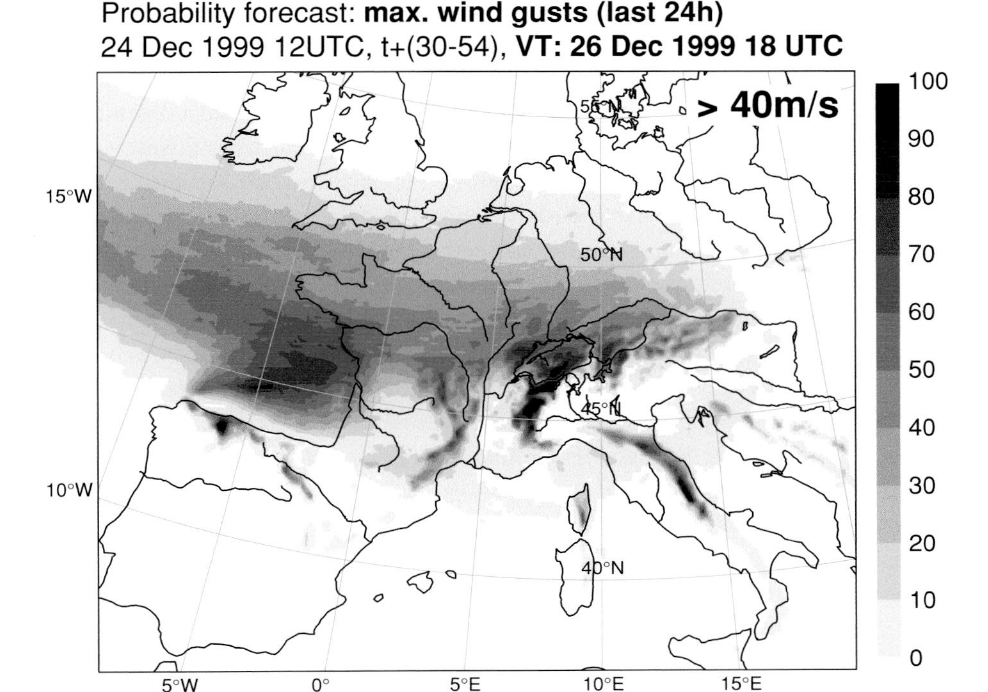

Figure 1.13 Probability that wind gusts exceed 40 m/s for the storm Lothar based on an ensemble (COSMO-LEPS) of limited-area model integrations using ECMWF ensemble boundary conditions. From Waliser *et al.* (2005).

1.5 Discussion

In this chapter, we have charted a revolution in the way weather and climate predictions are produced and disseminated as probability forecasts. The revolution started in studies of monthly predictability, spread to medium range and seasonal timescales, finally permeating extremes of meteorological predictions, the climate change and short-range weather forecast problems. The revolution is based on the notion that in many cases the most relevant information for some user is not necessarily what is most likely to happen, but rather a quantified probability of weather or climate events to which the user is sensitive. We introduced the case of Charlie, who wants to know if he can lay concrete. If Charlie's cost/loss ratio is less than 0.5, he may decide not to lay concrete even when frost is not likely to occur. If Charlie were one day to become minister for the environment, he might be faced with similar risk-based decisions on the adequacy of the current Thames barrier, one of London's key flood defences. A prediction of an 11 K warming associated with a doubling of CO_2 (Stainforth *et al.*, 2005) doesn't have to be likely in order for the replacement of the barrier to be in need of urgent consideration. However, the decision to replace may need better quantified estimates of uncertainty than we currently have.

Predicting the probability of occurrence of weather and climate events requires us to be able to quantify the sources of uncertainty in weather and climate prediction, and to estimate how these sources actually impact on the predictions themselves. In practice, these sources of uncertainty are not easy to quantify. This is not because we don't know the accuracy of instruments used to observe the atmosphere (and oceans). Rather, it is because the approximations used in making computational models of the essentially continuum multiphase fluid equations are themselves hard to quantify. Hence, for example, when model variables are compared with observations in data assimilation, the observation operator doesn't recognise the fact that the observation may be strongly influenced by scales of motion that the model is unable to resolve well. On many occasions and for certain types of observation (e.g. surface pressure) this may not be a serious problem, but occasionally and for other types of observation (e.g. humidity) it is. At present, this type of uncertainty is unquantified in operational data assimilation – from this perspective it is an example of an 'unknown unknown'. In the presence of such 'unknown unknowns', operational ensemble prediction systems run the danger of being underdispersive. This is potentially disastrous: if Charlie lays concrete when the risk of frost is predicted to be zero, and frost occurs, Charlie will never use ensemble prediction again! If, in his future career as politician, Charlie decides against replacing the Thames barrier on the basis of underdispersive ensemble climate forecasts, history may not be kind to him!

One specific conclusion of this chapter is that the development of accurate ensemble prediction systems on all timescales, hours to centuries, relies on a better quantification of model uncertainties. It has been argued that this may require a fundamental change in the way we formulate our models, from deterministic to stochastic dynamic. This change has been anticipated by Lorenz (1975) who said: 'I believe that the ultimate climatic models . . . will be stochastic, i.e. random numbers will appear somewhere in the time derivatives'. Stochastic representations of subgrid processes are particularly well suited to ensemble forecasting.

References

Barnett, T. P., D. W. Pierce, M. Latif, D. Dommenget and R. Saravanan (1999). Interdecadal interactions between the tropics and midlatitudes in the Pacific basin. *Geophys. Res. Lett.*, **26**, 615–18.

Broecker, W. (1995). Chaotic climate. *Sci. Am.*, November, 62–8.

Buizza, R. and T. N. Palmer (1995). The singular vector structure of the atmospheric global circulation. *J. Atmos. Sci.*, **52**, 1434–56.

Buizza, R., M. J. Miller and T. N. Palmer (1999). Stochastic simulation of model uncertainties in the ECMWF Ensemble Prediction System. *Quart. J. Roy. Meteor. Soc.*, **125**, 2887–908.

Buizza, R., P. L. Houtekamer, Z. Toth, G. Pellerin, M. Wei and Y. Zhu (2003). Assessment of the status of global ensemble prediction. *Proceedings of 2002 ECMWF Seminar on Predictability of Weather and Climate*. ECMWF.

Buizza, R., P. L. Houtekamer, Z. Toth, G. Pellerin, M. Wei and Y. Zhu (2005). A comparison of the ECMWF, MSC, and NCEP Global Ensemble Prediction Systems. *Mon. Wea. Rev.*, **133**, 1076–97.

Courtier, P., E. Andersson, W. Heckley, *et al.* (1998). The ECMWF implementation of three dimensional variational assimilation (3DVAR). I: Formulation. *Quart. J. Roy. Meteor. Soc.*, **124**, 1783–808.

Doblas-Reyes, F. J., M. Deque and J.-P. Piedlievre (2000). Multi-model spread and probabilistic seasonal forecasts in PROVOST. *Quart. J. Roy. Meteor. Soc.*, **126**, 2069–88.

Folland, C. K., T. N. Palmer and D. E. Parker (1986). Sahel rainfall and worldwide sea surface temperatures 1901–85. *Nature*, **320**, 602–7.

Frederiksen, J. S. and A. G. Davies (1997). Eddy viscosity and stochastic backscatter parameterizations on the sphere for atmospheric circulation models. *J. Atmos. Sci.*, **54**, 2475–92.

Griffies, S. M. and K. Bryan (1997). A predictability study of simulated North Atlantic multidecadal variability. *Clim. Dynam.*, **13**, 459–87.

Hoffman, R. and E. Kalnay (1983). Lagged average forecasting, alternative to Monte Carlo forecasting. *Tellus*, **35A**, 100–18.

Houtekamer, P., L. Lefaivre, J. Derome, H. Ritchie and H. Mitchell (1996). A system simulation approach to ensemble prediction. *Mon. Weather Rev.*, **124**, 1225–42.

Jung, T., T. N. Palmer and G. J. Shutts (2005). Influence of a stochastic parameterization on the frequency of occurrence of North Pacific weather regimes in the ECMWF model. *Geophys. Res. Lett.*, **32**, L23811.

Klinker, E. and P. D. Sardeshmukh (1992). The diagnosis of mechanical dissipation in the atmosphere from large-scale balance requirements. *J. Atmos. Sci.*, **49**, 608–27.

Khouider, B., A. J., Majda and M. A. Katsoulakis (2003). Coarse-grained stochastic models for tropical convection and climate. *Proc. Natl. Acad. Sci. USA*, **100**, 11941–6.

Landerer, F., J. Jungclaus, J. Marotzke (2006). Steric and dynamic change in response to the A1B scenario integration in the ECHAM5/MPI-OM coupled climate model. *J. Phys. Oceanogr.*, in press.

Leith, C. (1990). Stochastic backscatter in a sub-gridscale model: plane shear mixing layer. *Phys. Fluids A*, **2**, 297–9.

Lewis, J. M. (2005). Roots of ensemble forecasting. *Mon. Weather Rev.*, **133**, 1865–85.

Lorenz, E. N. (1963). Deterministic nonperiodic flow. *J. Atmos. Sci.*, **42**, 433–71.

(1965). A study of the predictability of the 28-variable atmospheric model. *Tellus*, **17**, 321–33.

(1975). Climatic predictability. In *The Physical Basis of Climate and Climate Modelling*. WMO GARP Publication Series No. 16. World Meteorological Organisation.

Meehl, G. A., G. J. Boer, C. Covey, M. Latif and R. J. Stouffer (2000). The Coupled model intercomparison project. *Bull. Am. Meteorol. Soc.*, **81**, 313.

Miyakoda, K., J. Sirutis and J. Ploshay (1986). One month forecast experiments – without anomaly boundary forcings. *Mon. Weather Rev.*, **114**, 1145–76.

Molteni, F. and T. N. Palmer (1993). Predictability and non-modal finite-time instability of the northern winter circulation. *Quart. J. Roy. Meteor. Soc.*, **119**, 269–98.

Molteni, F., R. Buizza, T. Petroliagis and T. N. Palmer (1996). The ECMWF ensemble prediction system: methodology and validation. *Quart. J. Roy. Meteor. Soc.*, **122**, 73–119.

Murphy, A. H. (1977). The value of climatological, categorical and probabilistic forecasts in the cost-loss ratio situation. *Mon. Weather Rev.*, **105**, 803–16.

Murphy, J. M. and T. N. Palmer (1986). Experimental monthly long-range forecasts for the United Kingdom. II: A real time long-range forecast by an ensemble of numerical integrations. *Meteorol. Mag.*, **115**, 337–49.

Murphy, J. M., D. M. H. Sexton, D. N. Barnett, *et al.* (2004). Quantifying uncertainties in climate change using a large ensemble of global climate model predictions. *Nature*, **430**, 768–72.

Nastrom, G. D. and K. S. Gage (1985). A climatology of atmospheric wavenumber spectra of wind and temperature observed by commercial aircraft. *J. Atmos. Sci.*, **42**, 950–60.

Palmer, T. N. (1988). Medium and extended range predictability and stability of the Pacific/North American mode. *Quart. J. Roy. Meteor. Soc.*, **114**, 691–713.

(2001). A nonlinear dynamical perspective on model error: a proposal for non-local stochastic-dynamic parametrisation in weather and climate prediction models. *Quart. J. Roy. Meteor. Soc.*, **127**, 279–304.

(2002). The economic value of ensemble forecasts as a tool for risk assessment: from days to decades. *Quart. J. Roy. Meteor. Soc.*, **128**, 747–74.

Palmer, T. N. and J. Shukla (2000). Editorial to DSP/PROVOST special issue, *Quart. J. Roy. Meteor. Soc.*, **126**, 1989–90.

Palmer, T. N. and J. Räisänen (2002). Quantifying the risk of extreme seasonal precipitation events in a changing climate. *Nature*, **415**, 512.

Palmer, T. N., C. Brankovic, F. Molteni, *et al.* (1990). The European Centre for Medium-Range Weather Forecasts (ECMWF) program on extended-range prediction. *Bull. Am. Meteorol. Soc.*, **71**, 1317–30.

Palmer, T. N., F. Molteni, R. Mureau, R. Buizza, P. Chapelet and J. Tribbia (1993). Ensemble prediction. In *ECMWF 1992 Seminar: Validation of Models over Europe*. ECMWF.

Palmer, T. N., C. Brankovic and D. S. Richardson (2000). A probability and decision-model analysis of PROVOST seasonal multimodel ensemble integrations. *Quart. J. Roy. Meteor. Soc.*, **126**, 2013–34.

Palmer, T. N., A. Alessandri, U. Andersen, *et al.* (2004). Development of a European multi-model ensemble system for seasonal to inter-annual prediction. *Bull. Am. Meteorol. Soc.*, **85**, 853–72.

Palmer, T. N., G. J. Shutts, R. Hagedorn, F. J. Doblas-Reyes, T. Jung and M. Leutbecher (2005). Representing model uncertainty in weather and climate prediction. *Ann. Rev. Earth Planet. Sci.*, **33**, 163–93.

Reynolds, C. and T. N. Palmer (1998). Decaying singular vectors and their impact on analysis and forecast correction. *J. Atmos. Sci.*, **55**, 3005–23.

Rodwell, M. J. and T. N. Palmer (2006). Assessing model physics with initial forecast tendencies: application to climate change uncertainty. *Quart. J. Roy. Meteor. Soc.*, Submitted.

Shukla, J. and J. M. Wallace (1983). Numerical simulation of the atmospheric response to equatorial Pacific sea surface temperature anomalies. *J. Atmos. Sci.*, **40**, 1613–30.

Shutts, G. J. (2005). A kinetic energy backscatter algorithm for use in ensemble prediction systems. *Quart. J. Roy. Meteor. Soc.*, **131**, 3079–102.

Shutts, G. J. and T. N. Palmer (2004). The use of high-resolution numerical simulations of tropical circulation to calibrate stochastic physics schemes. In *Proceedings of ECMWF Workshop on Intra-seasonal Variability*. ECMWF.

(2006). Convective forcing fluctuations in a cloud-resolving model: relevance to the stochastic parametrization problem. *J. Clim.*, in press.

Smith, D. M., A. W. Colman, S. Cusack, C. K. Folland, S. Ineson and J. M. Murphy (2006). Predicting surface temperature for the coming decade using a global climate model. *Nature*, in press.

Stainforth, D. A., T. Aina, C. Christensen, *et al.* (2005). Uncertainty in predictions of the climate response to rising levels of greenhouse gases. *Nature*, **433**, 403–6.

Stockdale, T. N., D. L. T. Anderson, J. O. S. Alves and M. A. Balmaseda (1998). Global seasonal rainfall forecasts using a coupled ocean-atmosphere model. *Nature*, **392**, 370–3.

Straus, D. M. and J. Shukla (2000). Distinguishing between the SST-forced variability and internal variability in mid-latitudes: analysis of observations and GCM simulations. *Quart. J. Roy. Meteor. Soc.*, **126**, 2323–50.

Toth, Z. and E. Kalnay (1993). Ensemble forecasting at NMC: the generation of perturbations. *Bull. Am. Meteorol. Soc.*, **74**, 2317–30.

Vitart, F. (2004). Monthly forecasting at ECMWF. *Mon. Weather Rev.*, **132**, 2761–79.

Waliser, A., M. Arpagaus, M. Leutbecher and C. Appenzeller (2006). The impact of moist singular vectors and horizontal resolution on short-range limited-area ensemble forecasts for two European winter storms. *Mon. Weather Rev.*, in press.

Wilson, S. (2000). Launching the Argo armada. *Oceanus*, **42**, 17–19.

Wolfram, S. (2002). *A New Kind of Science*. Wolfram Media Inc.

Zebiak, S. and M. Cane (1987). A model of the El Niño-Southern Oscillation. *Mon. Weather Rev.*, **115**, 2262–78.

2

Predictability from a dynamical meteorological perspective

Brian Hoskins
University of Reading

2.1 Introduction: origins of predictability

Predictability of weather at various timescales has its origins in the physics and dynamics of the system. The annual cycle is an example of very predictable behaviour with such an origin, though this predictability is of little other than general background use to the forecaster. The rapid rotation of the Earth with its shallow, generally stably stratified atmosphere leads to the dominance of phenomena with balance between their thermodynamic and dynamic structures. These structures evolve on timescales comparable to, or longer than, a day. This balanced motion is best described by consideration of two properties that are materially conserved under adiabatic conditions, potential temperature (θ, or equivalently entropy) and potential vorticity (PV). A feature of particular importance, leading to potentially predictable behaviour, is the ability of the atmosphere to support balanced, large-scale Rossby waves.

Atmospheric phenomena with recognised structures tend to exhibit characteristic evolutionary behaviour in time and thus to have a level of predictability. Particular examples of such phenomena are the midlatitude cyclone on synoptic timescales, the tropical Intra-Seasonal Oscillation and the El Niño–Southern Oscillation (ENSO) on annual timescales.

Slower parts of the climate system can leave an imprint on shorter timescales and hence give an element of predictability to them. Tropical sea surface temperature (SST) anomalies tend to persist and can lead to anomalous convective activity in

Predictability of Weather and Climate, ed. Tim Palmer and Renate Hagedorn. Published by Cambridge University Press.

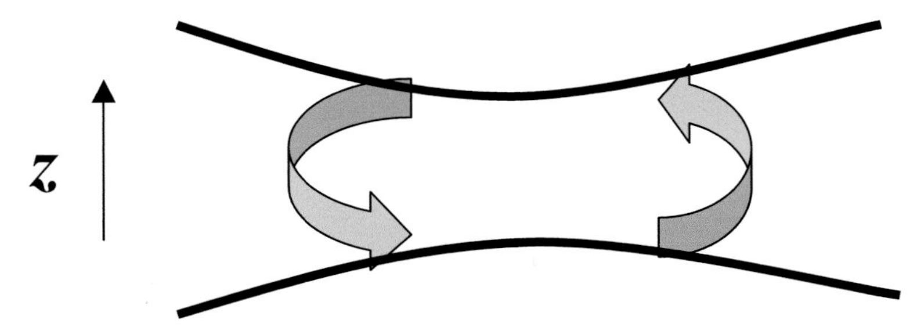

Figure 2.1 The balanced structure typically associated with a positive potential vorticity (PV) anomaly. Features shown are the cyclonic circulation in the region and the isentropes 'sucked' towards the anomaly in the vertical, consistent with high static stability there.

their region. This activity can trigger Rossby waves that communicate anomalous conditions to other regions of the globe. Soil moisture anomalies in tropical or extra-tropical regions can persist a month or more and similarly trigger both local and remote responses.

Here aspects of balanced motion and its description with PV and Rossby waves will be discussed in Section 2.2. The focus in Section 2.3 is on particular phenomena, midlatitude weather systems, blocking highs, and a particular mode of variability, the North Atlantic Oscillation (NAO). Section 2.4 gives some discussion of aspects of the summer of 2002 and their possible predictability. Some concluding comments are given in Section 2.5.

2.2 Balanced motion, potential vorticity and Rossby waves

Hydrostatic balance and geostrophic balance together lead to so-called thermal wind balance between the wind and temperature fields. The development of such motion is described by the quasi-geostrophic version of the conservation of PV. As discussed in detail in Hoskins *et al.* (1985), more general balanced motion is uniquely determined through an elliptic problem by the PV/θ distribution. Its development is described by the material conservation of PV on θ surfaces and θ on the lower boundary. Alternatively it is often convenient to summarise the upper tropospheric PV/θ distribution by the distribution of θ on the PV = 2PVU surface (here northern hemisphere signs are used for convenience), which away from the tropics can be considered to be the dynamical tropopause (Hoskins, 1997).

Because of the elliptic nature of the inversion problem, a positive PV anomaly is generally associated with both cyclonic motion and increased static stability (Figure 2.1). Similarly a negative PV anomaly is associated with anticyclonic motion and reduced static stability. As in Figure 2.2, the tip of PV 'trough' often elongates,

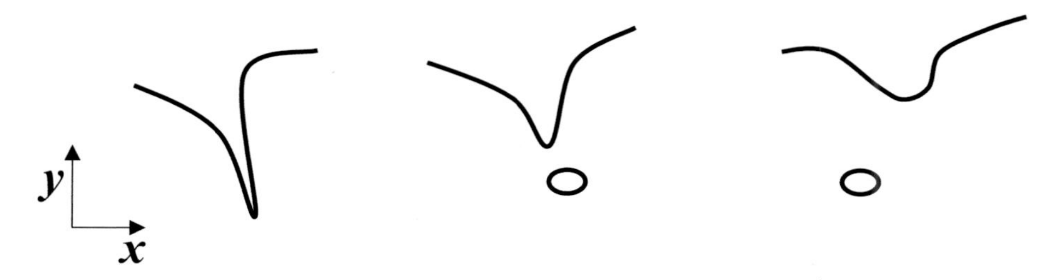

Figure 2.2 An upper air development of a geopotential trough, indicated by a PV contour on a θ surface or a θ contour on a PV surface in the tropopause region. Shown is the elongation of the trough, and the development and movement away of a cut-off.

develops its own cyclonic circulation and cuts itself off. Once it has done this, the cut-off low must continue to exist until either the PV anomaly is eroded by diabatic processes or it moves back into the higher PV region. This implies some extended predictability of such a cut-off low once it has formed.

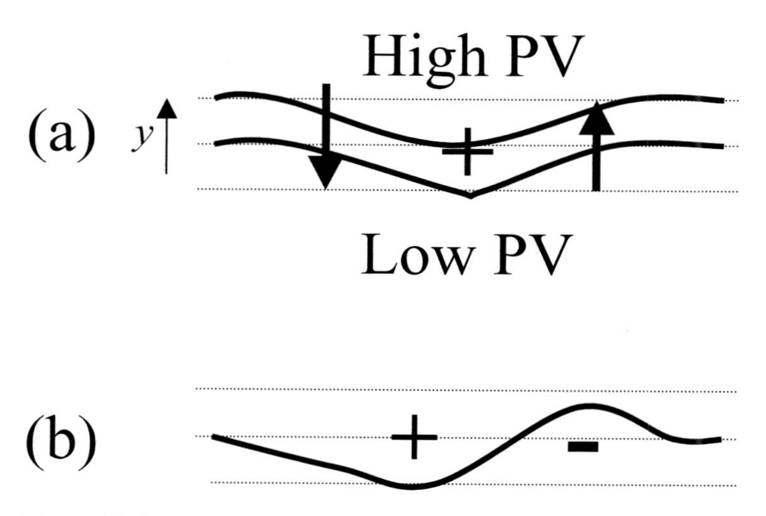

Figure 2.3 Rossby wave development. In (a) PV contours are shown displaced to the south, leading to a PV anomaly and cyclonic circulation. This circulation advects the PV meridionally, leading to the PV distribution and anomalies shown in (b).

If a contour is displaced equatorwards as in Figure 2.3(a), the associated cyclonic anomaly advects the PV distribution as shown in Figure 2.3(b). This gives a negative (anticyclonic) PV anomaly to the east and a positive (cyclonic) anomaly to the west. If there is a basic westerly flow, the net result is that the original PV anomaly will move to the east at less than the speed of the basic flow and could be stationary. However, the wave activity develops downstream, to the east, at a speed greater than that of the basic flow. The motion described is that of Rossby waves and the two speeds described respectively their phase and group speeds.

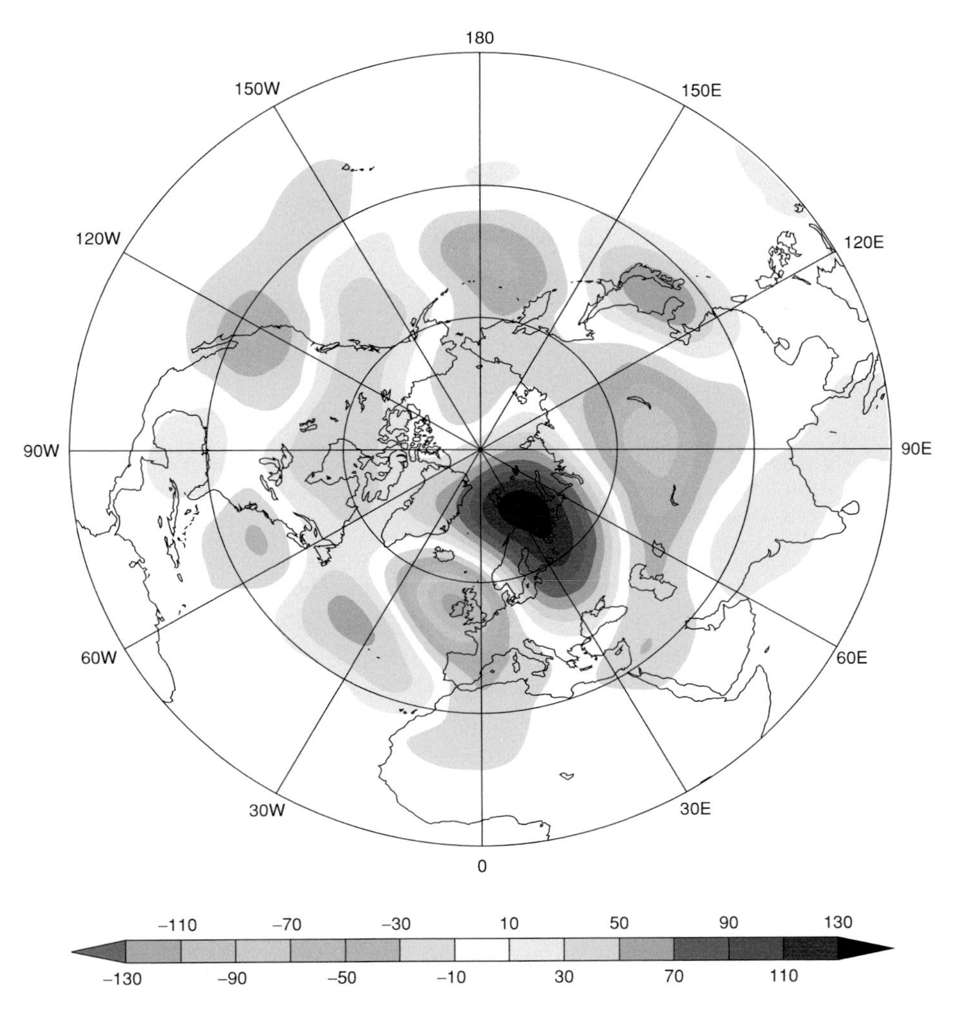

Figure 2.4 (See also colour plate section.) September–November 2000 300hPa geopotential height anomalies from climate. From Blackburn and Hoskins (personal communication).

On a β-plane Rossby wave activity spreads eastwards along regions of large meridional PV gradients as these provide the restoring mechanism for them. However, on the sphere, the path of long wavelength quasi-stationary Rossby wave activity is closer to great circles. Such Rossby waves generated in the tropics can arc polewards and eastwards into middle latitudes. Figure 2.4 (colour plate) shows the 300 hPa geopotential height anomalies for October 2000. As discussed by M. Blackburn and B. J. Hoskins (personal communication), the record rainfall over the UK is associated with the anomalous low over and to the west of the UK. It can be seen that this cyclone is preceded upstream, to the south-west, by an anticyclone. Streamfunction

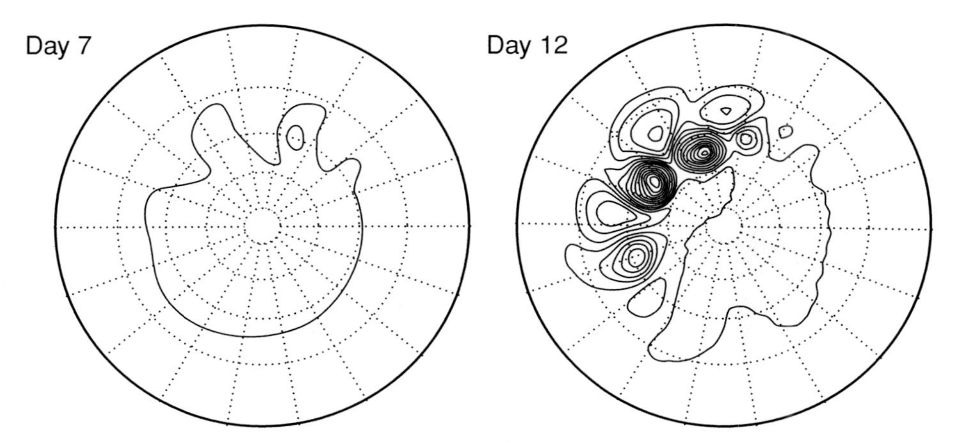

Figure 2.5 Surface pressure fields 7 days and 12 days after a baroclinically unstable zonal flow was disturbed near '3 o'clock'. The contours are drawn every 4 hPa. From Simmons and Hoskins (1979).

anomalies indicate a further cyclone south-west of this. The strong suggestion is that anomalous rainfall in the South American–Caribbean region triggered a stationary Rossby wave pattern that in turn led to the anomalous weather in the UK. If this hypothesis is correct, there is potential predictability of such anomalous midlatitude months. However, there must be the ability to represent the tropical anomalies in large-scale rainfall if it is to be realised.

2.3 Some phenomena

Middle latitude synoptic systems tend to have a characteristic structure including surface fronts and upper tropospheric troughs. They also have characteristic evolutions in time and therefore there is the implication of some predictability. These structural and evolutionary characteristics can be usefully interpreted in terms of theoretical descriptions in terms of normal mode baroclinic waves (Charney, 1947; Eady, 1949), non-linear life cycles (Simmons and Hoskins, 1978), the omega and vorticity equations (Hoskins *et al.*, 1978), and coupled mid-troposphere and surface Rossby waves (Heifetz *et al.*, 2004). A link with Rossby wave behaviour is shown by the ordered downstream development of baroclinic instability illustrated in Figure 2.5. In this numerical experiment from Simmons and Hoskins (1979) an unstable westerly flow had been perturbed at day zero at '3 o'clock'. Successive lows and highs have developed to the east and occluded, with the newest systems at day 12 near '6 o'clock'. Each system moves at some 10 m s^{-1}. However, the downstream development propagates at nearer 30 m s^{-1}, like a synoptic scale Rossby wave on the upper tropospheric jet. The orderly downstream propagation of baroclinic wave activity has been documented in a number of studies (e.g.

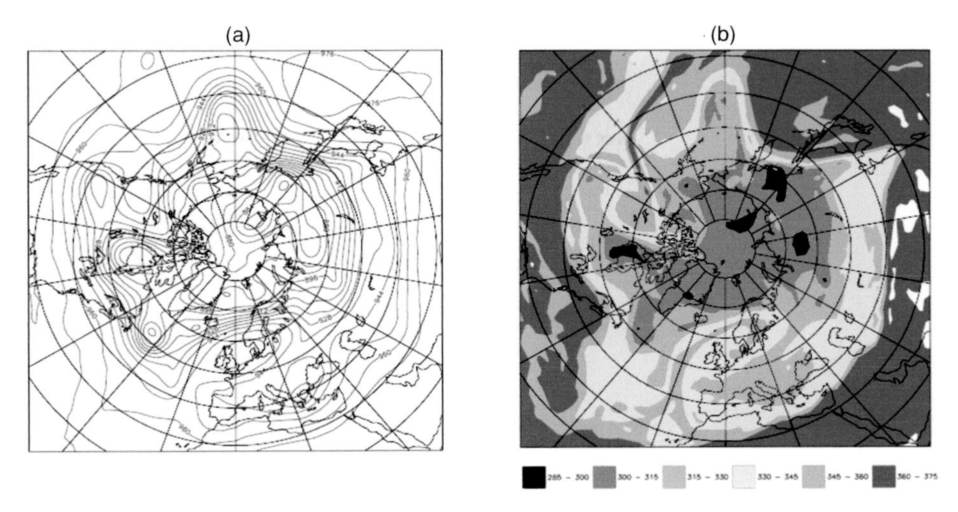

Figure 2.6 (See also colour plate section.) The block of 21 September 1998. Shown are the 250 hPa geopotential height field and the θ on PV2 field. From Pelly and Hoskins (2003a).

Chang, 1993). It gives the possibility that the development of new weather systems is predictable much beyond the synoptic timescale on which each individual system evolves.

The normal progression of middle latitude weather systems is sometimes interrupted by 'blocking highs'. An example of one of a block in the western European region is shown in Figure 2.6 (colour plate) in terms of its 300 hPa geopotential and θ on PV $= 2$ fields. The reversal of the zonal wind in the region of the block is associated with a reversal of the negative latitudinal gradient in θ. The formation of blocks, particularly in the European region, can often be viewed as a breaking of synoptic waves and, consistent with this, the timescale tends to be synoptic. However, once there is a low PV cut-off (here high θ cut-off) the decay is on the generally longer timescale of either diabatic processes or reabsorption into the subtropical region. Again there is associated enhanced predictability. These ideas are supported by Figure 2.7 from Pelly and Hoskins (2003a) which shows that on short timescales the decay of blocking-like features is on a timescale of about two days, but once a feature has lasted four days, and is probably associated with a PV cut-off, the decay time is about twice as long. It has indeed been found that the ECMWF Ensemble Prediction System has skill for the onset of blocking for about four days but for blocking events and the decay of them on timescales of about seven days (Pelly and Hoskins, 2003b).

A classic pattern of variability in the climate system is the North Atlantic Oscillation (NAO; Hurrell *et al.*, 2002) that describes the fluctuation of the surface westerly winds in that region. Alternative descriptions that have been used in recent years are the Arctic Oscillation (AO; Thompson and Wallace, 1998) that emphasises

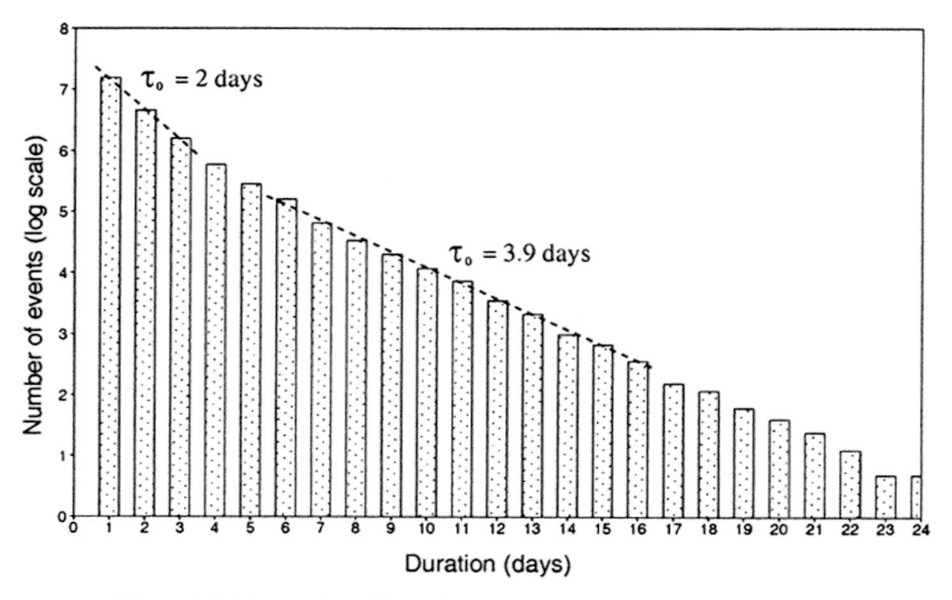

Figure 2.7 The number of blocking events in the period 1 August 2000 to 31 July 2001 lasting at least the number of days shown on the abscissa. The ordinate gives the logarithm of the number of sector blocking events and a straight line in the figure indicates a uniform decay rate of the events. For more details, see Pelly and Hoskins (2003a).

the fluctuations of the pressure in the polar cap and the Northern Annular Mode (NAM; Thompson and Wallace, 2000) that focuses on the polar vortex and the analogy with the southern hemisphere vortex. Figure 2.8 shows the autocorrelation timescales of the NAO and the northern hemisphere stratospheric vortex in winter (Ambaum and Hoskins, 2002). Synoptic timescale decay in the NAO changes to longer timescales beyond ten days. One hypothesis is that this reflects a link with the stratosphere and the longer timescales there, which are indicated in Figure 2.8.

The tropical Intra-Seasonal Oscillation (ISO), or Madden–Julian Oscillation (MJO), describes the large-scale organisation of tropical convection in the Indian Ocean that then migrates eastwards to the west Pacific, with its associated circulation changes. Consistent with the name, the timescale for the ISO is 30–60 days, although the word oscillation perhaps overemphasises its oscillatory nature. Once an ISO event has started, its typical evolution on the timescale of weeks is known. Since the convection associated with an ISO generates Rossby wave trains that lead to characteristic responses, particularly in the winter hemisphere (Matthews *et al.*, 2004), there is potential predictability on the timescale of weeks both in the tropics and higher latitudes. However, the current ability of models to simulate the ISO is generally poor and this potential predictability is yet to be realised.

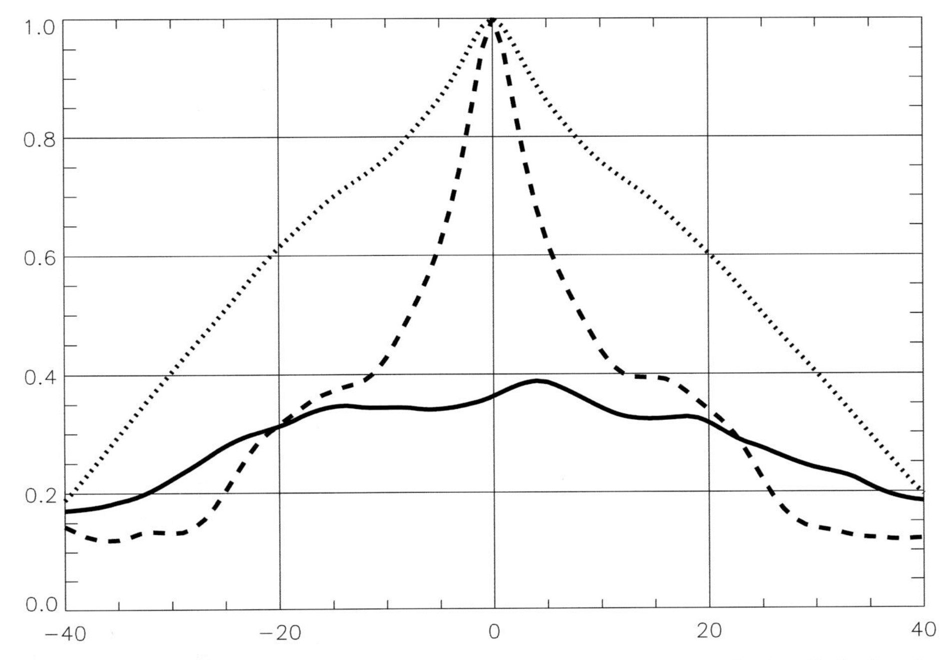

Figure 2.8 Lagged autocorrelations for the North Atlantic Oscillation (dashed) and the 500 K stratospheric vortex (dotted), and the lagged correlation between them (solid). The measures of the two patterns are the first principal components of daily mean sea-level pressure and 500 K PV, respectively. For more details, see Ambaum and Hoskins (2002).

2.4 Summer 2002

The northern summer of 2002 contained a number of climate features and anomalies that may have been linked. There were strong ISOs in the tropics. The surface westerly winds associated with these may have triggered the onset of an El Niño event that was observed to occur. Reduced Indian monsoon rainfall has been found to tend to occur in El Niño years and this certainly happened in 2002. The reduction in Indian monsoon rainfall was particularly dramatic in the middle of the season and was related to one very strong ISO event. The latent heat release associated with Indian monsoon rainfall leads to ascent in that region. It has been shown (Rodwell and Hoskins, 1996) that compensating descent occurs in the Mediterranean region and leads to the characteristic summer weather in the region. The reduction in the monsoon in 2002 is therefore consistent with the unusual occurrence in that year of weather systems moving into the Mediterranean region and then up into Europe, leading to flooding events there.

The possible linkages between all these events are explored by M. Blackburn *et al.* (personal communication). These linkages give the possibility of predictability.

However, they also indicate the range of processes and phenomena that may have to be modelled well in order to obtain this predictive power.

2.5 Concluding comments

A number of examples have been given in which phenomena and their structures give predictability. The dynamical perspective illustrated here provides a framework for consideration of the approach to prediction on various timescales and for the processes that need to be improved in models if potential predictability is to be found in practice.

References

Ambaum, M. H. P. and B. J. Hoskins (2002). The NAO troposphere-stratosphere connection. *J. Clim.*, **15**, 1969–78.

Chang, E. K. M. (1993). Downstream development of baroclinic waves as inferred from regression analysis. *J. Atmos. Sci*, **50**, 2038–53.

Charney, J. G. (1947). The dynamics of long waves in a baroclinic westerly current. *J. Meteorol.*, **4**, 135–62.

Eady, E. T. (1949). Long waves and cyclone waves. *Tellus*, **1**, 33–52.

Heifetz, E., C. H. Bishop, B. J. Hoskins and J. Methven (2004). The counter-propagating Rossby-wave perspective on baroclinic instability. I: Mathematical basis. *Quart. J. Roy. Meteor. Soc.*, **130**, 211–31.

Hoskins, B. J. (1997). A potential vorticity view of synoptic development. *Meteorol. Appl.*, **4**, 325–34.

Hoskins, B. J., I. Draghici and H. C. Davies (1978). A new look at the ω-equation. *Q. J. Roy. Meteor. Soc.*, **104**, 31–8.

Hoskins, B. J., M. E. McIntyre and A. W. Robertson (1985). On the use and significance of isentropic potential vorticity maps. *Quart. J. Roy. Meteor. Soc.*, **111**, 877–946.

Hurrell, J. W., Y. Kushnir, G. Ottersen and M. Visbeck (eds.) (2002). *The North Atlantic Oscillation: Climate Significance and Environmental Impact. Geophysical Monograph*, No. 134, American Geophysical Union.

Matthews, A. J., B. J. Hoskins and M. Masutani (2004). The global response to tropical heating in the Madden-Julian oscillation during the northern winter. *Quart. J. Roy. Meteor. Soc.*, **130**, 1991–2011.

Pelly, J. L. and B. J. Hoskins (2003a). A new perspective on blocking. *J. Atmos. Sci.*, **60**, 743–55.

(2003b). How well does the ECMWF Ensemble Prediction System predict blocking? *Quart. J. Roy. Meteor. Soc.*, **129**, 1683–702.

Rodwell, M. R. and B. J. Hoskins (1996). Monsoons and the dynamics of desserts. *Quart. J. Roy. Meteor. Soc.*, **122**, 1385–404.

Simmons, A. J. and B. J. Hoskins (1978). The life cycles of some non-linear baroclinic waves. *J. Atmos. Sci.*, **35**, 414–32.

(1979). The downstream and upstream development of unstable baroclinic waves. *J. Atmos. Sci.*, **36**, 1239–54.

Thompson, D. W. J. and J. M. Wallace (1998). The Arctic Oscillation signature in the wintertime geopotential height and temperature fields. *Geophys. Res. Lett.*, **25**, 1297–300.

(2000). Annular modes in the extratropical circulation. 1: Month-to-month variability. *J. Climate*, **13**, 1000–16.

3

Predictability – a problem partly solved

Edward N. Lorenz
Massachusetts Institute of Technology, Cambridge

Ed Lorenz, pioneer of chaos theory, presented this work at an earlier ECMWF workshop on predictability. The paper, which has never been published externally, presents what is widely known as the Lorenz 1996 model. Ed was unable to come to the 2002 meeting, but we decided it would be proper to acknowledge Ed's unrivalled contribution to the field of weather and climate predictability by publishing his 1996 paper in this volume.

The difference between the state that a system is assumed or predicted to possess, and the state that it actually possesses or will possess, constitutes the *error* in specifying or forecasting the state. We identify the rate at which an error will typically grow or decay, as the range of prediction increases, as the key factor in determining the extent to which a system is predictable. The long-term average factor by which an infinitesimal error will amplify or diminish, per unit time, is the leading Lyapunov number; its logarithm, denoted by λ_1, is the leading Lyapunov exponent. Instantaneous growth rates can differ appreciably from the average.

With the aid of some simple models, we describe situations where errors behave as would be expected from a knowledge of λ_1, and other situations, particularly in the earliest and latest stages of growth, where their behaviour is systematically different. Slow growth in the latest stages may be especially relevant to the long-range predictability of the atmosphere. We identify the predictability of long-term climate variations, other than those that are externally forced, as a problem not yet solved.

Predictability of Weather and Climate, ed. Tim Palmer and Renate Hagedorn. Published by Cambridge University Press.

3.1 Introduction

As I look back over the many meetings that I have attended, I recall a fair number of times when I have had the pleasure of being the opening speaker. It's not that this is necessarily a special honour, but it does allow me to relax, if not to disappear altogether, for the remainder of the meeting. On the present occasion, however, I find it is a true privilege to lead off. This is both because the subject of the seminar, predictability, is of special interest to me, and because much of the significant work in this field has taken place here at the European Centre.

Most of us who are here presumably have a special interest in the atmosphere, but the subject of predictability and the knowledge of it that we presently possess extend to much more general systems. By and large these systems fall into two categories, within which, to be sure, there are many subcategories. On the one hand there are real or realisable physical systems. On the other there are systems defined by mathematical formulas. The distinction between these categories is not trivial.

The former category includes the atmosphere, but also many much simpler systems, such as a pendulum swinging in a clock, or a flag flapping in a steady breeze. Instantaneous states of these systems cannot be observed with absolute precision, nor can the governing physical laws be expressed without some approximation. Exact descriptions of the dissipative processes are particularly elusive.

In the latter category, initial states may be prescribed exactly. Likewise, the defining formulas may be precisely written down, at least if the chosen finite-difference approximations to any differential equations, and the inevitable round-off procedures, are regarded as part of the system. In some instances the equations are of mathematical interest only, but in other cases they constitute models of real physical systems; that is, they may be fair, good, or even the best-known approximations to the equations that properly represent the appropriate physical laws. The relevance of mathematically defined systems cannot be too strongly emphasised; much of what we know, or believe that we know, about real systems has come from the study of models.

Systems whose future states evolve from their present states according to precise physical laws or mathematical equations are known as *dynamical systems*. These laws or equations encompass not only the internal dynamics of a system, but also any external factors that influence the system as it evolves. Often the concept of a dynamical system is extended to include cases where there may be some randomness or uncertainty in the evolution process, especially when it is believed that the general behaviour of the system would hardly be changed if the randomness could be removed; thus, in addition to mathematical models and abstractions, many real physical systems will qualify. Stochastic terms sometimes are added to otherwise deterministic mathematical equations to make them simulate real-system behaviour more closely.

In the ensuing discussion I shall frequently assume that our system is the atmosphere and its surroundings – the upper layers of the oceans and land masses – although I shall illustrate some of the points with rather crude models. By regularly calling our system the 'atmosphere' I do not mean to belittle the importance of the non-atmospheric portions. They are essential to the workings of the atmospheric portions, and, in fact, prediction of oceanic and land conditions can be of interest for its own sake, wholly apart from any coupling to the weather.

A procedure for predicting the evolution of a system may consist of an attempt to solve the equations known or believed to govern the system, starting from an observed state. Often, if the states are not completely observed, it may be possible to infer something about the unobserved portion of the present state from observations of past states; this is what is currently done, for example, in numerical weather prediction (see, for example, Toth and Kalnay, 1993). At the other extreme, a prediction procedure may be completely empirical. Nevertheless, whatever the advantages of various approaches may be, no procedure can do better than to duplicate what the system does. Any suitable method of prediction will therefore constitute, implicitly if not explicitly, an attempt at duplication – an attempt to reproduce the *result* of marching forward from the present state.

When we speak of 'predictability', we may have either of two concepts in mind. One of these is intrinsic predictability – the extent to which prediction is possible if an optimum procedure is used. The other is practical predictability – the extent to which we ourselves are able to predict by the best-known procedures, either currently or in the foreseeable future. If optimum prediction consists of duplication, it would appear that imperfect predictability must be due to one or both of two conditions – inability to observe the system exactly, and inability to formulate a perfect forward-extrapolation procedure. The latter condition is certainly met if the laws involve some randomness, or if future external influences cannot be completely anticipated.

When we cannot determine an initial state of a system precisely, there are two possible consequences. The system may be convergent; that is, two or more rather similar states, each evolving according to the same laws, may become progressively more similar. In this event, a precise knowledge of the true initial state is clearly not needed, and, in fact, the governing laws need not be known, since empirical methods will perform as well as any others. When we predict the oceanic tides, for example, which we can do rather well years in advance, we do not start from the observed present state of the ocean and extrapolate forward; we base our prediction on known periodicities, or on established relations between the tides and the computable motions of the sun, earth, and moon.

If, instead, the system is divergent, so that somewhat similar states become less and less similar, predictability will be limited. If we have no basis for saying which, if any, of two or more rather similar states is the true initial state, the governing laws cannot tell us which of the rather dissimilar states that would result from marching forward from these states will be the one that will actually develop. As will be noted in more detail in the concluding section, any shortcoming in the extrapolation procedure will have a similar effect. Systems of this sort are now known collectively as *chaos*. In the case of the atmosphere, it should be emphasised that it may be difficult to establish the absence of an intrinsic basis for discriminating among several estimates of an initial state, and the consequent intrinsic unpredictability; some estimates that now seem reasonable to us might, according to rules that we do not yet appreciate, actually be climatologically impossible and hence rejectable, while others might, according to similar rules, be incompatible with observations of earlier states.

3.2 First estimates of predictability

Two basic characteristics of individual chaotic dynamical systems are especially relevant to predictability. One quantity is the leading Lyapunov number, or its logarithm, the leading Lyapunov exponent. Let us assume that there exists a suitable measure for the difference between any two states of a system – possibly the distance between the points that represent the states, in a multidimensional phase space whose coordinates are the variables of the system. If two states are infinitesimally close, and if both proceed to evolve according to the governing laws, the long-term average factor by which the distance between them will increase, per unit time, is the first Lyapunov number. More generally, if an infinite collection of possible initial states fills the surface of an infinitesimal sphere in phase space, the states that evolve from them will lie on an infinitesimal ellipsoid, and the long-term average factors by which the axes lengthen or shorten, per unit time, arranged in decreasing order, are the Lyapunov numbers. The corresponding Lyapunov exponents are often denoted by $\lambda_1, \lambda_2, \ldots$; a positive value of λ_1 implies chaos (see, for example, Lorenz, 1993). Unit vectors in phase space pointing along the axes of the ellipsoid are the Lyapunov vectors; each vector generally varies with time.

Our interest in pairs of states arises from the case when one member of a pair is the true state of a system, while the other is the state that is believed to exist. Their difference is then the *error* in observing or estimating the state, and, if the assumed state is allowed to evolve according to an assumed law, while the true state follows the true law, their difference becomes the *error* in prediction. In the meteorological community it has become common practice to speak of the *doubling time* for small errors; this is inversely proportional to λ_1 in the case where the assumed and true laws are the same.

The other quantity of interest is the size of the attractor; specifically, the average distance ρ between two randomly chosen points of the attractor. The attractor is simply the set of points representing states that will occur, or be approximated arbitrarily closely, if the system is allowed to evolve from an arbitrary state, and transient effects associated with this state are allowed to die out. Estimation of these quantities is fairly straightforward for mathematically defined systems – ordinarily ρ^2 is simply twice the sum of the variances of the variables – but for real systems λ_1 may be difficult to deduce.

The third quantity that would seem to be needed for an estimate of the range of acceptable predictability is the typical magnitude of the error in estimating an initial state, ostensibly not a property of the system at all, but dependent upon our observing and inference techniques. For the atmosphere, we have a fair idea of how well we now observe a state, but little idea of what to expect in the years to come. Even though we may reject the notion of a future world where observing instruments are packed as closely as today's city dwellings, we do not really know what some undreamed-of remote-sensing technique may some day yield. However, assuming the size of an initial error, taking its subsequent growth rate to be given by λ_1, and recognising that the growth should cease when the predicted and actual states become as far apart as randomly chosen states – when the error reaches *saturation* – we can easily calculate the time needed for the prediction to become no better than guesswork.

How good are such naive estimates? We can demonstrate some simple systems where they describe the situation rather well, at least on the average. One system is one that I have been exploring in another context as a one-dimensional atmospheric model, even though its equations are not much like those of the atmosphere. It contains the K variables $X_1, \ldots, X_K$, and is governed by the K equations

$$dX_k/dt = -X_{k-2}X_{k-1} + X_{k-1}X_{k+1} - X_k + F, \tag{3.1}$$

where the constant F is independent of k. The definition of X_k is to be extended to all values of k by letting X_{k-K} and X_{k+K} equal X_k, and the variables may be thought of as values of some atmospheric quantity in K sectors of a latitude circle. The physics of the atmosphere is present only to the extent that there are external forcing and internal dissipation, simulated by the constant and linear terms, while the quadratic terms, simulating advection, together conserve the total energy $(X_1^2 + \cdots + X_K^2)/2$. We assume that $K > 3$; the equations are of little interest otherwise. The variables have been scaled to reduce the coefficients in the quadratic and linear terms to unity, and, for reasons that will presently appear, we assume that this scaling makes the time unit equal to 5 days.

For very small values of F, all solutions decay to the steady solution $X_1 = \ldots = X_K = F$, while, when F is somewhat larger, most solutions are periodic, but for still larger values of F (dependent on K) chaos ensues. For $K = 36$ and $F = 8.0$, for example, λ_1 corresponds to a doubling time of 2.1 days; if F is raised to 10.0, the time drops to 1.5 days.

Figures 3.1 and 3.2(a) have been constructed with $K = 36$, so that each sector covers 10 degrees of longitude, while $F = 8.0$. We first choose rather arbitrary values of the variables, and, using a fourth-order Runge–Kutta scheme with a time step Δt of 0.05 units, or 6 hours, we integrate forward for 14 400 steps, or 10 years. We then

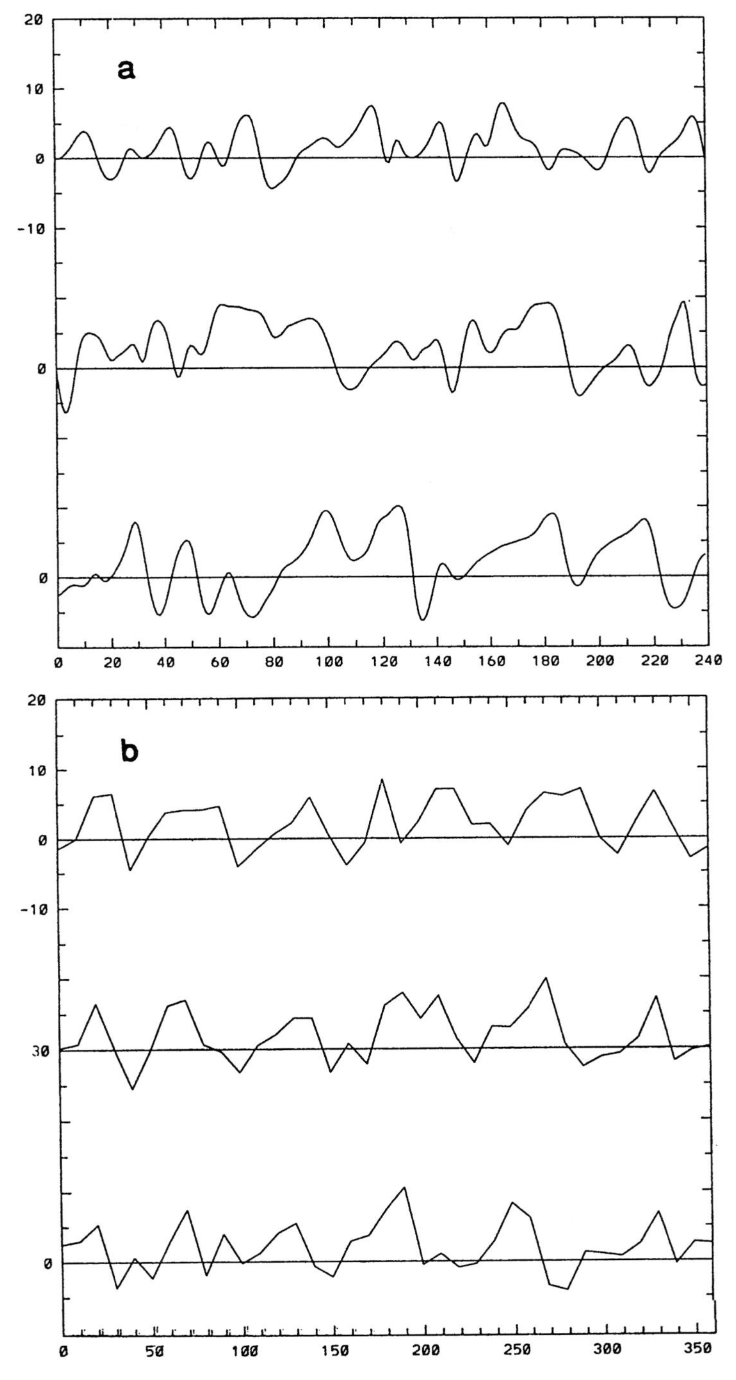

Figure 3.1 (a) Time variations of X_1 during a period of 180 days, shown as three consecutive 60-day segments, as determined by numerical integration of Eq. (3.1), with $K = 36$ and $F = 8.0$. Scale for time, in days, is at bottom. Scales for X_1 in separate segments are at left. (b) Longitudinal profiles of X_k at three times separated by 1-day intervals, determined as in (a). Scale for longitude, in degrees east, is at bottom. Scales for X_k in separate profiles are at left.

use the final values, which should be more or less free of transient effects, as new 'true' initial values, to be denoted by X_{k0}.

From Figure 3.1 we may gain some idea as to the resemblance or lack of resemblance between the *behaviour* of the model variables and some atmospheric variable such as temperature. Figure 3.1(a) shows the variations of X_1 during 720 time steps, or 180 days, beginning with the new initial conditions. The time series is displayed as three 60-day segments. There are some regularities – values lie mostly between -5 and $+10$ units, and about 12 maxima and minima occur every 60 days – but there is no sign of any true periodicity. Because of the symmetry of the model, all 36 variables should have statistically similar behaviour. Figure 3.1(b) shows the variations of X_k with k – a 'profile' of X_k about a 'latitude circle' – at the initial time, and one and two days later. The principal maxima and minima are generally identifiable from one day to the next, and they show some tendency to progress slowly westward, but their shapes are continually changing.

To produce the upper curve in Figure 3.2(a) we make an initial 'run' by choosing errors e_{k0} randomly from a distribution with mean 0 and standard deviation ε, here equal to 0.001, and letting $X'_{k0} = X_{k0} + e_{k0}$ be the 'observed' initial values of the K variables. We then use Eq. (3.1) to integrate forward from the true and also the observed initial state, for $N = 200$ steps, or 50 days, obtaining K sequences $X_{k0}, X_{k1}, \ldots, X_{kN}$ and K sequences $X'_{k0}, X'_{k1}, \ldots, X'_{kN}$, after which we let $e_{kn} = X'_{kn} - X_{kn}$ for all values of k and n.

We then proceed to make a total of $M = 250$ runs in the same manner, in each run letting the new values of X_{k0} be the old values of X_{kN} and choosing the values of e_{k0} randomly from the same distribution. Finally we let $e^2(\tau)$ be the average of the K values e^2_{kn}, where $\tau = n\Delta t$ is the prediction range, and let $\log E^2(\tau)$ be the average of the M values of $\log e^2(\tau)$, and plot $E(\tau)$ against the number of days (5τ), on a logarithmic scale. (The lower curve is the same except that the vertical scale is linear.)

For small n we see a nearly straight sloping line, representing uniform exponential growth, with a doubling time of 2.1 days, agreeing with λ_1, until saturation is approached. For large n we see a nearly straight horizontal line, representing saturation. It should not surprise us that the growth rate slackens before saturation is reached, rather than continuing unabated up to saturation and then ceasing abruptly.

The alternative procedure of simply letting $E^2(\tau)$ be the average value of $e^2(\tau)$, i.e. averaging the runs arithmetically instead of geometrically, would lead to a figure much like Figure 3.2(a), but with the sloping line in the upper curve indicating a doubling time of 1.7 days. Evidently the errors tend to grow more rapidly for a while in those runs where they have already acquired a large amplitude by virtue of their earlier more rapid growth, and it is these runs that make the major contribution to the arithmetic average. One could perhaps make equally good cases for studying geometric or arithmetic means, but only the former fits the definition of λ_1.

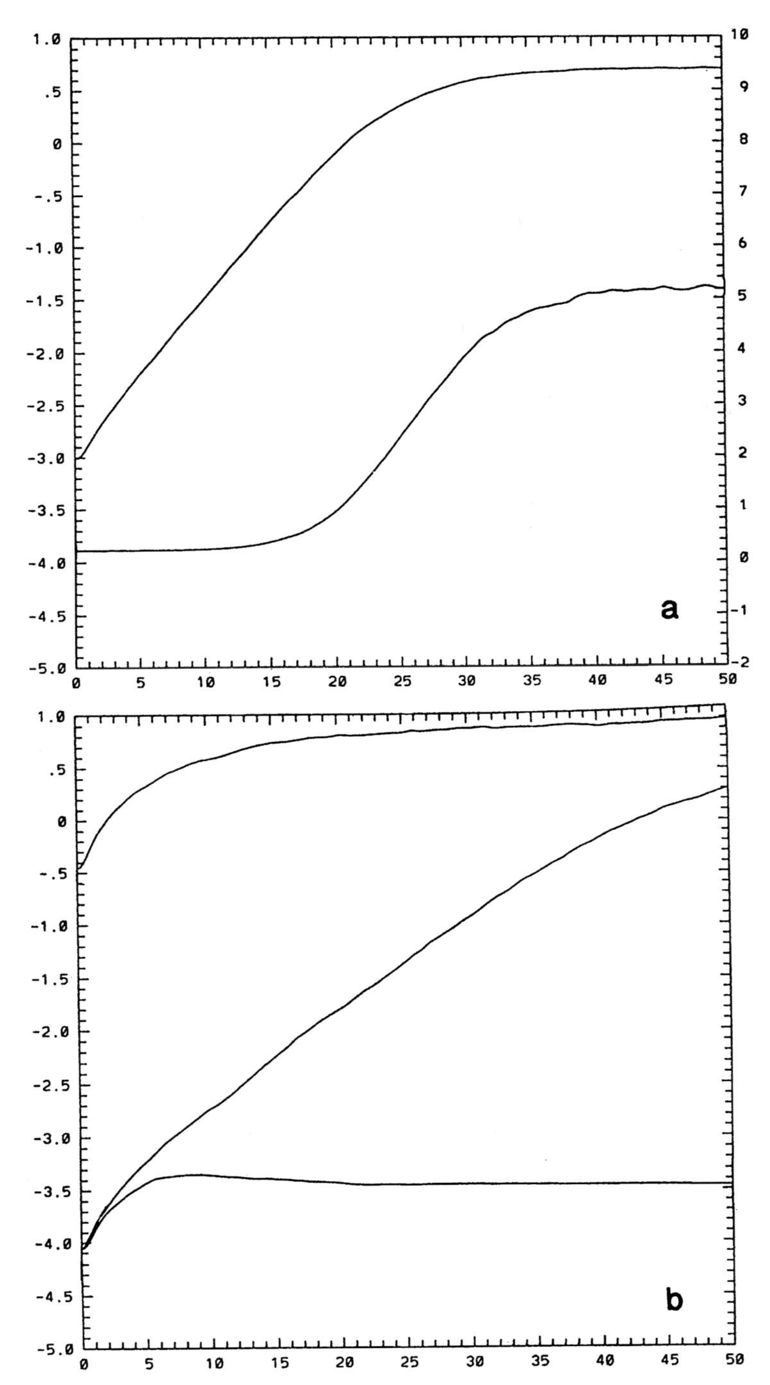

Figure 3.2 (a) Variations of average prediction error E (lower curve, scale at right) and $\log_{10} E$ (upper curve, scale at left) with prediction range τ (scale, in days, at bottom), for 50 days, as determined by 250 pairs of numerical integrations of Eq. (3.1), with $K = 36$ and $F = 8.0$ (as in Fig. 3.1). (b) The same as (a), but for variations of $\log_{10} E$ only, and as determined by 1000 pairs of integrations of Eq. (3.1), with $K = 4$, and with $F = 18.0$ (upper and middle curves, with different initial errors), and $F = 15.0$ (lower curve).

3.3 Atmospheric estimates

Some three decades ago a historic meeting, organised by the World Meteorological Organization, took place in Boulder, Colorado. The principal topic was long-range weather forecasting. At that time numerical modelling of the complete global circulation was just leaving its infancy; the three existing state-of-the-art models were those of Leith (1965), Mintz (1965), where A. Arakawa also played an essential role, and Smagorinsky (1965).

At such meetings the greatest accomplishments often occur between sessions. In this instance Jule Charney, who headed a committee to investigate the feasibility of a global observation and analysis experiment, persuaded the creators of the three models, all of whom were present, to use their models for predictability experiments, which would involve computations somewhat like those that produced Figure 3.2(a). On the basis of these experiments, Charney's committee subsequently concluded that a reasonable estimate for the atmosphere's doubling time was five days (Charney *et al.*, 1966). Taken at face value, this estimate offered considerable hope for useful two-week forecasts but very little for one-month forecasts.

The Mintz–Arakawa model that had yielded the five-day doubling time was a two-layer model. Mintz's graphs showed nearly uniform amplification before saturation was approached; presumably they revealed the model's leading Lyapunov exponent, although not, as we shall see, the leading exponent for the real atmosphere. As time passed by and more sophisticated models were developed, estimates of the doubling time appeared to drop. Smagorinsky's nine-level primitive-equation model, for example, reduced the time to three days (Smagorinsky, 1969).

Experiments more than a decade later with the then recently established operational model of ECMWF, based upon operational analyses and forecasts, suggested a doubling time between 2.1 and 2.4 days for errors in the 500-millibar height field (Lorenz, 1982). In the following years the model was continually modified, in an effort to improve its performance, and the newly accumulated data presently pushed the estimate below two days. There were small but significant variations of predictability with the season and the hemisphere, and quantities such as divergence appeared to be considerably less predictable than 500-m height.

One of the most recent studies (Simmons *et al.*, 1995), again performed with the ECMWF model, has reduced the estimate to 1.5 days. It is worth asking why the times should continually drop. Possibly the poorer physics of the earlier models overestimated the predictability, but it seems likely that a major factor has been spatial resolution. The old Mintz–Arakawa model used about 1000 numbers to represent the field of one variable at one level; the present ECMWF model uses about 45 000. Errors in features that formerly were not captured at all may well amplify more rapidly than those in the grossest features.

As with the Mintz–Arakawa model, the doubling times of the recent models appear consistent with the values of λ_1 for these models. Obviously not all of them can indicate the proper value of the exponent for the real atmosphere, and presumably none of them does.

Our reason for identifying the time unit in the model defined by Eq. (3.1) with five days of atmospheric time is now apparent. With $K = 36$ and $F = 8.0$ or 10.0, and indeed with any reasonably large value of K and these values of F, the doubling time for the model is made comparable to the times for the up-to-date global circulation models.

3.4 The early stages of error growth

Despite the agreement between the error growth in the simple model, and even in some global circulation models, with simple first estimates, reliance on the leading Lyapunov exponent, in most realistic situations, proves to be a considerable over-simplification. By and large this is so because λ_1 is defined as the long-term average growth rate of a very small error. Often we are not primarily concerned with averages, and, even when we are, we may be more interested in shorter-term behaviour. Also, in practical situations the initial error is often not small.

Sometimes, for example, we are interested in how well we can predict on specific occasions, or in specific types of situation, rather than in some general average skill. For any particular initial state, the initial growth rate of a superposed error will be highly dependent on the form of the error – on whether, for example, it assumes its greatest amplitude in synoptically active or inactive regions. In fact, there will be one error pattern – in phase space, it is an error vector – that will *initially* grow more rapidly than any other. The form and growth rate of this vector will of course depend upon the state on which the error is superposed.

Likewise, the *average* initial or early growth rate of *randomly* chosen errors superposed on a particular initial state will depend upon that state. Indeed, the identification of situations in which the atmosphere is especially predictable or unpredictable – the prediction of predictability – and even the identifiability of such situations – the predictability of predictability – have become recognised as suitable subjects for detailed study (see Kalnay and Dalcher, 1987; Palmer, 1988).

Assuming, however, that we are interested in averages over a wide variety of initial states, the value of λ_1 may still not tell us what we want to know, particularly in the earliest or latest stages of growth. In fact, in some systems the average *initial* growth rate of randomly chosen errors systematically exceeds the Lyapunov rate (see, for example, Farrell, 1989).

This situation is aptly illustrated by the middle curve in Figure 3.2(b), which has also been produced from Eq. (3.1), in the same manner as Figure 3.2(a), but with

K reduced to 4 and F increased to 18.0, and with $\varepsilon = 0.0001$. Also, because so few variables are averaged together, we have increased M to 1000. Between about 6 and 30 days the curve has a reasonably uniform slope, which agrees with λ_1, and indicates a doubling time of 3.3 days, but during the first 3.3 days the average error doubles twice. Systems exhibiting anomalously rapid initial error growth are in fact not uncommon. Certainly there are practical situations where we are mainly interested in what happens during the first few days, and here λ_1 is not always too relevant.

This phenomenon, incidentally, is in this case not related to the chaotic behaviour of the model. The lower curve in Figure 3.2(b) is like the middle one, except that F has been reduced to 15.0, producing a system that is not chaotic at all. Again the error doubles twice during the first six days, but then it levels off at a value far below saturation. If ε had been smaller, the entire curve would have been displaced downward by a constant amount.

When the initial error is not particularly small, as is often the case in operational weather forecasting, λ_1 may play a still smaller role. The situation is illustrated by the upper curve in Figure 3.2(b), which has been constructed exactly as the middle curve, except that $\varepsilon = 0.4$, or 5% of saturation, instead of 0.001. The rapid initial error growth is still present, but, when after four days it ceases, saturation is already being approached. Only a brief segment between 4 and 8 days is suggestive of 3.3-day doubling.

The relevance of the Lyapunov exponent is even less certain in systems, such as more realistic atmospheric models or the atmosphere itself, where different features possess different characteristic time scales. In fact, it is not at all obvious what the leading exponent for the atmosphere may be, or what the corresponding vector may look like. To gain some insight, imagine a relatively realistic model that resolves larger scales – planetary and synoptic scales – and smaller scales – mesoscale motions and convective clouds; forget about the fact that experiments with a global model with so many variables would be utterly impractical with today's computational facilities. Convective systems can easily double their intensity in less than an hour, and we might suppose that an initial error field consisting only of the omission of one incipient convective cloud in a convectively active region, or improperly including such a cloud, would amplify equally rapidly, and might well constitute the error pattern with the greatest *initial* growth rate.

Yet this growth rate need not be long-term, because the local instability responsible for the convective activity may soon subside, whereupon the error will cease to grow, while new instability may develop in some other location. A pattern with convective-scale errors distributed over many regions, then, would likely grow more steadily even if at first less rapidly, and might more closely approximate the leading Lyapunov vector.

Since this reasoning is highly speculative, I have attempted to place it on a slightly firmer basis by introducing another crude model which, however, varies with two distinct time scales. The model has been constructed by coupling two systems, each of which, aside from the coupling, obeys a suitably scaled variant of Eq. (3.1). There are K variables X_k plus JK variables $Y_{j,k}$, defined for $k = 1, \ldots, K$ and $j = 1, \ldots, J$, and the governing equations are

$$dX_k/dt = -X_{k-1}(X_{k-2} - X_{k+1}) - X_k - (hc/b)\sum_{j=1}^{J} Y_{j,k}, \tag{3.2}$$

$$dY_{j,k}/dt = -cbY_{j+1,k}(Y_{j+2,k} - Y_{j-1,k}) - cY_{j,k} + (hc/b)X_k. \tag{3.3}$$

The definitions of the variables are extended to all values of k and j by letting X_{k-K} and X_{k+K} equal X_k, as in the simpler model, and letting $Y_{j,k-K}$ and $Y_{j,k+K}$ equal $Y_{j,k}$, while $Y_{j-J,k} = Y_{j,k-1}$ and $Y_{j+J,k} = Y_{j,k+1}$. Thus, as before, the variables X_k can represent the values of some quantity in K sectors of a latitude circle, while the variables $Y_{j,k}$, arranged in the order $Y_{1,1}, Y_{2,1}, \ldots, Y_{J,1}, Y_{1,2}, Y_{2,2}, \ldots, Y_{J,2}, Y_{3,1}, \ldots$, can represent the values of some other quantity in JK sectors. A large value of J implies that many of the latter sectors are contained in one of the former, and we may think of the variables $Y_{j,k}$ as representing a convective-scale quantity, while, in view of the form of the coupling terms, the variables X_k should represent something that favours convective activity, possibly the degree of static instability.

In our computations we have let $K = 36$ and $J = 10$, so that there are ten small sectors, each one degree of longitude in length, in one large sector, while $c = 10.0$ and $b = 10.0$, implying that the convective scales tend to fluctuate ten times as rapidly as the larger scales, while their typical amplitude is $1/10$ as large. We have let h, the coupling coefficient, equal 1.0, and we have advanced the computations in time steps of 0.005 units, or 36 minutes. Our chosen value $F = 10.0$ is sufficient to make both scales vary chaotically; note that coupling replaces direct forcing as a driver for the convective scales.

Figure 3.3 reveals some of the typical behaviour of the model, by showing the distribution of X_k and $Y_{j,k}$ about a latitude circle, at times separated by 2 days. There are seven active areas (X_k large), generally 30 or 40 degrees wide, that fluctuate in width and intensity as they slowly propagate westward, while the convective activity, which is patently strongest in the active areas, tends to propagate eastward (note the signs in the subscripts in the non-linear terms in Eq. 3.3), but rapidly dies out as it leaves an active area.

Figure 3.4 presents separate error-growth curves for the large and small scales. For computational economy we have averaged 25 runs rather than 250. The small-scale errors begin to amplify immediately, doubling every 6 hours or so and approaching saturation by the third day. This growth rate is compatible with the computed value of λ_1 for the model. Meanwhile, the large-scale errors begin to grow at a similar rate once the small-scale errors exceed them by an order of magnitude, the growth evidently resulting from the coupling rather than the dynamics internal to the large scales. After the small-scale errors are no longer growing, the large-scale errors

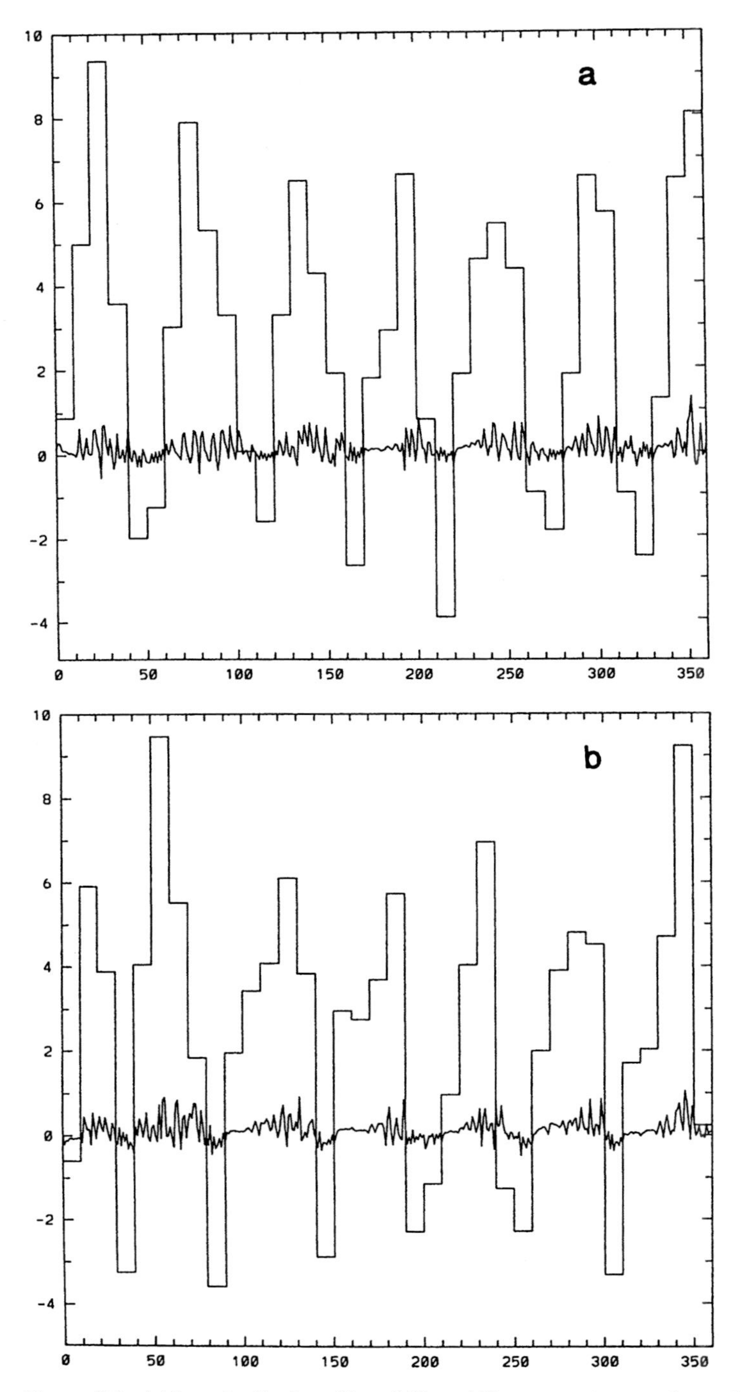

Figure 3.3 (a) Longitudinal profiles of X_k and $Y_{j,k}$ at one time, as determined by numerical integration of Eqs. (3.2) and (3.3), with $K = 36$, $J = 10$, $F = 10.0$, $c = 10.0$, $b = 10.0$, and $h = 1.0$. Scale for longitude, in degrees east, is at bottom. Common scale for X_k and $Y_{j,k}$ is at left. (b) The same as (a), but for a time two days later.

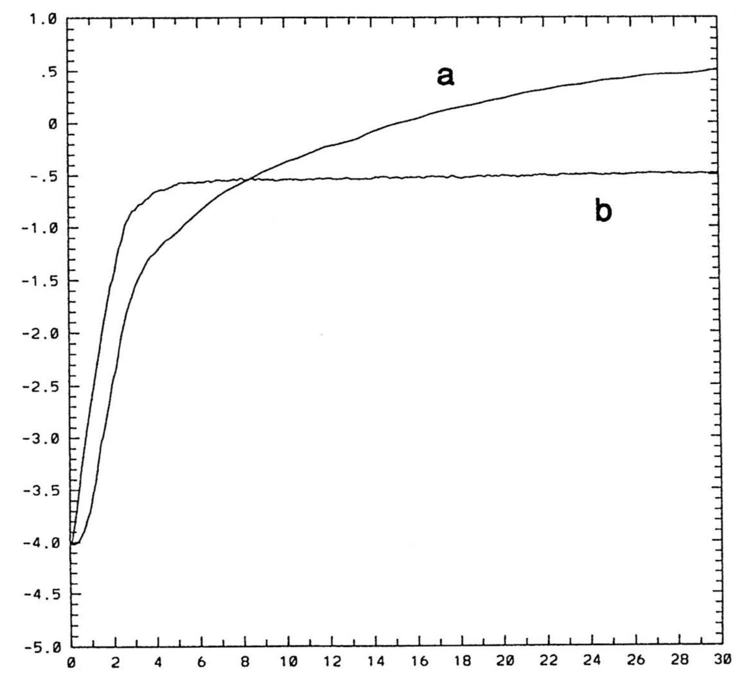

Figure 3.4 Variations of $\log_{10} E$ (scale at left) with prediction range τ (scale, in days, at bottom), shown separately for large scales (variables X_k, curve a) and small scales (variables $Y_{j,k}$, curve b), for 30 days, as determined by 25 pairs of integrations of Eqs. (3.2) and (3.3), with the parameter values of Figure 3.3.

continue to grow, at a slower quasi-exponential rate comparable to what appears in Figure 3.2(a), doubling in about 1.6 days. Finally they approach their own saturation level, an order of magnitude higher than that of the small-scale errors. Thus, after the first few days, the large-scale errors behave about as they would if the forcing were slightly weaker, and if the small scales were absent altogether.

In a more realistic model with many time scales or perhaps a continuum, we would expect to see the growth rate of the largest-scale errors subsiding continually, as, one after another, the smaller scales reached saturation. Thus we would not expect a large-scale-error curve constructed in the manner of Figure 3.4 to contain an approximate straight-line segment of any appreciable length.

We now see the probable atmospheric significance of the error doubling times of the various global circulation models. Each doubling time appears to represent the rate at which, in the *real* atmosphere, errors in predicting the features that are resolvable by the particular model will amplify, after the errors in unresolvable features have reached saturation. Of course, before accepting this interpretation, we must recognise the possibility that some of the small-scale features will not saturate rapidly; possibly they will act in the manner of coherent structures.

3.5 The late stages

As we have seen, prediction errors in chaotic systems tend to amplify less rapidly, on the average, as they become larger. Indeed, the slackening may become apparent long before the errors are close to saturation, and thus at a range when the predictions are still fairly good. For Eq. (3.1), and in fact for the average behaviour of some global atmospheric circulation models, we can construct a crude formula by assuming that the growth rate is proportional to the amount by which the error falls short of saturation. We obtain the equation

$$(1/E)dE/d\tau = \lambda_1(E^* - E)/E^*, \tag{3.4}$$

where E^* denotes the saturation value for E. Equation (3.4) possesses the solution

$$E = E^*\,(1 + \tanh(\lambda_1\tau))/2, \tag{3.5}$$

if the origin of τ is the range at which $E = E^*/2$. The well-known symmetry of the hyperbolic-tangent curve, when it is drawn with a linear vertical scale, then implies that the rate at which the error approaches saturation, as time advances, equals the rate at which it would approach zero, if time could be reversed. This relationship is evidently well approximated in the lower curve of Figure 3.2(a), and it has even been exploited to estimate growth rates for small errors, when the available data have covered only larger errors (see Lorenz, 1969b, 1982). It is uncertain whether the formula is more appropriate when E is the root-mean-square error or simply the mean-square error.

For many systems, however, Eq. (3.4) and hence Eq. (3.5) cannot be justified in the later stages. This may happen when, as in the case where the *early* growth fails to follow Eq. (3.4), the system possesses contrasting time scales. Here, however, the breakdown can occur because some significant feature varies more *slowly* than the features of principal interest – the ones that contribute most strongly to the chosen measure of total error.

Perhaps the feature most often cited as falling into this category is the sea surface temperature (SST), which, because of the ocean's high heat capacity, sometimes varies rather sluggishly. Along with the atmospheric features most strongly under its influence, the SST may therefore be expected to be somewhat predictable at a range when migratory synoptic systems are not. A slow final approach to saturation may thus be anticipated, particularly if the 'total error' includes errors in predicting the SST itself.

A perennial feature in which the SST plays a vital role is the El Niño–Southern Oscillation (ENSO) phenomenon. Phases of the ENSO cycle persist long enough for predictions of the associated conditions a few months ahead to be much better than guesswork, while some models of ENSO (e.g. Zebiak and Cane, 1987) suggest that the onsets of coming phases may also possess some predictability. Again, the phenomenon should lead to an ebbing of the late-stage growth rate.

Perhaps less important but almost certainly more predictable than the ENSO-related features are the winds in the equatorial middle-level stratosphere, dominated by the quasi-biennial oscillation (QBO). Even though one cannot be certain just when the easterlies will change to westerlies, or vice versa, nor how the easterlies or westerlies will vary from day to day within a phase, one can make a forecast with a fairly low expected mean-square error, for a particular day, a year or even several years in advance, simply by subjectively extrapolating the cycle, and predicting the average conditions for the anticipated phase. Any measure of the total error that gives appreciable weighting to these winds is forced to approach saturation very slowly in the latest stages.

Looking at still longer ranges, we come to the question, 'Is climate predictable?' Whether or not it is possible to predict climate changes, aside from those that result from periodic or otherwise predictable external activity, may depend on what is considered to be a climate change.

Consider again, for example, the ENSO phenomenon. To some climatologists, the climate changes when El Niño sets in. It changes again, possibly to what it had previously been, when El Niño subsides. We have already suggested that climatic changes, so defined, possess some predictability.

To others, the climate is not something that changes whenever El Niño arrives or leaves. Instead, it is something that often remains unchanged for decades or longer, and is characterised by the appearance and disappearance of El Niño at rather irregular intervals, but generally every two to seven years. A change of climate would be indicated if El Niño should start to appear almost every year, or only once in twenty years or not at all. Whether unforced changes of climate from one half-century or century to another, or one millennium to another, are at all predictable is much less certain.

Let us then consider the related question, 'Is climate a dynamical system?' That is, is there something that we can conscientiously call 'climate', determined by the state of the atmosphere and its surroundings, and undergoing significant changes over intervals of centuries but usually remaining almost unchanged through a single ENSO cycle or a shorter-period oscillation, whose future states are determined by its present and past according to some exact or approximate rule? To put the matter in perspective, let us first re-examine the justification for regarding the ever-changing synoptic pattern, and possibly the ENSO phenomenon, as dynamical systems.

Experience with numerical weather prediction has shown that we can forecast the behaviour of synoptic systems fairly well, far enough in advance for an individual storm to move away and be replaced by the next storm, without observing the superposed smaller-scale features at all, simply by including their influence in parametrised form. If instead of parametrising these features we omit them altogether, the models will still produce synoptic systems that behave rather reasonably, even though the actual forecasts will suffer from the omission. Evidently this is because the features that are small in scale are relatively small in amplitude, so that their influence acts much like small random forcing.

Moving to longer time scales, we find that some models yield rather good simulations of the behaviour of the ENSO phenomenon, even if not good forecasts of individual occurrences, without including the accompanying synoptic systems in any more than parametrised form. Here the synoptic systems do not qualify as being small in amplitude, but they appear to be rather weakly coupled to ENSO, so that again they may act like small random forcing.

Similarly, climatic fluctuations with periods of several decades or longer have more rapid oscillations superposed on them, ranging in timescale all the way from ENSO and the QBO to synoptic and small-scale features. Certainly these fluctuations are not small. Is their effect on the climate, if large, determined for the most part *by* the climate itself, so that climate can constitute a dynamical system? If this is not the case, are these features nevertheless coupled so weakly to the climate that they act like small random forcing, so that climate still constitutes a dynamical system? Or do they act more like strong random forcing, so that climate does not qualify as a dynamical system, and prospects for its prediction are not promising? At present the reply to these questions seems to be that we do not know.

3.6 Concluding remarks

In this overview I have identified the rate at which small errors will amplify as the key quantity in determining the predictability of a system. By an error we sometimes mean the difference between what is predicted and what actually occurs, but ordinarily we extend the concept to mean the difference, at any designated time, between two evolving states. We assume that there would be no prediction error if we could observe an initial state without error, and if we could formulate an extrapolation procedure without error, recognising that such formulation is not possible if the governing laws involve any randomness.

In my discussions and numerical illustrations I have found it convenient to consider the growth of errors that owe their existence to errors in the initial state, disregarding the additional influence of any inexactness in the extrapolation procedure. However, if the fault lies in the extrapolation and not in the initial state, the effect will be similar; after a reasonable time interval there will be noticeable errors in the *predicted* state, and these will proceed to grow about as they would have if they had been present initially. If the assumed and actual governing laws define systems with different leading Lyapunov exponents, the larger exponent will be the relevant one. Randomness in the governing laws will have the same effect as any other impediment to perfect extrapolation. In the case of the atmosphere, the inevitable small-scale features will work like randomness.

I have confined my quantitative discussions to results deduced from pairs or ensembles of numerical solutions of mathematical models with various degrees of sophistication, but alternative approaches have also been exploited. Some studies have been based on equations whose variables are ensemble averages of error magnitudes. These equations have been derived from conventional atmospheric models, but, to close the equations, i.e. to limit the number of variables to the number of equations, it has been necessary to introduce auxiliary assumptions of questionable validity (see, for example, Thompson, 1957; Lorenz, 1969a). Results agree reasonably well with those yielded by more conventional approaches.

There have also been empirical studies. Mediocre analogues – pairs of somewhat similar states – have been identified in northern-hemisphere weather data; their differences constitute moderate-sized errors, whose subsequent growth may be determined by noting how the states evolve (see Lorenz, 1969b). The growth rates of *small* errors may then be inferred from Eq. (3.4); again they are consistent with growth rates obtained from numerical integrations.

There are other aspects of the predictability problem that I have not touched upon at all, and I shall conclude by mentioning just one of these – the improvement in weather forecasting that may reasonably be expected in the foreseeable future. Recent experience, again with the ECMWF operational system, suggests that errors in present-day forecasting amplify more rapidly than they would if the continual error accumulation that results from imperfect extrapolation were not present, i.e. if all of the error growth resulted from amplification of already-present errors. There should therefore be room for improvement. Numerical estimates suggest that we may some day forecast a week in advance as well as we now forecast three days in advance, and two weeks ahead almost as well as we now forecast one week ahead.

References

Charney, J. G., R. G. Fleagle, V. E. Lally, H. Riehl and D. Q. Wark (1966). The feasibility of a global observation and analysis experiment. *Bull. Am. Meteorol. Soc.*, **47**, 200–20.

Farrell, B. F. (1989). Optimal excitation of baroclinic waves. *J. Atmos. Sci.*, **46**, 1193–206.

Kalnay, E. and A. Dalcher (1987). Forecasting forecast skill. *Mon. Weather Rev.*, **115**, 349–56.

Leith, C. E. (1965). Lagrangian advection in an atmospheric model. In *WMO-IUGG Symposium on Research and Development Aspects of Long-range Forecasting*, WMO Technical Note 66, 168–76.

Lorenz, E. N. (1969a). The predictability of a flow which possesses many scales of motion. *Tellus*, **21**, 289–307.

(1969b). Atmospheric predictability as revealed by naturally occurring analogues. *J. Atmos. Sci.*, **26**, 636–46.

(1982). Atmospheric predictability experiments with a large numerical model. *Tellus*, **34**, 505–13.

(1993). *The Essence of Chaos*. University of Washington Press.

Mintz, Y. (1965). Very long-term global integration of the primitive equations of atmospheric motion. In *WMO-IUGG Symposium on Research and Development Aspects of Long-range Forecasting*, WMO Technical Note 66, 141–67.

Palmer, T. N. (1988). Medium and extended range predictability and stability of the Pacific/North American mode. *Quart. J. Roy. Meteor. Soc.*, **114**, 691–713.

Simmons, A. J., R. Mureau and T. Petroliagis (1995). Error growth and estimates of predictability from the ECMWF forecasting system. *Quart. J. Roy. Meteor. Soc.*, **121**, 1739–1771.

Smagorinsky, J. (1965). Implications of dynamic modelling of the general circulation on long-range forecasting. In *WMO-IUGG Symposium on Research and Development Aspects of Long-range Forecasting*, WMO Technical Note 66, 131–7.

(1969). Problems and promises of deterministic extended range forecasting. *Bull. Am. Meteorol. Soc.*, **50**, 286–311.

Thompson, P. D. (1957). Uncertainty of initial state as a factor in the predictability of large-scale atmospheric flow patterns. *Tellus*, **9**, 275–95.

Toth, Z. and E. Kalnay (1993). Ensemble forecasting at NMC: the generation of perturbations. *Bull. Am. Meteorol. Soc.*, **74**, 2317–30.

Zebiak, S. E. and M. A. Cane (1987). A model El Niño-Southern Oscillation. *Mon. Weather Rev.*, **115**, 2262–78.

4

The Liouville equation and atmospheric predictability

Martin Ehrendorfer
Institut für Meteorologie und Geophysik,
Universität Innsbruck

4.1 Introduction and motivation

It is widely recognised that weather forecasts made with dynamical models of the atmosphere are inherently uncertain. Such uncertainty of forecasts produced with numerical weather prediction (NWP) models arises primarily from two sources: from imperfect knowledge of the initial model conditions and from imperfections in the model formulation itself. The recognition of the potential importance of accurate initial model conditions and an accurate model formulation dates back to times even prior to operational NWP (Bjerknes, 1904; Thompson, 1957). In the context of NWP, the importance of these error sources in degrading the quality of forecasts was demonstrated to arise because errors introduced in atmospheric models are, in general, growing (Lorenz, 1982, 1963, 1993, this volume), which at the same time implies that the predictability of the atmosphere is subject to limitations (Errico *et al.*, 2002). An example of the amplification of small errors in the initial conditions, or, equivalently, the divergence of initially nearby trajectories is given in Figure 4.1, for the system discussed by Lorenz (1984). The uncertainty introduced into forecasts through uncertain initial model conditions, and uncertainties in model formulations, has been the subject of numerous studies carried out in parallel with the continuous development of NWP models (e.g. Leith, 1974; Epstein, 1969; Palmer, 2000, this volume, Buizza, this volume).

Predictability of Weather and Climate, ed. Tim Palmer and Renate Hagedorn. Published by Cambridge University Press.

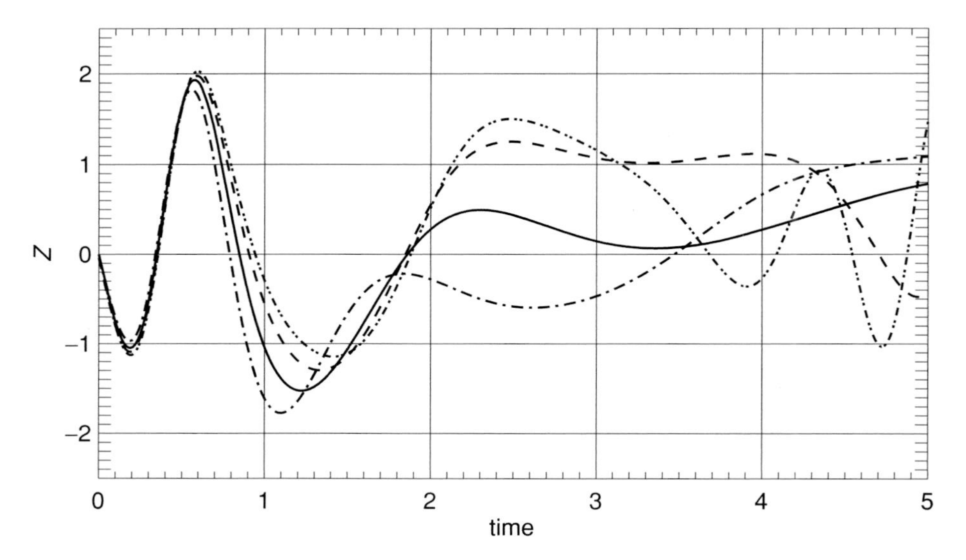

Figure 4.1 Time evolution of Z component of Lorenz (1984) model. Initial perturbation size is 0.04. The dashed curve indicates negative perturbation, dash–dot plus SV1 perturbation, and dash–dot–dot minus SV1 perturbation (see also Sections 4.4.2 and 4.5.2).

In addition to studying the intrinsic predictability of the atmosphere (e.g. Lorenz, 1969a, 1969b; Thompson, 1985a, 1985b), efforts have been directed at the quantification or prediction of forecast uncertainty that arises due to the sources of uncertainty mentioned above (see the review papers by Ehrendorfer, 1997; Palmer, 2000; and Ehrendorfer, 1999). In the context of predicting uncertainty, the Liouville equation (LE) arises as the general framework to describe in a probabilistic manner the time-dependent behaviour of an ensemble of solutions of a numerical model started from different initial conditions. It carries the name of Joseph Liouville (for an extensive biographical account, see Lützen, 1990), since its formulation can be traced back to a mathematical result that Liouville published in 1838 (see Section 4.2). The generality of the LE arises from the fact that the LE governs the time evolution of the probability density function (pdf) of the model's state vector $\mathbf{X}$ given the corresponding model dynamics. Closely analogous to the mass conservation equation in hydrodynamics, the LE is the mathematical formulation of the statement that realisations (i.e. members of the ensemble) cannot spontaneously appear or disappear; which, in turn, implies that the phase-space integral of the density of realisations is constant in time. These properties of the LE imply that it contains all (statistical) information about time-evolving model solutions, and as such is the basis for studying uncertainty prediction and state-dependent predictability of weather and climate phenomena.

The purpose of this chapter is to review the LE and its relation to atmospheric predictability. Following an overview of the LE, together with its various connections to mathematical physics (Section 4.2), the general solution to the LE is described

(Section 4.3). The relationship of various aspects of the LE to operational ensemble prediction, to the stochastic-dynamic equations, and to classical and statistical mechanics, is discussed in Section 4.4. The chapter concludes with a summarising outlook.

4.2 The Liouville equation: background

4.2.1 Historical background

The LE is apparently named after Joseph Liouville since it can be related to a result that he proved in 1838 on the material derivative of the 'Jacobian' of the transformation exerted by the solution of an ordinary differential equation on its initial condition (see Section 4.2.2). Subsequently, this result can be used in deriving the so-called transport theorem. Further, the LE, the Fokker–Planck equation and the Liouville theorem (LT) of statistical mechanics can all be viewed as special cases of the transport theorem. Comprehensive accounts on the historical details may be found in Lützen (1990), who mentions that Liouville's *Nachlass* contains 340 notebooks, consisting of more than 40 000 pages.

In view of the fundamental nature of Liouville's results, references to his name are found in various areas of physics, and in the atmospheric sciences. Some selected areas are briefly discussed here.

In *classical mechanics,* Liouville is referred to with regard to the LT (see Section 4.5.6) that expresses the conservation of phase space *volume* for Hamiltonian systems (see, for example, Abraham and Marsden, 1978; Goldstein, 1980; Marion and Thornton, 1988; Arnold, 1989; see also Section 4.5.6). The importance of the LT for *statistical mechanics* is emphasised, for example, by Landau and Lifschitz (1979) who state that *the pdf remains constant along trajectories* (for a Hamiltonian system; see also Eq. 4.108). Balescu (1991) points out that the LE is the most important equation of statistical mechanics and emphasises that the most important feature of the LE is its linearity. Lindenberg and West (1990) mention the LE as describing the conservation of phase-space points (their p. 21). Penrose (1989) refers to the LT as 'a very beautiful theorem, due to the distinguished French mathematician Joseph Liouville (1809–1882), that tells us that the *volume* of any region of the phase space must remain constant under any Hamiltonian evolution' (his p. 181; see also Section 4.5.6; see also Penrose, 1970). In the literature on *stochastic processes*, the LE plays an important role in its connection with the Fokker–Planck equation (Risken, 1989; Gardiner, 1990; van Kampen, 1992; see also Section 4.5.4). In that context, these equations are also referred to as the Chapman–Kolmogorov equation, or the Master equation (Gardiner, 1990). Also, as a further remark on terminology, Nicolis (1995) refers to a time-discrete version of a pdf-governing equation as the Frobenius–Perron equation.

In the meteorological and in the turbulence literature, the LE has been used as a starting point for various investigations, among them the problem of deriving equilibrium statistics of geophysical flows (see Holloway, 1986, for an excellent overview), described, for example, in Thompson (1972), Salmon *et al.* (1976), Thompson (1983), 1985b, and Merryfield *et al.* (2001) (see also Hasselmann, 1976; Stanišić, 1988; McComb, 1991; Penland and Matrosova, 1994; Hu and Pierrehumbert, 2001; Chu *et al.*, 2002; Weiss, 2003a). More specifically, with regard to the time evolution of uncertainty in atmospheric models, the LE was considered as a fundamental tool for approaching that question by Epstein (1969), Gleeson (1966), Tatarskiy (1969), and Fortak (1973).

Given the above brief (and necessarily incomplete) overview of some of the implications of Liouville's work, a more detailed inspection of his result is discussed in the next section.

4.2.2 Theoretical background

The result of J. Liouville in 1838

In the paper that is considered to contain the basic result for the LE and the LT, Liouville (1838) first discussed the general result for an *N*th order ordinary differential equation and then considered the particular case $N = 3$ by illustrating in terms of a third-order ordinary differential equation of the following form:

$$x''' = P(t, x, x', x''), \tag{4.1}$$

subsequently assuming the solution to be of the form

$$x = x(t, a, b, c). \tag{4.2}$$

He defined u, the determinant of the Jacobian of the mapping that the solution of the differential equation exerts on the initial condition, as

$$u = u(a, b, c, t) = \det \begin{pmatrix} \frac{\partial x(t,a,b,c)}{\partial a} & \frac{\partial x(t,a,b,c)}{\partial b} & \frac{\partial x(t,a,b,c)}{\partial c} \\ \frac{\partial x'(t,a,b,c)}{\partial a} & \frac{\partial x'(t,a,b,c)}{\partial b} & \frac{\partial x'(t,a,b,c)}{\partial c} \\ \frac{\partial x''(t,a,b,c)}{\partial a} & \frac{\partial x''(t,a,b,c)}{\partial b} & \frac{\partial x''(t,a,b,c)}{\partial c} \end{pmatrix}. \tag{4.3}$$

He then showed that it is true that

$$\frac{\partial u}{\partial t} = u \frac{\partial P}{\partial x''} \tag{4.4}$$

when the chain rule of differentiation is observed to rewrite, for example, the following partial derivative in the form

$$\frac{\partial x'''}{\partial a} = \frac{\partial P}{\partial a} = \frac{\partial P}{\partial x}\frac{\partial x}{\partial a} + \frac{\partial P}{\partial x'}\frac{\partial x'}{\partial a} + \frac{\partial P}{\partial x''}\frac{\partial x''}{\partial a}. \tag{4.5}$$

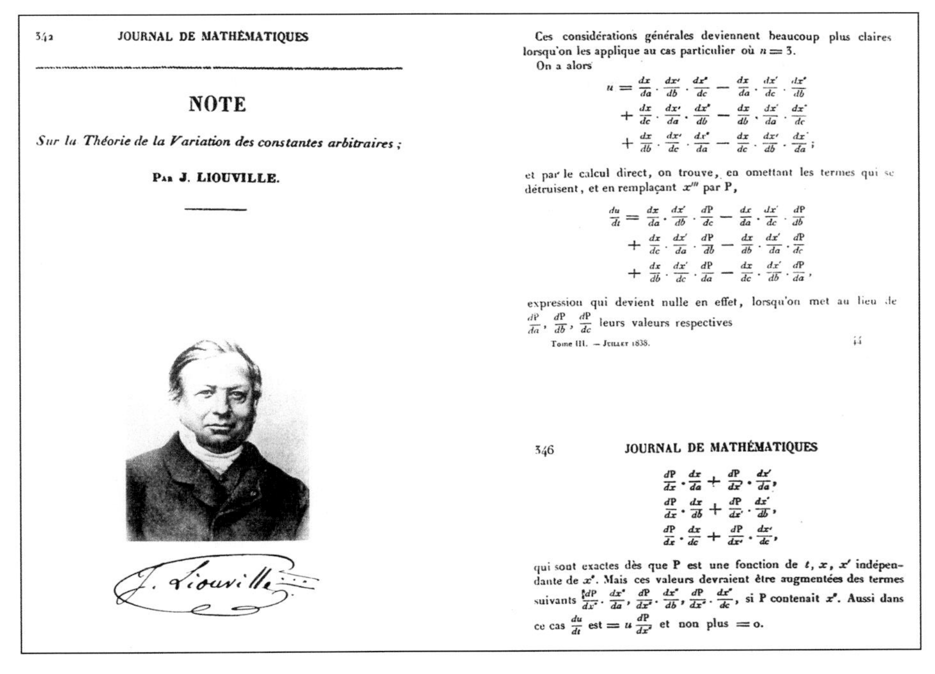

342 JOURNAL DE MATHÉMATIQUES

NOTE

Sur la Théorie de la Variation des constantes arbitraires ;

PAR J. LIOUVILLE.

Ces considérations générales deviennent beaucoup plus claires lorsqu'on les applique au cas particulier où $n = 3$.
On a alors

$$u = \frac{dx}{da}\cdot\frac{dx'}{db}\cdot\frac{dx''}{dc} - \frac{dx}{da}\cdot\frac{dx'}{dc}\cdot\frac{dx''}{db} + \frac{dx}{dc}\cdot\frac{dx'}{da}\cdot\frac{dx''}{db} - \frac{dx}{db}\cdot\frac{dx'}{da}\cdot\frac{dx''}{dc} + \frac{dx}{db}\cdot\frac{dx'}{dc}\cdot\frac{dx''}{da} - \frac{dx}{dc}\cdot\frac{dx'}{db}\cdot\frac{dx''}{da};$$

et par le calcul direct, on trouve, en omettant les termes qui se détruisent, et en remplaçant x''' par P,

$$\frac{du}{dt} = \frac{dx}{da}\cdot\frac{dx'}{db}\cdot\frac{dP}{dc} - \frac{dx}{da}\cdot\frac{dx'}{dc}\cdot\frac{dP}{db} + \frac{dx}{dc}\cdot\frac{dx'}{da}\cdot\frac{dP}{db} - \frac{dx}{db}\cdot\frac{dx'}{da}\cdot\frac{dP}{dc} + \frac{dx}{db}\cdot\frac{dx'}{dc}\cdot\frac{dP}{da} - \frac{dx}{dc}\cdot\frac{dx'}{db}\cdot\frac{dP}{da},$$

expression qui devient nulle en effet, lorsqu'on met au lieu de $\frac{dP}{da}, \frac{dP}{db}, \frac{dP}{dc}$ leurs valeurs respectives

Tome III. — JUILLET 1838. 44

346 JOURNAL DE MATHÉMATIQUES

$$\frac{dP}{dx}\cdot\frac{dx}{da} + \frac{dP}{dx'}\cdot\frac{dx'}{da},$$
$$\frac{dP}{dx}\cdot\frac{dx}{db} + \frac{dP}{dx'}\cdot\frac{dx'}{db},$$
$$\frac{dP}{dx}\cdot\frac{dx}{dc} + \frac{dP}{dx'}\cdot\frac{dx'}{dc},$$

qui sont exactes dès que P est une fonction de t, x, x' indépendante de x''. Mais ces valeurs devraient être augmentées des termes suivants $\frac{dP}{dx''}\cdot\frac{dx''}{da}$, $\frac{dP}{dx''}\cdot\frac{dx''}{db}$, $\frac{dP}{dx''}\cdot\frac{dx''}{dc}$, si P contenait x''. Aussi dans ce cas $\frac{du}{dt}$ est $= u\frac{dP}{dx''}$ et non plus $= 0$.

Figure 4.2 Reproduction of (parts of) original proof by J. Liouville. Photograph from Lützen (1990).

As already mentioned, the result was generalised to Nth-order equations and to systems of first-order equations. A part of the original paper from 1838 is reproduced in Figure 4.2.

Generalisation of Liouville's result

The generalisation of Liouville's result is known as the *material derivative of the Jacobian*, or as the *Euler expansion formula* (see, for example, chapter 13.4 in Lin and Segel, 1988). To illustrate that extension, an N-dimensional system of first-order ordinary differential equations:

$$\dot{\mathbf{X}} = \mathbf{\Phi}(\mathbf{X}, t) \tag{4.6}$$

is considered with the solution $\mathbf{X}$ being a function of both the initial condition $\mathbf{\Xi}$ and time t:

$$\mathbf{X} = \mathbf{X}(\mathbf{\Xi}, t). \tag{4.7}$$

Under the assumption of continuous dependence on the initial condition (e.g. Nicolis, 1995, p. 51), the Jacobian J may be defined as the determinant of the transformation that the solution exerts on the initial condition:

$$J = J(\Xi, t) = \det\left(\left. \frac{\partial \mathbf{X}(\Xi, t)}{\partial \Xi} \right|_{\Xi} \right). \tag{4.8}$$

At the same time it is implied that Ξ can be recovered from knowledge of $\mathbf{X}$ and t, through 'inverting' Eq. (4.7):

$$\Xi = \Xi(\mathbf{X}, t). \tag{4.9}$$

In that situation, two equivalent formulations of the generalisation of Liouville's result are

$$\frac{\partial J(\Xi, t)}{\partial t} = \left(\sum_{i=1}^{N} \left. \frac{\partial \Phi_i(\mathbf{X}, t)}{\partial X_i} \right|_{\mathbf{X}(\Xi, t)} \right) J(\Xi, t) \tag{4.10}$$

and

$$\left(\frac{\partial}{\partial t} + \sum_{i=1}^{N} \dot{X}_i \frac{\partial}{\partial X_i} \right) J(\mathbf{X}, t) = \left(\sum_{i=1}^{N} \frac{\partial \Phi_i(\mathbf{X}, t)}{\partial X_i} \right) J(\mathbf{X}, t). \tag{4.11}$$

Here, in the first formulation ('material' formulation), Eq. (4.10), J is considered as a function of Ξ and t, and the partial derivative with respect to time t is taken for fixed Ξ, and the divergence expression (in parentheses) must be written as a function of Ξ, too. In the second formulation ('spatial' formulation), Eq. (4.11), however, J is considered as a function of $\mathbf{X}$ and t, and an advective term appears. An illustration of these concepts appears in the next subsection.

A one-dimensional example

To illustrate the result on the material derivative of the Jacobian in the two versions (4.10) and (4.11), consider the following Riccati equation (see, e.g., Zwillinger, 1989):

$$\dot{X} = -X^2 \tag{4.12}$$

with the solution written in the form (4.7) as

$$X(\Xi, t) = \frac{\Xi}{1 + t\Xi} \tag{4.13}$$

and the 'inverse' mapping of the form (4.9) written as

$$\Xi(X, t) = \frac{X}{1 - tX}. \tag{4.14}$$

From (4.13), J is obtained as a function of Ξ and t through the defining equation (4.8) as

$$J(\Xi, t) = \frac{\partial X}{\partial \Xi} = \frac{1}{(1 + t\Xi)^2}. \tag{4.15}$$

Further, $J(X,t)$ may be found by expressing Ξ in (4.15) through its dependence on X as given in (4.14) in the form:

$$J(X,t) = \frac{1}{[1+t\Xi(X,t)]^2} = \frac{1}{\left(1+t\frac{X}{1-tX}\right)^2} = (1-tX)^2. \tag{4.16}$$

To see the correctness of the first formulation (4.10), the derivative of J using (4.15):

$$\frac{\partial J(\Xi,t)}{\partial t} = (-2)\underbrace{(1+t\Xi)^{-2}}_{=J(\Xi,t)}(1+t\Xi)^{-1}\Xi \tag{4.17}$$

is compared with the right-hand side of (4.10) written in the form:

$$\left.\frac{\partial \Phi(X,t)}{\partial X}\right|_{X(\Xi,t)} J(\Xi,t) = (-2)X(\Xi,t)J(\Xi,t). \tag{4.18}$$

Clearly, on the basis of (4.13), results (4.17) and (4.18) are the same, establishing Liouville's result in the formulation (4.10) for the example under consideration here. Similarly, for the second formulation, the left-hand side of (4.11) is, for the example under consideration here, found to be

$$\left(\frac{\partial}{\partial t} + \dot{X}\frac{\partial}{\partial X}\right) J(X,t) = (-2X)(1-tX) + (-X^2)(-2t)(1-tX), \tag{4.19}$$

or, equivalently:

$$\left(\frac{\partial}{\partial t} + \dot{X}\frac{\partial}{\partial X}\right) J(X,t) = (-2X)\underbrace{(1-tX)^2}_{=J(X,t)}, \tag{4.20}$$

which, on the basis of (4.16), is equal to the right-hand side of (4.17), thus establishing Liouville's result in the second formulation.

4.2.3 The transport theorem

The transport theorem (TT) describes how to rewrite the time derivative of the phase space integral of some quantity χ over a material region. This result is briefly derived below. In the derivation of the TT, Liouville's result is used in the formulation (4.10). Consequently, the TT may be viewed as one of the main implications of Liouville's result. In turn, all the consequences derivable from the TT, especially the LE and the LT, relate through this dependence back to the work of Liouville.

Consider the integral of χ over a material region $R(t)$ (see, for example, chapter 14.1 in Lin and Segel, 1988), with the material points subject to the dynamics (4.6):

$$I(t) \equiv \int_{R(t)} \chi(\mathbf{X},t)d\mathbf{X}. \tag{4.21}$$

The time derivative of $I(t)$ is

$$\frac{dI(t)}{dt} = \frac{d}{dt}\int_{R(t)} \chi(\mathbf{X}, t) d\mathbf{X}$$
$$= \frac{d}{dt}\int_{R(0)} \chi(\mathbf{X}(\Xi, t), t) J(\Xi, t) d\Xi, \tag{4.22}$$

where the integral has been transformed to the initial region $R(0)$ and $J(\Xi, t)$ is defined in Eq. (4.8). Since $R(0)$ does not depend on time, the time derivative can be taken inside the integral, noting that Ξ has to be held fixed:

$$\frac{dI(t)}{dt} = \int_{R(0)} \left.\frac{d}{dt}\right|_{\Xi} [\chi(\mathbf{X}(\Xi, t), t) J(\Xi, t)] d\Xi$$
$$= \int_{R(0)} \left\{ J(\Xi, t) \left.\frac{d\chi(\mathbf{X}(\Xi, t), t)}{dt}\right|_{\Xi} + \chi(\mathbf{X}(\Xi, t), t) \underbrace{\left.\frac{dJ(\Xi, t)}{dt}\right|_{\Xi}} \right\} d\Xi, \tag{4.23}$$

where the bracketed term has been differentiated according to the product rule. In the next step, the first term of the integrand is differentiated according to the chain rule, whereas result (4.10) is used to express the underbraced time derivative of the Jacobian:

$$\frac{dI(t)}{dt} = \int_{R(0)} J(\Xi, t) \left\{ \frac{\partial \chi(\mathbf{X}(\Xi, t), t)}{\partial t} \right.$$
$$\left. + \sum_{j=1}^{N} \frac{\partial \chi(\mathbf{X}(\Xi, t), t)}{\partial X_j} \left.\frac{dX_j(\Xi, t)}{dt}\right|_{\Xi} \right\} d\Xi$$
$$+ \int_{R(0)} \chi(\mathbf{X}(\Xi, t), t) \underbrace{\left(\sum_{i=1}^{N} \left.\frac{\partial \dot{X}_i}{\partial X_i}\right|_{\mathbf{X}=\mathbf{X}(\Xi,t)} \right) J(\Xi, t) d\Xi}. \tag{4.24}$$

At this point, the Jacobian appears as common factor in both terms of the integrand and is factored out:

$$\frac{dI(t)}{dt} = \int_{R(0)} \left\{ \frac{\partial \chi(\mathbf{X}(\Xi, t), t)}{\partial t} + \sum_{j=1}^{N} \frac{\partial \chi(\mathbf{X}(\Xi, t), t)}{\partial X_j} \left.\frac{dX_j(\Xi, t)}{dt}\right|_{\Xi} \right.$$
$$\left. + \chi(\mathbf{X}(\Xi, t), t) \left(\sum_{i=1}^{N} \left.\frac{\partial \dot{X}_i}{\partial X_i}\right|_{\mathbf{X}=\mathbf{X}(\Xi,t)} \right) \right\} J(\Xi, t) d\Xi. \tag{4.25}$$

Finally, the integration over $R(0)$ is reverted back to integration over $R(t)$, analogously to (4.22), through which step the dependencies on $\mathbf{X}$ reappear:

$$\frac{d}{dt}\int_{R(t)} \chi(\mathbf{X},t)d\mathbf{X} = \int_{R(t)} \left\{ \frac{\partial \chi(\mathbf{X},t)}{\partial t} + \sum_{j=1}^{N} \dot{X}_j \frac{\partial \chi(\mathbf{X},t)}{\partial X_j} + \chi(\mathbf{X},t) \sum_{i=1}^{N} \frac{\partial \dot{X}_i}{\partial X_i} \right\} d\mathbf{X}. \tag{4.26}$$

The result (4.26) is known as the TT. The above derivation has clearly shown the central role of Liouville's result (4.10) in order to arrive at the TT. The consequences of the TT are manifold. For example, if χ is set to the density of realisations in phase space $\rho(\mathbf{X}, t)$ and it is required that the phase space integral over any material region of that density cannot change (i.e. realisations cannot leave the material region or the probability mass in the material region must remain constant):

$$\frac{d}{dt}\int_{R(t)} \rho(\mathbf{X},t)d\mathbf{X} = 0, \tag{4.27}$$

which is the natural analogy to mass conservation in hydrodynamics, then the integrand in (4.26) has to vanish, which in turn is the LE (see Section 4.2.4). Equally important is the special case $\chi = 1$, which makes $I(t)$ equal to the *volume* $V(t)$ of a material region:

$$\chi \equiv 1 \quad \rightarrow \quad I(t) = \int_{R(t)} d\mathbf{X} \equiv V(t). \tag{4.28}$$

In this situation, the TT implies the result:

$$\frac{dV(t)}{dt} = \int_{R(t)} \left(\sum_{i=1}^{N} \frac{\partial \dot{X}_i}{\partial X_i} \right) d\mathbf{X}, \tag{4.29}$$

stating that the time derivative of the volume of a material region is equal to the integral of the phase-space divergence (see also Section 4.5.6). It is finally noted that setting $\chi \equiv \rho F$ in the TT, where ρ satisfies the requirement (4.27), leads, on the basis of (4.26), to the result:

$$\frac{d}{dt}\int_{R(t)} \rho(\mathbf{X},t)F(\mathbf{X},t)d\mathbf{X} = \int_{R(t)} \rho(\mathbf{X},t) \left\{ \frac{\partial F(\mathbf{X},t)}{\partial t} + \sum_{j=1}^{N} \dot{X}_j \frac{\partial F(\mathbf{X},t)}{\partial X_j} \right\} d\mathbf{X}, \tag{4.30}$$

which, similarly to the TT, gives a description on how to interchange the time derivative and the phase-space integration. It is noted in passing that combining the second and third of the terms on the right-hand side in Eq. (4.26) into a 'divergence' expression, and subsequently rewriting the integral over that divergence expression through the Gauss divergence theorem as a surface integral, leads to a reformulation of result (4.26) that is sometimes known as the *Reynolds transport theorem* (e.g. Lin and Segel, 1988).

4.2.4 Formulation of the Liouville equation

The LE is the continuity equation for the pdf ρ of the state vector $\mathbf{X}$ of a dynamical system. On the basis of the requirement (4.27), by setting $\chi = \rho$ in the TT, the LE is obtained as:

$$\frac{\partial \rho(\mathbf{X}, t)}{\partial t} + \sum_{k=1}^{N} \frac{\partial}{\partial X_k}[\rho(\mathbf{X}, t)\dot{X}_k(\mathbf{X}, t)] = 0, \tag{4.31}$$

where the flow $\dot{\mathbf{X}}$ of the dynamical system in phase space is given by (4.6); that is, the time evolution of the state $\mathbf{X}$ is governed by the (non-autonomous) dynamical system specified through $\mathbf{\Phi}$. Another formulation of the LE that is completely equivalent to (4.31) is given by expanding the phase-space derivative:

$$\frac{\partial \rho(\mathbf{X}, t)}{\partial t} + \sum_{k=1}^{N} \Phi_k(\mathbf{X}, t)\frac{\partial \rho(\mathbf{X}, t)}{\partial X_k} = -\psi(\mathbf{X}, t)\rho(\mathbf{X}, t), \tag{4.32}$$

where

$$\psi(\mathbf{X}, t) \equiv \sum_{k=1}^{N} \frac{\partial \Phi_k(\mathbf{X}, t)}{\partial X_k} \tag{4.33}$$

is the divergence of the flow in phase space. Evidently, since derived from (4.27), the LE expresses the conservation of the phase-space integral of the number density of realisations of the dynamical system (4.6). As a continuity equation, it is entirely analogous to the continuity equation in hydrodynamics. In the case of the LE, however, the phase space velocity is a known function at every point in phase space through (4.6), whereas when the continuity equation is considered in physical space in hydrodynamics the flow velocity has to be determined through the momentum equation.

Physically, the LE states that the local change of ρ – at a particular point in phase space – must be exactly balanced by the net flux of realisations across the faces of a small volume surrounding the point under consideration (Thompson, 1983). In the balance equation for the pdf, given by the LE (4.31), no source terms or non-convective fluxes appear. The absence of source terms is equivalent to the requirement that realisations are neither created nor destroyed in phase space. Inspection of the LE in the form (4.33) shows that the LE is a linear (in ρ), inhomogeneous partial differential equation with the single dependent variable being the pdf ρ. The independent variables in the LE are time t and the phase space coordinates $\mathbf{X}$. Finally, it is apparent that the solution to the LE is fundamentally dependent on the model dynamics $\mathbf{\Phi}(\mathbf{X}, t)$.

4.3 Solution of the Liouville equation

Given the one-to-one relationship between the initial state $\mathbf{X}(t=0) \equiv \Xi$ and the state $\mathbf{X}(t)$ of the dynamical system (4.6) at time t, as expressed through Eqs. (4.7) and (4.9) (see Section 4.2.2), it is possible to formulate an analytical solution to the LE. To obtain the solution to the LE, written in the form (4.32), the method of characteristics is used (e.g. Zwillinger, 1989) together with the fact that the LE is linear. For an initial condition prescribed through ρ_0:

$$\rho(\mathbf{X}, t=0) = \rho_0(\mathbf{X}), \tag{4.34}$$

the solution is

$$\rho(\mathbf{X}, t) = \underbrace{\rho_0(\Xi) \exp\left[-\int_0^t \psi(\mathbf{X}(\Xi, t'), t')dt'\right]}_{\equiv h(\mathbf{X}, t)}. \tag{4.35}$$

In this formulation of the solution (Ehrendorfer, 1994a, 1994b), for a given point $\mathbf{X}$ in phase space and time t, it is necessary to compute Ξ according to (4.9). Subsequently, Ξ is used for evaluating the solution (4.35) by computing $\rho_0(\Xi)$ and referring to the definition of ψ in Eq. (4.33). An illustration of such a computation is given in Section 4.4.1. It is therefore also evident that the function h is indeed a function of $\mathbf{X}$ and t; for a discussion of h, see Section 4.5, and especially Section 4.5.1. It is further apparent that the solution (4.35) satisfies the initial condition (4.34) and is both non-negative and – essentially through requirement (4.27) – correctly normalised. Solution (4.35) does indeed satisfy the LE (4.32) (Ehrendorfer, 1994a).

4.4 Illustrations

4.4.1 A one-dimensional autonomous example

As an illustration of the LE and its solution, as expressed in Eq. (4.35), the one-dimensional autonomous dynamical system

$$\frac{d}{dt}x = x - x^3 \tag{4.36}$$

is considered, with the functions Φ and ψ (see Equations 4.6 and 4.33) given as

$$\Phi(x) = x - x^3, \qquad \psi(x) = 1 - 3x^2. \tag{4.37}$$

System (4.36) has two stable equilibrium points at $x_s = \pm 1$ and one unstable equilibrium point at $x_u = 0$. The LE specific for this system is given by (see Eq. 4.32)

$$\frac{\partial \rho(x,t)}{\partial t} + (x - x^3)\frac{\partial \rho(x,t)}{\partial x} = -(1 - 3x^2)\rho(x,t). \tag{4.38}$$

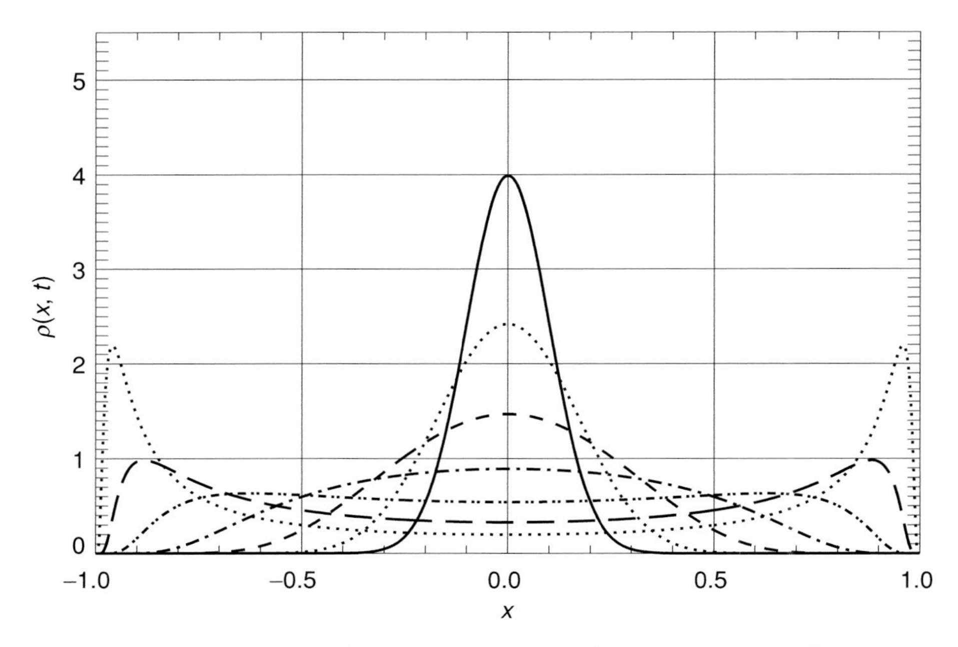

Figure 4.3 Solution of LE for system $\dot{x} = x - x^3$; initial pdf is $\mathtt{N}(0, 0.1^2)$; $\rho(x, t)$ is shown at time increments of 0.5. Initial pdf is plotted as a solid line; ordering of subsequent time (shown in different linestyles) is given by observing that value at $x = 0$ decreases monotonically with t.

The solution to the LE (4.38) for an arbitrary initial condition ρ_0 is obtained on the basis of (4.35) as:

$$\rho(x, t) = \rho_0(\xi[x, t]) \exp\left[- \int_0^t \underbrace{\{1 - 3(x\{\xi[x, t], t'\})^2\}\, dt'}_{=\psi(x)} \right]. \qquad (4.39)$$

Here, the function ψ is indicated for clarity; note that for the present autonomous system ψ possesses no *explicit* time dependence (time dependence enters only implicitly through the dependence of x on t). In the more general formulation of a non-autonomous system used in Sections 4.2.4 and 4.3, ψ possesses an explicit time dependence, as indicated in Eq. (4.33). For the present system, conditions (4.7) and (4.9) – explicitly solving (4.36) – are:

$$\begin{aligned} x = x(\xi, t) &= \xi e^t (1 - \xi^2 + \xi^2 e^{2t})^{-\frac{1}{2}} \\ &\Leftrightarrow \xi(x, t) = x e^{-t} (1 - x^2 + x^2 e^{-2t})^{-\frac{1}{2}}. \end{aligned} \qquad (4.40)$$

Illustrations of the functional form of the time evolution of the pdf, given the dynamics (4.36), are shown in Figures 4.3– 4.5. The initial pdf is taken as normal with the same variance in all three illustrations, but with different means. It is apparent that the pdf decreases at the unstable equilibrium point x_u as realisations are attracted to the stable equilibrium points x_s. Depending on how much initial probability mass lies on either

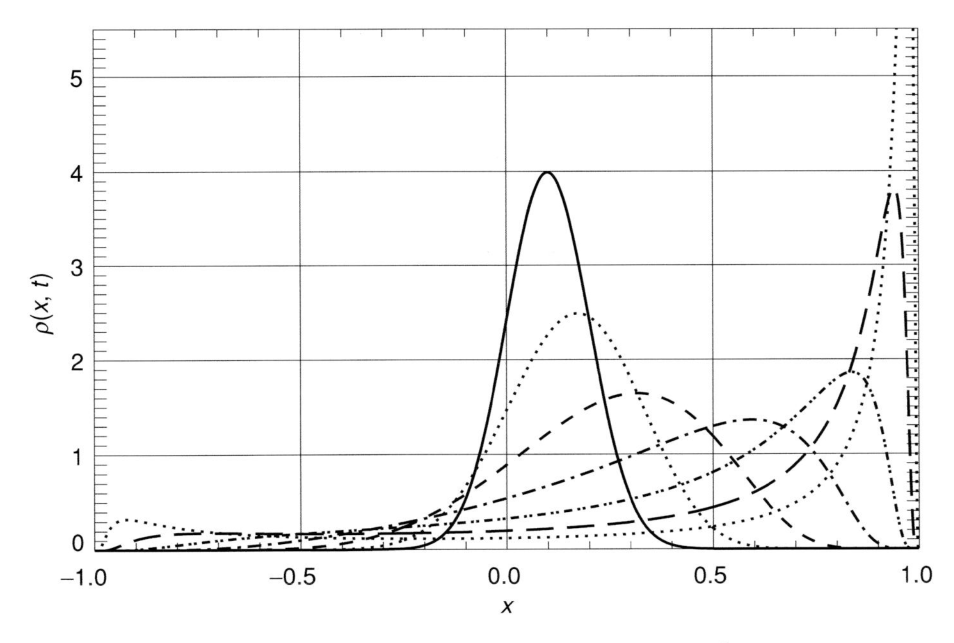

Figure 4.4 Same as Figure 4.3, except for initial pdf $\mathtt{N}(0.1, 0.1^2)$.

side of x_u, the pdf becomes differently asymmetric at later times; for example, in Figure 4.4 the pdf at the last time point shown (second dotted line) is much more strongly peaked at $x_s = +1$ than in Figure 4.5, since the initial pdf is centred at 0.1 in the former case, whereas it is centred at 0.05 in the latter case. The pdfs shown in Figure 4.3 are entirely symmetric, as the initial pdf is centred at zero.

An example code for carrying out the actual computation for evaluating the solution (4.39) is given below. Here `x` denotes x and `time` denotes t. Note that the time integral that appears in (4.39) is approximated in this code through an approximate trapezoidal rule (even though the integration could – presumably – be carried out analytically in the present situation).

```
xi = x * exp (-time) / sqrt (1. - x * x + x * x * exp (-2. * time))   ; find xi

a1 = (xi - smean)/sdev
y1(i) = 1./(sqrt(2.*!pi) * sdev) * exp(-0.5 * a1 * a1)   ; rho_0(xi)

xint = 0.0
for k = 1 ,  100 do begin
tp = k * time / 100.0
b = xi * exp (tp) /sqrt (1.0 - xi * xi + xi * xi * exp (2. * tp))   ; finding x(xi, t')
c = 1.0 - 3.0 * b * b   ; psi(x)
xint = xint + c & endfor

y1 (i) = y1 (i) * exp (-xint * time/100.0)   ; apply h(x, t)
```

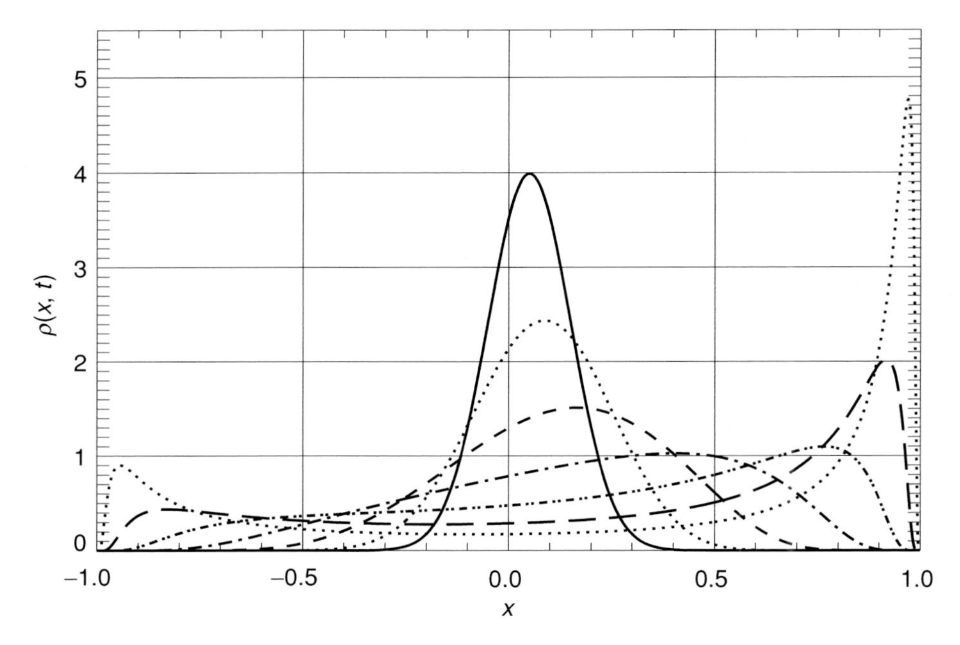

Figure 4.5 Same as Figure 4.3, except for inital pdf $\mathrm{N}(0.05, 0.1^2)$.

4.4.2 A three-dimensional autonomous example

To illustrate the LE in a somewhat higher-dimensional context, the prototypical chaotic model for atmospheric flow proposed by Lorenz (1984) is used. This model consists of three coupled, non-linear ordinary differential equations for the variables X, Y and Z, written as

$$\begin{pmatrix} \dot{X} \\ \dot{Y} \\ \dot{Z} \end{pmatrix} = \begin{pmatrix} -Y^2 - Z^2 - aX + aF \\ XY - bXZ - Y + G \\ bXY + XZ - Z \end{pmatrix}, \tag{4.41}$$

where the parameters a, b, F, and G take on the values $a = 0.25$, $b = 4.0$, $F = 8.0$, $G = 1.25$. Since this system will be used later for illustrating other properties of the LE, the tangent-linear system corresponding to (4.41) for primed perturbation variables is recorded here, too:

$$\begin{pmatrix} \dot{X}' \\ \dot{Y}' \\ \dot{Z}' \end{pmatrix} = \underbrace{\begin{pmatrix} -a & -2\bar{Y} & -2\bar{Z} \\ \bar{Y} - b\bar{Z} & \bar{X} - 1 & -b\bar{X} \\ b\bar{Y} + \bar{Z} & b\bar{X} & \bar{X} - 1 \end{pmatrix}}_{\mathsf{L}_{\overline{\mathbf{X}}(t)}} \begin{pmatrix} X' \\ Y' \\ Z' \end{pmatrix}, \tag{4.42}$$

where L denotes the tangent-linear model operator that depends on the time-dependent basic state $\overline{\mathbf{X}}(t)$ that is used for linearising the non-linear model (4.41). The tangent-linear system to any non-linear system is obtained by retaining the term linear in the

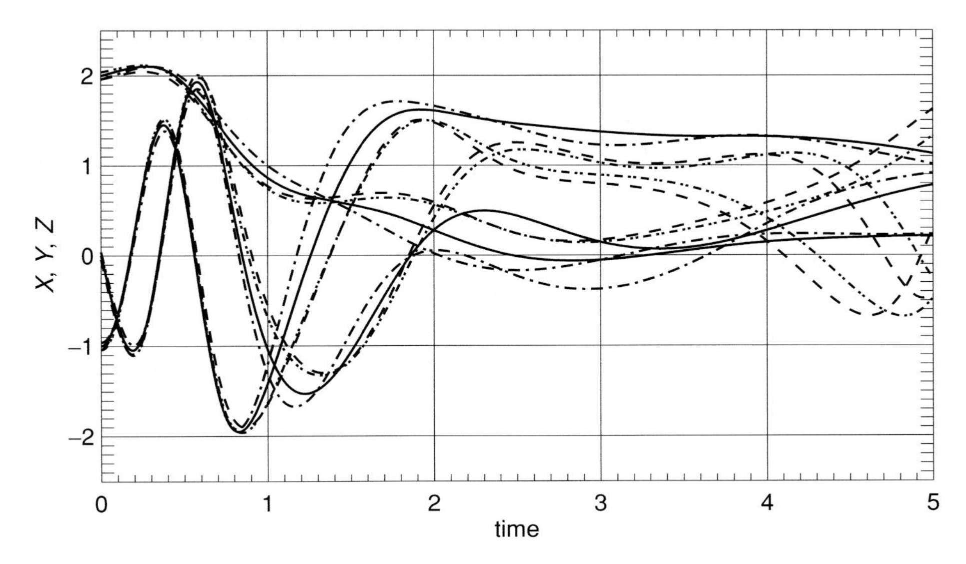

Figure 4.6 Non-linear evolution of states in the Lorenz (1984) system. Reference initial state (solid) is defined by $X = 2$, $Y = -1$, $Z = 0$. Perturbation sizes are 0.04, with dashed $(-, -, -)$ perturbation, dash–dot $(-, +, +)$ perturbation, and dash–dot–dot $(+, -, +)$ perturbation on reference initial state.

perturbation quantities in a Taylor series expansion of the non-linear model dynamics along $\overline{\mathbf{X}}(t)$:

$$\frac{d}{dt}\underbrace{(\mathbf{X}-\overline{\mathbf{X}})}_{\equiv \mathbf{X}'} = \mathbf{\Phi}(\mathbf{X}) - \mathbf{\Phi}(\overline{\mathbf{X}}) \approx \mathsf{L}_{\overline{\mathbf{X}}}\underbrace{(\mathbf{X}-\overline{\mathbf{X}})}_{\equiv \mathbf{X}'}, \tag{4.43}$$

where the tangent-linear model operator (illustrated here in Eq. 4.42) is given as

$$\mathsf{L}_{\overline{\mathbf{X}}} \equiv \left(\frac{\partial \Phi_i(\mathbf{X})}{\partial X_j}\right)_{\overline{\mathbf{X}}}, \tag{4.44}$$

and the tangent-linear resolvent $\mathsf{M}_{\Xi,t}$ connects an initial perturbed state to a perturbed state at time t:

$$\mathbf{X}'(t) = \mathsf{M}_{\Xi,t}\mathbf{X}'(t=0). \tag{4.45}$$

An approximation to the dynamics of perturbations made in the form of the linear model (4.43) will obviously be valid only to the degree that the perturbations are in some sense small. In the absence of analytical model solutions, numerical solutions to the non-linear and the tangent-linear Lorenz (1984) model, Eqs. (4.41) and (4.42), respectively, are obtained here with a predictor-corrector method to connect $\mathbf{X}$ and Ξ (see Eqs. 4.7 and 4.9). Illustrations of the behaviour of the non-linear system are

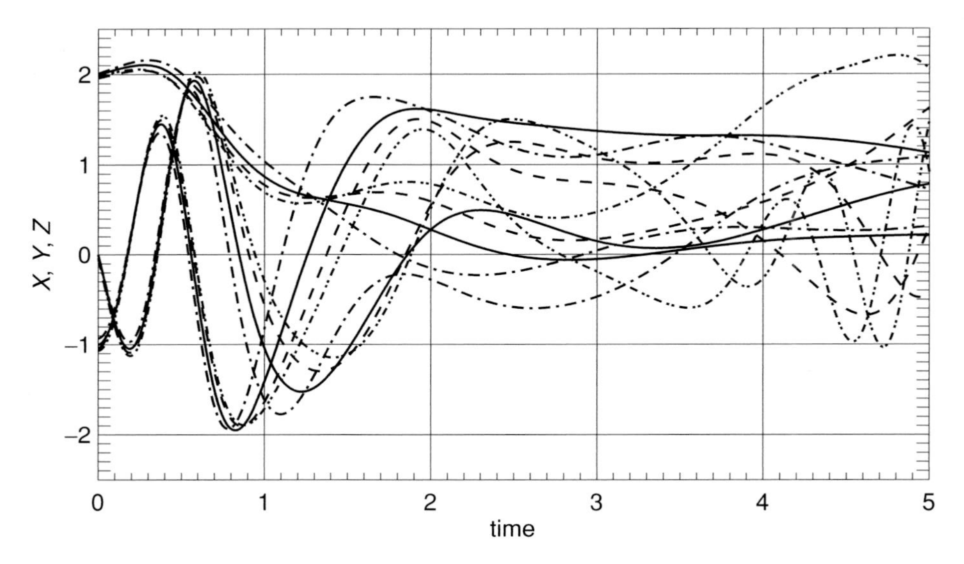

Figure 4.7 Same as Figure 4.6, except for dash–dot denoting positive SV1 perturbation, and dash–dot–dot negative SV1 perturbation. For perturbation sizes compare TLD/NLD picture in Figure 4.8.

shown in Figures 4.6 and 4.7. In Figure 4.8, the evolution of perturbations is shown, as obtained by both the fully non-linear difference (solid) in Eq. (4.43), as well as through (numerically) solving the tangent-linear model, as given in (4.45) (dashed). In the three panels in Figure 4.8, the size of the initial perturbation increases by a factor of two by going from panel to panel; it is apparent that the tangent-linear solution scales exactly, whereas the non-linear differences deviate from the tangent-linear solution earlier when the initial perturbation size becomes larger.

In the situation of system (4.41), the pdf with its time evolution described through the LE is a function of the three model variables and time, given as $\rho(X, Y, Z, t)$. Stating the LE in the form (4.32) is straightforward noting that ψ, defined in (4.33), takes on the form:

$$\psi(X, Y, Z) \equiv \frac{\partial \dot{X}}{\partial X} + \frac{\partial \dot{Y}}{\partial Y} + \frac{\partial \dot{Z}}{\partial Z} = -a + (X - 1) + (X - 1)$$
$$= 2X - 2 - a. \qquad (4.46)$$

As in Section 4.4.1, ψ contains no explicit time dependence as the dynamical system under consideration is autonomous. Referring back to the general solution of the LE

→

Figure 4.8 Evolution of perturbations in the Lorenz (1984) system. Initial perturbation size increases by a factor of 2 from panel to panel. Basic state is defined by (2, −1,0). Broken curves are non-linear difference (NLD) with *X* dotted, *Y* dashed, *Z* dash–dot. Solid curves are tangent-linear difference (TLD); variables may be identified by matching NLD at small *t*. Note that TLD scales exactly.

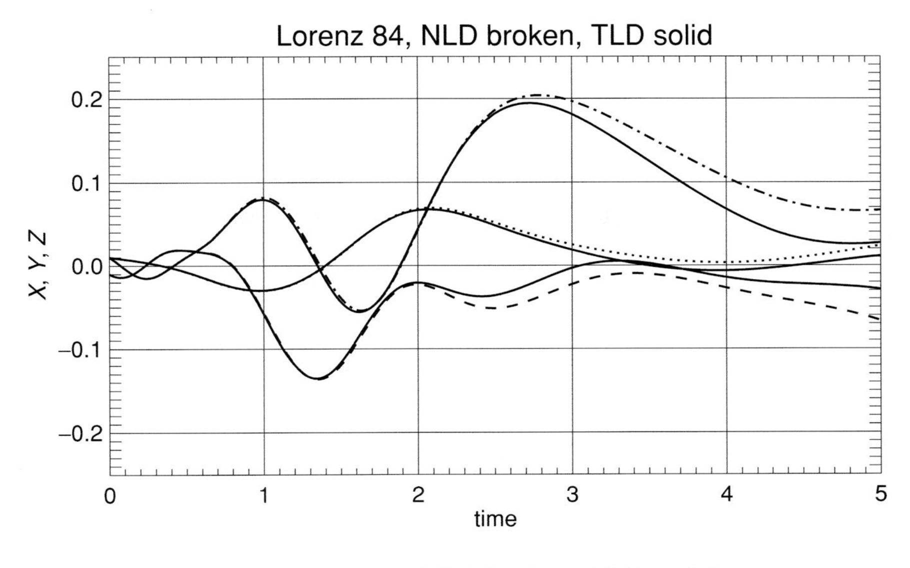
Lorenz 84, NLD broken, TLD solid
0.2
0.1
0.0
−0.1
−0.2
X, Y, Z
0
1
2
3
4
5
time

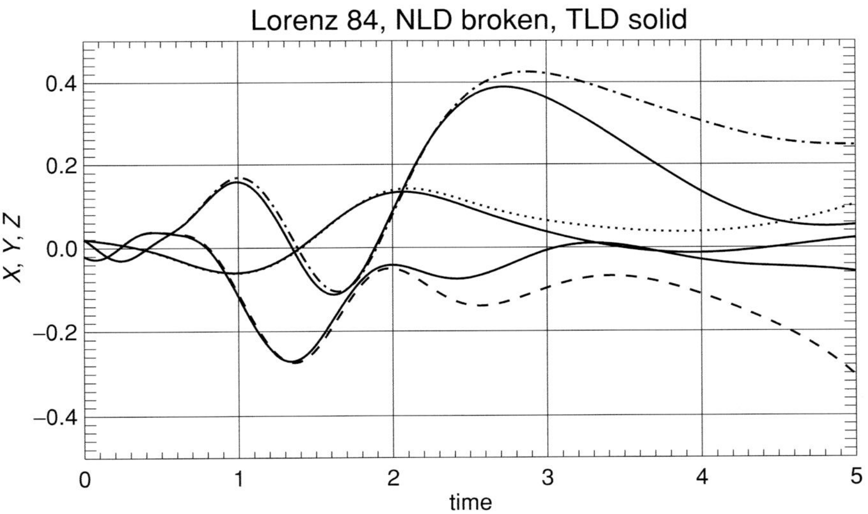
Lorenz 84, NLD broken, TLD solid
0.4
0.2
0.0
−0.2
−0.4
X, Y, Z
0
1
2
3
4
5
time

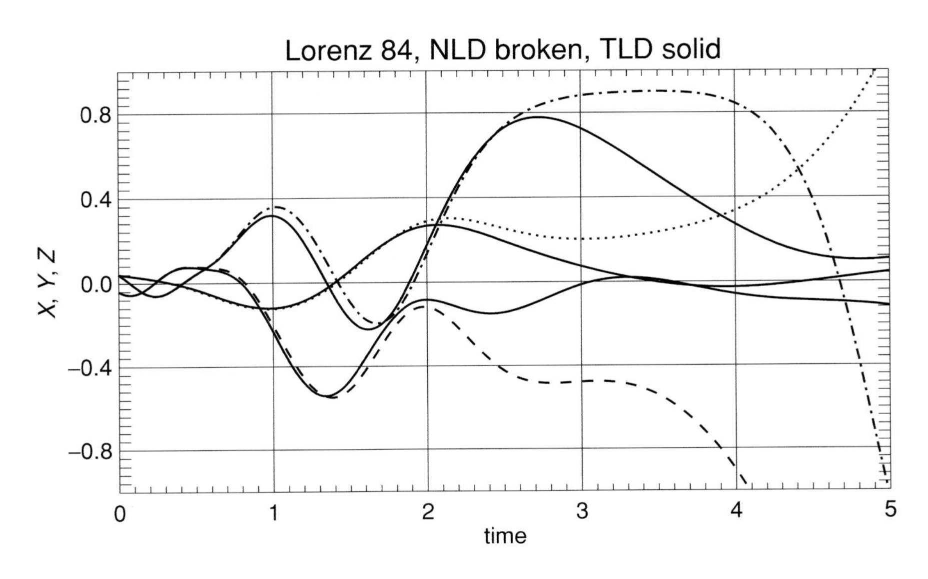
Lorenz 84, NLD broken, TLD solid
0.8
0.4
0.0
−0.4
−0.8
X, Y, Z
0
1
2
3
4
5
time

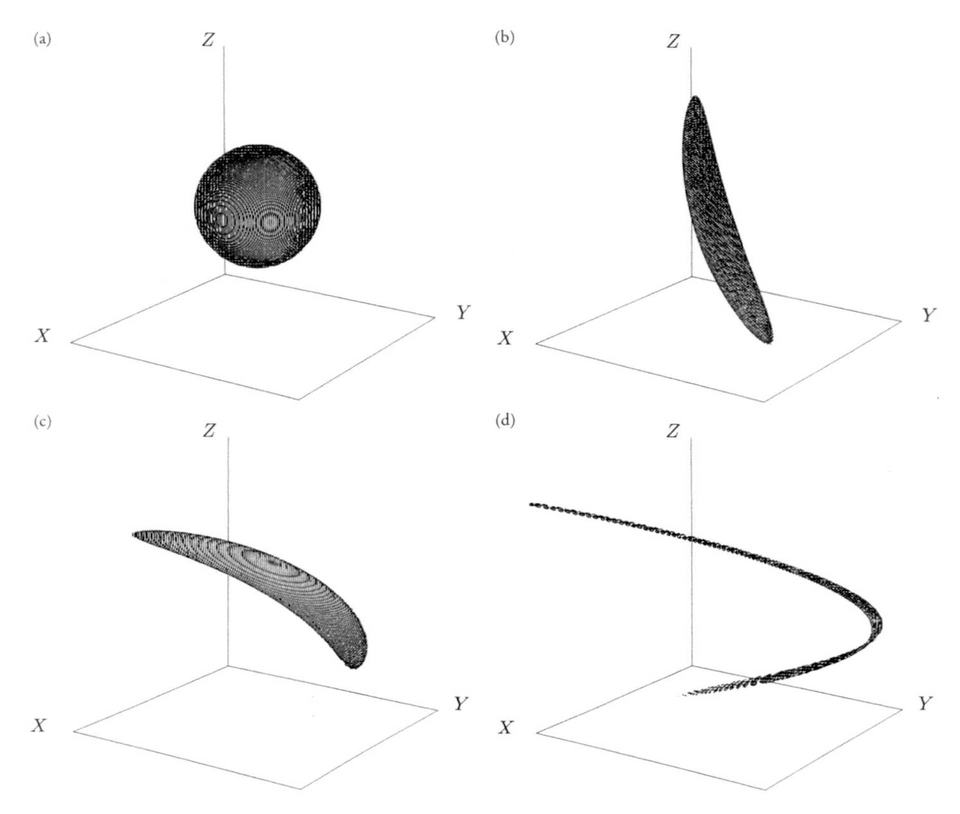

Figure 4.9 Solution of the LE for the Lorenz (1984) system displayed in phase space. From Ehrendorfer (1997).

given in Eq. (4.35), the solution of the LE relevant for the system (4.41) can be written in the form:

$$\rho(\mathbf{X}, t) = \underbrace{\rho_0(\Xi) \exp\left[-\int_0^t \psi(\mathbf{X}(\Xi, t'))dt'\right]}_{\equiv h(\mathbf{X},t)}. \tag{4.47}$$

In stating this form of the solution, the fact that ψ is not explicitly time dependent has been used. Also, as in Section 4.4.1, it is necessary to find Ξ through the (numerical) implementation of Eq. (4.9) appropriate for system (4.41) given $\mathbf{X}$ and t. The form of the solution stated in (4.47) is shown in Figure 4.9. Since, in the present situation, the pdf lives in a three-dimensional phase space, it is obviously necessary to somehow contour three-dimensional regions of the same values of the pdf. An attempt to visualise the pdf is shown in Figure 4.9 by showing the surface (as a contour in three-dimensional space) that encloses all pdf values that are greater than one at a given time. At the initial time (Figure 4.9a) that surface is a sphere, since the initial pdf ρ_0 is specified to be multivariate normal with the same variance for all three variables.

At a somewhat later time (Figure 4.9b) that surface becomes elongated along the direction of preferred error growth in the model (4.41), and still later (Figures 4.9c, d) acquires some curvature and clearly non-linear features. It is noted that this surface (defined through the equation 'ρ equal to some specified constant') is clearly not a material surface, since realisations on that surface obviously do not have to remain connected to that particular value of ρ (see also Ehrendorfer, 1997).

4.5 Comments on the Liouville equation

In this chapter various comments relating to the LE and atmospheric predictability are brought together.

4.5.1 A likelihood ratio

It is noted that the quantity h, defined in Eq. (4.35), may be referred to as a likelihood ratio, as it is the ratio of two pdfs:

$$h(\mathbf{X}, t) = \frac{\rho(\mathbf{X}, t)}{\rho_0(\Xi)} \equiv \exp\left[- \int_0^t \psi(\mathbf{X}(\Xi, t'), t') dt' \right]. \tag{4.48}$$

In contrast, since, on the basis of (4.7), $\mathbf{X}$ can be viewed as the time-dependent transform of Ξ, it is possible to ascertain that the pdf of $\mathbf{X}$ (viewed as the transformed Ξ) may be specified in terms of the pdf of Ξ in the form (see, for example, Theorem 4.6.18 in Dudewicz and Mishra, 1988):

$$\frac{\rho(\mathbf{X}, t)}{\rho_0(\Xi)} = \frac{1}{|J|}, \tag{4.49}$$

where J – already defined in Eq. (4.8) – is the determinant of the Jacobian of the transformation:

$$J \equiv \det\left(\frac{\partial \mathbf{X}(\Xi, t)}{\partial \Xi}\Big|_{\Xi} \right) = \det(\mathsf{M}_{\Xi,t}). \tag{4.50}$$

The second equality in Eq. (4.50) represents expressing the Jacobian in terms of the resolvent $\mathsf{M}_{\Xi,t}$ of the tangent-linear model, defined in Eq. (4.45) as the operator that maps an initial perturbation into a perturbation at time t. The identification used in Eq. (4.50) becomes clear by, referring to (4.7), taking the following Taylor series expansion:

$$\mathbf{X}(\Xi + \Xi', t) = \mathbf{X}(\Xi, t) + \frac{\partial \mathbf{X}(\Xi, t)}{\partial \Xi}|_{\Xi}\Xi' + hot. \tag{4.51}$$

It is evident from Eq. (4.51) that $\mathsf{M}_{\Xi,t}$ – the resolvent of the tangent-linear model that linearises the mapping $\mathbf{X} = \mathbf{X}(\Xi, t)$ around a trajectory starting at Ξ up to time t – is precisely the same as $\frac{\partial \mathbf{X}(\Xi,t)}{\partial \Xi}\Big|_{\Xi}$, as already indicated in Eq. (4.50).

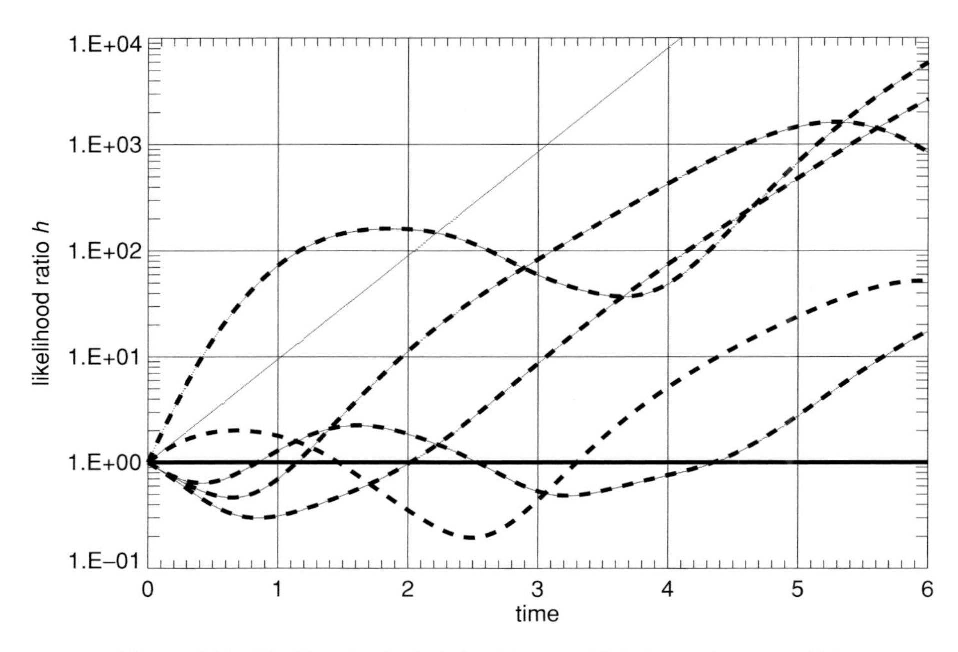

Figure 4.10 Likelihood ratio h, defined in Eq. (4.35), for the Lorenz (1984) system (4.41), for five different initial conditions. Overplotted curves show results computed by determinant of resolvent (thick dashed), and by evaluating ψ (thin dotted) in (4.52). The single dashed curve shows h for $\Xi = 0$; single thin dotted curve shows h when computed for a 'resting' basic state; in that case $h(t) = \exp[-(-a-2)t]$.

Consequently, on the basis of the results (4.48), (4.49), and (4.50), the likelihood ratio h can be expressed in the two different, but exactly equivalent formulations as follows:

$$h(\mathbf{X}, t) = \exp\left[-\int_0^t \psi(\mathbf{X}(\Xi, t'), t')dt'\right] = [|\det(\mathsf{M}_{\Xi,t})|]^{-1}. \tag{4.52}$$

The important result (4.52) will be referred to in subsequent sections. Figure 4.10 shows the (almost) exact agreement of computing h by the two methods indicated in Eq. (4.52), for five different initial conditions Ξ, for the Lorenz (1984) system (4.41).

4.5.2 The Liouville equation and singular vectors

To illustrate a connection between the LE and the computation of singular vectors (SVs), consider the linear non-autonomous system (see also, for example, Eq. 4.42):

$$\dot{\mathbf{X}} = \mathsf{L}(t)\mathbf{X} \tag{4.53}$$

as the dynamics (4.6) relevant for the LE (4.31). In that situation the phase space divergence ψ, defined in (4.33), is independent of the position in phase space, but remains to be a function of time t:

$$\psi(\mathbf{X}, t) = \psi(t) = \sum_{k=1}^{N} (\mathsf{L}(t))_{k,k} \equiv \sigma(t). \tag{4.54}$$

In the autonomous situation, L and σ become time independent, and, consequently, the eigendecomposition of L (assumed to exist) has the time-independent eigenvalues l_k. In that situation, ψ may be written as the sum of the eigenvalues l_k of L (e.g. Strang, 1993):

$$\psi = \text{trace}(\mathsf{L}) = \sum_{k=1}^{N} l_k. \tag{4.55}$$

Through the solution of the autonomous version of (4.53) in the form:

$$\mathbf{X}(t) = e^{\mathsf{L}t}\Xi \tag{4.56}$$

the resolvent M_t (see Eq. 4.45) is identified as

$$\mathsf{M}_t = e^{\mathsf{L}t}, \tag{4.57}$$

which yields the following relation between l_k and the eigenvalues m_k of M_t:

$$m_k = e^{l_k t}, \quad k = 1, \ldots, N. \tag{4.58}$$

Referring back to the definition of the likelihood ratio, Eq. (4.35), relation (4.52) becomes explicit in the present autonomous situation as follows:

$$\begin{aligned} h^{-1}(t) &= \exp\left(\int_0^t \sum_{k=1}^{N} l_k dt'\right) \\ &= \exp\left(t \sum_{k=1}^{N} l_k\right) = \prod_{k=1}^{N} e^{l_k t} = \prod_{k=1}^{N} m_k = \det \mathsf{M}_t. \end{aligned} \tag{4.59}$$

Equation (4.59) is an explicit illustration of the general result (4.52) for the special situation of the linear system (4.53) considered in autonomous form, where relation (4.58) was used, as well as the fact that the determinant of a matrix may be written as the product of its eigenvalues.

Referring back to the non-autonomous situation, it is evident that the LE (4.32) may be written on the basis of (4.53) and (4.54) as

$$\frac{\partial \rho}{\partial t} + \sum_{k=1}^{N} (\mathsf{L}(t)\mathbf{X})_k \frac{\partial \rho}{\partial X_k} + \sigma(t)\rho = 0, \tag{4.60}$$

with the explicit solution given on the basis of (4.35) and (4.54) as follows:

$$\rho(\mathbf{X}, t) = \rho_0(\Xi) \exp\left[-\int_0^t \sigma(t')dt'\right], \tag{4.61}$$

or, more explicitly, by denoting the resolvent of the non-autonomous system (4.53) as M_t:

$$\rho(\mathbf{X}, t) = \rho_0\big(\mathsf{M}_t^{-1}\mathbf{X}\big) \exp\left[-\int_0^t \sigma(t')dt' \right]. \tag{4.62}$$

Equation (4.62) is the solution to the LE (4.60) appropriate for the non-autonomous system (4.53) using the definition (4.54). It clearly allows for computing the time evolution of the pdf associated with a linear system for an arbitrary ρ_0, given knowledge about L and the associated resolvent M_t (see also Tribbia and Baumhefner, 1993).

As a next step the following SV problem using the same norm – described through $\mathsf{C}^\mathrm{T}\mathsf{C}$ – at initial and final times, is considered (see, for example, Ehrendorfer and Tribbia, 1997; Errico *et al.*, 2001):

$$(\mathsf{C}^\mathrm{T}\mathsf{C})^{-1}\mathsf{M}_t^\mathrm{T}(\mathsf{C}^\mathrm{T}\mathsf{C})\mathsf{M}_t\mathbf{z}_0 = \lambda\mathbf{z}_0 \quad \text{subject to:} \quad \mathbf{z}_0^\mathrm{T}(\mathsf{C}^\mathrm{T}\mathsf{C})\mathbf{z}_0 = 1. \tag{4.63}$$

Given the (squared) singular value spectrum λ_k, the product of the λ_k determines the likelihood ratio h through

$$\prod_{k=1}^{N} \lambda_k = \det\left[(\mathsf{C}^\mathrm{T}\mathsf{C})^{-1}\mathsf{M}_t^\mathrm{T}(\mathsf{C}^\mathrm{T}\mathsf{C})\mathsf{M}_t\right] = (\det \mathsf{M}_t)^2 = [h(\mathbf{X}, t)]^{-2}, \tag{4.64}$$

where (4.52) has been used (omitting the subscript Ξ) together with relations between determinants and eigenvalues of matrices (e.g. Strang, 1993). Result (4.64) implies the following relation between the likelihood ratio h and the eigenvalue spectrum λ_k:

$$h^2(\mathbf{X}, t) = \left(\prod_{k=1}^{N} \lambda_k \right)^{-1}. \tag{4.65}$$

Retracing the steps that lead to result (4.65), it is evident that (4.65) is valid for resolvents M_t that correspond to non-autonomous (and autonomous) linear systems. Clearly, result (4.65) may be used to re-express the solution (4.62) of the LE for system (4.53) in the form:

$$\rho(\mathbf{X}, t) = \rho_0\big(\mathsf{M}_t^{-1}\mathbf{X}\big) \left(\prod_{k=1}^{N} \lambda_k \right)^{-1/2}, \tag{4.66}$$

so that knowledge of the resolvent M_t and its singular value spectrum is sufficient to determine the time evolution of the PDF.

Referring back to (4.52), it is seen that a consequence of (4.65) is that for tangent-linear systems with vanishing phase space divergence (i.e. $\psi = 0$) the product of the squared singular values is one:

$$\psi = 0 \quad \Rightarrow \quad \prod_{k=1}^{N} \lambda_k = 1. \tag{4.67}$$

It is obvious that condition (4.67) does not necessarily imply that the singular value spectrum is symmetric in the sense that $\lambda_k = \lambda_{N-k+1}^{-1}$ $(k = 1, 2, \ldots, \frac{N}{2})$ even though it is interesting to ask under what circumstances such symmetry will exist; it should be expected that these circumstances will be closely related to properties of the basic state used to define the resolvent M_t. Further, it is clear that (4.67) implies that the averaged sum of the squared singular values is greater (or equal) than one:

$$\prod_{k=1}^{N} \lambda_k = 1 \quad \Rightarrow \quad \frac{1}{N}\sum_{k=1}^{N} \lambda_k \geq 1, \tag{4.68}$$

essentially, because the arithmetic mean is greater (or equal) than the geometric mean (with equality only if all λ_k are equal).

4.5.3 Lyapunov exponents

The Lyapunov exponents relevant for a non-linear dynamical system of the form (4.6) may be defined as (see, for example, Parker and Chua, 1989):

$$\gamma_k \equiv \lim_{t\to\infty} \underbrace{\frac{1}{t} \ln |m_k(\mathsf{M}_{\Xi,t})|}_{\equiv \hat{\gamma}_k}, \quad k = 1, 2, \ldots, N, \tag{4.69}$$

where the resolvent $\mathsf{M}_{\Xi,t}$ is defined in the context of Eq. (4.45), and m_k denotes the eigenvalues of $\mathsf{M}_{\Xi,t}$ (see Section 4.5.2). Clearly, through the relationship between h and M discussed in Section 4.5.1, especially through Eq. (4.52), the following relationship may be derived between the likelihood ratio h and the sum of the (unconverged) Lyapunov exponents $\hat{\gamma}_k$:

$$\begin{aligned}\sum_{k=1}^{N} \hat{\gamma}_k &= \frac{1}{t}\sum_{k=1}^{N} \ln |m_k| \\ &= \frac{1}{t} \ln \left|\prod_{k=1}^{N} m_k\right| = \frac{1}{t} \ln |\det \mathsf{M}_{\Xi,t}| = \frac{1}{t} \ln[h(\mathbf{X}, t)]^{-1}.\end{aligned} \tag{4.70}$$

On the basis of (4.70), it follows that the sum of the Lyapunov exponents may be expressed in terms of the likelihood ratio as

$$\sum_{k=1}^{N} \gamma_k = -\lim_{t\to\infty} \frac{1}{t} \ln h(\mathbf{X}, t). \tag{4.71}$$

Alternatively, by re-expressing h through the result (4.65) in terms of the product of λ_k, and inserting into (4.70), the following relationship between the sum of the (unconverged) Lyapunov exponents $\hat{\gamma}_k$ and the logarithms of the squared singular values λ_k is obtained:

$$\sum_{k=1}^{N} \hat{\gamma}_k = \sum_{k=1}^{N} \frac{1}{2t} \ln \lambda_k, \tag{4.72}$$

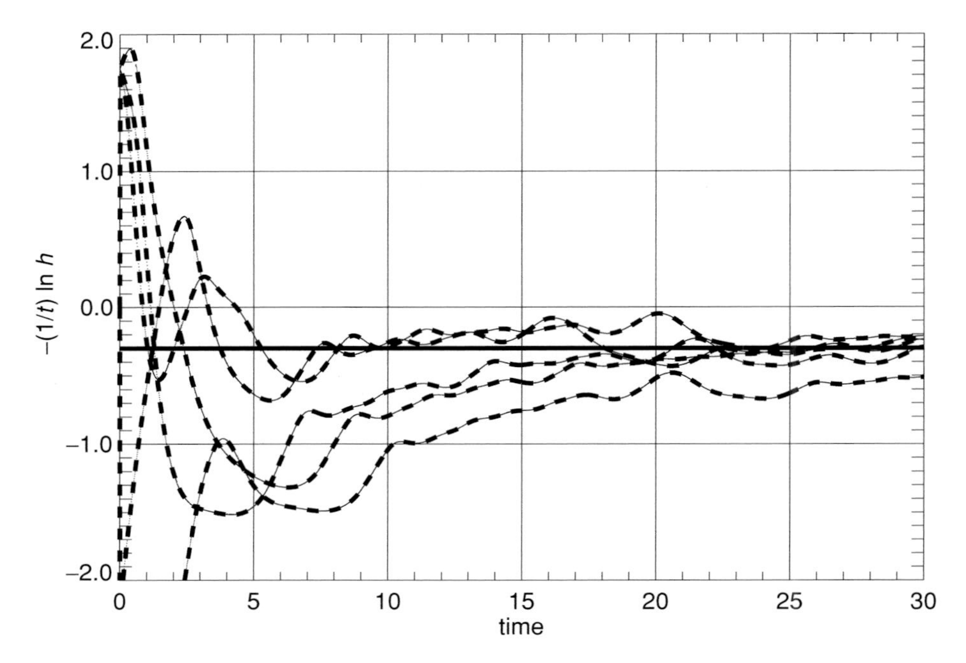

Figure 4.11 Sum of Lyapunov exponents as defined in (4.69), computed via the likelihood ratio h, defined in (4.35), for the Lorenz (1984) system (4.41), for five different initial conditions (as in Figure 4.10). Overplotted curves show results for $\sum_k \hat{\gamma}_k$ (see Eq. 4.70) obtained from h that is computed either through the determinant of resolvent (thick dashed), or by evaluating ψ (thin dotted) according to Eq. (4.52). The thick solid line is drawn at -0.3.

or, for the sum of the Lyapunov exponents themselves:

$$\sum_{k=1}^{N} \gamma_k = \lim_{t\to\infty} \frac{1}{2t} \sum_{k=1}^{N} \ln \lambda_k. \tag{4.73}$$

Even though both Eqs. (4.72) and (4.73) are written in terms of sums over k, the relation between an individual γ_k and λ_k should hold analogously. For the chaotic attractor of the Lorenz (1984) system, the sum of the three Lyapunov exponents is *negative* (about -0.3 for the parameters used; see Section 4.4.2, and Figure 4.11). Thus, in the limit, h *grows* exponentially (see Figure 4.10). Figure 4.11 illustrates how $-t^{-1} \ln h$ converges to $\sum_k \hat{\gamma}_k$ (Eq. 4.71) as a function of t for five different initial conditions.

4.5.4 The Fokker–Planck equation

Random processes added to the governing dynamics (4.6) appear in the equation governing the evolution of the pdf, namely the LE, as a diffusion term, since the transport of phase space points by random motions is a diffusive process (see, for example, Thompson, 1972, 1983; de Groot and Mazur, 1984). The LE written for such a dynamical system that includes random processes to account for, for example, model error, is generally referred to as the Fokker–Planck equation (FPE). As in the case of the LE, the FPE (see, for example, Risken, 1989; Gardiner, 1990; Soize, 1994) has only one dependent variable, namely the probability density function ρ, and it is a linear equation (Thompson, 1972).

Selected references to the FPE include the work by Hasselmann (1976), and Thompson (1983, 1985b), accounting by random processes for unknown forcings in the equations, and investigating the problem of determining equilibrium statistics of the atmosphere/climate system under such circumstances. It is worth pointing out that the FPE is – quite analogously to the LE – governing the time evolution of the pdf, and, as such, also contains all information about the equilibrium statistics of the forced dynamical system, as well as all relevant information regarding the predictability of that system (Hasselmann, 1976). The statistical equilibrium between forcing and dissipation has been referred to as a fluctuation–dissipation relation (see, for example, Penland, 1989 and Penland and Matrosova, 1994). Thompson (1983) has discussed the problem of finding analytical solutions to the stationary form of the FPE (see also Thompson, 1985b).

As an example, consider the one-dimensional dynamical system that includes δ-correlated (white) noise:

$$\dot{X} = \Phi(X) + \eta(t) \quad \text{with:} \quad \langle \eta(t)\eta(t') \rangle = \Gamma\delta(t - t'), \tag{4.74}$$

where η(t) is a normal random variable, with mean zero and variance Γ. The FPE for system (4.74) is obtained by inserting $\dot{X}$ into the LE (4.31) as:

$$\frac{\partial\rho(X,t)}{\partial t} + \frac{\partial}{\partial X}\left[\Phi(X)\rho(X,t) \underbrace{- \frac{\Gamma}{2}\frac{\partial\rho(X,t)}{\partial X}}_{=\eta\rho}\right] = 0, \tag{4.75}$$

where the first term in brackets results from the deterministic component in (4.74) and the second (diffusive) term results from the stochastic component in (4.74). Clearly, for a stationary (equilibrium) solution of (4.75) it is necessary that the bracketed term is a constant.

As an alternative to solving analytically for the stationary solution of (4.75), a numerical integration of the stochastic differential equation (4.74) may be carried out as, for example:

$$X(t + \Delta t) = X(t) + \Phi(X)\Delta t + \sqrt{\Gamma} * \sqrt{\Delta t} * g \quad \text{with:} \quad g \sim \mathrm{N}(0, 1), \tag{4.76}$$

where g is a standard-normal random variable (see, for example, Penland, 1989). Figure 4.12 shows the result of two integrations of Eq. (4.74) according to (4.76)

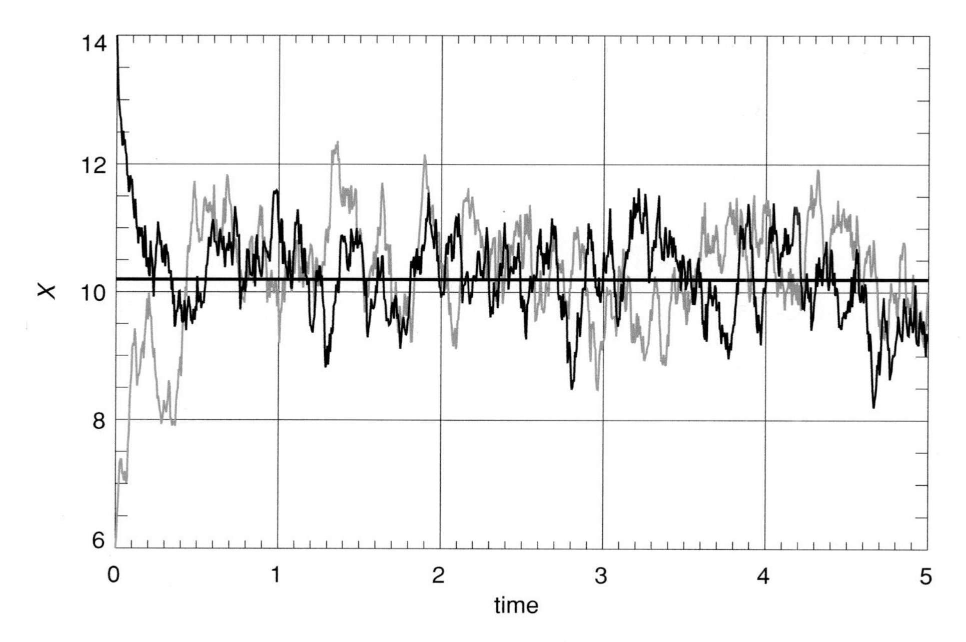

Figure 4.12 Numerical solution of one-dimensional stochastic differential Eq. (4.74) with $\Gamma = 10$ for system (4.77) and two different initial conditions. The equilibrium pdf is not normal; its mean is close to, but not equal to $X_{s,1} = 10.19615$ (black, solid).

with different initial conditions with the deterministic dynamics specified as

$$\Phi(X) = -X^2 + 10X + 2. \tag{4.77}$$

Note that the two stationary solutions of the *unforced* system (4.74), given the dynamics (4.77), are given by $X_{s,1} = 10.19615$ and $X_{s,2} = -0.19615$. It is noted that the equilibrium pdf of the forced system has a mean close to, but not equal to, $X_{s,1}$. For a detailed discussion of the impact of forcing on the equilibrium statistics of non-linear dynamical systems see Palmer (1999, 2000).

4.5.5 The stochastic-dynamic equations

The LE was used by Epstein (1969) as the starting point to derive the so-called stochastic-dynamic equations. The stochastic-dynamic equations describe the time evolution of the statistical moments of a state vector governed by a non-linear dynamical system such as Eq. (4.6). The statistical moments are the expectations of various functions of powers of the state vector.

To review the computation of expectations consider the random variable X with pdf $f(x)$, that is, $X \sim f(x)$, and the random variable Y that is related to X through the deterministic function r:

$$Y = r(X). \tag{4.78}$$

One possibility for the computation of the expectation of Y is given through determining the pdf $g(y)$ of Y (from the known pdf of X), essentially through Eq. (4.49), and then directly compute $E(Y)$ as (e.g. DeGroot, 1986, p. 184):

$$E[r(X)] = E[Y] = \int yg(y)dy. \tag{4.79}$$

In analogy to Eq. (4.79), the expectation of the model state $\mathbf{x}_t$ at time t is computed from the pdf ρ_t for the random variable $\mathbf{x}_t$, as

$$E[\mathbf{x}_t] = \int \mathbf{x}\rho_t(\mathbf{x})d\mathbf{x}. \tag{4.80}$$

As an extension, the expectation of any function s of $\mathbf{x}_t$ (e.g. any power of $\mathbf{x}_t$) is computed as

$$E[s(\mathbf{x}_t)] = \int s(\mathbf{x})\rho_t(\mathbf{x})d\mathbf{x}. \tag{4.81}$$

It is noted that result (4.81) accomplishes the computation of the expectation of the random variable $s(\mathbf{x}_t)$ *without* actually computing its associated pdf (see, for example, DeGroot, 1986), as is the case in Eq. (4.79). Equation (4.81), more abstractly rewritten as

$$E[r(X)] = \int r(x)f(x)dx, \tag{4.82}$$

is, in fact, the fundamental basis of ensemble prediction, in which the Monte Carlo approach is used to compute (among other things) the expectation of the state $\mathbf{x}_t$ at time t:

$$E[\mathbf{x}_t] = E[\mathsf{M}_t(\mathbf{x}_0)] = \int \mathsf{M}_t(\mathbf{x})\rho_0(\mathbf{x})d\mathbf{x}, \tag{4.83}$$

on the basis of the pdf ρ_0 of the initial state variable $\mathbf{x}_0$, where the non-linear model M_t takes on the role of the function r in Eq. (4.82) (see also Paegle and Robl, 1977).

Evidently, from (4.83), or (4.81), the expectation of the variable $s(\mathbf{x}_t)$ is time dependent. That time dependence is described by the stochastic-dynamic equations that govern the time evolution of expectations. An appropriate starting point for deriving these equations is given by taking the time derivative of Eq. (4.81), and subsequently using the transport theorem (4.26) to obtain the second line in Eq. (4.84), where ψ is defined in Eq. (4.33), and the operator $\frac{d}{dt}$ inside the integral is simply the sum of the first two operators inside the integral on the right-hand side of Eq. (4.26):

$$\begin{aligned}
\frac{d}{dt}E[s(\mathbf{x})] &\equiv \frac{d}{dt}\int s(\mathbf{x})\rho_t(\mathbf{x})d\mathbf{x} \\
&\overset{\text{TT}}{=} \int \left(\frac{d}{dt}(s\rho_t) + (s\rho_t)\psi\right) d\mathbf{x} \\
&= \int \left(\rho_t \frac{ds}{dt} + s\frac{d\rho_t}{dt} + s\rho_t\psi\right) d\mathbf{x} \\
&= \int \left(\rho_t \frac{ds}{dt} + s\left[\frac{d\rho_t}{dt} + \rho_t\psi\right]\right) d\mathbf{x} \\
&\overset{\text{LE}}{=} \int \rho_t \frac{ds}{dt} d\mathbf{x} \\
&= E\left[\frac{ds(\mathbf{x})}{dt}\right].
\end{aligned} \tag{4.84}$$

The third and fourth line in Eq. (4.84) are obtained by simple rearrangements, whereas the fifth line is obtained by observing that the bracketed term in the fourth line is zero on the basis of the LE in the form (4.31). The last line in Eq. (4.84) is obtained by recognising that the integral in the fifth line is simply the expectation of the time derivative of s. Note also the close analogy between result (4.30) and the equality expressed through the fifth line in Eq. (4.84). The important result Eq. (4.84) relates the time derivative of the expectation to the expectation of time derivatives and is therefore at the basis of deriving the stochastic-dynamic equations (see also Epstein, 1969, and Fortak, 1973):

$$\frac{d}{dt}E[s(\mathbf{x})] = E\left[\frac{ds(\mathbf{x})}{dt}\right]. \tag{4.85}$$

To illustrate the implications of Eq. (4.85) consider the following one-dimensional example with governing dynamics given by

$$\dot{X} = aX^2 + bX + c. \tag{4.86}$$

Consider the function s_1 as

$$s_1(X) = X, \tag{4.87}$$

with the notations

$$E[s_1(X)] = E[X] \equiv \mu, \quad \sigma^2 \equiv E[(X-\mu)^2]. \tag{4.88}$$

For the function s_1, result (4.85) implies

$$\frac{d\mu}{dt} = E[\dot{X}], \tag{4.89}$$

or, on the basis of (4.86):

$$\frac{d\mu}{dt} = E[aX^2 + bX + c]. \tag{4.90}$$

Using the identity

$$E(X^2) = \sigma^2 + \mu^2, \tag{4.91}$$

the stochastic-dynamic equation (4.90) for the mean μ may be written as

$$\dot{\mu} = a\mu^2 + b\mu + c + a\sigma^2. \tag{4.92}$$

It is obvious from Eq. (4.92) that the equation for the mean μ is *not* obtained by replacing X by μ in the governing dynamics (4.86). An additional term appears in the stochastic-dynamic equation (4.92) that contains the variance σ^2. Clearly, (4.92) can only be evaluated if σ^2 is available as a function of time.

The stochastic-dynamic equation for σ^2 can be derived by considering the function s_2:

$$s_2(X) = (X - \mu)^2, \tag{4.93}$$

with

$$E[s_2(X)] = \sigma^2. \tag{4.94}$$

For the function s_2, result (4.85) implies

$$\frac{d\sigma^2}{dt} = E\left[\frac{d}{dt}(X - \mu)^2\right], \tag{4.95}$$

which may be rewritten as

$$\frac{d\sigma^2}{dt} = E[2(X - \mu)(\dot{X} - \dot{\mu})], \tag{4.96}$$

or, due to Eqs. (4.88) and (4.89), as

$$\frac{1}{2}\frac{d\sigma^2}{dt} = E(X\dot{X}) - \mu\dot{\mu}. \tag{4.97}$$

Since, on the basis of (4.86) and (4.92), the two terms on the right-hand side of (4.97) can be rewritten as

$$\begin{aligned} E(X\dot{X}) &= aE(X^3) + bE(X^2) + c\mu \\ &= a(E(X - \mu)^3 + 3\mu\sigma^2 + \mu^3) + b\sigma^2 + b\mu^2 + c\mu, \\ \mu\dot{\mu} &= \mu(a\mu^2 + b\mu + c + a\sigma^2), \end{aligned} \tag{4.98}$$

the stochastic-dynamic equation (4.97) for σ^2 may be brought in the form:

$$\frac{1}{2}\frac{d\sigma^2}{dt} = aE(X - \mu)^3 + 2a\mu\sigma^2 + b\sigma^2. \tag{4.99}$$

Summarising, the stochastic-dynamic equations for the mean μ and the variance σ^2 for the dynamical system (4.86) are given as:

$$\begin{aligned}\frac{d\mu}{dt} &= a\mu^2 + b\mu + c + a\sigma^2,\\ \frac{d\sigma^2}{dt} &= 4a\mu\sigma^2 + 2b\sigma^2 + 2a\underbrace{E(X-\mu)^3}_{\equiv\vartheta}.\end{aligned} \tag{4.100}$$

As in the context of Eq. (4.92), the stochastic-dynamic equation for the variance σ^2 contains a higher-order moment term that needs to be specified somehow in order for Eqs. (4.100) to be useful. It is clearly possible to derive an equation for the evolution of $\vartheta \equiv E(X-\mu)^3$ by defining the function

$$s_3(X) = (X-\mu)^3, \tag{4.101}$$

and obtaining, on the basis of (4.85), and with the definition of ϑ:

$$\frac{d\vartheta}{dt} = E\left[\frac{d}{dt}(X-\mu)^3\right]. \tag{4.102}$$

Using steps analogous to those necessary to obtain Eq. (4.99), it is possible to rewrite (4.102) as

$$\frac{d\vartheta}{dt} = 6a\mu\vartheta + 3b\vartheta - 3a\sigma^4 + 3aE(X-\mu)^4. \tag{4.103}$$

As expected, the equation for ϑ is unclosed again, containing a fourth-order term. The basic principle of how to obtain the stochastic-dynamic equations is now evident. For a non-linear governing dynamical system, such as (4.86), these equations are non-linear themselves, as well as coupled, and unclosed. No finite closed hierarchy of stochastic-dynamic equations will be derivable for non-linear dynamics. This situation is very similar to the closure problem in turbulence work (e.g. Stanišić, 1988; McComb, 1991). This property is also very clearly stated by Epstein (1969): 'as long as the deterministic prognostic equations are nonlinear it is impossible to write a closed finite set of prognostic equations for the moments. In other words, to predict exactly the future behavior of even the mean of the distribution, all the moments of the distribution (or, equivalently, the entire distribution itself) must be known.' Nevertheless, when a closure assumption is utilised, the stochastic-dynamic equations provide the means to determine moments of the pdf without the need to integrate the governing dynamics multiple times, as is necessary when a Monte Carlo approach is taken.

4.5.6 Classical and statistical mechanics

As already mentioned in Sections 4.2.1 and 4.2.3 (see Eq. 4.29), the LE takes on a particularly simple form for Hamiltonian dynamics. In addition, Liouville's results imply conditions on the evolution of phase space volume for Hamiltonian dynamics that may be written as (e.g. Marion and Thornton, 1988)

$$\dot{q}_k = \frac{\partial H}{\partial p_k} \tag{4.104}$$

and

$$\dot{p}_k = -\frac{\partial H}{\partial q_k} \tag{4.105}$$

for coordinates q_k and momenta p_k. The formulation Eqs. (4.104) and (4.105) specify, for a given Hamiltonian H, a dynamical system of the form (4.6). Apparently it follows with Eqs. (4.104) and (4.105) that the phase space divergence ψ, defined in (4.33), vanishes, since

$$\psi = \frac{\partial \dot{q}_k}{\partial q_k} + \frac{\partial \dot{p}_k}{\partial p_k} = \frac{\partial}{\partial q_k}\frac{\partial H}{\partial p_k} + \frac{\partial}{\partial p_k}\left(-\frac{\partial H}{\partial q_k}\right) = 0. \tag{4.106}$$

The fact that the phase space divergence vanishes for Hamiltonian systems leads to the following two implications. First, the LE (4.32) takes on the simplified form

$$\frac{\partial \rho}{\partial t} + \sum_k \dot{q}_k \frac{\partial \rho}{\partial q_k} + \sum_k \dot{p}_k \frac{\partial \rho}{\partial p_k} = 0, \tag{4.107}$$

which may be expressed by saying, sometimes referred to as the *Liouville theorem*, that the pdf remains constant during (along) the motion (see, for example, Landau and Lifschitz, 1979, and Marion and Thornton, 1988, p. 235). Clearly, $\psi = 0$ implies, by (4.52), that the likelihood ratio h is one, so that the above statement can be seen in terms of the solution of the LE (4.35) as

$$\rho(\mathbf{X}, t) = \rho_0(\Xi). \tag{4.108}$$

Further, by inserting the dynamics into (4.107), Balescu (1991) expresses the LE in the form

$$\frac{\partial \rho}{\partial t} = \underbrace{\sum_k \left(-\frac{\partial H}{\partial p_k}\frac{\partial \rho}{\partial q_k} + \frac{\partial H}{\partial q_k}\frac{\partial \rho}{\partial p_k}\right)}_{\equiv [H,\rho]}, \tag{4.109}$$

or

$$\frac{\partial \rho}{\partial t} = [H, \rho] \tag{4.110}$$

by defining the *linear* operator $[H, \ldots]$ as the *Liouvillian* of the system (4.104) and (4.105). Referring to the central role of the LE in statistical mechanics, Balescu (1991) emphasises the need to exploit the linearity of the LE as thoroughly as possible (see also Figure 4.13).

It is very convenient to write Eq. (2.2.12) in differential form:

$$\frac{\partial F(q,p;t)}{\partial t} \equiv \partial_t F(q,p;t) = -[e^{-(H)t} F(q,p), H]_P$$

or else

$$\partial_t F(q,p;t) = [H(q,p), F(q,p;t)]_P \qquad (2.2.14)$$

This is called the *Liouville equation* and is, beyond any doubt, the most important equation of statistical mechanics, just as the Schrödinger equation is the central equation of quantum mechanics.

The most important feature of the Liouville equation is its *linearity*. This property introduces an important touch of simplicity into an otherwise very complex theory and should be exploited as thoroughly as possible. To stress this feature, it is often convenient to write Eq. (2.2.14) in a slightly different form*:

$$\partial_t F(t) = LF(t) \qquad (2.2.15)$$

where we introduced the linear operator L naturally defined as follows:

$$LF \equiv [H, F]_P = \sum_{n=1}^{N} \left\{ \frac{\partial H}{\partial q_n}\frac{\partial F}{\partial p_n} - \frac{\partial H}{\partial p_n}\frac{\partial F}{\partial q_n} \right\} \qquad (2.2.16)$$

The fundamental operator L will be called the *Liouvillian* of the system.

Figure 4.13 Reference from Balescu (1991, p. 41) concerning the LE for Hamiltonian dynamics.

Second, the property $\psi = 0$ for Hamiltonian systems implies on the basis of the general result (4.29) that the time derivative of a *material* volume in phase space is zero:

$$\frac{dV(t)}{dt} = 0. \qquad (4.111)$$

Consequently, for Hamiltonian evolution (with $\psi = 0$) one has $V(t) = \text{const}$, which is also referred to as the Liouville theorem (e.g. Arnold, 1989; Penrose, 1989; see also Section 4.2.1). The importance of the aspect of volume conservation for Hamiltonian systems has been discussed by Penrose (1989) (see also the quote in Section 4.2.1, as well as Figure 4.14). Penrose (1989) points out that in a way similar to how a small drop of ink placed in a large container of water spreads over the entire contents of the container (while preserving the volume of ink), an initially 'reasonably' shaped region in phase space will distort and stretch in a very complicated way and can get very thinly spread out over huge regions of the phase space (this effect is also illustrated in Figure 4.14).

In addition, through Eq. (4.71), implications exist for the Lyapunov exponents in case of Hamiltonian evolution (with $\psi = 0$).

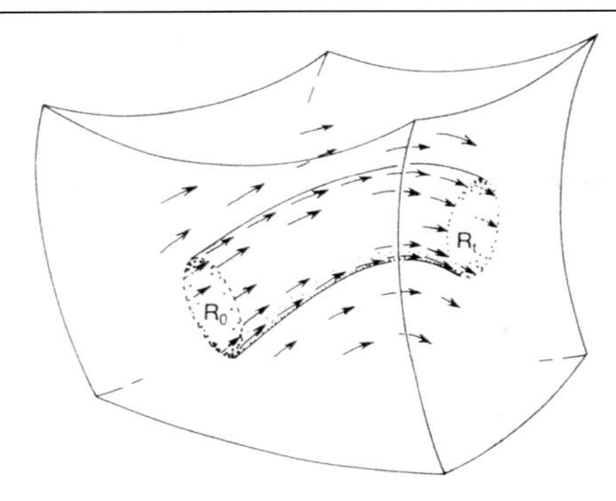

Fig. 5.13. As time evolves, a phase state region R_0 is dragged along by the vector field to a new region R_t. This could represent the time-evolution of a particular alternative for our device.

What can be said about Hamiltonian systems generally? Do regions in phase space tend to spread with time or do they not? It might seem that for a problem of such generality, very little could be said. However, it turns out that there is a very beautiful theorem, due to the distinguished French mathematician Joseph Liouville (1809–1882), that tells us that the *volume* of any region of the phase space must remain constant under any Hamiltonian evolution. (Of course, since our phase space has some high dimension, this has to be a 'volume' in the appropriate high-dimensional sense.) Thus the volume of each R_t must be the *same* as the volume of the original R_0. At first sight this would seem to answer our stability question in the affirmative. For the *size*—in the sense of this phase-space volume—of our region *cannot grow*, so it would seem that our region cannot spread itself out over the phase space.

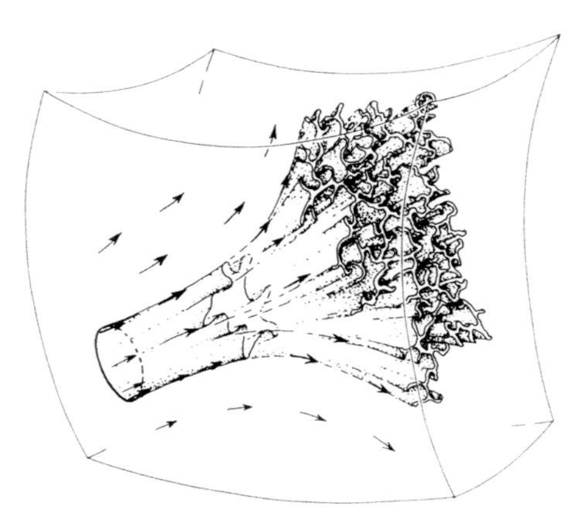

Fig. 5.14. Despite the fact that Liouville's theorem tells us that phase-space volume does not change with time-evolution, this volume will normally *effectively* spread outwards because of the extreme complication of this evolution.

Figure 4.14 Reference from Penrose (1989, p. 181) concerning the LT on phase space volume.

4.5.7 A spectral barotropic model

As a final illustration of the LE it will be demonstrated below that a spectral formulation of the barotropic model has zero divergence in phase space. The consequences for the LE are discussed.

The model equation in physical space is the barotropic quasi-geostrophic potential vorticity equation with free surface and bottom topography h_B:

$$\frac{\partial}{\partial t}\underbrace{[\nabla^2\psi - \Delta R^{-2}\psi]}_{\equiv \mathrm{L}\psi} + J\left(\psi, \underbrace{\nabla^2\psi + f - \Delta R^{-2}\psi + \Delta f_0 \frac{h_B}{\overline{h}}}_{\equiv q}\right) = 0, \qquad (4.112)$$

where ψ is the streamfunction for the geostrophic flow, R is the Rossby deformation radius, defined by $R^2 = g^*\overline{h}/f_0^2$, f is the Coriolis parameter indicating the rotation rate of the fluid, with reference value f_0, g^* is gravity, h_B is the bottom topography (positive above a $z = 0$ reference level), and $\overline{h}$ is the mean depth of the fluid. The operator J denotes the Jacobian operator defined as (**k** is the unit vector in the vertical):

$$J(\psi, q) = \mathbf{k} \cdot (\nabla\psi \times \nabla q), \qquad (4.113)$$

and the parameter Δ is included to allow for a divergent ($\Delta = 1$) or a non-divergent ($\Delta = 0$) model (in physical space). Equation (4.112) results from approximating the full shallow-water equations for a homogeneous rotating fluid assuming nearly geostrophic flow, with q being the geostrophically approximated shallow-water potential vorticity (for further details see Salmon, 1998). It is important to note that flow subject to (4.112) conserves (for appropriate boundary conditions) a form of total energy given as $\frac{1}{2}[(\nabla\psi)^2 + \Delta R^{-2}\psi^2]$. The spectral formulation of Eq. (4.112) for flow on a sphere with radius a, in spherical coordinates λ (longitude) and μ (sine of latitude), assuming an expansion of $\psi(\lambda, \mu, t)$ in terms of the spherical harmonics $Y_n^m(\lambda, \mu)$, may be written as

$$\dot{\psi}_n^m = \alpha_n J_n^m, \quad \text{with:} \quad \alpha_n \equiv \frac{a^2}{n(n+1) + \Delta(a/R)^2}, \qquad (4.114)$$

where $\psi_n^m(t)$ are the time-dependent expansion coefficients (observe the dot on top of ψ_n^m in Eq. (4.114) indicating time differentiation), the factor α_n results from inverting the linear operator L in Eq. (4.112), and the spectral Jacobian J_n^m is given as

$$J_n^m = \frac{1}{4\pi}\int_0^{2\pi}\int_{-1}^{+1} J(\psi, q)\left(Y_n^m\right)^* d\mu d\lambda, \qquad (4.115)$$

where $(Y_n^m)^*$ denotes the complex conjugate spherical harmonic. The spectral formulation (4.114) describes one of the simplest geophysical flow examples for an N-dimensional system of first-order autonomous differential equations of the form

(4.6), as it represents a set of non-linear coupled ordinary differential equations. The number N of equations depends on the type and degree of spectral truncation.

The special property of zero phase space divergence of system (4.114) results by first inserting the spectral expansions for ψ and q into the Jacobian formulation (4.115):

$$J_n^m = \frac{1}{4\pi} \int_0^{2\pi} \int_{-1}^{+1} J \left\{ \sum_{m'} \sum_{n'} \psi_{n'}^{m'}(t) Y_{n'}^{m'}(\lambda, \mu), \sum_{m''} \sum_{n''} q_{n''}^{m''}(t) Y_{n''}^{m''}(\lambda, \mu) \right\} \left(Y_n^m \right)^* d\mu d\lambda, \tag{4.116}$$

and subsequently rearranging integrals and summations as

$$J_i = \sum_j \sum_k \psi_j q_k \underbrace{\frac{1}{4\pi} \int_0^{2\pi} \int_{-1}^{+1} Y_i^* J(Y_j, Y_k) d\mu d\lambda}_{\equiv A_{ijk}} = \sum_{j,k} A_{ijk} \psi_j q_k. \tag{4.117}$$

In Eq. (4.117), the summation indices j and k (as well as the free index i) must be understood to stand for the index pair (n, m) in the previous equation (4.116). Note that the interaction coefficients A_{ijk}, defined in (4.117), are independent of the actual fields and may be precomputed; it is noted in passing that (4.117) forms the basis of the fully spectral (but computationally highly inefficient) interaction coefficient method (see Bourke, 1972, for a comparison of that method with the so-called transform method). The spectral expansion of q may be found from the spectral expansion of ψ as

$$q_n^m = -\alpha_n^{-1} \psi_n^m + g_n^m, \quad \text{with:} \quad g_n^m \equiv \left(f + \Delta f_0 \frac{h_B}{\overline{h}} \right)_n^m, \tag{4.118}$$

where g_n^m denotes the spectral coefficients of the expansion of the term in parenthesis in (4.118). In the more symbolic notation used in (4.117), Eq. (4.118) reads

$$q_k = -\alpha_k^{-1} \psi_k + g_k. \tag{4.119}$$

On the basis of (4.117) and (4.119), the spectral equation (4.114) may be written in the form:

$$\dot{\psi}_i = \underbrace{\alpha_i \sum_{j,k} A_{ijk} \psi_j \left(-\alpha_k^{-1} \psi_k + g_k \right)}_{\equiv f_i(\psi_l)}. \tag{4.120}$$

Obviously, a function f_i may be defined, as done in Eq. (4.120), that defines the right-hand side of Eq. (4.114), or, equivalently, Eq. (4.120), in terms of ψ_l and precomputed coefficients. Equation (4.120) most clearly demonstrates the analogy between the geophysical example Eq. (4.112) and the prototypical system Eq. (4.6).

At this point the property of the interaction coefficients to vanish if two indices are equal:

$$A_{iik} = A_{iji} = 0 \tag{4.121}$$

is observed that is basically a consequence of properties of the Jacobian operator (4.113) and the definition of the interaction coefficients in (4.117). It therefore follows that f_i, defined in (4.120), does not depend on ψ_i. This result clearly implies that the divergence in phase space (defined as ψ in Eq. 4.33) of system (4.120) is zero, even though the flow may evidently be divergent in physical space (for $\Delta = 1$). In other words, the dynamics (4.112) has no divergence in phase space (see also Salmon *et al.*, 1976). Using the symbolic notation that the vector $\mathbf{X}$ contains the ordered real and imaginary expansion coefficients:

$$\mathbf{X} = \begin{pmatrix} \dots \\ {}_r(\psi)_n^m \\ {}_i(\psi)_n^m \\ \dots \end{pmatrix}, \tag{4.122}$$

the dynamics (4.120) are written as

$$\dot{\mathbf{X}} = \mathbf{f}(\mathbf{X}), \tag{4.123}$$

(see Eq. 4.6) with divergence-free $\mathbf{f}$. As a consequence, the LE, as written in Section 4.2.4, takes on the form

$$\frac{\partial \rho}{\partial t} + \dot{\mathbf{X}} \cdot \nabla \rho = 0, \tag{4.124}$$

with solution (see Eq. 4.35) given by $\rho(\mathbf{X}, t) = \rho_0(\Xi)$; all the consequences discussed in Section 4.5.6 for a system with vanishing phase space divergence apply. Due to the phase-space non-divergence condition of the barotropic dynamics (4.112), the *stationary* solution ρ_s to the LE (4.124) is determined by:

$$\dot{\mathbf{X}} \cdot \nabla \rho_s = 0, \quad \text{or:} \quad \sum_k f_k(\mathbf{X}) \frac{\partial \rho_s}{\partial X_k} = 0, \tag{4.125}$$

with the definitions introduced above, in particular through Eq. (4.122) and (4.120). This linear last equation (4.125) determines the stationary pdf ρ_s (see also Thompson, 1983). For further discussion, reference is made to Holloway (1986), Salmon *et al.* (1976), and Salmon (1998).

4.6 Conclusions

The Liouville equation governs the time evolution of the pdf of the state of a dynamical system. Since such dynamical systems play a prominent role as prediction models in the atmospheric sciences in general, and in NWP in particular, the LE, governing the relevant pdfs, plays a fundamental role in atmospheric predictability. As the continuity equation for probability, the LE contains all the relevant information about the predictability of a (forced) dynamical system.

The LE is demonstrably linear and it is a partial differential equation. As the LE describes the entire pdf, closure problems typical for non-linear dynamical systems are avoided, as is the need for obtaining many realisations of the system, if specific values of the pdf are of interest. The LE (see also Ehrendorfer, 2003) provides the basis for making probabilistic statements about specific forecasts, as well as about atmospheric predictability in general. As such, the LE is fundamental for the areas of ensemble prediction (see, for example, Molteni *et al.*, 1996; Hamill *et al.*, 2000; Hamill *et al.*, 2003; Lewis, 2005) and atmospheric predictability (see, for example, Simmons and Hollingsworth, 2002; Weiss, 2003b; Tribbia and Baumhefner, 2004).

In the context considered here, an 'analytical' solution to the LE (see Eq. 4.35) was given for arbitrary governing dynamics. This solution avoids the fully numerical solution of an initial-value problem for a partial differential equation with very many independent variables. In the context of that solution, various considerations related to singular vectors, Lyapunov exponents, stochastic differential equations, the Fokker–Planck equation, the stochastic-dynamic equations, and statistical mechanics were considered in this chapter.

It is clear that a fundamental difficulty in dealing with the LE is related to the extremely high dimensionality of phase space that must be considered in contexts of realistic NWP models. Nevertheless, its theoretical attractiveness makes the LE a fundamental tool for exploring various aspects and relationships in the area of atmospheric predictability.

References

Abraham, R. and J. E. Marsden (1978). *Foundations of Mechanics*. Addison-Wesley.

Arnold, V. I. (1989). *Mathematical Methods of Classical Mechanics*. Springer.

Balescu, R. (1991). *Equilibrium and Nonequilibrium Statistical Mechanics*. Krieger Publishing Company.

Bjerknes, V. (1904). Das Problem der Wettervorhersage, betrachtet vom Standpunkte der Mechanik und der Physik. *Meteorol. Z.*, **21**, 1–7.

Bourke, W. (1972). An efficient, one-level, primitive-equation spectral model. *Mon. Weather Rev.*, **100**, 683–9.

Chu, P. C., L. M. Ivanov, T. M. Margolina and O. V. Melnichenko (2002). Probabilistic stability of an atmospheric model to various amplitude perturbations. *J. Atmos. Sci.*, **59**, 2860–73.

de Groot, S. R. and P. Mazur (1984). *Non-Equilibrium Thermodynamics*. Dover.

DeGroot, M. H. (1986). *Probability and Statistics*. Addison-Wesley.

Dudewicz, E. J. and S. N. Mishra (1988). *Modern Mathematical Statistics*. Wiley.

Ehrendorfer, M. (1994a). The Liouville equation and its potential usefulness for the prediction of forecast skill. I: Theory. *Mon. Weather Rev.*, **122**, 703–13.

(1994b). The Liouville equation and its potential usefulness for the prediction of forecast skill. II: Applications. *Mon. Weather Rev.*, **122**, 714–28.

(1997). Predicting the uncertainty of numerical weather forecasts: a review. *Meteorol. Z., N.F.*, **6**, 147–83.

(1999). Prediction of the uncertainty of numerical weather forecasts: problems and approaches. In *Proceedings ECMWF Workshop on Predictability, 20–22 October 1997*, pp. 27–99. ECMWF.

(2003). The Liouville equation in atmospheric predictability. In *Proceedings ECMWF Seminar on Predictability of Weather and Climate, 9–13 September 2002*, pp. 47–81. ECMWF.

Ehrendorfer, M. and J. J. Tribbia (1997). Optimal prediction of forecast error covariances through singular vectors. *J. Atmos. Sci.*, **54**, 286–313.

Epstein, E. S. (1969). Stochastic dynamic prediction. *Tellus*, **21**, 739–59.

Errico, R. M., M. Ehrendorfer and K. Raeder (2001). The spectra of singular values in a regional model. *Tellus*, **53A**, 317–32.

Errico, R. M., R. Langland and D. P. Baumhefner (2002). The workshop in atmospheric predictability. *Bull. Am. Meteorol. Soc.*, **83**, 1341–3.

Fortak, H. (1973). Prinzipielle Grenzen der deterministischen Vorhersagbarkeit atmosphärischer Prozesse. *Ann. Meteorol., N.F.*, **6**, 111–20.

Gardiner, C. W. (1990). *Handbook of Stochastic Methods for Physics, Chemistry and the Natural Sciences*. Springer.

Gleeson, T. A. (1966). A causal relation for probabilities in synoptic meteorology. *J. Appl. Meteorol.*, **5**, 365–8.

Goldstein, H. (1980). *Classical Mechanics*. Addison-Wesley.

Hamill, T. M., S. L. Mullen, C. Snyder, Z. Toth and D. P. Baumhefner (2000). Ensemble forecasting in the short to medium range: report from a workshop. *Bull. Am. Meteorol. Soc.*, **81**, 2653–64.

Hamill, T. M., C. Snyder and J. S. Whitaker (2003). Ensemble forecasts and the properties of flow-dependent analysis-error covariance singular vectors. *Mon. Weather Rev.*, **131**, 1741–58.

Hasselmann, K. (1976). Stochastic climate models. I: Theory. *Tellus*, **28**, 473–85.

Holloway, G. (1986). Eddies, waves, circulation, and mixing: statistical geofluid mechanics. *Annu. Rev. Fluid Mech.*, **18**, 91–147.

Hu, Y. and R. T. Pierrehumbert (2001). The advection-diffusion problem for stratospheric flow. I: Concentration probability distribution function. *J. Atmos. Sci.*, **58**, 1493–510.

Landau, L. D. and E. M. Lifschitz (1979). *Lehrbuch der Theoretischen Physik. Band V Statistische Physik, Teil 1*. Berlin: Akademie-Verlag.

Leith, C. E. (1974). Theoretical skill of Monte Carlo forecasts. *Mon. Weather Rev.*, **102**, 409–18.

Lewis, J. M. (2005). Roots of ensemble forecasting. *Mon. Weather Rev.*, **133**(7), 1865–85.

Lin, C. C. and L. A. Segel (1988). *Mathematics Applied to Deterministic Problems in the Natural Sciences*. Society for Industrial and Applied Mathematics.

Lindenberg, K. and B. J. West (1990). *The Nonequilibrium Statistical Mechanics of Open and Closed Systems*. VCH Publishers.

Liouville, J. (1838). Sur la théorie de la variation des constantes arbitraires. *J. Math. Pures Appl.*, **3**, 342–9.

Lorenz, E. N. (1963). Deterministic nonperiodic flow. *J. Atmos. Sci.*, **20**, 130–41.

(1969a). The predictability of a flow which possesses many scales of motion. *Tellus*, **21**, 289–307.

(1969b). Three approaches to atmospheric predictability. *Bull. Am. Meteorol. Soc.*, **50**, 345–9.

(1982). Atmospheric predictability experiments with a large numerical model. *Tellus*, **34**, 505–13.

(1984). Irregularity: a fundamental property of the atmosphere. *Tellus*, **36A**, 98–110.

(1993). *The Essence of Chaos*. University of Washington Press.

Lützen, J. (1990). *Joseph Liouville (1809–1882): Master of Pure and Applied Mathematics*. Springer.

Marion, J. B. and S. T. Thornton (1988). *Classical Dynamics*. Harcourt Brace Jovanovich.

McComb, W. D. (1991). *The Physics of Fluid Turbulence*. Oxford University Press.

Merryfield, W. J., P. F. Cummins and G. Holloway (2001). Equilibrium statistical mechanics of barotropic flow over finite topography. *J. Phys. Oceanog.*, **31**, 1880–90.

Molteni, F., R. Buizza, T. N. Palmer and T. Petroliagis (1996). The ECMWF ensemble prediction system: methodology and validation. *Quart. J. Roy. Meteor. Soc.*, **122**, 73–119.

Nicolis, G. (1995). *Introduction to Nonlinear Science*. Cambridge University Press.

Paegle, J. and E. Robl (1977). The time behavior of the probability density function of some simplified atmospheric flows. *J. Atmos. Sci.*, **34**, 979–90.

Palmer, T. N. (1999). A nonlinear dynamical perspective on climate prediction. *J. Climate*, **12**, 575–91.

(2000). Predicting uncertainty in forecasts of weather and climate. *Rep. Prog. Phys.*, **63**, 71–116.

Parker, T. S. and L. O. Chua (1989). *Practical Numerical Algorithms for Chaotic Systems*. Springer-Verlag.

Penland, C. (1989). Random forcing and forecasting using principal oscillation pattern analysis. *Mon. Weather Rev.*, **117**, 2165–85.

Penland, C. and L. Matrosova (1994). A balance condition for stochastic numerical models with application to the El Niño-Southern Oscillation. *J. Climate*, **7**, 1352–72.

Penrose, O. (1970). *Foundations of Statistical Mechanics*. Pergamon Press.

Penrose, R. (1989). *The Emperor's New Mind: Concerning Computers, Minds, and the Laws of Physics*. Penguin Books.

Risken, H. (1989). *The Fokker-Planck Equation: Methods of Solution and Applications*. Springer.

Salmon, R. (1998). *Lectures on Geophysical Fluid Dynamics*. Oxford University Press.

Salmon, R., G. Holloway and M. C. Hendershott (1976). The equilibrium statistical mechanics of simple quasi-geostrophic models. *J. Fluid Mech.*, **75**, 691–703.

Simmons, A. J. and A. Hollingsworth (2002). Some aspects of the improvement in skill in numerical weather prediction. *Quart. J. Roy. Meteor. Soc.*, **128**, 647–77.

Soize, C. (1994) *The Fokker-Planck Equation for Stochastic Dynamical Systems and its Explicit Steady State Solutions*. World Scientific.

Stanišić, M. M. (1988). *The Mathematical Theory of Turbulence*. Springer.

Strang, G. (1993). *Introduction to Linear Algebra*. Wellesley-Cambridge Press.

Tatarskiy, V. I. (1969). The use of dynamic equations in the probability prediction of the pressure field. *Izv. Atmos. Ocean. Physics*, **5**, 293–7.

Thompson, P. D. (1957). Uncertainty of initial state as a factor in the predictability of large scale atmospheric flow patterns. *Tellus*, **9**, 275–95.

(1972). Some exact statistics of two-dimensional viscous flow with random forcing. *J. Fluid. Mech.*, **55**, 711–17.

(1983). Equilibrium statistics of two-dimensional viscous flows with arbitrary random forcing. *Phys. Fluids*, **26**, 3461–70.

(1985a). Prediction of probable errors of predictions. *Mon. Weather Rev.*, **113**, 248–59.

(1985b). A statistical-hydrodynamical approach to problems of climate and its evolution. *Tellus*, **37A**, 1–13.

Tribbia, J. and D. Baumhefner (1993). On the problem of prediction beyond the deterministic range. In *Prediction of Interannual Climate Variations*, ed. J. Shukla, pp. 251–65. NATO ASI Series, Vol. 16. Springer.

(2004). Scale interactions and atmospheric predictability: an updated perspective. *Mon. Weather Rev.*, **132**, 703–13.

van Kampen, N. G. (1992). *Stochastic Processes in Physics and Chemistry*. North-Holland.

Weiss, J. B. (2003a). Coordinate invariance in stochastic dynamical systems. *Tellus*, **55A**, 208–18.

(2003b). Predictability and chaos. In *Handbook of Weather, Climate, and Water: Dynamics, Climate, Physical Meteorology, Weather Systems, and Measurements*, ed. T. D. Potter and B. R. Colman, pp. 83–93. Wiley.

Zwillinger, D. (1989). *Handbook of Differential Equations*. Academic Press.

5

Application of generalised stability theory to deterministic and statistical prediction

Petros J. Ioannou
Department of Physics,
National and Capodistrian University of Athens

Brian F. Farrell
Harvard University, Cambridge

Understanding of the stability of deterministic and stochastic dynamical systems has evolved recently from a traditional grounding in the system's normal modes to a more comprehensive foundation in the system's propagator and especially in an appreciation of the role of non-normality of the dynamical operator in determining the system's stability as revealed through the propagator. This set of ideas, which approach stability analysis from a non-modal perspective, will be referred to as generalised stability theory (GST). Some applications of GST to deterministic and statistical forecast are discussed in this review. Perhaps the most familiar of these applications is identifying initial perturbations resulting in greatest error in deterministic error systems, which is in use for ensemble and targeting applications. But of increasing importance is elucidating the role of temporally distributed forcing along the forecast trajectory and obtaining a more comprehensive understanding of the prediction of statistical quantities beyond the horizon of deterministic prediction. The optimal growth concept can be extended to address error growth in non-autonomous systems in which the fundamental mechanism producing error growth can be identified with the necessary non-normality of the system. In this review the influence of model error in both the forcing and the system is examined using the methods of stochastic dynamical systems theory. Deterministic and statistical prediction are separately discussed.

Predictability of Weather and Climate, ed. Tim Palmer and Renate Hagedorn. Published by Cambridge University Press.

5.1 Introduction

The atmosphere and ocean are constantly evolving and the present state of these systems, while notionally deterministically related to previous states, in practice becomes exponentially more difficult to predict as time advances. This loss of predictability of the deterministic state is described as sensitive dependence on initial conditions and quantified by the asymptotic exponential rate of divergence of initially nearby trajectories in the phase space of the forecast system (Lorenz, 1963) given by the first Lyapunov exponent (Lyapunov, 1907; Oseledets, 1968). Moreover, the optimality of the Kalman filter as a state identification method underscores the essentially statistical nature of the prediction problem (Ghil and Malanotte-Rizzoli, 1991; Berliner, 1996). The initial state is necessarily uncertain but so is the forecast model itself and the system is subject to perturbations from extrinsic and subgrid-scale processes. Given all these uncertainties the notion of a single evolving point in phase space is insufficient as a representation of our knowledge of forecast dynamics, and some measure of the uncertainty of the system state and the evolution of this uncertainty must be included in a comprehensive forecast system theory (Epstein, 1969; Ehrendorfer, this volume; Palmer, this volume).

The appropriate methods for studying errors in deterministic and statistical forecast are based on the system's propagator and proceed from advances in mathematics (Schmidt, 1906; Mirsky, 1960; Oseledets, 1968) and dynamical theory (Lorenz, 1963, 1965, 1985; Farrell, 1988, 1990; Lacarra and Talagrand, 1988; Molteni and Palmer, 1993; Penland and Magorian, 1993; Buizza and Palmer, 1995; Farrell and Ioannou, 1996a, 1996b; Moore and Kleeman, 1996; Kleeman and Moore, 1997; Palmer, 1999; DelSole and Hou, 1999a, 1999b; Ehrendorfer, this volume; Palmer, this volume; Timmermann and Jin, this volume).

We review recent advances in linear dynamical system and stability theory relevant to deterministic and statistical forecast. We begin with deterministic error dynamics in autonomous and non-autonomous certain systems and then address the problem of prediction of statistical quantities beyond the deterministic time horizon; finally, we study model error in certain and uncertain systems.

5.2 Deterministic predictability of certain systems

The variables in a certain forecast model are specified by the finite dimensional state vector $\mathbf{y}$ which is assumed to evolve according to the deterministic equation

$$\frac{d\mathbf{y}}{dt} = \mathbf{f}(\mathbf{y}). \tag{5.1}$$

Consider a solution of the forecast equations $\mathbf{y}(t)$ starting from a given initial state. Sufficiently small forecast errors $\mathbf{x} \equiv \delta\mathbf{y}$ are governed by the tangent linear equations

$$\frac{d\mathbf{x}}{dt} = \mathbf{A}(t)\,\mathbf{x}, \tag{5.2}$$

in which the Jacobian matrix

$$\mathbf{A}(t) \equiv \left.\frac{\partial \mathbf{f}}{\partial \mathbf{y}}\right|_{\mathbf{y(t)}}, \tag{5.3}$$

is evaluated along the known trajectory $\mathbf{y}(t)$.

The matrix $\mathbf{A}(t)$ is time dependent and in general its realisations do not commute, i.e. $\mathbf{A}(t_1)\mathbf{A}(t_2) \neq \mathbf{A}(t_2)\mathbf{A}(t_1)$. It follows that the evolution of the error field cannot be determined from analysis of the eigenvalues and eigenfunctions of $\mathbf{A}$, as would be the case for time independent normal matrices, but instead the analysis must be made using the methods of generalized stability theory (GST) (for a review see Farrell and Ioannou, 1996a, 1996b). GST concentrates attention on the behaviour of the propagator $\Phi(t, 0)$, which is the matrix that maps the initial error $\mathbf{x}(0)$ to the error at time t:

$$\mathbf{x}(t) = \Phi(t, 0)\,\mathbf{x}(0). \tag{5.4}$$

Once the matrix $\mathbf{A}(t)$ of the tangent linear system is available, the propagator is readily calculated. Consider a piecewise approximation of the continuous operator $\mathbf{A}(t)$: $\mathbf{A}(t) = \mathbf{A}_i$ where $\mathbf{A}_i$ is the mean of $\mathbf{A}(t)$ over $(i-1)\tau \leq t < i\tau$ for small enough τ. At time $t = n\tau$ the propagator is approximated by the time ordered product

$$\Phi(t, 0) = \prod_{i=1}^{n} e^{\mathbf{A}_i \tau}. \tag{5.5}$$

If $\mathbf{A}$ is autonomous (time independent) the propagator is the matrix exponential

$$\Phi(t, 0) = e^{\mathbf{A}t}. \tag{5.6}$$

Deterministic error growth is bounded by the optimal growth over the interval $[0, t]$:

$$\|\Phi(t, 0)\| \equiv \max_{\mathbf{x}(0)} \frac{\|\mathbf{x}(t)\|}{\|\mathbf{x}(0)\|}. \tag{5.7}$$

This maximisation is over all initial errors $\mathbf{x}(0)$. The optimal growth for each t is the norm of the propagator $\|\Phi(t, 0)\|$. The definition of the optimal implies a choice of norm. In many applications $\|\mathbf{x}(t)\|^2$ is chosen to correspond to the total perturbation energy.

We illustrate GST by applying it to the simple autonomous Reynolds[1] matrix $\mathbf{A}$:

$$\mathbf{A} = \begin{pmatrix} -1 & 100 \\ 0 & -2 \end{pmatrix}. \tag{5.8}$$

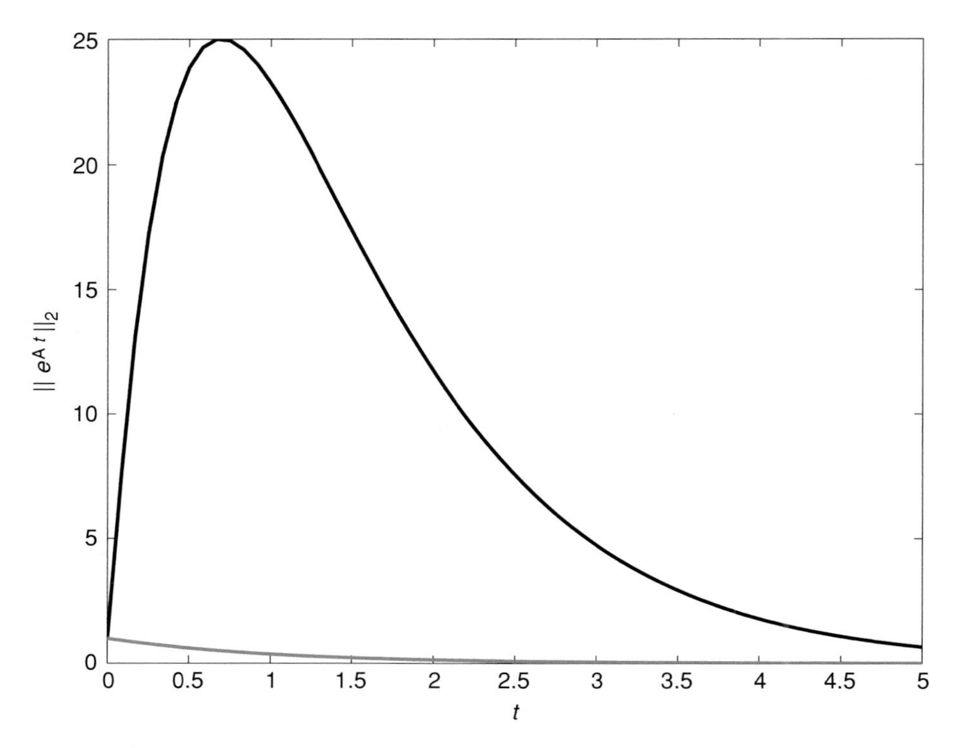

Figure 5.1 The upper curve gives the optimal growth as a function of time for the Reynold's matrix (5.8). The optimal growth is given by the norm of the propagator $e^{\mathbf{A}t}$. The lower curve shows the evolution of the amplitude of the least damped eigenmode which decays at rate -1.

Consider the model tangent linear system:

$$\frac{d\mathbf{x}}{dt} = \mathbf{A}\mathbf{x}. \tag{5.9}$$

Traditional stability theory concentrates on the growth associated with the most unstable mode, which in this example gives decay at rate -1, suggesting that the error decays exponentially at this rate. While this is indeed the case for very large times, the optimal error growth, shown by the upper curve in Figure 5.1, is much greater at all times than that predicted by the fastest growing mode (the lower curve in Figure 5.1). The modal prediction fails to capture the error growth because $\mathbf{A}$ is non-normal, i.e. $\mathbf{A}\mathbf{A}^\dagger \neq \mathbf{A}^\dagger\mathbf{A}$ and its eigenfunctions are not orthogonal.

The optimal growth is calculated as follows:

$$G = \frac{\|\mathbf{x}(t)\|^2}{\|\mathbf{x}(0)\|^2} = \frac{\mathbf{x}(t)^\dagger\mathbf{x}(t)}{\mathbf{x}(0)^\dagger\mathbf{x}(0)} = \frac{\mathbf{x}(0)^\dagger e^{\mathbf{A}^\dagger t} e^{\mathbf{A}t}\mathbf{x}(0)}{\mathbf{x}(0)^\dagger\mathbf{x}(0)}. \tag{5.10}$$

This Rayleigh quotient reveals that the maximum eigenvalue of the positive definite matrix $e^{\mathbf{A}^\dagger t} e^{\mathbf{A}t}$ determines the square of the optimal growth at time t. The corresponding eigenvector is the initial perturbation that leads to this growth, called the optimal perturbation (Farrell, 1988). Alternatively, we can proceed with a Schmidt decomposition (singular value decomposition) of the propagator:

$$e^{\mathbf{A}t} = \mathbf{U}\Sigma\mathbf{V}^\dagger \tag{5.11}$$

with $\mathbf{U}$ and $\mathbf{V}$ unitary matrices and Σ the diagonal matrix with elements the singular values of the propagator, σ_i, which give the growth achieved at time t by each of the orthogonal columns of $\mathbf{V}$. The largest singular value is the optimal growth and the corresponding column of $\mathbf{V}$ is the optimal perturbation. The orthogonal columns, $\mathbf{v}_i$, of $\mathbf{V}$ are called optimal vectors (or right singular vectors), and the orthogonal columns, $\mathbf{u}_i$, of $\mathbf{U}$ are the evolved optimal vectors (or left singular vectors) because from the Schmidt decomposition we have

$$\sigma_i \mathbf{u}_i = e^{\mathbf{A}t}\mathbf{v}_i. \tag{5.12}$$

The forecast system has typical dimension 10^7 so we cannot calculate the propagator directly as in Eq. (5.5) in order to obtain the optimal growth. Instead we integrate the system

$$\frac{d\mathbf{x}}{dt} = \mathbf{A}\mathbf{x} \tag{5.13}$$

forward to obtain $\mathbf{x}(t) = e^{\mathbf{A}t}\mathbf{x}(0)$ (or its equivalent in a time dependent system), and then integrate the adjoint system

$$\frac{d\mathbf{x}}{dt} = -\mathbf{A}^\dagger\mathbf{x} \tag{5.14}$$

backward in order to obtain $e^{\mathbf{A}^\dagger t}\mathbf{x}(t) = e^{\mathbf{A}^\dagger t}e^{\mathbf{A}t}\mathbf{x}(0)$. We can then find the optimal vectors (singular vectors) by the power method (Moore and Farrell, 1993; Molteni and Palmer, 1993; Errico, 1997). The leading optimal vectors are useful input for selecting the ensemble members in ensemble forecast (Buizza, this volume; Kalnay *et al.*, this volume) because they span and order in growth the initial error (Gelaro *et al.*, 1998). They also identify sensitive regions that can be targeted for further observation (Thorpe and Petersen, this volume).

We have remarked that optimal growth depends on the norm. The choice of norm is dictated by the physical situation; we are usually interested in growth in energy but other norms can be selected to concentrate on the perturbation growth in other physical measures such as growth in square surface pressure, or in square potential vorticity (for a discussion of the choice of the inner product see Palmer *et al.*, 1998; for a discussion of norms that do not derive from inner products see Farrell and Ioannou, 2000). Formally for autonomous operators there exist 'normal coordinates' in which the operator is rendered normal; however, this coordinate system is not usually physical in the sense that the inner product in these coordinates is not usually associated with a physically useful measure. But a more deeply consequential reason why the concept of 'normal coordinates' is not useful is that time dependent operators, such as the tangent linear forecast system, are inherently non-normal, in the sense that there is no transformation of coordinates that renders a general $\mathbf{A}(t)$ normal at all times. It follows that analysis of error growth in time dependent systems necessarily proceeds through analysis of the propagator as outlined above.

The tangent linear forecast system is generally assumed to be asymptotically unstable in the sense that the Lyapunov exponent of the tangent linear system is positive. Lyapunov showed that for a general class of time dependent but bounded matrices $\mathbf{A}(t)$ the perturbations $\mathbf{x}(t)$ grow at most exponentially so that $\| \mathbf{x}(t) \| \propto e^{\lambda t}$ as $t \to \infty$, where λ is the top Lyapunov exponent of the tangent linear system which can be calculated by evaluating the limit

$$\lambda = \overline{\lim}_{t\to\infty} \frac{\ln \| \mathbf{x}(t) \|}{t}. \tag{5.15}$$

This asymptotic measure of error growth is independent of the norm, $\| \cdot \|$.

It is of interest and of practical importance to determine the perturbation subspace that supports this asymptotic exponential growth of errors. Because this subspace has a much smaller dimension than that of the tangent linear system itself, a theory that characterises this subspace can lead to economical truncations of the tangent linear system. Such a truncation could be used in advancing the error covariance of the tangent linear system which is required for optimal state estimation. We now show that the inherent non-normality of time dependent operators is the source of the Lyapunov instability which underlies the exponential increase of forecast errors and that understanding the role of non-normality is key to understanding error growth.

Consider a harmonic oscillator with frequency ω. In normal coordinates (i.e. energy coordinates), $\mathbf{y} = [\omega x, v]^T$, where x is the displacement and $v = \dot{x}$, the system is governed by

$$\frac{d\mathbf{y}}{dt} = \mathbf{A}\mathbf{y}, \tag{5.16}$$

with

$$\mathbf{A} = \omega \begin{pmatrix} 0 & 1 \\ -1 & 0 \end{pmatrix}. \tag{5.17}$$

This is a normal system $\mathbf{A}\mathbf{A}^\dagger = \mathbf{A}^\dagger\mathbf{A}$ and the system trajectory lies on a constant energy surface, which is a circle. In these coordinates the perturbation amplitude is the radius of the circle and there is no growth.

Assume now that the frequency switches periodically between ω_1 and ω_2: so that for $T_1 = \pi/(2\omega_1)$ units of time the frequency is ω_1 and then for $T_2 = \pi/(2\omega_2)$ units of time the frequency is ω_2. There is no single transformation of coordinates that renders the matrix $\mathbf{A}$ normal when $\omega = \omega_1$ and also when $\omega = \omega_2$ so we revert to

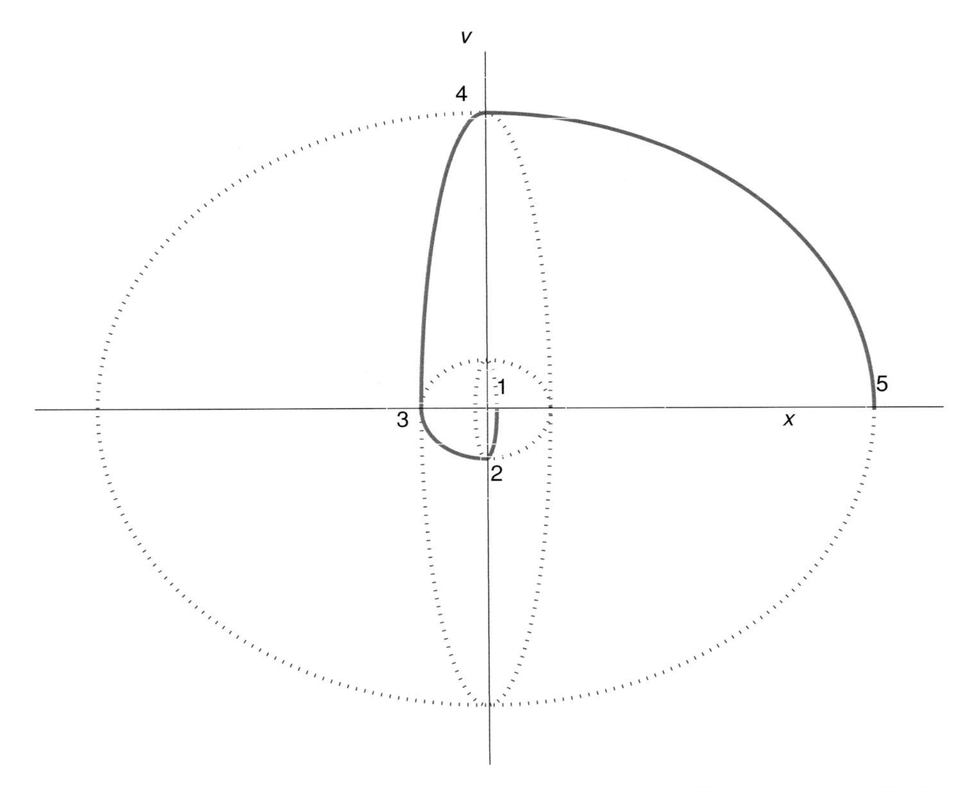

Figure 5.2 The parametric instability of the harmonic oscillator governed by (5.18) is caused by the non-normality of the time dependent operator. So long as there is instantaneous growth and the time dependent operators do not commute, asymptotic exponential growth occurs. For this example $\omega_1 = 1/2$ and $\omega_2 = 3$. See text for explanation of numbers 1 to 5.

the state $\mathbf{y} = [x, v]^T$ with dynamical matrices:

$$\mathbf{A}_{1,2} = \begin{pmatrix} 0 & 1 \\ -\omega_{1,2}^2 & 0 \end{pmatrix}. \tag{5.18}$$

When the frequency is ω_1 the state $\mathbf{y}$ traverses the ellipses of Figure 5.2 that are elongated in the direction of the x axis and when the frequency is ω_2 it traverses the ellipses of Figure 5.2 that are elongated in the v axis (both marked with dots). The dynamics of this system can be understood by considering the evolution of the initial condition $\mathbf{y}(0) = [1, 0]$ marked with 1 in Figure 5.2. Initial condition 1 goes to 2 at time $t = T_2$ under the dynamics of $\mathbf{A}_2$, then the dynamics switch to $\mathbf{A}_1$ taking the system from 2 to 3 at time $t = T_1 + T_2$; reverting back to $\mathbf{A}_2$, the system advances from 3 to 4 at $t = T_1 + 2T_2$, and then under $\mathbf{A}_1$, 4 goes to 5 at time $t = 2(T_1 + T_2)$ with coordinates $\mathbf{y}(2(T_1 + T_2)) = [36, 0]$. As time advances the trajectory clearly grows exponentially as this cycle is repeated despite the neutral stability of the system at each instant of time. How is this possible? The key lies in the inherent non-normality of the operator in time dependent systems. If the operator were time independent and stable, transient growth would necessarily give way to eventual decay. In contrast, a time dependent operator reamplifies the perturbations that would have otherwise decayed. This process of continual rejuvenation producing asymptotic destabilisation is generic and does not depend on the stability properties of the instantaneous operator state (Farrell and Ioannou, 1999).

As a further example consider harmonic perturbations $\Psi(y,t)e^{ikx}$ on a time dependent barotropic mean flow $U(y,t)$ in a β plane channel $-1 \le y \le 1$. The perturbations evolve according to

$$\frac{d\Psi}{dt} = \mathbf{A}(t)\Psi, \tag{5.19}$$

with time dependent operator:

$$\mathbf{A} = \nabla^{-2}\left(-ik\,U(y,t)\nabla^2 - ik\left(\beta - \frac{d^2U(y,t)}{dy^2}\right)\right), \tag{5.20}$$

in which discretised approximations of the operators on the right-hand side are implied. According to Rayleigh's theorem (Drazin and Reid, 1981) this flow cannot sustain growth unless the mean vorticity gradient $Q_y = \beta - U''$ changes sign. Let us consider only flows that are asymptotically stable at all times by Rayleigh's theorem, and for simplicity that the mean velocity switches periodically between the two flows shown in the left panels of Figure 5.3. The corresponding mean vorticity gradient is shown in the right panels of the same figure. Despite the asymptotic stability of each instantaneous flow implied by Rayleigh's theorem the periodically varying flow is asymptotically unstable. The Lyapunov exponent of the instability as a function of the switching period is shown in Figure 5.4.

This instability arises from sustaining the transient growth of the operator through time dependence. The same process is operative when the flow is varying continuously in time. In that case the Lyapunov exponent for given statistically stationary fluctuations in operator structure can be shown to depend on two parameters: the fluctuation amplitude and the autocorrelation time, T_c, of the fluctuations (Farrell and Ioannou, 1999). Snapshots of perturbation structure revealing the process of transient growth by the interaction of the perturbations with the time dependent operator in a continuously varying flow are shown in Figure 5.5. This mechanism of error growth predicts that the perturbation structure should project most strongly on the subspace of the leading optimal and evolved optimal (right and left singular) vectors. This is indeed the case, as can be seen in the example in Figure 5.6.

Study of the asymptotic error structure in more realistic tangent linear systems confirms the conclusions presented above that error structure in forecast systems projects strongly on the optimal vectors (Gelaro *et al.*, 2002). This result is key

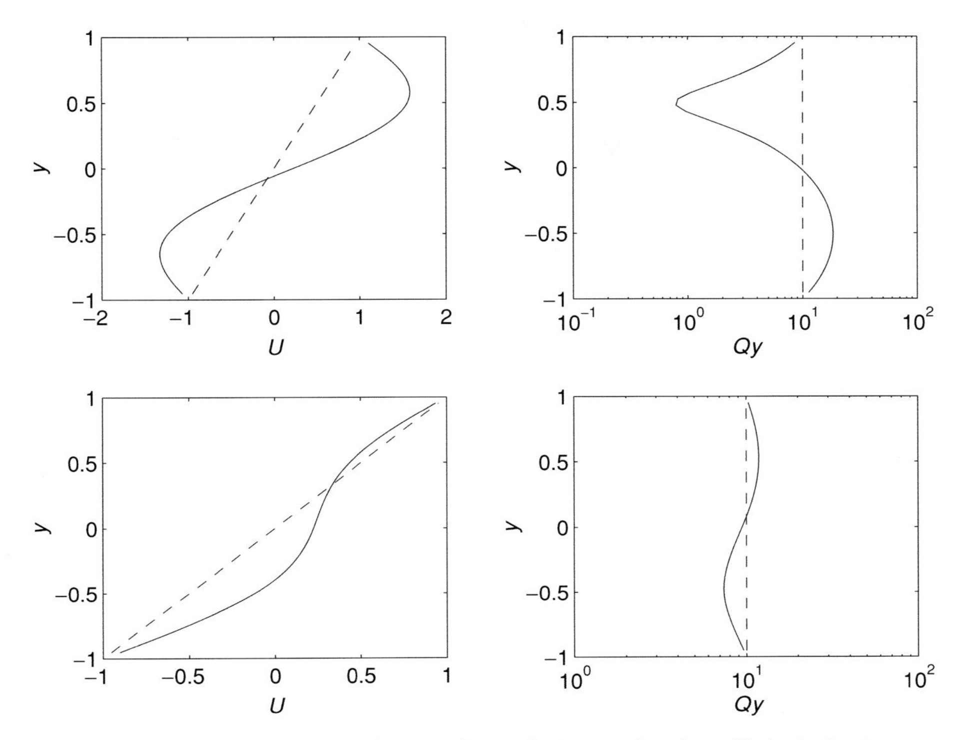

Figure 5.3 (Left panels) the mean flow velocity as a function of latitude for the Rayleigh stable example. (Right panel) the associated mean vorticity gradient, $Q_y = \beta - U''$, with $\beta = 10$.

for dynamical truncation of the error covariance which is required for optimal state estimation (Farrell and Ioannou, 2001).

5.3 Model error in deterministic forecast

We have discussed methods for determining the impact of uncertainties in the initial state on the forecast. However, as the initialisation of forecast models is improved with the advent of new data sources and the introduction of variational assimilation methods, the medium range forecast will become increasingly affected by uncertainties resulting from incomplete physical parametrisations and numerical approximations in the forecast model, and by the necessarily misrepresented subgrid-scale chaotic processes such as cumulus convection which act as temporally stochastic distributed forcing of the forecast error system. These influences, referred to collectively as model error, conventionally appear as an external forcing in the forecast error system (Allen *et al.*, this volume).

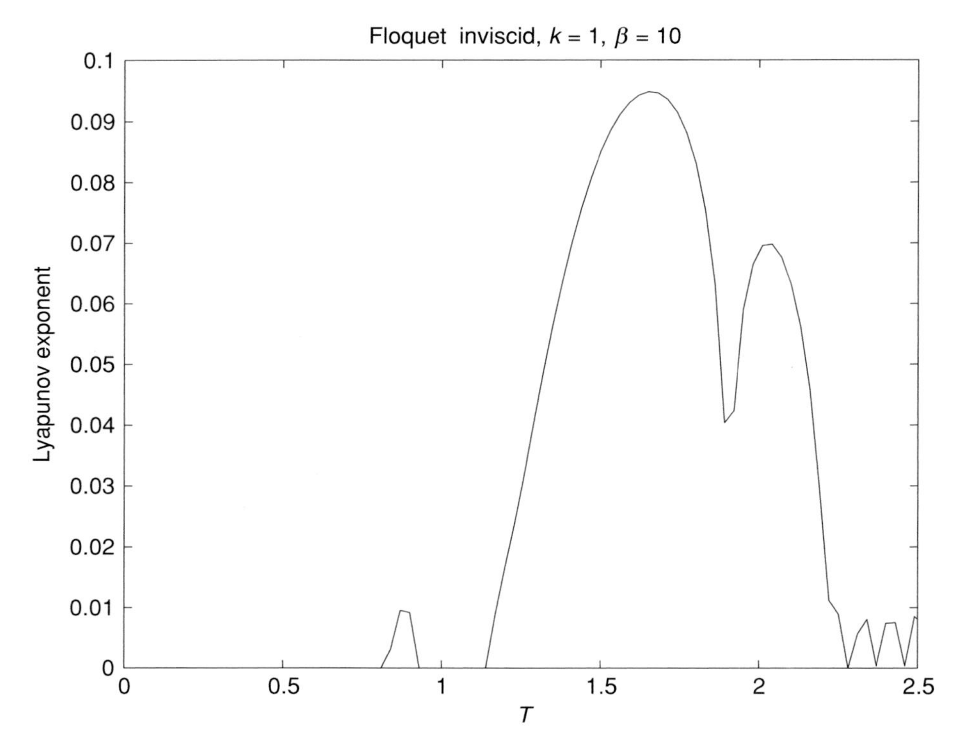

Figure 5.4 Lyapunov exponent as a function of switching period T for the example shown in Figure 5.3. The time dependent flow results from periodic switching every T time units between the Rayleigh stable flow profiles shown in Figure 5.3. The zonal wavenumber is $k = 1$ and $\beta = 10$.

Improving understanding of model error and specifically identifying forcings that lead to the greatest forecast errors are centrally important in predictability studies. In analogy with the optimal perturbations that lead to the greatest forecast error in the case of initial condition error, these continuous error sources will be called optimal distributed forcings. In an approach to this problem D'Andrea and Vautard (2000) obtained approximate optimal temporally distributed deterministic forcings of the forecast error system (which they refer to as forcing singular vectors) and Barkmeijer *et al.* (2003) obtained the optimal temporally distributed deterministic forcing of the forecast error system over fixed spatial structures. We here describe the method for determining the general optimal forcing in both the forecast and assimilation systems.

The underlying theory for determining the optimal forcing in the deterministic case is based on analysis of the dynamical system as a mapping from the space of input forcings to the space of states at later time. We seek the deterministic forcing $\mathbf{f}(t)$ of unit norm on $t \in [0, T]$ producing the greatest state norm at time T, i.e. that

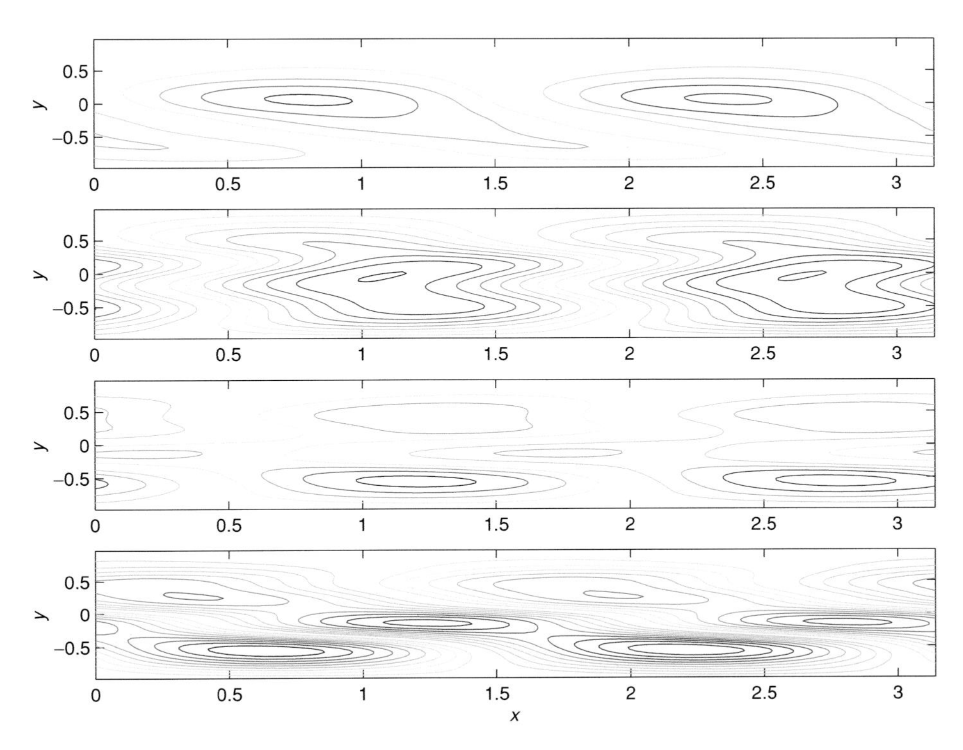

Figure 5.5 For a continuously varying barotropic flow, the structure of the state $\Psi(y, t)e^{ikx}$ in the zonal (x), meridional (y) plane at four consecutive times separated by an autocorrelation time T_c. The rms velocity fluctuation is 0.16 and the noise autocorrelation time is $T_c = 1$. The zonal wavenumber is $k = 2$, $\beta = 0$, and the Reynolds number is $Re = 800$. The Lyapunov exponent is $\lambda = 0.2$. At first (top panel) the Lyapunov vector is configured to grow, producing an increase over T_c of 1.7; in the next period the Lyapunov vector has assumed a decay configuration (second panel from top) and suffers a decrease of 0.7; subsequently (third panel from top) it enjoys a slight growth of 1.1; and finally (bottom panel) a growth by 1.8. Further details can be found in Farrell and Ioannou (1999).

maximises the square norm of the state $\|\mathbf{x}(T)\|^2$, assuming the state is initially zero, $\mathbf{x}(0) = 0$, and that $\mathbf{x}$ obeys the tangent linear forecast equation

$$\frac{d\mathbf{x}}{dt} = \mathbf{A}(t)\mathbf{x} + \mathbf{f}(t). \tag{5.21}$$

The forcing $\mathbf{f}(t)$ over the interval $[0, T]$ is measured in the square integral norm

$$\|\mathbf{f}\|^2_{L_2} = \int_0^T \mathbf{f}^\dagger(t)\mathbf{f}(t)dt, \tag{5.22}$$

while the state $\mathbf{x}$ is measured in the vector square norm

$$\|\mathbf{x}\|^2 = \mathbf{x}^\dagger \mathbf{x}. \tag{5.23}$$

The use of alternative inner products can be easily accommodated.

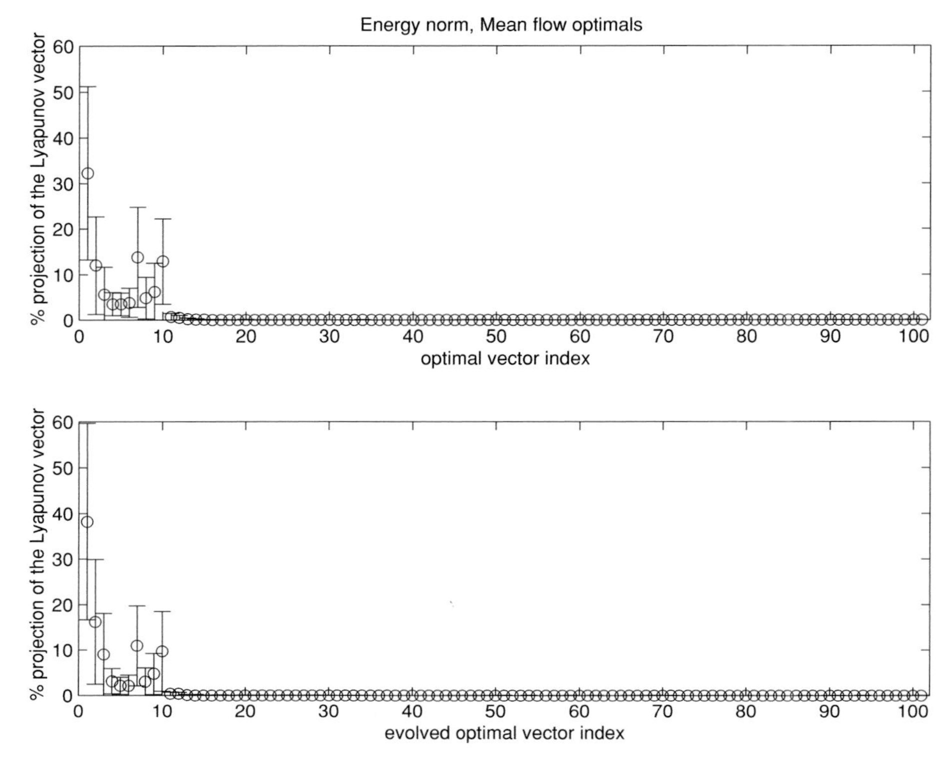

Figure 5.6 (Top panel) Mean and standard deviation of the projection of the Lyapunov vector on the optimal vectors of the mean flow calculated for a time interval equal to T_c in the energy norm. (Bottom panel) The mean projection and standard deviation of the Lyapunov vector on the T_c evolved optimal vectors of the mean flow in the energy norm.

The optimal forcing, $\mathbf{f}(t)$, is the forcing that maximises the forcing normalised final state norm at time T, i.e. that maximises the quotient

$$R_d = \frac{\| \mathbf{x}(T) \|^2}{\| \mathbf{f} \|_{L_2}^2}. \tag{5.24}$$

It can be shown (Dullerud and Paganini, 2000; Farrell and Ioannou, 2005) that this intractable maximisation over functions, $\mathbf{f}(t)$, can be transformed to a tractable maximisation over states, $\mathbf{x}(T)$. Specifically the following equality is true:

$$\max_{\mathbf{f}(t)} \frac{\| \mathbf{x}(T) \|^2}{\| \mathbf{f}(t) \|_{L_2}^2} = \max_{\mathbf{x}(T)} \frac{\| \mathbf{x}(T) \|^2}{\mathbf{x}(T)^\dagger \mathbf{C}^{-1} \mathbf{x}(T)}, \tag{5.25}$$

where $\mathbf{C}$ is the finite time state covariance matrix at time T under the assumption of temporally white noise forcing with unit covariance $\mathbf{I}$. The covariance can be obtained by integrating from $t = 0$ to $t = T$ the Lyapunov equation

$$\frac{d\mathbf{C}}{dt} = \mathbf{A}(t)\mathbf{C} + \mathbf{C}\mathbf{A}^{\dagger}(t) + \mathbf{I}, \tag{5.26}$$

with the initial condition $\mathbf{C}(0) = 0$. Note that the form of the second optimisation in Eq. (5.25) is reminiscent of the covariance or Mahalanobis metric used in predictability analysis (Palmer *et al.*, 1998), suggesting the interpretation of optimals weighted by the Mahalanobis metric as structures that are most easily forced.

Quotient (5.25) is maximised for unit forcing by the state

$$\mathbf{x}_{opt}(T) = \sqrt{\lambda_1}\mathbf{v}_1, \tag{5.27}$$

where λ_1 is the maximum singular value of $\mathbf{C}$ and $\mathbf{v}_1$ is the corresponding singular vector of $\mathbf{C}$ ($\mathbf{v}_1$ is conventionally called the top empirical orthogonal function (EOF) of $\mathbf{C}$). It can be shown (Farrell and Ioannou, 2005) that the optimal forcing and the associated state of the system can be obtained simultaneously by integrating the following coupled forward and adjoint systems backwards over the finite interval from time $t = T$ to the initial time $t = 0$:

$$\begin{aligned} \frac{d\mathbf{x}}{dt} &= \mathbf{A}(t)\mathbf{x} + \mathbf{f} \\ \frac{d\mathbf{f}}{dt} &= -\mathbf{A}^{\dagger}(t)\mathbf{f}, \end{aligned} \tag{5.28}$$

with $\mathbf{x}(T) = \sqrt{\lambda_1}\mathbf{v}_1$ and $\mathbf{f}(T) = \mathbf{v}_1/\sqrt{\lambda_1}$. The initial state $\mathbf{x}_{opt}(0) = 0$ is recovered as a consistency check.

5.4 Prediction of statistics of certain systems

Beyond the limit of deterministic forecast it is still possible to predict the statistical properties which constitute the climate of a system. Consider the perturbation structure, $\mathbf{x}$, produced by the forced equation

$$\frac{d\mathbf{x}}{dt} = \mathbf{A}\mathbf{x} + \mathbf{F}\mathbf{n}(t). \tag{5.29}$$

Here $\mathbf{A}$ may be the deterministic linear operator governing evolution of large-scale perturbations about the mean midlatitude flow, and $\mathbf{F}\mathbf{n}(t)$ an additive stochastic forcing with spatial structure $\mathbf{F}$, representing neglected non-linear and other forcing terms. For simplicity we assume that the components of $\mathbf{n}(t)$ are white noise with zero mean and unit variance. We wish to determine the perturbation covariance matrix

$$\mathbf{C}(t) = \langle \mathbf{x}\mathbf{x}^{\dagger} \rangle, \tag{5.30}$$

where $\langle \cdot \rangle$ denotes the ensemble average over the realisations of the forcing $\mathbf{F}\mathbf{n}(t)$. If a steady state is reached, $\langle \cdot \rangle$ is also the time mean covariance. We argue

(Farrell and Ioannou, 1993; DelSole, 1996, 1999, 2001, 2004)[2] that the midlatitude jet climatology can be obtained in this way because the transient climatology in the midlatitudes is the statistical average state resulting from random events of cyclogenesis. Because cyclogenesis is a rapid transient growth process primarily associated with the non-normality of **A**, its statistics are well approximated by the linear operator **A**. The diagonal elements of the steady state covariance **C** are the climatological variance of **x**, and they locate the storm track regions. All mean quadratic fluxes are also derivable from **C**, from which the observed climatological fluxes of heat and momentum can be obtained. In this way we obtain a theory for the climate and can address systematically statistical predictability questions such as how to determine the sensitivity of the climate, that is of **C**, to changes in the boundary conditions and physical parameters which are reflected in changes in the mean operator **A** and the forcing structure matrix **F**.

If $\mathbf{n}(t)$ is a white noise process it can be shown (Farrell and Ioannou, 1996a) that

$$\mathbf{C}(t) = \int_0^t e^{\mathbf{A}s}\mathbf{Q}e^{\mathbf{A}^\dagger s}\,ds, \tag{5.31}$$

where

$$\mathbf{Q} = \mathbf{F}\mathbf{F}^\dagger \tag{5.32}$$

is the covariance of the forcing. It can also be shown that the ensemble mean covariance evolves according to the deterministic equation

$$\frac{d\mathbf{C}}{dt} = \mathbf{AC} + \mathbf{CA}^\dagger + \mathbf{Q} \equiv \mathbf{HC} + \mathbf{Q}, \tag{5.33}$$

where **H** is a linear operator. It should be noted that the above equation is also valid for non-autonomous $\mathbf{A}(t)$. If **A** is time independent the solution of the above equation is

$$\begin{aligned}\mathbf{C}(t) &= e^{\mathbf{H}t}\,\mathbf{C}(0) + \left(\int_0^t e^{\mathbf{H}(t-s)}ds\right)\mathbf{Q}\\ &= e^{\mathbf{H}t}\,\mathbf{C}(0) + \mathbf{H}^{-1}(e^{\mathbf{H}t} - \mathbf{I})\mathbf{Q}.\end{aligned} \tag{5.34}$$

As $t \to \infty$ and assuming the operator **A** is stable a steady-state is reached, which satisfies the steady-state Lyapunov equation

$$\mathbf{AC}^\infty + \mathbf{C}^\infty\mathbf{A}^\dagger = -\mathbf{Q}. \tag{5.35}$$

This equation can be readily solved for $\mathbf{C}^\infty$, from which ensemble mean quadratic flux quantities can be derived.

Interpretation of **C** requires care. The asymptotic steady-state ensemble average, $\mathbf{C}^\infty$, is the same as the time averaged covariance and can be obtained from a single realisation of $\mathbf{x}(t)$ by averaging the covariance over a sufficient long interval. However, the time dependent $\mathbf{C}(t)$ cannot be associated with a time average[3] but rather is necessarily an ensemble average. With this consideration in mind, $\mathbf{C}(t)$ from Eq. (5.34) is appropriate for evolving the error covariance in ensemble prediction as will be discussed in the next section. In this section we consider a time independent and stable $\mathbf{A}$ and interpret the steady state $\mathbf{C}^{\infty}$ as the climatological covariance.

It has been demonstrated that such a formulation accurately models the midlatitude climatology (Farrell and Ioannou, 1994, 1995; DelSole, 1996, 1999, 2001, 2004; Whitaker and Sardeshmukh, 1998; Zhang and Held, 1999) and reproduces the climatological heat and momentum fluxes. The asymptotic covariance captures the distribution of the geopotential height variance of the midlatitude atmosphere as well as the distribution of heat and momentum flux in the extratropics.

The algebraic equation (5.35) gives $\mathbf{C}^{\infty}$ as an explicit functional of the forcing covariance $\mathbf{Q}$ and the mean operator $\mathbf{A}$, which is in turn a function of the mean flow and the physical process parameters. This formulation permits systematic investigation of the sensitivity of the climate to changes in the forcing and structure of the mean flow and parameters.

We first address the sensitivity of the climate to changes in the forcing under the assumption that the mean state is fixed.

We determine the forcing structure, $\mathbf{f}$, given by a column vector, that contributes most to the ensemble average variance $\langle E(t)\rangle$. This structure is the stochastic optimal (Farrell and Ioannou, 1996a; Kleeman and Moore, 1997; Timmermann and Jin, this volume).

The ensemble average variance produced by stochastically forcing this structure (i.e. introducing the forcing $\mathbf{f}n(t)$ in the right-hand side of Eq. (5.29) can be shown to be

$$\langle E(t)\rangle = \langle \mathbf{x}^{\dagger}\mathbf{x}\rangle = \mathbf{f}^{\dagger}\,\mathbf{B}(t)\,\mathbf{f}, \tag{5.36}$$

where $\mathbf{B}(t)$ is the stochastic optimal matrix

$$\mathbf{B}(t) = \int_0^t e^{\mathbf{A}^{\dagger}s} e^{\mathbf{A}s}\, ds. \tag{5.37}$$

The stochastic optimal matrix satisfies the time dependent back Lyapunov equation, analogous to Eq. (5.33):

$$\frac{d\mathbf{B}}{dt} = \mathbf{B}\mathbf{A} + \mathbf{A}^{\dagger}\mathbf{B} + \mathbf{I}. \tag{5.38}$$

If $\mathbf{A}$ is stable the statistical steady state $\mathbf{B}^{\infty}$ satisfies the algebraic equation

$$\mathbf{B}^{\infty}\mathbf{A} + \mathbf{A}^{\dagger}\mathbf{B}^{\infty} = -\mathbf{I}, \tag{5.39}$$

which can be readily solved for $\mathbf{B}^{\infty}$.

Having obtained $\mathbf{B}^{\infty}$ from (5.36) we obtain the stochastic optimal as the eigenfunction of $\mathbf{B}^{\infty}$ with the largest eigenvalue. The stochastic optimal determines the forcing structure, $\mathbf{f}$, that is most effective in producing variance. Forcings will have impact on the variance according to the forcing's projection on the top stochastic optimals (the top eigenfunctions of $\mathbf{B}^\infty$).

As another application the sensitivity of perturbation statistics to variations in the mean state can be obtained. Assume, for example, that the mean atmospheric flow U is changed by δU inducing the change $\delta\mathbf{A}$ in the mean operator. The statistical equilibrium that results satisfies the Lyapunov equation

$$(\mathbf{A}+\delta\mathbf{A})(\mathbf{C}^\infty+\delta\mathbf{C}^\infty)+(\mathbf{C}^\infty+\delta\mathbf{C}^\infty)+(\mathbf{A}+\delta\mathbf{A})^\dagger=-\mathbf{Q}\,, \qquad (5.40)$$

under the assumption that the forcing covariance $\mathbf{Q}$ has remained the same. Because $\mathbf{C}^\infty$ satisfies the equilibrium (5.35) the first order correction $\delta\mathbf{C}^\infty$ is determined from

$$\mathbf{A}\delta\mathbf{C}^\infty+\delta\mathbf{C}^\infty\mathbf{A}^\dagger \;=\; -(\delta\mathbf{A}\mathbf{C}^\infty+\mathbf{C}^\infty\delta\mathbf{A}^\dagger)\,. \qquad (5.41)$$

From this one can determine a bound on the sensitivity of the climate by determining the change in the mean operator that will result in the largest change $\delta\mathbf{C}^\infty$. This operator change leading to maximum increase in a specified quadratic quantity is called the optimal structural change. Farrell and Ioannou (2004) show that a single operator change fully characterises any chosen quadratic quantity tendency, in the sense that, if an arbitrary operator change is performed, the quadratic tendency, $\delta\mathbf{C}^\infty$, is immediately obtained by projecting the operator change on this single optimal structure change.

In this way the sensitivity of quadratic quantities such as variance, energy, and fluxes of heat and momentum, to change in the mean operator can be found. The mean operator change could include jet velocity, dissipation and other dynamical variables, and these jet structure changes, as well as the region over which the response is optimised, can be localised in the jet. The unique jet structure change producing the greatest change in a chosen quadratic quantity also completely characterises the sensitivity of the quadratic quantity to jet change in the sense that an arbitrary jet change increases the quadratic quantity in proportion to its projection on this optimal structure change. This result provides an explanation for observations that substantial differences in quadratic storm track quantities such as variance occur in response to apparently similar influences such as comparable sea surface temperature changes, and moreover provides a method for obtaining the optimal structural change.

5.5 Prediction of statistics of uncertain systems

The sensitivity of forecasts to various aspects of the model can be determined by performing parallel computations of the forecast system in which the uncertain aspects of the model are varied. These integrations produce an ensemble of forecasts (Palmer, Kalnay *et al.*, and Buizza, this volume). The ensemble mean of these predictions is for many systems of interest a best estimate of the future state (Gauss, 1809; Leith, 1974). These ensemble integrations also provide estimates of the probability density

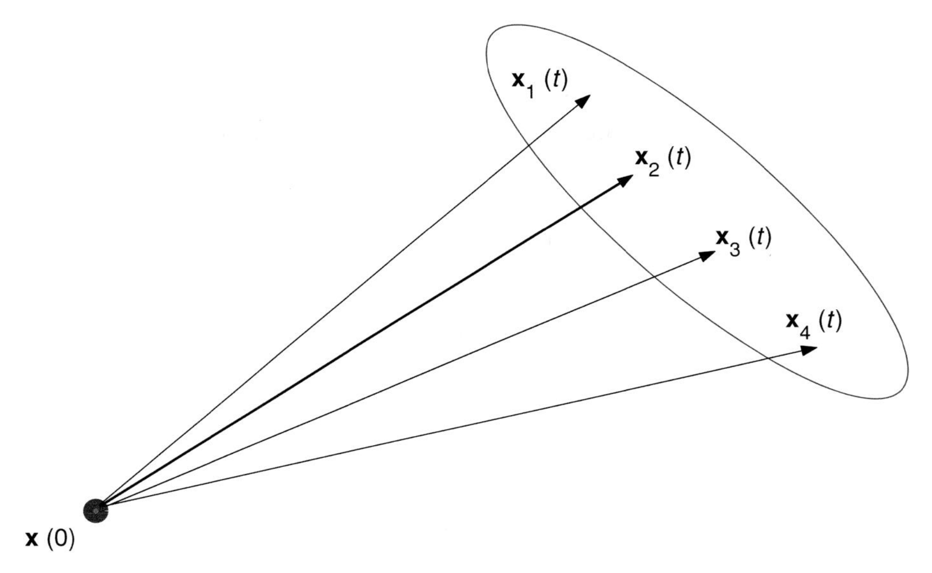

Figure 5.7 Schematic evolution of a sure initial condition $\mathbf{x}(0)$ in an uncertain system. After time t the evolved states $\mathbf{x}(t)$ lie in the region shown. Initially the covariance matrix $\mathbf{C}(0) = \mathbf{x}(0)\mathbf{x}^{\dagger}(0)$ is rank 1, but at time t the covariance matrix has rank greater than 1. For example if the final states were $\mathbf{x}_i(t)$ $(i = 1, \cdots, 4)$ with equal probability, the covariance at time t: $\mathbf{C}(t) = \frac{1}{4}\sum_{i=1}^{4} \mathbf{x}_i(t)\mathbf{x}_i^{\dagger}(t)$ would be rank 4, representing an entangled state. In contrast, in certain systems the degree of entanglement is invariant and a pure state evolves to a pure state.

function of the prediction. The covariance matrix of the predicted states $\mathbf{C} = \langle \mathbf{x}\mathbf{x}^{\dagger} \rangle$, where $\langle \cdot \rangle$ signifies the ensemble mean, provides the second moments of the probability density of the predictions and characterises the sensitivity of the prediction to variation in the model. We wish to determine bounds on the error covariance matrix resulting from such model uncertainties.

5.5.1 The case of additive uncertainty

Consider first a tangent linear system with additive model error. With the assumption that the model error can be treated as a stochastic forcing of the tangent linear system the errors evolve according to

$$\frac{d\mathbf{x}}{dt} = \mathbf{A}(t)\mathbf{x} + \mathbf{F}\mathbf{n}(t), \tag{5.42}$$

where $\mathbf{A}(t)$ is the tangent linear operator which is considered certain, $\mathbf{F}$ the structure matrix of the uncertainty which is assumed to be well described by zero mean and unit covariance white noise processes $\mathbf{n}(t)$. Such systems are uncertain and as a result a single initial state maps to a variety of states at a later time depending on the realisation of the stochastic process $\mathbf{n}(t)$. This is shown schematically in Figure 5.7 for the case of four integrations of the model.

Let us assume initially that the ensemble had model error covariance $\mathbf{C}(0)$. At a time t later the error covariance is given by Eq. (5.34). The homogeneous part of (5.34) is the covariance resulting from the deterministic evolution of the initial $\mathbf{C}(0)$ and represents error growth associated with uncertainty in specification of the initial state. Predictability studies traditionally concentrate on this source of error growth. The inhomogeneous part of (5.34) represents the contribution of additive model error.

The deterministic part of the growth of the error covariance at any time is bounded by the growth produced by the optimal perturbation. The forced error growth at any time, by contrast, is bounded by the error covariance at time t forced by the quite different stochastic optimal that is determined as the eigenfunction with largest eigenvalue of the stochastic optimal matrix

$$\mathbf{B}(t) = \int_0^t e^{\mathbf{A}^\dagger s} e^{\mathbf{A}s}\, ds \tag{5.43}$$

at time t. Given an initial error covariance, $\mathbf{C}(0)$, and a forcing covariance, $\mathbf{Q}$, it is of interest to determine the time at which the accumulated covariance from the model error exceeds the error produced by uncertainty in the initial conditions. As an example consider the simple system (5.8). Assume that initially the state has error such that $\mathrm{trace}(\mathbf{C}(0)) = 1$ and that the additive model error has covariance $\mathrm{trace}(\mathbf{Q}) = 1$. The growth of errors due to uncertainty in the initial conditions is plotted as a function of time in Figure 5.8. After approximately unit time the error covariance is dominated by the accumulated error from model uncertainty.

From this simple example it is realised that in both stable and unstable systems as the initial state is more accurately determined error growth will inevitably be dominated by model error. At present the success of the deterministic forecasts and increase in forecast accuracy obtained by decreasing initial state error suggest that improvements in forecast accuracy are still being achieved by reducing the uncertainty in the initial state.

5.5.2 The case of multiplicative uncertainty

Consider now a forecast system with uncertain parametrisations (Palmer, 1999; Sardeshmukh *et al.*, 2001, 2003) so that the tangent linear system operator itself is uncertain and for simplicity takes the form

$$\mathbf{A}(t) = \mathbf{A} + \varepsilon \mathbf{B} \eta(t), \tag{5.44}$$

where $\eta(t)$ is a scalar stochastic process with zero mean and unit variance and $\mathbf{B}$ is a fixed matrix characterising the structure of the operator uncertainty, and ε is a scalar amplitude factor.

An important property of these multiplicative uncertain systems is that different realisations produce highly disparate growths. Fix the inner product with which the perturbation magnitude is measured and concentrate on calculation of the error

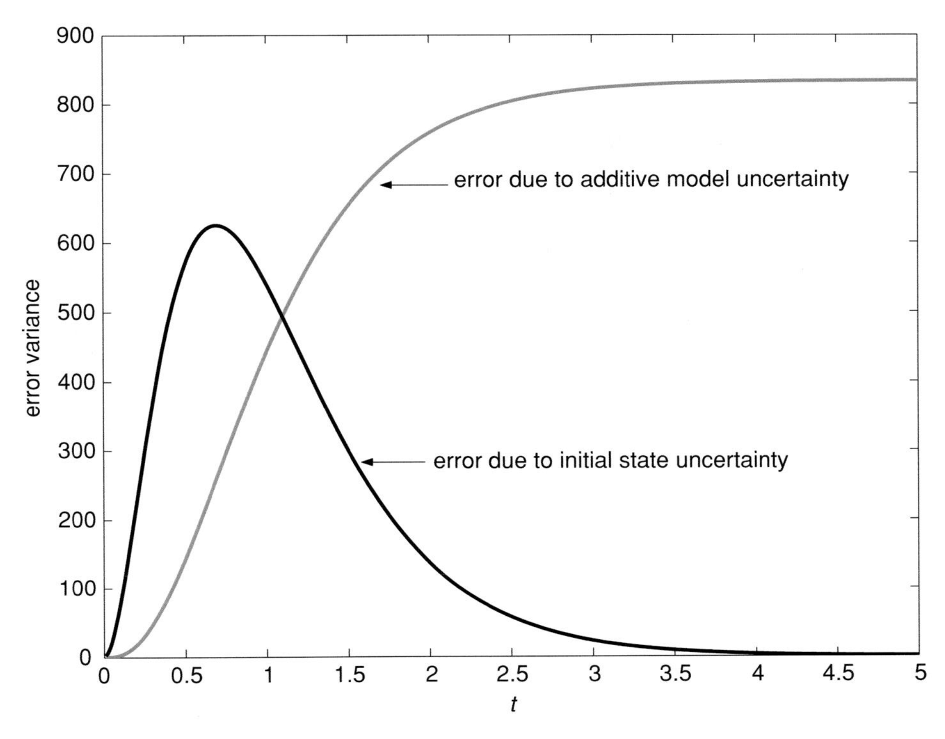

Figure 5.8 Error variance resulting from free evolution of the optimal unit variance, initial covariance and evolution of error variance forced by additive uncertainty with unit forcing variance. The operator **A** is the simple 2×2 Reynolds operator given in (5.8).

growth. Because of the uncertainty in the operator, different realisations, η, will result in different perturbation magnitudes, and because the probability density function of η is known, perturbation amplitudes can be ascribed a probability. A measure of error growth is the expectation of the growths:

$$\langle g \rangle = \int P(\omega))\, g(\omega) d\omega, \tag{5.45}$$

where ω is a realisation of η, $P(\omega)$ is the probability of its occurrence, and $g(\omega)$ is the error growth associated with this realisation of the operator. Because of the convexity of the expectation the root-mean-square second moment error growth exceeds the amplitude error growth, i.e.

$$\sqrt{\langle g^2 \rangle} \geq \langle g \rangle. \tag{5.46}$$

It follows that in uncertain systems different moments generally have different growth rates and the Lyapunov exponent of an uncertain system may be negative while higher moments are unstable. This emphasises the fact that rare or extreme events that are weighted more by the higher moment measure, while difficult to predict, are potentially highly consequential.

As an example consider two classes of trajectories which with equal probability give growth in unit time of $g = 2$ or $g = 1/2$. What is the expected growth in unit time? While the Lyapunov growth rate is 0 because

$$\lambda = \frac{\langle \log g \rangle}{t} = \frac{\log 2}{2} + \frac{\log(1/2)}{2} = 0, \tag{5.47}$$

the second moment growth rate is positive:

$$\lambda_2 = \frac{\log \langle g^2 \rangle}{t} = \log\left(\frac{1}{2} \times 4 + \frac{1}{2} \times \frac{1}{4}\right) = 0.75, \tag{5.48}$$

proving that the error covariance increases exponentially fast and showing that uncertain systems may be Lyapunov (sample) stable, while higher order moments are unstable. The second moment measures include the energy, so that this example demonstrates that multiplicative uncertain systems can be Lyapunov stable while expected energy grows exponentially with time. In fact if the uncertainty is Gaussian there is always a higher moment that grows exponentially (Farrell and Ioannou, 2002a).

One implication of this property is that optimal error growth in multiplicative uncertain systems is not derivable from the norm of the ensemble mean propagator. To obtain the optimal growth it is necessary to first determine the evolution of the covariance $\mathbf{C} = \langle \mathbf{x}\mathbf{x}^{\dagger} \rangle$ under the uncertain dynamics and then determine the optimal $\mathbf{C}(0)$ of unit trace that leads to greatest trace($\mathbf{C}(t)$) at later times.

Consider the multiplicative uncertain tangent linear system:

$$\frac{d\mathbf{x}}{dt} = \mathbf{A}\mathbf{x} + \varepsilon n(t)\mathbf{B}\mathbf{x}, \tag{5.49}$$

where $\mathbf{A}$ is the sure mean operator and $\mathbf{B}$ is the structure of the uncertainty in the operator and $n(t)$ is its time dependence. Take $n(t)$ to be a Gaussian random variable with zero mean, unit variance and autocorrelation time t_c. Define $\Phi(t, 0)$ to be the propagator for a realisation of the operator $\mathbf{A} + \varepsilon\eta(t)\mathbf{B}$. For that realisation the square error at time t is

$$\mathbf{x}(t)^{\dagger}\mathbf{x}(t) = \mathbf{x}(0)^{\dagger}\Phi^{\dagger}(t, 0)\Phi(t, 0)\,\mathbf{x}(0) \tag{5.50}$$

where $\mathbf{x}(0)$ is the initial error. The optimal initial error, i.e. the initial error that leads to the greatest variance at time t, for this realisation is the eigenvector of

$$\mathbf{H}(t) = \Phi^{\dagger}(t, 0)\Phi(t, 0) \tag{5.51}$$

with largest eigenvalue.

For uncertain dynamics we seek the greatest expected variance at t by determining the ensemble average

$$\langle \mathbf{H}(t) \rangle = \left\langle \Phi^{\dagger}(t, 0)\Phi(t, 0) \right\rangle. \tag{5.52}$$

The optimal initial error is identified as the eigenvector of $\langle \mathbf{H}(t) \rangle$ with largest eigenvalue. This eigenvalue determines the optimal ensemble expected error growth. This gives constructive proof of the remarkable fact that there is a single sure initial error that maximises expected error growth in an uncertain tangent linear system; that is, the greatest ensemble error growth is produced when all ensemble integrations are initialised with the same state.

To quantify the procedure one needs to obtain an explicit form of the ensemble average $\langle \mathbf{H}(t) \rangle$ in terms of the statistics of the uncertainty. It turns out that this is possible for Gaussian fluctuations (Farrell and Ioannou, 2002b, 2002c), in which case $\langle \mathbf{H}(t) \rangle$ evolves according to the exact equation

$$\frac{d\langle \mathbf{H} \rangle}{dt} = (\mathbf{A} + \varepsilon^2 \mathbf{E}(t)\mathbf{B})^\dagger \langle \mathbf{H}(t) \rangle + \langle \mathbf{H}(t) \rangle (\mathbf{A} + \varepsilon^2 \mathbf{E}(t)\mathbf{B}) \tag{5.53}$$

$$+ \varepsilon^2 (\mathbf{E}^\dagger(t) \langle \mathbf{H} \rangle \mathbf{B} + \mathbf{B}^\dagger \langle \mathbf{H} \rangle \mathbf{E}(t)) \tag{5.54}$$

where

$$\mathbf{E}(t) = \int_0^t e^{-\mathbf{A}s} \mathbf{B} e^{\mathbf{A}s} e^{-\nu s} \, ds. \tag{5.55}$$

For autocorrelation times of the fluctuations which are small compared with the time scales of $\mathbf{A}$, the above equation reduces to

$$\frac{d\langle \mathbf{H} \rangle}{dt} = \left(\mathbf{A} + \frac{\varepsilon^2}{\nu} \mathbf{B}^2 \right)^\dagger \langle \mathbf{H}(t) \rangle + \langle \mathbf{H}(t) \rangle \left(\mathbf{A} + \frac{\varepsilon^2}{\nu} \mathbf{B}^2 \right)$$
$$+ \frac{2\varepsilon^2}{\nu} \mathbf{B}^\dagger \langle \mathbf{H} \rangle \mathbf{B}. \tag{5.56}$$

As an example application of this result, the ensemble for an uncertain tangent linear system arising from a forecast system with Gaussian statistical distribution of parameter value variation could be constructed from this basis of optimals, i.e. the optimals of $\langle \mathbf{H} \rangle$.

5.6 Conclusions

Generalised stability theory (GST) is required for a comprehensive understanding of error growth in deterministic autonomous and non-autonomous systems. In contrast to the approach based on normal modes in GST applied to deterministic systems, attention is concentrated on the optimal perturbations obtained by singular value analysis of the propagator or equivalently by repeated forward integration of the system followed by backward integration of the adjoint system. The optimal perturbations are used to understand and predict error growth and structure and for such tasks as building ensembles for use in ensemble forecast. In addition this approach provides theoretical insight into the process of error growth in both autonomous and non-autonomous systems. In the case of non-autonomous systems the process of error growth is identified with the intrinsic non-normality of a time dependent system and the unstable error is shown to lie in the subspace of the leading optimal and evolved optimal (right and left singular) vectors.

Beyond the deterministic time horizon, GST can be used to address questions of predictability of statistics and of sensitivity of statistics to changes in the forcing and changes in the system operator. As an example of the power of these methods, the sensitivity of a statistical quantity is found to be related to a single structured change in the mean operator.

Finally, we have seen how GST addresses error growth in the presence of both additive and multiplicative model error. In the case of multiplicative model error the role of rare trajectories is found to be important for the stability of higher statistical moments, including quadratic moments which relate sample stability to stability in energy.

Acknowledgements

This work was supported by NSF ATM-0123389 and by ONR N00014-99-1-0018.

Notes

1. It is called the Reynolds matrix because it captures the emergence of rolls in three-dimensional boundary layers that are responsible for transition to turbulence.
2. See also the related linear inverse model perspective (Penland and Magorian, 1993; Penland and Sardeshmukh, 1995).
3. Under certain conditions it can be associated with a zonal mean; for discussion and physical application of this interpretation see Farrell and Ioannou (2003).

References

Barkmeijer, J., T. Iversen and T. N. Palmer (2003). Forcing singular vectors and other sensitive model structures. *Quart. J. Roy. Meteor. Soc.*, **129**, 2401–24.

Berliner, L. M. (1996). Hierarchical Baysian time series models. In *Maximum Entropy and Bayesian Methods*, ed. K. M. Hanson and R. N. Silver, pp. 15–22. Kluwer Academic Publishers.

Buizza, R. and T. N. Palmer (1995). The singular-vector structure of the atmospheric global circulation. *J. Atmos. Sci.*, **52**, 1434–56.

D'Andrea, F. and R. Vautard (2000). Reducing systematic errors by empirically correcting model errors. *Tellus*, **52A**, 21–41.

DelSole, T. (1996). Can quasigeostrophic turbulence be modelled stochastically? *J. Atmos. Sci.*, **53**, 1617–33.

(1999). Stochastic models of shear-flow turbulence with enstrophy transfer to subgrid scales. *J. Atmos. Sci.*, **56**, 3692–703.

(2001). A theory for the forcing and dissipation in stochastic turbulence models. *J. Atmos. Sci.*, **58**, 3762–75.

(2004). Stochastic models of quasigeostrophic turbulence. *Surv. Geophys.*, **25**, 107–49.

DelSole, T. and A. Y. Hou (1999a). Empirical stochastic models for the dominant climate statistics of a general circulation model. *J. Atmos. Sci.*, **56**, 3436–56.

(1999b). Empirical correction of a dynamical model. I: Fundamental issues. *Mon. Weather Rev.*, **127**, 2533–45.

Drazin, P. G. and W. H. Reid (1981). *Hydrodynamic Stability*. Cambridge University Press.

Dullerud, G. E. and F. Paganini (2000). *A Course in Robust Control Theory: a Convex Approach*. Springer Verlag.

Epstein, E. S. (1969). Stochastic-dynamic prediction. *Tellus*, **21**, 739–59.

Errico, R. M. (1997). What is an adjoint mode? *Bull. Am. Meteorol. Soc.*, **78**, 2577–91.

Farrell, B. F. (1988). Optimal excitation of neutral Rossby waves. *J. Atmos. Sci.*, **45**, 163–72.

(1990). Small error dynamics and the predictability of atmospheric flows. *J. Atmos. Sci.*, **47**, 2409–16.

Farrell, B. F. and P. J. Ioannou (1993). Stochastic dynamics of baroclinic waves. *J. Atmos. Sci.*, **50**, 4044–57.

(1994). A theory for the statistical equilibrium energy and heat flux produced by transient baroclinic waves. *J. Atmos. Sci.*, **51**, 2685–98.

(1995). Stochastic dynamics of the mid-latitude atmospheric jet. *J. Atmos. Sci.*, **52**, 1642–56.

(1996a). Generalized stability. I: Autonomous operators. *J. Atmos. Sci.*, **53**, 2025–40.

(1996b). Generalized stability. II: Non-autonomous operators. *J. Atmos. Sci.*, **53**, 2041–53.

(1999). Perturbation growth and structure in time dependent flows. *J. Atmos. Sci.*, **56**, 3622–39.

(2000). Perturbation dynamics in atmospheric chemistry. *J. Geophy. Res.*, **105**, 9303–20.

(2001). State estimation using reduced order Kalman filter. *J. Atmos. Sci.*, **58**, 3666–80.

(2002a). Perturbation growth and structure in uncertain flows: Part I. *J. Atmos. Sci.*, **59**, 2629–46.

(2002b). Perturbation growth and structure in uncertain flows: Part II. *J. Atmos. Sci.*, **59**, 2647–64.

(2002c). Optimal perturbation of uncertain systems. *Stoch. Dynam.*, **2**, 395–402.

(2003). Structural stability of turbulent jets. *J. Atmos. Sci.*, **60**, 2101–18.

(2004). Sensitivity of perturbation variance and fluxes in turbulent jets to changes in the mean jet. *J. Atmos. Sci.*, **61**, 2645–52.

(2005). Optimal excitation of linear dynamical systems by distributed forcing. *J. Atmos. Sci.*, **62**, 460–75.

Gauss, K. F. (1809). *Theory of the Motion of the Heavenly Bodies Moving about the Sun in Conic Sections*. C. H. Davis, Translation, republished by Dover Publications, 1963.

Gelaro, R., R. Buizza, T. N. Palmer and E. Klinker (1998). Sensitivity analysis of forecast errors and the construction of optimal perturbations using singular vectors. *J. Atmos. Sci.*, **55**, 1012–37.

Gelaro, R., C. Reynolds and R. M. Errico (2002). Transient and asymptotic error growth in a simple model. *Quart J. Roy. Meteor. Soc.*, **128**, 205–28.

Ghil, M. and P. Malanotte-Rizzoli (1991). Data assimilation in meteorology and oceanography. *Adv. Geophys.*, **33**, 141–266.

Kleeman, R. and A. M. Moore (1997). A theory for the limitations of ENSO predictability due to stochastic atmospheric transients. *J. Atmos. Sci.*, **54**, 753–67.

Lacarra, J. and O. Talagrand (1988). Short range evolution of small perturbations in a barotropic model. *Tellus*, **40A**, 81–95.

Leith, C. E. (1974). Theoretical skill of Monte Carlo forecasts. *Mon. Weather Rev.*, **102**, 409–18.

Lorenz, E. N. (1963). Deterministic non-periodic flow. *J. Atmos. Sci.*, **20**, 130–41.

(1965). A study of the predictability of a 28-variable atmospheric model. *Tellus*, **17**, 321–33.

(1985). The growth of errors in prediction. In *Turbulence and Predictability in Geophysical Fluid Dynamics and Climate Dynamics*, ed. M. Ghil, pp. 243–65. North-Holland.

Lyapunov, M. A. (1907). Problème général de la stabilité du mouvement. Annales Fac. Sciences Toulouse 9. Reprinted by Princeton University Press, 1949.

Mirsky, L. (1960). Symmetric gage functions and unitarily invariant norms. *Q. J. Math*, **11**, 50–9.

Moore, A. M. and R. Kleeman (1996). The dynamics of error growth and predictability in a coupled model of ENSO. *Quart. J. Roy. Meteor. Soc.*, **122**, 1405–46.

Molteni, F. and T. N. Palmer (1993). Predictability and finite time instability of the northern winter circulation. *Quart. J. Roy. Meteor. Soc.*, **119**, 269–98.

Moore, A. M. and B. F. Farrell (1993). Rapid perturbation growth on spatially and temporally varying oceanic flows determined using an adjoint method: application to the Gulf Stream. *J. Phys. Ocean.*, **23**, 1682–702.

Oseledets, V. I. (1968). A multiplicative ergodic theorem. Lyapunov characteristic exponents. *Trans. Moscow Math. Soc.*, **19**, 197–231.

Palmer, T. N. (1999). A nonlinear dynamical perspective on climate prediction. *J. Climate*, **12**, 575–91.

Palmer, T. N., R. Gelaro, J. Barkmeijer and R. Buizza (1998). Singular vectors, metrics, and adaptive observations. *J. Atmos. Sci.*, **55**, 633–53.

Penland, C. and T. Magorian (1993). Prediction of Niño 3 sea surface temperatures using linear inverse modeling. *J. Climate*, **6**, 1067–76.

Penland, C. and P. D. Sardeshmukh (1995). The optimal growth of tropical sea surface temperature anomalies. *J. Climate*, **8**, 1999–2004.

Sardeshmukh, P. D., C. Penland and M. Newman (2001). Rossby waves in a stochastically fluctuating medium. *Prog. Probability*, **49**, 369–84.

(2003). Drifts induced by multiplicative red noise with application to climate. *Europhys. Lett.*, **63**, 498–504.

Schmidt, E. (1906). Zur Theorie der linearen und nichtlinearen Intgralgleighungen. *Math. Annalen* **63**, 433–476.

Whitaker, J. S. and P. D. Sardeshmukh (1998). A linear theory of extratropical synoptic eddy statistics. *J. Atmos. Sci.*, **55**, 237–58.

Zhang, Y. and I. Held (1999). A linear stochastic model of a GCM's midlatitude storm tracks. *J. Atmos. Sci.*, **56**, 3416–35.

6

Ensemble-based atmospheric data assimilation

Thomas M. Hamill
Physical Sciences Division NOAA Earth System Research Laboratory, Boulder

Ensemble-based data assimilation techniques are being explored as possible alternatives to current operational analysis techniques such as three- or four-dimensional variational assimilation. Ensemble-based assimilation techniques utilise an ensemble of parallel data assimilation and forecast cycles. The background-error covariances are estimated using the forecast ensemble and are used to produce an ensemble of analyses. The background-error covariances are flow dependent and often have very complicated structure, providing a very different adjustment to the observations than are seen from methods such as three-dimensional variational assimilation. Though computationally expensive, ensemble-based techniques are relatively easy to code, since no adjoint nor tangent linear models are required, and previous tests in simple models suggest that dramatic improvements over existing operational methods may be possible.

A review of the ensemble-based assimilation is provided here, starting from the basic concepts of Bayesian assimilation. Without some simplification, full Bayesian assimilation is computationally impossible for model states of large dimension. Assuming normality of error statistics and linearity of error growth, the state and its error covariance may be predicted optimally using Kalman filter (KF) techniques. The ensemble Kalman filter (EnKF) is then described. The EnKF is an approximation to the KF in that background-error covariances are estimated from a finite ensemble of forecasts. However, no assumptions about linearity of error growth are made. Recent algorithmic variants on the standard EnKF are also described, as well

Predictability of Weather and Climate, ed. Tim Palmer and Renate Hagedorn. Published by Cambridge University Press.

as methods for simplifying the computations and increasing the accuracy. Examples of ensemble-based assimilations are provided in simple and more realistic dynamical systems.

6.1 Introduction

The purpose of this chapter is to introduce the reader to promising new experimental methods for atmospheric data assimilation involving the use of ensemble forecasts (e.g. Evensen, 1994; Evensen and van Leeuwen, 1996; Houtekamer and Mitchell, 1998; Burgers *et al.*, 1998; Tippett *et al.*, 2003; Anderson, 2003; Evensen, 2003; Lorenc, 2003). There is a natural linkage between data assimilation and ensemble forecasting. Ensemble forecasts (Toth and Kalnay, 1993, 1997; Molteni *et al.*, 1996; Houtekamer *et al.*, 1996a) are designed to estimate the flow-dependent uncertainty of the forecast, while data assimilation techniques require accurate estimates of forecast uncertainty in order to optimally blend the prior forecast(s) with new observations. Ensemble-based assimilation methods integrate the two steps; the ensemble of forecasts is used to estimate forecast-error statistics during the data assimilation step, and the output of the assimilation is a set of analyses. This process is cycled, the short-term ensemble forecasts from the set of analyses providing the error statistics again for the next assimilation cycle.

Rather than starting with the specifics of recently proposed ensemble-based assimilation techniques, in this chapter we will take a step back and try to motivate their use by quickly tracing them from first principles, noting the approximations that have been made along the way. This will take us from Bayesian data assimilation (Section 6.2), which is conceptually simple but computationally prohibitive, to the Kalman filter (Section 6.3), a simplification assuming normality and linearity of error growth, to ensemble-based data assimilation methods (Section 6.4), which may be more computationally tractable and robust. This review will include a description of stochastic and deterministic ensemble update algorithms, a simple pictorial example, discussions of model error and covariance localisation, and some pseudocode of an ensemble filter. Important ongoing research issues are discussed (Section 6.5) and conclusions provided (Section 6.6).

Several other useful review papers on ensemble-based data assimilation are available. Evensen (2003) provides a review of most of the proposed ensemble-based assimilation approaches, a more theoretical examination of the treatment of model errors, and a wide array of references to ensemble-based assimilation in the atmospheric and oceanographic literature. Lorenc (2003) also reviews ensemble methods, and in particular provides some thoughts on the potential relative strengths and weaknesses compared with the current state-of-the-art assimilation algorithm, four-dimensional variational analysis (4D-Var). Tippett *et al.* (2003) discusses the similarities and differences between a number of the proposed algorithms, and Anderson

(2003) discusses a way of interpreting ensemble-based techniques using simple linear regression terminology.

To keep the size of this chapter manageable, several topics will be omitted. We will not describe the full variety of ensemble filters nor Kalman filters, in particular leaving out a discussion of reduced-order Kalman filters (e.g. Farrell and Ioannou, 2001). Related subjects such as atmospheric predictability will be discussed only in relevance to the assimilation problem, and applications of ensemble filters to problems like adaptive observations will not be included.

In subsequent discussion, the atmosphere state, which is of course a continuum, is assumed to be adequately described in discretised fashion, such as by the values of winds, temperature, humidity, and pressure at a set of grid points.

6.2 Bayesian data assimilation

Conceptually, the atmospheric data assimilation problem is a relatively simple one. The task at hand is to estimate accurately the probability density function (pdf) for the current atmospheric state given all current and past observations. Much of the material in this section follows Anderson and Anderson (1999). If the reader is interested in further background material on the subject, Lorenc (1986) provides a formulation of data assimilation in a Bayesian context, and Cohn (1997) provides a more rigorous statistical formulation of the problem.

When considering Bayesian assimilation, there are two general steps to the assimilation. Assume that a pdf of the state of the atmosphere is available (in the lack of any knowledge, this may be the climatological pdf). The first step is to assimilate recent observations, thereby sharpening the pdf. The second step is to propagate the pdf forward in time until new observations are available. If the pdf is initially sharp (i.e. the distribution is relatively narrow), then chaotic dynamics and model uncertainty will usually broaden the probability distribution. The update and forecast steps are then repeated. We will describe each of these steps separately, starting with the assimilation of new observations.

6.2.1 Bayesian updating

Assume that an estimate of the pdf has been propagated forward to a time when observations are available. The state can be estimated more specifically by incorporating information from the new observations. This will be termed the 'update'.

The following notational convention is used. Boldface characters will denote vectors or matrices, while use of the italicised font denotes a scalar. $\mathbf{x}^{\mathrm{t}}_{\mathrm{t}-1}$ will denote the n-dimensional true model state at time $t-1$: $\mathbf{x}^{\mathrm{t}}_{\mathrm{t}-1} = [x^{\mathrm{t}}_{\mathrm{t}-1_{(1)}}, \ldots, x^{\mathrm{t}}_{\mathrm{t}-1_{(n)}}]$. Also, assume a collection of observations ψ_{t}. This vector includes observations $\mathbf{y}_{\mathrm{t}}$ at the most recent time as well as observations at all previous times $\psi_{\mathrm{t}} = [\mathbf{y}_{\mathrm{t}}, \psi_{\mathrm{t}-1}]$, where

$\psi_{t-1} = [\mathbf{y}_{t-1}, \ldots, \mathbf{y}_0]$. There are M_t observations at time t, i.e. $\mathbf{y}_t = [y_{t(1)}, \ldots, y_{t(M_t)}]$. Let $P(\mathbf{x}_t^t)$ be a multivariate probability density function, defined such that $Pr(\mathbf{a} \leq \mathbf{x}_t^t \leq \mathbf{b}) = \int_{\mathbf{a}}^{\mathbf{b}} P(\mathbf{x}_t^t) d\mathbf{x}_t^t$, and probability density integrates to 1.0 over the entire phase space.

Formally, the update problem is to accurately estimate $P(\mathbf{x}_t^t \mid \psi_t)$, the probability density estimate of the current atmospheric state, given the current and past observations. Bayes' rule tells us that this quantity can be re-expressed as

$$P(\mathbf{x}_t^t \mid \psi_t) \propto P(\psi_t \mid \mathbf{x}_t^t) P(\mathbf{x}_t^t). \tag{6.1}$$

Bayes' rule is usually expressed with a normalisation constant in the denominator on the right-hand side of Eq. (6.1); for simplicity, the term in the denominator will be dropped here, and it is assumed that when coded, the developer will ensure that probability density integrates to 1.0.

One hopefully minor assumption is made: observation errors are independent from one time to the next. Hence, $P(\psi_t \mid \mathbf{x}_t^t) = P(\mathbf{y}_t \mid \mathbf{x}_t^t) P(\psi_{t-1} \mid \mathbf{x}_t^t)$. This may not be true for observations from satellites, where instrumentation biases may be difficult to remove. Also, errors of observation representativeness (Daley, 1993) may be flow dependent and correlated in time. But under this assumption, (6.1) is equivalent to

$$P(\mathbf{x}_t^t \mid \psi_t) \propto P(\mathbf{y}_t \mid \mathbf{x}_t^t) P(\psi_{t-1} \mid \mathbf{x}_t^t) P(\mathbf{x}_t^t). \tag{6.2}$$

By Bayes' rule again, $P(\psi_{t-1} \mid \mathbf{x}_t^t) P(\mathbf{x}_t^t) \propto P(\mathbf{x}_t^t \mid \psi_{t-1})$. Hence, (6.2) simplifies to

$$P(\mathbf{x}_t^t \mid \psi_t) \propto P(\mathbf{y}_t \mid \mathbf{x}_t^t) \; P(\mathbf{x}_t^t \mid \psi_{t-1}). \tag{6.3}$$

In principle, Eq. (6.3) is elegantly simple. It expresses a recursive relationship: the 'posterior', the pdf for the current model state, given all the observations, is a product of the probability distribution for the current observations $P(\mathbf{y}_t \mid \mathbf{x}_t^t)$ and the 'prior', $P(\mathbf{x}_t^t \mid \psi_{t-1})$, also known as the 'background'. The prior is the pdf of the model state at time t given all the past observations up to time $t-1$. Typically, the prior will have been estimated in some fashion from a cycle of previous data assimilations and short-term forecasts up to the current time; approximations of how this may be computed will be discussed in Section 6.2.2.

Let's now demonstrate the update step of Bayesian assimilation with a simple example. $P(\mathbf{x}_t^t \mid \psi_{t-1})$ is an estimate of the prior for a two-dimensional model state. This was produced by assimilating all prior observations up to and including time $t-1$ and estimating in some manner how that pdf has evolved in the time interval between $t-1$ and t. Consider how to update the pdf given a new scalar observation y, which in this example is observing the same quantity as the first component of the state vector measures. The pdf for the observation $P(y_t \mid \mathbf{x}_t^t)$ is assumed to be distributed normally about the actual observation, $\sim N(y_t, \sigma^2)$. Here, let $y_t = 58$ and $\sigma^2 = 100$.

Selected contours of the prior are plotted in Figure 6.1(a); as shown, the prior is bimodal. The shape of the marginal prior distributions $P(x_{t(1)} \mid \psi_{t-1})$ and

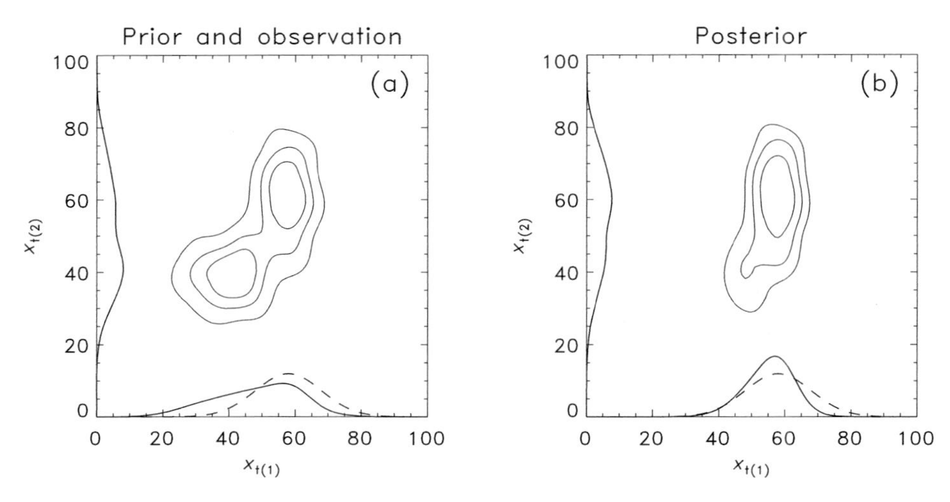

Figure 6.1 Example of Bayesian data assimilation update. Here the model state is two dimensional and a single observation is assimilated. This observation measures the same variable as the first component of the model state. (a) Probability density for joint and marginal prior distributions (solid) and observation distribution (dashed). The three contours enclose 75%, 50%, and 25% of the probability density, respectively. (b) Probability density for posterior distributions. Contours levels set as in (a).

$P(x_{t(2)} \mid \psi_{t-1})$ are plotted along each axis in solid lines. The dashed line denotes the observation probability distribution $P(y_t \mid \mathbf{x}_t^t)$. This probability varies with the value $x_{t(1)}$, but given $x_{t(1)}$ is the same for any value of $x_{t(2)}$. The updated posterior distribution is computed using Eq. (6.3) and is shown in Figure 6.1(b). Note that the assimilation of the observation enhanced the probability in the lobe overlapping the observation distribution and decreased it in the other lobe. Overall, the posterior distribution is more sharp (specific) than the prior, as is expected.

6.2.2 Forecasting of probability density

With an updated model pdf, a method for forecasting the evolution of this pdf forward in time is needed. Assume that we have an (imperfect) non-linear forecast model operator $\mathcal{M}$ so that the time evolution of the state can be written as a stochastic differential equation:

$$d\mathbf{x}_t^t = \mathcal{M}(\mathbf{x}_t^t)\,dt + \mathcal{G}(\mathbf{x}_t^t)d\mathbf{q} \tag{6.4}$$

where $d\mathbf{q}$ is a Brownian-motion process with covariance $\mathbf{Q}_t dt$ and $\mathcal{G}$ is the model-error forcing. Conceptually, the time evolution of the pdf can be modelled with the Fokker–Planck equation (e.g. Gardiner, 1985, section 5.3):

$$\frac{\partial P(\mathbf{x}_t^t)}{\partial t} = -\nabla . \left[\mathcal{M}\left(\mathbf{x}_t^t\right)P(\mathbf{x}_t^t)\right] + \sum_{i,j} \frac{\partial^2}{\partial x_{t(i)}^t \partial x_{t(j)}^t} \left(\frac{\mathcal{G}\mathbf{Q}_t\mathcal{G}^T}{2}\right)_{ij} P(\mathbf{x}_t^t) \tag{6.5}$$

If $\mathcal{G}\mathbf{Q}_t\mathcal{G}^T$ is zero, then only the first term remains, and the Fokker–Planck equation reduces to the Liouville equation (Ehrendorfer, 1994a, 1994b), a continuity equation for the conservation of probability. Probability thus diffuses with time according to the chaotic dynamics of the forecast model. The second term includes the effects of model error, including the increased diffusion of probability due to model uncertainty as well as noise-induced drift (Sardeshmukh *et al.*, 2001).

6.2.3 Limitations of Bayesian data assimilation

Unfortunately, neither the update nor the forecast steps in Bayesian data assimilation can be applied directly to real-world numerical weather prediction (NWP) applications. For the update step, one problem with modelling a complicated pdf in higher dimensions is the 'curse of dimensionality' (e.g. Hastie *et al.*, 2001, pp. 22–7). Were one to try to estimate the probability density in a higher-dimensional space using a small ensemble, one would find that the model of probability was very poor unless simplifying assumptions about the form of the distribution were made. Even were this problem surmountable, the computational cost would be extravagant. In the prior example the probability density was evaluated on a 100×100 grid. Suppose a similarly complicated structure for the prior existed in 100 dimensions. Then if the joint probabilities were monitored on a similar grid for each dimension, this would involve evaluating and modifying 100^{100} density estimates. Such computations are already prohibitive for a 100-dimensional model state; the problem becomes incomprehensible for model states of $O(10^7)$. Similarly, the Fokker–Planck equation cannot be integrated in high-dimensional systems using Eq. (6.5) due to computational constraints.

Consequently, *Monte Carlo* techniques are typically applied. Suppose we cannot explicitly compute the sharpening of the pdf from updating to new observations, nor the subsequent diffusion of probability in the forecast due to chaos and model error. As an approximation, let's randomly sample the initial probability distribution $P(\mathbf{x}_t^t \mid \psi_t)$. Thereafter, let's simulate the effects of chaos, model error and observations. Ensemble forecast techniques will be used to model the growth of errors due to the initial condition uncertainty, and some additional random noise will be added to each member to correct for the uncertainty contributed by model error. Monte Carlo data assimilation methods will be used that draw the ensemble of model states toward the observations in a process that recognises the uncertainty inherent in the observations. Hopefully, with a large enough random sample, probabilities estimated from the ensemble relative frequency will converge to the probabilities that would be calculated explicitly, were that computationally feasible.

6.3 Kalman filters

The methods underlying ensemble-based data assimilation come in part from Monte Carlo techniques, but the underlying concepts also are derived from a method known as the *Kalman filter* (Kalman, 1960; Kalman and Bucy, 1961; Jazwinski, 1970 section 7.3; Gelb, 1974 section 4.2; Maybeck, 1979 section 5.3; Ghil, 1989; Daley, 1991 section 13.3; Cohn, 1997; Talagrand, 1997; Daley, 1997). We review the Kalman filter first. The Kalman filter is an approximation to Bayesian state estimation which assumes linearity of error growth and normality of error distributions. There are two components of the Kalman filter, an update step where the state estimate and an estimate of the forecast uncertainty are adjusted to new observations, and a forecast step, where the updated state and the uncertainty estimate are propagated forward to the time when the next set of observations becomes available.

6.3.1 The extended Kalman filter

We now consider an implementation of the Kalman filter called the extended Kalman filter, or 'EKF' (Jazwinski, 1970; Gelb, 1974; Ghil and Malanotte-Rizzoli, 1991; Gauthier *et al.*, 1993; Bouttier, 1994). The EKF assumes that background and observation error distributions are Gaussian: $\mathbf{x}_{\mathrm{t}}^{\mathrm{b}} = \mathbf{x}_{\mathrm{t}}^{\mathrm{t}} + e$, where $e \sim N(0, \mathbf{P}_{\mathrm{t}}^{\mathrm{b}})$. That is, the probability density of the prior is distributed as a multivariate normal distribution with known $n \times 1$ mean background $\mathbf{x}_{\mathrm{t}}^{\mathrm{b}}$ and $n \times n$ background-error covariance matrix $\mathbf{P}_{\mathrm{t}}^{\mathrm{b}}$. Similarly, $\mathbf{y} = \mathcal{H}(\mathbf{x}_{\mathrm{t}}^{\mathrm{t}}) + \varepsilon$, where $\varepsilon \sim N(0, \mathbf{R})$ and $\mathcal{H}$ is the $M_t \times n$ 'forward' operator that maps the state to the observations. Let $\mathbf{H}$ represent the $m \times n$ Jacobian matrix of $\mathcal{H}$: $\mathbf{H} = \frac{\partial \mathcal{H}}{\partial \mathbf{x}}$ (see Gelb, 1974, section 6.1). Also, let $\mathcal{M}$ represent the non-linear model forecast operator. $\mathbf{M}$ is the $n \times n$ Jacobian matrix of $\mathcal{M}$, $\mathbf{M} = \frac{\partial \mathcal{M}}{\partial \mathbf{x}}$. $\mathbf{M}$ is often called the *transition matrix* between times t and $t+1$. $\mathbf{M}^{\mathrm{T}}$ is its adjoint (see Le Dimet and Talagrand, 1986, and Lacarra and Talagrand, 1988). $\mathbf{Q}$ will represent the $n \times n$ covariance of model errors accumulated between update cycles.

The EKF equations are

$$\mathbf{x}_{\mathrm{t}}^{\mathrm{a}} = \mathbf{x}_{\mathrm{t}}^{\mathrm{b}} + \mathrm{K}\left(\mathbf{y}_{\mathrm{t}} - \mathcal{H}\left(\mathbf{x}_{\mathrm{t}}^{\mathrm{b}}\right)\right) \tag{6.6a}$$

$$\mathbf{K} = \mathbf{P}_{\mathrm{t}}^{\mathrm{b}}\mathbf{H}^{\mathrm{T}}\left(\mathbf{H}\mathbf{P}_{\mathrm{t}}^{\mathrm{b}}\mathbf{H}^{\mathrm{T}} + \mathrm{R}\right)^{-1} \tag{6.6b}$$

$$\mathbf{P}_{\mathrm{t}}^{\mathrm{a}} = (\mathbf{I} - \mathbf{K}\mathbf{H})\mathbf{P}_{\mathrm{t}}^{\mathrm{b}} \tag{6.6c}$$

$$\mathbf{x}_{\mathrm{t+1}}^{\mathrm{b}} = \mathcal{M}\left(\mathbf{x}_{\mathrm{t}}^{\mathrm{a}}\right) \tag{6.6d}$$

$$\mathbf{P}_{\mathrm{t+1}}^{\mathrm{b}} = \mathbf{M}\mathbf{P}_{\mathrm{t}}^{\mathrm{a}}\mathbf{M}^{\mathrm{T}} + \mathbf{Q} = \mathbf{M}\left(\mathbf{M}\mathbf{P}_{\mathrm{t}}^{\mathrm{a}}\right)^{\mathrm{T}} + \mathbf{Q}. \tag{6.6e}$$

Equations (6.6a–6.6c) describe the update step. The optimal analysis state $\mathbf{x}_{\mathrm{t}}^{\mathrm{a}}$ is estimated by correcting the background $\mathbf{x}_{\mathrm{t}}^{\mathrm{b}}$ toward the 'observation increment' $\mathbf{y}_{\mathrm{t}} - \mathcal{H}(\mathbf{x}_{\mathrm{t}}^{\mathrm{b}})$, weighted by the Kalman-gain matrix $\mathbf{K}$. The effect of $\mathbf{K}$ is to apply observation increments to correct the background at relevant surrounding grid points. Equation (6.6c) indicates how to update the background-error covariance to reflect the reduction in uncertainty from assimilating the observations. Equations (6.6d–6.6e) propagate the resulting analysis and error covariance forward in time to when observations are next available. The expected analysis state is propagated forward with the full non-linear forecast model. Model errors are assumed to be uncorrelated with the growth of analysis errors through the tangent-linear forecast dynamics.

The conceptual appeal of the Kalman filter relative to an analysis scheme like three-dimensional variational assimilation (3D-Var; Lorenc, 1986; Parrish and Derber, 1992) is that the error covariances of the forecast and subsequent analysis are explicitly prognosed. The analysis reduces error variances in locations where accurate observations are plentiful, and the error covariances are also forecast forward in time, growing at a rate proportional to the local error growth. Consequently, the structure of the background-error covariances and hence the adjustment to observations $\mathbf{x}_t^a - \mathbf{x}_t^b$ can be quite complicated and flow and time dependent (e.g. Bouttier, 1994).

6.3.2 Considerations in the use of Kalman filters

What approximations may limit the accuracy of the EKF? First, Kalman filters assume linear growth and normality of errors, for the assimilation problem becomes somewhat more tractable when these assumptions are made. Non-normality of the prior such as the bimodality in Figure 6.1(a) is typically assumed to be uncommon in atmospheric data assimilation. These linear and normal assumptions may be inappropriate for atmospheric data assimilations of moisture, cloud cover, and other aspects of the model state that may be very sensitive to motions at small scales, where the timescale of predictability is small and errors grow and saturate rapidly. Similarly, if observations are not regularly available, error covariances estimated with tangent linear dynamics may grow rapidly without bound (Evensen, 1992; Gauthier *et al.*, 1993; Bouttier, 1994).

Second, error statistics must be carefully estimated and monitored; in particular, it is important that the background-error covariance matrix be estimated accurately. For example, if background error variances are underestimated, the EKF will assume the error statistics are indicating that the background is relatively more accurate than the nearby observations and thus will not correct the background to the observations to the extent it should (Daley, 1991, p. 382). Estimating $\mathbf{Q}$ may be particularly difficult. In practice, accurately determining even the time-averaged statistics of $\mathbf{Q}$ may be quite complicated (Cohn and Parrish, 1991; Daley, 1992; Dee, 1995; Blanchet *et al.*, 1997). For both the Kalman filter and ensemble-based methods, the accuracy of the assimilation is likely to depend strongly on this assumed model for $\mathbf{Q}$. Methods for estimating $\mathbf{Q}$ will be discussed for ensemble-based methods in Section 6.4.4.

Another disadvantage of the Kalman filters for atmospheric data assimilation is their computational expense. Though Kalman filters provide a dramatic reduction in the computational cost relative to full Bayesian data assimilation, for a highly dimensional state vector, the computational costs in weather prediction models may still be impossibly large. Consider the last line in Eq. (6.6). For an n-dimensional model state vector, it will require $2n$ applications of $\mathbf{M}$ to forecast the error covariances.

Some reductions of computational expense may be possible. For example, there have been suggestions that this computation may be more practical if the tangent linear calculations are performed in a subspace of the leading singular vectors (Fisher, 1998; Farrell and Ioannou, 2001).

Much more can be said about the Kalman filter, such as its equivalence to 4D-Var under certain assumptions (Li and Navon, 2001), the manner of computing **M**, iterated extensions of the basic extended Kalman filter (Jazwinski, 1970; Gelb, 1974; Cohn, 1997), and the properties of its estimators (which, in the case of the discrete filter, if assumptions hold, provide the Best Linear Unbiased Estimate, or BLUE; see Talagrand, 1997).

6.4 Ensemble-based data assimilation

Ensemble-based assimilation algorithms use Monte Carlo techniques and may be able to provide more accurate analyses than the EKF in situations where non-linearity is pronounced and pdfs exhibit some non-normality. If these assimilation algorithms can work accurately with many fewer ensemble members than elements in the state vector, then they will be computationally much less expensive as well.

Many researchers have proposed a variety of ensemble-based assimilation methods. Despite the many differences between the various ensemble-based algorithms, all comprise a finite number (perhaps ten to a few hundred) of parallel data assimilation and short-range forecast cycles. Background-error covariances are modelled using the ensemble of forecasts, and an ensemble of analyses are produced, followed by an ensemble of short-term forecasts to the next time observations are available. Ensemble-based assimilation algorithms also have the desirable property that if error dynamics are indeed linear and the error statistics Gaussian, then as the ensemble size increases, the state and covariance estimate from ensemble algorithms converge to those obtained from the extended Kalman filter (Burgers *et al.*, 1998).

The concepts behind ensemble assimilation methods have been used in engineering and aerospace applications as far back as the 1960s (Potter, 1964; Andrews, 1968; Kaminski *et al.*, 1971; Maybeck, 1979, ch. 7). Leith (1983) sketched the basic idea for atmospheric data assimilation. The idea was more completely described and tested in an oceanographic application by Evensen (1994) and in atmospheric data assimilation by Houtekamer and Mitchell (1998).

For notational simplicity, the t time subscript used in previous sections is dropped; it is assumed unless noted otherwise that we are interested in estimating the state pdf at time t. We start off by assuming that we have an ensemble of forecasts that randomly sample the model background errors at time t. Let's denote this ensemble as $\mathbf{X}^{\mathrm{b}}$, a matrix whose columns comprise ensemble members' state vectors:

$$\mathbf{X}^{\mathrm{b}} = \left(\mathbf{x}_1^{\mathrm{b}}, \ldots, \mathbf{x}_m^{\mathrm{b}}\right), \tag{6.7}$$

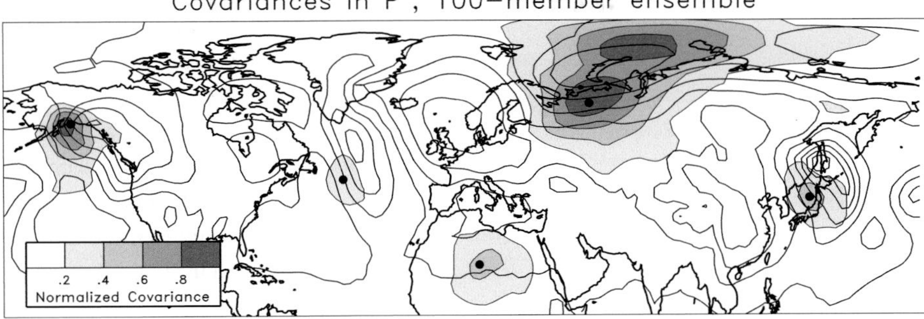

Figure 6.2 Background-error covariances (grey shading) of sea-level pressure in the vicinity of five selected observation locations, denoted by dots. Covariance magnitudes are normalised by the largest covariance magnitude on the plot. Solid lines denote ensemble mean background sea-level pressure contoured every 8 hPa.

The subscript now denotes the ensemble member. The ensemble mean $\overline{\mathbf{x}}^{\mathrm{b}}$ is defined as

$$\overline{\mathbf{x}}^{\mathrm{b}} = \frac{1}{m}\sum_{i=1}^{m}\mathbf{x}_i^{\mathrm{b}}. \tag{6.8}$$

The perturbation from the mean for the ith member is $\mathbf{x}_i'^{\mathrm{b}} = \mathbf{x}_i^{\mathrm{b}} - \overline{\mathbf{x}}^{\mathrm{b}}$. Define $\mathbf{X}'^{\mathrm{b}}$ as a matrix formed from an ensemble of perturbations

$$\mathbf{X}'^{\mathrm{b}} = \left(\mathbf{x}_1'^{\mathrm{b}}, \ldots, \mathbf{x}_m'^{\mathrm{b}}\right) \tag{6.9}$$

and let $\hat{\mathbf{P}}^{\mathrm{b}}$ represent an estimate of $\mathbf{P}^{\mathrm{b}}$ from a finite ensemble

$$\hat{\mathbf{P}}^{\mathrm{b}} = \frac{1}{m-1}\mathbf{X}'^{\mathrm{b}}\mathbf{X}'^{\mathrm{b}^{\mathrm{T}}}. \tag{6.10}$$

Unlike the Kalman filter or 3D-Var, the background-error covariance estimate is generated from a specially constructed ensemble of non-linear forecasts. The finite sample will introduce errors (see, for example, Casella and Berger, 1990, section 5.4, and Hamill *et al.*, 2001, section 2) relative to the EKF. However, estimating the covariances using an ensemble of non-linear model forecasts may provide a powerful advantage over the EKF. Envision a situation where errors grow rapidly but saturate at low amplitude; the linear assumption of error growth in the EKF will result in an overestimate of background error variance, but the differences among ensemble members will not grow without bound and thus should provide a more accurate model of the actual background-error statistics. Unlike data assimilation algorithms such as 3D-Var (in most operational implementations), the background-error covariances can vary in time and space. If this error covariance model is relatively accurate, it will thus provide a better adjustment to the observations.

Figure 6.2 illustrates the potential benefit from estimating background-error covariances using an ensemble-based data assimilation system. Here we see a snapshot of sea-level pressure background-error covariances with five locations around the northern hemisphere, estimated from a 100-member ensemble. The data were taken from the ensemble data assimilation experiment of Whitaker *et al.* (2004), which tested the efficacy of assimilating only a sparse network of surface pressure observations concentrated over the USA, Europe, and east Asia. A covariance localisation with a correlation length of approximately 2700 km was applied (see Section 6.4.5). Notice that the magnitude and the spatial structure of the background-error covariances change from one location to the next, with larger covariances for the point south of Alaska and northern Russia and smaller covariances at other locations. The horizontal extent of the positive covariance also changed markedly from one location to the next. The background-error covariances control the magnitude of the adjustment to the observation, drawing more to observations when background errors are large. Hence, observations will affect the analysis very differently around each of the five locations, which is the essence of why ensemble-based algorithms may outperform methods assuming fixed background-error covariances.

We will first consider the update step in two general classes of ensemble filters, stochastic (fully Monte Carlo) and deterministic. Both classes propagate the ensemble of analyses with non-linear forecast models; the primary difference is whether or not random noise is applied during the update step to simulate observation uncertainty. A brief pictorial example of the update step is then provided, followed by a discussion of the ensemble forecast process and how model error may be treated. A description of an important algorithmic modification, covariance localisation, is then provided. Finally, some pseudocode for a simple deterministic filter is provided.

6.4.1 Stochastic update algorithms

The most well-known stochastic ensemble-based data assimilation algorithm is the *ensemble Kalman filter*, or 'EnKF' (Houtekamer and Mitchell, 1998, 1999, 2001; Burgers *et al.*, 1998; Keppenne, 2000; Mitchell and Houtekamer, 2000; Hamill and Snyder, 2000; Hamill *et al.*, 2001; Heemink *et al.*, 2001; Keppenne and Rienecker, 2002; Mitchell *et al.*, 2002; Hamill and Snyder, 2002; Houtekamer *et al.*, 2005). This algorithm updates each member to a different set of observations perturbed with random noise. Because randomness is introduced every assimilation cycle, the update is considered stochastic.

The EnKF performs an ensemble of parallel data assimilation cycles, $i = 1, \ldots, m$, with each member updated to a somewhat different realisation of the observations:

$$\mathbf{x}_i^{\mathrm{a}} = \mathbf{x}_i^{\mathrm{b}} + \hat{\mathbf{K}}\big(\mathbf{y}_i - \mathcal{H}(\mathbf{x}_i^{\mathrm{b}})\big). \tag{6.11}$$

In (6.11), the $\mathbf{y}_i = \mathbf{y} + \mathbf{y}'_i$ are 'perturbed observations', defined such that $\mathbf{y}'_i \sim N(0, \mathbf{R})$, and

$$\frac{1}{m}\sum_{i=1}^{m} \mathbf{y}'_i = 0. \tag{6.12}$$

The m sets of perturbed observations are thus created to update the m different background fields. Here, in (6.11),

$$\hat{\mathbf{K}} = \hat{\mathbf{P}}^{\mathrm{b}}\mathbf{H}^{\mathrm{T}}(\mathbf{H}\hat{\mathbf{P}}^{\mathrm{b}}\mathbf{H}^{\mathrm{T}} + \mathbf{R})^{-1}, \tag{6.13}$$

similar to the Kalman gain of the EKF gain in (6.6b), but using the ensemble to estimate the background-error covariance matrix as in (6.10).

Notice that the EnKF assimilates perturbed observations in Eq. (6.11) rather than the observations themselves. To understand this, let $\mathbf{X}'^{\mathrm{a}}$ be a matrix of analysis ensemble member deviations from the analysis mean state, as (6.9) defined background deviations. Let $\hat{\mathbf{P}}^{\mathrm{a}}$ be formed from the ensemble of analyses assimilating perturbed observations using (6.11). Then as the ensemble size approaches infinity and if the dynamics are linear, $\hat{\mathbf{P}}^{\mathrm{a}} = \frac{1}{m-1}\mathbf{X}'^{\mathrm{a}}\mathbf{X}'^{\mathrm{a}\mathrm{T}} \rightarrow \mathbf{P}^{\mathrm{a}}$, where $\mathbf{P}^{\mathbf{a}}$ is the extended Kalman filter analysis-error covariance from (6.6c) (Burgers *et al.*, 1998). If unperturbed observations are assimilated in (6.11) without other modifications to the algorithm, the analysis-error covariance will be underestimated, and observations will not be properly weighted in subsequent assimilation cycles.

Adding noise to the observations in the EnKF can introduce spurious observation-background error correlations that can bias the analysis-error covariances, especially when the ensemble size is small (Whitaker and Hamill, 2002). Pham (2001) proposed an alternative to perturbing the observations, adding noise to background forecasts in a manner that also ensures analysis-error covariances are equal to those produced by the EKF. Anderson (2003) proposed a sequential observation processing method that minimises this effect. Houtekamer and Mitchell (1998) proposed the use of a 'double' EnKF with two parallel sets of ensembles, each set used to estimate background-error covariances to update the other set. See van Leeuwen (1999), Houtekamer and Mitchell (1999), and Whitaker and Hamill (2002) for a discussion of covariance biases in the single and double EnKFs.

Several algorithms have been proposed for simplifying and parallelising the coding of the EnKF. One technique that is uniformly used is to form the Kalman gain (6.13) from the ensemble without ever forming the actual background-error covariance matrix. For a complex numerical weather prediction model with a high-dimensional state vector, explicitly forming $\hat{\mathbf{P}}^{\mathrm{b}}$ as in (6.10) would be computationally prohibitive; for example, in a model with 10^7 elements in its state, storing and readily accessing the 10^{14} elements of $\hat{\mathbf{P}}^{\mathrm{b}}$ is not possible. However, in ensemble-based methods, $\hat{\mathbf{K}}$ can be formed without ever explicitly computing the full $\hat{\mathbf{P}}^{\mathrm{b}}$ (Evensen, 1994; Houtekamer and Mitchell, 1998). Instead, the components of $\hat{\mathbf{P}}^{\mathrm{b}}\mathbf{H}^{\mathrm{T}}$ and $\mathbf{H}\hat{\mathbf{P}}^{\mathrm{b}}\mathbf{H}^{\mathrm{T}}$ of $\hat{\mathbf{K}}$ are computed separately. Define

$$\overline{\mathcal{H}(\mathbf{x}^{\mathrm{b}})} = \frac{1}{m}\sum_{i=1}^{n}\mathcal{H}\big(\mathbf{x}_{\mathrm{i}}^{\mathrm{b}}\big),$$

which represents the mean of the estimate of the observation interpolated from the background forecasts. Then

$$\hat{\mathbf{P}}^{\mathrm{b}}\mathbf{H}^{\mathrm{T}} = \frac{1}{m-1}\sum_{i=1}^{m}\big(\mathbf{x}_{i}^{\mathrm{b}} - \overline{\mathbf{x}^{\mathrm{b}}}\big)\big(\mathcal{H}(\mathbf{x}_{i}^{\mathrm{b}}) - \overline{\mathcal{H}(\mathbf{x}^{\mathrm{b}})}\big)^{\mathrm{T}}, \tag{6.14}$$

and

$$\mathbf{H}\hat{\mathbf{P}}^{\mathrm{b}}\mathbf{H}^{\mathrm{T}} = \frac{1}{m-1}\sum_{i=1}^{m}\big(\mathcal{H}(\mathbf{x}_{i}^{\mathrm{b}}) - \overline{\mathcal{H}(\mathbf{x}^{\mathrm{b}})}\big)\big(\mathcal{H}(\mathbf{x}_{i}^{\mathrm{b}}) - \overline{\mathcal{H}(\mathbf{x}^{\mathrm{b}})}\big)^{\mathrm{T}}. \tag{6.15}$$

Of course, if the number of observations is as large as the elements in the model state, $\hat{\mathbf{P}}^{\mathrm{b}}\mathbf{H}^{\mathrm{T}}$ and $\mathbf{H}\hat{\mathbf{P}}^{\mathrm{b}}\mathbf{H}^{\mathrm{T}}$ will be as large as $\hat{\mathbf{P}}^{\mathrm{b}}$, negating this advantage. However, another possible coding simplification is *serial processing*. If observations have independent errors uncorrelated with the background, they can be assimilated simultaneously or serially (sequentially), producing the same result (Kaminski *et al.*, 1971; Gelb, 1974 p. 304; Bishop *et al.*, 2001). The analysis ensemble after the assimilation of the first observation is used as the background ensemble for the assimilation of the second, and so on. When observations are assimilated serially, for each observation that is assimilated, $\mathbf{H}\hat{\mathbf{P}}^{\mathrm{b}}\mathbf{H}^{\mathrm{T}}$ and $\mathbf{R}$ become scalars. Thus, the inverse $(\mathbf{H}\hat{\mathbf{P}}^{\mathrm{b}}\mathbf{H}^{\mathrm{T}} + \mathbf{R})^{-1}$ in the gain matrix is trivial to compute. Also, the application of the covariance localisation, discussed later, is much more straightforward to apply. Serial stochastic ensemble filters have been demonstrated in Houtekamer and Mitchell (2001), Hamill *et al.* (2001), Hamill and Snyder (2002), and Anderson (2003).

The equivalence of serial and simultaneous processing is only true if observations have independent errors (Kaminski *et al.*, 1971). Practically, however, many observations may have vertically or horizontally correlated errors. Consider two alternatives to deal with this. First, if the size of a batch of observations with correlated errors is relatively small, these correlated batches can be processed simultanteously without much more computational expense (Houtekamer and Mitchell, 2001; Mitchell *et al.*, 2002; Houtekamer *et al.*, 2005); the matrix inverse of $(\mathbf{H}\hat{\mathbf{P}}^{\mathrm{b}}\mathbf{H}^{\mathrm{T}} + \mathbf{R})^{-1}$ should not be prohibitively expensive. Another option is to transform the observations and the forward operator so that the observations are effectively independent (Kaminski *et al.*, 1971).

Several investigators have proposed speeding up the performance of the stochastic EnKF by separately updating different grid points independently on different processors. Keppenne and Rienecker (2002) designed an algorithm whereby all observations in the region of a particular set of grid points are simultaneously assimilated to update those grid points, while other distinct sets of grid points are updated independently.

Houtekamer and Mitchell (2001) propose a method that uses both serial processing of observations and processing different regions separately from one another. They also discuss other ways of minimising the amount of information that needs to be swapped between processors on a parallel computer. Reichle *et al.* (2002, 2003) and Reichle and Koster (2004) demonstrate a parallelised EnKF algorithm applied to soil-moisture state estimation.

6.4.2 Deterministic update algorithms

Several methods have been proposed to correct the background ensemble to new observations so that $\hat{\mathbf{P}}^{\mathrm{a}} \rightarrow \mathbf{P}^{\mathrm{a}}$ without adding random noise. Algorithms that do not add stochastic noise are called *deterministic* algorithms, so named because if the background ensemble and the associated error statistics are known, the ensemble of analysis states will be completely known as well. These algorithms (e.g. Lermusiaux and Robinson, 1999; Bishop *et al.*, 2001; Anderson, 2001; Whitaker and Hamill, 2002; Lermusiaux, 2002; Hunt *et al.*, 2004) update in a way that generates the same analysis-error covariance update that would be obtained from the Kalman filter, assuming that the Kalman filter's background-error covariance is modelled from the background ensemble. Tippett *et al.* (2003) describe the similarities and differences between several of these algorithms. In each, the background-error covariances are never explicitly formed, with manipulations being performed using the matrix square root (i.e. Eq. (6.9), the matrix of ensemble member deviations from the mean). As pointed out in Tippett *et al.*, since $\hat{\mathbf{P}}^{\mathrm{b}} = \frac{1}{m-1}\mathbf{X}'^{\mathrm{b}}\mathbf{X}'^{\mathrm{b}\mathrm{T}}$, given a matrix $\mathbf{U}$ representing any $n \times n$ orthogonal transformation such that $\mathbf{U}\mathbf{U}^{\mathrm{T}} = \mathbf{U}^{\mathrm{T}}\mathbf{U} = \mathbf{I}$, then $\hat{\mathbf{P}}^{\mathrm{b}}$ can also be represented as $\hat{\mathbf{P}}^{\mathrm{b}} = \frac{1}{m-1}(\mathbf{X}'^{\mathrm{b}}\mathbf{U})(\mathbf{X}'^{\mathrm{b}}\mathbf{U})^{\mathrm{T}}$. Hence, many square-root filters can be formulated that produce the same analysis-error covariance.

Since Tippett *et al.* (2003) review many of these methods, we will explicitly describe only one of these, a particularly simple implementation, the 'ensemble square-root filter', or 'EnSRF', described by Whitaker and Hamill (2002), which is mathematically equivalent to the filter described in Anderson (2001). The EnSRF algorithm has been used for the assimilation at the scale of thunderstorms by Snyder and Zhang (2003), Zhang *et al.* (2004) and Dowell *et al.* (2004). Whitaker *et al.* (2004) used the algorithm for the global data assimilation of surface pressure observations. Like the EnKF, the EnSRF conducts a set of parallel data assimilation cycles. It is convenient in the EnSRF to update the equations for the ensemble mean (denoted by an overbar) and the deviation of the ith member from the mean separately:

$$\overline{\mathbf{x}}^{\mathrm{a}} = \overline{\mathbf{x}}^{\mathrm{b}} + \hat{\mathbf{K}}\big(\mathbf{y} - \mathcal{H}(\overline{\mathbf{x}}^{\mathrm{b}})\big), \tag{6.16}$$

$$\mathbf{x}'^{\mathrm{a}}_i = \mathbf{x}'^{\mathrm{b}}_i - \tilde{\mathbf{K}}\mathcal{H}(\mathbf{x}'^{\mathrm{b}}_i). \tag{6.17}$$

Here, $\hat{\mathbf{K}}$ is the traditional Kalman gain as in Eq. (6.13), and $\tilde{\mathbf{K}}$ is the 'reduced' gain used to update deviations from the ensemble mean.

When sequentially processing independent observations, $\hat{\mathbf{K}}$, $\tilde{\mathbf{K}}$, $\mathbf{H}\hat{\mathbf{P}}^{\mathrm{b}}$ and $\hat{\mathbf{P}}^{\mathrm{b}}\mathbf{H}^{\mathrm{T}}$ are all n-dimensional vectors, and $\mathbf{H}\hat{\mathbf{P}}^{\mathrm{b}}\mathbf{H}^{\mathrm{T}}$ and $\mathbf{R}$ are scalars. Thus, as first noted by Potter (1964), when observations are processed one at a time,

$$\tilde{\mathbf{K}} = \left(1 + \sqrt{\frac{\mathbf{R}}{\mathbf{H}\hat{\mathbf{P}}^{\mathrm{b}}\mathbf{H}^{\mathrm{T}} + \mathbf{R}}}\right)^{-1} \hat{\mathbf{K}}. \tag{6.18}$$

The quantity multiplying $\hat{\mathbf{K}}$ in Eq. (6.18) thus becomes a scalar between 0 and 1. This means that, in order to obtain the correct analysis-error covariance with unperturbed observations, a modified Kalman gain that is reduced in magnitude relative to the traditional Kalman gain is used to update deviations from the ensemble mean. Consequently, deviations from the mean are reduced less in the analysis using $\tilde{\mathbf{K}}$ than they would be using $\hat{\mathbf{K}}$. In the stochastic EnKF, the excess variance reduction caused by using $\hat{\mathbf{K}}$ to update deviations from the mean is compensated for by the introduction of noise to the observations.

In the EnSRF, the mean and departures from the mean are updated independently according to Eqs. (6.16) and (6.17). If observations are processed one at a time, the EnSRF requires about the same computation as the traditional EnKF with perturbed observations, but for moderately sized ensembles and processes that are generally linear and Gaussian, the EnSRF produces analyses with significantly less error (Whitaker and Hamill, 2002). Conversely, Lawson and Hansen (2003) suggest that if multimodality is typical and ensemble size is large, the EnKF will perform better.

Another deterministic update algorithm is the ensemble transform Kalman filter (ETKF) of Bishop *et al.* (2001). The ETKF finds the transformation matrix $\mathbf{T}$ such that $\hat{\mathbf{P}}^{\mathrm{a}} = \frac{1}{m-1}(\mathbf{X}'^{\mathrm{b}}\mathbf{T})(\mathbf{X}'^{\mathrm{b}}\mathbf{T})^{\mathrm{T}} \rightarrow \mathbf{P}^{\mathrm{a}}$ (see Bishop *et al.* for details on the computation of $\mathbf{T}$). Compared with the EnSRF, an advantage of the ETKF is its computational speed; a disadvantage is that the ETKF cannot apply covariance localisations (Section 6.5), which may make the analyses very inaccurate unless large ensembles are used. The ETKF has been successfully demonstrated for generating perturbed initial conditions for ensemble forecasts about a mean state updated using 3D-Var (Wang and Bishop, 2003), and computationally efficient hybrid ETKF-variational schemes are being explored (Etherton and Bishop, 2004), which may have an advantage in situations with significant model errors.

6.4.3 A simple demonstration of stochastic and deterministic update steps

Consider again the Bayesian data assimilation problem illustrated in Figure 6.1. There, a bimodal two-dimensional probability distribution was updated to an observation of one component. Let's explore the characteristics of the EnKF and EnSRF update applied to this problem.

A 100-member random sample was first generated from the bimodal pdf in Figure 6.1(a). These samples are denoted by the black dots in Figure 6.3(a). Let's keep track of the assimilation for one particular member, denoted by the larger black dot.

The EnKF and EnSRF adjust the background to the observations with weighting factors that assume the distributions are normal. Estimated from this random sample, the background-error covariance is

$$\hat{\mathbf{P}}^{\mathrm{b}} = \begin{pmatrix} \sigma^2\left(x^{\mathrm{b}}_{(1)}\right) & Cov\left(x^{\mathrm{b}}_{(1)}, x^{\mathrm{b}}_{(2)}\right) \\ Cov\left(x^{\mathrm{b}}_{(1)}, x^{\mathrm{b}}_{(2)}\right) & \sigma^2\left(x^{\mathrm{b}}_{(2)}\right) \end{pmatrix} \simeq \begin{pmatrix} 150.73 & 109.70 \\ 109.70 & 203.64 \end{pmatrix}.$$

The shape of this distribution is illustrated by the black contours in Figure 6.3(a). Here, the observation measures the same aspect as the first component of our state variable: $\mathcal{H} = [1,\ 0]$. As in Figure 6.1, assume $\mathbf{R} = 100$, so $\mathbf{H}\hat{\mathbf{P}}^{\mathrm{b}}\mathbf{H}^{\mathrm{T}} + \mathbf{R} \simeq 150.73 + 100.00 = 250.73$. $\hat{\mathbf{P}}^{\mathrm{b}}\mathbf{H}^{\mathrm{T}} \simeq [150.73,\ 109.70]^{\mathrm{T}}$, and hence $\hat{\mathbf{K}} = \mathbf{P}^{\mathrm{b}}\mathbf{H}^{\mathrm{T}}(\mathbf{H}\mathbf{P}^{\mathrm{b}}\mathbf{H}^{\mathrm{T}} + \mathbf{R})^{-1} \simeq [0.60,\ 0.44]^{\mathrm{T}}$.

For the EnKF, perturbed observations were then generated, denoted by the short vertical lines along the abscissa in Figure 6.3(a). Equation (6.11) was then applied, updating background samples to their associated perturbed observations, generating analysis samples. For example, the enlarged black dot in Figure 6.3(a) was updated to the perturbed observation marked with the '*'. The resulting analysis sample is the enlarged black dot in Figure 6.3(b). For the noted sample, the first component of the background state was much less than the mean, and the perturbed observation was greater than the mean background state. The resulting analysis nudged the posterior state toward the mean in both components. Other dots in Figure 6.3(b) denote other updated EnKF member states.

In the EnSRF, the ensemble background mean state $\sim[47.93,\ 50.07]^{\mathrm{T}}$ was updated to the mean observed value 58.0 using $\hat{\mathbf{K}}$ computed above and Eq. (6.16), resulting in a mean analysed state of $\sim[53.55,\ 54.16]$. As with the EnKF, given the positive observation increment and the positive correlation of the background-error covariances between the two components, both components of the mean state were adjusted upward. EnSRF perturbations from the mean were updated using Eq. (6.17) and the reduced gain, here $\tilde{\mathbf{K}} \simeq 0.613\,\hat{\mathbf{K}}$.

Compare the EnKF and EnSRF random samples of the posterior from Figures 6.3(b–c) and their fitted distribution (thin lines) with the correct Bayesian posterior (bold lines). The samples from both distributions do not appear to sample randomly the correct posterior. The EnKF and EnSRF posterior distributions are shifted slightly toward lower values in both components. The EnSRF posterior samples preserve the original shape from the prior, though their values are shifted in mean and compressed together. In comparison, the EnKF samples are randomised somewhat through the assimilation of the perturbed observations, and in this case, its distribution is rather more diffuse than that of the EnSRF. The EnKF samples appear to overlap more with the correct distribution than the samples from the EnSRF.

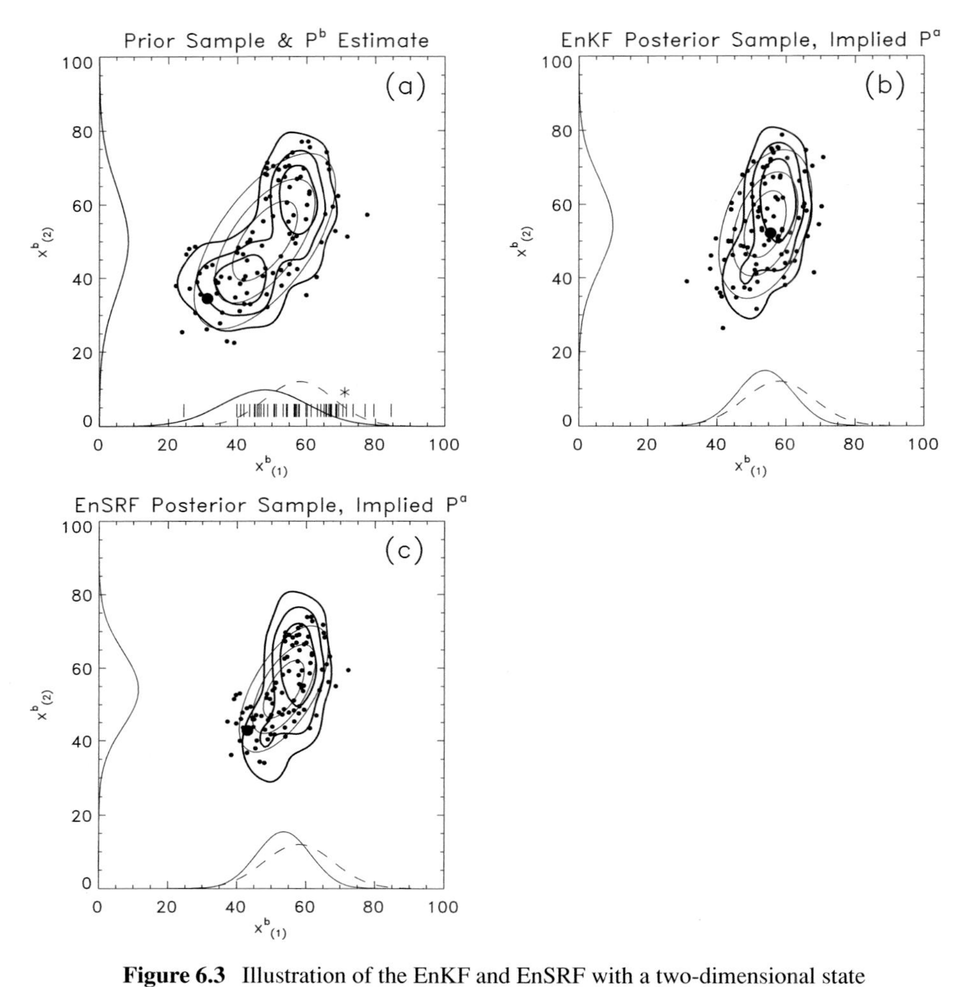

Figure 6.3 Illustration of the EnKF and EnSRF with a two-dimensional state variable and observations of the first component of the model state. (a) Random samples (black dots) from the probability distribution in Figure 6.1(a), and the original prior pdf, contoured in bold lines. Implied bivariate normal probability background distribution estimated from the sample ensemble contoured in thin lines, and the observation sampling distribution (dashed). Solid vertical lines along abscissa denote individual perturbed observations sampled from this distribution. The one large black dot and the perturbed observation marked with a star denote the sample discussed in the text. (b) Analysed samples from the EnKF assimilation scheme (dots), the implied analysis-error bivariate normal distribution from this sample (thin contours), and the true posterior pdf from Figure 6.1 (bold contours). (c) Analysed samples from EnSRF (dots), implied bivariate normal pdf (thin contours) and the true posterior pdf (bold contours). In each panel, the three contours enclose 75%, 50%, and 25% of the probability density, respectively.

Why can't ensemble-based methods correctly adjust the prior ensemble to the new observations so that the samples reflect a random draw from the Bayesian posterior? The reason is that ensemble-based methods implicitly assume a second-moment closure; that is, the distributions are assumed to be fully described by means and covariances. The example shown above demonstrates that some inaccuracies can be expected in these analyses if indeed there are higher-moment details in these distributions (Lawson and Hansen, 2004). Hopefully, highly non-normal distributions are not frequently encountered, as radically more expensive and unproven techniques than those discussed here may then be required (e.g. Gordon *et al.*, 1993).

6.4.4 Ensemble propagation of the pdf and model-error parametrisation

In real-world applications, background-error covariances cannot simply be estimated at the next assimilation cycle by conducting an ensemble of deterministic forecasts forward from the current cycle's analyses. Because of model deficiencies, even if the true state of the atmosphere is perfectly known, the resulting forecast will be imperfect: $\mathbf{x}^{t}_{(t+1)} = \mathcal{M}\left(\mathbf{x}^{t}_{(t)}\right) + \eta$, where here we denote the time index in parentheses and $\mathcal{M}$ is again the non-linear forecast operator. Let's first assume that our forecast model is unbiased $\langle \eta \rangle = 0$, again with model-error covariance $\langle \eta\eta^{\mathrm{T}} \rangle = \mathbf{Q}$ (here the angle brackets denote a statistical expected value). In practice, the assumption of no bias is probably not justified, and if the bias can be determined, the forecasts ought to be corrected for this bias (Dee and Todling, 2000; Evensen, 2003), or more ideally, the forecast model ought to be improved. In any case, consider the error covariance at the next assimilation time. Assume again that forecast error due to initial-condition uncertainty and model error are uncorrelated $\langle (\mathcal{M}(\overline{\mathbf{x}}^{a}_{(t)}) - \mathcal{M}(\mathbf{x}^{t}_{(t)}))\eta^{\mathrm{T}} \rangle = 0$, and assume linearity of the error growth $\mathcal{M}(\overline{\mathbf{x}}^{a}_{(t)}) - \mathcal{M}(\mathbf{x}^{t}_{(t)}) \simeq \mathbf{M}(\overline{\mathbf{x}}^{a}_{(t)} - \mathbf{x}^{t}_{(t)})$. Then the true background-error covariance at the next assimilation time is

$$
\begin{aligned}
&\left\langle \left(\overline{\mathbf{x}}^{b}_{(t+1)} - \mathbf{x}^{t}_{(t+1)}\right)\left(\overline{\mathbf{x}}^{b}_{(t+1)} - \mathbf{x}^{t}_{(t+1)}\right)^{\mathrm{T}} \right\rangle \\
&\quad = \left\langle \left(\mathcal{M}\left(\overline{\mathbf{x}}^{a}_{(t)}\right) - \mathcal{M}\left(\mathbf{x}^{t}_{(t)}\right) - \eta\right)\left(\mathcal{M}\left(\overline{\mathbf{x}}^{a}_{(t)}\right) - \mathcal{M}\left(\mathbf{x}^{t}_{(t)}\right) - \eta\right)^{\mathrm{T}} \right\rangle \\
&\quad \simeq \left\langle \mathbf{M}\left(\overline{\mathbf{x}}^{a}_{(t)} - \mathbf{x}^{t}_{(t)}\right)\left(\overline{\mathbf{x}}^{a}_{(t)} - \mathbf{x}^{t}_{(t)}\right)^{\mathrm{T}} \mathbf{M}^{\mathrm{T}} \right\rangle + \left\langle \eta\eta^{\mathrm{T}} \right\rangle \\
&\quad = \mathbf{M}\mathbf{P}^{a}_{(t)}\mathbf{M}^{\mathrm{T}} + \mathbf{Q}
\end{aligned}
\tag{6.19}
$$

where $\mathbf{M}$ is again the Jacobian of the non-linear operator. Consider what happens when covariances are estimated directly from an ensemble of forecasts propagated forward from an ensemble of $i = 1, \ldots, m$ analyses using the fully non-linear forecast model

$$
\mathbf{x}^{b}_{i(t+1)} = \mathcal{M}\left(\mathbf{x}^{a}_{i(t)}\right). \tag{6.20}
$$

Calculating the expected covariance, we get

$$\begin{aligned}
&\left\langle\left(\mathbf{x}^{\mathrm{b}}_{\mathrm{i}\,(t+1)} - \overline{\mathbf{x}}^{\mathrm{b}}_{(t+1)}\right)\left(\mathbf{x}^{\mathrm{b}}_{\mathrm{i}\,(t+1)} - \overline{\mathbf{x}}^{\mathrm{b}}_{(t+1)}\right)^{\mathrm{T}}\right\rangle \\
&\quad = \left\langle\left(\mathcal{M}\left(\mathbf{x}^{\mathrm{a}}_{\mathrm{i}(t)}\right) - \mathcal{M}\left(\overline{\mathbf{x}}^{\mathrm{a}}_{(t)}\right)\right)\left(\mathcal{M}\left(\mathbf{x}^{\mathrm{a}}_{\mathrm{i}(t)}\right) - \mathcal{M}\left(\overline{\mathbf{x}}^{\mathrm{a}}_{(t)}\right)\right)^{\mathrm{T}}\right\rangle \\
&\quad \simeq \left\langle \mathrm{M}\left(\mathbf{x}^{\mathrm{a}}_{\mathrm{i}(t)} - \overline{\mathbf{x}}^{\mathrm{a}}_{(t)}\right)\left(\mathbf{x}^{\mathrm{a}}_{\mathrm{i}(t)} - \overline{\mathbf{x}}^{\mathrm{a}}_{(t)}\right)^{\mathrm{T}} \mathrm{M}^{\mathrm{T}}\right\rangle \\
&\quad \simeq \mathbf{M}\hat{\mathbf{P}}^{\mathrm{a}}_{(\mathrm{t})}\mathbf{M}^{\mathrm{T}}.
\end{aligned} \tag{6.21}$$

Comparing (6.19) and (6.21), it is apparent that an ensemble of analyses that are simply propagated forward with the non-linear forecast model will have too small an expected amount of spread, missing the extra model-error covariance $\mathbf{Q}$. Let us define some hypothetical set of background forecasts at time $t+1$ that *do* have the correct covariance, i.e. define $\breve{\mathbf{x}}^{\mathrm{b}}_{\mathrm{i}(t+1)}$ such that $\langle(\breve{\mathbf{x}}^{\mathrm{b}}_{\mathrm{i}(t+1)} - \overline{\breve{\mathbf{x}}^{\mathrm{b}}}_{(t+1)})(\breve{\mathbf{x}}^{\mathrm{b}}_{\mathrm{i}(t+1)} - \overline{\breve{\mathbf{x}}^{\mathrm{b}}}_{(t+1)})^{\mathrm{T}}\rangle = \mathbf{M}\hat{\mathbf{P}}^{\mathrm{a}}_{(\mathrm{t})}\mathbf{M}^{\mathrm{T}} + \mathbf{Q}$. Such an ensemble is possible if we add noise to our existing ensemble:

$$\breve{\mathbf{x}}^{\mathrm{b}}_{\mathrm{i}(t+1)} = \mathbf{x}^{\mathrm{b}}_{\mathrm{i}\,(t+1)} + \xi_{\mathrm{i}}, \tag{6.22}$$

where $\langle\xi_{\mathrm{i}}\xi_{\mathrm{i}}^{\mathrm{T}}\rangle = \mathbf{Q}$, $\langle\xi_{\mathrm{i}}\rangle = 0$, and $\langle\mathbf{x}^{\mathrm{b}}_{\mathrm{i}(t+1)}\,\xi_{\mathrm{i}}^{\mathrm{T}}\rangle = 0$.

Several methods have been proposed for incorporating noise into the ensemble of forecasts so that they account for model error. First, the forecast model could be stochastic-dynamic instead of deterministic, with additional terms in the prognostic equations to represent interactions with unresolved scales and/or misparameterised effects; in essence, $\mathcal{M}$ is changed so that the ensemble of forecasts integrates random noise in addition to the deterministic forecast dynamics, as in Eq. (6.4). Over an assimilation cycle, this additional variance added to the ensemble as a result of integrating noise should be designed to increase the covariance by the missing $\mathbf{Q}$. A second possibility is that one may choose to run a forecast model without integrating noise but to add noise to each member at the data assimilation time so as to increase the ensemble variance appropriate to the missing $\mathbf{Q}$. Third, it may be possible to use a multimodel ensemble to estimate covariances, or to achieve satisfactory results by inflating the deviations of ensemble members about their mean.

Little work has yet been done on the first of these three approaches. Buizza *et al.* (1999) demonstrated a simple technique for integrating noise to account for deterministic subgrid-scale parametrisations. Under their methodology, the parametrised terms in the prognostic equations were multiplied by a random number. Shutts (2004) describes an updated stochastic backscatter approach. Penland (2003) outlines a more general approach for integrating system noise in numerical models. To date, however, a comprehensive noise integration scheme has not yet been demonstrated in an operational weather prediction model. Palmer (2001) discusses the potential appeal of such an approach.

The second general approach is to augment the ensemble-estimated model of covariances during the update step with noise representing the missing model error covariances. Mitchell and Houtekamer (2000) describe one such approach whereby innovation statistics were used to develop a simple model-error covariance model. More recently, Houtekamer *et al.* (2005) have tested an additive-error filter with operational data. Hamill and Whitaker (2005) have recently attempted to use differences between high- and low-resolution model forecasts to parametrise the additive errors.

A third approach, use of multiple forecast models for generating the ensemble of background forecasts (e.g. Houtekamer *et al.*, 1996b; Harrison *et al.*, 1999; Evans *et al.*, 2000; Ziehmann, 2000; Richardson, 2000; Hou *et al.*, 2001), is appealing for its simplicity. A wider range of forecasts is typically generated when different weather forecast models are used to forecast the evolution of different ensemble members. Unfortunately, it is not clear whether or not the differences between members are actually representative of model errors; initial experimentation has shown that the multimodel ensembles tend to produce unrealistic estimates of error covariances. Forecast errors at larger scales ought to be mostly in balance, but when estimated from multimodel ensembles, preliminary results suggest that the errors can be greatly out of balance, with detrimental effects on the subsequent assimilation (M. Buehner, personal communication). See also Hansen (2002) for a discussion of the use of multimodel ensembles in data assimilation in a simple model.

A last approach is to modify the observation- or background-error covariances in some manner so they draw more to the observations. Pham (2001) proposes reducing $\mathbf{R}$ with a 'forgetting factor' to achieve this. Another approach is 'covariance inflation', discussed in Anderson and Anderson (1999). Ensemble members' deviations about their mean are inflated by an amount r (slightly greater than 1.0) before the first observation is assimilated:

$$\mathbf{x}_{\mathrm{i}}^{\mathrm{b}} \leftarrow r\left(\mathbf{x}_{\mathrm{i}}^{\mathrm{b}} - \overline{\mathbf{x}}^{\mathrm{b}}\right) + \overline{\mathbf{x}}^{\mathrm{b}}. \tag{6.23}$$

Here, the operation $\leftarrow$ denotes a replacement of the previous value of $\mathbf{x}_{\mathrm{i}}^{\mathrm{b}}$. Application of a moderate inflation factor has been found to improve the accuracy of assimilations (Hamill *et al.*, 2001; Whitaker and Hamill, 2002; Whitaker *et al.*, 2004). Note that inflation increases the spread of the ensemble, but it does not change the subspace spanned by the ensemble. Hence, if model error projects into a substantially different subspace, this parametrisation may not be effective.

6.4.5 Covariance localisation

In ensemble assimilation methods, the accuracy of error covariance models is especially important. Unlike 3D-Var, the effects of a misspecification of error statistics can affect the analysis-error covariance, which is then propagated forward in time. Hence, if the analysis errors are underestimated in one cycle, the forecast errors may be underestimated in the following cycle, underweighting the new observations. The process can feed back on itself, the ensemble assimilation method progressively ignoring observational data more and more in successive cycles, leading eventually to a useless ensemble. This is known as *filter divergence* (e.g. Maybeck, 1979, p. 337; Houtekamer and Mitchell, 1998).

One of the most crucial preventatives is to model background-error covariances realistically (Hamill *et al.*, 2001). Of course, an adequate parametrisation of model error will be necessary in all but perfect model simulations (see previous section). However, filter divergence can occur even in simulations where the forecast model is perfect, for background-error covariances will incur large sampling errors when estimated from small ensembles. While more ensemble members would be desirable to reduce these sampling errors, more members requires more computational expense.

One common algorithmic modification to improve background-error covariance estimates from small ensembles is *covariance localisation*. The covariance estimate from the ensemble is multiplied point by point with a correlation function that is 1.0 at the observation location and zero beyond some prespecified distance. Houtekamer and Mitchell (1998) and Evensen (2003) simply use a cut-off radius so that observations are not assimilated beyond a certain distance from the grid point. This may be problematic in situations where observations are sparse, for then there will be grid points affected by the observation adjacent to grid points unaffected by the observation, potentially introducing spurious discontinuities.

A preferable approach is to use a correlation function that decreases monotonically with increasing distance (Houtekamer and Mitchell, 2001). Mathematically, to apply covariance localisation, the Kalman gain $\hat{\mathbf{K}} = \hat{\mathbf{P}}^{\mathrm{b}}\mathbf{H}^{\mathrm{T}}(\mathbf{H}\hat{\mathbf{P}}^{\mathrm{b}}\mathbf{H}^{\mathrm{T}} + \mathbf{R})^{-1}$ is replaced by a modified gain

$$\hat{\mathbf{K}} = (\rho_S \circ \hat{\mathbf{P}}^{\mathrm{b}})\mathbf{H}^{\mathrm{T}}\,(\mathbf{H}(\rho_S \circ \hat{\mathbf{P}}^{\mathrm{b}})\mathbf{H}^{\mathrm{T}} + \mathbf{R})^{-1}, \tag{6.24}$$

where the operation $\rho_S \circ$ in (6.24) denotes a Schur product (an element-by-element multiplication) of a correlation matrix $\mathbf{S}$ with local support with the covariance model generated by the ensemble. For horizontal localisation, one such correlation matrix can be constructed using an approximately Gaussian-shaped function that is actually a compactly supported, fourth-order piece-wise polynomial, described in Gaspari and Cohn (1999). The Schur product of matrices $\mathbf{A}$ and $\mathbf{B}$ is a matrix $\mathbf{C}$ of the same dimension, where $c_{ij} = a_{ij}\ b_{ij}$. When covariance localisation is applied to smaller ensembles, it can actually result in more accurate analyses than would be obtained from larger ensembles without localisation (Houtekamer and Mitchell, 2001). Mathematically, localisation increases the effective rank of the background-error covariances (Hamill *et al.*, 2001). In the extreme, if the correlation matrix $\mathbf{S}$ were the identity matrix, the covariance model would consist of grid points with variances and zero covariance and the rank of the covariance matrix after localisation would increase from $m - 1$ to n, the dimension of the state vector. In practice, such an extreme localisation would harm the quality of the analysis, destroying the mass-wind balance (Mitchell and Houtekamer, 2002; Lorenc, 2003) and prohibiting the observation from changing the analysis at nearby grid points. Hence, broader localisations are typically used. Generally, the larger the ensemble, the broader the optimum correlation length scale of the localisation function (Houtekamer and Mitchell, 2001; Hamill *et al.*, 2001). See Whitaker *et al.* (2004) and Houtekamer *et al.* (2005) for examples of ensemble assimilations that also include a vertical covariance localisation.

As a concrete example of horizontal covariance localisation, consider Figure 6.4. This used the same data set as in Figure 6.2, a global ensemble-data assimilation scheme utilising only sea-level pressure observations (Whitaker *et al.*, 2004). Part (a) of Figure 6.4 (colour plate) provides a map of sea-level pressure correlations at grid points around the northern hemisphere with a grid point in the western Pacific Ocean on 0000 UTC 14 December 2001. When directly estimated using the 25-member ensemble subsampled from the 200-member ensemble (Figure 6.4b), correlations for grid points in the region around the observation are positive. The shape of the correlation function was anisotropic, with positive correlations generally limited to a region east of the axis of the cyclone. Background errors for regions in the eastern Pacific and near the Greenwich meridian also appeared to be highly correlated with background errors at the observation location. However, when the correlations are estimated from a 200-member ensemble, it is apparent that these distant correlations in the 25-member ensemble were artefacts of the limited sample size. The errors in the eastern Pacific and along the Greenwich meridian were not dynamically interconnected with the errors in the western Pacific. When the covariance localisation function (Figure 6.4c) was applied to the 25-member ensemble, the resulting correlation model (Figure 6.4d) more closely resembles that from the larger ensemble.

In applying the covariance localisation, distant grid points are forced to be statistically independent. Should they be? As a thought experiment, consider a two-member ensemble. Dynamically, there is no a-priori reason to expect that, say, the growth of spread over Japan is dynamically interconnected to the growth of spread over Africa, and neither interconnected with the growth of differences over South America. This two-member ensemble may identify many distinct regions where rapid growth of differences is occurring, but with a covariance model estimated from only two members, the ensemble assumes they are all intimately coupled. Covariance localisation is thus an heuristic attempt to modify the model of background-error covariances so that a limited-size ensemble will not represent distant, distinct features as dynamically interrelated when in fact they only appear to be so due to limited sample size. If indeed distant regions are dynamically coupled, the localisation will cause the loss of this information. The effect on the data assimilation will be that observations will not be able to change the analysis and reduce the analysis-error variance in distant regions; local observations will have to be relied upon instead. This is judged to be less detrimental than the opposite, to let observations affect distant regions when this is inappropriate.

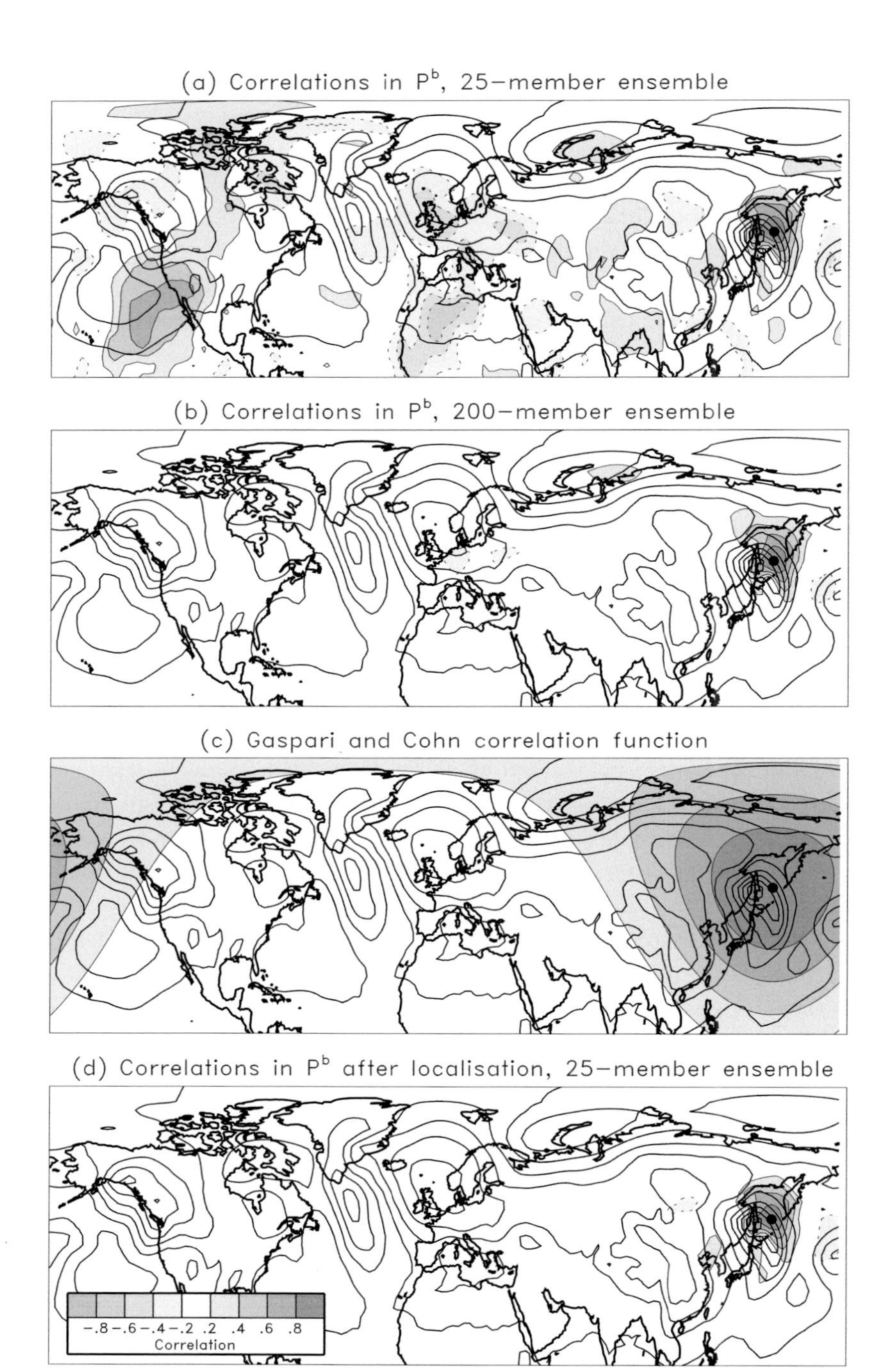

Figure 6.4 (See also colour plate section.) Illustration of covariance localisation. (a) Correlations of sea-level pressure directly estimated from 25-member ensemble with pressure at a point in the western Pacific (colours). Solid lines denote ensemble mean background sea-level pressure contoured every 8 hPa. (b) As in (a), but using 200-member ensemble. (c) Covariance localisation correlation function. (d) Correlation estimate from 25-member ensemble after application of covariance localisation.

6.4.6 Pseudocode for an ensemble Kalman filter

The previous detail on ensemble-based assimilation algorithms may make them appear more complex than they are. In many circumstances, the basic algorithm is extremely easy to code. Here some pseudocode is provided for the EnSRF filter discussed in Section 6.4.2. Assume that model error is treated through the introduction of additive error noise, and assume that observations have independent errors, so that they can be serially processed. The steps are:

1. Construct an ensemble of arbitrary initial conditions with a large amount of spread, perhaps by taking random samples from the model's climatology.
2. Perform the forecast step; integrate an ensemble of short-range forecasts forward to the time when the next set of observations are available (Eq. 6.20).
3. Perform the EnSRF update step:
   (a) Add samples of model error to all members (6.22).
   (b) Loop through all the available observations.
      (i) Determine the ensemble-mean background (6.8) and the matrix of ensemble perturbations (6.9).
      (ii) Determine the subcomponents of the estimated Kalman gain (6.14 and 6.15), applying covariance localisation (6.24) if desired.
      (iii) Form the Kalman gain (6.13) and reduced Kalman gain (6.18).
      (iv) Update the ensemble mean (6.16) and the individual perturbations (6.17).
      (v) Set the background mean and perturbations for the assimilation of the next observation to the newly updated analysis mean and perturbations.
4. Add the updated mean and the perturbations together to reform the ensemble of analysed states.
5. Go back to step 2.

6.5 Discussion

6.5.1 Major research questions

Researchers are just beginning to test ensemble-based atmospheric data assimilation methods in full numerical weather prediction modelling systems using real observations. From these and other studies, we can make an educated guess at some of the major issues that will need to be resolved before operational implementation is practical.

As discussed previously, in order to ensure a high-quality analysis, great care must be taken to ensure that the error-covariance models are realistic in ensemble methods. These methods cycle the covariance estimates. Thus, for example, if observation errors are assumed to be unbiased and independent but in fact are biased or correlated (Liu and Rabier, 2003), these errors will cause the analysis-error covariance to be misestimated, later affecting the subsequent background-error estimates and subsequent fit to the observations. Accurate estimation of model error in particular is likely to be crucial, as was discussed in Section 6.4.4.

Practically, an ensemble-based assimilation method ought to be self-correcting, able to detect when the system is not appropriately fitting the observations. Theoretically, this can be done by monitoring the innovation statistics ($\mathbf{y} - \mathcal{H}(\mathbf{x}^{\mathrm{b}})$), which ought to be white noise with zero mean and covariance ($\mathbf{H}\hat{\mathbf{P}}^{\mathrm{b}}\mathbf{H}^{\mathrm{T}} + \mathbf{R}$) (Maybeck, section 5.4; Dee, 1995). Perhaps the influence of model error can then be increased or decreased so the innovation statistics have the correct properties (Mitchell and Houtekamer, 2000).

Other problems may be more subtle. For instance, initial tests with real observations (Houtekamer *et al.*, 2005) suggest that when many observations are frequently assimilated, the errors due to chaotic effects may not grow rapidly after the analysis, as expected. The reasons for this are not yet fully apparent. It is known that the more observations that are assimilated, the spectrally whiter and more random are the analysis errors (Hamill *et al.*, 2002); consequently, it may take longer than the time between updates for the dynamics to organise the perturbations into growing structures. The slow growth of analysis errors may also be exacerbated by the addition of random model error to the background forecasts, because of imbalances introduced by covariance localisation, and/or because the computational costs require the use of reduced-resolution models with unrealistically slow error growth characteristics.

A final major concern is the computational expense. The cost of most ensemble methods scales as the number of observations times the dimension of the model state times the number of ensemble members. In the coming years, observations will increase in number faster than computer processing speed. It may be possible to mitigate this problem in one of several ways. Perhaps computations can be speeded up through parallelisation (Houtekamer and Mitchell, 2001; Keppenne and Rienecker, 2002), perhaps the method can be cast in a variational framework where the costs do not scale with the number of observations (Hamill and Snyder, 2000; Etherton and Bishop, 2004), or perhaps many high-density observations can be combined into fewer 'superobservations' (Lorenc, 1981).

6.5.2 Comparisons with 4D-Var

An important question is whether, for a given amount of computer time, a better analysis could be produced by an ensemble-based assimilation or by the current state-of-the art, four-dimensional variational analysis (4D-Var; Le Dimet and Talagrand, 1986; Courtier *et al.*, 1994; Rabier *et al.*, 1998, 2000). Such direct comparisons of ensemble assimilation methods and 4D-Var in realistic scenarios have yet to be performed and ideally should wait until ensemble methods have been given a chance to mature.

Some intelligent guesses can be made regarding their relative advantages and disadvantages (for another view, see Lorenc, 2003). Ensemble-based methods are much easier to code and maintain, for neither a tangent linear nor an adjoint of the forecast model is required, as they are with 4D-Var. Ensemble-based methods produce an ensemble of possible analysis states, providing information on both the mean analysis and its uncertainty. Consequently, the ensemble of analysis states can be used directly to initialise ensemble forecasts without any additional computations.

Another advantage is that if the analysis uncertainty is very spatially inhomogeneous and time dependent, in ensemble-based methods this information will be fed through the ensemble from one assimilation cycle to the next. In comparison, in 4D-Var, the assimilation typically starts at each update cycle with the same stationary model of error statistics. Hence, the influence of observations may be more properly weighted in ensemble-based methods than in 4D-Var. Ensemble-based methods also provide a direct way to incorporate the effects of model imperfections directly into the data assimilation. In comparison, in current operational implementations of 4D-Var, the forecast model dynamics are a strong constraint (Courtier *et al.*, 1994; but see Bennett *et al.*, 1996 and Zupanski, 1997 for possible alternatives). If the forecast model used in 4D-Var does not adequately represent the true dynamics of the atmosphere, model error may be large, and 4D-Var may fit a model trajectory that was significantly different from the trajectory of the real atmosphere during that time window.

Ensemble-based techniques may have disadvantages relative to 4D-Var, including some that will only be discovered through further experimentation. Most ensemble-based techniques are likely to be at least as computationally expensive as 4D-Var, and perhaps significantly more expensive when there are an overwhelmingly large number of observations (though see Hamill and Snyder, 2000 and Etherton and Bishop, 2004 for more computationally efficient alternatives). Ensemble approaches may be difficult to apply in limited-area models because of difficulty of specifying an appropriate ensemble of lateral boundary conditions, and the method is very sensitive to misestimation of the error covariances.

6.5.3 Applications of ensemble-based assimilation methods

Ensemble data assimilation techniques offer the potential of generating calibrated analyses that may be useful for a variety of applications. Anderson (2001) showed that the ensemble techniques can be used for parameter estimation. Hamill and Snyder (2002) showed that ensemble assimilation techniques facilitate the calculation of regions where adaptive observations are necessary. Snyder and Zhang (2003), Zhang *et al.* (2004), and Dowell *et al.* (2004) demonstrate the feasibility of ensemble filters for mesoscale data assimilation of radar observations. Reichle *et al.* (2002, 2003) apply ensemble filters to estimation of soil moisture. Hamill *et al.* (2003) demonstrate how analysis-error covariance singular vectors, the most rapidly growing forecast structures consistent with analysis errors, can be diagnosed using ensemble filters.

6.6 Conclusions

This chapter presented a brief tutorial of ensemble-based atmospheric data assimilation. The technique is being explored by a rapidly growing number of researchers as a possible alternative to other atmospheric data assimilation techniques such as three- and four-dimensional atmospheric data assimilation. The technique is appealing for its comparative algorithmic simplicity and its ability to deal explicitly with model error. Testing of ensemble filters has progressed rapidly over the past few years from perfect-model experiments in toy dynamical systems to the assimilation of real observations into global NWP models. Recent results are both suggestive of the potential, though substantial continued development may be necessary for these methods to become competitive with or superior to the existing four-dimensional variational techniques.

Acknowledgements

Deszo Devenyi (NOAA/FSL), Chris Snyder (NCAR/MMM), Jeff Whitaker (NOAA/CDC), Ryan Torn (University of Washington), Brian Etherton (UNC/Charlotte), Sharanya Majumdar (University of Miami), Jim Hansen (MIT) and three anonymous reviewers are thanked for their informal reviews of early versions of this manuscript. This was originally prepared as a paper for ECMWF's 2002 Predictability Workshop. The many participants who offered interesting questions and constructive criticism at this workshop are thanked as well. Preparation of this manuscript was supported by National Science Foundation grants ATM-0112715, -0205612, and 0130154.

References

Anderson, J. L. (2001). An ensemble adjustment filter for data assimilation. *Mon. Weather Rev.*, **129**, 2884–903.

(2003). A local least squares framework for ensemble filtering. *Mon. Weather Rev.*, **131**, 634–42.

Anderson, J. L. and S. L. Anderson (1999). A Monte Carlo implementation of the nonlinear filtering problem to produce ensemble assimilations and forecasts. *Mon. Weather Rev.*, **127**, 2741–58.

Andrews, A. (1968). A square-root formulation of the Kalman covariance equations. *AIAA J.*, **6**, 1165–8.

Bennett, A. F., B. S. Chua and L. M. Leslie (1996). Generalized inversion of a global numerical weather prediction model. *Met. Atmos. Phys.*, **60**, 165–78.

Bishop, C. H., B. J. Etherton and S. J. Majumdar (2001). Adaptive sampling with the ensemble transform Kalman filter. 1: Theoretical aspects. *Mon. Weather Rev.*, **129**, 420–36.

Blanchet, I., C. Frankignoul and M. A. Cane (1997). A comparison of adaptive Kalman filters for a tropical Pacific ocean model. *Mon. Weather Rev.*, **125**, 40–58.

Bouttier, F. (1994). A dynamical estimation of forecast error covariances in an assimilation system. *Mon. Weather Rev.*, **122**, 2376–90.

Buizza, R., M. Miller and T. N. Palmer (1999). Stochastic representation of model uncertainties in the ECMWF ensemble prediction system. *Quart. J. Roy. Meteor. Soc.*, **125**, 2887–908.

Burgers, G., P. J. van Leeuwen and G. Evensen (1998). Analysis scheme in the ensemble Kalman filter. *Mon. Weather Rev.*, **126**, 1719–24.

Casella, G. and R. L. Berger (1990). *Statistical Inference*. Duxbury Press.

Cohn, S. E. (1997). An introduction to estimation theory. *J. Meteorol. Soc. Jpn.*, **75(1B)**, 257–88.

Cohn, S. E. and D. F. Parrish (1991). The behavior of forecast error covariances for a Kalman filter in two dimensions. *Mon. Weather Rev.*, **119**, 1757–85.

Courtier, P., J.-N. Thépaut and A. Hollingsworth (1994). A strategy for operational implementation of 4D-Var, using an incremental approach. *Quart. J. Roy. Meteor. Soc.*, **120**, 1367–87.

Daley, R. (1991). *Atmospheric Data Analysis*. Cambridge University Press.

(1992). Estimating model-error covariances for applications to atmospheric data assimilation. *Mon. Weather Rev.*, **120**, 1735–46.

(1993). Estimating observation error statistics for atmospheric data assimilation. *Ann. Geophys.*, **11**, 634–47.

(1997). Atmospheric data assimilation. *J. Meteorol. Soc. Jpn.*, **75(1B)**, 319–29.

Dee, D. P. (1995). On-line estimation of error covariance parameters for atmospheric data assimilation. *Mon. Weather Rev.*, **123**, 1128–45.

Dee, D. P. and R. Todling (2000). Data assimilation in the presence of forecast bias: the GEOS moisture analysis. *Mon. Weather Rev.*, **128**, 3268–82.

Dowell, D. C., F. Zhang, L. J. Wicker, C. Snyder and N. A. Crook (2004). Wind and thermodynamic retrievals in the 17 May 1981 Arcadia, Oklahoma supercell: ensemble Kalman filter experiments. *Mon. Weather Rev.*, **132**, 1982–2005.

Ehrendorfer, M. (1994a). The Liouville equation and its potential usefulness for the prediction of forecast skill. I: Theory. *Mon. Weather Rev.*, **122**, 703–13.

(1994b). The Liouville equation and its potential usefulness for the prediction of forecast skill. II: Applications. *Mon. Weather Rev.*, **122**, 714–28.

Etherton, B. J. and C. H. Bishop (2004). Resilience of hybrid ensemble / 3DVAR analysis schemes to model error and ensemble covariance error. *Mon. Weather Rev.*, **132**, 1065–80.

Evans, R. E., M. S. J. Harrison and R. J. Graham (2000). Joint medium-range ensembles from the Met. Office and ECMWF systems. *Mon. Weather Rev.*, **128**, 3104–27.

Evensen, G. (1992). Using the extended Kalman filter with a multilayer quasi-geostrophic ocean model. *J. Geophys. Res.*, **97**, 17905–24.

(1994). Sequential data assimilation with a nonlinear quasi-geostrophic model using Monte Carlo methods to forecast error statistics. *J. Geophys. Res.*, **99(C5)**, 10143–62.

(2003). The ensemble Kalman filter: theoretical formulation and practical implementation. *Ocean Dynam.*, **53**, 343–67.

Evensen, G. and P. J. van Leeuwen (1996). Assimilation of Geosat altimeter data for the Agulhas current using the ensemble Kalman filter with a quasigeostrophic model. *Mon. Weather Rev.*, **124**, 85–96.

Farrell, B. F. and P. J. Ioannou (2001). State estimation using a reduced-order Kalman filter. *J. Atmos. Sci.*, **58**, 3666–80.

Fisher, M. (1998). *Development of a Simplified Kalman Filter*. ECMWF Research Department Technical Memorandum 260. European Centre for Medium-Range Weather Forecasts. 16 pp. Available from Library, *ECMWF*, Shinfield Park, Reading, Berkshire, RG2 9AX, England.

Gardiner, C. W. (1985). *Handbook of Stochastic Methods*, 2nd edn. Springer.

Gaspari, G. and S. E. Cohn (1999). Construction of correlation functions in two and three dimensions. *Quart. J. Roy. Meteor. Soc.*, **125**, 723–57.

Gauthier, P., P. Courtier and P. Moll (1993). Assimilation of simulated lidar data with a Kalman filter. *Mon. Weather Rev.*, **121**, 1803–20.

Gelb, A. (ed.) (1974). *Applied Optimal Estimation*. MIT Press.

Ghil, M. (1989). Meteorological data assimilation for oceanography. Part 1: Description and theoretical framework. *Dyn. Atmos. Oceans*, **13**, 171–218.

Ghil, M. and P. Malanotte-Rizzoli (1991). Data assimilation in meteorology and oceanography. *Adv. Geophys.*, **33**, 141–266.

Gordon, N. J., D. J. Salmond and A. F. M. Smith (1993). Novel approach to nonlinear/non-Gaussian Bayesian state estimation. *IEEE Proceedings – F*, **140**, 107–13.

Hamill, T. M. and C. Snyder (2000). A hybrid ensemble Kalman filter / 3d-variational analysis scheme. *Mon. Weather Rev.*, **128**, 2905–19.

(2002). Using improved background-error covariances from an ensemble Kalman filter for adaptive observations. *Mon. Weather Rev.*, **130**, 1552–72.

Hamill, T. M. and J. S. Whitaker (2005). Accounting for the error due to unresolved scales in ensemble data assimilation: a comparison of different approaches. *Mon. Weather Rev.*, **133**, 3132–47.

Hamill, T. M., J. S. Whitaker and C. Snyder (2001). Distance-dependent filtering of background-error covariance estimates in an ensemble Kalman filter. *Mon. Weather Rev.*, **129**, 2776–90.

Hamill, T. M., C. Snyder and J. S. Whitaker (2003). Ensemble forecasts and the properties of flow-dependent analysis-error covariance singular vectors. *Mon. Weather Rev.*, **131**, 1741–58.

Hansen, J. A. (2002). Accounting for model error in ensemble-based state estimation and forecasting. *Mon. Weather Rev.*, **130**, 2373–91.

Harrison, M. S. J., T. N. Palmer, D. S. Richardson and R. Buizza (1999). Analysis and model dependencies in medium-range ensembles: two transplant case studies. *Quart. J. Roy. Meteor. Soc.*, **125**, 2487–515.

Hastie, T., R. Tibshirani and J. Friedman (2001). *The Elements of Statistical Learning*. Springer.

Heemink, A. W., M. Verlaan and A. J. Segers (2001). Variance-reduced ensemble Kalman filtering. *Mon. Weather Rev.*, **129**, 1718–28.

Hou, D., E. Kalnay and K. K. Droegemeier (2001). Objective verification of the SAMEX-98 ensemble forecasts. *Mon. Weather Rev.*, **129**, 73–91.

Houtekamer, P. L. and H. L. Mitchell (1998). Data assimilation using an ensemble Kalman filter technique. *Mon. Weather Rev.*, **126**, 796–811.

(1999). Reply to comment on "Data assimilation using an ensemble Kalman filter technique." *Mon. Weather Rev.*, **127**, 1378–9.

(2001). A sequential ensemble Kalman filter for atmospheric data assimilation. *Mon. Weather Rev.*, **129**, 123–37.

Houtekamer, P. L., L. Lefaivre and J. Derome (1996a). The RPN ensemble prediction system. In *Proceedings ECMWF Seminar on Predictability, Vol. II, Reading*, pp. 121–146. [Available from ECMWF, Shinfield Park, Reading, Berkshire RG2 9AX, United Kingdom].

Houtekamer, P. L., J. Derome, H. Ritchie and H. L. Mitchell (1996b). A system simulation approach to ensemble prediction. *Mon. Weather Rev.*, **124**, 1225–42.

Houtekamer, P. L., H. L. Mitchell, G. Pellerin, *et al.* (2005). Atmospheric data assimilation with the ensemble Kalman filter: results with real observations. *Mon. Weather Rev.*, **133**, 604–20.

Hunt, B. R., E. Kalnay, E. J. Kostelich, *et al.* (2004). Four-dimensional ensemble Kalman filtering. *Tellus*, **56A**, 273–7.

Jazwinski, A. H. (1970). *Stochastic Processes and Filtering Theory.* Academic Press.

Kalman, R. E. (1960). A new approach to linear filtering and prediction problems. *T. ASME – J. Basic Eng.*, **82D**, 35–45.

Kalman, R. E. and R. S. Bucy (1961). New results in linear filtering and prediction theory. *T. ASME – J. Basic Eng.*, **83D**, 95–108.

Kaminski, P. G., A. E. Bryson, Jr and S. F. Schmidt (1971). Discrete square root filtering: a survey of current techniques. *IEEE T. Automat. Control*, **AC-16**, 727–36.

Keppenne, C. L. (2000). Data assimilation into a primitive equation model with a parallel ensemble Kalman filter. *Mon. Weather Rev.*, **128**, 1971–81.

Keppenne, C. L. and M. M. Rienecker (2002). Initial testing of a massively parallel ensemble Kalman filter with the Poseidon isopycnal ocean general circulation model. *Mon. Weather Rev.*, **130**, 2951–65.

Lacarra, J. F. and O. Talagrand (1988). Short-range evolution of small perturbations in a barotropic model. *Tellus*, **40A**, 81–95.

Lawson, G. and J. A. Hansen (2003). Implications of stochastic and deterministic filters as ensemble-based data assimilation methods in varying regimes of error growth. *Mon. Weather Rev.*, **132**, 1966–81.

Le Dimet, F.-X. and O. Talagrand (1986). Variational algorithms for analysis and assimilation of meteorological observations: theoretical aspects. *Tellus*, **38A**, 97–110.

Leith, C. E. (1983). Predictability in theory and practice. In *Large-scale Dynamical Processes in the Atmosphere,* ed.B. J. Hoskins and R. P. Pearce, chapter 13. Academic Press.

Lermusiaux, P. F. J. (2002). On the mapping of multivariate geophysical fields: sensitivities to size, scales, and dynamics. *J. Atmos. Ocean. Tech.*, **19**, 1602–37.

Lermusiaux, P. F. J. and A. R. Robinson (1999). Data assimilation via error subspace statistical estimation. 1: Theory and schemes. *Mon. Weather Rev.*, **127**, 1385–407.

Li, Z. and I. M. Navon (2001). Optimality of variational data assimilation and its relationship with the Kalman filter and smoother. *Quart. J. Roy. Meteor. Soc.*, **127**, 661–83.

Liu, Z.-Q. and F. Rabier (2003). The potential of high-density observations for numerical weather prediction: a study with simulated observations. *Quart. J. Roy. Meteor. Soc.*, **129**, 3013–35.

Lorenc, A. C. (1981). A global three-dimensional multivariate statistical interpolation scheme. *Mon. Weather Rev.*, **109**, 701–21.

(1986). Analysis methods for numerical weather prediction. *Quart. J. Roy. Meteor. Soc.*, **112**, 1177–94.

(2003). The potential of the ensemble Kalman filter for NWP – a comparison with 4D-Var. *Quart. J. Roy. Meteor. Soc.*, **129**, 3183–203.

Maybeck, P. S. (1979). *Stochastic Models, Estimation, and Control.* Vol. 1 Academic Press.

Mitchell, H. L. and P. L. Houtekamer (2000). An adaptive ensemble Kalman filter. *Mon. Weather Rev.*, **128**, 416–33.

Mitchell, H. L., P. L. Houtekamer and G. Pellerin (2002). Ensemble size, balance, and model-error representation in an ensemble Kalman filter. *Mon. Weather Rev.*, **130**, 2791–808.

Molteni, F., R. Buizza, T. N. Palmer and T. Petroliagis (1996). The ECMWF ensemble prediction system: methodology and validation. *Quart. J. Roy. Meteor. Soc.*, **122**, 73–119.

Palmer, T. N. (2001). A nonlinear dynamical perspective on model error: a proposal for non-local stochastic-dynamic parametrization in weather and climate prediction models. *Quart. J. Roy. Meteor. Soc.*, **127**, 279–304.

Parrish, D. F. and J. C. Derber (1992). The National Meteorological Center's Spectral Statistical Interpolation Analysis System. *Mon. Weather Rev.*, **120**, 1747–63.

Penland, C. (2003). A stochastic approach to nonlinear dynamics: a review. (Electronic supplement to 'Noise out of chaos and why it won't go away'), *Bull. Am. Meteorol. Soc.*, **84**, 921–5.

Pham, D. T. (2001). Stochastic methods for sequential data assimilation in strongly nonlinear systems. *Mon. Weather Rev.*, **129**, 1194–207.

Potter, J. (1964). *W matrix Augmentation*. M.I.T. Instrumentation Laboratory Memo SGA 5–64, Massachusetts Institute of Technology, Cambridge, Massachusetts.

Rabier, F., J.-N. Thepaut and P. Courtier (1998). Extended assimilation and forecast experiments with a four-dimensional variational assimilation system. *Quart. J. Roy. Meteor. Soc.*, **124**, 1861–87.

Rabier, F., H. Järvinen, E. Klinker, J.-F. Mahfouf and A. Simmons (2000). The ECMWF operational implementation of four-dimensional variational assimilation. I: Experimental results with simplified physics. *Quart. J. Roy. Meteor. Soc.*, **126**, 1143–70.

Reichle, R. H. and R. D. Koster (2004). Assessing the impact of horizontal error correlations in background fields on soil moisture estimation. *J. Hydrometeorol.*, **4**, 1229–42.

Reichle, R. H., D. B. McLaughlin and D. Entekhabi (2002). Hydrologic data assimilation with the ensemble Kalman filter. *Mon. Weather Rev.*, **130**, 103–14.

Reichle, R. H., J. P. Walker, R. D. Koster and P. R. Houser (2003). Extended versus ensemble Kalman filtering for land data assimilation. *J. Hydrometeorol.*, **3**, 728–40.

Richardson, D. S. (2000). Ensembles using multiple models and analyses. *Quart. J. Roy. Meteor. Soc.*, **127**, 1847–64.

Sardeshmukh, P. D., C. Penland and M. Newman (2001). Rossby waves in a stochastically fluctuating medium. In *Stochastic Climate Models*, ed. P. Imkeller and J.-S. von Storch, pp. 369–84, Progress in Probability 49. Birkhaueser, Basel.

Snyder, C. and F. Zhang (2003). Assimilation of simulated Doppler radar observations with an ensemble Kalman filter. *Mon. Weather Rev.*, **131**, 1663–77.

Shutts, G. (2004). A Stochastic Kinetic-energy Backscatter Algorithm for Use in Ensemble Prediction Systems. ECMWF Technical Memorandum 449. Available from ECMWF, Shinfield Park, Reading, Berkshire RG2 9AX, United Kingdom.

Talagrand, O. (1997). Assimilation of observations, an introduction. *J. Meteorol. Soc. Jpn.*, **75(1B)**, 191–209.

Tippett, M. K., J. L. Anderson, C. H. Bishop, T. M. Hamill and J. S. Whitaker (2003). Ensemble square root filters. *Mon. Weather Rev.*, **131**, 1485–90.

Toth, Z. and E. Kalnay (1993). Ensemble forecasting at NMC: the generation of perturbations. *Bull. Am. Meteorol. Soc.*, **74**, 2317–30.

(1997). Ensemble forecasting at NCEP and the breeding method. *Mon. Weather Rev.*, **12**, 3297–319.

van Leeuwen, P. J. (1999). Comment on "Data assimilation using an ensemble Kalman filter technique". *Mon. Weather Rev.*, **127**, 1374–7.

Wang, X. and C. H. Bishop (2003). A comparison of breeding and ensemble transform Kalman filter ensemble forecast schemes. *J. Atmos. Sci.*, **60**, 1140–58.

Whitaker, J. S. and T. M. Hamill (2002). Ensemble data assimilation without perturbed observations. *Mon. Weather Rev.*, **130**, 1913–24.

Whitaker, J. S., G. P. Compo, X. Wei and T. M. Hamill (2004). Reanalysis without radiosondes using ensemble data assimilation. *Mon. Weather Rev.*, **132**, 1190–200.

Zhang, F., C. Snyder and J. Sun (2004). Impacts of initial estimate and observation availability on convective-scale data assimilation with an ensemble Kalman filter. *Mon. Weather Rev.*, **132**, 1238–53.

Ziehmann, C. (2000). Comparison of a single-model EPS with a multi-model ensemble consisting of a few operational models. *Tellus*, **52a**, 280–99.

Zupanski, D. (1997). A general weak constraint applicable to operational 4DVAR data assimilation systems. *Mon. Weather Rev.*, **125**, 2274–92.

7

Ensemble forecasting and data assimilation: two problems with the same solution?

Eugenia Kalnay, Brian Hunt, Edward Ott, Istvan Szunyogh
University of Maryland, College Park

7.1 Introduction

Until 1991, operational numerical weather prediction (NWP) centres used to run a single computer forecast started from initial conditions given by the analysis, which is the best available estimate of the state of the atmosphere at the initial time. In December 1992, both the US National Centers for Environmental Prediction (NCEP) and ECMWF started running *ensembles* of forecasts from slightly perturbed initial conditions (Molteni and Palmer, 1993; Toth and Kalnay, 1993; Tracton and Kalnay, 1993; Toth and Kalnay, 1997; Buizza *et al.*, 1998; Buizza, this volume).

Ensemble forecasting provides human forecasters with a range of possible solutions, whose average is generally more accurate than the single deterministic forecast (e.g. Figures 7.3 and 7.4), and whose spread gives information about the forecast errors. It also provides a quantitative basis for probabilistic forecasting.

Schematic Figure 7.1 shows the essential components of an ensemble: a control forecast started from the analysis, two additional forecasts started from two perturbations to the analysis (in this example the same perturbation is added and subtracted from the analysis so that the ensemble mean perturbation is zero), the ensemble average, and the 'truth', or forecast verification, which becomes available later. The first schematic shows an example of a 'good ensemble' in which 'truth' looks like a member of the ensemble. In this case, the ensemble average is closer to the truth than the control due to non-linear filtering of errors, and the ensemble spread is related

Predictability of Weather and Climate, ed. Tim Palmer and Renate Hagedorn. Published by Cambridge University Press.

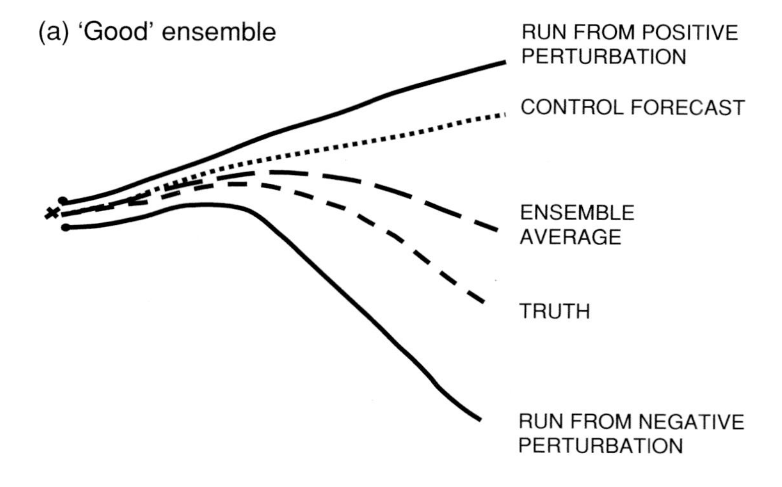

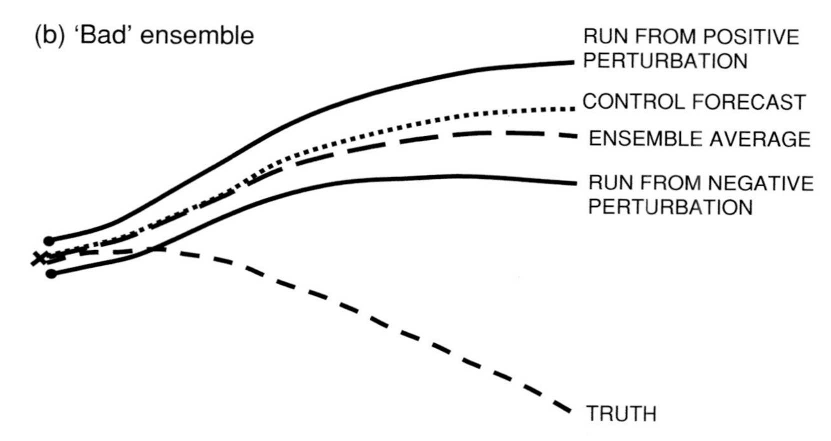

Figure 7.1 Schematic of the essential components of an ensemble of forecasts. The analysis (denoted by a cross) constitutes the initial condition for the control forecast (dotted); two initial perturbations (dots around the analysis), chosen in this case to be equal and opposite; the perturbed forecasts (full line); the ensemble average (long dashes); and the verifying analysis or truth (dashed). (a) is a 'good ensemble' in which the truth is a plausible member of the ensemble. (b) is an example of a bad ensemble, quite different from the truth, pointing to the presence of deficiencies in *the forecasting system* (in the analysis, in the ensemble perturbations and/or in the model).

to the forecast error. The second schematic is an example of a 'bad ensemble': due to a poor choice of initial perturbations and/or to model deficiencies, the forecasts are not able to track the verifying truth, and remain relatively close to each other compared with the truth. In this case the ensemble is not helpful to the forecasters at all, since the lack of ensemble spread would give them unjustified overconfidence in the erroneous forecast. Nevertheless, for NWP development, the 'bad' ensemble is still very useful: after verification time, the poor performance of the ensemble clearly indicates that there is a deficiency in the *forecasting system*. A single failed 'deterministic' forecast, by contrast, would not be able to distinguish between a deficiency in the system and growth of errors in the initial conditions as the cause of failure.

Ideally, the initial perturbations should sample well the analysis 'errors of the day' and the spread among the ensemble members should be similar to that of the forecast errors. The two essential problems in the design of an ensemble forecasting system are how to create effective initial perturbations, and how to handle model deficiencies, which, unless included in the ensemble, result in the forecast error being larger than the ensemble spread.

In this chapter we give a brief historic review of ensemble forecasting, current methods to create perturbations, and discuss ensemble Kalman filtering methods designed to perform efficient data assimilation, but which can, at the same time, provide optimal initial ensemble perturbations and estimate the model errors. We compare the properties of ensemble Kalman filters with those of 4D-Var, the only operational data assimilation method that currently includes the effect of the 'errors of the day'.

7.2 Ensemble forecasting methods

Human forecasters have always performed *subjective ensemble forecasting* by either checking forecasts from previous days, and/or comparing forecasts from different centres, approaches similar to lagged forecasting and multiple systems forecasting. The consistency among these forecasts at a given verification time provided a level of confidence in the forecasts, confidence that changed from day to day and from region to region.

7.2.1 Early methods

Epstein (1969) introduced the idea of stochastic-dynamic forecasting (SDF), and pointed out that it could be also used in the analysis cycle to provide the forecast error covariance. Epstein designed SDF as a shortcut to estimate the true probability distribution of the forecast uncertainty, given by the Liouville equation (Ehrendorfer, this volume), which Epstein approximated running a huge (500) number of perturbed (Monte Carlo) integrations for the three-variable Lorenz (1963) model. However, since SDF involves the integration of forecast equations for each element of the covariance matrix, this method is still not computationally feasible for models with a large number of degrees of freedom.

Leith (1974) suggested the direct use of a Monte Carlo forecasting approach (MCF), where random perturbations sampling the *estimated* analysis error covariance are added to the initial conditions. He indicated that a relatively small number of integrations (of the order of eight) is enough to approximate an important property of an infinite ensemble. In a large ensemble the average forecast error variance at long time leads converges to the climatological error variance, whereas individual forecast errors have an average error variance that is *twice* as large. Since the estimation of the analysis error covariance was constant in time, the MCF method did not include the effects of 'errors of the day'. Errico and Baumhefner (1987) applied this method to realistic global models, using perturbations that represented a realistic (but constant) estimation of the error statistics in the initial conditions. Hollingsworth (1980) showed that for atmospheric models, random errors in the initial conditions took too long to spin up into growing 'errors of the day', making MCF an inefficient approach for ensemble forecasting.

Hoffman and Kalnay (1983) suggested, as an alternative to MCF, the lagged averaged forecasting (LAF) method, in which forecasts from earlier analyses were included in the ensemble. Since the ensemble members are forecasts of different length, they should be weighted with weights estimated from the average forecast error for each lead time. Hoffman and Kalnay found that, compared with MCF, LAF resulted in a better prediction of skill (a stronger relationship between ensemble spread and error), presumably because LAF includes effects from 'errors of the day'. The main disadvantage of LAF, namely that 'older' forecasts are less accurate and should have less weight, was addressed by the scaled LAF (SLAF) approach of Ebisuzaki and Kalnay (1991), in which the LAF perturbations (difference between the forecast and the current analysis) are scaled by their 'age', so that all the initial SLAF perturbations have errors of similar magnitude. They also suggested that the scaled perturbations should be both added and subtracted from the analysis, thus increasing the ensemble size and the probability of 'encompassing' the true solution within the ensemble. SLAF can be easily implemented in both global and regional models, including the impact of perturbed boundary conditions (Hou *et al.*, 2001).

7.2.2 Operational Ensemble Forecasting methods

In December 1992 two methods to create perturbations became operational at NCEP and at ECMWF. They are based on bred vectors and singular vectors respectively, and like LAF, they include 'errors of the day'. These and other methods that have since become operational or are under consideration in operational centres are briefly discussed. More details are given in Kalnay (2003).

Singular vectors

Singular vectors (SVs) are the linear perturbations of a control forecast that grow fastest within a certain time interval (Lorenz, 1965), known as 'optimisation period', using a specific norm to measure their size. SVs are strongly sensitive to the length of the interval and to the choice of norm (Ahlquist, 2000). Ehrendorfer and

Tribbia (1997) showed that *if the initial norm used to derive the singular vectors is the analysis error covariance norm, then the initial singular vectors evolve into the eigenvectors of the forecast error covariance at the end of the optimisation period.* This indicates that if the analysis error covariance is known, then singular vectors based on this specific norm are ideal perturbations.

ECMWF implemented an ensemble system with initial perturbations based on singular vectors using a total energy norm (Molteni and Palmer, 1993; Molteni *et al.*, 1996; Buizza *et al.*, 1997; Palmer *et al.*, 1998; Buizza, this volume).

Bred vectors

Breeding is a non-linear generalisation of the method to obtain leading Lyapunov vectors, which are the sustained fastest growing perturbations (Toth and Kalnay, 1993, 1997; Szunyogh and Toth, 2002). Bred vectors (BVs) (like leading Lyapunov vectors) are independent of the norm and represent the shapes of the instabilities growing upon the evolving flow. In areas where the evolving flow is very unstable (and where forecast errors grow fast), the BVs tend to align themselves along very low dimensional subspaces (the locally most unstable perturbations). An example of such a situation is shown in Figure 7.2, where the forecast uncertainty in a 2.5-day forecast of a storm is very large, but the subspace of the ensemble uncertainty lies *within a one-dimensional space*. In this extreme (but not uncommon) case, a single observation at 500 hPa would be able to identify the best solution. The differences between the forecasts are the bred vectors. The non-linear nature of BVs allows for the saturation of fast-growing instabilities such as convection, or, in the case of EL Niño–Southern Oscillation ENSO coupled instabilities, the weather noise (Peña and Kalnay, 2004).

In unstable areas of fast growth, BVs tend to have shapes that are independent of the forecast length or the norm, and depend only on the verification time. This suggests that forecast errors, to the extent that they reflect instabilities of the background flow, should have shapes similar to bred vectors, and this has been confirmed with model simulations (Corazza *et al.*, 2003).

NCEP implemented an ensemble system based on breeding in 1992, and the US Navy, the National Centre for Medium Range Weather Forecasting in India, and the South African Meteorological Weather Service implemented similar systems. The Japanese Meteorological Agency implemented an ensemble system based on breeding, but they impose a partial global orthogonalisation among the bred vectors, reducing the tendency of the bred vectors to converge towards a low dimensional space of the most unstable directions (Kyouda and Kusunoki, 2002).

Multiple data assimilation ensembles

Houtekamer *et al.* (1996) developed a system based on running an ensemble of data assimilation systems using perturbed observations, implemented in the Canadian

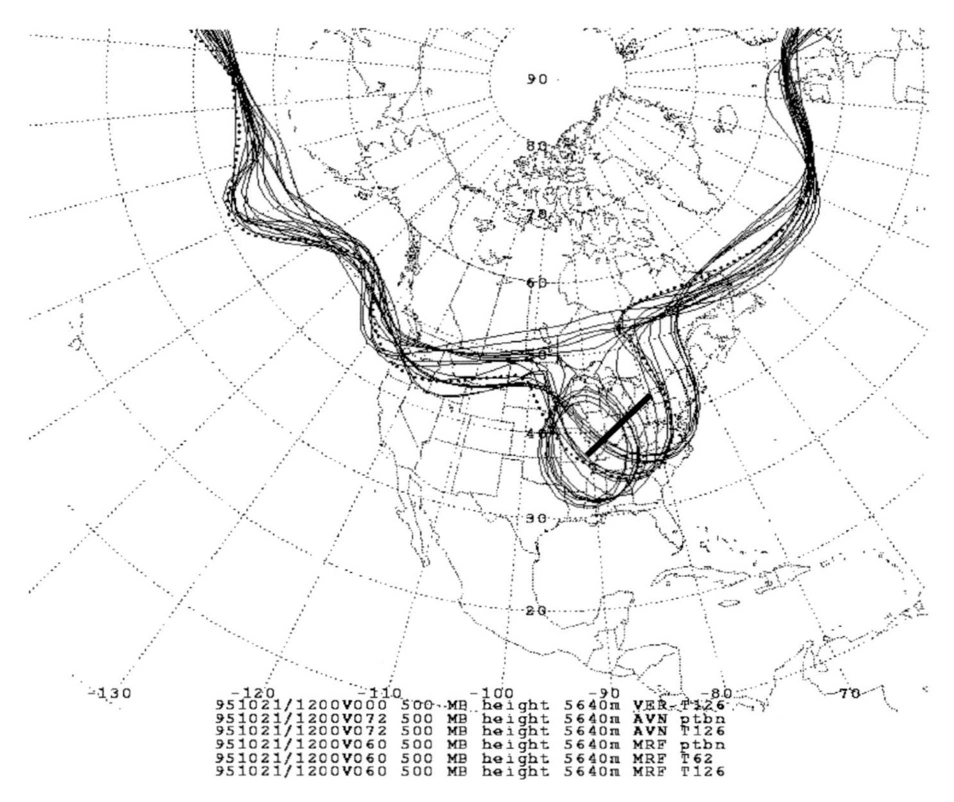

Figure 7.2 'Spaghetti plots' showing a 2.5 day ensemble forecast verifying on 95/10/21. Each 5640 gpm contour at 500 hPa corresponds to one ensemble forecast, and the dotted line is the verifying analysis. Note that the uncertainty in the location of the centre of the predicted storm in the Midwest of the USA is very large, but that it lies on a one-dimensional space (thick line).

Weather Service. Hamill *et al.* (2000) showed that in a quasi-geostrophic system, a multiple data assimilation system performs better than the singular vectors and the breeding approaches. With respect to the computational cost, the multiple data assimilation system and the singular vector approach are comparable, whereas breeding is essentially cost free.

Perturbed physical parametrisations

The methods discussed above only include perturbations in the initial conditions, assuming that the error growth due to model deficiencies is small compared with that due to unstable growth of initial errors. Several groups have also introduced changes in the physical parametrisations to allow for the inclusion of uncertainties in the model formulation (Houtekamer *et al.*, 1996; Stensrud *et al.*, 2000). Buizza *et al.* (1999) developed a perturbation approach that introduces a stochastic perturbation of the impact of subgrid-scale physical parametrisations by multiplying the time

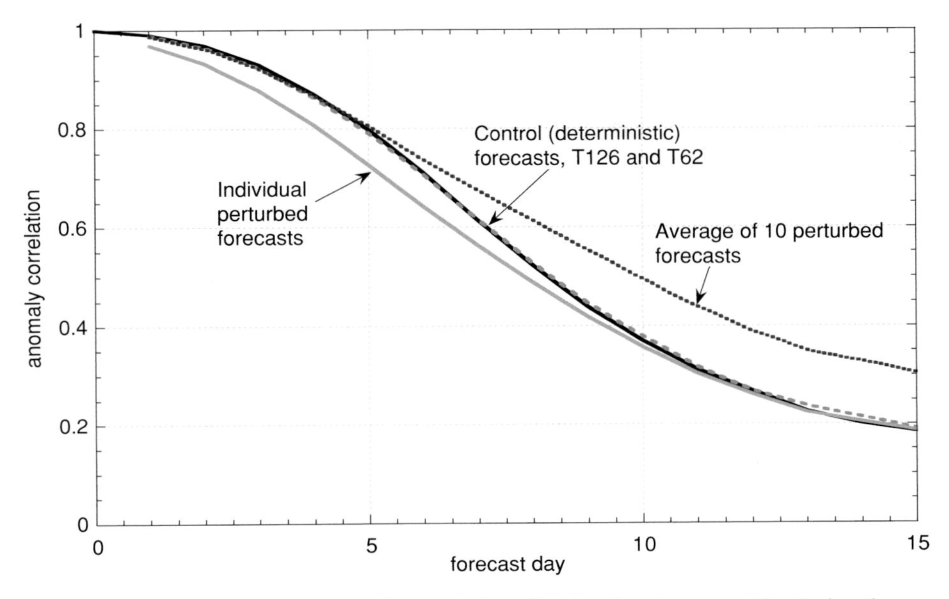

Figure 7.3 Average anomaly correlation of the bred vector ensembles during the winter of 1997/1998 (data courtesy Jae Schemm, of NCEP). Note that, on the average, the individual perturbed forecasts are worse than the control, but their ensemble average is better.

derivative of the 'physics' by a random number normally distributed with mean 1 and standard deviation 0.2. This simple approach resulted in a clear improvement of the performance of the ensemble system.

Both the perturbations of the initial conditions and of the subgrid-scale physical parametrisations have been shown to be successful towards achieving the goals of ensemble forecasting. However, since they both introduce changes to the best estimate of the initial conditions and the model, which are used for the control forecast, it is not surprising that on the average, the individual forecasts are worse than the unperturbed control (see example in Figure 7.3). Nevertheless, the ensemble average is an improvement over the control, especially after the perturbations grow into a non-linear regime that tends to filter out some of the errors because the ensemble solutions tend to diverge in the most uncertain components of the forecast.

Multiple system ensembles

An alternative to the introduction of perturbations is the use of multiple systems. Different operational or research centres, each aiming to be the best, choose different competitive approaches to data assimilation and forecasting systems. In principle, a combination of these different systems should sample well the uncertainty in both the models and the initial conditions. It has been known that the ensemble average of multiple centre forecasts is significantly better than even the very best individual

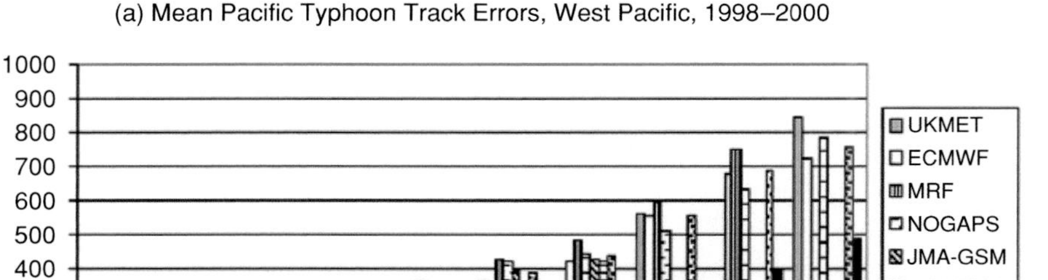

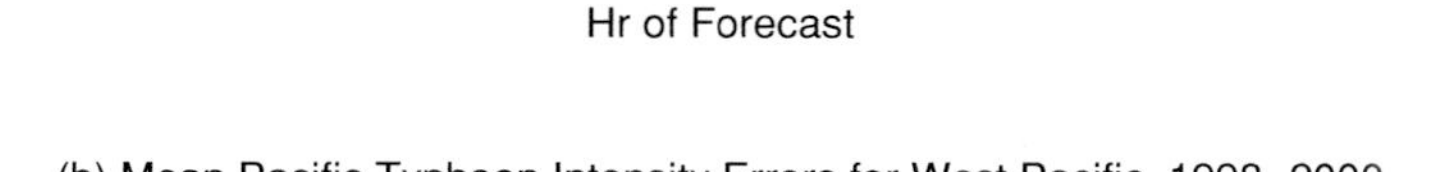

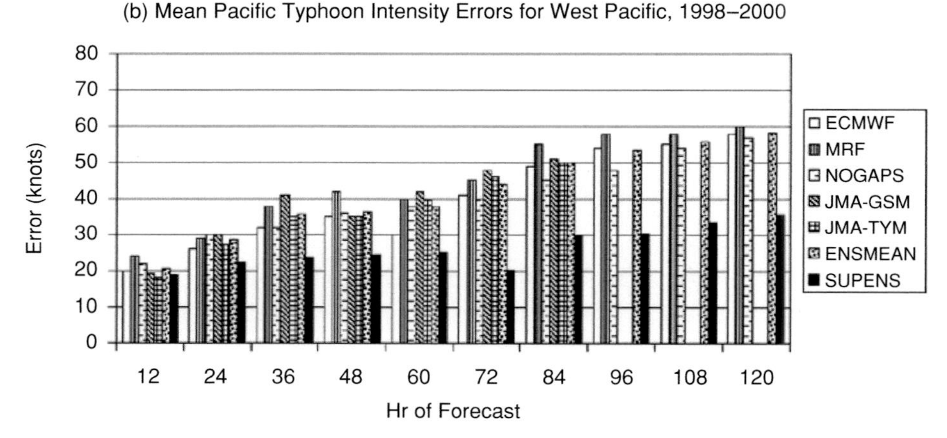

Figure 7.4 Mean typhoon track and intensity errors for the west Pacific, 1998–2000, for several global models (ECMWF, UK MetOffice, US MRF, US Navy NOGAPS, Japan JMA global and typhoon models, the ensemble mean of all these models), and a superensemble obtained by linear regression trained on the first half of the same season. From Kumar *et al.* (2003).

forecasting system (e.g. Kalnay and Ham, 1989; Fritsch *et al.*, 2000; Arribas *et al.*, 2005). This has also been shown to be true for regional models (Hou *et al.*, 2001). Krishnamurti *et al.* (1999), this volume) introduced the concept of 'superensemble', a multiple system ensemble where linear regression is used to correct the bias of each of the operational systems from past performance, and the predictors are combined to minimise the ensemble average prediction errors. This results in remarkable forecast improvements (e.g. Figure 7.4). This method is also called 'poor person' ensemble approach to reflect that it does not require running a forecasting system (Arribas *et al.*, 2005).

Other methods

This field is changing quickly, and improvements and changes to the operational systems are under development. For example, ECMWF has implemented changes in the length of the optimisation period for the SVs, a combination of initial and final or evolved SVs (which are more similar to BVs), and the introduction of a stochastic element in the physical parametrisations, all of which contributed to improvements in the ensemble performance (Buizza *et al.*, 2000). NCEP is considering the implementation of the ensemble transform Kalman filter (Bishop *et al.*, 2001) to replace breeding since the BVs have a tendency to converge to the leading Lyapunov vectors, providing insufficient spread. A recent comparison of the ensemble performance of the Canadian, US and ECMWF systems (Buizza *et al.*, 2005) suggests that the ECMWF SV perturbations behave well beyond the two-day optimisation period, at which time the model advantages of the ECMWF system are also paramount, giving the best performance. The BV perturbations of NCEP are somewhat better at short ranges, and the multiple analysis method performs well at all ranges. See other chapters in this volume for further discussions on applications of ensemble forecasting at all ranges: Hagedorn *et al.*, Lalaurette and van der Grijn, Mylne, Palmer, Shukla and Kinter, Tibaldi *et al.*, Timmermann and Jin, Latif *et al.*, Toth *et al.*, Waliser, and Webster *et al.*

7.3 Ensemble Kalman filtering for data assimilation and ensemble forecasting

As indicated before (e.g. Ehrendorfer and Tribbia, 1997), 'perfect' initial perturbations for ensemble forecasting should sample well the analysis errors. Thus, ideal initial perturbations $\delta\mathbf{x}_i$ in an ensemble with K members should have a covariance that spans well the analysis error covariance $\mathbf{A}$:

$$\frac{1}{K-1}\sum_{i=1}^{K}\delta\mathbf{x}_i\delta\mathbf{x}_i^T \approx \mathbf{A} \tag{7.1}$$

Until recently, the problem has been the lack of knowledge of $\mathbf{A}$, which changes substantially from day to day and from region to region due to instabilities of the background flow. These instabilities, associated with the 'errors of the day', are not taken into account in data assimilation systems, except for 4D-Var and Kalman filtering (KF), methods that are computationally very expensive. 4D-Var has been implemented at ECMWF, Météo-France, the Canadian Meteorological Service (CMS) and Japan Meteorological Agency (Rabier *et al.*, 2000; Desroziers *et al.*, 2003; Andersson *et al.*, 2004; Gauthier, 2004). The implementation of 4D-Var at ECMWF required some cost-saving simplifications such as reducing the resolution of the analysis from $\sim$40 km in the forecast model to $\sim$120 km in the assimilation model. Even at the lower resolution, the 4D-Var clearly outperformed the previous operational 3D-Var. Versions of 4D-Var are also under development in other centres.

The original formulations of Kalman filtering and extended Kalman filtering are prohibitive because they would require the equivalent of N model integrations, where N is the number of degrees of freedom of the model, of the order of 10^6 or more. Considerable work has been done on finding simplifying assumptions to reduce the cost of KF (e.g. Fisher *et al.*, 2003), but so far they have been successful only under special circumstances.

An alternative approach to Kalman filter which is much less expensive is ensemble Kalman filter (EnKF) suggested first by Evensen (1994) and now under development by several groups (Hamill, this volume). In the original extended KF the background error covariance is updated by using the linear tangent model and its adjoint (Cohn, 1997; Ghil *et al.*, 1981) equivalent to running the model about N times, where N is the number of degrees of freedom. By contrast, EnKF attempts to estimate the evolution of the background error covariance from an ensemble of K forecasts, with $K \ll N$. In the formulation of EnKF of Evensen (2003) and Houtekamer and Mitchell (1998), ensembles of data assimilation are driven by *perturbed observations* and used to derive the background error covariance from the ensemble of forecasts. A more recent class of EnKF is known as *square-root filters* (Anderson and Anderson, 1999; Bishop *et al.*, 2001; Anderson, 2001; Tippett *et al.*, 2002; Whitaker and Hamill, 2002; Ott *et al.*, 2002, 2004; Whitaker *et al.*, 2004). The ensemble forecasts are used to obtain a most likely forecast (the ensemble mean $\overline{\mathbf{x}}_b = \frac{1}{K}\sum_{i=1}^{K}\mathbf{x}_i$) and the background error covariance (obtained from the perturbations of the ensemble with respect to the mean $\mathbf{B} = \frac{1}{K-1}\sum_{i=1}^{K}(\mathbf{x}_i - \overline{\mathbf{x}}_b)(\mathbf{x}_i - \overline{\mathbf{x}}_b)^T$) at the time of the analysis. The full Kalman filter equations and the new observations are then used to obtain the most likely analysis $\mathbf{x}_a$, with the analysis increment (difference between the single analysis and the forecast) lying within the subspace of the forecast ensemble perturbations. After this is completed, the new initial analysis perturbations $\delta\mathbf{x}_i$ for the next analysis cycle are obtained by solving the square root problem $\frac{1}{K-1}\sum_{i=1}^{K}\delta\mathbf{x}_i\delta\mathbf{x}_i^T = \mathbf{A}$, where $\mathbf{A}$ is the analysis error covariance estimated by the Kalman filter, and K is the number of ensemble members. The solution of this problem is not unique and different square-root filters have adopted different solutions to the square-root problem (Tippett *et al.*, 2002; Ott *et al.*, 2002). An advantage of the square-root filter approach is that there is no need to add perturbations to the observations, which reduces the sampling error (Whitaker and Hamill, 2002).

Several of the EnKF approaches reduce the computational cost by assimilating the observations one at a time, for the whole physical domain, a method known as sequential assimilation of observations. This is done using a localisation of the error covariance in the horizontal and in the vertical, to avoid spurious long-distance correlations due to sampling. Although the sequential assimilation of observations is very efficient with limited observations (such as those available before 1979), it becomes less practical with the abundant satellite observations currently available (about 3.5 million data per assimilation cycle, about 40 times more than a decade ago) and planned for the next decade.

A different approach, also within the class of the square-root filters, is the local ensemble Kalman filter (LEKF) method (Ott *et al.*, 2002, 2004; Szunyogh *et al.*, 2005). In the LEKF the square root filter problem is solved *locally in physical space*, not in observation space. For each grid point a local three-dimensional volume of the order of 700 km by 700 km by a few vertical layers is used to perform the analysis. The Kalman filter equations are solved exactly in the subspace locally spanned by the global ensemble members, using *all the observations* available within the volume. This localisation in space results in a further reduction of the number K of ensemble members needed to obtain an accurate solution, so that matrix operations are done in a very low dimensional space. The analysis is carried out independently at each grid point, leading to a completely parallel algorithm.

In schematic Figure 7.5 we compare ensemble Kalman filtering (EnKF) with the 3D-Var approach, in which the background error covariance is estimated as an average over many cases. Figure 7.5(a) shows how the 3D-Var analysis maximises the joint probability defined by the observations error covariance and the background error covariance, both of which are high dimensional. Since the 3D-Var background error covariance is a statistical average, the 3D-Var analysis does not know about 'errors of the day'. Figure 7.5(b) shows that in the EnKF, the ensemble perturbations define a very low-dimensional subspace within which the forecast errors lie, and the KF analysis maximises the joint probability within that subspace. Because the computations are performed within this subspace, the rank of the matrices involved is low, and the Kalman filter equations providing the analysis and analysis error covariance can be solved directly, not iteratively.

Figure 7.6 (colour plate) shows the background errors and the analysis corrections based on a given set of noisy observations in a quasi-geostrophic data assimilation system (Morss *et al.*, 2000; Hamill and Snyder, 2000; Corazza *et al.*, 2003). The top panel corresponds to 3D-Var, with a background error covariance constant in time. Because the system does not know about the dynamical stretching produced by the 'errors of the day', the analysis increments introduced by the new observations tend to be isotropic. The bottom panel shows that the local ensemble Kalman filter (Ott *et al.*, 2004; Szunyogh *et al.*, 2005) is much more efficient in correcting the background errors. The large improvements made on the analysis are also apparent in forecasts (not shown).

Performing the EnKF locally in space substantially reduces the number of ensemble members required for the analysis, as shown in Figure 7.7, from Ott *et al.*, 2004; obtained using the Lorenz (1996) model. When performed globally, the number of ensemble members required for the EnKF to converge to the optimal value is proportional to the size of the model. When done locally, the number of ensemble members

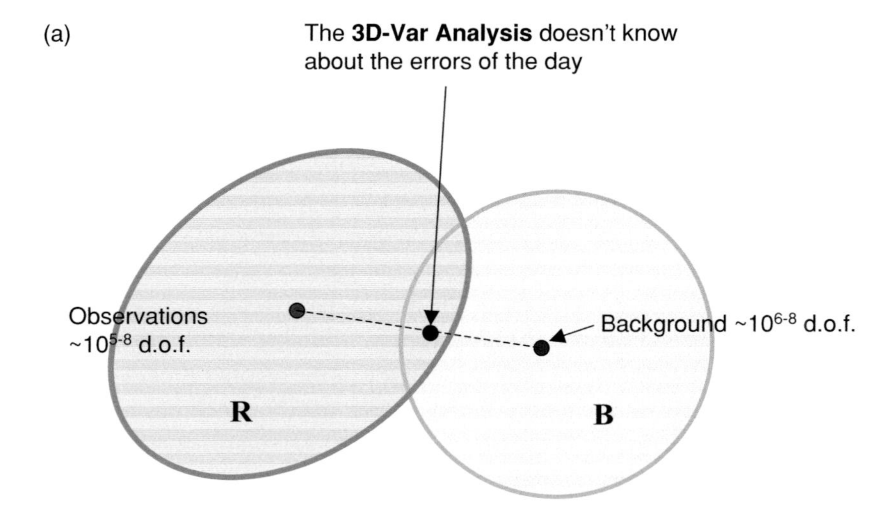

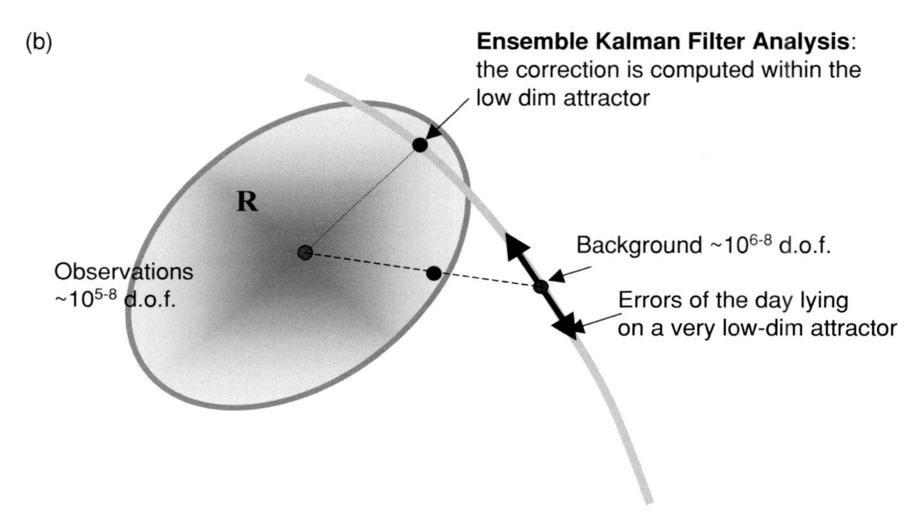

Figure 7.5 Schematic of the analysis given the background forecast in a very large dimensional space, the background error covariance **B** (which in the case of 3D-Var is isotropic and constant in time), the vector of observations, in a very large dimensional space, with observations error covariance **R**. The analysis estimate of the true state of the atmosphere maximises the joint probability distribution. (a) 3D-Var. (b) EnKF, in which the ensemble forecast members define a subspace within which the analysis lies.

is reduced from 27 to 8, and it does not increase with the size of the model. In addition, the analysis for different grid points can be carried out in parallel, since they are independent of each other.

The efficiency of localisation is schematically shown in Figure 7.8, where three independent unstable regions a, b, c, can each have wave number 1 and wave number

Background error (shaded) and 3D- Var analysis increments

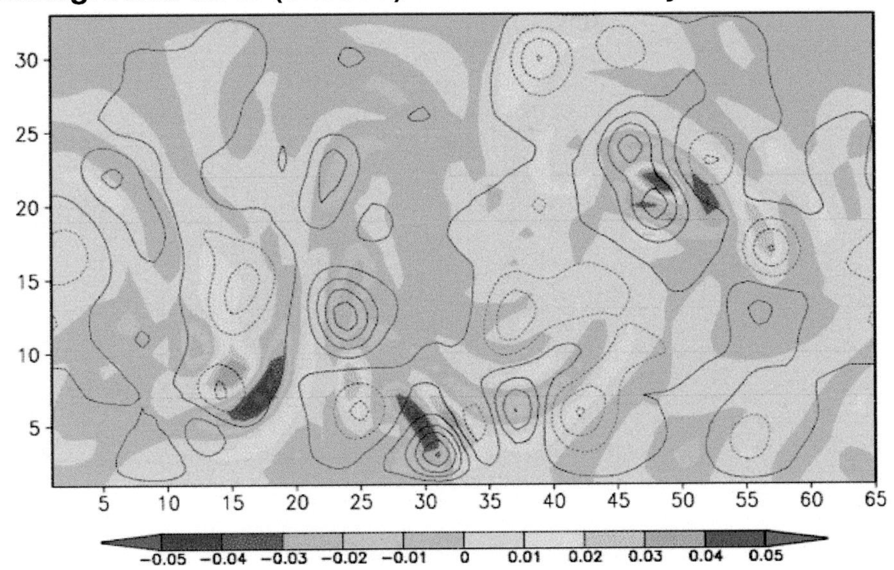

Background error (shaded) and LEKF analysis increments

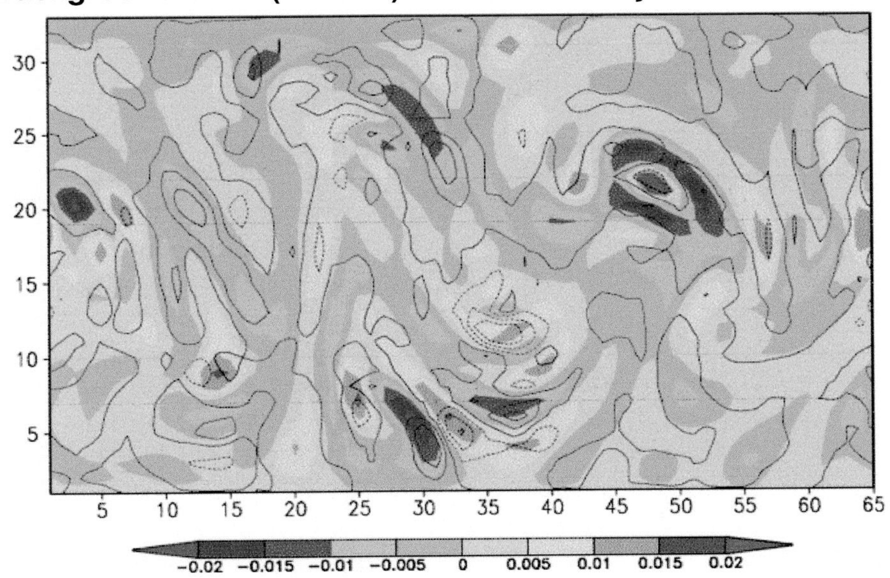

Figure 7.6 (See also colour plate section.) Simulation of data assimilation in a quasi-geostrophic model, assimilating potential vorticity observations at a particular day (15 June). The shades represent the 12 hr forecast (background) error and the contours the analysis corrections. (Top) 3D-Var. (Bottom) local ensemble Kalman filter. Figures courtesy of Matteo Corazza.

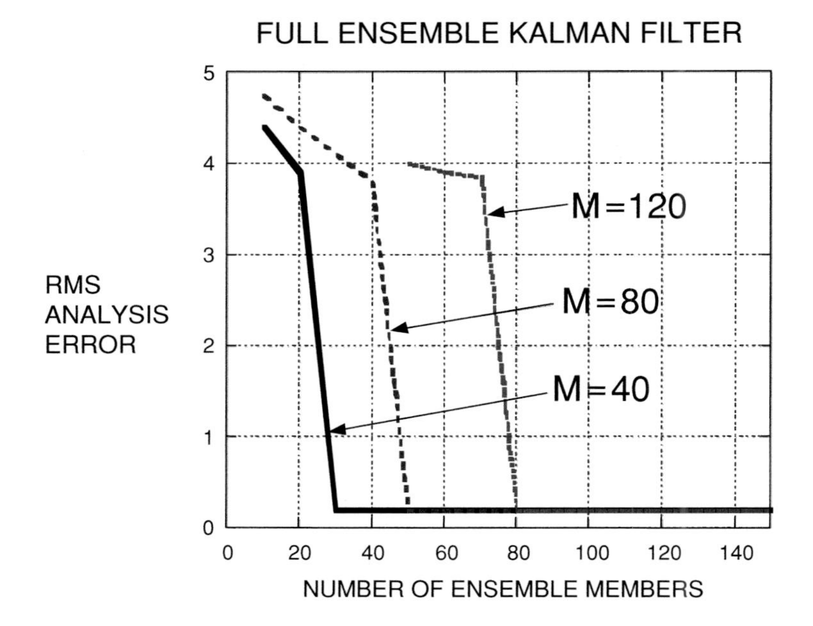

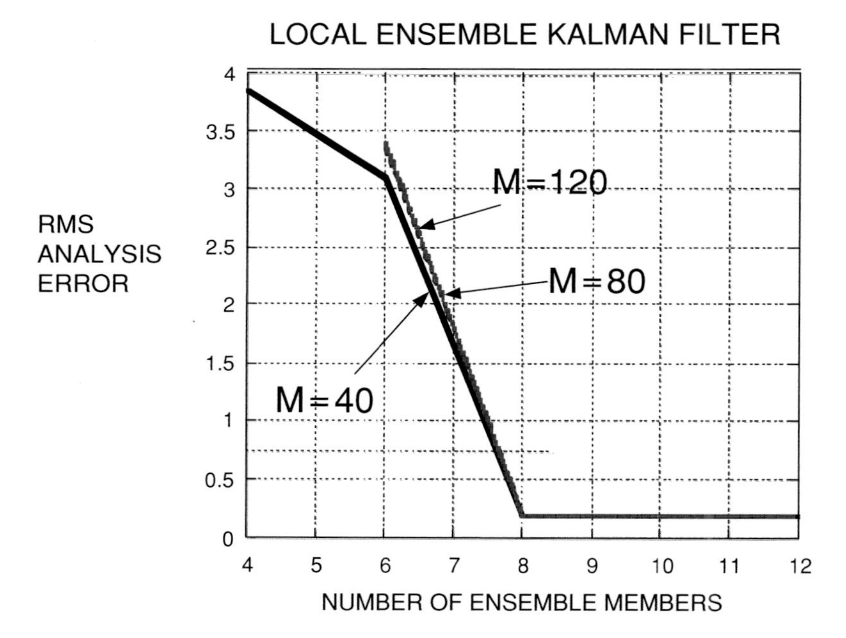

Figure 7.7 Number of ensemble members required for convergence to the optimal solution in a Lorenz (1996) model. (Top) using a full global ensemble Kalman filter. (Bottom) using a local ensemble Kalman filter. The size of the domain, M, is either 40, 80 or 120. Note that the Kaplan–Yorle dimension is about 27 for the 40-variable model and increases linearly with size. Figure adapted from Ott *et al.* (2004).

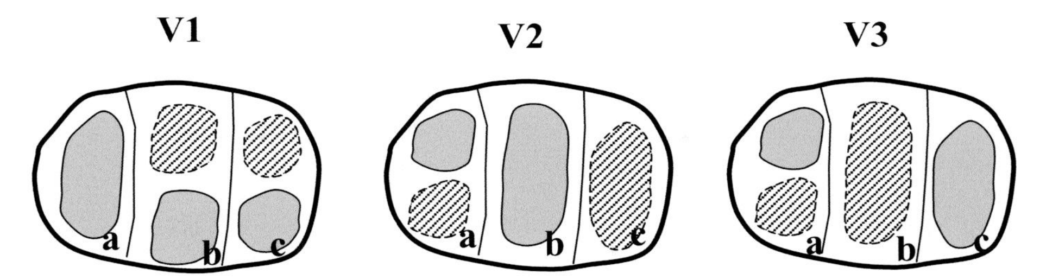

Figure 7.8 Schematic showing the advantage of performing a local rather than a global analysis. The domain is composed of three regions, **a**, **b**, **c**, each of which has possible instabilities with wave numbers 1 and 2. From a local point of view, the ensemble perturbations **V1** and **V2** are sufficient to represent all possible unstable perturbations, whereas from a global point of view, **V3** is independent from **V2**, and there are many more independent perturbations.

2 instabilities, in the same way that different areas of the world can develop baroclinic instabilities that evolve independently from each other. Three of the possible ensemble perturbations are depicted. From a local point of view, the first two perturbations are enough to represent all possible combinations of wave numbers 1 and 2 instabilities, whereas from a global point of view the third perturbation and many others are linearly independent of the first two.

The LEKF has been tested in a 'perfect model' mode, using the operational NCEP Global Forecasting System at a resolution of T62/28 levels, with excellent results (Szunyogh *et al.*, 2005). Figure 7.9 shows the evolution of the analysis of surface pressure, when about 11% of the points (separated by about 200 km) have 'rawinsonde' observations. The analysis errors for temperature and winds show similar quick convergence to values much smaller than the observational errors. Figure 7.10 shows the vertical root-mean-square analysis errors for temperature and zonal wind for several levels of observational density, including 11% of the grid points, a density similar to that of the rawinsondes in the northern hemisphere extratropics, and 2%, a density similar to rawinsondes in the southern hemisphere. The ability of the LEKF to extract information through the knowledge of the 'errors of the day' is very encouraging. Figure 7.11 shows that despite the local nature of the analysis, in the perfect model simulation, the LEKF, with just 2% observations, is able to reach 'superbalance', being able to reproduce the evolution of not only slow synoptic waves but also that of a gravity wave present in the 'nature' run.

7.4 Prospects for operational implementation

At the time of this writing (December 2004), tests of the LEKF with real observations have not yet been carried out. Tests with other square-root filters have been performed only with rawinsonde upper air observations and cloud tracked winds, and

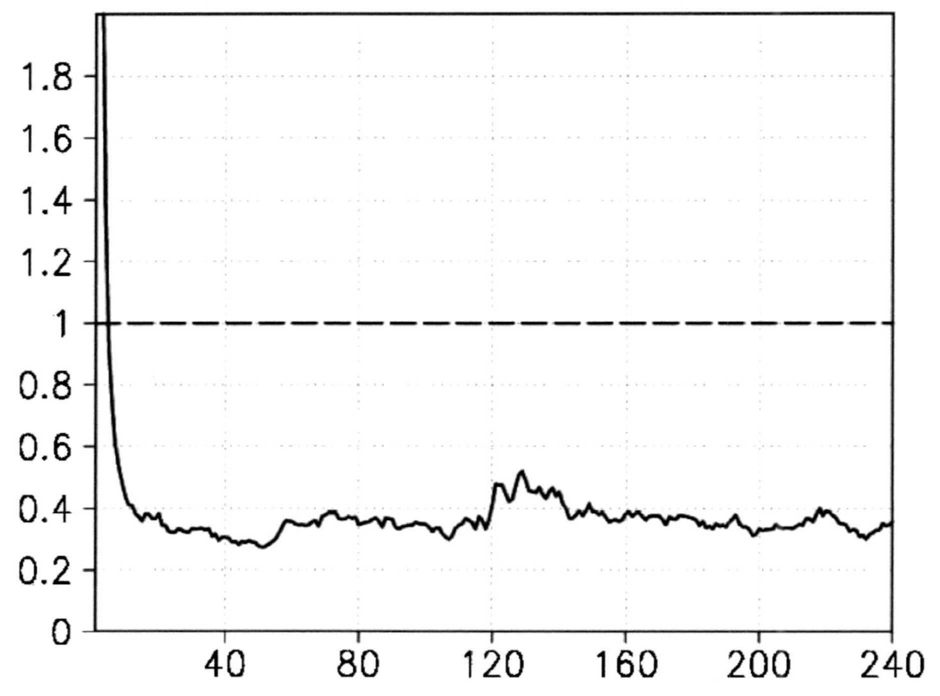

Figure 7.9 Evolution of the LEKF analysis error in surface pressure in hPa as a function of assimilation step (in units of 6 hr). The rms error of the observations is shown by the dashed line. Observations are made at 11% of the grid points, and the model has T62 horizontal resolution (about 200 km). From Szunyogh *et al.* (2005).

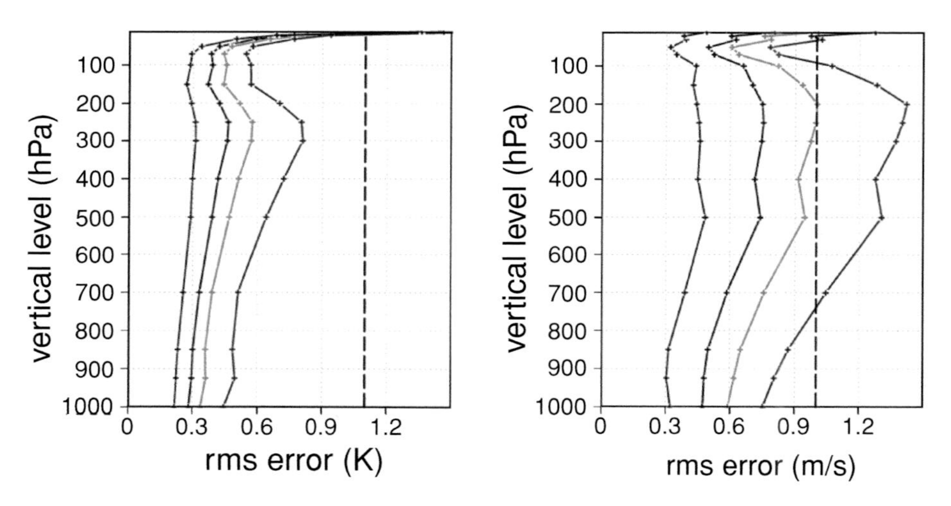

Figure 7.10 Root-mean-square global analysis error for temperatures (left, °C) and tropical analysis error for zonal winds (right, m/sec). The dashed line is the rms of observations. From left to right, the following percentage of the grid points have 'rawinsonde' data: 100%, 11%, 5%, 2%. Since the grid resolution is about 200 km, the second is similar to the current rawinsonde density in the northern hemisphere, and the fourth to the southern hemisphere and tropics. From Szunyogh *et al.* (2005).

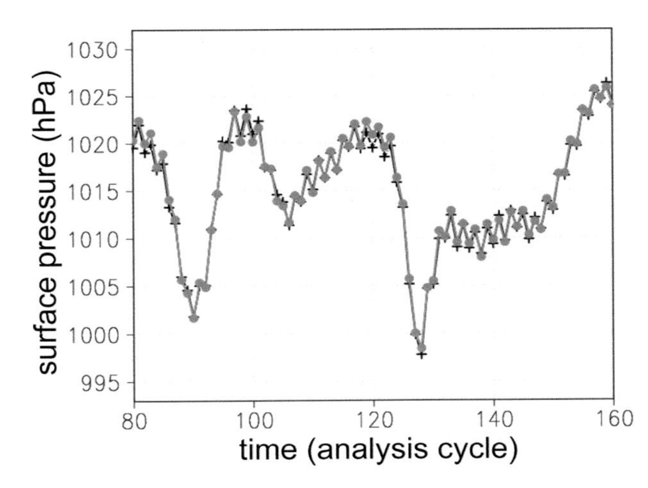

Figure 7.11 Comparison of the true (crosses) and analysed (circles) gravity wave observed at 30N 150W. The observing network has density similar to that of the southern hemisphere. From Szunyogh *et al.* (2005).

yield results comparable in the northern hemisphere to the 3D-Var used in the NCEP Reanalysis, which also used satellite retrievals (Jeff Whitaker, personal communication, 2004). The perturbed observations approach has been tested at the Canadian Meteorological Centre (CMC) in an operational environment and also yields results comparable to those of the very mature operational 3D-Var in the Canadian Meteorological Service (Houtekamer *et al.*, 2005). By contrast, 4D-Var has been shown to be superior to 3D-Var in several centres (ECMWF, Météo-France, CMC). This relatively disappointing performance of the EnKF approach, although it is to be expected in the still early stages of testing, is in contrast to the excellent results obtained with perfect model simulations.

From this experience we cannot say at this point whether EnKF will be able to compete with or replace 4D-Var in operational centres as the next data assimilation system of choice. However, EnKF has a number of very attractive advantages that hold promise once the new systems are tested and tuned. Because of familiarity we will describe the advantages of the LEKF system that has been developed at the University of Maryland (Ott *et al.*, 2002, 2004; Hunt *et al.*, 2004; Szunyogh *et al.*, 2005), comparing with the corresponding characteristics of 4D-Var, but some of the advantages are generic to other EnKFs.

(a) The LEKF is very efficient, due to its complete parallelism and relatively few required ensemble members, and can use data simultaneously. In the original formulation it takes only 15 minutes to assimilate 1.5 million observations using 40 ensemble members on a cluster of 25 dual processor 2.8 GH PCs. This includes the 6-hour 40-member ensemble forecast, which takes about 6 minutes (and which would be free within an operational centre that performs ensemble forecasting). B. R. Hunt (personal communication, 2003) has developed an alternative algorithm (local ensemble transform Kalman filter, LETKF) not based on singular value decomposition, which is about three times faster than the LEKF while yielding essentially identical results (I. Szunyogh, and E. Kostelich, personal communication, 2004). The efficiency of the LEKF ensures that it can be used operationally with a resolution at least the same as that of the operational ensemble forecasting system.

(b) Like other EnKF methods, the LEKF does not require the development and maintenance of the model's linear tangent or adjoint models, saving the large effort needed for these models, and avoiding the need for linear approximations, since the full non-linear model is used for every operation.

(c) Similarly, like other EnKF, the LEKF does not require the Jacobian or the adjoint of the observation operator H, another important advantage (Houtekamer and Mitchell, 1998; Szunyogh *et al.*, 2005). Basically, if we define a matrix $\mathbf{E}$ of perturbation vectors, so that $\mathbf{B} = \mathbf{E}\mathbf{E}^T; \mathbf{H}\mathbf{B}\mathbf{H}^T = (\mathbf{H}\mathbf{E})(\mathbf{H}\mathbf{E})^T$, one can replace the linear $\mathbf{HE}$ with its fully non-linear expression,

$$\mathbf{HE} = \mathbf{H}\left[\delta\mathbf{x}_1 | \ldots | \delta\mathbf{x}_K\right] \approx \left[H(\mathbf{x}_1) - H(\overline{\mathbf{x}}) | \ldots | H(\mathbf{x}_1) - H(\overline{\mathbf{x}})\right],$$

which is simpler and more accurate.

(d) It can be easily extended to four dimensions (4DLEKF), so that observations can be assimilated at their time of observation, in between analysis times (Hunt *et al.*, 2004). This is performed at a relatively low computational cost, by using the ensemble to 'transport' observational increments from the time of observation to the analysis time. This is important since experience suggests that the ability to assimilate observations at the right time is the main advantage of 4D-Var (rather than the evolution of the covariance).

(e) The 4DLEKF can be used as a smoother, taking advantage of future observations, as would be possible in a Reanalysis mode, at the cost of doubling the computations. At a given time t_0 a preliminary (operational) analysis and analysis error covariance is performed using only past observations, then ensemble forecasts between this time and the next analysis time t_1 are used to bring the observations in that interval back to t_0, and a final, more accurate analysis and analysis error covariance can be obtained at t_0. The problem of using the data twice (which would erroneously reduce the analysis error covariance) can be handled by increasing the observation error covariance (e.g. doubling it for observations used both at t_0 and t_1).

(f) The LEKF provides diagnostic tools that can be used to tune the system in the future. One of them is the local effective dimension ED (Patil *et al.*, 2001) obtained from the singular values σ_i of the ensemble

$$ED = \frac{\left(\sum_{i=1}^{k} \sigma_i\right)^2}{\sum_{i=1}^{k} \sigma_i^2}$$

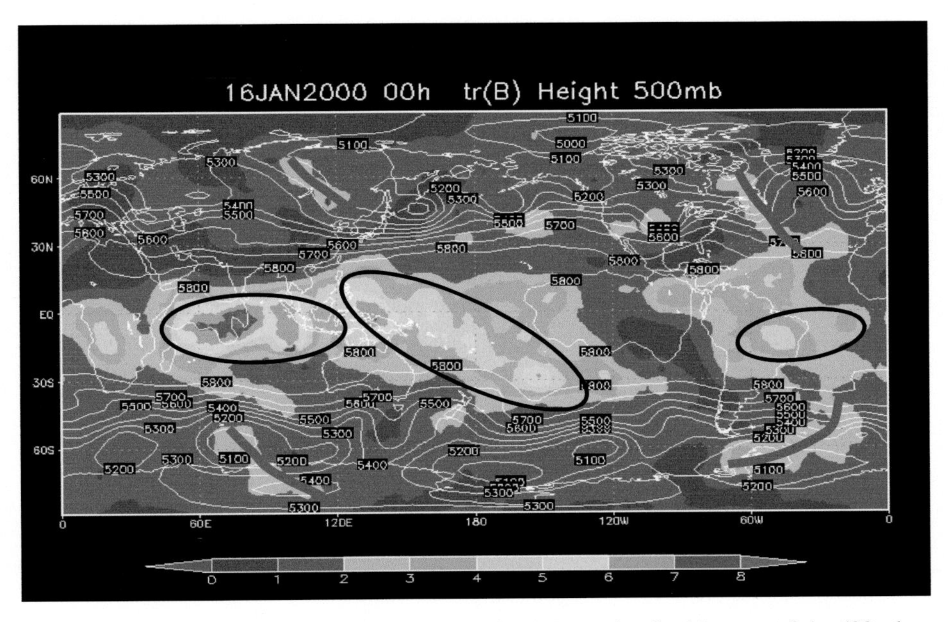

Figure 7.12 (See also colour plate section.) Example of a 6 hr trace of the 500 mb height forecast error covariance showing the potential use of LEKF for adaptive observations. Regions in blue and purple do not need immediate observations. Midlatitude areas marked with red have large errors but a low effective ensemble dimension, so that they are prime areas for targeting. Tropical regions with large errors (ovals), by contrast, have also large effective ensemble dimension, presumably because the error growth is dominated by convection.

Szunyogh *et al.* (2005) showed that there is a strong relationship between the projection of the forecast error on the ensemble perturbations (not available in real data assimilation experiments) and the effective dimension. Their Figure 7.7 shows that in regions where the projection of the error on the ensemble is close to 100%, as in most of the midlatitudes, the effective dimension (about 10) is much smaller than the actual ensemble dimension (40). In the tropics, where the projection is only about 40%, the effective dimension is larger, of the order of 30 or more. Other diagnostics that can be used for tuning are comparisons of observations minus forecast and observations minus analysis, which can be compared with their predicted values.

(g) EnKF can be used to obtain best estimates of model bias as part of the data assimilation (e.g. Anderson, 2001).

(h) Finally, the LEKF provides estimates of the background and the analysis error covariances (and thus ideal initial perturbations), although they are based on a space with limited dimension given by the ensemble size. This information has other important applications such as adaptive observations, a new area of great growth in recent years (Thorpe and Petersen, this volume). Figure 7.12 (colour plate) suggests that EnKF makes it possible to have interactive targeted observations, with remote sensing instruments dwelling only on regions identified by the ensemble as having low accuracy forecasts.

7.5 Final comments

Although the EnKF approach in general, and the LEKF/LETKF in particular, seem to have many advantages, so far there has been no direct comparison between LEKF and 4D-Var in a realistic system. The EnKF is based on a reduced dimension, albeit computed with full non-linearity, and it provides an estimate of the background and analysis covariances. 4D-Var requires an estimate of the background error covariance **B** at the initial time of the assimilation window, and it is common to start with a 3D-Var **B**. Obtaining a better evolving estimate of **B** with a reduced Kalman filter has been difficult (Fisher *et al.*, 2003). Mike Fisher showed in a recent presentation that 4D-Var is equivalent to a full Kalman smoother (which at the end of the interval yields the same estimate as the Kalman filter) if the assimilation window is sufficiently long, because the lack of an initial **B** is 'forgotten'. Fisher proposed that a 4D-Var with a 3–10 day window and including model errors (weak constraint) will provide a solution equivalent to a *full rank* Kalman filter, and it could perhaps be made computationally affordable. This is an attractive alternative to EnKF and only experience will tell whether one is better than the other.

An important remaining problem is that of model deficiencies, leading to model systematic errors and to problems such as those suggested in Figure 7.1(b). A successful approach to start addressing this problem for ensemble forecasting applications is using multiple models (Krishnamurti *et al.*, 1999; Hou *et al.*, 2001). Other approaches (DelSole and Hou, 1999; Kaas *et al.*, 1999) rely on empirical methods to reduce the errors using past observations. Yet another method, known as 'dressing', adds random perturbations to the ensemble forecasts in order to reproduce the observed error covariance with the ensemble (Roulston and Smith, 2003; Wang and Bishop, 2004). It is possible that the ensemble Kalman filtering approach will also be able to handle efficiently model errors by augmenting the model variables with a relatively small number of parameters associated with model errors, and using the observations to estimate the optimal value of their time-varying coefficients (Danforth *et al.*, 2006).

If current research in ensemble Kalman filtering methods achieves its promise, then the problems of data assimilation and ensemble forecasting may indeed have a unified solution.

References

Ahlquist, J. (2000): Almost Anything can be a Singular Vector. http://www.met.fsu.edu/ftp/ahlquist/singvect.ps

Anderson, J. L. (2001). An ensemble adjustment filter for data assimilation. *Mon. Weather Rev.,* **129**, 2884–903.

Anderson, J. L. and S. L. Anderson (1999). A Monte Carlo implementation of the nonlinear filtering problem to produce ensemble assimilations and forecasts. *Mon. Weather Rev.*, **127**, 2741–58.

Andersson, Erik, C. Cardinali, M. Fisher, *et al.* (2004). Developments in ECMWF's 4D-Var system. In *Symposium on Forecasting the Weather and Climate of the Atmosphere and Ocean*, 20th Conference on Weather Analysis and Forecasting, 16th Conference on Numerical Weather Prediction, Seattle, Washington, January 2004. Paper J1.4, American Meteorological Society.

Arribas, A., K. B. Robertson and K. R. Mylne (2005). Test of a poor man's ensemble prediction system for short-range probability forecasting. *Mon. Weather Rev.*, **133**, 1825–39.

Bishop, C. H., B. J. Etherton and S. J. Majumdar (2001). Adaptive sampling with the ensemble transform Kalman filter. I: Theoretical aspects. *Mon. Weather Rev.*, **129**, 420–36.

Buizza, R. (1997). Potential forecast skill of ensemble prediction, and spread and skill distributions of the ECMWF Ensemble Prediction System. *Mon. Weather Rev.*, **125**, 99–119.

Buizza, R., Petroliagis, T., Palmer, T. N., *et al.* (1998). Impact of model resolution and ensemble size on the performance of an ensemble prediction system. *Quart. J. Roy. Meteor. Soc.*, **124**, 1935–60.

Buizza, R., M. Miller and T. N. Palmer (1999). Stochastic representation of model uncertainties in the ECMWF Ensemble Prediction System. *Quart. J. Roy. Meteor. Soc.*, **125**, 2887–908.

Buizza, R., Barkmeijer, J., Palmer, T. N. and O. S. Richardson (2000). Current status and future developments of the ECMWF Ensemble Prediction System. *Meteorol. Appl.*, **7**, 163–75.

Buizza, R., P. L. Houtekamer, Z. Toth, G. Pellerin, M. Wei, Y. Zhu (2005). A comparison of the ECMWF, MSC and NCEP Global Ensemble Prediction Systems. *Mon. Weather Rev.*, **133**, 1076–97.

Cohn, S. (1997). An introduction to estimation theory. *J. Meteorol. Soc., Jpn.* **75** (1B), 257–88.

Corazza, M., E. Kalnay, D. J. Patil, *et al.* (2003). Use of the breeding technique to estimate the structure of the analysis "errors of the day". *Nonlinear Proc. Geoph.*, **10**, 233–43.

Danforth, C., E. Kalnay and T. Miyoshi (2006). Estimating and correcting global weather model error. *J. Atmos. Sci.*, **63**, in press.

DelSole, T. and A. Y. Hou (1999). Empirical stochastic models for the dominant climate statistics of a general circulation model. *J. Atmos. Sci.*, **56**, 3436–3456.

Desroziers, G., G. Hello and J. N. Thépaut (2003). A 4D-VAR re-analysis of the FASTEX experiment. *Quart. J. Roy. Meteor. Soc.*, **129**, 1301–16.

Ebisuzaki, W. and E. Kalnay (1991). Ensemble experiments with a new lagged average forecasting scheme. In *Research Activities in Atmospheric and Oceanic Modeling*, pp. 6.31–6.32. WMO Report 15. [Available from WMO, C.P. No 2300, CH1211, Geneva, Switzerland].

Ehrendorfer, M. and J. J. Tribbia (1997). Optimal prediction of forecast error covariances through singular vectors. *J. Atmos. Sci.*, **54**, 286–313.

Epstein, E. S. (1969). Stochastic-dynamic prediction. *Tellus*, **21**, 739–59.

Errico, R. and D. P. Baumhefner (1987). Predictability experiments using a high-resolution limited-area model. *Mon. Weather Rev.*, **115**, 488–504.

Evensen, G. (1994). Sequential data assimilation with a nonlinear quasi-geostrophic model using Monte Carlo methods to forecast error statistics. *J. Geophys. Res.*, **99** (C5), 10143–62.

(2003). The ensemble Kalman filter: theoretical formulation and practical implementation. *Ocean Dynam.*, **53**, 343–67.

Fisher, M., L. Isaksen, M. Ehrendorfer, A. Beck and E. Andersson (2003). A critical evaluation of the reduced-rank Kalman filter (RRKF) approach to flow-dependent cycling of background error covariances. ECMWF Technical Memorandum.

Fritsch, J. M., J. Hilliker, J. Ross and R. L. Vislocky (2000). Model consensus. *Weather Forecast*, **15**, 571–82.

Gauthier, P. (2004). THORPEX activities in Canada. In *THORPEX workshop, Boulder, CO, June 2004.* (http://box.mmm.ucar.edu/uswrp/recent_meetings/ThorpexWorkshop/Canada_Gauthier.pdf)

Ghil, M., S. Cohn, J. Tavantzis, K. Bube and E. Isaacson (1981). Applications of estimation theory to numerical weather prediction. In *Dynamic Meteorology: Data Assimilation Methods*, ed. L. Bengtsson, M. Ghil and E. Kallen. Springer-Verlag.

Hamill, T. M. and C. Snyder (2000). A hybrid ensemble Kalman filter-3D variational analysis scheme. *Mon. Weather Rev.*, **128**, 2905–919.

Hoffman, R. N. and E. Kalnay (1983). Lagged average forecasting, an alternative to Monte Carlo forecasting. *Tellus*, **35A**, 100–18.

Hollingsworth, A. (1980). An experiment in Monte Carlo forecasting. In *ECMWF Workshop on Stochastic-Dynamic Forecasting*, pp. 65–85. ECMWF, Shinfield Park, Reading, UK, RG2 9AX.

Hou, D., E. Kalnay and K. K. Droegemeier (2001). Objective verification of the SAMEX '98 ensemble forecasts. *Mon. Weather Rev.*, **129**, 73–91.

Houtekamer, P. L. and H. L. Mitchell (1998). Data assimilation using an ensemble Kalman filter technique. *Mon. Weather Rev.*, **126**, 796–811.

(2001). A sequential ensemble Kalman filter for atmospheric data assimilation. *Mon. Weather Rev.*, **129**, 123–37.

Houtekamer, P. L., L. Lefaivre and J. Derome (1996). A system simulation approach to ensemble prediction. *Mon. Weather Rev.*, **124**, 1225–42.

Houtekamer, P. L., H. L. Mitchell, G. Pellerin, *et al.* (2005). Atmospheric data assimilation with the ensemble Kalman filter: Results with real observations. *Mon. Weather Rev.*, **133**, 604–20.

Hunt, B. R., E. Kalnay, E. J. Kostelich, J. *et al.* (2004). Four-dimensional ensemble Kalman filtering. *Tellus*, **56A**, 273–7.

Kaas, E., A. Guldberg, W. May and M. Decque (1999). Using tendency errors to tune the parameterization of unresolved dynamical scale interactions in atmospheric general circulation models. *Tellus*, **51A**, 612–29.

Kalnay, E. (2003). *Atmospheric Modeling, Data Assimilation, and Predictability*. Cambridge University Press.

Kalnay, E. and M. Ham (1989). Forecasting forecast skill in the Southern Hemisphere. In *Preprints of the 3rd International Conference on Southern Hemisphere Meteorology and Oceanography, Buenos Aires, 13–17 November 1989*. Boston, MA: American Meteorological Society.

Krishnamurti, T. N., C. M. Kishtawal, T. E. LaRow, *et al.* (1999): Improved weather and seasonal climate forecasts from multimodel superensemble. *Science*, **285**(5433), 1548–50.

Kumar, T. S. V., T. N. Krishnamurti, M. Fiorino and M. Nagata (2003). Numerical prediction of typhoon tracks and intensity using a multimodel superensemble. *Mon. Weather Rev.*, **131**, 574–83.

Kyouda, M. and S. Kusunoki (2002). Ensemble Prediction System. In *Outline of the Operational Numerical Weather Prediction at the Japan Meteorological Society*, pp. 59–63. Appendix to the WMO Numerical Weather Prediction Progress Report.

Leith, C. E. (1974). Theoretical skill of Monte Carlo forecasts. *Mon. Weather Rev.*, **102**, 409–18.

Lorenz, E. N. (1963). Deterministic nonperiodic flow. *J. Atoms. Sci.*, **20**, 130–41.

Lorenz, E. N. (1965). A study of the predictability of a 28-variable atmospheric model. *Tellus*, **17**, 321–33.

Lorenz, E. N. (1996). Predictability: a problem partly solved. In *Proceedings of Seminar on Predictability, Vol. 1*. European Centre for Medium-Range Weather Forecasts, Shinfield Park, Reading, Berkshire, RG2 9AX, United Kingdom.

Molteni, F. and T. N. Palmer (1993). Predictability and finite time instability of the northern winter circulation. *Quart. J. Roy. Meteor. Soc.*, **119**, 269–98.

Molteni, F., R. Buizza, T. N. Palmer and T. Petroliagis (1996). The ECMWF ensemble prediction system: methodology and validation. *Quart. J. Roy. Meteor. Soc.*, **122**, 73–119.

Morss, R. E., K. A. Emanuel and C. Snyder (2001). Idealized adaptive observation strategies for improving numerical weather prediction. *J. Atmos. Sci.*, **58**, 210–34.

Ott, E., B. H. Hunt, I. Szunyogh, *et al.* (2002). Exploiting local low dimensionality of the atmospheric dynamics for efficient Kalman filtering. arXiv:archive/paper 0203058, http://arxiv.org/abs/physics/0203058.

Ott, E., B. H. Hunt, I. Szunyogh, *et al.* (2004). A local ensemble Kalman filter for atmospheric data assimilation. *Tellus*, **56A**, 415–28.

Palmer, T. N., R. Gelaro, J. Barkmeijer and R. Buizza (1998). Singular vectors, metrics and adaptive observations. *J. Atmos. Sci.*, **55**, 633–53.

Patil, D. J., B. R. Hunt, E. Kalnay, J. A. Yorke and E. Ott (2001). Local low dimensionality of atmospheric dynamics. *Phys. Rev. Lett.*, **86**, 5878–81.

Peña, M. and E. Kalnay (2004). Separating fast and slow modes in coupled chaotic systems. *Nonlinear Proc. Geoph.*, **19**, 319–27.

Rabier, F., H. Järvinen, E. Klinker, J. F. Mahfouf and A. Simmons (2000). The ECMWF operational implementation of four-dimensional variational assimilation. I: Experimental results with simplified physics. *Quart. J. Roy. Meteor. Soc.*, **126**, 1143–70.

Roulston, M. S. and L. A. Smith (2003). Combining dynamical and statistical ensembles. *Tellus*, **55A**, 16–30.

Stensrud, D. J., J.-W. Bao and T. T. Warner (2000). Using initial condition and model physics perturbations in short-range ensembles. *Mon. Weather Rev.*, **128**, 2077–107.

Szunyogh, I. and Z. Toth (2002). The effect of increased horizontal resolution on the NCEP global ensemble mean forecasts. *Mon. Weather Rev.*, **130**, 1125–43.

Szunyogh, I., E. J. Kostelich, G. Gyarmati, *et al.* (2005). Assessing a local ensemble Kalman filter: perfect model experiments with the NCEP global model. *Tellus*, **57A**, 528–45.

Tippett, M. K., Anderson, J. L., Bishop, C. H., Hammill, T. M. and Whitaker, J. S. (2002). Ensemble square-root filters. *Mon. Weather Rev.*, **131**, 1485–90.

Toth, Z. and E. Kalnay (1993). Ensemble forecasting at NMC: the generation of perturbations. *Bull. Am. Meteorol. Soc.*, **74**, 2317–30.

(1997). Ensemble forecasting at NCEP and the breeding method. *Mon. Weather Rev.*, **125**, 3297–319.

Tracton, M. S. and E. Kalnay (1993). Ensemble forecasting at NMC: practical aspects. *Weather Forecast.*, **8**, 379–98.

Wang, X. and C. H. Bishop (2004). Ensemble augmentation with a new dressing kernel. In *20th AMS Conference on Weather Analysis and Forecasting, Seattle, January 2004.* American Meteorological Society.

Whitaker, J. S. and T. H. Hamill (2002). Ensemble data assimilation without perturbed observations. *Mon. Weather Rev.*, **130**, 1913–24.

Whitaker, J. S., G. P. Compo, X. Wei and T. H. Hamill (2004). Reanalysis without radiosondes using ensemble data assimilation. *Mon. Weather Rev.*, **132**, 1190–200.

8

Approximating optimal state estimation

Brian F. Farrell
Harvard University, Cambridge

Petros J. Ioannou
Department of Physics, National and Capodistrian University of Athens

Minimising forecast error requires accurately specifying the initial state from which the forecast is made by optimally using available observing resources to obtain the most accurate possible analysis. The Kalman filter accomplishes this for linear systems and experience shows that the extended Kalman filter also performs well in non-linear systems. Unfortunately, the Kalman filter and the extended Kalman filter require computation of the time-dependent error covariance matrix which presents a daunting computational burden. However, the dynamically relevant dimension of the forecast error system is generally far smaller than the full state dimension of the forecast model which suggests the use of reduced order error models to obtain near optimal state estimators. A method is described and illustrated for implementing a Kalman filter on a reduced order approximation of the forecast error system. This reduced order system is obtained by balanced truncation of the Hankel operator representation of the full error system. As an example application a reduced order Kalman filter is constructed for a time-dependent quasi-geostrophic storm track model. The accuracy of the state identification by the reduced order Kalman filter is assessed and comparison made with the state estimate obtained by the full Kalman filter and with the estimate obtained using an approximation to 4D-Var. The accuracy assessment is facilitated by formulating the state estimation methods as observer systems. A practical approximation to the reduced order Kalman filter that utilises 4D-Var algorithms is examined.

Predictability of Weather and Climate, ed. Tim Palmer and Renate Hagedorn. Published by Cambridge University Press.

8.1 Introduction

An important component of forecast error is error in the analysis of the initial state from which the forecast is made. Analysis error can be reduced by taking more observations, by taking more accurate observations, by taking observations at locations chosen to better constrain the forecast, and by extracting more information from the observations that are available. The last of these, obtaining the maximum amount of information from observations, is attractive because it makes existing observations more valuable and because, at least for linear systems, there is a solution to the problem of extracting the maximum information from a given set of observations: under appropriate assumptions the problem of extracting the maximum amount of information from a set of observations of a linear system in order to minimise the uncertainty in the state estimate is solved by the Kalman filter (KF) (Kalman, 1960; Ghil and Malanotte-Rizzoli, 1991; Wunsch, 1996; Ide *et al.*, 1997; Lermusiaux and Robinson, 1999). Moreover, application of the Kalman filter to the local tangent error equations of a non-linear system provides an approximation to the optimal data assimilation method. This non-linear extension of the KF is referred to as the extended Kalman filter (EKF) (Ghil *et al.*, 1981; Miller *et al.*, 1994; Ide and Ghil, 1997; Ghil, 1997).

Unfortunately, the Kalman filter and the extended Kalman filter require statistical description of the forecast error in the form of the error covariance; obtaining the required error covariance involves integrating a system with dimension equal to the square of the dimension of the forecast system. Direct integration of a system of such high dimension is not feasible. Attempts to circumvent this difficulty (see review of Ghil, 1997) have involved various approximations to the error covariance (Tippett *et al.*, 2000; Bishop *et al.*, 2001) and approximate integration methods (Evensen, 1994; Dee,1995; Fukumori and Malanotte-Rizzoli, 1995; Cohn and Todling, 1996; Verlaan and Heemink, 1997; Houtekamer and Mitchell, 1998; Hamill, this volume; Kalnay *et al.*, this volume).

While the formal dimension of the forecast error system obtained by linearising the forecast model about a base trajectory is the same as that of the forecast system itself, there are reasons to believe that the effective dimension is far lower. The trajectory of the system state in a high dimensional dynamical system typically lies on a small dimensional subspace of the entire phase space. In chaotic systems all initial conditions approach this attractor, which can be embedded in a space of dimension at most $2d + 1$, where d is the attractor dimension (Takens, 1981). An estimate of the attractor dimension can be made from the number of positive Lyapunov exponents (the Kaplan–Yorke dimension; Kaplan and Yorke, 1979), but in any case the attractor dimension is bounded above by the number of Lyapunov exponents associated with positive volume growth along the system trajectory in phase space (Illyashenko, 1983). While this is useful conceptually for bounding the dimension of the embedding space, identifying the subspace itself is more difficult in the case of non-linear and time-dependent systems. However, in the case of stochastically forced linear normal systems the analogous subspace to which the solution is primarily confined can be easily found by eigenanalysis of the covariance matrix of the system forced white in space and time. The resulting empirical orthogonal function (EOF) spectrum typically falls off rapidly in physical models. The eigenvectors may be identified with the modes of the normal operator, and the corresponding eigenvalues are the variance accounted for by the modes (North, 1984; Farrell and Ioannou, 1996 (henceforth FI96)). The fact that a restricted number of EOFs account for nearly all of the variance in normal systems shows that the effective dynamical dimension of these systems is small compared with the dimension of their phase space. This notion of quantifying the effective dimension of normal linear systems can be extended to bound the effective dimension of non-normal systems (Farrell and Ioannou, 2001a (henceforth FI01)).

In the case of the tangent linear forecast error system, the spectrum of optimal perturbations of the error propagator over the forecast interval typically comprises a few hundred growing structures (Buizza and Palmer, 1995) and Lyapunov spectra for error growth have shown similar numbers of positive exponents (Palmer *et al.*, 1998) which suggests from the above considerations that the effective dimension of the error system for scales resolved by forecast models is $O(10^3)$.

The problem of reducing the order of a linear dynamical system can be cast mathematically as that of finding a finite dimensional representation of the dynamical system so that the Eckart–Schmidt–Mirsky (ESM) theorem (Stewart and Sun, 1990) can be applied to obtain an approximate truncated system with quantifiable error. The ESM theorem states that the optimal k order truncation of an n dimensional matrix in the euclidean or Frobenius norm is the matrix formed by truncating the singular value decomposition of the matrix to its first k singular vectors and singular values. A method for exploiting the ESM theorem to obtain a reduced order approximation to a dynamical system was developed in the context of controlling lumped parameter engineering systems and is called balanced truncation (Moore, 1981; Glover, 1984; Zhou and Doyle, 1998). Balanced truncation was applied to the set of ordinary differential equations approximating the partial differential equations governing perturbation growth in time-independent atmospheric flows by FI01.

We first review the method of balanced truncation and then illustrate it with a simple matrix example. Then we apply it to a storm track model (Farrell and Ioannou, 2001b). We then review some salient aspects of optimal state estimation using an analysis method based on an observer model of the assimilation system and discuss the structure of the gain matrix in the presence of model error and the asymptotic behaviour of the assimilation error as the number of observations increases. We finally construct a reduced order Kalman filter based on balanced truncation and apply it to a time-dependent Lyapunov unstable quasi-geostrophic model of a forecast tangent linear error system with which we examine the approach of an approximation to the optimal observer based on 4D-Var.

8.2 The storm track model

Consider an idealised model of the midlatitude storm track consisting of a Boussinesq atmosphere with constant stratification and constant shear in thermal wind balance on a β-plane channel with periodic boundary conditions in the zonal, x, direction; solid walls located at two latitudes in the meridional, y, direction and a solid lid at height $z = H$, simulating the tropopause. The observed zonal localisation of a midlatitude storm track is simulated in the model by terminating the channel with a linear damping modelling the storm track exit region. The stability properties of such a storm track model are discussed in FI96.

Zonal and meridional lengths are non-dimensionalised by $L = 1200$ km; vertical scales by $H = fL/N = 12$ km; velocity by $U_0 = 50$ m/s; and time by $T = L/U_0$, so that a time unit is approximately 6.7 h. The Brunt–Vaisala frequency is $N = 10^{-2}\,\mathrm{s}^{-1}$, and the Coriolis parameter is $f = 10^{-4}\,\mathrm{s}^{-1}$. The corresponding non-dimensional value of the planetary vorticity gradient is $\beta = 0.46$.

The non-dimensional linearised equation which governs evolution of streamfunction perturbations is

$$\frac{\partial \nabla^2 \psi}{\partial t} = -U(z)\nabla^2 D\psi - \left(\beta - \frac{d^2U(z)}{dz^2}\right) D\psi - \nabla(r(x)\nabla\psi), \tag{8.1}$$

in which the perturbation is assumed to be in the form $\psi(x, z, t)e^{ily}$, where l is the meridional wave number; $\nabla^2\psi$ is the perturbation potential vorticity, with $\nabla^2 \equiv \partial^2/\partial x^2 + \partial^2/\partial z^2 - l^2$; and $D \equiv \partial/\partial x$. The perturbation potential vorticity damping rate $r(x)$ is taken to vary smoothly in the zonal direction with form:

$$r(x) = \frac{\mu}{2}\left[2 - \tanh\left(\frac{x - \pi/4}{\delta}\right) + \tanh\left(\frac{x - 7\pi/2}{\delta}\right)\right], \tag{8.2}$$

in which parameters controlling the maximum damping rate and the width of the damping region have been chosen to be $\mu = 5$ and $\delta = 1.5$, respectively. The mean velocity profile is $U(z) = 0.2 + z$. The zonal extent of the re-entrant channel is $0 < x < 4\pi$; latitudinal walls are located at $y = 0$ and $y = 1$, and the ground and tropopause boundaries are located at $z = 0$ and $z = 1$, respectively. In the following we consider perturbations with $l = 1$. A cross-section of the idealised storm track at a given latitude is shown in Figure 8.1. Conservation of potential temperature at the ground and tropopause provides the boundary conditions

$$\frac{\partial^2 \psi}{\partial t \partial z} = -U(0)D\frac{\partial \psi}{\partial z} + U'(0)D\psi - r(x)\frac{\partial \psi}{\partial z} - \Gamma_g(D^2 - l^2)\psi \quad \text{at } z = 0, \tag{8.3}$$

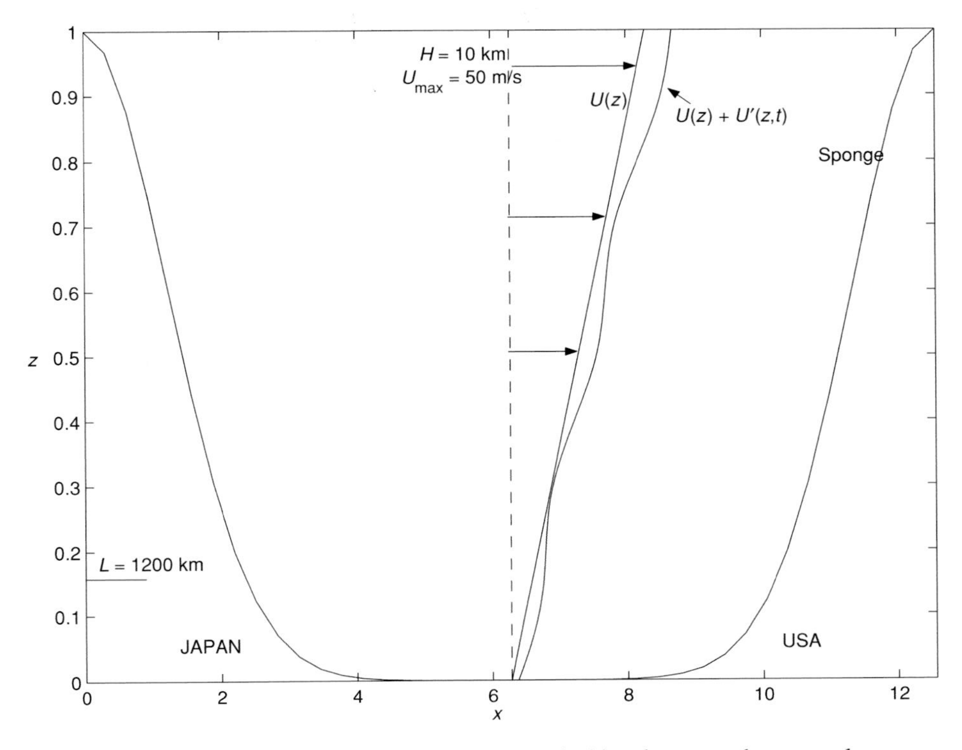

Figure 8.1 The cross-section of the storm tack. Also shown are the sponge layers.

and

$$\frac{\partial^2 \psi}{\partial t \partial z} = -U(1) D \frac{\partial \psi}{\partial z} + U'(1) D \psi - r(x) \frac{\partial \psi}{\partial z} \quad \text{at } z = 1, \tag{8.4}$$

where $U'(0)$ and $U'(1)$ denote the velocity shear at $z = 0$ and $z = 1$ respectively. The coefficient of Ekman damping

$$\Gamma_g \equiv \frac{N}{U_0} \sqrt{\frac{\nu}{2f}}$$

is given the value $\Gamma_g = 0.0632$ corresponding to a vertical eddy momentum diffusion coefficient $\nu = 20$ m^2/s in the boundary layer.

The waves evolve with nearly zero damping in the middle third of the channel (a length of $2\pi L \approx 7500$ km) which models the core of the storm track. Because in this model absolute instabilities do not exist in flows that are westerly everywhere, the storm track is asymptotically stable for all meridional wave numbers because all perturbations are eventually absorbed on entering the highly dissipative sponge (FI96).

Two scenarios are investigated. In the first a transiently growing disturbance excited near the western boundary of the storm track is modelled using the reduced order system, the purpose being to illustrate the accuracy of the reduced order model approximation of the autonomous dynamics. In the second, time dependence is added to produce a Lyapunov unstable model of a tangent linear forecast error system, the time mean operator remaining stable, with the purpose of evaluating the accuracy of the Kalman filter obtained by the reduced order model in an unstable time dependent system. Such an unstable time dependent system provides an even more stringent test of the state estimator than does the time independent stable and unstable model error systems studied by Todling and Ghil (1994), Ghil and Todling (1996) and Cohn and Todling (1996).

The perturbation dynamics of the time mean storm track are governed by

$$\frac{d\psi}{dt} = \mathbf{A}\psi, \tag{8.5}$$

where

$$\mathbf{A} = (\nabla^2)^{-1}(-(0.2 + z)D\nabla^2 - \beta D - \nabla(r(x)\nabla), \tag{8.6}$$

in which the Helmholtz operator, ∇^2, has been made invertible by incorporating the boundary conditions[1].

The dynamical operator is approximated spectrally in the zonal direction and with finite differences in the vertical. With 40 zonal harmonics and 10 levels in the vertical the resulting dynamical system has $N = 400$ degrees of freedom.

8.3 Reducing the model order by balanced truncation

Although this storm track model is of small enough dimension for direct numerical solution, we are interested in using it to explore the accuracy of approximate solutions obtained using reduced order models that could be implemented in far larger systems such as arise in numerical forecast.

Before proceeding with the order reduction we must first choose the norm that will be used to measure the accuracy of the approximation. The accuracy is measured by the norm of the Euclidean length of the errors incurred in a chosen variable. This norm is the square root of the euclidean inner product in this variable. If another norm is selected to measure the accuracy of the approximation, the most direct method of accounting for this choice is to transform the variable used to represent the state of the system so that the Euclidean inner product in the transformed variable corresponds to the new norm. The reduced order approximate system resulting from balanced transformation will in general depend on the norm chosen. As discussed in FI01, optimal order reduction of dissipative stable normal systems is immediate: it is Galerkin projection of the dynamics onto the least damped modes. Difficulties in the reduction process arise when the system is non-normal in the variable corresponding to the chosen norm. The Galerkin projection on the least damped modes is suboptimal and the reduction must proceed by including in the retained subspace the distinct subspaces of the preferred excitations and preferred responses of the system. Throughout this chapter we have chosen streamfunction as the error variable, the root-mean-square of which is to be minimised in the construction of the model order reduction. However, we find that the results do not change qualitatively if the energy norm is chosen instead.

The preferred structures of response of a general non-normal system are revealed by stochastically forcing the system with spatially and temporally uncorrelated unitary forcing and calculating the eigenfunctions of the resulting mean covariance matrix $\mathbf{P} = \langle \psi \psi^\dagger \rangle$ (the brackets denote an ensemble average, and $\dagger$ the Hermitian transpose of a vector or a matrix). The covariance matrix under such forcing is given by

$$\mathbf{P} = \int_0^\infty e^{\mathbf{A}t} e^{\mathbf{A}^\dagger t} dt, \tag{8.7}$$

and this integral is readily calculated by solving the Lyapunov equation (FI96)

$$\mathbf{AP} + \mathbf{PA}^\dagger = -\mathbf{I}, \tag{8.8}$$

which $\mathbf{P}$ satisfies, as can be easily verified. The Hermitian and positive definite matrix $\mathbf{P}$ characterises the response of the system, and its orthogonal eigenvectors, ordered in decreasing magnitude of their eigenvalue, are the empirical orthogonal functions (EOFs) of the system under spatially and temporally uncorrelated forcing.

In control literature the covariance matrix $\mathbf{P}$ is called the controllability Gramian and it is given an alternative interpretation as a measure of the efficiency by which forcings place the system in a given state. This alternative deterministic interpretation of $\mathbf{P}$ is very useful in predictability and stems from the observation that if we force the system

$$\frac{d\psi}{dt} = \mathbf{A}\psi + \mathbf{f}(t) \tag{8.9}$$

from $t = -\infty$ to $t = 0$, with initial condition $\psi(-\infty) = 0$ then all the states that can be reached at $t = 0$ with square integrable forcings satisfying

$$\int_{-\infty}^0 \mathbf{f}^\dagger \mathbf{f} dt \leq 1 \tag{8.10}$$

are exactly

$$\mathbf{P}^{1/2}\psi \tag{8.11}$$

with $||\psi|| \leq 1$ (for a proof see Dullerud and Paganini (2000) or Farrell and Ioannou (2005)). For example, let $\mathbf{v}_i$ be the eigenfunctions of $\mathbf{P}$ with eigenvalues λ_i ordered in descending order in their magnitude, i.e. $\lambda_1 > \lambda_2 > \ldots$. Then the top eigenfunction $\mathbf{v}_1$, in which as we have seen most of the response of the system is concentrated, is also the state that can be most easily forced in the sense that, given forcings of unit amplitude, the largest state that can result at $t = 0$ is $\sqrt{\lambda_1}\mathbf{v}_1$. In this way the eigenfunctions of the covariance matrix split the state space into an orthonormal basis that identifies the likelihood of occurrence of a given state. This is the reason that in predictability studies in which the impact of uncertain initial conditions is investigated, the initial states are conditioned by the covariance or Mahalanobis metric (Palmer *et al.*, 1998):

$$\| \psi \|_M^2 \equiv \psi^\dagger \mathbf{P}^{-1} \psi. \tag{8.12}$$

A geometric interpretation of these states is offered by the controllability ellipsoid

$$\psi^\dagger \mathbf{P}^{-1} \psi = 1, \tag{8.13}$$

which has semi-major axes in the directions of $\mathbf{v}_i$ with length $\sqrt{\lambda_i}$.

The preferred structures of excitation of the system are determined from the stochastic optimal matrix

$$\mathbf{Q} = \int_0^\infty e^{\mathbf{A}^\dagger t} e^{\mathbf{A}t} dt, \tag{8.14}$$

the orthogonal eigenvectors of which, when ordered in decreasing magnitude of their eigenvalue, rank the forcing structures according to their effectiveness in producing the statistically maintained variance (for a deterministic interpretation of **Q** see FI01). The eigenvectors of **Q** are called the stochastic optimals (SOs) and because of the non-normality of the system are distinct from the EOFs. The stochastic optimal matrix **Q** satisfies the back Lyapunov equation

$$\mathbf{A}^\dagger \mathbf{Q} + \mathbf{Q}\mathbf{A} = -\mathbf{I}. \tag{8.15}$$

The stochastic optimal matrix **Q** is called the observability Gramian in control literature and is given an alternative deterministic interpretation. Let the system

$$\frac{d\psi}{dt} = \mathbf{A}\psi, \tag{8.16}$$

with $t = 0$, be at state ψ_0. The states of the system

$$\psi(t) = e^{\mathbf{A}t} \psi_0, \tag{8.17}$$

produced by this initial condition have square integral norm:

$$\int_0^\infty \psi^\dagger(t)\psi(t) dt = \psi_0^\dagger \mathbf{Q} \psi_0. \tag{8.18}$$

The eigenvectors of **Q**, $\mathbf{u}_i$ when ordered in decreasing magnitude of their eigenvalue, μ_i, rank the initial conditions in effectiveness in producing square integrable output, and the observability ellipsoid

$$\psi^\dagger \mathbf{Q}^{-1} \psi = 1, \tag{8.19}$$

which has semi-major axes in the directions of $\mathbf{u}_i$ with length $\sqrt{\mu_i}$, identifies the initial conditions that produce maximum square integrable output, or equivalently, orders the initial states in the degree that they can be identified from observations of the system $\psi(t)$.

Lyapunov equations (8.8) and (8.15) have unique positive definite solutions $\mathbf{P}$ and $\mathbf{Q}$ if $\mathbf{A}$ is stable. If the operator $\mathbf{A}$ is not stable, $\mathbf{P}$ and $\mathbf{Q}$ can be obtained by stabilising the operator by adding Rayleigh friction. Alternatively, finite time horizon $\mathbf{P}$ and $\mathbf{Q}$ matrices can be used to obtain a reduced order system that best approximates the dynamics over a finite time interval. In any case the covariance matrix $\mathbf{P}$ and stochastic optimal matrix $\mathbf{Q}$ or an approximation to these matrices need to be determined or approximated in order to proceed with order reduction by balanced truncation.

For general non-normal systems the observability and controllability ellipsoids are distinct. A successful order reduction must accurately approximate the dynamics by including in the truncation both the directions of the system's response (the dominant eigenfunctions of $\mathbf{P}$) and also the directions in which, when the system is forced, it most effectively responds (the dominant eigenfunctions of $\mathbf{Q}$). The fact that the observability and the controllability ellipsoids are distinct is an indication that the directions of greatest response of the system are different from the directions in which it is most effectively forced. If we can identify a coordinate transformation in which the controllability and observability ellipsoids are the same, then in these balanced coordinates, reduction of the order of the system can proceed by retaining the dominant directions of the common ellipsoid. The semi-major axes of this common ellipsoid in the balanced representation are the Hankel singular vectors, and the lengths of the semi-major axes are the Hankel singular values which turn out to be the square root of the eigenvalues of the product of the covariance and stochastic matrix, $\mathbf{PQ}$. The balanced truncation thus transforms the internal coordinates of the system so that the transformed covariance matrix $\mathbf{P}$ and stochastic optimal matrix $\mathbf{Q}$ become identical and diagonal (while preserving the inner product of the physical variables). The dynamical system is then truncated in these transformed balanced coordinates. The balanced truncation retains a leading subset of empirical orthogonal functions and stochastic optimals of the dynamical system and preserves the norm. Balanced truncation preserves the stability of the full system and provides an approximation with known error bounds which is found in practice to be nearly optimal (Moore, 1981; Glover, 1984; FI01) as will now be shown.

A successful order reduction must accurately approximate the dynamics of the system, which can be expressed as the mapping of all past (square integrable) forcings to all future responses. This linear mapping of inputs to outputs is called the Hankel operator. Application of the ESM theorem to the Hankel operator provides the optimal low order truncation of the dynamics. Remarkably, because of the separation between past forcings and future responses in the Hankel operator representation of the dynamics, this operator has finite rank equal to the order of the system; and its singular values, denoted by h, turn out to be the lengths of the semi-major axes of the balanced controllability–observability ellipsoid.

The procedure used to implement balanced truncation is now briefly reviewed. Consider a general k order truncation of the N dimensional system (8.5):

$$\frac{d\tilde{\psi}_k}{dt} = \mathbf{A}_k \tilde{\psi}_k, \tag{8.20}$$

where $\mathbf{A}_k$ is the reduced $k \times k$ dynamical matrix, with $k < N$, and $\tilde{\psi}_k$ the associated reduced order k-dimensional state vector which is related to the full state vector by the transformation $\tilde{\psi} = \mathbf{X}\tilde{\psi}_k$. Similarly, the reduced state vector $\tilde{\psi}_k$ is related to the full state vector by $\tilde{\psi}_k = \mathbf{Y}^\dagger \tilde{\psi}$ (the dagger denotes the hermitian transpose of a matrix), which implies that $\mathbf{Y}^\dagger \mathbf{X} = \mathbf{I}_k$, where $\mathbf{I}_k$ is the k-order identity matrix. Matrices $\mathbf{Y}$ and $\mathbf{X}$ determine the transformation from the full system to the reduced system. The matrix $\mathbf{A}_k$, governing the dynamics in (8.20), is

$$\mathbf{A}_k = \mathbf{Y}^\dagger \mathbf{A} \mathbf{X}. \tag{8.21}$$

Details of the construction on the biorthogonal matrices $\mathbf{X}$ and $\mathbf{Y}$ are given in Farrell and Ioannou (2001b).

A measure of the accuracy of the truncation is the maximum difference that can occur between the full system response, $\psi(t)$, and the reduced order system response, $\tilde{\psi}(t)$. This measure is the H_∞ norm of the error system:

$$\|\mathbf{A} - \mathbf{A}_k\|_\infty = \sup_\omega \|\mathbf{R}(\omega) - \tilde{\mathbf{R}}(\omega)\|_2, \tag{8.22}$$

in which the resolvent of the full system, $\mathbf{R}(\omega)$, is defined as $\mathbf{R}(\omega) = (i\omega\mathbf{I} - \mathbf{A})^{-1}$ and the resolvent of the full order projection of the reduced system is $\tilde{\mathbf{R}}(\omega) = \mathbf{X}(i\omega\mathbf{I}_k - \mathbf{A}_k)^{-1}\mathbf{Y}^\dagger$. It is to be recalled that the L_2 norm of a matrix, denoted as $\|\cdot\|_2$, is equal to its largest singular value.

Assuming the Hankel singular values have been ordered decreasing in magnitude, it can be shown that the error in the H_∞ norm Eq. (8.22) of the balanced approximation of the full system by any k order system $\mathbf{A}_k$ satisfies the inequality:

$$h_{k+1} \le \|\mathbf{A} - \mathbf{A}_k\|_\infty \le 2 \sum_{i=k+1}^{N} h_i, \tag{8.23}$$

where h_{k+1} is the first neglected Hankel singular value (Zhou and Doyle, 1998). Although h_{k+1} is only a lower bound on the error, we have found in examples that this lower bound is nearly attained.

8.3.1 A simple example of balanced truncation

Consider the 3×3 dynamical system

$$\frac{d\psi}{dt} = \mathbf{A}\psi, \tag{8.24}$$

with

$$\mathbf{A} = \begin{pmatrix} -0.1 & 100 & 0 \\ 0 & -0.2 & 0 \\ 0 & 0 & -0.01 \end{pmatrix}. \tag{8.25}$$

This system is stable but highly non-normal in the first coordinates whilst its third coordinate, which when excited decays the slowest, does not interact with the other coordinates. The non-normality of the system leads to substantial optimal growth as revealed by the norm of the propagator, $\|e^{\mathbf{A}t}\|$, which measures the maximum state norm that can be produced at time t, by initial states of unit norm. The optimal growth as a function of time is shown in Figure 8.2. We wish to obtain a 2×2 system that best approximates the dynamics of the original system.

We first obtain a 2×2 order reduction by Galerkin projection on the two least damped eigenmodes $\mathbf{e}_1, \mathbf{e}_2$ of $\mathbf{A}$. This is achieved as follows: form the 3×2 matrix

$$\mathbf{E} = [\mathbf{e}_1, \mathbf{e}_2] \tag{8.26}$$

and the 2×2 diagonal matrix $\mathbf{D}_2$ with diagonal elements the first two least damped eigenvalues of $\mathbf{A}$. The 2×2 reduced order matrix in coordinates that preserve the original state norm is

$$\mathbf{A}_2 = \mathbf{M}^{1/2}\mathbf{D}_2\mathbf{M}^{-1/2}, \tag{8.27}$$

where $\mathbf{M} = \mathbf{E}^\dagger\mathbf{E}$. The performance of the norm of the propagator of the modally reduced system as a function of time is shown in Figure 8.2 (curve 5), and clearly this truncation provides a very poor representation of the dynamics.

We obtain next a 2×2 order reduction by Galerkin projection on the top two optimal vectors of the propagator for time $t = 5$ when the global optimal growth is achieved. This reduction is achieved as follows. Form the matrix

$$\mathbf{V} = [\mathbf{v}_1, \mathbf{v}_2], \tag{8.28}$$

where $\mathbf{v}_1$, $\mathbf{v}_2$ are the optimal vectors, and then the matrix $\mathbf{A}_2 = \mathbf{V}^\dagger\mathbf{A}\mathbf{V}$, which is the matrix representation in the coordinates ϕ, in which $\psi = \mathbf{V}\phi$. Then the 2×2 reduction in this basis is $\mathbf{A}_2$. Because the basis vectors are orthonormal the square norm of the transformed states ϕ is equal to the square norm of the original states. This procedure can be followed to obtain order reduction by Galerkin projection on any set of orthonormal vectors. The performance of the reduced system in this basis is poor, as shown in Figure 8.2 (curve 2). Note that the reduced order system is unstable. Selecting as a basis the singular vectors for time $t = 3$ leads to even worse performance (curve 3 in Figure 8.2). The same results would have been obtained if we had used as a basis the corresponding evolved optimals (the left singular vectors). In general it is found that if the singular vectors are used for order reduction it is best to use the singular vectors or the evolved optimals for a sufficiently long time. Short time optimals can be very suboptimal as a basis for truncation, because

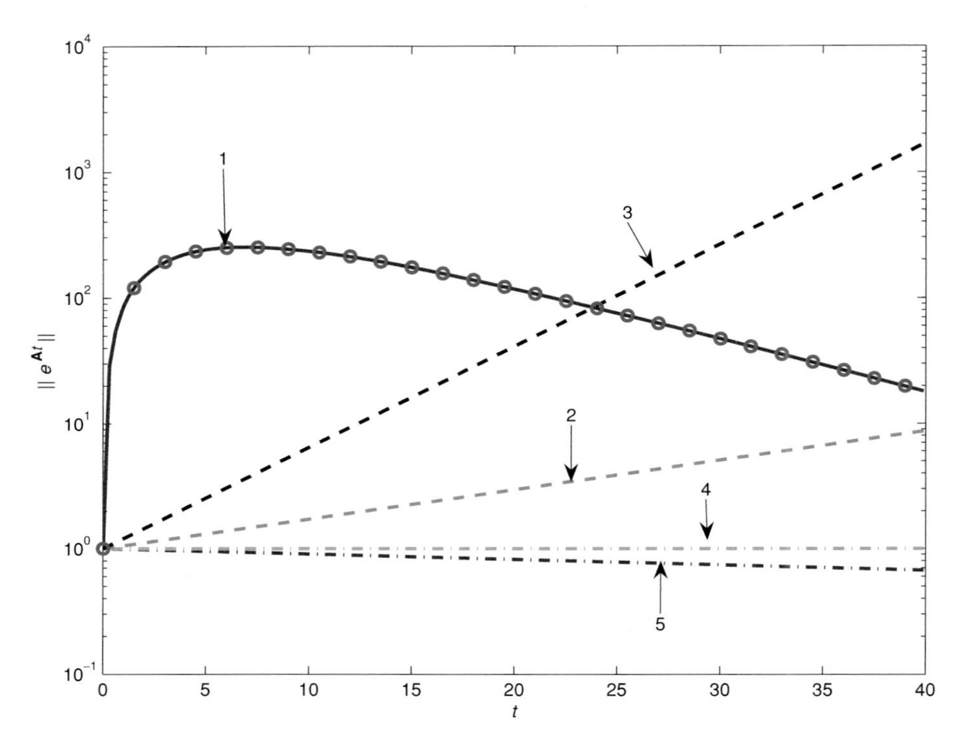

Figure 8.2 The norm of the propagator $\|e^{\mathbf{A}t}\|$ as a function of time for various order 2 approximations of the dynamics. Curve 1: the optimal growth of the original system (8.24). The circles are the optimal growth of the order 2 system obtained with balanced truncation. This reduced system captures accurately the optimal growth of the full 3×3 system. Curve 2: the optimal growth obtained from a 2×2 truncation obtained by Galerkin projection on the two optimal vectors associated with the two largest singular values of the propagator for time $t = 5$. The order 2×2 operator that results is unstable. Curve 3: same as 2 but the optimal vectors are obtained for time $t = 3$. Curve 4: the optimal growth obtained from 2×2 truncations obtained by Galerkin projections on the two eigenvectors associated with the two largest eigenvalues of the covariance matrix **P** *or the matrix* **Q** (both truncations give identical growths). Curve 5: the optimal growth obtained from a 2×2 truncation obtained by a Galerkin projection on the two least damped eigenmodes of the matrix **A**. The performance of all the truncated systems, except the one obtained by balanced truncation, is very poor.

these vectors are often associated with directions of rapid growth that does not persist.

We reduce the order of the system by Galerkin projection on the top two eigenvectors of the covariance matrix **P**. The performance is very poor (see Figure 8.2, curve 4). The same poor performance is obtained if we use the top two eigenvectors of the covariance matrix **Q**. The reason for the failure is that the controllability ellipsoid $\mathbf{x}^T \mathbf{P}^{-1} \mathbf{x} = 1$ is elongated in the direction of $\psi = [1, 10^{-3}, 0]^T$ with semi-major axis of about 900 (Figure 8.3). The second largest direction is the normal direction $\psi = [0, 0, 1]^T$ with semi-major axis length of about 7. This second direction is the

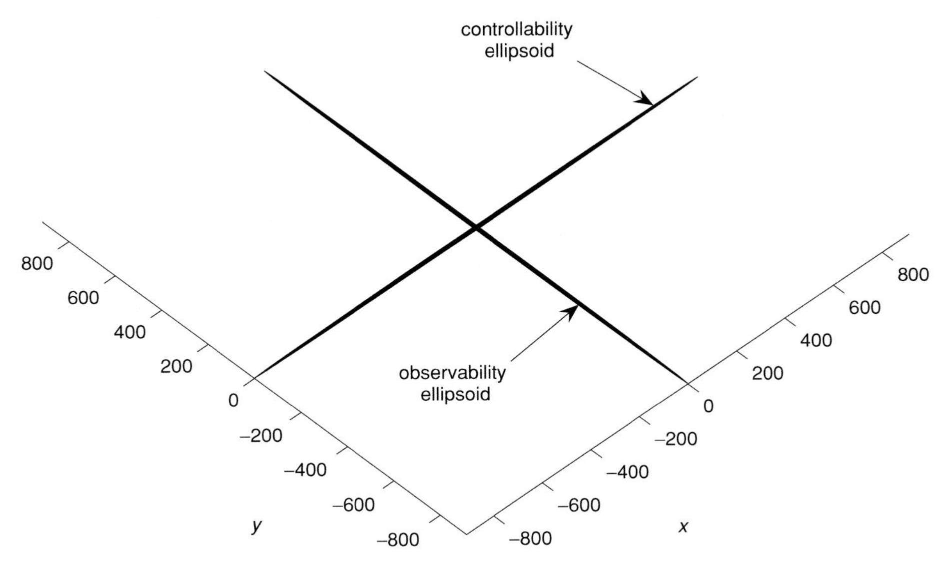

Figure 8.3 The two ellipsoids associated with the covariance matrix **P** and the stochastic optimal matrix **Q**. The ellipsoids are respectively $\mathbf{x}^T\mathbf{P}^{-1}\mathbf{x} = 1$ (the controllability ellipsoid) and $\mathbf{x}^T\mathbf{Q}^{-1}\mathbf{x} = 1$ (the observability ellipsoid) and have semimajor axes proportional to the square roots of the eigenvalues of the matrices **P** and **Q**. The ellipsoids are like needles. The elongated direction of the **P** ellipsoid indicates that the dominant response of the system is in that direction, while the elongated direction of the **Q** ellipsoid identifies the forcing directions in which the system responds most readily. These two directions are different. A good reduction of the order of the dynamics must include both of these directions. This is systematically achieved with balanced truncation.

normal direction, which because of the relatively slow decay persists for a long time when it is excited. The observability ellipsoid $\mathbf{x}^T\mathbf{Q}^{-1}\mathbf{x} = 1$ is elongated in the direction of $\psi = [2 \times 10^{-3}, 1, 0]^T$ with semi-major axis of about 900 (Figure 8.3). The second largest direction is again the normal direction $\psi = [0, 0, 1]^T$ with semi-major axis length of about 7. It is thus clear that retaining the two dominant directions of either the **P** or the **Q** matrix doesn't retain the two dominant directions of both the **P** and the **Q** matrix that are both necessary for a good description of the dynamics.

The transformation

$$\mathbf{X} = \begin{pmatrix} -15.6 & -15.6 & 0 \\ -0.02 & 0.04 & 0 \\ 0 & 0 & 1 \end{pmatrix}, \tag{8.29}$$

and its associated biorthogonal

$$\mathbf{Y} = \begin{pmatrix} -0.04 & -0.020 & 0 \\ -15.6 & 15.5 & 0 \\ 0 & 0 & 1 \end{pmatrix}, \tag{8.30}$$

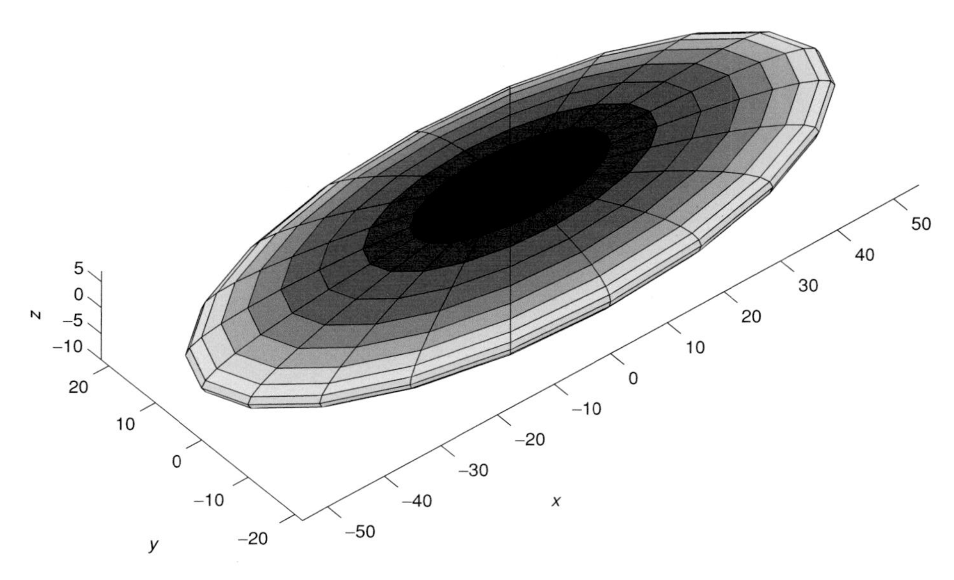

Figure 8.4 The ellipsoid of the **P** and **Q** matrices in the balanced coordinates. These coordinates are constructed so that the transformed covariance and the optimal matrix ellipsoids become identical. In these coordinates the system can be most effectively truncated because the directions of the system's dominant response coincide with the directions in which the system is most easily forced. The length of the semimajor axes of this ellipsoid are the square roots of the Hankel singular values.

renders both the covariance matrix $\tilde{\mathbf{P}} = \mathbf{Y}^\dagger \mathbf{P} \mathbf{Y}$ and the stochastic optimal matrix $\tilde{\mathbf{Q}} = \mathbf{X}^\dagger \mathbf{Q} \mathbf{X}$ diagonal and equal to each other. This associated common ellipsoid (Figure 8.4) in the balanced coordinates has semi-major axes equal to the square roots of the Hankel singular values, namely approximately equal to 54, 21, 7. Balanced order reduction proceeds in this coordinate system by retaining the two directions that are associated with the two top Hankel singular values, which are to a very good approximation the first and second direction. In this way the reduced order 2×2 balanced system includes the dominant directions of both the controllability and observability ellipsoids and is expected to be a near optimal reduction of order. This is shown in Figure 8.2 in which the circles that give the optimal growth of the propagator of the balanced system reproduce exactly the optimal growth of the original system. However, the accuracy of the reduced system can be best examined by considering the difference of the exact and reduced systems by evaluating the norm of the difference of the resolvents of the two systems as a function of frequency ω. This difference is plotted as a function of frequency in Figure 8.5. The maximum of this difference over all frequencies defines the H_∞ norm of the error which has been shown to satisfy inequality (8.23), which in this case becomes

$$50 \leq \| \mathbf{A} - \mathbf{A}_2 \|_\infty \leq 100, \tag{8.31}$$

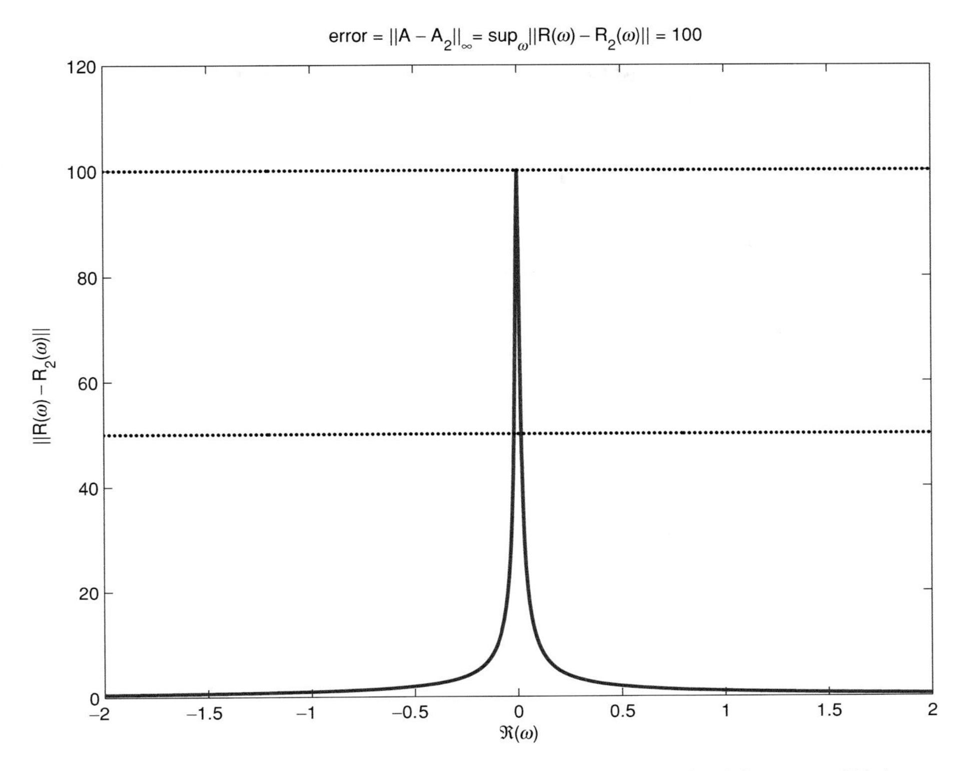

Figure 8.5 The norm of $\| \mathbf{R}(\omega) - \mathbf{R}_2(\omega) \|$ as a function of real frequency, $\Re(\omega)$, where $\mathbf{R}_2(\omega) = \mathbf{X}(i\omega\mathbf{I}_2 - \mathbf{A}_2)^{-1}\mathbf{Y}^\dagger$ is the resolvent of the 2×2 system obtained from balanced truncation. This maximum of this norm over all frequencies defines the H_∞ norm of the error system, which provides the greatest error that can be produced in the truncated system when it is forced by monochromatic sinusoidal forcing. The two horizontal lines indicate the bounds that the balanced truncation error must satisfy in this measure. The lower dotted line is the first neglected Hankel singular value $h_3 = 50$, the top is $2h_3$. The H_∞ norm of the error assumes the upper bound of inequality (8.31). This error is realised when the system is forced at zero frequency ($\omega = 0$) with the structure of the most persistent mode of the system $[0, 0, 1]^T$.

given that the neglected Hankel singular value is $h_3 = 50$. The error system is found in this way to perform worst at $\omega = 0$ when forced with the structure of the neglected persistent mode $[0, 0, 1]^T$.

8.4 Applying balanced truncation to the mean storm track perturbation model

In order to obtain a balanced truncation of the storm track model governed by operator (8.6) we first obtain the covariance matrix, $\mathbf{P}$, and the stochastic optimal matrix, $\mathbf{Q}$, by solving Lyapunov equations (8.8) and (8.15) respectively. The eigenfunction of $\mathbf{P}$

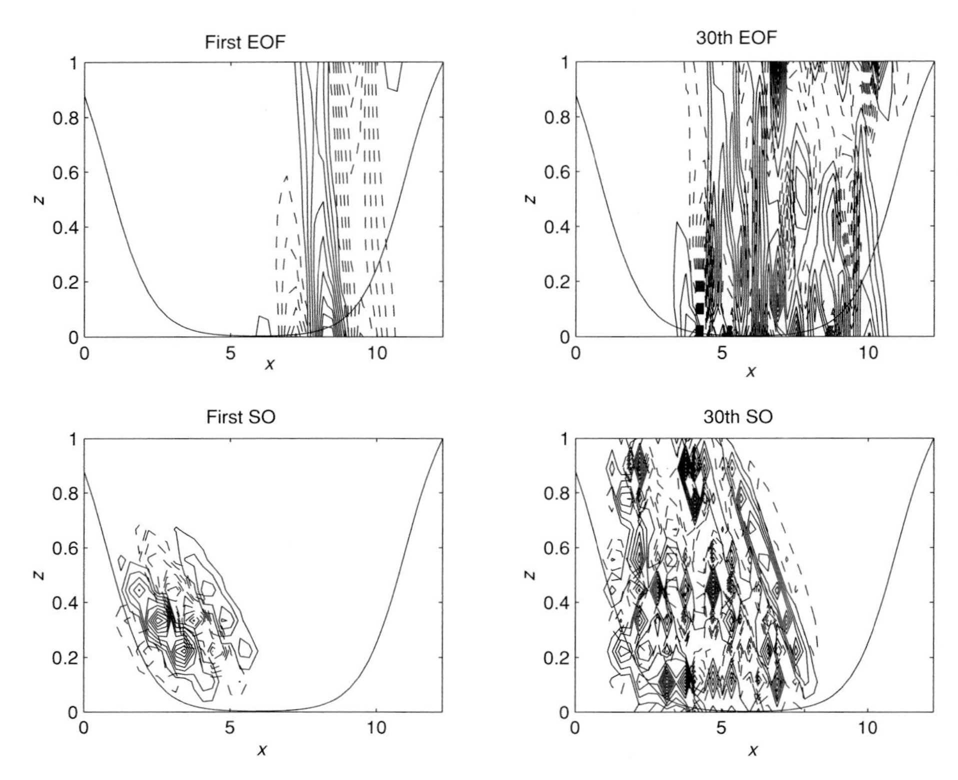

Figure 8.6 For the stable time mean storm track model. Top panels: the stream function of the first and the thirtieth EOF. The first EOF accounts for 23% of the maintained variance, the thirtieth EOF accounts for 0.35% of the variance. Bottom panels: the structure of the streamfunction of the first and thirtieth stochastic optimal (SO). The first SO is responsible for producing 19.7% of the maintained variance; the thirtieth SO is responsible for producing 0.48% of the maintained variance.

associated with the largest eigenvalue is the first EOF of the perturbation field, and the eigenfunction of $\mathbf{Q}$ associated with the largest eigenvalue is the first SO of the perturbation field. The structure of the first EOF, which accounts for 23% of the streamfunction perturbation variance, is concentrated in the exit region of the storm track, as can be seen in Figure 8.6 (top left panel). By contrast, the first SO, which is responsible for generating 19.7% of the streamfunction perturbation variance, is concentrated at the entrance region of the storm track and is nearly orthogonal to the first EOF, as can also be seen in Figure 8.6 (bottom left panel). This near orthogonality between the EOF structures and SO structures remains even at order 30. Balanced truncation accomplishes an accurate representation of the dynamics by retaining both the structure of the dominant EOFs and of the SOs. It is clear from Figure 8.6 that truncations based on projections on the leading EOFs will be very suboptimal as the leading EOFs span well only perturbations concentrated in the exit region of the storm track, leaving the dynamically important entry region of the storm track, where

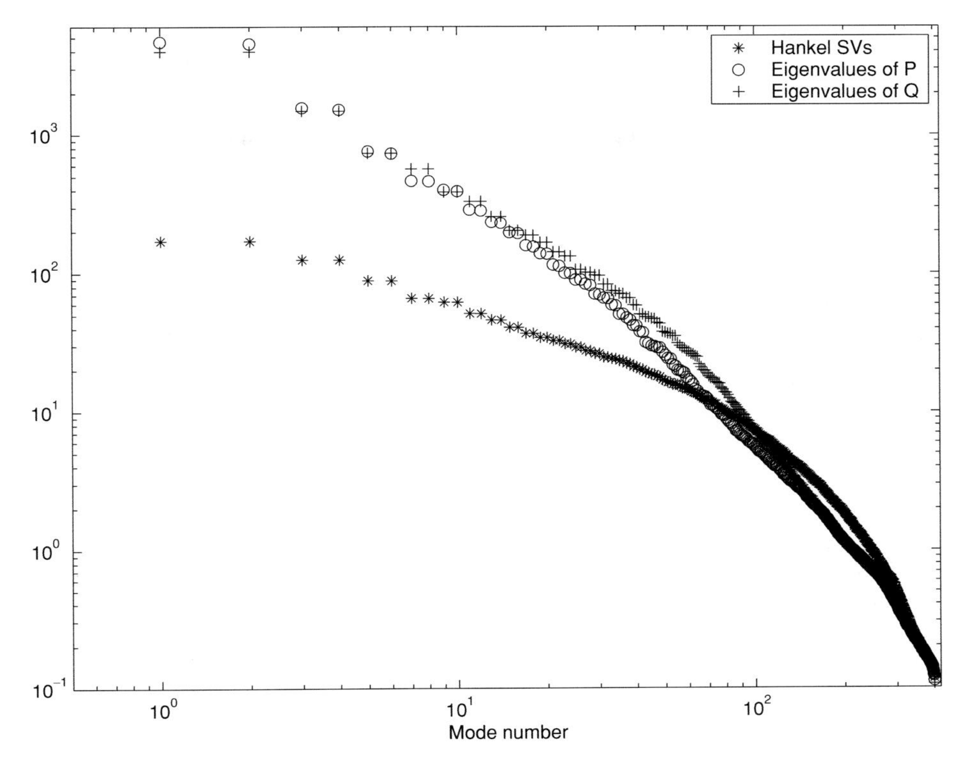

Figure 8.7 The Hankel singular values (stars) compared with the eigenvalues of the covariance matrix **P** (circles), and the eigenvalues of the stochastic optimal matrix **Q** (crosses). The Hankel singular values are the square roots of the eigenvalues of the product **PQ**. Note that the EOFs (the eigenvalues of **P**) and the SOs (the eigenvalues of **Q**) fall much more rapidly with mode number than do the Hankel singular values.

perturbations have greatest potential growth, virtually without support in the span of the retained basis.

Although the error in the frequency response of a balanced truncation (cf. 8.31) is bounded above by twice the sum of the neglected Hankel singular values and below by the first neglected Hankel singular value, experience shows balanced truncation of tangent linear forecast error systems results in truncation errors close to the lower bound. The Hankel singular values and the eigenvalues of **P** and the **Q** for the storm track model are shown in Figure 8.7. Note that the decrease with mode number of the eigenvalues of **P** and of **Q** is more rapid than that of the Hankel singular values. But this more rapid decrease with mode number of the eigenvalues of **P** and **Q** does not indicate the order required for an accurate approximation; this is instead determined by the first neglected Hankel singular value which falls more slowly with mode number.

It is often assumed that a system can be well approximated by Galerkin projection onto a subspace of its EOFs, with the effectiveness of the truncation being judged

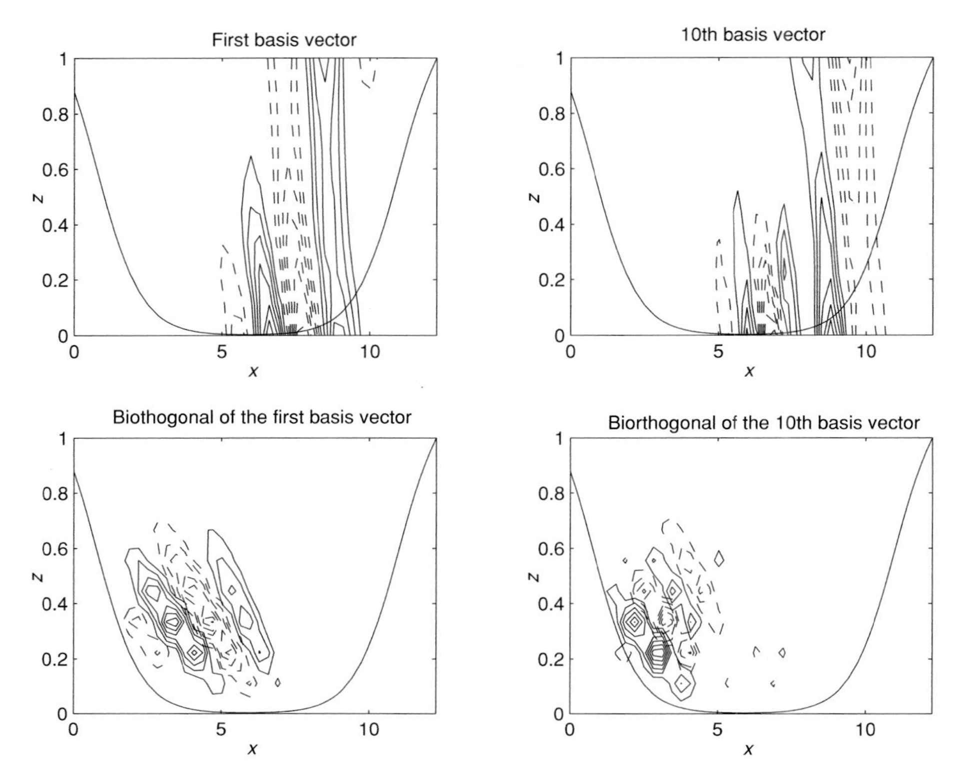

Figure 8.8 For the stable time mean storm track model. (Top left panel) the streamfunction of the first basis vector of the expansion for the balanced truncation of the system. It is given by the first column of $\mathbf{X}$. (Top right panel) the streamfunction of the tenth basis vector of the expansion for the balanced truncation of the system. It is given by the tenth column of $\mathbf{X}$. (Bottom left panel) the streamfunction of the biorthogonal of the first basis vector. It is given by the first column of $\mathbf{Y}$. (Bottom right panel) the streamfunction of the tenth basis vector. It is given by the tenth column of $\mathbf{Y}$.

from the magnitude of the eigenvalues of the neglected EOFs. This is valid only for normal systems and we see here that for non-normal systems the decrease with mode number of the eigenvalues of the covariance matrix is misleading and generally optimistic even as an estimate of the order of the system required for an accurate approximation.

A subset of the columns of $\mathbf{X}$ is retained in the balanced truncation. This non-orthogonal basis and its biorthogonal, the columns of $\mathbf{Y}$, are constructed so as to capture the structures supporting the dynamics most efficiently, simultaneously accounting for the preferred responses (EOFs) and the preferred excitations (SOs) of the dynamics. The first and the tenth structure retained in the dynamics (the first and the tenth column of $\mathbf{X}$) and their biorthogonal structures (the first and tenth column of $\mathbf{Y}$) are shown in Figure 8.8.

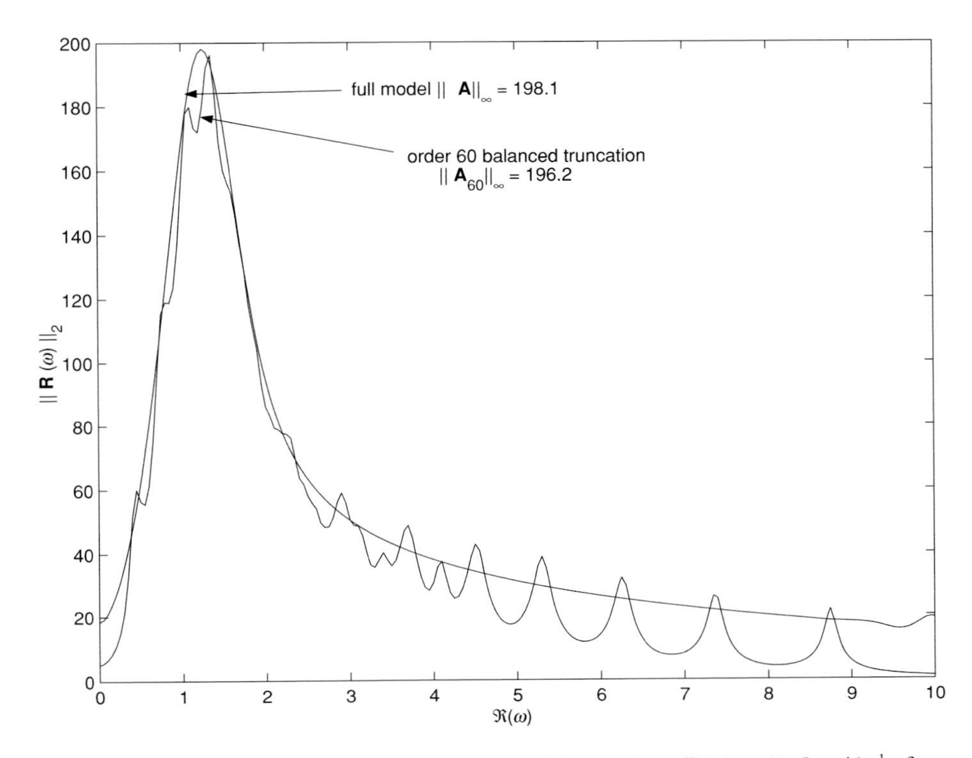

Figure 8.9 The maximum singular value of the resolvent $\mathbf{R}(\omega) = (i\omega I - \mathbf{A})^{-1}$ of the full system $\mathbf{A}$ as a function of the real frequency, $\Re(\omega)$. The maximum of this curve as a function of ω is the H_∞ norm of $\mathbf{A}$ which is found here to be 198.1. Also plotted is the maximum singular value of the resolvent associate with $\mathbf{A}_{60}$, which is the operator obtained from an order 60 balanced truncation of $\mathbf{A}$. The maximum of this curve is the H_∞ norm of $\mathbf{A}_{60}$ which is found to be 196.2.

The storm track model and its reduced order approximate have very different eigenvalue spectra. The eigenvalue spectrum of the reduced order approximate is such that the frequency response of the approximate system is as close as possible to that of the original system, which is shown in Figure 8.9. This results both from a decrease in the stability of the reduced system compared with that of the full system and from the increase in growth due to the non-normality in the reduced system.

The accuracy of the approximation is measured by the H_∞ norm of the error dynamical system $\|\mathbf{A} - \mathbf{A}_{60}\|_\infty$, which, as discussed in the previous section, lies between the lower bound given by the first neglected Hankel singular value, $h_{61} = 13.8$, and the upper bound:

$$2\sum_{i=61}^{400} h_i = 1.8 \times 10^3.$$

The largest singular value of the error system resolvent as a function of frequency is shown in Figure 8.10, where it can be seen that $\|\mathbf{A} - \mathbf{A}_{60}\|_\infty = 28.5$, which shows

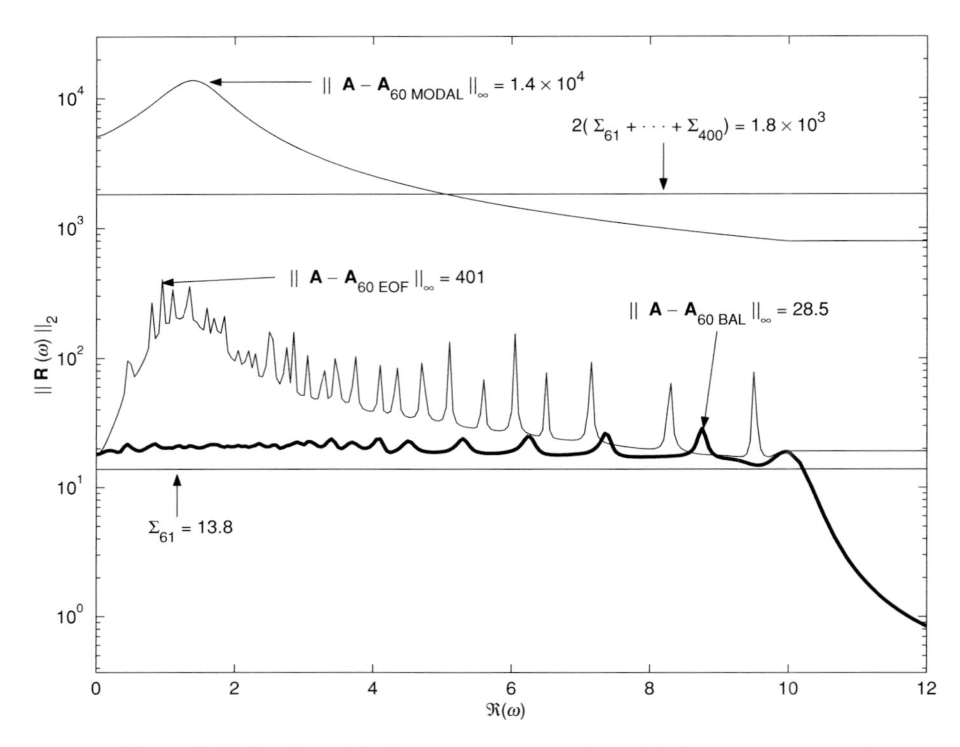

Figure 8.10 For the stable time mean storm track model: the maximum singular value of the error system $\mathbf{A} - \mathbf{A}_{60}$ as a function of real frequency, $\Re(\omega)$. The system $\mathbf{A}_{60}$ is an order 60 approximation obtained from $\mathbf{A}$ by balanced truncation. The maximum of this curve is the H_∞ error of the order 60 balanced truncation which is found here to be 28.5. Also indicated with a straight line is the theoretical minimum error of an order 60 truncation, which equals the first neglected Hankel singular value $\Sigma_{61} = 13.8$. The balanced truncation is seen to be nearly optimal.

that the balanced truncation error in this example is only approximately twice its lower bound. The error is nearly white for the span of frequencies that correspond to the frequencies of the system eigenmodes. For comparison, the error incurred in an order 60 Galerkin projection of the dynamics onto the first 60 EOFs and the error incurred in an order 60 Galerkin projection onto the first 60 least damped modes, are also shown in Figure 8.10. It can be seen that the EOF projection performs apprecia-bly worse than the balanced truncation, while the modal truncation at this order is useless.

The optimal growth[2] as a function of optimising time attained by the full system and by the following: the order 60 balanced truncation; the order 60 system obtained by Galerkin projection on the first 60 EOFs; the order 60 system obtained by Galerkin projection on the first 60 SOs; and the order 60 system obtained by Galerkin pro-jection on the first 60 least damped modes, are all shown in Figure 8.11. Note that the balanced truncation performs very well, reproducing the optimal growth nearly perfectly up to $t = 5$, corresponding to about two days. By comparison the EOF

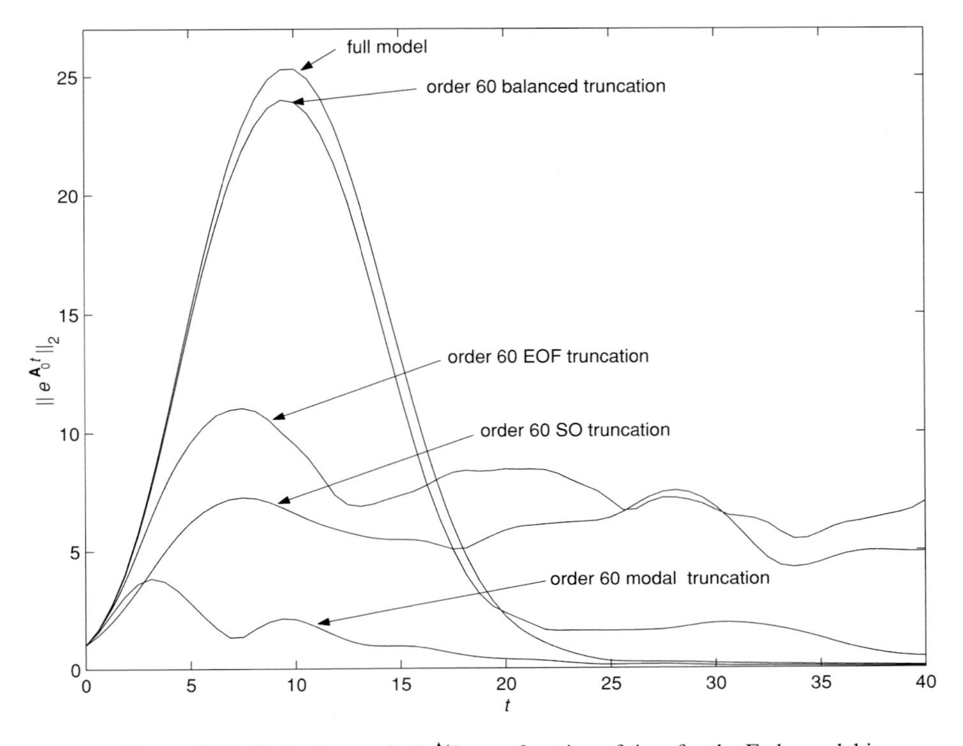

Figure 8.11 Optimal growth, $\|e^{\mathbf{A}t}\|$, as a function of time for the Eady model in a channel with sponge layers and meridional wave number $l = 1$. Shown is the optimal growth for the full system with 400 degrees of freedom and the optimal growth produced by an order 60 approximate system obtained by balanced truncation of the full system. Shown also for comparison is the optimal growth attained by the order 60 approximate system obtained by Galerkin projection on the first 60 EOFs, the first 60 SOs and the first 60 least damped modes.

and SO truncations perform appreciably worse and the modal truncation gives even poorer results.

The structure of the initial perturbation that leads to greatest square streamfunction growth at $t = 10$ in the full system, together with the resulting structure, is shown in Figure 8.12; for comparison these structures as obtained by the truncated system are also shown. The structures are well captured by the order 60 reduced system.

We have demonstrated how to obtain balanced truncation of a stable time independent system but the method of balanced truncation can be extended to unstable systems (Sznaier *et al.*, 2002) and to time dependent systems in which balancing is performed sequentially over finite time intervals (Van Dooren, 2000).

In forecast applications we seek an accurate reduction of the dynamics of the time dependent tangent linear operator calculated on the system trajectory over a limited time interval (24 or 48 hours). One choice is to balance on the time mean operator over this interval. Another choice is to balance on the time dependent

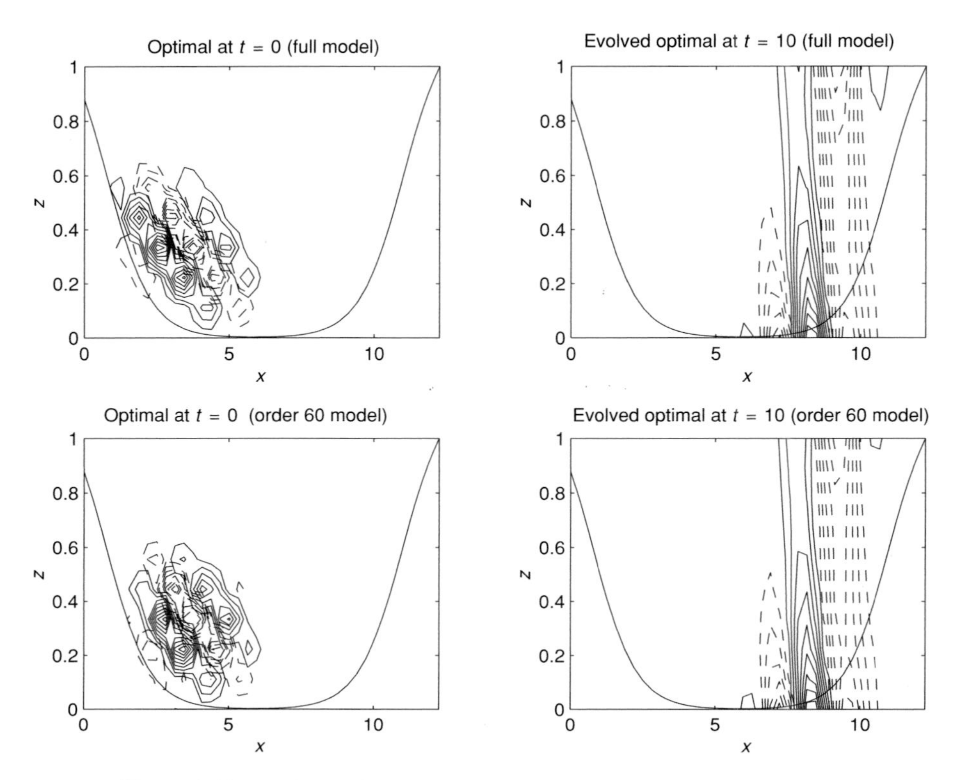

Figure 8.12 For the stable time mean storm track model. The structure of the streamfunction of the optimal perturbation that leads to the greatest energy growth at $t = 10$ (left panels), and the evolved optimal streamfunction, which is the structure that these optimals evolve into at the optimising time $t = 10$ (right panels). The top panels are for the full system while the bottom panels are for the order 60 balanced truncation.

version of the tangent linear operator over this or an extended interval about the assimilation time, obtaining approximation of the **P** and **Q** matrices on this interval. Both procedures have been tested using the time dependent version of our storm track model and found to produce accurate truncations. We examine below results obtained from a reduced order Kalman filter in which the truncation is made on the time dependent tangent linear operator over 48 hours centred on the assimilation time.

The time mean tangent linear operator (the mean being calculated over an interval) is generally asymptotically stable. This is because realistic states of the atmosphere support primarily instabilities with positive group velocities and do not support absolute instabilities (unstable modes with zero group velocity) (Farrell, 1982; Lin and Pierrehumbert, 1993; DelSole and Farrell, 1994). The asymptotic instability of the tangent linear system arises primarily from the continual instigation of transient growth which occurs in non-periodic time dependent systems in the same way that the Mathieu instability arises in time periodic systems. This mechanism is discussed in Farrell and Ioannou (1999) and has been verified in the context of a forecast system by Reynolds and Errico (1999) and Gelaro *et al.* (2000). The stability of the mean operator allows balancing to be performed on a stable operator, although the error system itself is non-autonomous and asymptotically unstable. However, it is not necessary to balance on the mean operator, and as remarked above comparable results can be obtained by balancing on the time dependent tangent linear operator over an appropriate interval; experiment suggests approximately 48 hours.

8.5 Assimilation as an observer system

Consider assimilating data taken from truth, $\mathbf{x}_\mathrm{t}$. The forecast error $\mathbf{e}_\mathrm{f} = \mathbf{x}_\mathrm{f} - \mathbf{x}_\mathrm{t}$ obeys the equation

$$\frac{d\mathbf{e}_\mathrm{f}}{dt} = \mathbf{A}\mathbf{e}_\mathrm{f} + \mathbf{Q}^{1/2}\mathbf{w}_\mathrm{m}, \tag{8.32}$$

in which $\mathbf{A}$ is the unstable tangent linear operator, $\mathbf{Q}$ is the model error covariance, and $\mathbf{w}_\mathrm{m}$ is assumed to be a vector of temporally uncorrelated noise processes.

Introduce n observations, $\mathbf{y}_\mathrm{ob}$, obtained from truth $\mathbf{x}_\mathrm{t}$ as

$$\mathbf{y}_\mathrm{ob} = \mathbf{H}\mathbf{x}_t + \mathbf{R}^{1/2}\mathbf{w}_\mathrm{o}, \tag{8.33}$$

where $\mathbf{H}$ is the observation matrix, $\mathbf{R}$ is the observational error covariance and $\mathbf{w}_o$ is an n vector of white noise processes.

Assimilate these observations to obtain an analysis, x_a, with analysis error $\mathbf{e}_\mathrm{a} = \mathbf{x}_\mathrm{a} - \mathbf{x}_\mathrm{t}$ satisfying the Luenberger observer system:

$$\begin{aligned}\frac{d\mathbf{e}_\mathrm{a}}{dt} &= \mathbf{A}\mathbf{e}_\mathrm{a} + \mathbf{K}(\mathbf{y}_\mathrm{ob} - \mathbf{H}\mathbf{x}_\mathrm{a}) + \mathbf{Q}^{1/2}\mathbf{w}_\mathrm{m} \\ &= (\mathbf{A} - \mathbf{K}\mathbf{H})\mathbf{e}_\mathrm{a} + \mathbf{K}\mathbf{R}^{1/2}\mathbf{w}_\mathrm{o} + \mathbf{Q}^{1/2}\mathbf{w}_\mathrm{m}.\end{aligned} \tag{8.34}$$

The gain, $\mathbf{K}$, is chosen to minimise the analysis error variance trace $(\langle\mathbf{e}_\mathrm{a}\mathbf{e}_\mathrm{a}^\dagger\rangle)$. Unlike the forecast error system, a Luenberger observer system is asymptotically stable. Any gain, $\mathbf{K}$, that stabilises the tangent linear operator results in an observer with bounded error, this error being forced by a combination of model error $\mathbf{Q}$ and observational error $\mathbf{R}$ (cf. Eq. 8.34). Good gains do not just stabilise the operator but simultaneously reduce the non-normality of the tangent linear operator so that the minimum of trace $(\langle\mathbf{e}_\mathrm{a}\mathbf{e}_\mathrm{a}^\dagger\rangle)$ is maintained by the combination of observational and model error.

Just as generalised stability of the tangent linear forecast system reveals the potential for forecast failures due to transient growth of initialisation error or unresolved forcings distributed over the forecast interval, so also does generalised stability analysis of the observer system reveal how model error and initialisation error contribute to analysis failures.

8.5.1 The case of an optimal observer

The gain $\mathbf{K}$ that minimises the statistical steady analysis error variance trace $(\langle \mathbf{e}_a \mathbf{e}_a^\dagger \rangle)$ is the Kalman gain. For simplicity of presentation we take as our example an operator $\mathbf{A}$ that is time independent and observations taken continuously in time. A stationary error system with continuous observations is chosen for heuristic reasons although in forecast systems the tangent linear operator is time dependent and observations are introduced at discrete intervals. However, the statistical properties of optimal state estimation are general and results are qualitatively similar across observer systems.

The asymptotic Kalman gain resulting from continual assimilation of observations with observation matrix $\mathbf{H}$ is

$$\mathbf{K} = \mathbf{P}\mathbf{H}^\dagger \mathbf{R}^{-1}, \tag{8.35}$$

with $\mathbf{P}$ the stabilising solution of the algebraic Riccati equation

$$\mathbf{A}\mathbf{P} + \mathbf{P}\mathbf{A}^\dagger - \mathbf{P}\mathbf{H}^\dagger \mathbf{R}^{-1}\mathbf{H}\mathbf{P} + \mathbf{Q} = 0. \tag{8.36}$$

It is a property of the Kalman filter that the matrix $\mathbf{P}$ obtained as a solution of the algebraic Riccati equation is also the asymptotic error covariance of the observer system (8.34).

8.5.2 4D-Var as an observer system

4D-Var data assimilation with assimilation window T can be viewed as a special case of an observer in which a climatological background error covariance $\mathbf{B}$ is advanced for T units of time. In our autonomous model system the error covariance is advanced according to

$$\mathbf{P} = e^{\mathbf{A}T} \mathbf{B} e^{\mathbf{A}^\dagger T}, \tag{8.37}$$

from which we obtain the gain:

$$\mathbf{K}_{\text{4D-Var}} = \mathbf{P}\mathbf{H}^\dagger (\mathbf{H}\mathbf{P}\mathbf{H}^\dagger + \mathbf{R})^{-1}. \tag{8.38}$$

This gain produces a stabilised observer if enough observations are made.

The asymptotic error in the observer (8.34) is obtained by calculating the covariance, $\mathbf{P}$, that solves the equation

$$(\mathbf{A} - \mathbf{K}_{\text{4D-Var}}\mathbf{H})\,\mathbf{P} + \mathbf{P}\,(\mathbf{A} - \mathbf{K}_{\text{4D-Var}}\mathbf{H})^\dagger + \mathbf{K}_{\text{4D-Var}}\mathbf{R}\mathbf{K}_{\text{4D-Var}}^\dagger + \mathbf{Q} = 0. \tag{8.39}$$

8.6 Effect of the number of observations on the performance of the assimilation

Consider convergence of the assimilated state to truth as more observations are taken in the presence of model error. To fix ideas assume that repeated independent observations are made at each of the grid points of our model.

If the state of the assimilation system has dimension N and n observations are taken at each grid point, the observation matrix for these n observations, $\mathbf{H}_n$, is an $nN \times N$ matrix

$$\mathbf{H}_n = \mathbf{I}_N \bigotimes e, \tag{8.40}$$

where $\mathbf{I}_N$ is the identity N^2 dimensional matrix, $\bigotimes$ denotes the Kronecker product and e is the unit column $e = [1, \ldots, 1]^T$ of dimension n.

Consider an observation error covariance matrix $\mathbf{R} = r\mathbf{I}_N \bigotimes \mathbf{I}_n$, where $\mathbf{I}_n$ is the n^2 dimensional identity matrix and let $\mathbf{K}_n$ be the Kalman gain that results from these n observations. The Kalman gain is

$$\mathbf{K}_n = \mathbf{P}_n \mathbf{H}_n^\dagger \mathbf{R}^{-1} = \frac{1}{r} \mathbf{P}_n \left(\mathbf{I}_N \bigotimes e^\dagger \right), \tag{8.41}$$

with $\mathbf{P}_n$ the stabilising solution of the algebraic Ricatti equation

$$\mathbf{A}\mathbf{P}_n + \mathbf{P}_n \mathbf{A}^\dagger - \mathbf{P}_n \mathbf{H}_n^\dagger \mathbf{R}^{-1} \mathbf{H}_n \mathbf{P}_n + \mathbf{Q} = 0, \tag{8.42}$$

where $\mathbf{Q}$ is the model error covariance. On substitution of the specific expressions above for the observation matrix $\mathbf{H}_n$ and the observational error covariance matrix $\mathbf{R}$, (8.42) assumes the simplified form

$$\mathbf{A}\mathbf{P}_n + \mathbf{P}_n \mathbf{A}^\dagger - \frac{n}{r} \mathbf{P}_n^2 + \mathbf{Q} = 0, \tag{8.43}$$

from which we conclude that the analysis error in the observer system resulting from assimilation of n observations at each grid point with each observation having observational error variance r is equal to the analysis error that results from observing the same system with a single isolated observation with observational error variance r/n. It remains to determine how the error covariance P_n scales with n.

In the absence of model error ($\mathbf{Q} = 0$) the answer is immediate:

$$\mathbf{P}_n = \frac{\mathbf{P}}{n}, \tag{8.44}$$

where $\mathbf{P}$ is the assimilation error covariance associated with a single observation which satisfies the algebraic Riccati equation

$$\mathbf{A}\mathbf{P} + \mathbf{P}\mathbf{A}^\dagger - \frac{1}{r} \mathbf{P}^2 = 0. \tag{8.45}$$

So in the absence of model error the assimilation square error tends to zero as more observations are taken at the expected rate of n^{-1}.

Consider now the case in which model error exists. In that case we may expand P_n in an asymptotic series:

$$\mathbf{P}_n = \frac{p_o}{\sqrt{n}} + \frac{p_1}{n} + \cdots. \tag{8.46}$$

The leading term in this expansion is given by

$$p_o = \sqrt{r}\mathbf{Q}^{1/2}, \tag{8.47}$$

and consequently the asymptotic error covariance in the presence of model error has the leading behaviour

$$\mathbf{P}_n = \sqrt{\frac{r}{n}}\mathbf{Q}^{1/2}. \tag{8.48}$$

We conclude that in the presence of model error the assimilation square error of the Kalman filter in our example tends to zero at rate $n^{-1/2}$ as more observations are made.

It is instructive to compare this with the behaviour of analysis error in a 4D-Var data assimilation as the number of observations increases. In the absence of model error the 4D-Var analysis square error also tends to zero at rate n^{-1}, but in the presence of model error if the background covariance $\mathbf{B}$ is not rescaled as more observations are taken the analysis error asymptotes to a non-zero constant value.

In order to understand this behaviour consider the asymptotic error as $n \to \infty$ in the unstable stochastically forced scalar system with growth rate a:

$$\frac{de}{dt} = ae + q^{1/2}w. \tag{8.49}$$

The associated algebraic Riccati equation is

$$2ap_n - \frac{n}{r}p_n^2 + q = 0, \tag{8.50}$$

with stabilising solution

$$p_n = a\frac{r}{n} + \sqrt{a^2\left(\frac{r}{n}\right)^2 + q\frac{r}{n}}. \tag{8.51}$$

This stabilising solution is also the error in the observer system after assimilation of n observations. Note that in the absence of model error and for all n:

$$p_n = \frac{2ar}{n} \quad \text{if } q = 0, \tag{8.52}$$

and that the Kalman gain is

$$\mathbf{K}_n = \frac{2a}{n}[1, 1, \ldots, 1, 1], \tag{8.53}$$

and that the weight given each in the assimilation is

$$\mathbf{K}_n\mathbf{H}_n = 2a, \tag{8.54}$$

indicating that the weight given to observations is proportional to the error growth rate and is independent of the number of observations.

With model error and as $n \to \infty$:

$$p_n \approx \sqrt{\frac{qr}{n}} \quad \text{if } q \neq 0, \tag{8.55}$$

and the Kalman gain is

$$\mathbf{K}_n \approx \sqrt{\frac{q}{rn}}[1, 1, \ldots, 1, 1], \tag{8.56}$$

so that the weight given to observations is

$$\mathbf{K}_n\mathbf{H}_n = \sqrt{\frac{nq}{r}}, \tag{8.57}$$

independent of error growth rate and indicating that as the number of observations tends to infinity in the presence of model error the model is increasingly discounted and the observations accepted. A comparison of the error as a function of the number of observations in the scalar system is shown in Figure 8.13.

Regardless of model error, the error in the optimal observer vanishes if enough observations are assimilated, a result that holds in higher dimensions, as we have seen.

8.7 Approach of 4D-Var to the Kalman filter as the assimilation interval increases

In the absence of model error 4D-Var is equivalent to the extended Kalman filter if the assimilation window is extended to infinity. Present implementations of 4D-Var employ assimilation windows of 12 hours and it may appear that these implementations must be suboptimal and that the assimilation could be improved by lengthening the assimilation window.

Consider the asymptotic gain arising from a single observation in the time independent storm track model with and without model error. The asymptotic gain is shown in Figure 8.14 (top panel). It is evident that in the absence of model error the gain is not localised: the gain identifies the unstable structures of the forecast model and provides loadings designed to destroy these structures which have the character of a global mode. As shown in Figure 8.14 (bottom panel), in the presence of model error the gain becomes localised to the neighbourhood of the observation, because the model error that is distributed in the system produces incoherent responses far from the observation location that cancel when the ensemble average response of the system is taken so that the gain in the presence of model error is localised.

Because 4D-Var calculates the gains without model error the gain associated with a 4D-Var assimilation as the assimilation window is increased extends into the far field. This evolution of the gain associated with an initial climatological background

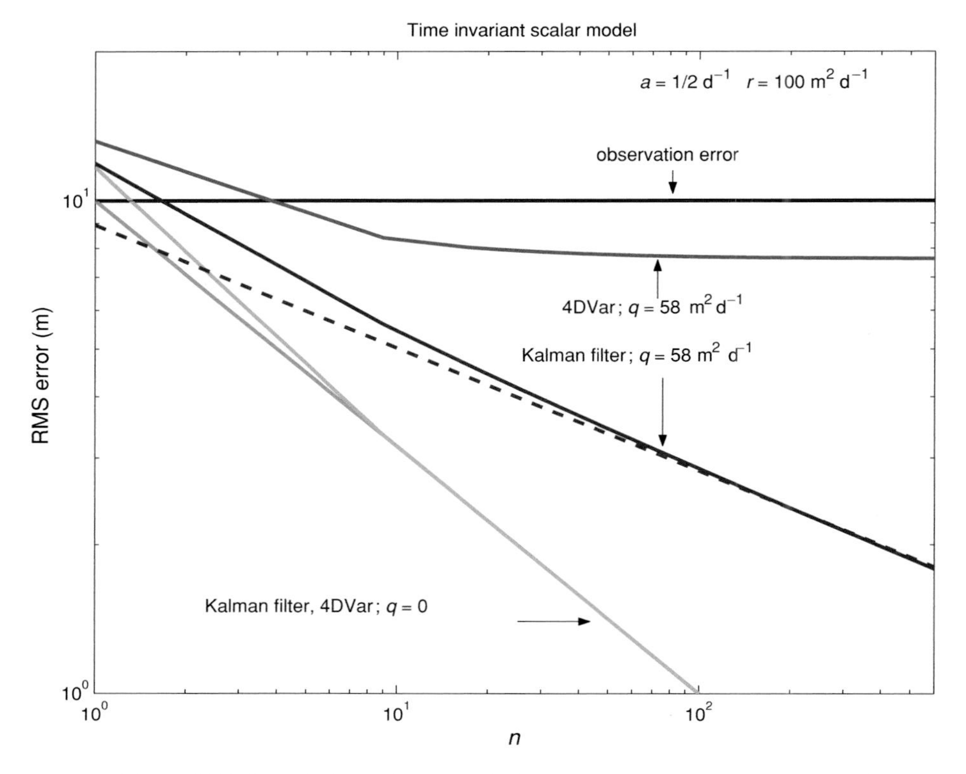

Figure 8.13 Error in the scalar optimal observer system and a scalar system with an equivalent 4D-Var observer as a function of the number of observations. The gain in the optimal observer is the asymptotic Kalman gain. The growth rate is $a = 1/2d^{-1}$, the observational error is 10 m. The model error variance is $q = 58m^2\,d^{-1}$ resulting in a model-induced error of 10 m after a day. With $q = 0$ the error in both the observer system with the Kalman filter and the 4D-Var falls as $n^{-1/2}$. With $q \neq 0$ the error in the 4D-Var observer asymptotes to a constant value while in the observer with the Kalman filter falls as $n^{-1/4}$.

B in a 4D-Var assimilation is shown in Figure 8.15. With time the climatological gain associated with the background error covariance assumes a global structure.

In the absence of model error the gain as the assimilation interval increases approaches the structure of the gain of the Kalman filter and the analysis error of 4D-Var asymptotes to the analysis error obtained by a Kalman filter. The convergence of 4D-Var assimilation error to that of the Kalman filter is shown for the time dependent version of the storm track model in Figure 8.16.

However, the perfect model assumption is physically unrealistic, and the 4D-Var assimilation scheme produces gains that have global structure as the assimilation window is increased. We find in our model storm track that 4D-Var performs best with an assimilation window that is large enough to allow the gain to be affected by the flow but short enough so that far-field loadings do not have time to form. An

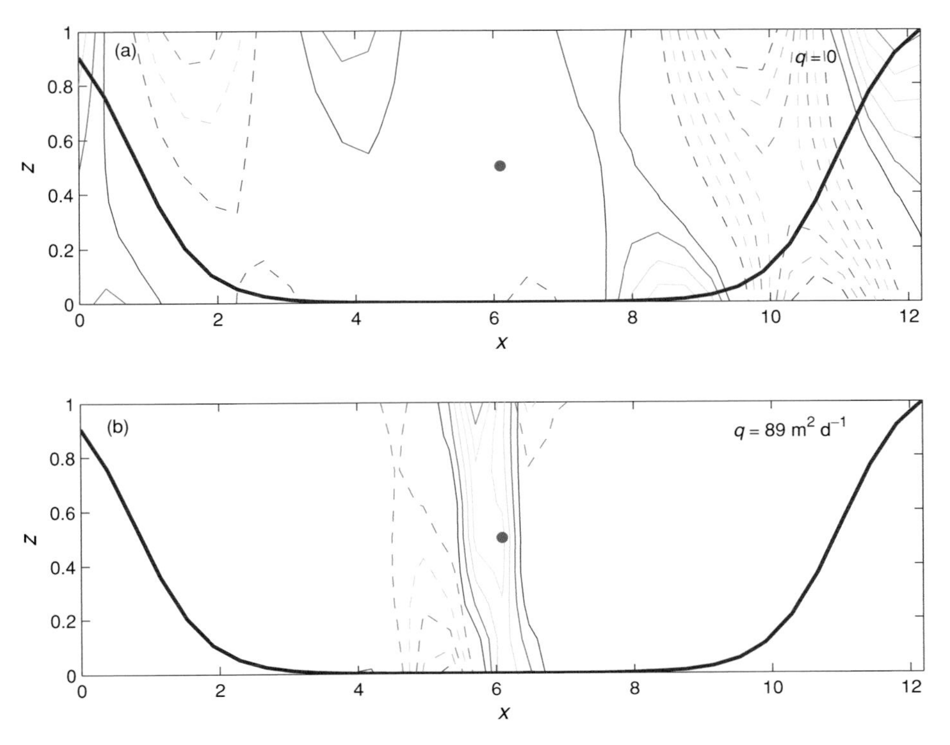

Figure 8.14 The asymptotic Kalman gain for observation at the centre of the channel in the storm track model. Top panel: the gain for the case of no model error. Bottom panel: the gain for the case with model error. The model error q produces an rms model error of 5 m in a day. The rms observational error is 10 m. The asymptotic Kalman gain has been calculated for the time mean flow. Note that the model error leads to localisation of the gain in the neighbourhood of the observations.

example of 4D-Var analysis error as a function of the assimilation interval is shown in Figure 8.17. In this example the optimal assimilation interval is 36 hours.

We conclude that neglect of model error in the formulation of 4D-Var makes 4D-Var operate best for rather short assimilation intervals. Model error must be introduced to make 4D-Var an optimal observer. In the next section we propose a method for introducing model error into 4D-Var.

8.8 Reduced order error covariance estimate

We now formulate the observer system in which the error covariance is advanced in the truncated space to obtain a reduced order Kalman gain. The resulting observer system in reduced coordinates is

$$\frac{d\mathbf{e}_k}{dt} = (\mathbf{A}_k - \mathbf{K}_k\mathbf{H}_k)\,\mathbf{e}_k + \mathbf{K}_k\mathbf{R}_k^{1/2}\mathbf{w}_\mathrm{o} - \mathbf{Q}_k^{1/2}\mathbf{w}_\mathrm{m}, \tag{8.58}$$

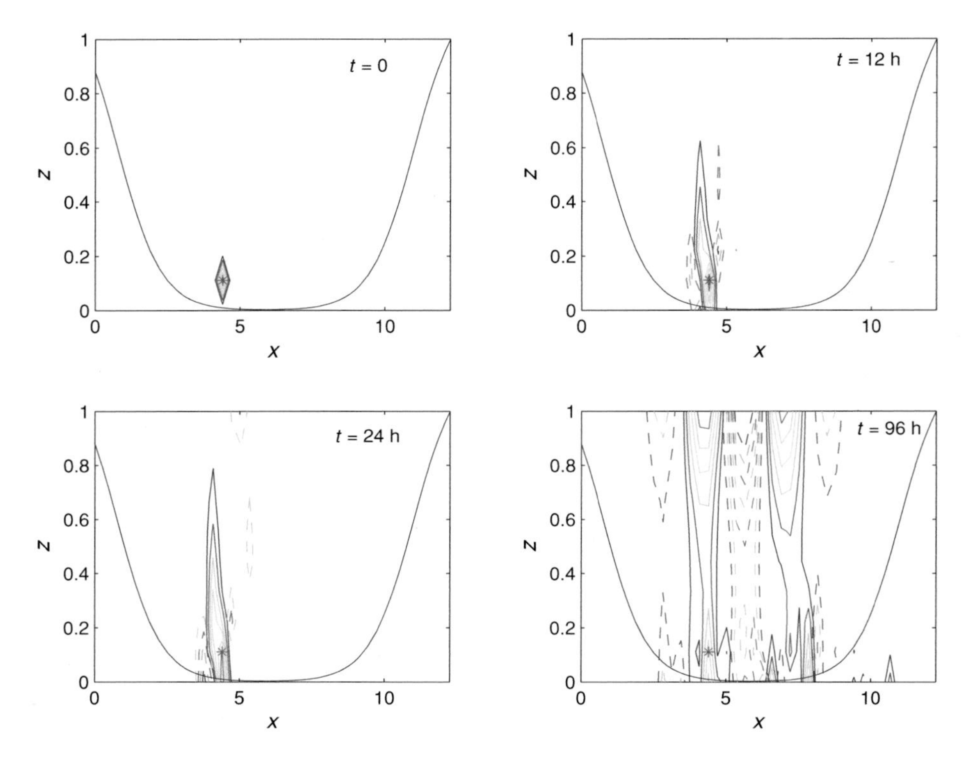

Figure 8.15 Evolution of the gain associated with the observation marked with a star in 4D-Var as a function of the assimilation interval in the unstable time mean storm track error model. The background **B** matrix is the identity. As the assimilation interval increases 4D-Var gains extend into the far field.

where the reduced analysis is $\mathbf{e}_k = \mathbf{Y}^\dagger \mathbf{e}_\mathrm{a}$ for $k \ll N$ and the reduced $k \times k$ operator is

$$\mathbf{A}_k = \mathbf{Y}^\dagger \mathbf{A} \mathbf{X}. \tag{8.59}$$

The n observations, y_ob, are assimilated in the reduced space according to

$$\mathbf{y}_\mathrm{ob} = \mathbf{H}_k \mathbf{x}_k + \mathbf{R}^{1/2} \mathbf{w}_\mathrm{o}, \tag{8.60}$$

where the reduced order observation matrix is

$$\mathbf{H}_k = \mathbf{H}\mathbf{X}. \tag{8.61}$$

The error system in the reduced space is used to obtain the Kalman gain $\mathbf{K}_k$ and to propagate the error covariance,

$$\mathbf{P}_k = \langle \mathbf{e}_k \mathbf{e}_k^T \rangle. \tag{8.62}$$

The error covariance of the full system is then approximated from that of the reduced covariance $\mathbf{P}_k$ by

$$\mathbf{P} = \mathbf{X}\mathbf{P}_k \mathbf{X}^\dagger. \tag{8.63}$$

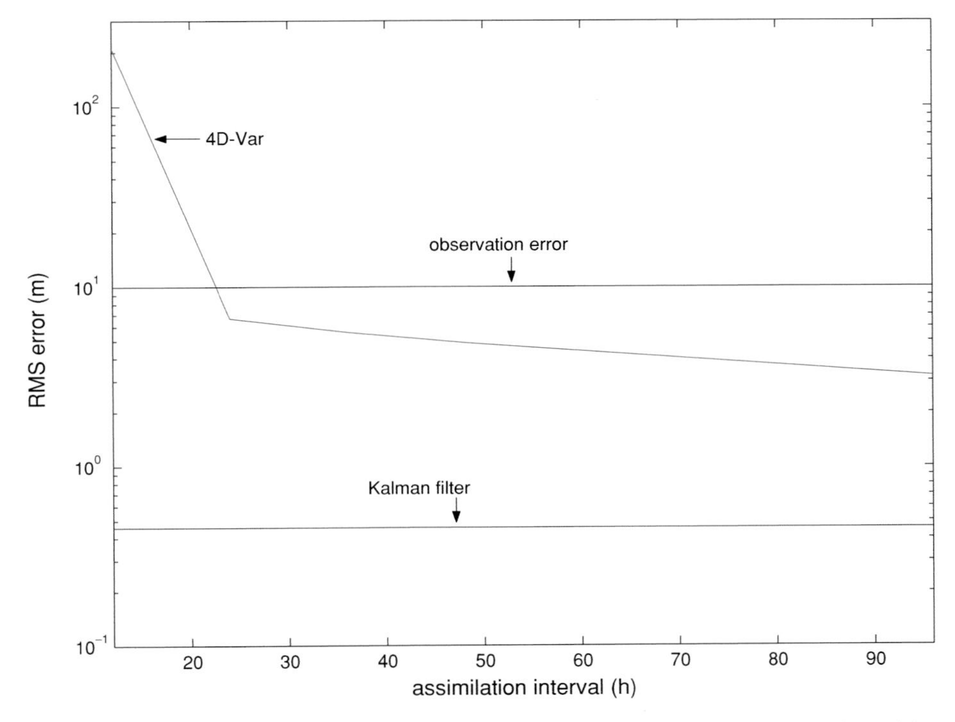

Figure 8.16 Error in 4D-Var assimilations in the time dependent storm track model with no model error as a function of assimilation interval. Also shown is the error obtained with sequential application of a Kalman filter. Sixteen observations are assimilated with rms observational error of 10 m. As the assimilation interval tends to infinity the 4D-Var error approaches that of the Kalman filter.

This error covariance is used in our 4D-Var model. By introducing this covariance in 4D-Var we evolve the error covariance and simultaneously also introduce model error. Introduction of this reduced order covariance in 4D-Var makes the 12-hour 4D-Var perform nearly optimally. Analysis of the performance of this filter is shown in Figure 8.18. Using the reduced order covariance obtained without model error leads to degradation of the 4D-Var assimilation due to unrealistic far field loadings in the gains.

8.9 Conclusions

A data assimilation system combines observations and dynamics expressed through a numerical forecast model to obtain an estimate of the state of the atmosphere. An optimal data assimilation system combines observations and dynamics to obtain the statistically best state estimate. Statistical optimality requires information about the observation error and about the error in the numerical forecast. This latter is difficult to obtain because of the high dimension of the error system so that approximations

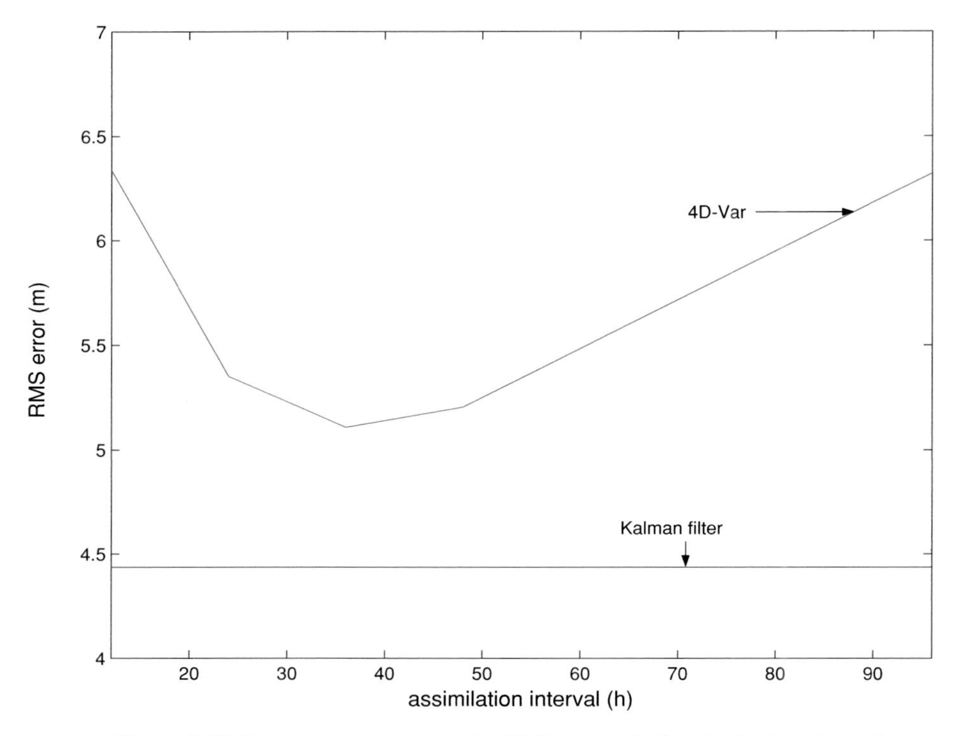

Figure 8.17 Root-mean-square error in 4D-Var assimilations in the time dependent storm track model with model error as a function of assimilation interval. The best 4D-Var performance is achieved in this example for assimilation over the interval 36 h. Also shown is the error obtained with the Kalman filter. Forty observations are assimilated with rms observational error of 10 m; the model error variance is $q = 12\, m^2\, d^{-1}$, so that a model error of 5 m accumulates in one day.

to the forecast error have to be made to implement practical applications of optimal state estimation. A promising method for obtaining an approximation to forecast error is to advance the error covariance in a state space of reduced dimension compared with that of the full forecast error system. The error covariance in the reduced space can then be used in an approximate optimal state estimation method such as 4D-Var or the extended Kalman filter. Such a reduction is possible because the significantly unstable subspace of the error system is of much lower dimension than the complete state dimension.

Assimilation systems can be usefully modelled as observer systems in which any gain matrix that stabilises the analysis error system is an observer and the gain that results in minimum analysis error is the optimal observer. This perspective on assimilation provides insight by allowing generalised stability analysis of the observer system to be performed, revealing for instance the distributed error sources that serve to most effectively degrade the analysis (Farrell and Ioannou, 2005).

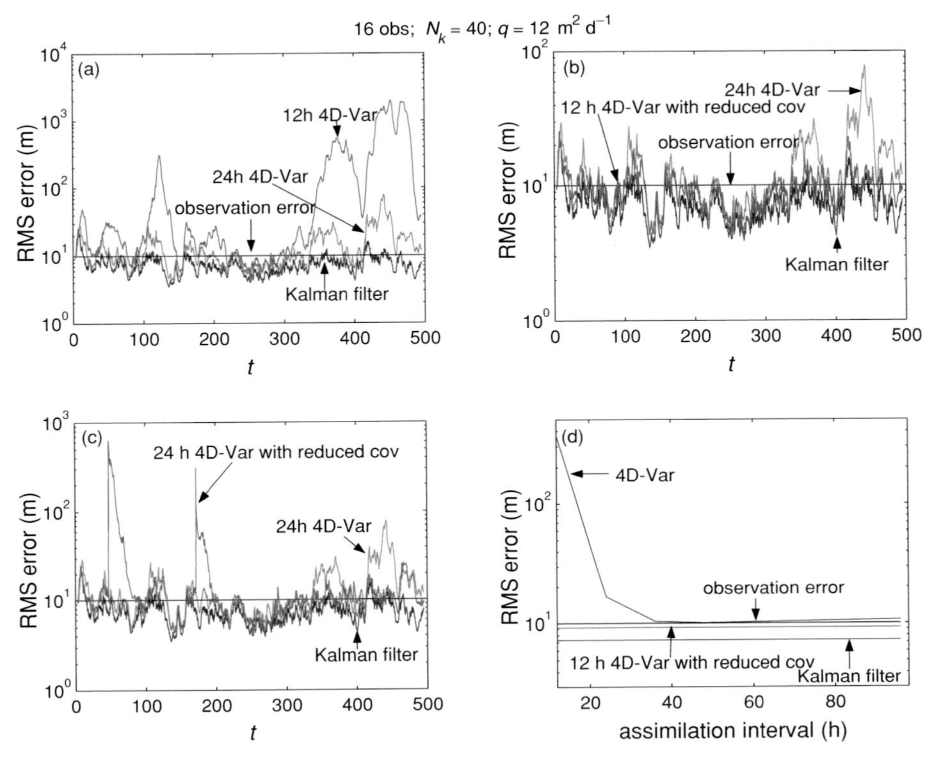

Figure 8.18 Error in a simulation of the time dependent storm track model with model error. (a) Comparison of the errors in a 12-h and 24-h 4D-Var with the error in the full Kalman filter. Panel (b): Comparison of the error in a 24-h 4D-Var with the error in a 12-h 4D-Var in which the isotropic static **B** has been preconditioned with the error covariance obtained from a reduced rank Kalman filter with balanced truncation. The reduced rank Kalman filter has been obtained with model error. In the truncated system 40 degrees of freedom (dof) have been retained out of the 400 dof of the system. The isotropic **B** introduced to the reduced rank covariance has amplitude equal to the smallest eigenvalue of the reduced rank covariance. Also shown is the error resulting from the Kalman filter. The 12-h 4D-Var performance is nearly optimal. (c) Comparison of the error in a 24-h 4D-Var with the error in a 24-h 4D-Var in which the isotropic static **B** has been preconditioned with the error covariance obtained from the reduced Kalman filter. The 24-h 4D-Var preconditioned with the covariance from the reduced Kalman filter propagates the covariance without model error longer and its performance is worse than that of the corresponding 12-h 4D-Var. (d) Root-mean-square error in 4D-Var assimilations in the time dependent storm track model with model error as a function of assimilation interval. Also shown is the error obtained with sequential application of a Kalman filter and the error from the 12-h 4D-Var which was preconditioned with the reduced rank covariance. Sixteen observations are assimilated with rms observational error of 10 m. The model error variance coefficient is $q = 12\, m^2\, d^{-1}$, so that a model error of 5 m accumulates after a day.

Analysis of the observer system modelling 4D-Var and the Kalman filter reveals that the number of observations assimilated increases the analysis error asymptotes to a finite value comparable to observational error and independent of the number of observations, unless the forecast error covariance is systematically adjusted to account for the increase of observations. One way this adjustment can be accomplished is by advancing the forecast error covariance in the dynamically relevant reduced order system that supports the growing error structures.

The result from using this accurate forecast covariance is that as the number of observations n increases, the associated Kalman filter obtains assimilation error $O(n^{-1/4})$ (with model error present) while the 4D-Var simulation fails to systematically reduce the estimation error. Assuming that redundancy of observation in the restricted subspace of significantly growing error structures has or soon will be available it is important to systematise the error covariance calculation in order to take advantage of these observations.

The gain under the assumption of a perfect model develops far field loadings that degrade the assimilation because the model error is in fact non-vanishing. The error covariance obtained by introducing model error into the reduced system suppresses these far field loadings. The error covariance calculated in the reduced system provides a method for introducing model error into 4D-Var, thus reducing the deleterious effects of the perfect model assumption and allowing accurate equivalent gains to be realised on short assimilation intervals.

Acknowledgements

This work was supported by NSF ATM-0123389 and by ONR N00014-99-1-0018.

Notes

1. For waves with a constant meridional wavenumber l, the operator ∇^2 is invertible even for homogeneous boundary conditions.
2. The optimal growth at time, t, is defined as the maximum perturbation growth that can occur over time t. For an autonomous system, governed by $\mathbf{A}$, the optimal growth at t is given by the largest singular value of $e^{\mathbf{A}t}$ or by $\|e^{\mathbf{A}t}\|_2$.

References

Bishop, C. H., B. J. Etherton, S. J. Majumdar and J. Sharanya (2001). Adaptive sampling with the ensemble transform Kalman filter. I: Theoretical aspects. *Mon. Weather Rev.*, **129**, 420–36.

Buizza, R. and T. Palmer (1995). The singular vector structure of the atmospheric general circulation. *J. Atmos. Sci.*, **52**, 1434–56.

Cohn, S. E. and R. Todling (1996). Approximate Kalman filters for stable and unstable dynamics. *J. Meteorol. Soc. Jpn.*, **75**, 257–88.

Dee, D. P. (1995). On-line estimation of error covariance parameters for atmospheric data assimilation. *Mon. Weather Rev.* **123**, 1128–45.

Delsole, T. M. and B. F. Farrell (1994). Nonlinear equilibration of localized instabilities on a baroclinic jet. *J. Atmos. Sci.*, **51**, 2270–84.

Dullerud, G. E. and F. Paganini (2000). *A Course in Robust Control Theory: a Convex Approach*, Springer Verlag.

Evensen, G. (1994). Sequential data assimilation with a nonlinear quasi-geostrophic model using Monte-Carlo methods to forecast error statistics. *J. Geophys. Res.*, **99** (C5), 10143–62.

Farrell, B. F. (1982). Pulse asymptotics of the Charney baroclinic instability problem. *J. Atmos. Sci.*, **39**, 507–17.

Farrell, B. F. and P. J. Ioannou (1996). Generalized stability. I: Autonomous operators. *J. Atmos. Sci.*, **53**, 2025–41.

(1999). Perturbation growth and structure in time dependent flows. *J. Atmos. Sci.*, **56**, 3622–39.

(2001a). Accurate low dimensional approximation of the linear dynamics of fluid flow. *J. Atmos. Sci.*, **58**, 2771–89.

(2001b). State estimation using a reduced-order Kalman filter. *J. Atmos. Sci.*, **58**, 3666–80.

(2005). Optimal excitation of linear dynamical systems by distributed forcing. *J. Atmos. Sci.*, **62**, 460–75.

Fukumori, I. and P. Malanotte-Rizzoli (1995). An approximate Kalman filter for ocean data assimilation; an example with an idealized Gulf Stream model. *J. Geophys. Res. Oceans*, **100**, 6777–93.

Gelaro, R., C. Reynolds and R. M. Errico (2000). Transient and asymptotic perturbation growth in a simple model. *Quart. J. Roy. Meteor. Soc.*, **128**, 205–28.

Ghil, M. (1997). Advances in sequential estimation for atmospheric and oceanic flows. *J. Meteorol. Soc. Jpn.*, **75**, 289–304.

Ghil, M. and P. Malanotte-Rizzoli (1991). Data assimilation in meteorology and oceanography. *Adv. Geophys.*, **33**, 141–266.

Ghil, M. and R. Todling (1996). Tracking atmospheric instabilities with the Kalman filter. II: Two layer results. *Mon. Weather Rev.*, **124**, 2340–52.

Glover, K. (1984). An optimal Hankel-norm approximation of linear multivariable systems and their $\mathbf{L}^{\infty}$-error bounds. *Int. J. Control*, **39**, 1115–93.

Houtekamer, P. L. and H. L. Mitchell (1998). Data assimilation using an ensemble Kalman filter technique. *Mon. Weather Rev.*, **126**, 796–811.

Ide, K. and M. Ghil (1997). Extended Kalman filtering for vortex systems. I: Methodology and point vortices. *Dyn. Atmos. Oceans*, **27**, 301–32.

Ide, K., P. Courtier, M. Ghil and A. C. Lorenc (1997). Unified notation for data assimilation: operational, sequential, and variational. *J. Meteorol. Soc. Jpn.*, **75**, 181–9.

Illyashenko, Yu. S. (1983). On the dimension of attractors of k-contracting systems in infinite-dimensional space. *Vestn. Mosk. Univ. Ser. 1 Mat. Mekh.*, **3**, 52–8.

Kalman, R. E. (1960). A new approach to linear filtering and prediction problems. *J. Basic Eng.*, **82D**, 35–45.

Kaplan, J. L. and J. A. Yorke (1979). Preturbulence: a regime observed in a fluid flow model of Lorenz. *Commun. Math. Phys.*, **67**, 93–108.

Lermusiaux, P. F. J. and A. R. Robinson (1999). Data assimilation via error statistical estimation. I: Theory and schemes. *Mon. Weather Rev.*, **127**, 1385–407.

Lin, S. J. and R. T. Pierrehumbert (1993). Is the midlatitude zonal flow absolutely unstable? *J. Atmos. Sci.*, **50**, 505–17.

Miller, R. N., M. Ghil and F. Gauthiez (1994). Advanced data assimilation in strongly nonlinear dynamical systems. *J. Atmos. Sci.*, 51, 1037–56.

Moore, B. C. (1981). Principal component analysis in linear systems: controllability, observability, and model reduction. *IEEE T. Automat. Contr.*, **AC-26**, 17–31.

North, G. (1984). Empirical orthogonal functions and normal modes. *J. Atmos. Sci.*, **41**, 879–87.

Palmer, T. N., R. Gelaro, J. Barkmeijer and R. Buizza (1998). Singular vectors, metrics, and adaptive observations. *J. Atmos. Sci.*, **55**, 633–53.

Reynolds, C. A. and R. M. Errico (1999). Convergence of singular vectors towards Lyapunov vectors. *Mon. Weather Rev.*, **127**, 2309–23.

Stewart, G. W. and J-G. Sun (1990). *Matrix Perturbation Theory*, Academic Press.

Sznaier, M., A. C. Doherty, M. Barahona, J. C. Doyle and H. Mabuchi (2002). A new bound of the $L_2([0, T])$ induced norm and applications to model reduction. In *Proceedings 2002 American Control Conference, Anchorage*. IEEE.

Takens, F. (1981). Detecting strange attractors in turbulence. In *Dynamical Systems and Turbulence*, ed. D. Rand and L.-S. Young, pp. 366–381. Lecture Notes in Mathematics 898. Springer.

Tippett, M. K., S. E. Cohn, R. Todling and D. Marchesin (2000). Low-dimensional representation of error covariance. *Tellus*, **52**, 533–53.

Todling, R. and M. Ghil (1994). Tracking atmospheric instabilities with the Kalman filter. I: Methodology and one-layer results. *Mon. Weather Rev.*, **122**, 183–204.

Zhou, K. and J. C. Doyle (1998). *Essentials of Robust Control.* Upper Saddle River, NJ. Prentice Hall.

Van Dooren, P. (2000). Gramian based model reduction of large-scale dynamical systems. In *Numerical Analysis 1999*, ed. D. F. Griffiths and G. A. Watson, pp. 231–47. Chapman and Hall/CRC Press.

Verlaan, M. and A. W. Heemink (1997). Tidal flow forecasting using reduced rank square root filters. *Stoch. Hydrol. Hydraul.*, **11**, 349–68.

Wunsch, C. (1996). *The Ocean Circulation Inverse Problem.* Cambridge University Press, Cambridge.

9

Predictability past, predictability present

Leonard A. Smith
Centre for the Analysis of Time Series,
London School of Economics and Pembroke College, Oxford

Maybe we oughta help him see,
The future ain't what it used to be.
Tom Petty

9.1 Introduction

Predictability evolves. The relation between our models and reality is one of similarity, not identity, and predictability takes form only within the context of our models. Thus predictability is a function of our understanding, our technology and our dedication to the task. The imperfection of our models implies that theoretical limits to predictability in the present may be surpassed; they need not limit predictability in the future. How then are we to exploit probabilistic forecasts extracted from our models, along with observations of the single realisation corresponding to each forecast, to improve the structure and formulation of our models? Can we exploit observations as one agent of a natural selection and happily allow our understanding to evolve without any ultimate goal, giving up the common vision of slowly approaching the Perfect Model? This chapter addresses these questions in a rather applied manner, and it adds a fourth: Might the mirage of a Perfect Model actually impede model improvement?

Given a mathematical dynamical system, a measurement function that translates between states of this system and observations, and knowledge of the statistical

Predictability of Weather and Climate, ed. Tim Palmer and Renate Hagedorn. Published by Cambridge University Press.

characteristics of any observational noise, then in principle we can quantify predictability quite accurately. But this situation describes the perfect model scenario (PMS), not the real world. In the real world we define the predictability of physical systems through our mathematical theories and our *in silico* models. And all our models are wrong: useful but imperfect. This chapter aims to illustrate the utility of existing ensemble prediction systems, not their imperfections. We will see that economic illustrations are of particular value, and investigate the construction of probability forecasts of observables from model simulations. General arguments and brief illustrations are given below; mathematical details and supporting statistics can be found in the references. While the arguments below are often couched in terms of economic users, their implications extend to the ways and means of meteorology as a physical science. Just as it is important not to confuse utility with empirical adequacy, it is also important to accept both as means of advancing any physical science.

In the next three sections we make a quick tour of useful background issues in forecasting, economics and predictability. When considering socioeconomic value it is helpful not to confuse severe weather and high-impact weather: the value of a weather forecast depends not only on its information content but also on our ability to take some mitigating action; a great deal of the unclaimed value in current operational products lies in their ability to yield useful information regarding unremarkable weather which carries significant economic impact. Then in Section 9.5 we consider the question of comparing forecasts and the notion of 'best'. This continues in Section 9.6 with a number of issues at the interface of meteorology and statistics, while illustrations of their economic relevance are noted in Section 9.7. It is quite popular nowadays to blame forecast busts on 'uncertainty in initial condition' (or chaos), we discuss what this phrase might possibly mean in Section 9.8, before concluding in Section 9.9. In reality predictability evolves and, as shown in Section 9.4, 'the future' evolves even within the mathematical fiction of a perfect model scenario where predictability does not.

9.2 Contrasting 1995 and 2002 perspectives on predictability

What has changed in the short time since the 1995 ECMWF Seminar on Predictability? Since I cannot avoid directly criticising what was happening in 1995, we will focus mostly on my contribution to the seminar (Smith, 1996).

Ensemble formation for systems of chaotic differential equations was a major topic of discussion in 1995; in contrast this chapter does not contain a single differential equation. In fact, it contains only one equation and, as it turns out, that equation is ill posed. In 1995 my focus was on constructing perfect ensembles, while below we will be more concerned with interpreting operational ensembles. The 1995 paper quantifies the difference between some forecast probability density function (pdf) and a perfect pdf obtained by propagating current uncertainty forward in time under a perfect model, while below I am content to discuss how to change an ensemble of simulations into a pdf forecast in the first place. There is also a question as to how one should evaluate any forecast pdf, given that we never have access to the 'perfect pdf', if such a thing exists, but only observations of a single realisation of weather. That is, we have only measurements of the one thing that happened, a target often called the *verification*. In general, it seems to me that the 1995 discussion focused on doing maths within the perfect model scenario (PMS), whereas we are now more interested in quantifying information content and debating resource allocation.

Smith (1996) discussed quantifying model error, while now I have been reduced to pondering model error, which I now refer to as *model inadequacy* (following Kennedy and O'Hagan, 2001). Any particular model can be thought of as one member of a model class. As a very simple example, consider different models that share the same structural form but have different parameter values, these are in the same model class; or consider the collection of all one-dimensional maps consisting of finite-order polynomials. Model inadequacy reflects the fact that not only is the best model we have imperfect, but there is no member of the available model class which is perfect. This is a much deeper flaw than having incorrect parameter values: in this case there are no 'Correct' parameter values to be had. And this case is ubiquitous within physical science.

The concept of *i*-shadowing was introduced in the 1995 Predictability Seminar, as was the notion of an accountable probability forecast. A model is said to *i-shadow* over a given period in time if there exists a model trajectory that is consistent with the observations, given the observational noise, over that period. For historical reasons, meteorologists call the model state that corresponds to the operational best guess of current atmospheric conditions *the analysis*. The question of quantifying just how long operational models can shadow either the observations or even the corresponding time series of analyses remains of key interest. The notion of an accountable ensemble forecast was also introduced in the 1995 Seminar (see also Smith, 2001) as a generalisation of Popper's idea of accountability in the single forecast scenario. Popper (1956) realised that forecasts would fail due to uncertainty in the initial condition even if the model was perfect; he called a model accountable if it correctly specified the accuracy of measurement required to obtain a given forecast accuracy. For an accountable ensemble forecast the size of the ensemble will accurately reflect the resolution of the probability forecast. The relevant point here is that any forecast product extracted from an accountable probability forecast will suffer *only* from the fact that the forecast ensemble has a finite number of members: we could never reject the null hypothesis that both the members of the forecast ensemble and the verification ('Truth') were drawn from the same distribution.

The distribution of shadowing times is arguably the best measure we have for contrasting various non-linear models and quantifying the relevance of uncertainty in the initial condition (as opposed to model inadequacy). I hope that in the pages that follow methods which reflect the quality of a simulation model (i.e. shadowing times) are clearly distinguished from methods which reflect the quality of a complete probabilistic forecast system (i.e. ignorance as defined in Good (1952) and Section 9.5.5). Recall that with a non-linear model, a probabilistic forecasting system can only be evaluated as a whole: non-linearity links data assimilation, ensemble formation, model structure and the rest.

In general, I would identify two major changes in my own work from 1995 to 2002. The first is a shift from doing mathematics to doing physics; more specifically, of trying to identify when very interesting mathematics is taking us to a level of detail that cannot be justified given the limited ability of our model to reflect the phenomena we are modelling. Indeed, I now believe that model inadequacy prevents accountable probability forecasts in a manner not dissimilar to that in which uncertainty in the initial condition precludes accurate best first guess (BFG) forecasting in the root-mean-square sense. In fact we may need to replace our current rather naïve concept of probability forecasts with something else; something which remains empirically relevant when no perfect model is in hand. The second change reflects a better understanding of the role of the forecast user as the true driver for real-time weather forecasting. Economic users can play particularly vital roles both as providers of valid empirical targets, the ultimate test of mathematical modelling (at least within mathematical physics), and also as a valuable source of data for assimilation. In the next section, we will develop an ensemble of users with which to illustrate this interaction.

9.3 An ensemble of users

Tim Palmer's chapter in this volume (see also Palmer, 2002) introduced his golf buddy, Charlie the contractor. Charlie is forced by the nature of his work to make binary decisions, for example whether or not to pour concrete on a given afternoon. The weather connection comes in as another binary event: if it freezes then the cement will not set properly. By using cost-loss analysis (Angstrom, 1919; Murphy, 1977; Richardson, 2000), Charlie can work out the probability threshold for freezing at which he should take the afternoon off and go play golf. Of course, if Charlie is presented with a definitive forecast ('The low temperature tonight will be 4 degrees C' or 'No ground frost tonight'), then pours the concrete and it does freeze, he is likely to be somewhat disappointed. As Tim noted, he may look for someone to sue. There are, of course, no definitive forecasts and to sell any forecast as unequivocal is to invite lawsuits. When a forecaster has foreknowledge of the uncertainty of a forecast and yet still presents an unequivocal forecast to the public, justifying it as being 'for their own good', she is inviting such a law suit. Arguably, the public has the right to expect a frank appraisal of the forecaster's belief in the forecast. As it turns out, Charlie also plays the horses; he knows much about odds. He does not even hold the naïve expectation that the corresponding implied 'probabilities' (of each horse winning) should sum up to one!

But there is more to the world than binary decisions (and golf). While I do not know Charlie, at a recent London School of Economics alumni dinner I met Charles. Charles now works in the financial futures market; while he no doubt plays golf, he does not see himself as making binary decisions; he is interested in 'How much' questions rather than 'Yes or No' questions. This is because Charles buys and sells large quantities of petroleum products (heating oil, gas, various flavours of crude, jet fuel and so on) always being careful not to take delivery of any of it. He has also started pricing weather derivatives, a wide variety of weather derivatives in fact. He is fluent in stochastic calculus and knows a bit of probability theory, enough to know that in order to gauge his risk he wants more than a single probability threshold.

Charles has an interesting view of what constitutes a good four-week forecast. He doesn't care at all about the average temperature in week four, nor whether the Monday two weeks hence is in fact going to be a very cold day. While Charlie is concerned as to whether or not cement will set tonight, Charles is not concerned about any particular day. Instead, Charles is very concerned about the number of days between now and the end of the month on which cement will not set, inasmuch as this is the kind of variable that weather derivatives near expiry might be based upon. Charles knows how to place his bets given a good probability forecast; his question is whether a given probability forecast is a good one! For better or for worse, one of the advantages of providing a probability forecast is that no single probability forecast need ever be judged 'wrong'; finding oneself overly worried about what might happen in any one forecast suggests we have not fully accepted this basic lesson of first-year Statistics. Nevertheless, if Charles only bets when the forecast probability of winning is over 90% and, after many bets, he finds he has won only half the time, then he will have a strong case against the forecast vendor.

The financial markets are inundated with vendors of various forecasts, and Charles is familiar with forecasts that fail to provide value. He already knows how to sue, of course, but never does so; life is too short. Rather, he relies on natural selection in the marketplace: if the forecasts do not contain the information he needs, or are not presented in a manner such that he can extract that information (even if it is there), then he simply stops buying them and speaks badly of them in London wine bars.

And then there is Charlotte, another LSE graduate now working in the energy sector. Charlotte's goal is not to make money *per se*, but rather to generate electricity as efficiently as possible, using a mix of wind power alongside a set of combined cycle gas turbine (CCGT) generators.[1] Her definition of 'efficiently' is an economic one, but includes issues of improving air quality and decreasing carbon dioxide production. Ideally, she wants a forecast which includes the full probability density function (pdf) of future atmospheric states, or at least the pdf for temperature, pressure, wind speed and humidity at a few dozen points on the surface of the Earth. And she would like to forecast weather dependence of demand as well, especially where it is sensitive to variables that also impact the efficiency of generation (Altalo and Smith, 2004; Smith *et al.*, 2005). The alternative of using only marginal distributions, or perhaps a bounding box (Smith, 2001; Weisheimer *et al.*, 2005), may prove an operational alternative. Charlotte's aim is not to be perfect but rather simply to do better, so she is happy to focus on a single site if that is required by the limited information content in the forecast.

A quick calculation[2] shows that even with an accountable forecast the ensemble size required in order to resolve conditional probabilities will remain prohibitive for quite some time. Of course, the future may allow flow-dependent resource allocation, including the distribution of ensemble members over multiple computer sites on those days when such resources are justified. A somewhat longer and rather more dubious calculation suggests that generating this style of weather forecasting might feed back on the weather which is being forecast. Certainly the effect would far exceed that of the flapping of a seagull's wing, unless the forecasters were relocated to some remote location, say on the moon.

Like Charles, Charlotte is also interested in the number of cold days in the remainder of this month, or this season. This is especially true if they are likely to be consecutive cold days. The consumption of natural gas, and hence its price, depends on such things. And it is impossible to deduce them from knowing that there is a 10% chance of a cold day on four consecutive days in week two: thinking of the ensemble members as scenarios, Charlotte wants to know whether that 10% is composed of the same ensemble members each day (in which case there is a 10% chance of a four-day cold spell) or whether it was the case that 10% of the total ensemble was cold on each of the four days, but the cold temperatures corresponded to different members each day, in which case the chance of an extended cold spell could be very low (depending on the size of the ensemble).

Why does she care about the four consecutive cold days problem? By law (of Parliament), natural gas will be diverted from industrial users to domestic users in times of high demand. If she can see that there is a moderate probability of such a period in advance, she can fill her reserve tanks before the start of the cold spell (and take a forward position in gas and electricity markets as well). This is a fairly low cost action, because if the cold spell fails to materialise, she can simply decrease purchase of natural gas next week. The carrying costs of an early purchase are small, the profit loss of running low is huge: she is happy to overstock several times a year, in order to have full reserves during the cold spells that do occur.

And she can mix this weather information in with a variety of other indicators and actions; from scheduling (or postponing) preventive maintenance, to allowing optional leave of absence, so as to embed probabilistic weather information naturally into a scheme of seamless forward planning.

Decisions similar to Charlotte's are made across the economy. We consider two more examples of such decisions: one faced by the owner of a small corner shop and one faced by a multinational energy giant. By overstocking soft drinks whenever the probability of a heat wave exceeds a relatively low threshold, shop owners can hedge against the significant impact of running out of stock when a heat wave finally materialises. The marginal costs of the extra stock are relatively low, as in Charlotte's case, and the shop owner will happily make this investment several times when no heat wave materialises. We find something like the inverse of the understock problem in offshore oil wells. In deep water, floating oil wells store the extracted oil onsite, until a tanker can come and collect it; for a variety of reasons, it is costly for the local storage tanks to get too full. To offload the oil, a tanker must not only reach the well, but seas must be calm enough for it to dock; obviously it can make sense to send a tanker early if there is a good chance of sustained heavy seas in the vicinity of the well near the originally scheduled time of arrival.

Cost functions targeting Charles' and Charlotte's desires could prove very valuable to modellers; inasmuch as we have not already fit our models to such targets, they provide a fresh viewpoint from which we can detect hidden shortcomings of our models. But even beyond this, Charlotte and her colleagues are not only collecting traditional meteorological observations at various points scattered about the country (power station locations), they also collect real-time data on weather-related demand integrated over spatial regions comparable with the grid resolution of a weather model and on timescales of seconds: the assimilation of such observations might also prove of value. This value will hinge on the information content of the observations, not their number: a huge number of uninformative or redundant observations may tell us much less than a few relevant measurements.

Charlie, Charles and Charlotte each aim to extract as much relevant information as possible from the forecast, but no more. Each of them realises that, in the past, weather forecasts have been presented as if they contained much more information than even a casual verification analysis would support. The five-day forecasts for Oxford at www.metoffice.com present the day 5 forecast with the same air of authority given to the day 2 forecast. The Weather Channel, which provides 10-day point forecasts at www.weather.com, and other vendors are concerned to present the uncertainty they know is associated with their current apparently unequivocal forecasts. Yet a decade after ensemble forecasting became operational, it is still not clear how to do so. And the situation is getting worse: there is talk of commercially available point forecasts out to 364 days, each lead time presented as if it were as reliable as any other (in this last case, no doubt, the vast majority are equally reliable). Questions of how best to communicate uncertainty information to numerate users and how to rapidly communicate forecast uncertainty to the general public are central to programmes like THORPEX. Answering these questions will require improving the research interface between social psychology, meteorology and mathematics. Our current progress in this direction can be observed in the forecasts posted at www.dime.lse.ac.uk (see also http:// lsecats.org and www.meteo.psu.edu/~roulston).

So we now have our ensemble of three users, each with similar but distinct interests in weather forecasts and different resources available to evaluate forecast information.

Charlie is primarily interested in binary choices. Charles' main interests lie both in a handful of meteorological standards (where he only cares about the legal value of the standard, not what the weather was) and in very broad meteorologically influenced demand levels. Charlotte is interested in accurate, empirically falsifiable forecasts; she doesn't care what the analysis was nor what the official temperature at Heathrow was. She has temperature records of her own in addition to measurements of the efficiency of her CCGT generators and wind turbines. In a very real sense, her 'economic variables' are more 'physical' than any model-laden variable intended to reflect the 500 mb pressure height field within some model's analysis.

There are, of course, important societal uses of weather forecasts beyond economics; many of these societal applications are complicated by the fact that the psychological reactions figure into the effectiveness of the forecast (for an example, see Roulston and Smith, 2004). For most of what follows, however, we will consider weather forecasts from the varying viewpoints of Charlie, Charles and Charlotte. Obviously, I aim to illustrate how probabilistic forecasts derived from operational ensemble prediction systems (EPS) compare, in terms of economic relevance, with forecasts derived from a best first guess (BFG) approach, and we shall see that probabilistic forecasts are more physically relevant as well. But I will also argue that accountable probability forecasts may well lie forever beyond our grasp, and that we must be careful not to mislead our 'users' or ourselves in this respect. To motivate this argument, we will first contrast the Laplacian view of predictability with a twenty-first century view that accounts for uncertainty in the initial condition, if not model inadequacy.

9.4 Contrasting nineteenth vs. twenty-first century perspectives on predictability

Imagine (for a moment) an intelligence that knew the True Laws of Nature and had accurate but not exact observations of a chaotic system extending over an arbitrarily long time. Such an agent, even if sufficiently powerful to subject all this data to (exact) mathematical analysis, could not determine the current state of the system and thus the present, as well as the future, would remain uncertain in her eyes. Yet the future would hold no surprises for her, she could make accountable probability forecasts, and low probability events would occur only as frequently as expected. The degree of perfection that meteorologists have been able to give ensemble weather forecasting reflects their aim to approximate the intelligence we have just imagined, although we will forever remain infinitely remote from such intelligence.[3]

It is important to distinguish determinism and predictability (see Earman, 1986; Bishop, 2003 and the references therein). Using the notion of indistinguishable states (Judd and Smith, 2001, 2004) we can illustrate this distinction with our twenty-first century demon, which has a perfect model and infinite computational capacity but access to only finite resolution observations. If the model is chaotic, then we can prove that, in addition to the 'True' state, there exists a set of states that are indistinguishable from the particular trajectory that actually generated the data. More precisely, we have shown that given many realisations of the observational noise, there is not one but, in each case, a collection of trajectories that cannot be distinguished from the 'True' trajectory given only the observations. The system is deterministic, the future trajectory of each state is well defined and unique, but uncertainty in the initial condition limits even the demon's prediction to the provision of probability distributions.

Note that the notion of shadowing is distinct from that of indistinguishable states (or indistinguishable trajectories); *i*-shadowing contrasts a trajectory of our mathematical model with a time series of targets usually based on observations of some physical system. This is often cast as a question of existence: does the model admit one or more trajectories that are consistent with the time series of observations given the noise model? The key point here is that we are contrasting our model with the observations. This is very different from the case of indistinguishable states, where we are contrasting various model trajectories with each other and asking whether or not we are likely to be able to distinguish one from another given only noisy observations. In this case, one considers all possible realisations of the observational noise. Thus when working with indistinguishable states we consider model trajectories and the statistics of the observational noise, whereas with shadowing we contrast a model trajectory and the actual set of observations in hand. Shadowing has a long history in non-linear dynamical systems dating back to the early 1960s; a discussion of the various casts of shadow can be found in Smith (2001).

In terms of actually constructing an operational ensemble, the relevant members of the indistinguishable sets are those that have, in fact, shadowed given the particular realisation of the observational noise in the recent past. More clearly: recall that each set of indistinguishable states is determined by integrating over all possible realisations of the observational noise; even when we wish to pick our ensemble members from this set, we will weight them with respect to the observations obtained in the one realisation of the noise which we have access to (our observations).

Given the arguments above, it follows that within the perfect model scenario an infinite number of distinct, infinitely long shadowing model trajectories would exist, each trajectory shadowing observations from the beginning of time up until the present. These trajectories are easily distinguished from each other within the model state space, but the noisy observations do not contain enough information to identify which one was used to generate the data. The contents of this set of indistinguishable states will depend on the particular realisation of the observational noise, but the set will always include the generating trajectory (also known as 'Truth'). This fact implies that even if granted all the powers of our twenty-first century demon, we would still have to make probabilistic forecasts. Epistemologically, one could argue that the 'true state' of the system is simply not defined at this point in time, and that the future is no more than a probability distribution. Accepting this argument implies that after each new observation is made, the future really ain't what it used to be.

Of course, even this restriction to an ever-changing probabilistic future is a difficulty to which we can only aspire. We do not have a perfect model; we do not even have a perfect model class, meaning that there is no combination of model parameters (or available parametrisations) for which any model accessible to us will provide a shadowing trajectory. Outside the perfect model scenario (PMS), the set of indistinguishable states tends to the empty set (Judd and Smith, 2004). In other words, the probability that the data was generated from *any* model in the model class approaches zero as we collect more observations, an awkward fact for the applied Bayesian.[4] But accepting this fact allows us to stop aiming for the perfect model, just as accepting uncertainty in the initial condition freed us from the unattainable goal of a single accurate forecast trajectory from 'inaccurate' starting conditions. When the model is very good (that is, the typical *i*-shadowing times are long compared with the forecast time) then the consideration of indistinguishable states suggests a new approach to several old questions.

9.5 Indistinguishable states and the fair valuation of forecasts

In this section we will consider the aims of model building and the evaluation of ensemble forecasts. Rather than repeat arguments in Smith (1997, 2001) on the importance of distinguishing model-variables from observed variables, we will consider the related question of the 'best' parameter value for a given model class. Arguably there is no such thing outside of PMS, and the insistence on finding a best can degrade forecast performance. We will then consider the game of weather roulette, and its use as an illustration for the economic decisions Charles and Charlotte make every day. A fair comparison of the value of different forecasts requires contrasting like with like, for example we must 'dress' both BFG and EPS simulations in order to obtain a fair evaluation of the probabilistic forecasts available from each method. Weather roulette also allows us to illustrate Charles' favoured cost function for economic forecasts, the logarithmic skill score called *ignorance* (Roulston and Smith, 2002). Relative ignorance can also be used to obtain insight on operational questions such as the division of computational resource between ensemble size and model resolution for a given target, as illustrated with Heathrow temperatures below (see also Smith *et al.*, 2001). Two worked economic examples are discussed in Section 9.7.

9.5.1 Against best

What parameter values should best be used in a forecast model? If the system that is generating the data corresponds to a particular set of parameters in our forecast model, then we have a *perfect model class*; obviously that set of parameters would be a candidate for 'best'. Outside PMS, at least, the question of best cannot be decoupled from the question of why we are building the model. We will duck both questions, and simply consider a single physically well-understood parameter, the freezing point of water,[5] within three (imperfect) modelling scenarios.

At standard pressure, it is widely accepted that liquid water freezes at zero degrees C. In a weather model with one millimetre resolution, I would be quick to assign this value to the model-freezing point. In a weather model with one-angstrom resolution, I would hope the value of zero degrees C would 'emerge' from the model all by itself. And at 40 kilometre resolution? Well at 40 km resolution I have no real clue as to the relation between model-temperature and temperature. I see no defensible argument for setting this parameter to anything other than that value which yields the best distribution of shadowing trajectories (that is, the distribution which, in some sense, reflects the longest shadowing times; a definition we will avoid here).

Of course, the physical relevance of the mathematical form used for the model parametrisation assumes that the parameter value lies in some range; internal consistency (and relevance) of the model parametrisation itself places some constraints on the values of the model parameters within it. This suggests, for example, that the freezing point of water should be somewhere around zero, but does not suggest any precise value.

This confusion between model-parameters and their physical analogues, or even better between model-variables and the observations (direct empirical measurements), is common. The situation is not helped by the fact that both are often given the same name; to clarify this we will distinguish between temperature and model-temperature where needs be.

Translating between model variables and observables is also related to representation error. Here we simply note that representation error is a shortcoming of the model (and the model-variables) not the observations. A reanalysis is a useful tool, but its value derives from the observations. In fifty years' time the temperature observations recorded today at Heathrow airport will still be important bits, whereas no one will care about model-temperature at today's effective grid resolution. The *data*, not the model-state space, endure.

Outside PMS, it is not clear how to relate model-variables to variables. To be fair, one should allow each model its own projection operator. Discussion of the difficulties this introduces will be pursued elsewhere, but there is no reason to assume that this operator should be one-to-one; it is likely to be one-to-many, as one model state almost surely corresponds to many atmospheric states if it corresponds to any. It may even be the case that there are many atmospheric states for each model state, and that each atmospheric state corresponds to more than one model state: a many-to-many map. But again, note that it might be best to avoid the assumption that there is 'an' atmospheric state altogether. We do not require this assumption. All we ever have are model states and integer observations.

9.5.2 Model inadequacy: the biggest picture

Let us now reconsider the issues of forecasting within the big picture. Once again, suppose for the moment that there exists some True state of the atmosphere, call it **X**(t). This is the kind of thing Laplace's nineteenth-century demon would need to know to make perfect forecasts, and given **X**(t) perfect BFG forecasts would follow even if the system were chaotic. As Laplace noted, mere mortals can never know **X**, rather we make direct measurements (that is, observations) *s*, which correspond to some projection of **X** into the space of our observing system. Given *s*, and most likely employing our model **M** as well, we project all the observations we have into our model-state space to find a distribution for *x*(t). Traditionally attention has focused on one model state, often called the analysis, along with some uncertainty information regarding the accuracy of this state. This is nothing more than data assimilation, and the empirically relevant aim of data assimilation is an ensemble (or probability distribution function), not any single state. Using our model, we then iterate *x*(t) forward in time to some future state, where we reverse the process (via downscaling or model output statistics) to extract some observable *w*. In the meantime, the atmosphere has evolved itself to a new state, and we compare our forecast of *w* with our new observation and the corresponding observed projection from the new **X**.

Although pictures similar to the one drawn in the preceding paragraph are commonplace, the existence of Truth, that is the existence of **X**, is mere hypothesis. We have no need of that hypothesis, regardless of how beautiful it is. All we ever have access to are the observations, which are mere numbers. The existence of some 'True' atmospheric state, much less some mathematically perfect model under which it evolves, is untestable. While such questions of belief are no doubt of interest to philosophers and psychologists, how might they play a central role within science itself? In questions of resource allocation and problem selection, they do.

Accepting that there is no perfect model is a liberating process; perhaps more importantly, it might allow better forecasts of the real atmosphere. Doing so can impact our goals in model improvement, for example, by suggesting the use of ensembles of models and improving their utility, as opposed to exploring mathematical approximations which will prove valid only for some even better model which, supposing it exists, is currently well beyond our means. The question, of course, is where to draw the line; few would question that there is significant information content in the 'Laws of Physics' however they lie, yet we have no reliable guide for quantifying this information content. In which cases will PMS prove to be a productive assumption? And when would we do better by maintaining two (or seven) distinct but plausible model structures and *not* trying to decide which one was best? What is the aim of data assimilation in the multimodel context? Indeed, is there a single aim of data assimilation in any context: might the aim for nowcasting differ from that of medium-range forecasting?

Charlie, Charles and Charlotte, along with others in the economy rather than in economics, seem untroubled by many of these issues; they tend to agree on the same cost function, even if they agree on nothing else. So we will leave these philosophical issues for the time being, and turn to the question of making money.

9.5.3 Weather roulette

A major goal of this chapter is to convince you that weather roulette is not only a reasonable illustration of the real-world economic decisions that Charles and Charlotte deal with on a daily basis, but that it also suggests a relevant skill score for probabilistic forecast systems. Weather roulette (M. S. Roulston and L. A. Smith, unpublished data) is a game in which you bet a fixed stake (or perhaps your entire net worth) on, say, the temperature at Heathrow. This is repeated every day. You can place a wager on each number between −5 and 29, where 29 will include any temperature greater than 29 and −5 any value below −5. How should you spread your bet? First off, we can see that it would be foolish not to put something on each and every possibility, just to be sure that we never go bust. The details of the distribution depend on your attitudes toward risk and luck, among other things; we will consider only the first. In fact we will initially take a risk neutral approach where we believe in our forecast probabilities: in this case we distribute our stake proportionally according to the predicted probability of each outcome. We imagine ourselves playing against a house that sets probabilistically fair odds using a probability distribution different from ours. Using this approach we can test the performance of different probability forecast systems in terms of how they fare (either as house or as punter).

As a first step, let's contrast how the ECMWF ensemble forecast would perform betting against climatological probabilities. Given a sample-climatology based on 17 years of observations, we know the relative frequency with which each option has been observed (given many centuries of data, we might know this distribution for each day of the year). But how do we convert an ensemble forecast of about 50 simulations into a probability forecast for the official observed temperature at Heathrow airport? There are a number of issues here.

9.5.4 Comparing like with like

How can Charlotte contrast the value of two different probability forecasting systems, say one derived from a high-resolution BFG forecast and the other from an EPS forecast? Or perhaps two EPS forecasts which differ either in the ensemble formation method or as to the ensemble size? There are two distinct issues here: how to turn simulations into forecasts and how to agree on the verification.

The first question is how to turn a set of model(s) simulations into a probability weather forecast. In the case of ensembles there are at least three options: fitting some parametric distribution to the entire ensemble, or dressing the individual ensemble's members (each with an appropriate kernel), or treating the entire forecast ensemble (and the corresponding verification) as a single point in a higher dimensional space and then basing a forecast upon analogues in this product space. We will focus on dressing the ensemble, which treats each individual member as a possible scenario. The product space approach treats the statistics of the ensemble as a whole; examples in the context of precipitation and wind energy can be found in Roulston *et al.* (2001, 2003).

Treating the singleton BFG as a point prediction (that is, a delta function) does not properly reflect its value in practice. To obtain a fair comparison with an EPS, we 'dress' the forecast values by replacing each value with a distribution. For the singleton BFG, one can construct a useful *kernel*[6] simply by sampling the distribution of historical 'error' statistics. The larger the archive of past forecast-verification pairs is, the more specific the match between simulation and kernel. Obviously we expect a different kernel for each forecast lead-time, but seasonal information can also be included if the archive span is long enough. Dressing the BFG in this way allows a fair comparison with probability forecasts obtained from ensemble forecasts. If nothing else, doing so places the same shortcoming due to counting statistics on each case. To see the unfairness in treating a BFG forecast as if it were an ensemble which had forecast the same value 50 times, recall the game of roulette. Think of the advantage one would have in playing roulette if, given an accountable probability forecast, you could place separate bets with 50 one dollar chips on each game, rather than having to place one 50 dollar chip on each game. To make a fair comparison, it is crucial that the single high-resolution simulation is dressed and not treated as an unequivocal forecast.

Suppose we have in hand a projection operator that takes a single model state into the physical quantity we are interested in, in this case the temperature at Heathrow. Each ensemble member could be translated into a specific forecast; because the ensemble has a fixed number of members, we will want to dress these delta functions to form a probability distribution that accounts both for our finite sample and for the impact of model error. We can use a best member approach to dressing, which is discussed in Roulston and Smith (2003) or simply use Gaussian kernels. Why don't we dress the ensemble members with the same kernel used for the BFG? Because to do so would double count for uncertainty in the initial condition. The only accounting for this uncertainty in the BFG is the kernel itself, while the distribution of the ensemble members aims to account for some of the uncertainty in the initial condition explicitly. Thus, other things being equal, the BFG kernel is too wide for ensemble members. Which kernel is appropriate for the ensemble members? Just as the error in the high-resolution forecast is too wide, the error in the ensemble mean is irrelevant. By using a kernel based on the best member of the ensemble, we aim to use a kernel that is as narrow as possible, but not more so. We do not, of course, know which member of the ensemble will turn out to provide the best forecast, but we do know that one of them will. Of course one must take some care in identifying the best member: considering the distant future of a trajectory will obliterate the more relevant information in the recent past. The critical issue here is to dress the simulations, the ideal kernel may well depend upon the circumstances, the size of the forecast archive, for instance.

The necessity of dressing emphasises the role of the forecast archive in the proper valuation validation of EPS forecasts. It also shows that the market value of an EPS will be diminished if a suitable EPS forecast archive is not maintained. This remains the case when we use simple kernels, such as a Gaussian distribution, and use the archive to determine any free parameters.

The second question of model verification addresses how we come to agree on what actually happened. This is not as clear-cut as it might first appear; a cyclist may arrive home drenched only to learn that, officially, it had not rained. Often, as in the case of the cyclist, what 'happened' is decided by decree: some agency is given the power to define what officially happened. Charles is happy with this scenario as long as the process is relatively fair: he wants to close his book at the end of each day and go home. Charlotte may be less pleased if the 'official' forecast says her generators were running at 77%, while a direct calculation based on the amount of gas burned yields 60%. In part, this difference arises because Charles really is playing a game, while Charlotte is trying to do something real. Outside the perfect model scenario, there appears to be no unique optimal solution.

9.5.5 Ignorance

Roulston and Smith (2002) discuss an information theoretical skill score that reflects expected performance at weather roulette. Called *ignorance*,[7] this skill score reflects the expected wealth doubling time of two gamblers betting against each other using probability forecasts, each employing Kelly systems.[8] For instance, a difference in ignorance scores of one bit per game would suggest that, on average, the players with lower ignorance score would double their stake on each iteration. The typical wealth doubling (or halving) time of a balanced roulette table is about 25 games, favouring the house, while scientific roulette players have claimed a doubling time on the order of three and favouring the player. How do these values arise?

Prior to the release of the ball, the probability of each number on a balanced roulette wheel can be taken to be equal, that is 1 in 37 in Europe (and 1 in 38 in the United States). If the house offers odds of 36-for-1, then the expected value of a unit stake after a single bet on a single number is 36/37. Raising this to the 25th power yields the reduction of the initial stake to one half; hence the expected wealth halving time is, in this case, roughly 25 games.

Roulette is a particularly appropriate analogy in that bets can be placed after the ball has been released. If, using observations taken after the ball is released, one could predict the quarter of the wheel on which the ball would land, and if such predictions were correct one third of the time, then one would expect a pay-off of roughly 36/9 once in every three games. This is comparable to reports of empirical returns averaging 30% or the expected stake doubling time of about three noted above.

There are several important things to note here. First, the odds offered by the house are not fair, even if the wheel is. Specifically, the implied 'probabilities' (the reciprocal of the odds) over the available options do not sum up to one: $37 \times (1/36) > 1$.

In practice it is rarely, if ever, the case that this sum is one, although this assumption is built into the most common definition of fair odds.[9]

Second, ignorance contrasts probability forecasts given only a single realisation as the verification. That is, of course, the physically relevant situation. In the 1995 ECMWF Seminar, we discussed the difference between the forecast distribution and some 'true' distribution as quantified via the Kolmogorov–Smirnov (KS) distance; a more elegant measure of the difference between two probability distributions is provided by the *relative entropy* (see Kleeman, 2002). What these and similar approaches fail to grasp, however, is that there is no 'true' distribution to contrast our forecast distribution with. Outside PMS, our forecast distribution rarely even asymptotes to the climatology unless forced to (and if forced, it asymptotes to a sample-climatology).

In the case of roulette, each spin of the wheel yields a single number. Before the ball is released, a uniform prior is arguably fair; but the relevant distribution is defined at the point when all bets must be placed, and this is *not* uniform. And there's the rub. What is a fair distribution at this point? Even if we assume it exists, it would depend on the size of your computer, and your knowledge of mechanics, and the details of this particular wheel, and this particular ball. And so on. Computing the relative entropy (or the KS distance) requires knowledge of both the forecast distribution and the true distribution conditioned on the observations. The equation that defines the relative entropy has two terms. The first term reflects the relative frequency with which certain forecast probabilities are verified; the second term reflects the relative frequency with which certain probabilities are forecast by a perfect model. Thus outside PMS the second term is unknowable, and hence the relative entropy is unavailable. The first term reflects the ignorance skill score. A shortcoming of ignorance is that it provides only a relative skill score ranking alternatives. Its advantage, of course, is that it is deployable in practice. As we have only the data, and data are but numbers, only limited skill scores like ignorance are available in practice. Similarly, our real-time performance will depend on the single future we experience, not the expectation over some many-worlds collection of 'possibilities'.

9.5.6 Heathrow temperature

So back to weather roulette: How does the dressed ensemble fare against climatology? Rather well. Consider daily bets, each with a stake of £100, placed over the period of a year starting on 23 December, 1999. The three-day forecast based on the ECMWF ensemble made almost £5000 in this period, while the ten-day forecast made about £1000 over the same period. As expected even in a perfect model, the value of the ensemble relative to climatology decreases with lead time. And there is more. The dressed ECMWF ensemble can also be used to bet against house odds based on the (dressed) ECMWF high-resolution best first guess (BFG) forecast. In this case the relative information gain is greater at a lead time of 10 days than at three, with winnings of over £5000 and over £2000, respectively. This also makes sense, inasmuch as the value of the ensemble forecast relative to climatology is greatest in the short range, while its value relative to a high-resolution forecast is greater at longer lead times (M. S. Roulston and L. A. Smith, unpublished data).

But there is still more, we can contrast the 51 member ensemble with the 10 member ensemble (all ensembles are dressed before use; the best-member kernel will, of course, vary with the size of the ensemble); is there statistically significant value added in the 51 member case relative to the case of randomly selecting 10 members? Yes. Relative to the 12-member case? Yes. And the 17-member case? No. Arguably, if Charles were betting only on Heathrow temperatures he would have done as well dressing 17 members as with using (that is, buying) all of them. He would have taken them for free (why not?) but not paid much for them. There is nothing universal about the number 17, in this context. The relevant ensemble size will depend on the details of the target variable as well as the model.

Of course any meteorological sales person worth their salt would immediately point out that users like Charlotte are interested in conditional probabilities of multiple variables at multiple locations. Charlotte is interested in the efficiency of each of her CCGT plants, as well as some integrated measure of electricity demand. But Charles is happy to stick with Heathrow, as long as his probability forecasts are making him money and costing him as little as possible. Making/pricing complicated multi-site derivatives is hard work; and he knows that it is dangerous as well: the curse of dimensionality implies tight constraints on the conditional probability forecasts that can be pulled out of even the best Monte Carlo ensemble with only a few dozen, or a few thousand, members. Charles simply need not take these risks, if the market in trading Heathrow temperatures is both sufficiently profitable and sufficiently liquid. Charlotte, by contrast, is exposed to these risks; the best she can do is understand the limits placed on the available forecasts by current technology.

There are three important take-home messages here: first that, like it or not, the ideal distribution between ensemble size and resolution will be target (that is, user) dependent; second that the marginal value of the $n + 1$st ensemble member will go to zero at different times for different uses; and third that the value of the EPS as a whole will depend critically upon the provision of an archive that allows users to convert these simulations into forecasts of the empirical quantities of interest.

9.6 Lies, damn lies, and the perfect model scenario

For the materialist, science is what teaches us what to believe. For the empiricist, science is more nearly what teaches us how to give up our beliefs.

Bas van Fraassen

Philosophers might call the meteorologist striving for PMS a realist; he believes in the physical reality of the model states and in the truth, or approximate truth, of the model. Alternatively, one modern variety of empiricist aims only for empirical adequacy. Such labels might seem inappropriate outside a philosophy department, if they were not relevant for the allocation of resources and hence the progress of science.

A model is empirically adequate if the dynamics of the model are in agreement with the observations. In this context, the word prediction often refers to prophecies as well as forecasts (see Smith, 1997). To judge empirical adequacy requires some accounting for observational noise and the fair use of a projection operator between the model-variables and the observables; *i*-shadowing would be a necessary condition for empirical adequacy of a dynamical system. There is still much to be understood in this direction, nevertheless it is not clear to me that *any* of our dynamical models are empirically adequate when applied to dynamic (non-transient) physical systems. There are a number of points, however, where the decision to work within PMS impacts the relevance of the statistics used to judge our models and the choice of how to 'improve' them.

9.6.1 Addressing model inadequacy and multiple model ensembles

The philosophical foundations of theories for objective probability distributions are built about the notion of equally likely cases or events (see Gillies, 2000). Within PMS, the perfect ensemble is an invocation of this ideal for chaotic dynamical systems that are perfectly modelled but imperfectly observed. This view of the world is available only to our twenty-first century demon who has access to various sets of indistinguishable states. Given a collection of good but imperfect models, we might try and use them simultaneously to address the issue of model inadequacy. But once we employ multiple imperfect models, the epistemological foundations that justify empirical probability forecasting turn to sand. While we can tolerate uncertainty in the parameter values of a model that comes from the correct model class by invoking what are effectively Bayesian methods, it appears we cannot find an internally consistent framework to support *objective* (empirically relevant) probability forecasts[10] when using multiple models drawn from distinct model classes, the union of which is known to be imperfect. Of course one can draw comfort in those aspects of a forecast in which models from each model class independently assign similar probabilities; but only comfort, not confidence (see Smith, 2002). Both systematic and flow dependent differences in the skill scores between the probabilistic forecasts from each model class may help us identify which physical phenomena deserve the most attention in each model class. Thus with time we can improve each of the models in the ensemble, while our probabilistic forecasts remain infinitely remote from the accountable probability forecasts of our twenty-first century demon.

9.6.2 Model inadequacy and stochastic parametrisations

Every attempt at model improvement is an attempt to reduce either parameter uncertainty or model inadequacy. *Model inadequacy* reflects the fact that there is no model within the class of models available to us that will remain consistent with all the data. For example, there may be no set of parameter values that enable a current model to shadow the observations *out-of-sample*[11] (or even in-sample?). Finding the (metric dependent) 'best' parameters, or distributions of parameters, and improving the data assimilation scheme are attempts at minimising the effects of model inadequacy within a given class of models. But model inadequacy is that which remains even when the best model within the model class is in hand; it affects both stochastic models and deterministic models.

Historically, physicists have tended to employ deterministic models, and operational numerical weather prediction models have been no exception to this trend. There are at least two good reasons why our forecast models should be stochastic in theory. The first comes from recent results (Judd and Smith, 2004) which establish that, given an imperfect non-linear chaotic model of a deterministic system, better state estimation (and perhaps, even better probabilistic forecasts) can almost certainly be obtained by using a stochastic model even when the system which generated the data really is deterministic! The second is the persuasive argument that, given current model resolution, it makes much more sense (physically) to employ stochastic subgrid-scale parametrisations than to employ dogmatically some mean value, even a good estimate of the expected value (Palmer, 2001; Smith, 2002). And in addition to these theoretical arguments, stochastic parametrisations have been shown to be better in practice (Buizza *et al.*, 1999). It is useful to separate arguments for improving a forecast based on each of these two reasons; we should maintain the distinction between methods which improve our model class (say, by adding stochastic physics) and those that deal with residual model inadequacy (which will always be with us).

While adopting stochastic parametrisations will make our models fundamentally stochastic, it neither removes the issue of model inadequacy nor makes our model class perfect. Consider what is perhaps the simplest stochastic model for a time series: independent and identically distributed (IID) normal (Gaussian) random variables of mean zero and standard deviation one. Data are numbers. *Any* data set has a finite (non-zero) probability of coming from this trivial IID model. Adjusting the mean and the standard deviation of the model to equal the observed sample-mean and sample-deviation will make it more difficult to reject the null hypothesis that our IID 'model' generated the data, but not much more difficult. We are soon faced with probabilities so low that, following Borel (1950), we could say *with certainty* that even a stochastic model does not shadow the observations. The possibility to resemble differs from the ability to shadow.

One often overlooked point here is that whenever we introduce a stochastic element into our models we also introduce an additional constraint: namely that to be a shadowing trajectory, the innovations must be consistent with the stochastic process specified in the model. Stochastic parametrisations may prove a tremendous improvement, but they need not yield *i*-shadowing trajectories even if there exist *some* series of innovations that would produce trajectories similar to a time series of analysis states. To be said to shadow, the particular series of innovations must have a reasonable probability given the stochastic process. Experience suggests that it makes little difference if we require a 95% isopleth, or 99%, or 99.999% for that matter. Model inadequacy manifests itself rather robustly. Without this additional constraint the application of the concept of *i*-shadowing to stochastic models would be trivial; the concept would be useless as any rescaled IID process could be said to shadow any set of observations. With this constraint, the introduction of stochastic terms does not guarantee shadowing, and their contribution to improved probabilistic forecasts can be fairly judged.

Even within PMS we must be careful that any critical, theoretically sound assumptions hold if they are relevant in the particular case in question (Hansen and Smith, 2000; Gilmour *et al.* 2001). Outside PMS a model's inability to shadow holds implications for operational forecasting, and for the Bayesian paradigm in applied science if not in mathematics. First of all, it is demonstrable that we can work profitably with imperfect models full of theory-laden (or better still, *model-laden*[12]) variables, but we can also be badly misled to misallocate resource in the pursuit of interesting mathematics, which assumes an unjustified level of perfection in our models. Ultimately, only observations can adjudicate this argument – regardless of what we 'know' must be the case.

9.6.3 Dressing ensemble forecasts and the so-called 'superensemble'

The superensemble method introduced by Krishnamurti *et al.* (1999) is a very interesting method for extracting a single 'locally optimised' BFG forecast from a multimodel ensemble forecast. In short, one finds the optimal weights (in space, time, and target variable) for recombining an ensemble of multimodel forecasts so as to optimise some root-mean-square skill score of the resulting BFG forecast.

The localised relative skill statistics generated within this 'superensemble' approach must contain a wealth of data of value in understanding the shortcomings of each component model and in addressing these model inadequacies. Nevertheless the 'superensemble' approach aspires only to form a single BFG forecast, and thus it might be more aptly called a 'super ensemble-mean' approach. How might we recast the single forecast output from the 'superensemble' approach, in order to make a 'like with like' comparison between the 'superensemble' output and a probability forecast? The obvious approach would be to dress it, either with some parametric distribution or in the same way we dressed the high-resolution forecasts in Section 9.5 above. Hopefully the shortcoming of either approach is clear: by first forming a single super ensemble-mean, we have discarded any information in the original distribution of the individual model states. Alternatively, dressing the individual ensemble members retains information from their distribution. So, while it is difficult to see how any 'superensemble' approach could outperform either a dressing approach or a product space method (or any other method which retains the distribution information explicitly), it would be interesting to see if in fact there is any relevant information in this distribution!

9.6.4 Predicting the relevance of indistinguishable states

The indistinguishable states approach suggests interesting alternatives both to current methods of ensemble formation and to the optimised selection of additional observations (Judd and Smith, 2001, 2004). The second are often called *adaptive observations* since the additional observation that is suggested will vary with the current state of the atmosphere (see Lorenz and Emanuel, 1998; Hansen and Smith, 2000).

Ensemble formation via indistinguishable states avoids the problems of adding finite perturbations to the current best guess analysis. The idea is to direct computational resources towards maintaining a very large ensemble. Rather than discarding ensemble members from the last initialisation some would simply be reweighted as more observations are obtained (see also Beven, 2002). It would relax (that is, discard) the assumptions of linearised uncertainty growth, for example that the observational uncertainty was small relative to the length scale on which the linearisation is relevant (see Hansen and Smith, 2000; Judd, 2003), or that the uncertainty distributions are Gaussian. And, by making perturbations as far into the distant past as possible, the ensemble members are as consistent with the long-term dynamics as possible; there are no unphysical 'balance' issues. Perhaps most importantly, an indistinguishable states approach appears to generalise beyond the assumption that near shadowing trajectories of reasonable duration do, in fact, exist when it is difficult to see how any of the current alternative approaches might function in that case. Much work remains to be done in terms of quantifying state dependent systematic model error (such as drift, discussed by Orrell *et al.* 2001) and detecting systematic differences between the behaviour of the analysis and that of the ensemble and its members.

When required, the ensemble would be reseeded from additional trajectories initiated as far back in time as practical, unrealistic perturbations would be identified and discarded without ever being included in a forecast. Reweighting evolved ensemble members given additional data (rather than discarding them), allows larger ensembles to be maintained, and becomes more attractive both as the period between initialising weather ensembles decreases and in the case of seasonal ensembles where many observations may be collected between forecast launches. In the former case at least, we can form lagged ensemble ensemble forecasts (LEEPS) by reweighting (and perhaps changing the kernel of) older ensemble members if they either remain relevant in light of the current observations or contribute to the probability forecast at any lead-time. Of course, outside PMS the relative weighting and the particular kernel assigned to the older members can differ from that of the younger members, and it would be interesting to use the time at which this weighting went to zero in estimating an upper bound for a reasonable maximum forecast lead time. And outside PMS, one must clearly distinguish between the ensemble of model simulations and a probabilistic forecast of weather observables.

Seasonal forecasts as studied within DEMETER (see www.ecmwf.int/demeter) provide ensembles over initial conditions and model structure. While it may prove difficult to argue for maintaining multiple models within PMS, the need to at least sample *some* structural uncertainties outside PMS provides an *a-priori* justification for multimodel ensembles, as long as the various models are each plausible. Indeed, it is the use of a single model structure that can only be justified empirically in this case, presumably on the grounds that, given the available alternatives, one model structure is both significantly and consistently better than all the others.

Within PMS, ensembles of indistinguishable states based on shadowing trajectories aims to yield nearly accountable probability forecasts, while operational methods based on singular vector or on bred vector perturbations do not have this aim, even in theory. The indistinguishable states framework also suggests a more flexible approach to adaptive observations if one model simulation (or a set of simulations) was seen to be of particular interest. To identify adaptive observations one can simply divide the current trajectories into two groups, one group in which each member has the interesting property (for example, a major storm) and the other group in which the simulations do not; call these groups red and blue. One could then identify which observations (in space, time, and model-variable) are most likely to provide information on distinguishing the distribution of red trajectories from the distribution of blue, or better said: which observations are most likely to give members of one of the groups high probability and those of the other low probability. As additional regular observations are obtained, the main computational overhead in updating our method is to reweight the existing trajectories, a relatively low computational cost and an advantage with respect to alternative approaches (see Hansen and Smith, 2001, and references thereof). Given a multimodel multi-initial condition ensemble, the question of adaptive observations shifts from complex assumptions about the growth of uncertainty under imperfect models, to a question of how to best distinguish between two subsets of known trajectories.

With multimodel forecasts we can also use the indistinguishable states framework to select observations that are most likely to 'falsify' the ensembles from one of two models on a given day. To paraphrase John Wheeler: each of our models is false; the trick is to falsify them as quickly as possible. Here only the observations can adjudicate; while what we decide to measure is constrained by what we can think of measuring, the measurement obtained is not. Le Verrier was as confident of his prediction of Vulcan as he was of his prediction of Neptune, and while both planets were observed for some time only Neptune is with us today.

Just as the plural of datum is not information, the plural of good idea is not theoretical framework. The indistinguishable states approach to forecasting and predictability has significant strengths over competing strategies, but its operational relevance faces a number of hurdles that have yet to be cleared. This statement should not be taken to indicate that the competition has cleared them cleanly!

9.6.5 A short digression toward longer timescales: the in-sample science

Noting the detailed discussion in the chapters by Tim Palmer and Myles Allen, we will not resist a short digression towards longer timescales (additional discussion can be found in Allen, 1999, Smith 2002, Stainforth *et al.*, 2005 and the references therein). An operational weather model makes many 7-day forecasts over its short lifetime; contrast this case with that of a climate model used to make 50-year forecasts, yet considered obsolete within a year or two. The continuous feedback from making forecasts on new unseen (out-of-sample) data is largely denied the climate modeller, who is constrained by the nature of the problem forever to violate one of the first maxims of undergraduate statistics: never restrict your analysis to in-sample statistics. By construction, climate modelling is an in-sample science. And the fundamentally transient nature of the problem makes it harder still.

As argued elsewhere (Smith, 2002), an in-sample science requires a different kind of consistency constraint. If model inadequacy foils our attempts at objective probability forecasts within the weather scenario, there is little if any chance of recovering these in the climate scenario.[13] We can, however, interpret multiple models in a different way. While approaches like best-member dressing can take into account the fact that different models will perform better in different conditions in the weather context, a climate modeller cannot exploit such observations. In the weather scenario we can use all information available at the time the models are launched when interpreting the distribution of model simulations, and as we launch (at least) once a day, we can learn from our mistakes.

Inasmuch as climate forecasting is a transient experiment, we launch only once. It is not clear how one might combine a collection of single runs of different climate models into a sensible probability forecast. But by studying the in-sample behaviour of ensembles under a variety of models, we can tune each model until an ensemble of initial conditions under each and every individual model can at least bound the in-sample observations, say from 1950 to 2000. If ensembles are then run into the future, we can look for the variables (space and time scales) on which the individual model ensembles bound the past and agree (in distribution) regarding the future.

Of course agreement does not ensure relevance to the future; our collection of models can share a common flaw. But if differences in the subtle details of different models that have similar (plausible) in-sample performance are shown to yield significantly different forecast distributions, then no coherent picture emerges from the overall ensemble. Upon adding a new model to this mix, we should not be surprised by a major change to the overall forecast distribution. This may still occur even if 'each and every one' of the current models has similar forecast distributions, but in this case removing (and hopefully, adding) one model is less likely to do so. In any event, we can still use these differences in the forecast distributions to gain physical insight, and improve each model (individually) using the in-sample data yet again (Smith, 2002). But as long as the details can be shown to matter significantly, we can form no coherent picture.

9.7 Socio-economic relevance: why forecast everyday?

From a scientific point of view, it is interesting to ask why we make forecasts every day? Why not spend all available funds making detailed observations in, say, January, and then spend the rest of the year analysing them, using the same computational resources but with a higher resolution model than possible operationally? Once we got the physics of the processes for January down pat, we could move on to February. And so on. There are a number of reactions to this question, but the most relevant here is the simple fact that numerical weather forecasting is more than just a scientific enterprise; real-time forecasting is largely motivated by the socioeconomic benefits it provides. One of the changes we will see in this century is an increase in the direct involvement of users, both in the consideration of their desires and in the exploitation of their data sets. Closing this loop can benefit both groups: users will employ forecast products more profitably while modellers will have to leave the 500 mb model-pressure height field behind, along with the entire model state space, and again give more consideration to empirically accessible variables.

Electricity demand provides real-time observations reflecting a number of environmental variables, updated on the timescale of a model time step, and spatially integrated over areas similar to model grid spacing. Might not assimilating this data (or using it to reweight the trajectories of what were indistinguishable) be more likely to yield relevant information than a single thermometer accurate to 16 bits? Charlotte uses demand observations every day; she would certainly be willing to sell (or barter) these bits for cheaper (or better) forecasts.

It is easily observed that many talented meteorologists dismiss the idea of a two-way exchange with socio-economics out of hand. They state, rather bluntly, that meteorologists should stick to the 'real science' of modelling the atmosphere, even dismissing the comparison of forecast values with observations as mere 'post-processing'.[14] Interestingly, if only coincidentally, these same scientists are often those most deeply embedded within PMS. Of course I do not really care about forecasting today's electricity demand *per se* any more than I am interested in benthic foraminifera; but I do care very much about empirical adequacy and our ability to forecast things we can actually measure: any empirically accessible quantity whatever its origin. I am not overly concerned whether a quantity is called 'heating degree days' or 'temperature'. As long as they correspond to something we can measure, not a model variable, I'll take as many as I can get. Probabilistic forecasts for both temperature and heating degree days, as well as cumulative heating degree days, are posted daily on www.dime.lse.ac.uk.

9.7.1 Ensembles and wind power

So let us consider three examples of economic meteorology. The first is a study of a hypothetical wind farm using real wind data, real forecasts, real electricity prices and a non-existent wind farm just south of Oxford. (Detailed results are available in Roulston *et al.*, 2003.) The economic constraints vary with changes in regulation, which occur almost as frequently as changes in an operational weather model. In our study, the wind farm must contract for the amount of electricity it will produce in a given half hour a few days in advance; it will not be paid for any overproduction, and will have to supply any shortfall by buying in electricity at the spot price on the day. What we can do in this example is to contrast several different schemes for setting the size of the contract, and then evaluate them in terms of the income of our fictional wind farm. Contrasting the use of climatology, the dressed ECMWF high-resolution forecast and the dressed ECMWF ensemble forecast shows, for instance, that at day 4 there is a clear advantage in using the ensemble. Would Charlotte buy the ECMWF ensemble? That question involves the cost of the ensemble relative to its benefits, the cost of the high-resolution run, the size of the current forecast archive and the availability of alternative, less expensive probability forecasts. But it is the framework that is important here: she can now make an economic choice between probability forecasts, even if some of those probability forecasts are based on single BFG model runs (singleton ensembles). As shown in our next example, it is likely that in some cases the dressed ensemble forecast may not contain significantly more information than the dressed high-resolution forecast. Still, the framework allows us to see when this is the case, and often why. Figure 9.1 shows the daily income from the wind farm: the upper panel uses climatology forecasts; the lower panel uses the dressed ensemble (for more detail, see Roulston *et al.*, 2003). The ensemble forecast provides increased profit at times of unseasonable strong winds (March 2000) while avoiding loss at times of unseasonably weak winds (January 2000). It is not perfect, of course, just better. Presenting the impact of weather forecasts in this format allows Charlotte to make her own decisions based on her company's attitude to risk.

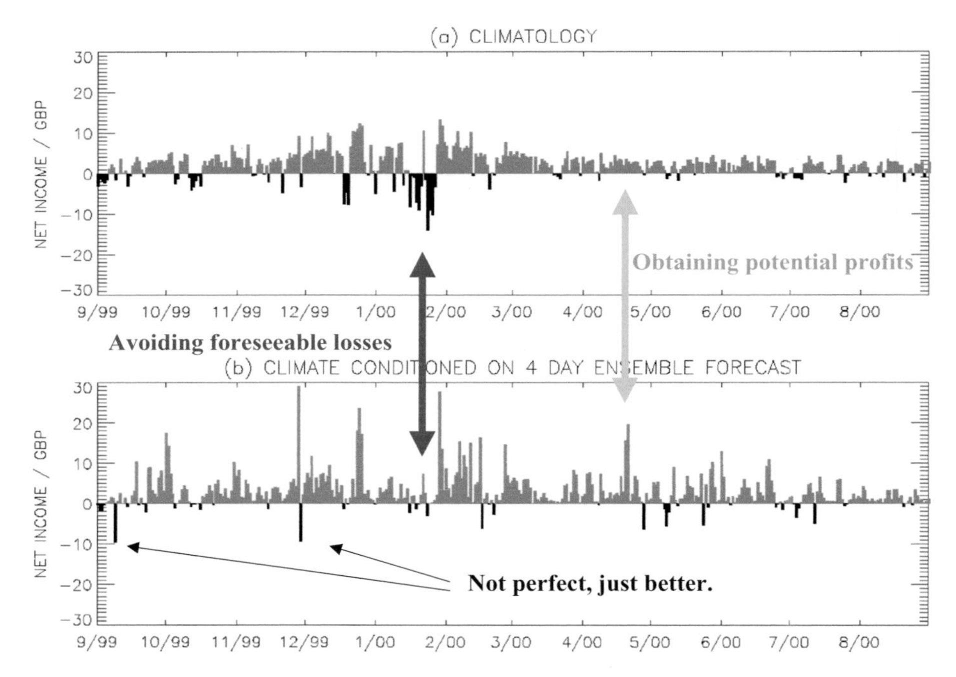

Figure 9.1 The daily income from a hypothetical wind farm based on observed winds, real electricity prices and ECMWF forecasts. (a) The profit when the estimated production use is based on climatology. (b) The same when based on the ECMWF ensemble. For more detail, see Roulston *et al.* (2003).

9.7.2 Ensembles and wave risk

Our second example comes from a research project led by Mark Roulston in cooperation with Jerome Ellepola of Royal Dutch Shell. Shell is interested in waves at offshore platforms all over the world, both fixed rigs (for example, oil well platforms) and floating platforms (Floating Production Storage and Offloading Vessels or FPSOVs). The ECMWF ensemble system includes a wave model (Jansen *et al.* 1997). Our aim is to evaluate the relative ignorance both of dressed ensemble forecasts and the dressed high-resolution forecast. In this case buoy data supplied by Shell provide the target observations and (out-of-sample) verification. The results here differ significantly at Bonga, a floating FPSOV off the west coast of Africa, and at Draugen, a fixed platform in the North Sea. Details of this study can be found in Roulston *et al.* (2005).

At Bonga, we find no statistically significant advantage in using the ensemble forecasts of significant wave height in the frequency bands of interest to Shell, even at day 10. Physically, one might argue that the waves arriving at Bonga now tend to have originated far away in space–time: having a good atmospheric forcing in day 1 and 2 yields a low ignorance wave forecast even at day 10. In that case, the wind that generates the relevant waves has already 'hit the water' well before the waves reach the FPSOV, suggesting that the current forecasts may contain useful information at Bonga well beyond day 10. Alternatively, one might argue that the current ECMWF ensemble does not target the tropics, reducing the relevance of the ensemble at Bonga, and that doing so would increase the value of the wave ensemble forecast in week one. Either way, after checking for statistically insignificant but economically important extreme events, Shell might argue that there was no reason to buy more than the dressed BFG for Bonga. Of course, it is also possible that increasing the size of the forecast archive might increase the relative skill of the ensemble-based forecast.

At Draugen the situation is quite different, relatively fast-growing nearby weather events sampled in the ensemble result in a significant information advantage for the ensemble wave height forecasts. In the North Sea the probability forecasts based on the ensembles have a clear advantage over those from the high-resolution simulations. These results suggest that significant thought should go into setting the price structure for BFG-based probability forecasts and EPS-based forecasts (and yet other multimodel options which dress the union of BFG and EPS simulations). Such issues are relevant to the economics of forecasting, if not economic forecasts; Charlotte is interested in both.

9.7.3 Electricity demand: probabilistic forecasts or probabilities?

The third example involves forecasting electricity demand in California, a nearly ideal 'how much' question. Modern economies run on electricity; to maintain a reliable electricity grid one must first forecast likely demand and then arrange enough generation to meet that demand. Producing excess generation is expensive, while not having enough generation can be catastrophic (Altalo and Smith, 2004). This asymmetry of impact between positive and negative forecast errors is common in industry; it brings the difficulties of using 'probability' forecasts into sharp focus. A case study for the California electricity grid can be found in Smith *et al.* (2005).

If we do not expect our probability forecasts to be accountable, then we should not be surprised when traditional methods for using these forecasts, such as maximising the expected utility, fail miserably in practice. It is clear that we have extracted useful probabilistic information from our multi-initial condition, multimodel ensembles; it is not at all clear that from this information we can extract a probability forecast which is useful as such. A method to do so would be of great interest scientifically, of great value economically, and of great use socially. In the meantime, however, alternative ad hoc methods for using these predictive distributions are sometimes found to yield more statistically robust results in practice.

9.8 So what is 'uncertainty in initial condition', really?

This may seem a trivial question. The standard answer to this trivial question would be the equation

$$e = x - \mathbf{X}$$

where e is the uncertainty in the initial condition, x is the model state we used at $t = 0$, and $\mathbf{X}$ is the true state of the real system at time $t = 0$. There is a small difficulty here. Perhaps unexpectedly[15] it is with the symbol '$-$', which is undefined as when contrasting apples and oranges, making our one equation ill-posed. While x sits in our model-state space (without doubt one large integer in this digital age), $\mathbf{X}$ is, at best, the true state of the atmosphere. These vectors, if they both exist, exist in different spaces even if we confuse model-temperature with temperature (see Smith, 1997). There is an easy mathematical way out here, simply replacing $\mathbf{X}$ with $\mathbf{P}(\mathbf{X})$ where $\mathbf{P}$ is some projection operator that takes the true state space into our model-state space (this is touched upon in Smith, 2001, Orrell *et al.*, 2001 and Judd and Smith, 2004). Introducing the projection operator shifts the ambiguity in the minus sign to the projection operator, $\mathbf{P}$. It is not clear that we can nail down $\mathbf{P}$ beyond simply saying that it is many-to-many, which may prove more troubling than it seems at first.

The main difficulty in interpreting this equation, however, may be of a different sort: if there is no model state that shadows the time series of each consecutive analysis to within the observational uncertainty, then the shortcomings of the forecast simply cannot be attributed to uncertainty in the initial condition. Why? Because in this case, there was no model initial condition to be uncertain of. It was not that we did not know which value of x to use, but that there was no value of x which would have given an accurate (useful) point forecast. There is no clear definition of uncertainty in the initial condition outside the perfect model scenario.

Both the projection operator and tests of empirical adequacy are bound up with the definition of observational 'noise', while the identification of shadowing trajectories requires projecting observational noise into the model state space. Although most scientists believe that they can recognise it when they see it, much remains to be said about the concept of noise. Turning to a more practical matter: how might we proceed in practice other than by developing the best model structure our technology can support, inserting physically motivated parametrisations with empirically estimated parameter values, insisting that water freeze at exactly zero degrees, and trusting that in time our model will slowly approach Truth?

There is an alternative. Its implications are not yet clear, but if I am lucky enough to be invited to Predictability 2009 then I hope to learn more at that time. The alternative is to embrace model inadequacy while relinquishing the twenty-first century Laplacian dream of accountable probability forecasts. To adopt instead a goal which is less attractive but conceivably attainable: using ensembles of initial conditions evolved under a collection of imperfect models, aiming to say as much as is justified by our models. We should expect our forecasts to be blatantly wrong as rarely as possible, but not less rarely than possible.

9.9 Conclusions and prospects

Predictability present: There is no doubt that the current operational ensemble systems have value beyond that recognised in industry; this is an opportunity. The question should be seen as one of how to exploit this information content, not as to whether or not it 'exists'. Each of our three users can base better economic decisions and reduce their exposure to weather risk by using probabilistic forecasts based on existing ensemble prediction systems. This is not to say that current probability forecasts are accountable, but rather that current ensemble forecasts are valuable. Charlie, who deals for the most part with binary decisions, can determine the ideal probability thresholds at which he should act and interpret forecasts more profitably. Charles can translate the forecast probabilities both better to gauge the likely behaviour of weather derivatives in the near term, and to extract the likely impact of weather on the futures markets. And Charlotte, our most numerate user, can incorporate probabilistic weather forecasts in a variety of applications from hours to weeks, with the aim of including probabilistic weather information to allow seamless forward planning for weather impacts. Current ensemble prediction systems have demonstrable economic value.

Two obvious questions arise. First, what can be done to raise the level of exploitation of these forecasts? And second, how can we best move forward to increase their value? The answer to the first question involves education, technology transfer, and both the production and advertisement of case studies illustrating the value of current forecast products in realistic economic examples. Answering the question of how best to move forward would, no doubt, benefit from a better understanding of what constitutes the notion of 'forward', but the improvement of forecast models and their associated data assimilation, the improved generation and retention of members in initial condition ensembles, the wider use of multimodel ensembles and improved methods for translating ensembles of simulations into weather forecasts will each play a role.

Almost a century ago, L. F. Richardson began the first numerical weather forecast by hand, while envisioning the use of parallel computing in numerical weather forecasting. Today, the electronic computer plays two rather distinct roles in physical science. First it allows us to calculate approximate solutions to a wide variety of equations at speeds Richardson could only dream of. Second, and perhaps even more importantly, it allows us to record and access data that, in turn, reveal just how imperfect our models are. Having accepted that probabilistic forecasts are here to stay, it will be interesting to see when it proves profitable to shift from trying to make our model perfect towards trying to make our forecasts better. Accepting that our best dynamical models are not and never need be empirically adequate will open new avenues toward understanding the physics of the Earth System, and may allow us to achieve predictability past the limitations we face at present.

Acknowledgements

It is difficult to write the acknowledgements of a paper that covers a decade. The more important of these insights were obtained through joint work with Mark Roulston, Kevin Judd, Jim Hansen, David Orrell and Liam Clarke while Jost van Hardenburg supplied critical calculations and insights. Myles Allen, Judy Curry, Milena Cuellar, Tim Palmer, Dave Stainforth, Alan Thorpe and Antje Weisheimer made useful comments on earlier drafts, and I am grateful to numerous others for discussions during and after the ECMWF Seminar. Contributions by Isla Gilmore, Pat McSharry and Christine Ziehmann helped lay the foundations on which this iteration was built, clarified by recent discussions with Jim Berger, Jochen Broecher, Devin Kilminster, Ele Montuschi and Naomi Oreskes.

I happily acknowledge personal and philosophical debts to Nancy Cartwright, Donald Giles and Robert Bishop, each of whom I hope to repay. Also, I again thank Tim Palmer for the invitation to the 1995 Predictability Seminar, which had a significant if unpredictable impact on my research; I hope he does not regret the results. LSE CATS's Faraday Partners, NG Transco, EDF, London Electricity and Risk Management Solutions have contributed to my understanding of economics of weather forecasting; I am particularly grateful to Mary Altalo, Melvin Brown, Neil Gordon, Steve Jewson, Shanti Majithia and Dave Parker. Roulston, Judd and myself have each benefited from the Predictability DRI under grant N00014-99-1-0056. Kilminster and myself have also been supported by the USWRP and NOAA. I gratefully acknowledge the continuing support of Pembroke College, Oxford.

Notes

1. The use of CCGT generators comes from the fact that their efficiency in converting fuel to electricity varies with temperature, pressure and humidity; any generation method with weather dependent efficiency would suffice here. For details on forecasting for CCGT generators, see Gordon and Smith (2005).
2. The construction of state-dependent conditional probability distributions from ensembles requires having enough members to estimate a *distribution* of the $n + 1$st variable, given particular values of the first n variables. This is just another guise of the curse of dimensionality, made worse both by the need for a distribution of the target variable and by Charlotte's particular interest in the tails of (each of the many distinct conditional) distributions.
3. After Laplace, see Nagel (1961). Note that the distinction between uncertainty in initial conditions and model inadequacy was clear to Laplace (although perhaps not to Bayes); Laplace distinguished uncertainty of the current state of the universe from 'ignorance of true causes'.
4. It is not clear how well the Bayesian paradigm (or any other) can cope with model inadequacy, specifically whether or not it can yield empirically relevant probability

forecasts given an imperfect model class. I am grateful to Jim Berger for introducing me to a Bayesian research programme that strives for a systematic approach to extracting approximately accountable probability forecasts in real time.

5. This section has generated such varied and voluminous feedback that I am loath to alter it at all. A few things might be clarified in this footnote. First I realise that ground frost may occur when the 2-metre temperature is well above zero, that the freezing point of fresh water might be less relevant to the Earth System than the freezing point of sea water, and (I've learned) that the freezing point of deliquescent haze is quite variable. Yet each of these variables is amenable to physical measurement; my point is simply that their 40 km resolution model-variable namesakes are not, except inasmuch as they affect the model trajectories. Granted, in the lab 'zero degrees C' may be argued exact by definition; any reader distracted by this fact should read the phrase as '32 degrees F' throughout. I am also well aware that the value given to one parameter will affect others, but I would avoid the suggestion that 'two wrongs make a right': outside PMS there is no 'right'. Whenever different combinations of parameter values shadow, then we should keep (sample) all such combinations until such time as we can distinguish between them given the observations. This is one reason why I wish to avoid, as far as possible, the notion of shadowing anything other than the observations themselves, since comparisons with averages and the like might yield the 'two wrongs' without the 'better'. Lastly I realise that it is *always* possible that out-of-sample, the parameters may fail to yield a reasonable forecast however they have been tuned: all forecasts must be interpreted within the rosy scenario, as discussed in Smith (2002).
6. A kernel is a distribution function used to smooth the single value from each simulation, either by substituting the kernel itself or by sampling from it. Note that different kernels can be applied to different ensemble members, as long as they can be distinguished without reference to any future verification (for example, one would expect the ECMWF EPS control member to have a different kernel and be more heavily weighted than the perturbation members at short lead-times; but even the rank order of members may prove useful at longer lead times).
7. As far as I know, ignorance was first introduced in this context by Good (1952) who went so far as to suggest that the wages of British meteorologists be based on the score of their forecasts.
8. In each round, the stake is divided across all options, the fraction on each option proportional to the predicted probability of that option. Kelly (1956) was in fact interested in interpreting his result in terms of information theory; see Epstein (1977).
9. A typical definition of 'fair odds' would be those odds on which one would happily accept either side of the bet; it is not clear that this makes sense outside PMS. Operational fair odds should allow the house offering those odds some opportunity of remaining solvent even if it does not have access to some non-existent perfect probability forecasts. One might define fair odds outside PMS as those set by a not-for-profit house aiming only to maintain its endowment while providing a risk management resource (L. A. Smith *et al.*, unpublished data).

10. A necessary condition for 'objective' as used here is that the forecasts are accountable. This can be tested empirically; we have found no dynamic physical system for which accountable forecasts are available. This is a stronger constraint than other useful meanings of objective probabilities; for example that all 'rational men' would converge to the same probability forecast given the same observations and background information. Subjective probability forecasts can, of course, be obtained quite easily from anyone.
11. The phrase *out-of-sample* reflects the situation where the data used in evaluation were not known before the test was made, in particular that the data were not used in constructing the model or determining parameters. If the same data were used in building the model and testing it, then the test is *in-sample*. Just as passing an in-sample test is less significant than passing an out-of-sample test, failing an in-sample test is more damning. Note that data are only out-of-sample once.
12. Our theories often involve variables that cannot be fully observed; philosophers would call variables like a temperature field *theory-laden*. Arguably, this temperature field is something rather different from its *in silico* realisation in a particular computational model. Thus the model variables composing, say, the T42 temperature 'field' might be called *model-laden*.
13. I would have said no chance, but Myles Allen argues effectively that climate models *might* provide information on 'climate variables' without accurately resolving detailed aspects of the Earth System.
14. 'Post-what?' one might ask, as a computer simulation is not a forecast until expressed in terms of an observable weather variable.Verification against the analysis may be a necessary evil, but its limitations make it inferior to verification against observations.
15. As with an embarrassing number of insights I initially felt were unexpected, Ed Lorenz provided a clear discussion of this issue several decades ago; in this case, for example, he explicitly discussed models that were similar enough to be 'subtractable' (Lorenz, 1985).

References

Allen, M. R. (1999). Do-it-yourself climate modelling. *Nature*, **401**, 642.

Altalo M. G. and L. A. Smith (2004). Using ensemble weather forecasts to manage utilities risk. *Environ. Finance*, **20**, 8–9.

Angstrom, A. K. (1919). Probability and practical weather forecasting. *Centraltryckeriet Tecknologforeningens Forlag*.

Beven, K. J. (2002). Towards an alternative blueprint for a physically-based digitally simulated hydrologic response modelling system. *Hydrol. Process.*, **16(2)**, 189–206.

Bishop, R. (2003). On separating prediction from determinism. *Erkenntnis*, **58**, 169–88.

Borel, E. (1950). *Probability and Certainty*. New York: Walker.

Buizza, R., Miller, M. J. and T. N. Palmer (1999). Stochastic simulation of model uncertainties in the ECMWF Ensemble Prediction System. *Quart. J. Roy Meteor. Soc.*, **125**, 2887–908.

Earman, J. (1986). *A Primer on Determinism*. D. Reidel.

Epstein, R. A. (1977). *The Theory of Gambling and Statistical Logic*. Academic Press.

Fraassen, B. C. van (2002). *The Empirical Stance*. Yale University Press.

Gillies, D. (2000). *Philosophical Theories of Probability*. Routledge.

Gilmour, I., L. A. Smith and R. Buizza (2001). Linear regime duration: is 24 hours a long time in synoptic weather forecasting? *J. Atmos. Sci.*, **22**, 3525–39.

Good, I. (1952). Rational decisions. *J. Roy. Stat. Soc. B*, **14**.

Gordon N. and L. A. Smith (2005). Weather forecasting for combined Cycle Gas Turbine (CCGT). *Proceedings of the First THORPEX Science Symposium* (in press). American Meteorological Society.

Hansen, J. A. and L. A. Smith (2000). The role of operational constraints in selecting supplementary observations. *J. Atmos. Sci.* **57**(17), 2859–71.

Hansen, J. A. and L. A. Smith (2001). Probabilistic noise reduction. *Tellus*, **5**, 585–98.

Janssen, P. A. E. M., B. Hansen and J. R. Bidlot (1997). Verification of the ECMWF wave forecasting system buoy and altimeter data. *Weather Forecast.*, **12**, 763–84.

Judd, K. (2003). Nonlinear state estimation, indistinguishable states and the extended Kalman filter, *Physica D*, **183**, 273–81.

Judd, K. and L. A. Smith (2001). Indistinguishable states. I: The perfect model scenario. *Physica D*, 125–41.

(2004). Indistinguishable states. II: Imperfect model scenario. Physica *D*, **196**, 224–42.

Kelly, J. L. Jr. (1956). A new interpretation of information rate. *Bell System Technical J.*, **35**(4), 917–26.

Kennedy, M. and A. OHagan (2001). Bayesian calibration of computer codes. *J. Roy. Stat. Soc. B, 63*, 425–64.

Kleeman, R. (2002). Measuring dynamical prediction utility using relative entropy. *J. Atmos. Sci.*, **50**, 2057–72.

Krishnamurti, T. N., *et al.* (1999). Improved skills for weather and seasonal climate forecasts from multimodel superensemble. *Science*, Sept 3.

Lorenz, E. N. (1985). The growth of errors in prediction. In *Turbulence and Predictability in Geophysical Fluid Dynamics and Climate Dynamics*, ed. M. Ghil, pp. 243–65. North Holland.

Lorenz, E. N. and K. Emanuel (1998). Optimal sites for supplementary weather observations. *J. Atmos. Sci.*, **55**, 399–414.

Murphy, A. H. (1977). The value of climatological, categorical and probabilistic forecasts in the cost-loss ratio situation. *Mon. Weather Rev.*, **105**, 803–16.

Nagel, E. (1961). *The Structure of Science*. Harcourt, Brace and World.

Orrell, D., L. A. Smith, T. Palmer and J. Barkmeijer (2001). Model error and operational weather forecasts. *Nonlinear Proc. Geoph.*, **8**, 357–71.

Palmer, T. N. (2001). A nonlinear perspective on model error. *Quart. J. Roy. Meteor. Soc.*, **127**, 279–304.

(2002). The economic value of ensemble forecasts as a tool for risk assessment: from days to decades. *Quart. J. Roy. Meteor. Soc.*, **128**, 747–74.

Popper, K. (1956). *The Open Universe*. Routledge. Reprinted 1982.

Richardson, D. S. (2000). Skill and relative economic value of the ECMWF ensemble prediction system. *Quart. J. Roy. Meteor. Soc.*, **126**, 649–68.

Roulston, M. S. and Smith, L. A. (2002). Evaluating probabilistic forecasts using information theory. *Mon. Weather Rev.*, **130**(6), 1653–60.

(2003). Combining dynamical and statistical ensembles. *Tellus*, **55A**, 16–30.

(2004). The boy who cried wolf revisited: the impact of false alarm intolerance on cost-loss scenarios. *Weather Forecast.*, **19**(2), 391–7.

Roulston, M. S., C. Ziehmann and L. A. Smith (2001). *A Forecast Reliability Index from Ensembles: A Comparison of Methods*. Report. Deutscher Wetterdienst.

Roulston, M. S., Kaplan, D. T., Hardenberg, J. and Smith, L. A. (2003). Using medium range weather forecasts to improve the value of wind energy production. *Renewable Energy*, **28**(4), 585–602.

Roulston, M. S., J. Ellepola, J. von Hardenberg and L. A. Smith (2005). Forecasting wave height probabilities with numerical weather prediction models. *Ocean Eng.*, **32**, 1841–63.

Smith, L. A. (1996). Accountability and error in forecasts. In *Proceedings of the 1995 Predictability Seminar*, ECMWF.

(1997). The Maintenance of Uncertainty. In *Proceedings of the International School of Physics* (*Enrico Fermi*), Course CXXXIII, ed. G. Cini (Societa Italiana di Fisica, Bologna), pp. 177–368.

(2001). Disentangling uncertainty and error: on the predictability of nonlinear systems. In *Nonlinear Dynamics and Statistics*, ed. A. I. Mees, pp. 31–64, Boston: Birkhauser.

(2002). What might we learn from climate forecasts? *Proc. Nat. Acad. Sci.*, **99**, 2487–92.

Smith, L. A., M. Roulston and J. von Hardenberg (2001). *End to End Ensemble Forecasting: Towards Evaluating the Economic Value of the Ensemble Prediction System*. Technical Memorandum 336. ECMWF.

Smith, L. A., M. Altalo and C. Ziehmann (2005). Predictive distributions from an ensemble of weather forecasts: extracting California electricity demand from imperfect weather models. *Physica D* (in press).

Stainforth, D. A., T. Aina, C. Christensen, *et al.* (2005). Evaluating uncertainty in the climate response to changing levels of greenhouse gases. *Nature* **433**(7024), 403–6.

Weisheimer, A., L. A. Smith and K. Judd (2005). A new view of seasonal forecast skill: bounding boxes from the DEMETER ensemble forecasts. *Tellus*, **57A**, 265–79.

10

Predictability of coupled processes

Axel Timmermann
IPRC, SOEST, University of Hawai'i, Honolulu

Fei-Fei Jin
*Department of Meteorology,
Florida State University, Tallahassee*

The predictability of coupled multiple-timescale dynamical systems is investigated. New theoretical concepts will be presented and discussed that help to quantify maximal prediction horizons for finite amplitude perturbations as well as optimal perturbation structures. Several examples will elucidate the applicability of these new methods to seasonal forecasting problems.

10.1 Introduction

One of the key elements of all natural sciences is forecasting. In some situations it is the existence of newly predicted particles that decides on the validity of physical theories; in other situations it is the predicted trajectory of a dynamical system that can falsify scientific hypotheses. In the seminal Lorenz (1963) paper, E. Lorenz discovered that certain low-dimensional non-linear dynamical systems bear an interesting and at that time unexpected property: two initially close trajectories will diverge very quickly in phase space, eliminating the possibility for long-term forecasting. This paper, strongly inspired by the theoretical considerations of Barry Saltzman, gave birth to chaos theory. It is this particular dependence on the initial conditions that is intrinsic to all weather, ocean and climate prediction efforts.

In a uni-timescale chaotic system we have the following situation. The rate of divergence of initially infinitesimally close trajectories can be quantified roughly in

Predictability of Weather and Climate, ed. Tim Palmer and Renate Hagedorn. Published by Cambridge University Press.

terms of the Lyapunov exponent λ that can be expressed as

$$\lambda = \lim_{t\to\infty}\lim_{\delta x(0)\to 0} \frac{1}{t}\ln\frac{\delta x(t)}{\delta x(0)} \tag{10.1}$$

where $\delta x(0)$ represents the initial error. With a given error tolerance level of Δ the future state of the system can be predicted T_p ahead, with

$$T_\mathrm{p} \sim \frac{1}{\lambda}\ln\left(\frac{\Delta}{\delta x(0)}\right). \tag{10.2}$$

If the dynamical system has different timescales, the Lyapunov exponent is proportional to the smallest characteristic timescale, irrespective of the variance carried by fluctuations on these timescales. By contrast, for geophysical flows, predictions can be made far beyond the Lyapunov timescale, which characterises small-scale fluctuations associated with turbulence. In particular for the atmosphere it is the large-scale dynamics that determines long-term predictability (Lorenz, 1969). Apparently, the Lyapunov exponent and its association with predictability horizons are of very limited use in multiple-timescale systems. A similar paradoxical situation can occur when we plan to compute the singular vectors for coupled two-timescale systems, such as the coupled atmosphere–ocean system. In order to determine these optimal perturbations using a coupled general circulation model, one needs to compute the Jacobian and its adjoint along the non-linear phase-space trajectory. An important question is now, what is the associated timescale of the linearised and adjoint coupled models? Do these models really capture the long-term dynamics, associated for example with the ENSO phenomenon, or are they mostly capturing the growth of weather perturbations in an adiabatically slow varying oceanic background? Before we dwell into these problems of how to determine optimal perturbation patterns of coupled systems let us return to the concept of Lyapunov exponents.

In an attempt to overcome the difficulties associated with the infinitesimally small amplitude perturbations Aurell *et al.* (1997) have generalised the Lyapunov exponent concept to non-infinitesimal finite amplitude perturbations. This generalisation has the advantage that both the non-linear dynamical evolution of these perturbations as well as the predictability of multiple-timescale systems can be treated appropriately (Bofetta *et al.*, 1998). The so-called *finite-size Lyapunov* is defined as

$$\lambda(\delta) = \frac{1}{T_r(\delta)}\ln r, \tag{10.3}$$

with δ denoting the finite amplitude perturbation of the initial conditions, and T_r the time it takes for the initial perturbation to grow by a factor of r. In Bofetta *et al.* (1998) it is shown that this is a suitable quantity to describe the predictability of multiple-timescale systems. Bofetta *et al.* (1998) illustrate this new concept by studying the predictability of two coupled sets of non-linear Lorenz (1963) equations, one characterised by a slow timescale and the other by a fast timescale. It is shown that for small perturbations $\lambda(\delta)$ is almost equal to the maximum Lyapunov exponent

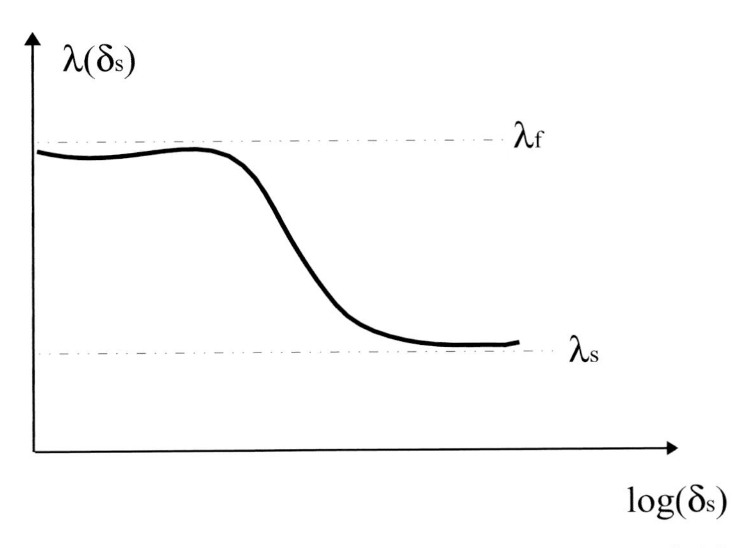

Figure 10.1 Finite size Lyapunov exponent of two coupled (slow–fast) Lorenz models computed from the slow variables. The two horizontal lines represent the uncoupled Lyapunov exponents. Figure schematised after Bofetta *et al.* (1998).

associated with the shortest timescale, whereas for large initial perturbations $\lambda(\delta)$ asymptotically approaches the Lyapunov exponent of the slow decoupled system. This is illustrated in Figure 10.1.

Furthermore, the predictability time T_{p} of the slow component of the coupled fast–slow Lorenz (1963) system becomes orders of magnitude larger than the one based on the classical Lyapunov estimate, when large error tolerances are allowed for, as illustrated in Figure 10.2.

An interesting example for a coupled slow–fast system, and a challenging one for seasonal forecasters, is the tropical Pacific coupled atmosphere–ocean system with its primary oscillatory instability, the so-called El Niño–Southern Oscillation (ENSO) phenomenon. ENSO variability arises from the so-called Bjerknes positive air–sea feedback and a negative feedback provided by slow ocean adjustment processes (Jin, 1997). In some ENSO models (e.g. the Zebiak and Cane 1987 model and hybrid coupled models) the atmospheric dynamics is treated in a special way (e.g. Gill, 1980; Kleeman, 1991): the linearised atmospheric shallow water equations on an equatorial β plane for zonal, meridional velocities and the geopotential height anomaly can be written as

$$\frac{\partial u}{\partial t} - \beta y\, v + \frac{\partial \Phi}{\partial x} = -ru \tag{10.4}$$

$$\frac{\partial v}{\partial t} + \beta y\, u + \frac{\partial \Phi}{\partial y} = -rv \tag{10.5}$$

$$\frac{\partial \Phi}{\partial t} + c_a^2 \left(\frac{\partial u}{\partial x} + \frac{\partial v}{\partial y} \right) = \alpha Q(x, y) - r\Phi \tag{10.6}$$

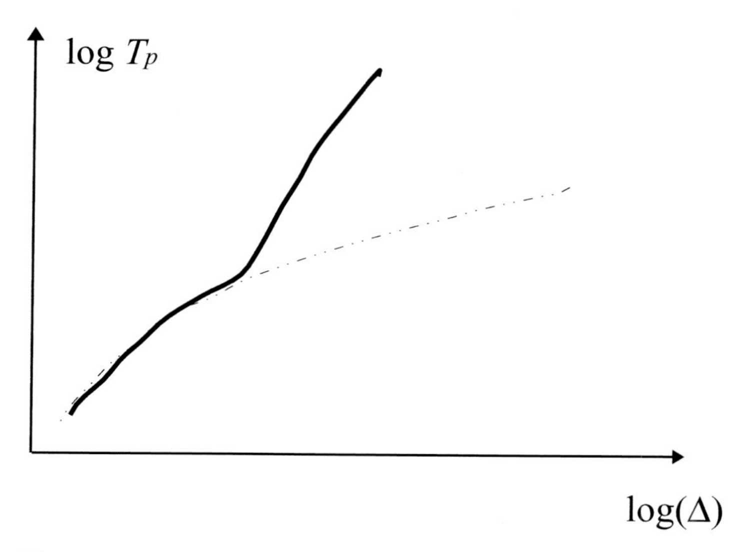

Figure 10.2 Predictability time for the slow component of the two coupled Lorenz models as a function of error tolerance. The dashed line represents the classical Lyapunov estimate. Figure schematised after Bofetta *et al.* (1998).

where $c_a = 60\ \mathrm{m/s^2}$ represents the equatorial wave speed in the atmosphere. Given a clear timescale separation between atmospheric adjustment and oceanic adjustment, these equations can be simplified such that all acceleration terms are dropped and only the stationary solution (u, v, Φ) for a given diabatic forcing Q is determined. $Q(x, y)$ is a function of the actual SST pattern. This technique, which has been quite successful in studying ENSO dynamics and predictability, can be compared with the so-called slaving principle of Haken (1978). Fast variables are assumed to relax very rapidly (or instantaneously) to a state that is a function of the slow variable state vector. This basic principle can be described mathematically as follows. Consider the fast variable x_f that is governed by

$$\frac{dx_\mathrm{f}}{dt} = F(x_\mathrm{f}, x_\mathrm{s}, \zeta(t)) \tag{10.7}$$

with x_s representing a slow variable, F a non-linear function and ζ noise with variance Σ. The dynamics of the slow variable is assumed to be given by

$$\frac{dx_\mathrm{s}}{dt} = x_\mathrm{f}. \tag{10.8}$$

For illustrative purposes we focus on the simplified system

$$F(x_\mathrm{f}, x_\mathrm{s}, \zeta(t)) = -\beta x_\mathrm{f} - \frac{dV(x_\mathrm{s})}{dx_\mathrm{s}} + \sqrt{2\beta\sigma}\zeta(t).$$

In case of large β one expects a very fast relaxation of the dynamics of x_f towards a quasi-stationary state in which $dx_\mathrm{f}/dt \to 0$. For large enough β, i.e. for quick relaxation times, we obtain

$$x_{\mathrm{f}} = -\frac{1}{\beta}\left[\frac{dV(x_{\mathrm{s}})}{dx_{\mathrm{s}}} - \sqrt{2\beta\sigma}\zeta(t)\right]. \tag{10.9}$$

Substituting this equation into the dynamical equation (10.8) yields

$$\frac{dx_{\mathrm{s}}}{dt} = -\frac{1}{\beta}\frac{dV(x_{\mathrm{s}})}{dx_{\mathrm{s}}} + \sqrt{2\sigma/\beta}\zeta(t). \tag{10.10}$$

It should be noted here that in comparison with Eqs. (10.7) and (10.8) the order of the stochastic differential equation is reduced by one, which might have severe consequences for the dynamics and the predictability. For illustrative purposes let us discuss a climate-relevant example. Locally the heat budget in a fixed mixed layer slab ocean with depth H can be expressed in terms of

$$\frac{dT}{dt} = -\lambda T + \frac{Q}{c_{\mathrm{p}}\rho H}, \tag{10.11}$$

where T represents the mixed layer temperature and $-\lambda T$ the first term in a Taylor expansion that describes the adjustment of ocean and atmospheric temperatures as well as entrainment processes at the base of the mixed layer. According to Hasselmann (1976) and Frankignoul and Hasselmann (1977), Q can be parametrised in high latitudes in terms of stochastic weather fluctuations. Here we choose the following more general ansatz:

$$\frac{dQ}{dt} = -\alpha Q + \sigma\zeta(t) + F_1 T, \tag{10.12}$$

which represents a red noise process and a coupling of the atmospheric fluctuations to the mixed layer temperature. This dependence parametrises symbolically the fact that changed meridional temperature gradients have an influence on the eddy momentum transport and heat fluxes in the atmosphere, through changes in baroclinicity. Without adiabatic elimination the equations can be written in terms of a stochastically driven linear oscillator equation:

$$\frac{d^2T}{dt^2} + (\alpha+\lambda)\frac{dT}{dt} + \left(\alpha\lambda - \frac{F_1}{c_{\mathrm{p}}\rho H}\right)T - \frac{\sigma}{c_{\mathrm{p}}\rho H}\zeta(t) = 0, \tag{10.13}$$

whereas in the case of adiabatic elimination we obtain the following red noise equation:

$$\frac{dT}{dt} + \left(\lambda - \frac{F_1}{\alpha c_{\mathrm{p}}\rho H}\right)T - \frac{\sigma}{c_{\mathrm{p}}\rho H\alpha}\zeta(t) = 0. \tag{10.14}$$

The associated spectra look quite different. The full spectrum of the process exhibits a spectral peak at a particular frequency. This frequency depends mainly on the air–sea coupling F_1 as well as on the damping timescales of the atmosphere and the ocean. The adiabatic elimination leads to a temperature spectrum that does not exhibit any spectral peak. Generally also, the associated prediction times are different. Hence,

Figure 10.3 Westerly wind burst (WWB) over the West Pacific Ocean seen from satellite in May 2002.

the usage of the adiabatic elimination has to be done with care, in particular since the order of the differential equations is reduced by one.

This pedagogic example of the adiabatic elimination procedure illustrates that a rather straightforward but also 'brutal' way of getting rid of the two-timescale problem in predictability studies is to use a quasi-stationarity assumption for the fast variables. This assumption is implicitly made in intermediate and hybrid ENSO prediction models. However, predicting ENSO is more complicated and it has become questionable (Syu and Neelin, 2000) whether the adiabatic elimination approach is successful. The question was raised (Syu and Neelin, 2000) as to whether fast variables can really be eliminated by slaving principles, or whether they also play a fundamental role in triggering ENSO events. Fast, so-called intraseasonal oscillations in the atmosphere (Figure 10.3) can kick off Kelvin waves that, under appropriate preconditioning of the warm pool, can lead to the onset of El Niño or La Niña events.

Furthermore the amplitude of westerly wind bursts (WWBs) depends crucially on the existing SST boundary conditions. In terms of a dynamical systems approach this amplitude dependence of the WWB noise can be treated as a multiplicative noise source, which provides effectively an additional non-linearity (as can be seen from variable transformations).

Once such oceanic equatorial Kelvin waves are triggered, the oceanic evolution is highly predictable to at least a season in advance. Kelvin waves can be compared with a Jack-in-the box. Nobody knows exactly when the box is opened, but once it is opened the future is well determined. Of course, this particular feature has strong implications for ENSO prediction, since the predictability time of the individual WWBs is very small, whereas the predictability time for the Kelvin wave propagation and the subsequent generation of sea surface temperature (SST) anomalies by local air–sea interactions is much larger. Is there a possibility of overcoming the initial uncertainty in predicting WWBs? This is a question of utmost importance since the living conditions of about 30% of the world population are significantly influenced by ENSO-related climate anomalies.

We have already discussed the possibility of eliminating fast fluctuations from the equations of the tropical Pacific coupled atmosphere–ocean system by adiabatic elimination. In a sense the resulting equations can be viewed as a kind of time-averaged set of equations, in which only slow oceanic timescales are retained. There are now a few ways to improve on this. One is to take into account the fast fluctuations empirically (see Section 10.2); another way is to derive effective equations of motion directly from the data under consideration (see Sections 10.3 and 10.4).

10.2 Stochastic optimals

Tropical Pacific climate variability can be decomposed to a first order into variability $\vec{\Phi}$ associated with the ENSO phenomenon and non-ENSO related stochastic fluctuations $\vec{\theta}$ of variance Σ. Anomalies will be computed with respect to the annual cycle basic state. The stochastically forced dynamical equations for the tropical atmosphere–ocean system can be written symbolically in the form

$$\frac{d\vec{\Phi}}{dt} = \vec{F}(\vec{\Phi}) + \Sigma\vec{\theta}. \tag{10.15}$$

For a Zebiak–Cane type intermediate ENSO model (Zebiak and Cane, 1987) the vector $\vec{\Phi}$ might consist of the SST anomalies and the expansion coefficients for the different equatorial wave modes. Given a stable (linearly damped) ENSO mode, this equation can be linearised around a non-linear trajectory $\vec{\Phi}(t)$. We obtain a dynamical equation for the perturbations (denoted by primes) which reads

$$\frac{d\vec{\Phi}'}{dt} = \frac{\partial\vec{F}}{\partial\vec{\Phi}}\vec{\Phi}' \mid_{\vec{\Phi}(t)} + \Sigma\vec{\theta}' = \mathbf{A}\vec{\Phi}' + \Sigma\vec{\theta}'. \tag{10.16}$$

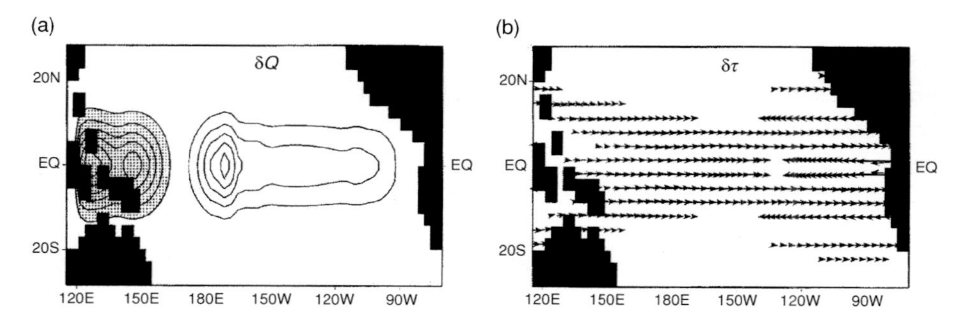

Figure 10.4 First stochastic optimal for (a) heat flux and (b) wind stress anomalies as computed from an intermediate ENSO model. Figure from Moore and Kleeman (1999).

Bold face expressions denote matrices. Integral solutions of this equation can be obtained, once the propagator $\mathbf{R(t_1, t_2)} = \mathbf{e^{A(t_2-t_1)}}$ is known. Then the solutions can be expressed in terms of

$$\vec{\Phi}'(t_2) = \mathbf{R(t_1, t_2)}\vec{\Phi}'(t_1) + \int_{t_1}^{t_2} \mathbf{R(t_1, t_2)}\Sigma\vec{\theta}'dt. \tag{10.17}$$

The first term on the right-hand side represents the predictable part of the dynamics, whereas the second part belongs to the unpredictable stochastic part. It is the ratio between these terms that strongly determines the overall predictability of the system (Grötzner *et al.*, 1999). It has been shown by Kleeman and Moore (1997) that the total variance at a given time τ can be obtained from the stochastic forcing terms $\vec{\theta}$ and the so-called stochastic optimals. Assuming Gaussian white noise, the stochastic optimals are the eigenvectors of the operator

$$\mathbf{Z} = \int_{t_1}^{t_2} \mathbf{R^{\dagger}(t, t_1)XR(t, t_1)dt}, \tag{10.18}$$

where $\mathbf{X}$ is the kernel defining the variance norm of interest (Farrell and Ioannou, 1996). By contrast, the optimal perturbations (singular vectors) (Trefethen *et al.*, 1993) $\vec{\alpha}$ can be computed from the eigenequation

$$\mathbf{R^{\dagger}XR}\vec{\alpha} = \lambda\mathbf{X}\vec{\alpha}. \tag{10.19}$$

They represent the fastest growing perturbations of the coupled system before nonlinearities become important. The eigenvectors of Z (stochastic optimals) corresponding to the maximal eigenvalues are the spatial structures that the stochastic forcing must possess in order to maximise the stochastically induced variance in the model. They are an important quantity for predictability studies of coupled processes. Moore and Kleeman (1999) computed the stochastic optimals for an intermediate ENSO model. The result is shown in Figure 10.4. The heat flux anomalies as well as the wind anomalies are reminiscent of anomalies associated with intraseasonal variability (Hendon and Glick, 1997). In fact it turned out (Moore and Kleeman, 1999) that the optimal perturbations and stochastic optimals have quite similar structures, indicating that for initial error growth as well as for the establishment of variance throughout its evolution, ENSO is highly sensitive to intraseasonal atmospheric variability. If for a certain time period the atmospheric anomalies do not project well onto the stochastic optimals, the resulting ENSO variance is expected to be small (given a stable ENSO mode). Hence, the pattern of fast atmospheric fluctuations governs the generation of El Niño and La Niña anomalies. It should be emphasised here that this concept holds only for a stable oscillatory ENSO mode, otherwise the amplitude of ENSO is mainly governed by non-linearities. This stability assumption is questioned, however, by many climate researchers. At least there is the possibility that ENSO operates close to a Hopf bifurcation point, meaning that ENSO is stable during some decades whereas it is unstable during other decades. This hybrid hypothesis has been confirmed by the study of An and Jin (2000). In addition to the stability properties of ENSO, Moore and Kleeman (1999) found also that the phase of the annual cycle as well as the non-linearities and the integrated past noise history are important factors in controlling the sensitivity of ENSO towards stochastic forcing.

Furthermore, it has to be noted here that linear systems cannot transfer energy from weather scales to ENSO scales. Hence, the argument that intraseasonal variability forces ENSO is strictly speaking incorrect in a linear framework. What can force ENSO in a stochastic environment are low-frequency variations which probably project onto the patterns of intraseasonal variability. Another possibility is that intraseasonal variability rectifies ENSO through non-linear interactions. This dilemma has been nicely illustrated in Roulston and Neelin (2000).

10.3 Non-linear probabilistic methods

In this section an empirical non-linear method is described that allows for the determination of optimal perturbation structures associated with low and high predictability. The principal idea of this method is to compute transition probability densities between areas in state space. From these probability densities we learn how quickly certain initial probability density functions are broadened. This dispersion of the probability densities is associated with information loss and hence entropy production. The dispersion rates for different state space elements can be averaged to determine a maximum predictability horizon. The method is closely linked to the empirical derivation of a so-called *Markov chain*. Let us consider a non-linear discretised dynamical system

$$x_{t+1} = f(x_t, \zeta_t), \tag{10.20}$$

where ζ_t represents a stochastic component. For simplification the equations are written in univariate form. The generalisation to multivariate systems is straightforward.

Another way to write down the dynamical equation is given by

$$x_{t+1} = \int dy \delta[x - f(y)]. \tag{10.21}$$

The integrand $\delta[x - f(y)]$ is called Frobenius Perron operator and will be abbreviated by $m(x, y)$. Now all 'reasonable' functions $p(x)$ can be written in terms of

$$p(x_{t+1}) = \int dy \delta[x - f(y)] p(y_t). \tag{10.22}$$

If p belongs to the family of probability density functions, i.e. fulfilling $p \leq 1$ and $\int dy p(y) = 1$ this equation describes how a probability distribution at time t is transformed into a probability distribution at time $t + 1$ under the action of the non-linear 'transition probability kernel' (Frobenius Perron operator) $m(x, y)$. This kernel can be approximated by discretising the state space of the system into N cells. The state space cells are denoted by x^i and the corresponding probability density by $\vec{p} = (p(x^1), \ldots p(x^i), \ldots p(x^N))$. The evolution equation of the discretised probability density can be written as

$$p^i(t+1) = \sum_{j=1}^{N} m_t(i, j) p^j(t). \tag{10.23}$$

It becomes apparent that a good approximation of the Frobenius Perron matrix $m(i, j)$ (transition probability matrix) depends crucially on an intelligent discretisation of the state space. Here we use an equinumber (equal-residence time) partition in state space rather than an equidistant partition of the phase space variables.

The advantage of this equinumber partition is that we can exploit some beautiful mathematical properties of the equinumber partition such as the double stochasticity of the transition matrices (meaning that $\sum_i m(i, k) = 1$ and $\sum_k m(i, k) = 1$). The disadvantage is that resolution for extreme state space directions is low. For high dimensional systems the requirements for the amount of available data needed to compute the transition probabilities become very large (Pasmanter and Timmermann, 2003). Here we assume that ENSO can be described by relatively few degrees of freedom. In this mathematical context predictability can be easily quantified by the rate at which information about the state of the system is lost – i.e. predictability is associated with the broadening of the probability density. A convenient way of measuring how quickly information is lost is to compute the 'distance' between the highly localised initial probability density p_0 and the broadened probability density function at time t using the so-called entropy measure

$$I(t) = \sum_{i=1}^{N} p^i(t) \ln \frac{p^i(t)}{p_0^i}. \tag{10.24}$$

Using this equation an averaged loss of information in one time step can be computed from

$$\langle \Delta I \rangle = -N^{-1} \sum_{i=1}^{N} \sum_{j=1}^{N} m(i,k) \ln m(i,k). \qquad (10.25)$$

Notice here that the information loss or equivalently the entropy production depends strongly upon the chosen partition. The rate of information loss determines how far ahead we can predict the state of a system; in other words when the probability density becomes too broad, the state of the system is becoming less and less defined. We can compute entropy production for all possible initial conditions in state space, denoted by the cell number j. An interesting problem is to find the initial conditions $\delta_{j,k}$ that are associated with the largest entropy production for a given forecast time τ. These physical initial conditions expressed in terms of a localised δ function in probability space are associated with the eigenvectors of the Markov transition matrices. Since each cell number corresponds to an N-dimensional vector in physical space, we can compute those initial structures that are associated with the slowest or the largest entropy production for a given lead time τ. Using this method, optimal initial perturbations can be determined empirically for a non-linear coupled system, given a long multivariate data set that covers the state space densely enough. This method is applied to ENSO data obtained from a 100-year-long present-day control simulation performed with the Coupled General Circulation Model ECHAM4/OPYC3 (Timmermann *et al.*, 1999a).

A one-dimensional cyclic Markov chain is constructed from the monthly simulated NINO3 anomalies. An equinumber partition of the state space spanned by this time series is chosen and the annual cyclostationarity of the climatic background state is taken into account by an extension of the Markov chain method to cyclostationary non-linear processes (Pasmanter and Timmermann, 2003). Figure 10.5 shows the entropy production for different seasons averaged over all initial conditions. We observe a relatively fast entropy growth, and hence a low predictability, during the spring season as compared with the autumn and winter seasons. This well-known feature of the so-called spring persistence barrier has puzzled many researchers and is still a major obstacle for long-lead ENSO forecasting. It can be partly attributed to the seasonally varying atmosphere–ocean instability, which attains largest values in spring and which contributes also to the phase-locking of ENSO to the annual cycle. In this section we have discussed how to deal empirically with the long-term predictability of coupled processes, without considering the fast timescale explicitly.

10.4 Non-linear deterministic methods

This section describes a relatively new method (Breiman and Friedman, 1985; Timmermann *et al.*, 2001) that allows for the empirical extraction of equations of motions

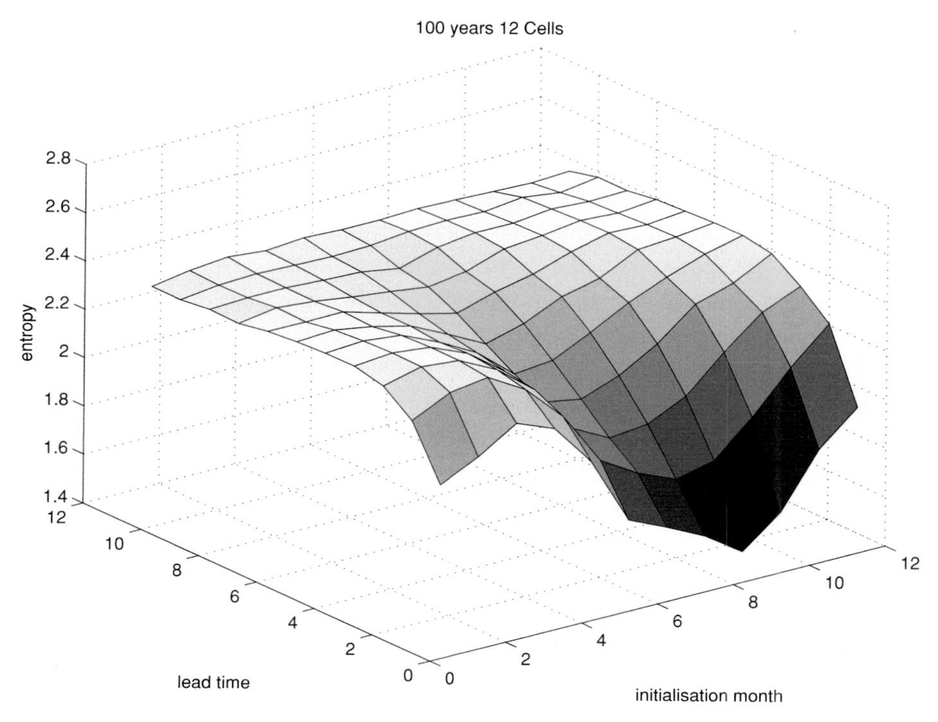

Figure 10.5 Seasonal dependence of the entropy production: entropy as a function of lead time and of the initialisation month.

from data. Let us begin with the 'mother' deterministic dynamical system, prognosing the variables x_j:

$$\dot{x}_i = F_i(\vec{x}). \tag{10.26}$$

The question arises now whether it is possible to reconstruct the 'mother' function F from the data x_j under consideration. This is a very difficult task and requires large amounts of data. A simple trick, however, facilitates the reconstruction enormously. If the dynamical equations can be written in terms of a sum of non-linear functions

$$\dot{x}_i = \sum_j \Phi_i^j(x_j) + \text{Residual}, \tag{10.27}$$

the functions $\Phi_i^j(x_j)$ can be determined statistically by performing a multiple non-parametric regression analysis between the variables x_j and their numerically computed derivatives. This multiple non-parametric regression exercise can be solved by using the so-called Alternating Conditional Expectation (ACE) Value algorithm. This iterative technique that provides look-up tables for the estimated functions $\tilde{\Phi}_i^j(x_j)$ was developed by Breiman and Friedman (1985). It was applied for the first time in a climate context by Timmermann *et al.* (2001).

The ACE algorithm converges in a consistent way to optimal transformations, as has been shown by Breiman and Friedman (1985). (For details on the numerical implementation of the ACE algorithm see Voss, 2001; Voss and Kurths, 1997.) The modified ACE algorithm works as follows. Globally optimal solutions (in a least square sense, i.e. $\langle(\dot{x}^j - \sum_{i=1} \tilde{\Phi}_i^j(x^i))^2\rangle = \min$) can be obtained iteratively from the sequence

$$\tilde{\Phi}_{i,0}^j(x^i) = \langle \dot{x}^j | x^i \rangle, \tag{10.28}$$

$$\tilde{\Phi}_{i,k}^j(x^i) = \left\langle \dot{x}^j - \sum_{p \neq i} \tilde{\Phi}_{p,k^*}^j(x^p) \middle| x^i \right\rangle. \tag{10.29}$$

The index j corresponds to the component of the differential equation, k to the iteration step ($k > 0$) and p to the sum over the predictor components. The index k^* equals k for $p < i$ and $k - 1$ for $p > i$, and $\langle \ldots | \ldots \rangle$ denotes the conditional expectation value. The so-called optimal transformations $\tilde{\Phi}_{i,k}^j(x^i)$ produced by this algorithm are given in the form of numerical tables. In the expression (10.29) only scalar quantities are involved, and in contrast to Eq. (10.26) only one-dimensional conditional probabilities (or, equivalently, two-dimensional joint probabilities) have to be estimated. These can be interpreted in terms of the time transition probabilities or as the dynamical contribution to the componentwise Frobenius Perron operator of the underlying dynamical system.

Minimising $\langle(\dot{x}^j - \sum_{i=1} \Phi_i^j(x^i))^2\rangle$ is equivalent to the maximisation of the correlation

$$\Psi(\dot{x}^j, x^1, \ldots, x^N) = \frac{\left\langle \dot{x}^j \sum_i^N \tilde{\Phi}_i^j(x^i) \right\rangle}{\left(\langle \dot{x}^{j2} \rangle \left\langle \left[\sum_i^N \tilde{\Phi}_i^j(x^i) \right]^2 \right\rangle \right)^{1/2}}, \tag{10.30}$$

where it is assumed that all variables have zero mean. Hence, this technique to solve the non-linear regression problem is also called maximal correlation approach.

Once derived empirically, the functions $\Phi_i^j(x_j)$ can be used in order to build a numerical (forecasting) model for the variables under consideration. It is useful to concentrate a priori on a set of dynamically relevant variables. Here, we show how this empirical non-linear technique can be applied to the ENSO prediction problem. We used a 240-year-long climate simulation performed with the Coupled General Circulation Model (CGCM) ECHAM4/OPYC3 (Timmermann *et al.*, 1999a). The simulated ENSO activity (pattern, phase-locking, amplitude) is quite realistic, although the frequency of the simulated ENSO is somewhat too short (2.3 years). We performed an empirical orthogonal function (EOF) analysis of the simulated sea level depth anomalies (in a 1.5 layer model equivalent to thermocline depth anomalies) as well as of the SST anomalies. According to the recharge oscillator concept for ENSO (Jin, 1997), these two variables are key to explaining ENSO. The leading principal components of two SST (T_1, T_2) and two sea level EOFs

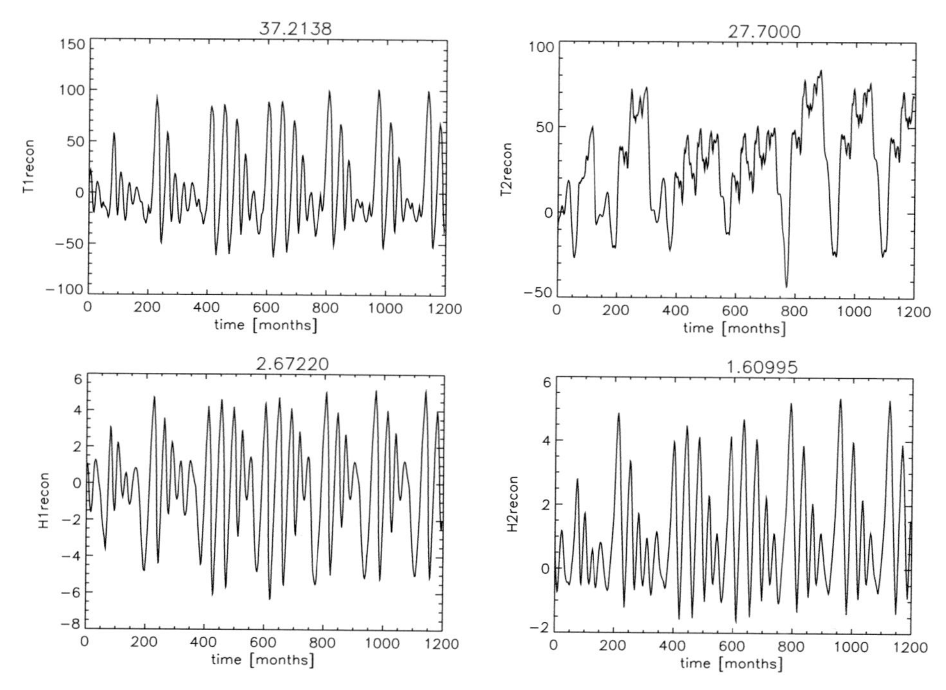

Figure 10.6 Simulated principal components for the two leading temperature and sea level anomaly EOFs. The simulation is based on a four-dimensional non-linear ENSO model that has been derived empirically from a CGCM simulation (Timmermann *et al.*, 2001) using non-parametric multiple regression techniques.

(H_1, H_2) were taken, their derivatives were computed and the ACE algorithm was applied to these variable pairs. For computational convenience the resulting functions $\Phi_i^j(x_j)$ were fitted by higher order polynomials and the resulting ordinary differential equation system was integrated in time using a Runge–Kutta scheme of fourth order. As can be seen from Figure 10.6 the principal components simulated by this four-dimensional non-linear low order ENSO model captures most essential features of the CGCM ENSO. An interesting reconstructed feature is that ENSO exhibits decadal amplitude modulations. These decadal amplitude modulations of ENSO can be explained by the theory of homoclinic orbits (Timmermann, 2003). How can such a reduced dynamical model be used in order to assess the predictability of ENSO?

This empirical model, which was derived from monthly anomalies, captures the dynamics of effective, averaged variables. Short weather fluctuations are not explicitly resolved, but their effect on the dynamics of ENSO is captured empirically. The full CGCM model is far too complex to compute optimal perturbations for ensemble prediction experiments. Besides, the above-mentioned problem of fast–slow timescale systems prevents a straightforward linearisation of the coupled atmosphere–ocean model code and the computation of its adjoint. The simplified

ENSO model, however, based on the simulated dynamics of the CGCM, can be linearised around its non-linear trajectory, by a Taylor expansion of the functions $\Phi_i^j(x_j)$. The resulting 4×4 matrix can be transposed and singular vectors can be determined for any given lead time. The same procedure can be performed for different variables for higher dimensional systems and different 'mother' CGCMs. If the resulting optimal perturbations are robust with respect to reasonable changes in the dimensions etc., these patterns can be used in order to perturb the CGCM in seasonal forecasting ensemble simulations. Also here, empirical modelling might help to circumvent the multiple-timescale problem sketched in the introduction.

10.5 Beyond error growth: predictability of the third kind

In weather and climate prediction we distinguish two different types of prediction (Lorenz, 1975).

- Predictability of the first kind characterises initial value problems, exemplified by conventional weather forecasting practice. It measures how uncertainties in the initial conditions evolve during the forecasting period. Different techniques, such as the breeding vector and the singular vector techniques, have been developed that help to find those initial perturbations that are associated with the maximal growth of initial errors with respect to an a-priori chosen norm.
- By contrast, for predictions of the second kind, an attempt is made to forecast how a system will respond to prescribed changes in its determining parameters. The response of the climate system to a doubling of the atmospheric carbon dioxide concentrations is a well-known example.

An important question arises now: is there a possibility to determine the state of a non-linear dynamical system beyond error growth timescales, even when the external 'parameter' forcing is constant? And if yes, can we exploit this kind of information for long-term forecasting? The answer to this question is a preliminary yes. Owing to non-linearities, certain chaotic systems possess global phase-space topologies that can be exploited using statistical techniques. The following system of equations which constitutes the normal form of a triple instability (Arnéodo *et al.* 1991) and which describes the dynamics of the chemical Beloushov–Zhabotinsky reaction is a suitable and illustrative example. The dynamical equations are:

$$\dot{x} = y \tag{10.31}$$

$$\dot{y} = z \tag{10.32}$$

$$\dot{z} = -\eta z - \nu y - \mu x - k_1 x^2 - k_2 y^2 - k_3 xy - k_4 xz - k_5 x^2 y \tag{10.33}$$

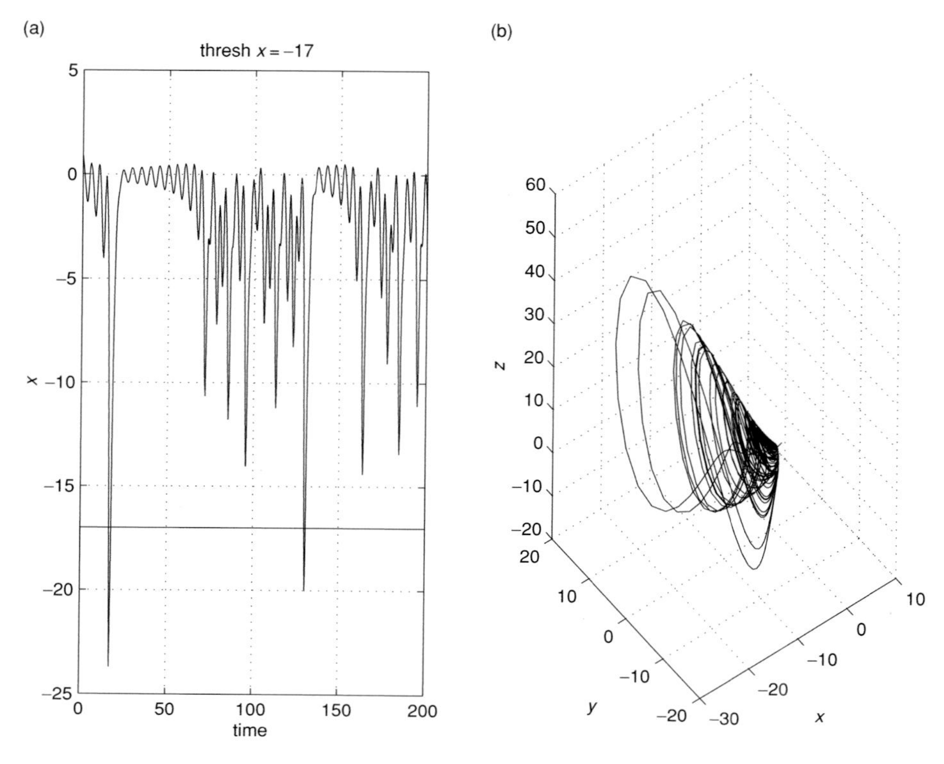

Figure 10.7 (a) Time series of the x-variable of the triple instability dynamical system. A threshold level of $x = -17$ is indicated. (b) Phase-space plot (x,y,z) of the dynamical equation. The underlying dynamical system is believed to capture important dynamics of the chemical Beloushov–Zhabotinsky reaction.

with $k_1 = -1$, $k_2 = 1.425$, $k_3 = 0$, $k_4 = -0.2$, $k_5 = 0.01$, $\eta = 1$, $\mu = 1.38$, $\nu = 1.3$. The dynamics is shown in Figure 10.7.

We observe a fast chaotic oscillation with a period of a few time steps and a peculiar bursting behaviour with a much longer timescale (50–100 time steps). The phase-space view of the system illustrates that the trajectory spirals out of the neighbourhood of a saddle node and returns to it via a large amplitude excursion. This kind of behaviour is reminiscent of near-homoclinic dynamics. In fact our parameter values are chosen such that the system operates close (in parameter space) to a homoclinic orbit. Due to the strong skewness of the probability distribution of the variable x, large amplitude excursions become less likely for large x values. As expected for such a skewed probability distribution, the averaged return times of large negative excursions grow with the size of the excursions. What is, however, peculiar for this system, in contrast to skewed white noise or other surrogates, is that there are lower bounds of return times, in particular for large amplitude bursting events. The reason is the global topology of the attractor that follows the shadow of a homoclinic orbit. The system, after a large negative excursion, has to return to the neighbourhood of

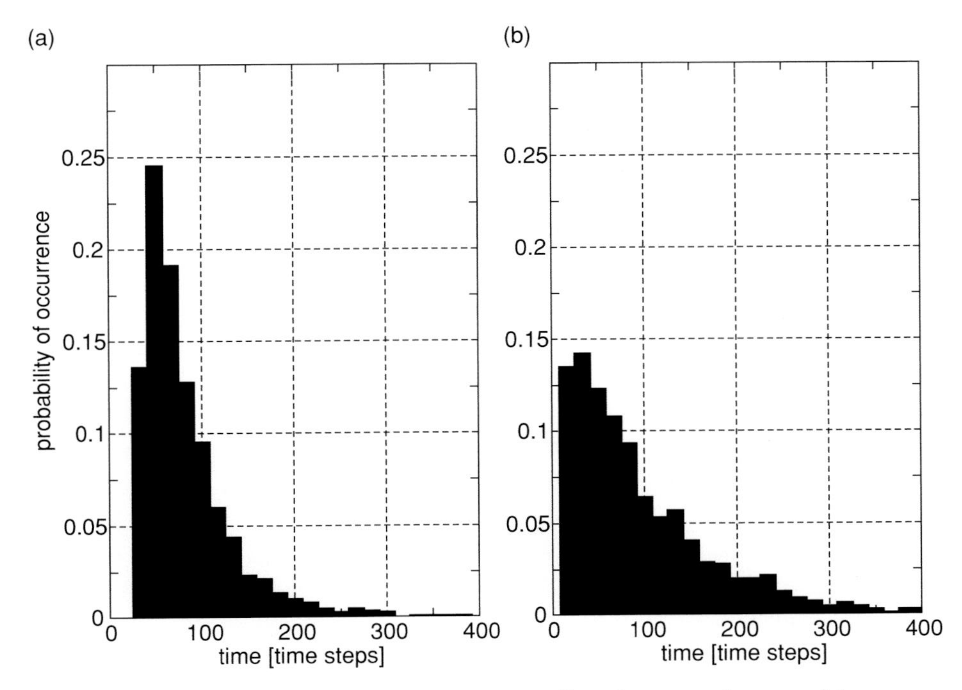

Figure 10.8 Histogram of the return times of bursting events that exceed the threshold level of $x < -17$ for (a) the non-linear time series and (b) the linear surrogate time series.

the saddle node, and spiral around it a couple of times before a new extremum of x can be attained. There are now several interesting questions which can be addressed:

- If there was a bursting event n time steps back, what is the probability of occurrence for a similar event, m time steps ahead?
- If the distance between the last two observed bursting events was n time steps, what is the probability that the next bursting events are separated by m time steps?
- Are there forbidden return times of extreme events?
- If yes, what is the dynamical (physical) reason for such forbidden return times?

Let us assume for a moment, that a 'bursting event' stands for a very strong El Niño event. It is obvious in this case that answers to these questions would be of utmost importance and are beyond the reach of 'classical' ENSO prediction schemes. We have integrated forward in time the system of Equations (10.31, 10.32, 10.33) using a Runge–Kutta method of fourth order. In order to shed light on the peculiarities of the non-linear attractor topology we have also computed surrogate data from the time series using the method of Schreiber and Schmitz (1996). These surrogates have the same spectrum as the non-linear time series but randomised phases. Figure 10.8

shows the probability distribution of return times of events which cross the -17 threshold for the x-variable for the non-linear time series (a) and the corresponding surrogate time series (b). For the fully non-linear trajectory we clearly see 'forbidden return times' for large amplitude events. Certain shortcuts in phase space are not allowed.

Conditional return-time probability maps can be computed which address the first two questions posed above. These maps are distinct from those of the surrogate time series, owing to the global properties of the bursting attractor.

Using a low-dimensional non-linear dynamical system we have shown that predictive information can be extracted from the system far beyond the typical growth timescales associated with initial condition errors. Our approach exploits the bursting behaviour of certain dynamical systems. We suggested a new kind of predictability (the so-called *third kind of predictability*) which is based on the global properties of the attractor, rather than on the local properties (such as for the first kind of predictability).

Inspired by these considerations, let us ask the question: is there any example in climate research for which we can exploit this kind of global predictability? The answer is: maybe. It is well known that ENSO exhibits amplitude modulations on a decadal timescale. Large El Niño events do not occur in a close sequence. There seems to be a certain systematics in their return, which can be explained physically using, for example, the non-linear recharge model (Timmermann *et al.*, 2003). In fact it can be shown (Timmermann, 2003) that several ENSO models of different complexity exhibit bursting behaviour similar to the observations and that is due to the presence of homoclinic or heteroclinic connections between stationary points. A proof, however, that the real tropical Pacific climate system falls into this category has not yet been provided. The following considerations are very preliminary, but they might help to illustrate the philosophy of our approach. Part (b) of Figure 10.9 (colour plate) depicts the observed eastern equatorial Pacific SST anomaly time series and its smoothed interannual wavelet energy. Computing the wavelet energy is a means to obtain a time series that represents the envelope of the original SST anomaly time series. We clearly observe that eras of high ENSO variance alternate with eras of low ENSO variance on a decadal timescale. If the physical mechanism for these amplitude modulations is similar to the one found in different ENSO models (Timmermann, 2003; Timmermann *et al.*, 2003), we sense the temptation to predict the next high ENSO variance era, maybe 10 years ahead.

In order to predict the evolution of ENSO variance we fit polynomials (Casdagli, 1989; Hegger *et al.* 1999) to the embedded tendency time series of simulated and observed ENSO variance (Figure 10.9a and b). The respective estimated dynamical model is used to make out-of-sample forecasts. For the CGCM simulation an embedding dimension of three and a delay of 48 months is used. For the observations we choose a delay of 36 months. The 10-year lead anomaly correlation skill (not shown) obtained for the envelope curves shown in Figure 10.9 is about 0.8 and 0.6,

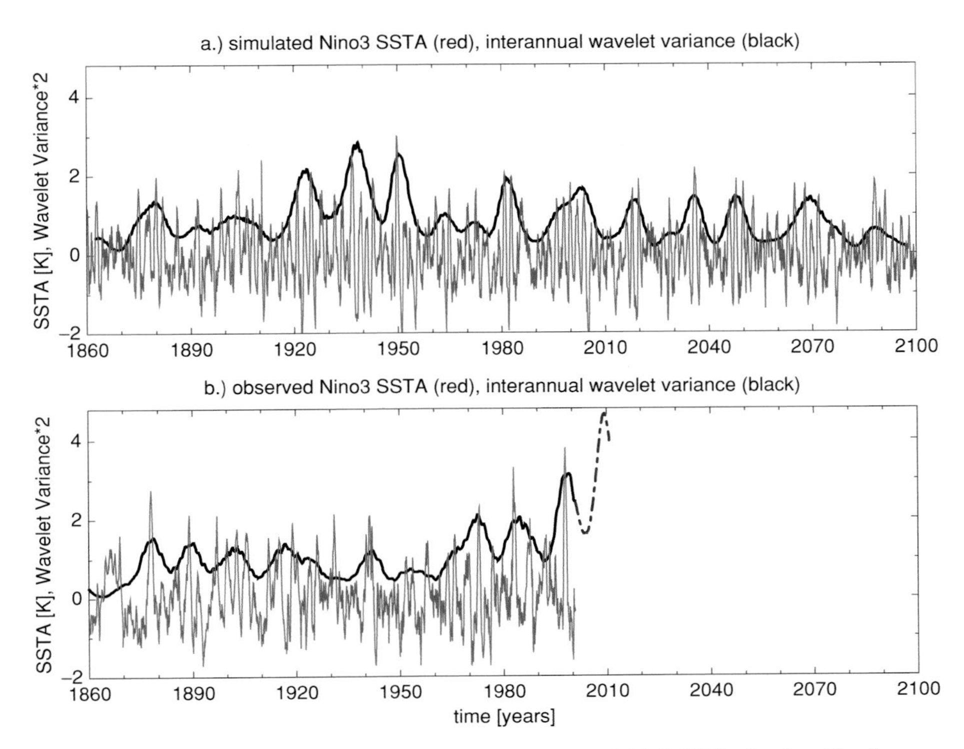

Figure 10.9 (See also colour plate section.) (a) NINO-3 SSTA simulated by the ECHAM4/OPYC3 model (red) and its interannual wavelet energy (black). (b) Observed NINO-3 SSTA time series (red) and its interannual wavelet energy (black), 10-year forecast of interannual wavelet energy (dashed) using non-linear prediction techniques.

respectively. Motivated by this result we perform a real forecast of the next ENSO regime. Our prediction for the envelope curve of ENSO is shown in Figure 10.9(b) (dash-dotted). According to our forecasting procedure an increase of ENSO variability is expected in the coming decade.

It has to be noted here that there is no physical justification for the mathematical ansatz chosen here other than the idea that ENSO bursting is governed by non-linear dynamics.

What has not been taken into consideration here is how errors in the estimation of the wavelet variance that naturally emerge due to the shadowing effect of continous wavelets (Torrence and Compo, 1998) affect the forecasts of the envelope. The associated question is: how well can we estimate the interannual wavelet variance for the SST anomaly on 12 September 2004. Different estimates can be obtained with different buffering techniques of the future time series such as zero-tapering or AR(2) surrogates (Timmermann *et al.*, 1999a). According to this 'illustrative' forecast, the positive trend increases and a super ENSO regime is predicted for the coming 6–8 years. A possibility to improve on this deterministic forecasting ansatz is to take into account the errors in estimating the initial ENSO variance, errors in estimating the right coefficients in the fitted three-dimensional polynomial as well as the small sample size (140 years of monthly data). With these refinements, it will be possible to make Monte Carlo ensemble forecasts, which will yield probabilities for the next very active ENSO regime to occur in say 5–10 years. These very preliminary results are presented here in order to illustrate the methodology and to document the predictive information which might be contained in the bursting behaviour of certain dynamical systems, which has not yet been exploited for seasonal or longer forecasting purposes.

10.6 Some remarks on ENSO prediction models

If we take a look into the *COLA Experimental long lead forecast bulletin* (http://grads.iges.org/ellfb/home.html) we will find many statistical ENSO prediction models and, as has been shown by Landsea and Knaff (2000), these statistical models that can be run on PCs can compete with the most sophisticated numerical coupled atmosphere–ocean models used in seasonal forecasting (e.g. Stockdale *et al.*, 1998). However, it has to be noted here that the statistical models predict only indicators such as the Southern Oscillation index or the NINO3 SST anomaly index and not the full weather patterns and their ENSO-related changes. Why are these simple models, which are based on a few lines of Fortran code and a random number generator, so successful in predicting ENSO indicators? Are the sophisticated numerical models just expensive but physically consistent random number generators? Do we really need to resolve global small-scale weather patterns, in order to predict ENSO? It is obvious that the timing and amplitude of individual westerly wind bursts (WWBs), kicking off oceanic Kelvin waves that might initiate an El Niño or a La Niña event, cannot be predicted six months ahead either by coupled atmosphere–ocean models or by statistical models, which do not take into account WWBs explicitly. Hence, in both cases long-term ENSO prediction is hampered by the difficulty in predicting individual WWBs. The predictable part of, for example, an El Niño event is mainly due to the 'Jack-in-the-Box' part associated with the Kelvin wave propagation plus the SST adjustment part and the slow discharging of the equatorial heat content due to the zonally integrated Sverdrup transport. An important question now is: do we really need to resolve atmospheric weather patterns, oceanic instability waves and diurnal cycles in order to predict the warming of the eastern equatorial Pacific several months or seasons ahead? Let us assume we have a perfect model formulated in terms of non-linear partial differential equations (PDE) and a very bright *mathematician* who is capable of averaging these PDEs in time analytically. The result would be a new set of averaged equations $\frac{1}{T}\int_0^T dt\text{PDE(t)}$. Now let us assume that we have a perfect model based on the original PDEs that simulates ENSO most realistically and a clever *statistician* who instead of averaging the equations averages the ensemble output of the model. Based on the central limit theorem and assuming independent variables he would argue that the distributions of the simulated variables would approach a Gaussian distribution for longer averaging periods, given independent samples. The longer the averaging period, the more linear the process appears to be. The statistician would be able to derive a quasi-linear model from the simulated data. It would in fact be surprising if the *mathematician* and the *statistician* obtain very different results. Hence, it appears that the key for the success of linear statistical ENSO prediction schemes is the central limit theorem. Averaging reduces the noise but still basic features of the ENSO mode are retained (see Eq. 10.17). There is, however, one important statistical feature that cannot be captured by linear statistical models that operate on a monthly basis: the skewness of the observed monthly Niño 3 SST probability distribution (Burgers and Stephenson, 1999). The role of the non-linear temperature advection in generating this skewed probability distribution has been explained by Jin *et al.* (2003). It is in particular this non-linearity that is associated also with the propagation characteristics of SST anomalies that determines the occurrence of very large El Niño events. This skewness cannot be captured by linear statistical models, and hence they exhibit errors in the amplitude either of El Niño or La Niña events.

We have seen that both numerical ENSO models as well as statistical models have some value in predicting ENSO anomalies several months ahead. Unfortunately the advantages of these two modelling philosophies are rarely combined. Despite the fact that the success of combined multimodel forecasts is well documented (Metzger *et al.*, 2004) only a few examples exist where weather and climate prediction centres employ both approaches in an *optimal* way (see Krishnamurti *et al.*, this volume). The idea of a combined multimodel ensemble (superensemble) forecast is simple. Let us assume we have N ENSO prediction models, denoted by $M_i, i\varepsilon\{1, \ldots N\}$, each providing a six month forecast, say, for the NINO3 SST anomaly index X_i. The basic idea of combined forecasts is to find an optimal linear combination such that residual errors with respect to the observations and an a-priori chosen norm are minimised. The optimal forecast is obtained from $X = \sum_{i=1,N} \alpha_i X_i(M_i)$. Basically the models are weighted by their hindcast and/or forecast skills. This strategy can be refined in many ways. One is to take into account also the time derivatives; another is to choose the α_i in such a way that they are dependent on the physical situation, representing the fact that some models work well, for example for a basic state that favours westward-propagating SST anomalies; others perform well for eastward-propagating anomalies, etc. Hence, with very little statistical efforts (one just has to compute a multiple linear regression) it would be easy to improve ENSO forecasts. A combined multimodel ensemble climate community ENSO forecast adds extra value to the zoo of existing forecasts because it can compensate for the weaknesses of some models by the strengths of other models, depending on the existing situation in the tropical Pacific.

The value of an ENSO prediction model is often measured by the difference between the anomaly correlation skill (or root-mean-square) of the model under consideration and a damped persistence forecast. Why do we choose a damped persistence forecast, which corresponds to a red noise process, if ENSO is an oscillatory mode? In fact if the performance of the ENSO prediction model is tested against a physically more justified null hypothesis, such as a noise driven oscillator, the difference between model anomaly correlation skill and AR(2) anomaly correlation skill would be even less, as shown by Burgers and Stephenson (1999). Fairness requires that we have to compare ENSO forecasts against a reasonable physical null hypothesis, not an unreasonable statistical one.

Acknowledgements

A. T. wishes to thank his friend R. A. Pasmanter for a wonderful collaboration on the topics of climate prediction and ENSO dynamics. This paper was initiated during A. T.'s second visit to Hawai'i in winter 2001.

References

An, S. I. and F.-F. Jin (2000). An Eigen analysis of the interdecadal changes in the structure and frequency of the ENSO mode. *Geophys. Res. Lett.*, **27**, 2573–6.

Arnéodo, A., J. Elezgaray, J. Pearson and T. Russo (1991). Instabilities of front patterns in reactin-diffusion systems. *Physica D.*, **49**, 141–60.

Aurell, E. *et al.* (1997). Predictability in the large: an extension of the concept of Lyapunov exonent. *J. Phys. A*, **30**, 1.

Bofetta, G., P. Giuliani, G. Paladin and A. Vulpiani (1998). An extension of the Lyapunov analysis for the predictability problem. *J. Atmos. Sci.*, **55**, 3409–16.

Breiman, L. and J. H. Friedman (1985). Estimating optimal transformations for multiple regression and correlation (with discussion). *J. Am. Stat. Assoc.*, **80**, 580.

Burgers, G. and D. B. Stephenson (1999). The normality of El Niño. *Geophys. Res. Lett.*, **26**, 1027–30.

Casdagli, M. (1989). Nonlinear prediction of chaotic time series. *Physica D*, **35**, 335–56.

Farrell, B. F. and P. J. Ioannou (1996). Generalized stability theory. I: Autonomous operators. *J. Atmos. Sci.*, **53**, 2025–40.

Frankignoul, C. and K. Hasselmann (1977). Stochastic climate models. II: Application to sea-surface temperature anomalies and thermocline variability. *Tellus*, **29**, 289–305.

Gill, A. (1980). Some simple solutions for the heat-induced tropical circulation. *Quart. J. Roy. Meteor. Soc.*, **106**, 447–62.

Grötzner, A., M. Latif, A. Timmermann and R. Voss (1999). Interannual to decadal predictability in a coupled ocean–atmosphere general circulation model. *J. Climate*, **12**, 2607–24.

Haken, H. (1978). *Synergetics: An Introduction*, 2nd edn. Springer.

Hasselmann, K. (1976). Stochastic climate models. 1: Theory. *Tellus*, **28**, 473–85.

Hegger, R., H. Kantz and T. Schreiber (1999). Practical implementation of nonlinear time series methods: the TISEAN package. *Chaos*, **9**, 413.

Hendon, H. H. and J. Glick (1997). Intraseasonal air-sea interactions in the tropical Indian and Pacific Oceans. *J. Climate*, **10**, 647–61.

Jin, F.-F. (1997). An equatorial ocean recharge paradigm for ENSO. I: Conceptual model. *J. Atmos. Sci.*, **54**, 811.

Jin, F.-F., S. I. An, A. Timmermann and X. Zhang (2003). Strong El Niño events and nonlinear dynamical heating. *Geophys. Res. Lett.*, **30**, 1120.

Kleeman, R. (1991). A simple model of the atmospheric response to ENSO sea surface temperature anomalies. *J. Atmos. Sci.*, **48**, 3–18.

Kleeman, R. and A. M. Moore (1997). A theory for the limitation of ENSO predictability due to stochastic atmospheric transients. *J. Atmos. Sci.*, **54**, 753–67.

Landsea, C. W. and J. A. Knaff (2000). How much skill was there in forecasting the very strong 1997–98 El Niño? *Bull. Am. Meteorol. Soc.*, **81**, 2107–19.

Lorenz, E. N. (1963). Deterministic non-periodic flow. *J. Atmos. Sci.*, **20**, 130.

(1969). The predictability of a flow which possesses many scales of motion. *Tellus*, **21**, 289.

(1975). Climate predictability. In *The Physical Bases of Climate and Climate Modeling*, pp. 132–6. GARP Publication Series 16. World Meteorological Organization.

Metzger, S. M., M. Latif and K. Fraedrich (2004). Combining ENSO forecasts: a feasibility study. *Mon. Wea. Rev.*, **132**, 456–72.

Moore A. M. and R. Kleeman (1999). Stochastic forcing of ENSO by the intraseasonal oscillation. *J. Climate*, **12**, 1199–220.

Pasmanter, R. A. and A. Timmermann (2003). Cyclic Markov chains with an application to an intermediate ENSO model. *Nonlinear Proc. Geoph.*, **203**, 197–210.

Roulston, M. S. and J. D. Neelin (2000). The response of an ENSO model to climate noise, weather noise and intraseasonal forcing. *Geophys. Res. Lett.*, **27**, 3723–6.

Schreiber, T. and A. Schmitz (1996). Improved surrogate data for nonlinear tests. *Phys. Rev. Lett.*, **77**, 635.

Stockdale, T. N., D. L. T. Anderson, J. O. S. Alves and M. A. Balmsadea (1998). Global seasonal rainfall forecasts using a coupled ocean-atmosphere model. *Nature*, **392**, 370–3.

Syu, H.-H. and J. D. Neelin (2000). ENSO in a hybrid coupled model. I: sensitivity to physical parametrizations. *Clim. Dynam.*, **16**, 19–34.

Timmermann, A., M. Latif, A. Bacher, J. Oberhuber and E. Roeckner (1999a). Increased El Niño frequency in a climate model forced by future greenhouse warming. *Nature*, **398**, 694–6.

Timmermann, A., M. Latif and R. Voss (1999b). Modes of climate variability as simulated by the coupled atmosphere-ocean model ECHAM3/LSG. I: ENSO-like climate variability and its low-frequency modulation. *Clim. Dynam.*, **15**, 605–18.

Timmermann, A., H. Voss and R. Pasmanter (2001). Empirical dynamical system modeling of ENSO using nonlinear inverse techniques. *J. Phys. Oceanogr.*, **31**, 1579–98.

Timmermann, A. (2003). Decadal ENSO amplitude modulations: a nonlinear mechanism. *Global Planet. Change*, 135–56.

Timmermann, A., F.-F. Jin and J. Abshagen (2003). A nonlinear theory for El Niño bursting. *J. Atmos. Sci.*, **60**, 152–65.

Torrence, C. and G. P. Compo (1998). A practical guide to wavelet analysis. *Bull. Am. Meteorol. Soc.*, **79**, 61–78.

Trefethen, L. N., A. E. Trefethen, S. C. Reddy and T. A. Driscoll (1993). Hydrodynamic stability without eigenvalues. *Science*, **261**, 578–84.

Voss, H. and J. Kurths (1997). Reconstruction of nonlinear time delay models from data by the use of optimal transformations. *Phys. Lett. A*, **234**, 336–44.

Voss, H. U. (2001). Analyzing nonlinear dynamical systems with nonparametric regression. In *Nonlinear Dynamics and Statistics*, ed. A. I. Mees, pp. 413–34. Boston: Birkhäuser.

Zebiak, S. and M. Cane (1987). A model of the El Niño-Southern Oscillation. *Mon. Weather Rev.*, **115**, 2262–78.

11

Predictability of tropical intraseasonal variability

Duane E. Waliser
Jet Propulsion Laboratory, California Institute of Technology, Pasadena

11.1 Introduction

Not long after the development of numerical weather forecasting in the 1950s, predictability studies emerged with the desire to determine the theoretical limits associated with deterministic weather forecasting (e.g. Thompson, 1957; Lorenz, 1965, 1982; this volume; Palmer, this volume). Estimating these limits helped to better quantify the capabilities and skill level of operational weather forecast models and to determine how far and fast the community should press the embryonic field of numerical weather forecasting. Numerical predictability studies expanded to include the ocean and the climate scale with the advent of seasonal-to-interannual forecasting based on the El Niño–Southern Oscillation (ENSO) (e.g. Cane *et al.*, 1986; Graham and Barnett, 1995; Kirtman *et al.*, 1997; Barnston *et al.*, 1999; Anderson, this volume; Hagedorn *et al.*, this volume; Shukla and Kinter, this volume). In this case, it was of interest to understand the theoretical limits for predicting tropical Pacific Ocean sea surface temperature (SST) anomalies, and then in turn their implications for predicting monthly or seasonal anomalies of midlatitude circulation, temperature and rainfall.

Very recently, predictability at the intraseasonal timescale (i.e. lead times of about 2 weeks to 2 months) has garnered great interest (Schubert *et al.*, 2002; Waliser *et al.*, 2003a; ECMWF, 2004). This evolution of research and operations in regard to specific prediction regimes (i.e. weather, seasonal and then intraseasonal) has

Predictability of Weather and Climate, ed. Tim Palmer and Renate Hagedorn. Published by Cambridge University Press.

mimicked quite remarkably that anticipated by John von Neumann (1955; relevant excerpt can be found in Waliser, 2005). His foresight followed from traditional mathematical approaches and was thus based on the expectation that the simplifying extremes of the prediction problem would be tackled first. In this case, the short lead time, 'initial-value problem' of weather forecasting, followed by the 'boundary-value problem' associated with seasonal-to-interannual forecasting, and then finally the regime in between (i.e. intraseasonal) where neither extreme holds. Given the long-standing maturity of weather forecasting and the more recent establishment of operational seasonal-to-interannual forecasting, there is now a fairly well defined gap in prediction capability at the intraseasonal timescale. While the simple existence of this gap has certainly contributed to the community's growing interest in this timescale, attention has also been stimulated by the recognition that a number of noteworthy phenomena and processes have the potential to lend predictability at this intervening timescale. These phenomena and processes include the Arctic Oscillation (AO)/North Atlantic Oscillation (NAO), the Pacific North American (PNA) pattern, the Madden–Julian Oscillation (MJO), soil moisture and SST variability, and the intermittent occurrence of midlatitude blocking. Improved predictions of these phenomena and processes have the potential to provide significant practical benefits which include useful low-frequency weather forecasts over much of the tropics – an area where forecasting has typically been exceptionally challenging (Waliser *et al.*, 1999b; 2003b; Wheeler and Weickmann, 2001; Newman *et al.*, 2003; Barlow *et al.*, 2005; Wheeler and McBride, 2005; Hoskins, this volume), skilful forecasts of active and break monsoon conditions (Waliser *et al.*, 2003c; Webster and Hoyos, 2004; Liess *et al.*, 2005; Webster *et al.*, this volume), and improved extratropical surface temperature and precipitation predictions (Higgins *et al.*, 2000; Kirtman *et al.*, 2001; Thompson and Wallace, 2001; Whitaker and Weickmann, 2001; Baldwin *et al.*, 2003; Bond and Vecchi, 2003; Koster *et al.*, 2004; Vecchi and Bond, 2004).

While the discussion and associated studies cited above indicate that a number of intraseasonal phenomena and processes have bearing on the intraseasonal (aka subseasonal) prediction problem, the MJO in particular has been singled out as one of the most underexploited in terms of lending potential for near-term gains in forecast skill (Schubert *et al.*, 2002; Waliser *et al.*, 2003a, 2005; ECMWF, 2004). This focus on the MJO is not only due to the characteristics of the phenomenon itself and the direct impact it has on a broad region of the tropics at the intraseasonal timescale but also because of the influences the MJO has on timescales of variability outside this band. For the case of weather, the MJO – through its relatively slow modulations of tropical diabatic heating – offers the hope for extending (at least occasionally) the range of useful forecasts of weather and/or weather statistics (e.g. Ferranti *et al.*, 1990; Jones *et al.*, 2004a), including tropical storms and hurricanes (Maloney and Hartmann, 2000a, 2000b; Mo, 2000; Goswami *et al.*, 2003) and extreme United States west coast rainfall events (Mo, 1999; Higgins *et al.*, 2000; Jones, 2000). For the seasonal-to-interannual timescale, the MJO represents an intermittent yet important component of atmospheric forcing (e.g. westerly wind bursts and the development of El Niño) as well as a key component atmospheric 'noise' that can limit the skill associated with forecasts at this timescale (e.g. McPhaden, 1999; Moore and Kleeman, 1999; Kessler and Kleeman, 2000; Zhang *et al.*, 2001).

Given the importance of the MJO to considerations of weather and climate predictability as outlined above, this chapter reviews predictability issues associated with tropical intraseasonal variability, with a particular emphasis on the MJO. In the following section, a brief observational description of the MJO is presented, including its seasonal and interannual modulations. In Section 11.3, the physical theory underlying this variability, particularly as it relates to the phenomenon's predictability, is briefly reviewed. In Section 11.4, an assessment of our present-day understanding of the predictability of the MJO is given. In Section 11.5, a number of practical considerations of MJO prediction are discussed. Section 11.6 concludes with a discussion of the outstanding issues and questions regarding future research and progress in this area. Additional reviews of this and related material can be found in Lau and Waliser (2005); Waliser (2006) and Zhang (2005).

11.2 Physical description

The dominant form of intraseasonal atmospheric variability, particularly in terms of rainfall generation and global reach of influence, is most often referred to as the Madden–Julian Oscillation (MJO; also known as the 30–60 day, 40–50 day, and intraseasonal oscillation (ISO)) after its discoverers (Madden and Julian, 1971, 1994, 2005). The left panels of Figure 11.1 (colour plate) illustrate the canonical space–time structure of rainfall and low-level winds in the tropics associated with an MJO 'event' during boreal winter, with the interval between maps being 12.5 days. These maps illustrate its eastward propagation and equatorially trapped character. The left panels of Figure 11.2 show similar information but for mid-tropospheric geopotential heights and upper-level winds. Comparison of the corresponding upper and lower tropical wind fields emphasises the baroclinic nature of its wind anomalies, with upper tropospheric divergence (convergence) occurring in conjunction with positive (negative) rainfall anomalies and vice versa for the lower troposphere. In addition, it can be seen that the MJO has a global scale. At upper levels, wind anomalies are primarily characterised by wave number 1. At lower levels, wind and rain anomalies are primarily characterised by wave number 2, with a significant modulation by the relatively warmer (cooler) eastern (western) hemisphere background state. For example, over the Indian and west Pacific Oceans, there is evidence of considerable interaction between the wind and rainfall anomalies. In these regions, where the coupling between the convection and warm surface waters is strong, the oscillation propagates rather slowly, about 5–10 m/s. However, once the disturbances reach the vicinity of the Date Line, and thus cooler eastern Pacific Ocean equatorial waters,

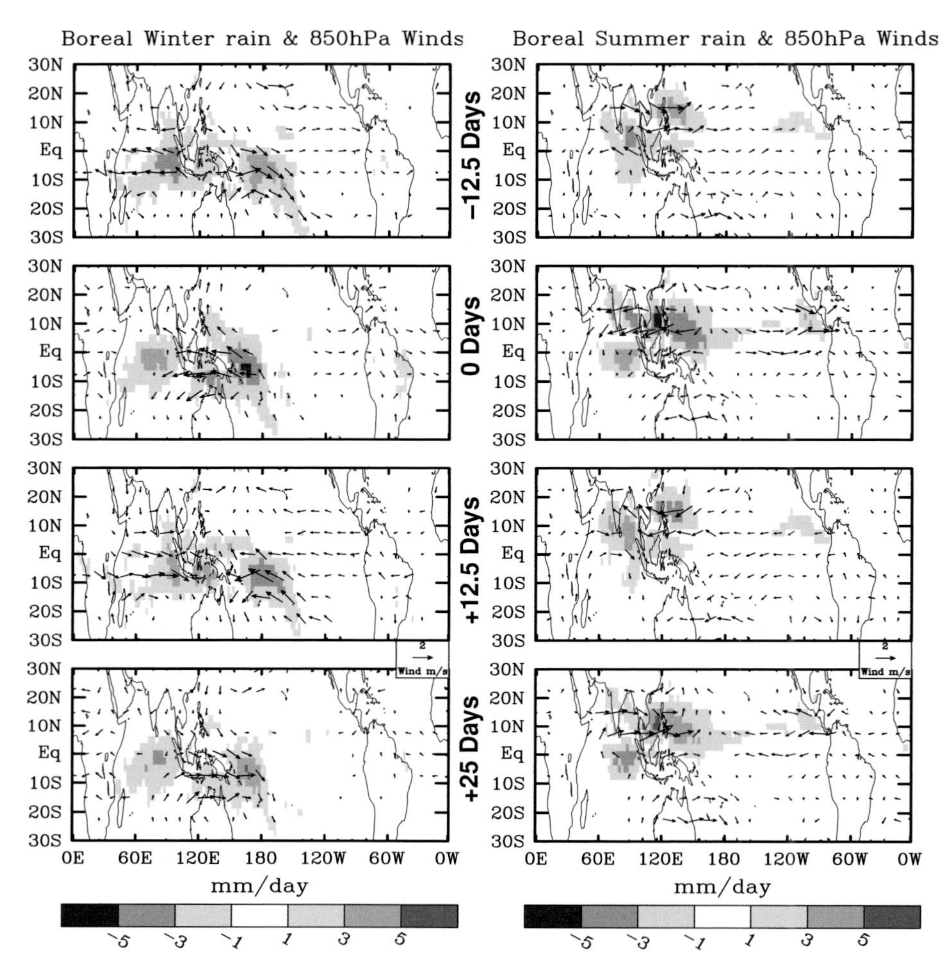

Figure 11.1 (See also colour plate section.) Canonical structure of an MJO event based on 5-day average (i.e. pentad) NCEP/NCAR Reanalysis (Kalnay *et al.*, 1996) and CMAP rainfall data (Xie and Arkin, 1997) from 1979 to 2000. Data were bandpassed filtered with a 30–90 day filter and then separated into boreal winter (Nov–Apr) and summer (May–Oct). Extended EOF (EEOF) analysis with +/–5 pentad lags was performed on tropical rainfall (30N to 30S, 30E to 180E) to identify the dominant 'mode' for the winter and summer separately. Composite events were constructed by selecting events if the EEOF amplitude time series exceeded 1 standard deviation ($N = 43$ (49) for winter (summer)). The resulting composites have dimensions lag (-5 to $+5$ pentads), latitude and longitude. In the plots above, only 4 panels of the boreal winter composite are shown, each separated by 2.5 pentads (i.e. 12.5 days). Plots show composite rainfall and 850 hPa wind vectors for (left) boreal winter and (right) boreal summer. Only values that exceed the 90% confidence limit are shown.

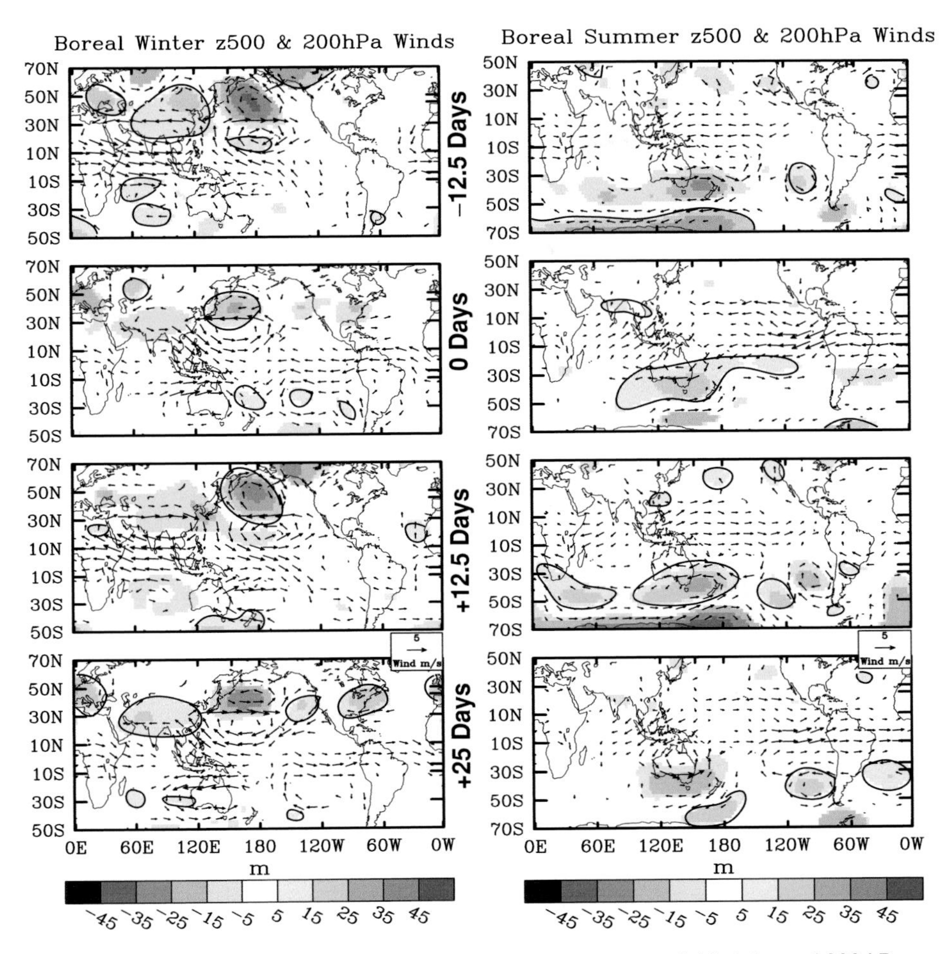

Figure 11.2 Same as above except for 500 hPa geopotential heights and 200 hPa wind. In this case, encircled shading denotes positive anomalies.

the convection tends to subside and propagate southeastward into the South Pacific Convergence Zone. Beyond the Date Line, the disturbance is primarily evident only in the near-equatorial wind field with characteristics similar to a dry Kelvin wave with a speed of about 15–20 m/s or greater (Hendon and Salby, 1994).

Another important feature associated with the MJO, especially in relation to its connections to midlatitudes, is its off-equatorial structure and variability. From the left panels of Figure 11.1 and Figure 11.2 there is evidence of off-equatorial Rossby wave gyres that straddle the near-equatorial rainfall anomalies. For example, in the composite maps at lag +12.5 days, the positive rainfall (i.e. heating) anomaly is located over the Maritime continent. Associated with this are upper-level cyclonic (anticyclonic) gyres to the north-east and south-east (north-west and south-west) centred at latitudes of about 20°. These gyres are more easily identified in the life-cycle analysis of the MJO by Hendon and Salby (1994) and are consistent with the circulation that is expected in association with a near-equatorial tropospheric heating anomaly (Matsuno, 1966; Gill, 1980). One of the important manifestations of these tropical heating and subtropical streamfunction anomalies is that they act as Rossby wave sources for midlatitude variability (e.g. Weickmann, 1983; Liebmann and Hartmann, 1984; Weickmann *et al.*, 1985; Lau and Phillips, 1986; Sardeshmukh and Hoskins, 1988; Berbery and Noguespaegle, 1993). For example, the +12.5-day lag map of Figure 11.2 shows evidence of a wave train emanating from the tropics and extending poleward and eastward over the Pacific Ocean and North America. Such connections with the extratropics have important ramifications for midlatitude weather variability, regime changes and forecasting capabilities (e.g. Ferranti *et al.*, 1990; Higgins *et al.*, 2000; Jones *et al.*, 2004a).

The intraseasonal variability characteristics discussed above tend to be most strongly exhibited during the boreal winter and spring when the Indo-Pacific warm pool is centred at or near the equator. In the boreal summer, the MJO is still present although its spatial variability and propagation characteristics are modified by the changes in the circulation associated with the annual cycle. The right panels of Figure 11.1 and Figure 11.2 illustrate the canonical space–time structure of the MJO in boreal summer (for more in-depth observational descriptions see recent reviews by Goswami, 2005; Hsu, 2005; Waliser, 2006). Note that the summertime manifestation of the MJO is often referred to as the intraseasonal oscillation (ISO), the boreal summer ISO, or monsoon ISO (MISO). Examination of the rainfall map at lag 0 days shows that positive rainfall anomalies in the western and central Indian Ocean for the boreal summer case occur in conjunction with negative rainfall anomalies over a region extending between India and the western equatorial Pacific. This system then appears to propagate both eastward – similar to the boreal winter case – and northward (Yasunari, 1979; Lau and Chan, 1986; Lawrence and Webster, 2002; Hsu, 2005). As with the boreal winter case, the associated midlatitude variability occurs primarily in the winter hemisphere.

Most relevant to the present discussion of predictability is the large spatial scale and slow evolution of the rainfall patterns in Figure 11.1 relative to typical synoptic-scale weather features. These characteristics suggest a measure of predictability at a timescale on the order of weeks (e.g. Waliser *et al.*, 1999b). This feature, along with their somewhat regular occurrence, their impact on the Asian–Australian monsoons, as well as their influence on extratropical weather patterns motivate the need to develop the capability to predict these 'events' and improve our understanding of their predictability. To gain an appreciation for their dramatic impact on the Asian and Australian monsoons, Figure 11.3 shows the annual cycle of rainfall and the anomalous evolution of unfiltered and filtered rainfall over India and northern Australia for a sample of three years. These time series emphasise the overall dominance, apart from the annual variation, of the intraseasonal timescale on these monsoon systems, including its obvious role in dictating active and break phases. In addition, the

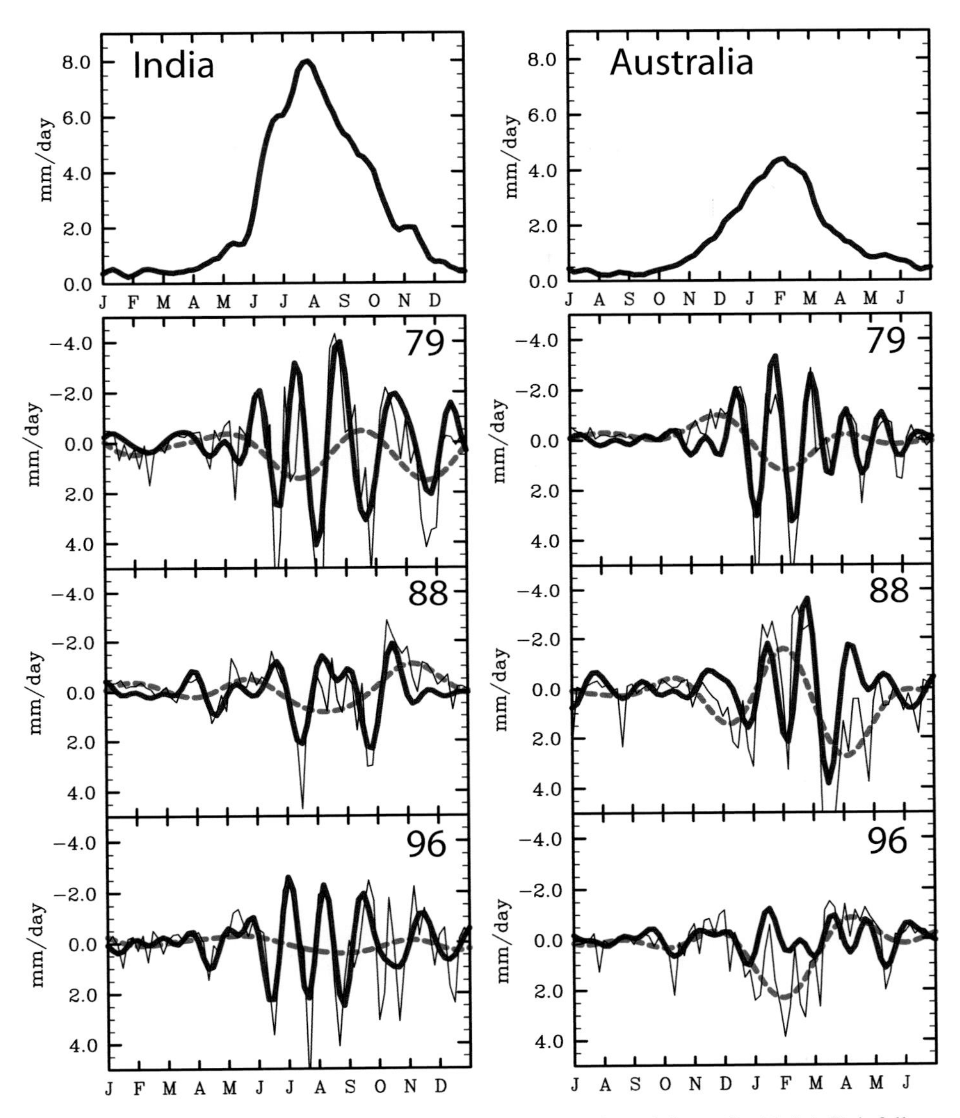

Figure 11.3 Time series of rainfall over India (left) and Australia (right). Rainfall data is based on pentad values of the satellite and *in-situ* merged CMAP product of Xie and Arkin (1997) from 1979 to 1999. The data plotted for India (Australia) are the domain averages of the grid points lying within India (Australia, lying north of 25°S). (Top) Mean 73-pentad annual cycle. (Lower three panels) The thin black lines are pentad anomaly values, the thick black lines are 30–90 day bandpassed values, and the thick dashed lines are 90 day low-pass values for the years 1979, 1988 and 1996.

three years sampled is enough to illustrate that the MJO exhibits a considerable amount of year-to-year variability (e.g. Ferranti *et al.*, 1997; Hendon *et al.*, 1999; Slingo *et al.*, 1999; Lawrence and Webster, 2001; Teng and Wang, 2003).

While the diagrams in Figure 11.1 and Figure 11.2 illustrate what might be considered typical winter and summer MJO events, it is important to recognise that these events have considerably more complexity in reality. For example, the study by Wang and Rui (1990), and later by Jones *et al.* (2004b), have further diagnosed the 'synoptic climatology' of tropical MJO events, including their seasonal modulation. These studies show for example that boreal winter events display considerable variation in the longitude that the convection develops and/or dissipates. Moreover, it is well known that the convection associated with MJO events typically propagates further east during El Niño events (e.g. Kessler, 2001). For the boreal summer case, Kemball-Cook and Wang (2001) show that there is a systematic intraseasonal change in the spatial structure and propagation characteristics of the MJO. In the early part of the summer (e.g. May–June), the off-equatorial variability is generally found west of South-east Asia and the Maritime continent, while in the later part of the summer, it expands to include much of the north-western tropical Pacific.

In addition to the above complexities, there are also finer scale structures embedded in the MJO that deserve mention. For example, studies by Nakazawa (1988), Lau *et al.* (1991), and Chen *et al.* (1996) have shown that the convective variability organises on a wide range of time and space scales within the large-scale anomalies that are emphasised in Figure 11.1. In addition, the latitudinal asymmetry of the boreal summer MJO makes its evolution and physical description more complicated than the boreal winter case, where the latter is generally thought of as an eastward-propagating, convectively coupled, equatorially trapped wave complex. These aspects of the physical description also hold for the boreal summer case, particularly in the Indian Ocean and far western Pacific. However, the summertime hemisphere in the boreal summer case experiences a relatively larger increase in SST and surface moisture, and an accompanying enhancement in the large-scale easterly vertical shear, than for the boreal winter case. These features, along with the land–sea distribution in the area, promote the emanation and growth of Rossby waves that are forced by the near-equatorial convection anomalies (Li and Wang, 1994; Wang and Xie, 1996, 1997; Kemball-Cook and Wang, 2001; Lawrence and Webster, 2002). The overall eastward propagation of the large-scale, near-equatorial convective anomaly, combined with the inherent westward-propagation of these Rossby waves (Matsuno, 1966) and a number of mechanisms that promote the northward propagation of the latter (e.g. Jiang *et al.*, 2004), largely account for the appearance of the eastward-propagating, north-west-south-east tilted, large-scale 'rainband' evident in Figure 11.1. Other noteworthy features associated with the boreal summer MJO include what is known as the climatological ISO (CISO; e.g. Wang and Xu, 1997) and significant variability at the 10–20 day timescale. Elaboration on the above detailed features of the MJO as well as more comprehensive aspects of MJO theory are beyond the scope of this chapter and the reader is referred to a number of recent reviews on these subjects (Chang, 2004; Goswami, 2005; Hsu, 2005; Wang, 2005; Wheeler and McBride, 2005; Waliser, 2006).

11.3 Predictability

By the late 1980s, many characteristics of the MJO were fairly well documented and it was clear that it was a somewhat well defined phenomenon with a number of reproducible features from one event to another as well as in events from one year to the next. Given this, and the degree that research had shown a number of important interactions of the MJO with other features of our weather and climate system, it was an obvious step to consider MJO forecasting in more earnest. Since numerical weather and climate models typically had, and still have, a relatively poor representation of the MJO (e.g. Slingo *et al.*, 1996; Waliser *et al.*, 2003e; Slingo *et al.*, 2005; Allen *et al.*, this volume), a natural avenue to consider was the development of empirical models. Along with the possibility of providing more skilful forecasts than numerical methods available at the time, this avenue also provided a means to establish an initial estimate of the predictability limit for the MJO – at least that which could be ascertained from the observations alone.

There were a number of different approaches and data sets used in these empirical studies. For example, von Storch and Xu (1990) examined Principal Oscillating Patterns of equatorial 200 mb velocity potential anomalies with an emphasis on boreal winter. The model of Waliser *et al.* (1999b) was based on a field-to-field Singular Value Decomposition that used previous and present pentads of outgoing longwave radiation (OLR) to predict future pentads of OLR with separate models developed for boreal winter and summer conditions. Lo and Hendon (2000) developed a lag regression model that used as predictors the first two and first three principal components of spatially filtered OLR and 200 hPa streamfunction, respectively, to predict the evolution of the OLR and 200 hPa streamfunction anomalies associated with the boreal winter MJO. A similar strategy was used by Jones *et al.* (2004c). Mo (2001) utilised empirical basis functions in time by using a combination of singular spectrum analysis for the filtering and identification of the principal modes of variability and the maximum entropy method for the forecasting component. The procedure was applied to monitor and forecast OLR anomalies in the intraseasonal band over the Indian–Pacific sector as well as the pan-American region. In a quite different approach, Wheeler and Weickmann (2001) utilised tropical wave theory (Matsuno, 1966; Wheeler and Kiladis, 1999) as the basis for their filtering and forecasting technique. In order to monitor and predict the evolution of a given mode of interest, near-equatorial time-longitude sections were Fourier analysed in two dimensions; the specific zonal wave numbers and frequencies associated with the mode(s) of interest (e.g. MJO) were retained, and then the modified spectrum was inverse Fourier analysed. Goswami and Xavier (2003) identified all active and break phases associated with boreal summer MJO events and then as a means for prediction calculated the typical (i.e. ensemble average) transition from active to break (and break to active) conditions as a function of lead time. Most recently, Webster and Hoyos (2004) have developed a physically based, multi-predictor, Bayesian model to predict regional rainfall and river discharge associated with the Asian monsoon, with a particular emphasis on intraseasonal variations over India. A number of additional empirical schemes that have relevance to real-time prediction will be discussed in Section 11.4.

The above discussion gives a flavour of the types of empirical MJO models that have been developed to date. For the most part, each of the above studies developed their model on a given portion of the observed record and then tested it on an independent portion. Glossing over the details, the upshot of these studies is that empirical models demonstrate useful predictive skill for the MJO on the order of 15–25 days or more, depending on the spatial scale and quantity being predicted. However, as with any empirical model, these models are limited in the totality of the weather and climate system they can predict, their ability to adapt to arbitrary conditions, and their ability to take advantage of known physical constraints. Thus one might conclude that if dynamical models had a realistic representation of the MJO, this limit might be extended somewhat. While the majority of dynamical models to date still exhibit significant shortcomings in terms of their MJO simulation – particularly if pressed to do operational prediction, there have been a few models, or versions of models, that have demonstrated success at representing a number of the principal features of the MJO (Slingo *et al.*, 1996; Sperber *et al.*, 1997; Waliser *et al.*, 1999a; Kemball-Cook *et al.*, 2002; Maloney, 2002; Fu *et al.*, 2003; Zheng *et al.*, 2004). This degree of model success at least provides the means to perform 'perfect-model', or so-called 'twin-predictability', experiments to ascertain an estimate of the theoretical limits of prediction for the MJO. In this case, 'forecasts' are verified against others that only differ in the initial conditions (e.g. Lorenz, 1965; Shukla, 1985; Palmer, this volume). This approach was taken in two recent studies by Waliser *et al.* (2003b, 2003c). In this case, the experiments were performed with the NASA Goddard Laboratory for Atmospheres (GLA) general circulation model (GCM) (Kalnay *et al.*, 1983; Sud and Walker, 1992). In a number of studies, this model has been shown to exhibit a relatively realistic MJO (Slingo *et al.*, 1996; Sperber *et al.*, 1997; Waliser et al., 2003d) with reasonable amplitude, propagation speed, surface flux characteristics, seasonal modulation, and interannual variability (Waliser *et al.*, 2001). One of its principal deficiencies is its relatively weak variability in the equatorial Indian Ocean, a problem quite common in atmospheric GCMs (AGCMs, Waliser *et al.*, 2003d).

For these studies, a 10-year control simulation using specified annual cycle SSTs was performed in order to provide initial conditions from which to perform an ensemble of twin predictability experiments. Note that this analysis was performed separately on boreal winter and summer MJO activity (e.g. left and right panels of

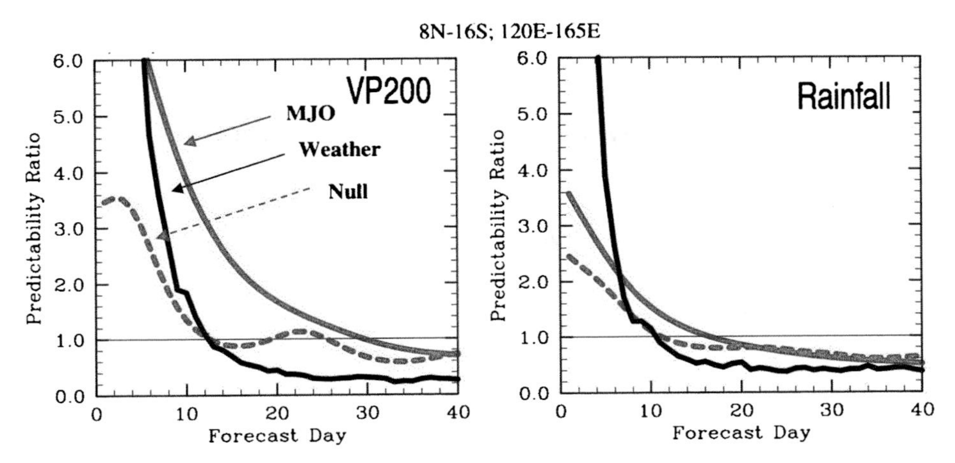

Figure 11.4 Predictability measure, defined as the ratio of the MJO signal and the MJO forecast error (see Waliser *et al.*, 2003b), versus lead time based on 120 northern hemisphere winter MJO twin-predictability forecast cases for VP200 (left) and rainfall (right) from the NASA/GLA model for 120 active/strong boreal winter MJO cases (solid black), 30 weak/null boreal winter MJO cases (dashed grey) and for unfiltered 'weather' variations (using the 120 active MJO cases; solid grey) for the region 8° N–16° S and 120–165° E.

Figure 11.1, respectively). The following discussion describes the boreal winter study (Waliser *et al.*, 2003b) but the methods are quite similar for the boreal summer analogue (Waliser *et al.*, 2003c) of which a few results are also mentioned. Initial conditions were taken from periods of strong MJO activity identified via extended empirical orthogonal function (EOF) analysis of 30–90 day bandpassed tropical rainfall during the October through April season. From the above analysis, 15 cases were chosen when the MJO convection was located over the Indian Ocean, Maritime continent, western Pacific Ocean, and central Pacific Ocean, respectively, making 60 cases in total. In addition, 15 cases were selected which exhibited very little to no MJO activity. Two different sets of small random perturbations, determined in a rather ad hoc and simplistic manner, were added to these 75 initial states. Simulations were then performed for 90 days from each of these 150 perturbed initial conditions (Buizza, this volume; Kalnay *et al.*, this volume).

A measure of potential predictability was constructed based on a ratio of the signal associated with the MJO, in terms of bandpassed (30–90 day filter) rainfall or 200 hPa velocity potential (VP200), and the mean square difference between sets of twin (bandpassed) forecasts. Predictability was considered useful if this ratio was greater than one, and thus if the mean square error was less than the signal associated with the MJO. The results, shown in Figure 11.4, indicate that useful predictability for this model's MJO extends out to about 20 to 30 days for VP200 and to about 10 to 15 days for rainfall. This is in contrast to the timescales of useful predictability for the model's weather, or for cases in which the MJO is absent. In these latter cases, the predictability limit is roughly 12 days for VP200 and 7 days for rainfall. Note that these latter two regimes are related, in that when the MJO is quiescent, the model lacks a low-frequency component that might help it retain predictability over long timescales and is in a regime where the processes and timescales of weather are the only phenomena left to provide predictability. Additional support for this conclusion was demonstrated from a predictability analysis on EOF decompositions of the model data (Waliser *et al.*, 2003b). This analysis shows more definitively that the enhanced predictability derives from the (about two) EOF modes that represent the MJO variability in the model. In addition to the above, the predictability measure exhibits modest dependence on the phase of the MJO, with greater predictability for the convective phase at short ($< \sim 5$ days) lead times and for the suppressed phase at longer ($> \sim 15$ days) lead times. This result appears consistent with the empirical model results of Goswami and Xavier (2003) that showed break monsoon phases to be more predictable than active phases.

Additional experiments have been carried out to assess the sensitivity of the results above to changes in background state. For example, the study by Waliser *et al.* (2003c) was performed in an analogous fashion, although with more cases ($N = 168$), to examine the predictability limits associated with the boreal summer MJO. The left panels of Figure 11.5 show maps that depict the size of the MJO 'signal' in terms of VP200 at lead times of 5, 15 and 25 days. Due to the relatively long timescale of the MJO, this signal remains roughly constant over this period. The right panels show the associated mean squared error at the same lead times. Evident is the fact that the error is generally less than the signal even up to 25 days, indicating predictability at these lead times. As expected, the maps show that the predictable VP200 signal is limited to the tropical regions where the MJO has an impact on this quantity. A similar diagram for rainfall (not shown here), indicates useful predictability out to about 15 days, with the geographic extent being even more limited. The plots in Figure 11.6 provide a more quantitative illustration of the above for a select region within the area of high MJO variability. The lead time at which the error and signal intersect can roughly be equated to the limit of predictability (i.e. where the predictability ratio discussed above becomes one). Similar to the boreal winter results, the limit of predictability for the upper level circulation (i.e. VP200) and rainfall for the boreal summer MJO is about 25–30 days and 15–20 days, respectively. Consistent with the discussion above, this figure also demonstrates that when the MJO cases analysed are divided into strong versus weak cases, predictability associated with the strong cases is enhanced.

Predictability measures were also examined under El Niño and La Niña conditions. In this case, analogous experiments to those described above for the boreal winter case were performed but with imposed El Niño and La Niña SST anomalies. To construct these anomalies, the observed SST anomalies between September and the following August for the El Niño years of 1957, 1972, 1982, 1986, 1991 and 1997 were averaged together. This 12-month 'raw' anomaly was then subject to a Fourier

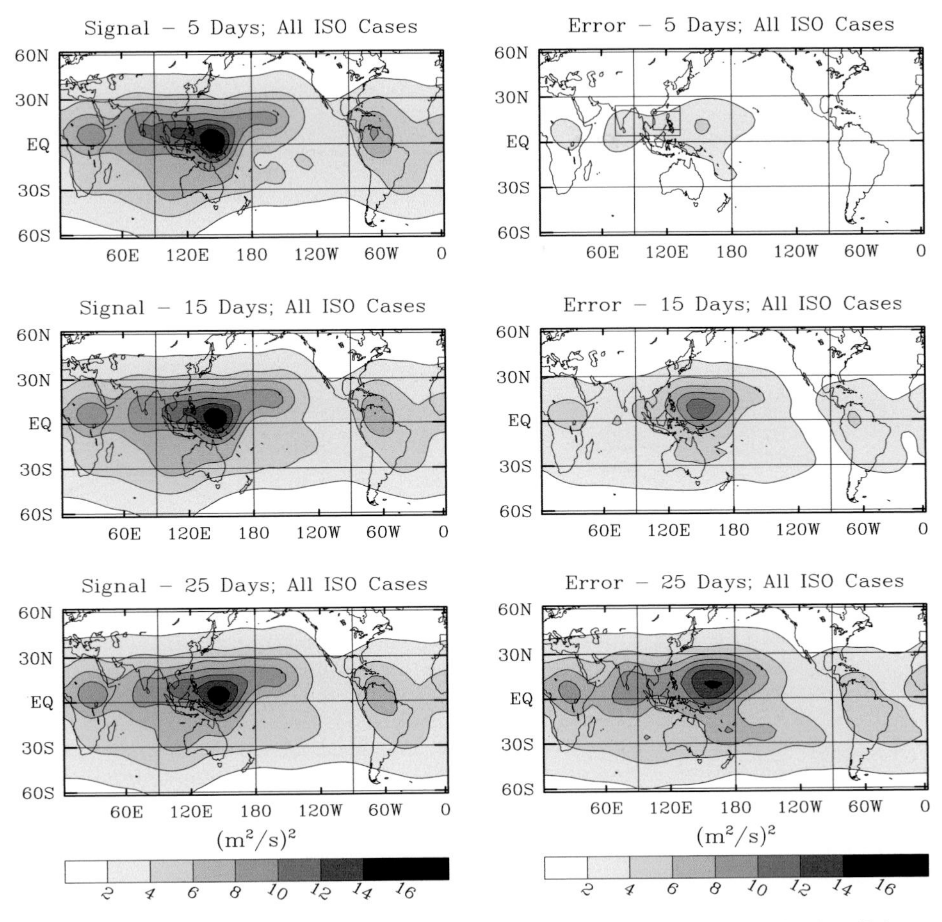

Figure 11.5 The MJO mean signal (left) and mean forecast error (right) for all the boreal summer MJO cases ($N = 168$) at lead times of 5 (top), 15 (middle), and 25 (bottom) days for filtered (30–90 days) 200 hPa velocity potential (VP200). VP200 values have been scaled by 10^{-12}. From Waliser *et al.* (2003c).

analysis, retaining only the lowest three harmonics. This step ensured that the imposed anomaly consisted of only low-frequency and, more importantly, periodic variations. To account for signal loss in the compositing and filtering procedure, the resulting anomaly was multiplied by a factor of 2, and then added to the climatological SSTs to provide a perpetual 12-month evolving El Niño condition. The procedure was performed exactly the same for the La Niña case, except using June through the following May anomalies for the La Niña years 1950, 1954, 1956, 1956, 1970, 1973, 1974, 1988. The 12-month means of the anomalous El Niño and La Niña SST patterns are shown in the upper panels of Figure 11.7. For each case, 10-year simulations were performed and then analysed in the same manner as described above.

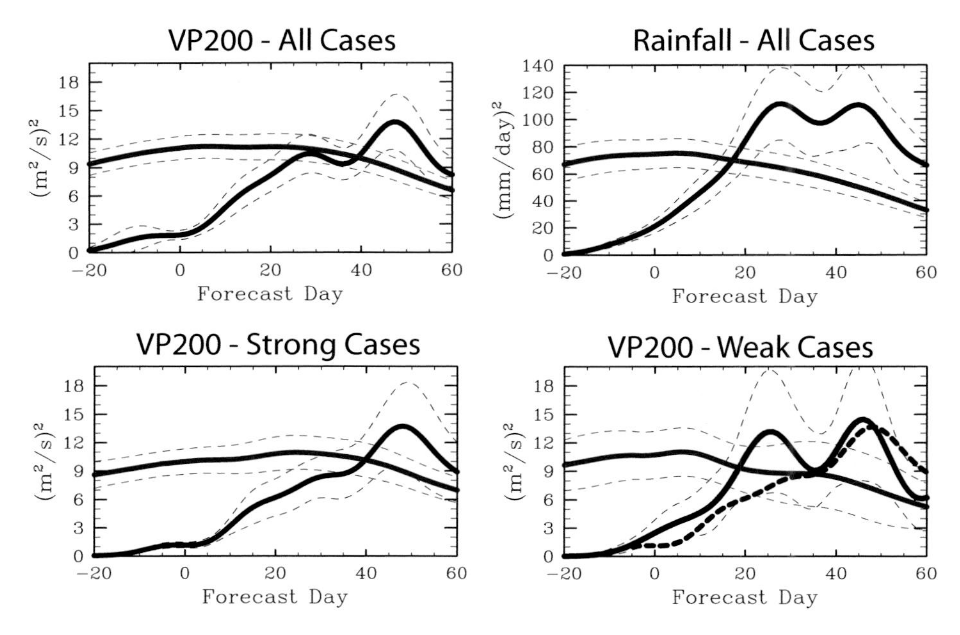

Figure 11.6 (Upper) The thick solid black lines which increase with forecast time are the mean squared forecast error for the (30–90 day) filtered (left) 200 hPa velocity potential (VP200) and (right) rainfall, over the region 12° N–16° N and 117.5° E–122.5° E (model grid point at centre of smaller black box in Figure 11.5) for all the boreal summer MJO cases ($N = 168$). The thick solid black lines which are roughly constant with forecast time are the mean MJO signal for the same quantities, and over the same region and cases. The thin dotted lines depict the 95% confidence limits for the above quantities using a student *t*-test. VP200 values have been scaled by 10^{-12}. (Lower) Same as upper left panel, except that the left panel is based on forecasts using the strongest, and the right the weakest, MJO cases ($N = 80$ in each case). The thick dashed line in the lower right panel is a re-plotting of the mean-squared error from the lower left panel to allow for easier comparison between the strong and weak MJO cases. VP200 values have been scaled by 10^{-12}. Adapted from Waliser *et al.* (2003c).

The middle and lower panels of Figure 11.7 illustrate that the predictability of the model's MJO is considerably enhanced (diminished) for the imposed El Niño (La Niña) conditions. Examination of the results shows that part of these changes derives from the changes in the SST in the central Pacific and the associated extension (contraction) of the MJO propagation path for the El Niño (La Niña) case. More substantial is the fact that overall the MJO signal, meaning the amplitude of the typical event analysed, is considerably larger (smaller) for the El Niño (La Niña) case than for the control (i.e. climatological SST) case. Given a somewhat similar error growth rate, this change in signal also promotes the changes observed to the model's MJO predictability. This latter aspect raises an interesting question relative to the studies performed to date that examine the relation between interannual SST variability and MJO activity (e.g. Gualdi *et al.*, 1999; Hendon *et al.*, 1999; Slingo

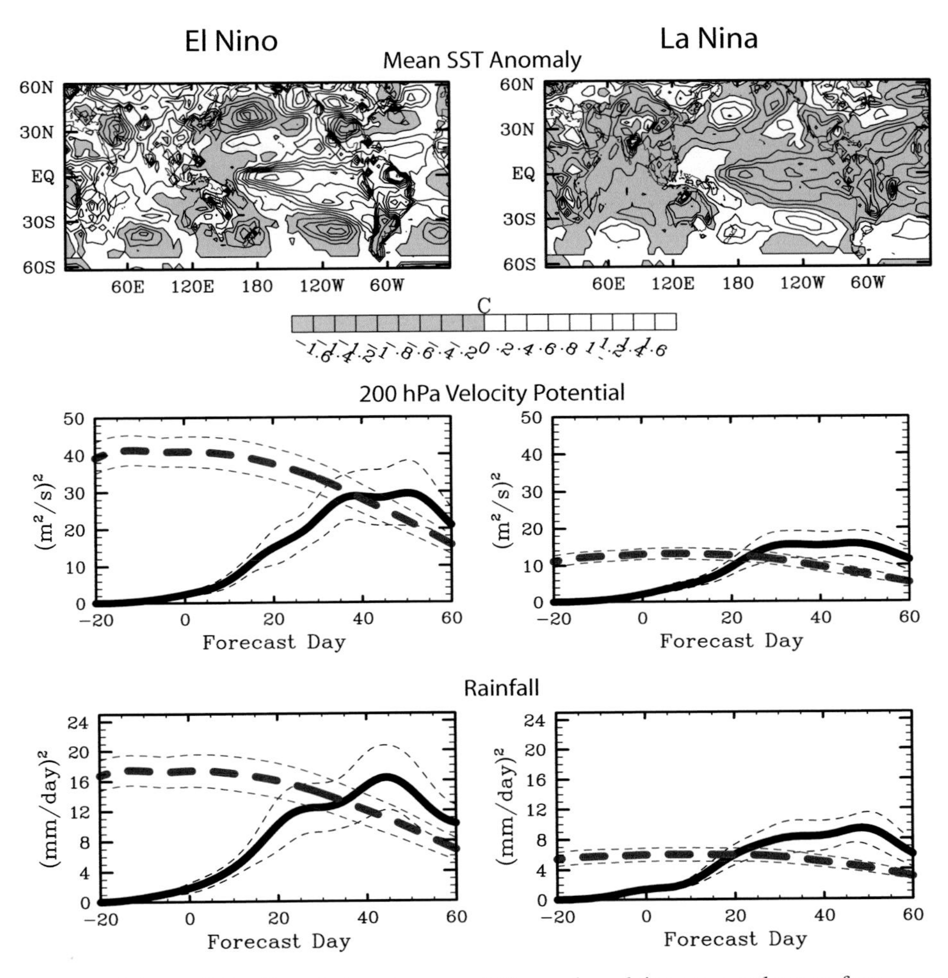

Figure 11.7 (Upper) Annual mean of 12-month evolving perpetual sea surface temperature (SST) anomaly applied to the set of predictability experiments associated with El Niño (left) and La Niña (right) conditions. (Middle) The thick solid black lines are the mean squared forecast error for the (30–90 day) filtered 200 hPa velocity potential (VP200) over the western Pacific Ocean (4° N–12° S; 147.5° E–162.5° E) for all the boreal winter El Niño (left) and La Niña (right) MJO cases (each have $N = 120$). The thick dashed grey lines are the mean MJO signal for the same quantities, and over the same region and cases. The thin dotted lines depict the 95% confidence limits for the above quantities using a student t-test. VP200 values have been scaled by 10^{-12}. (Lower) Same as middle, except for rainfall.

et al., 1999), which for the most part (even for this same model, Waliser *et al.*, 2001) have found little or no relation, particularly for the boreal winter case. The results here suggest that perpetually warm or cool anomalous SST conditions might have a considerable impact on the MJO (e.g. Slingo *et al.*, 1999; Zveryaev, 2002) while the observed intermittent interannual SST variability does not.

While the results from these numerical studies are encouraging from the viewpoint of intraseasonal prediction, and are not entirely inconsistent with the sorts of complimentary empirical studies mentioned above, there are a number of issues to consider that might impact the limit of predictability estimate they provide. First, the GLA model employed has been shown to have too much high frequency, low wave-number activity (Slingo *et al.* 1996). Relative to the MJO, this variability would be considered to be unorganised, errant convective activity that may erode the relatively smooth evolution of the MJO and thus diminish its predictability. Second, these simulations were carried out with fixed climatological SST values. A previous study with this model showed that coupled SSTs tend to have an enhancing and organising influence on the MJO, making it stronger and more coherent (Waliser *et al.*, 1999a). Thus the exclusion of SST coupling may lead to an underestimate of the predictability as well (Timmermann and Jin, this volume).

There are also a number of aspects associated with the model and/or analysis to suggest that the above results might overestimate the predictability of the MJO. The first is that the model's coarse resolution and inherent reduced degrees of freedom relative to the true atmosphere may limit the amount of small-scale variability that would typically erode large time and space scale variability. However, it is important to note in this regard that the low-order EOFs of intraseasonally filtered model output typically do not capture as much variability as analogous EOFs of observed quantities. Thus the model's MJO itself still has room to be more robust and coherent, which would tend to enhance predictability. In addition to model shortcomings, the simple manner that perturbations were added to the initial conditions may also lead to an overestimate of the predictability. The perturbation structure and the size of the perturbations may be too conservative and may not adequately represent the type of initial condition error that would be found in an operational context. However, even if that is the case, it would seem that adequate size 'initial' errors would occur in the forecast in a matter of a day or two and thus one would expect this aspect to overestimate the predictability by only a couple of days, if at all.

In order to address some of the uncertainties mentioned above, an analogous study for boreal summer conditions using the ECHAM AGCM has recently been undertaken (Liess *et al.*, 2005). The modelling and analysis framework is similar to that described above with two important exceptions. First, rather than select a large number of events (i.e. $\sim$15–20) for each of four phases of the boreal summer MJO (i.e. convection in Indian Ocean, Maritime continent, South-east Asia, north-west tropical Pacific) and performing only a few (i.e. two) perturbation experiments with each, this study has selected the three strongest events in a 10-year simulation and then performed a larger ensemble of forecasts for each of the four phases (i.e. 15). In addition, rather than use simply determined perturbations, this study uses the breeding method (Toth and Kalnay, 1993; Cai *et al.*, 2003). Figure 11.8 shows the combined results from all twelve 15-member ensemble MJO forecasts using the ECHAM5 AGCM. The data for the figure are taken from 90° E to 120° E and 10° N

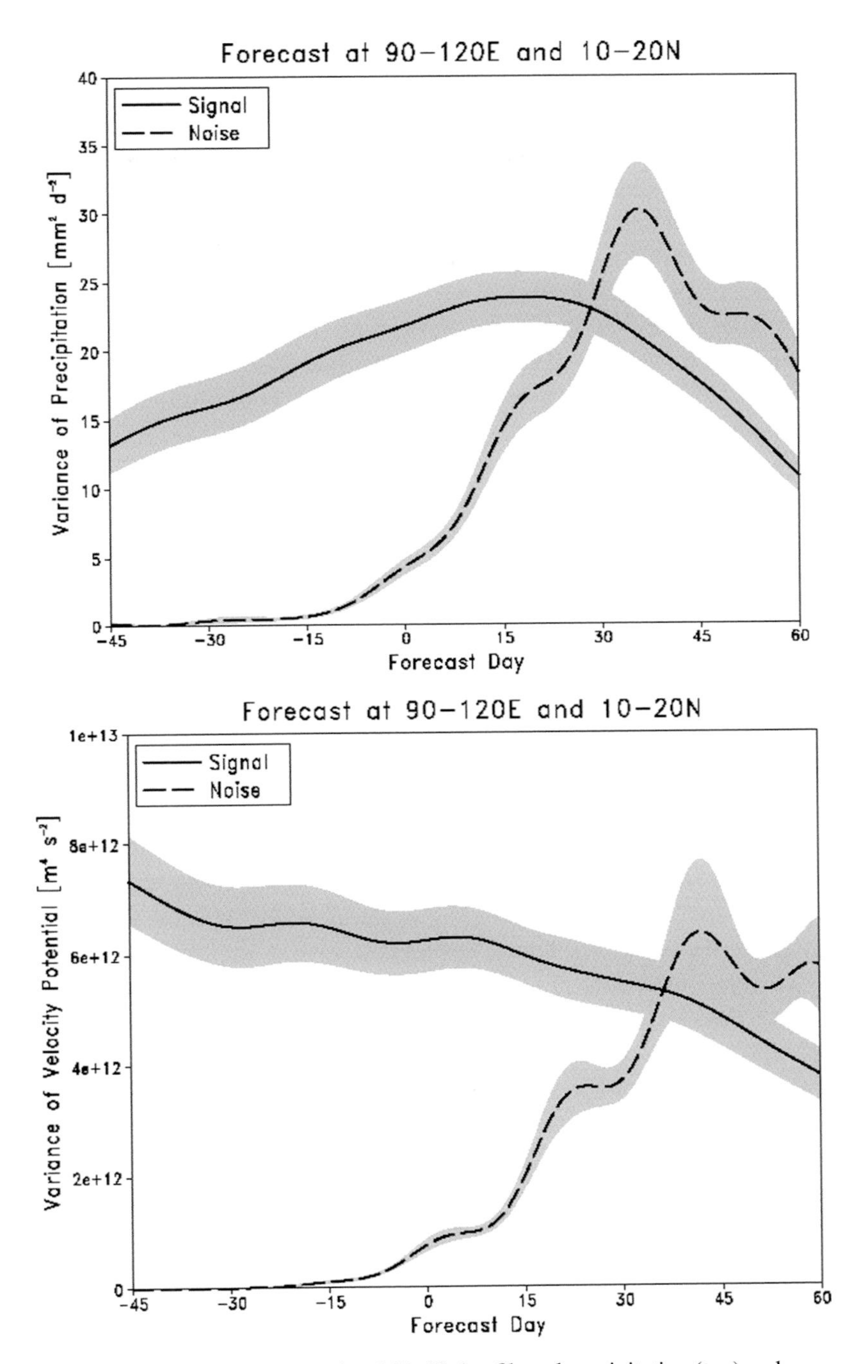

Figure 11.8 Signal-to-noise ratio of 30–90 day filtered precipitation (top) and 200 hPa velocity potential (bottom) predictions averaged over all four phases of three MJO events. Shadings represent the significance at the 95% interval based on all twelve 15-member ensemble forecasts. All values are averaged over the region 90 to 120° E and 10 to 20° N. Adapted from Liess *et al.* (2005).

to 20° N for 30–90 day bandpass filtered rainfall (upper) and VP200 (lower) anomalies. These results suggest that the boreal summer MJO has dynamical predictability with lead times potentially up to and beyond 30 days. These lead times are at least as large, if not larger, than those found in the Waliser *et al.* studies highlighted above. However, it should be noted that the event analysed here is a particularly robust and strong one for the model, and those above were based on both strong and moderate sized events which could account for the difference. In any case, even though the above results do not take into account systematic model bias relative to the observations, they, along with many of the other studies discussed above, indicate that a promising avenue and time scale of operational prediction lies ahead.

11.4 Practical considerations

Based on the motivating factors presented in the Introduction and the promising results derived from the empirical forecast and dynamical predictability studies discussed in the previous section, there is ample reason to push towards an operational MJO predictive capability. Ideally, it would be most convenient if our present-day numerical weather forecast models could demonstrate skill at MJO simulation and prediction. If this was the case, our medium- to long-range forecasts could simply be extended to provide useful subseasonal predictions of the MJO and the ancillary weather and circulation systems it interacts with (e.g. tropical storms, midlatitude flows). Unfortunately, due to the poor representation of the MJO by most GCMs, this avenue cannot be readily or fully exploited (Allen *et al.*, this volume; Palmer, this volume). This has been found to be particularly true in the few studies carried out to test the predictive skill of the MJO in operational weather forecast models. For example, the studies by Chen and Alpert (1990), Lau and Chang (1992), Jones *et al.* (2000), Hendon *et al.* (2000) were all performed on the most recent or previous versions of the National Oceanic and Atmospheric Administration's (NOAA's) National Centers for Environmental Prediction (NCEP) (or NMC) medium range forecast (MRF) model's Dynamic Extended Range Forecasts (DERFs). In general these studies only found useful skill out to about 7–10 days for MJO-related variability, and were simply hampered by MJO variability that was too weak and/or that propagated too fast. Probably the most optimistic set of forecast skill experiments for the MJO were a set of Asian monsoon MJO case studies performed by Krishnamurti *et al.* (1990, 1992, 1995). The novel approach in these cases was that an attempt was made to filter out the 'weather' time and space scales from the initial conditions and leave only the 'low-frequency modes'. In this case, the results demonstrated useful forecast skill out to 3–4 weeks; however, there are some uncertainties associated with making such a technique operational as well with how the boundary-layer forcing (i.e. SST) was handled. For a more thorough discussion of the above studies, as well as the real-time efforts highlighted below, see Waliser (2005).

Given the need for forecast capability at the intraseasonal timescale, along with the poor representation of the MJO in dynamical models, a number of real-time efforts have been developed based on empirical methods. These include those based on a number of the schemes mentioned above, such as Wheeler and Weickmann (2001), Jones *et al.* (2004c), Lo and Hendon (2000) as improved on by Wheeler and Hendon (2004), and Webster and Hoyos (2004). In addition, van den Dool and Saha (2002) have recently applied the empirical wave propagating (EWP) technique developed by van den Dool and Qin (1996) to forecast the MJO. EWP is a 'phase-shifting' technique that allows one, in the diagnostic step, to determine the amplitude-weighted-average climatological phase speed of anomaly waves (e.g. equatorial MJO), where the waves are represented as either zonal or spherical harmonics. The diagnostic step results in a table of phase speeds for waves in the anomaly field as a function of zonal wave number, calendar month and latitude, based on the observed data. This technique has shown to be particularly well suited for empirically forecasting the large-scale upper-level anomalies (e.g. VP200) associated with the MJO.

In quite a different approach, stemming from somewhat different and/or more comprehensive objectives, Newman *et al.*(2003) have developed and implemented a real-time forecasting scheme that has applicability to the MJO based on what is often referred to as the Linear Inverse Model (LIM; Winkler *et al.*, 2001). The LIM is based on NCEP/NCAR reanalysis data (Kalnay *et al.*, 1996) that has had the annual cycle removed, been smoothed with a 7-day running mean filter, gridded to T21 spatial resolution, and been reduced by EOF decomposition. The specific fields used include global 250 and 750 hPa streamfunction and tropical column-integrated diabatic heating. For the boreal winter (summer) model, the first 30 (30) streamfunction and 7 (20) diabatic heating EOFs are used. In this model, historical data are used to define the relationship between a given state (i.e. a weekly average) and conditions one week later, with the process being iterated to produce multi-week forecasts. The advantage of the model is that it includes both tropical and extratropical quantities in the forecasts. In this way, the interaction between the two can be more readily examined and diagnosed. The results in Figure 11.9 show that for tropical forecasts of diabatic heating, the LIM slightly outperforms a research version of the (dynamic) NCEP MRF model at lead times of two weeks, for both northern hemisphere summer and winter, particularly in regions where the MJO is most strongly associated with the diabatic heating field.

Based on the sorts of activities and preliminary successes described above, along with the need to take a more systematic approach to diagnosing problems in dynamical forecasts of the MJO, an experimental MJO prediction programme has recently been implemented (Waliser *et al.*, 2006). The formal components of this programme arose from two parallel streams of activity. The first was the occurrence of the

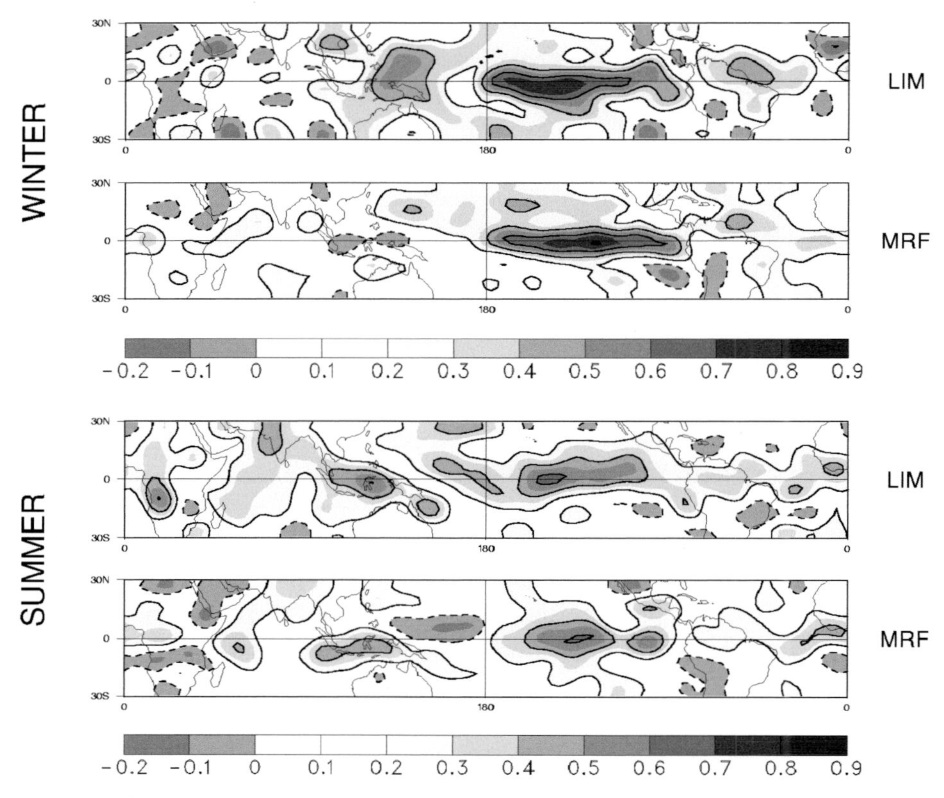

Figure 11.9 Anomaly correlations between forecast and verification column-integrated diabatic heating using the LIM forecast model (Winkler *et al.*, 2001; Newman *et al.*, 2003) and a research version of the NCEP MRF model (i.e. MRF98) for both the northern hemisphere winter (top) and summer (bottom). Forecasts were made for June–August periods for the years 1979–2000. Solid (dashed) contours indicate positive (negative) values.

intraseasonal workshop mentioned in the Introduction (Schubert *et al.*, 2002) and the recognition of the importance of the MJO in regard to the potential skill to be had from intraseasonal predictions. The second stream of activity ensued from the priorities and recommendations of the US CLIVAR Asian–Australian Monsoon Working Group. These streams of activity led to the identification of forecast contributors (which include many of the efforts described above), the formulation of an initial framework for such a programme, the identification of a sponsor that could provide scientific and technical support as well as serve as the data host/server (i.e. NOAA's Climate Diagnostics Center), and a more formal implementation meeting (Waliser *et al.*, 2003a).

The motivation for the above experimental programme involves not only the obvious objective of forecasting MJO variability but also to serve as a basis for model intercomparison studies. The latter includes using the forecasts and biases in model error growth as a means to learn more about, and possibly rectify, model shortcomings but also includes using the empirical models to provide some measure of the expectations that should be attributed to the dynamical models in terms of MJO predictive skill. In addition, it is hoped that this programme and its forecasts will provide a modelling resource to those trying to diagnose interactions between the MJO and other aspects of weather and intraseasonal variability (e.g. PNA, AO). While the immediate goal of the program has been to assemble and provide what is readily available from the community in terms of 2–4 week forecasts of the MJO, there are a number of challenges faced by such an effort that are worth highlighting. The most notable involve how to deal with forecast models that have yet to have or routinely do not have a lead-dependent forecast climatology which is necessary to remove a model's systematic biases, the manner the MJO signal(s) are to be extracted from the heterogeneous set of models (e.g. empirical and numerical), the degree that coupled models and ensembles need to be or can be incorporated into the project, and of course the general logistical problems of dealing with assembling a very non-uniform set of forecast products from different agencies and researchers in near real-time and streamlining them for the purpose of this project.

In terms of ocean coupling, a number of recent studies (Wu *et al.*, 2002; Fu and Wang, 2004; Zheng *et al.*, 2004) have indicated that accurate MJO predictions can only be produced if SST coupling is accounted for in dynamical forecasts. For example, the plots in Figure 11.10, taken from Zheng *et al.* (2004), show that the observed (i.e. quadrature) phase relationship between MJO-related convection and SST anomalies is properly represented in their coupled GCM (CGCM). However, the relationship becomes incorrectly represented (i.e. nearly in phase) in the corresponding AGCM simulations that use specified SSTs taken from the CGCM simulations. These results hold for boreal summer and winter, the Indian Ocean and western Pacific Ocean, as well as another GCM configuration (Fu and Wang, 2004). One of the most important implications of this result is that if specified SSTs are used in a prediction environment, phase errors in tropical convection on the order of 5–10 days (or 5–20° longitude) will occur. This is substantial when considering the local tropical prediction but also problematic when considering the impact on the extratropics. Thus, subseasonal (e.g. MJO) predictions must include ocean coupling – i.e. even a 'two-tier' prediction framework is inadequate.

11.5 Discussion

The review of the studies examined in this chapter was meant to summarise what is known regarding the predictability of tropical intraseasonal variability as well as the current state of affairs of our ability to predict it. Notable is the fact that nearly all the studies presented were primarily based on the MJO, which although

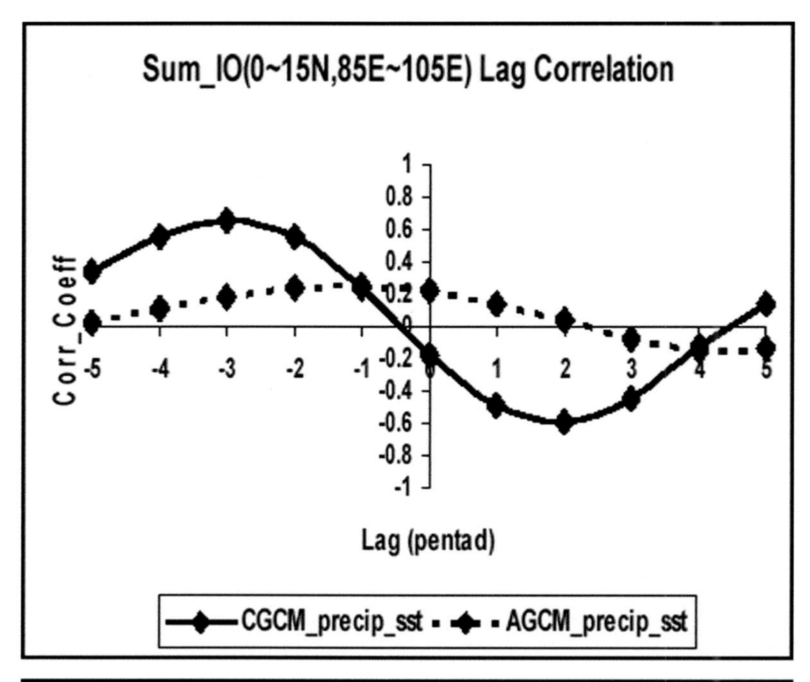

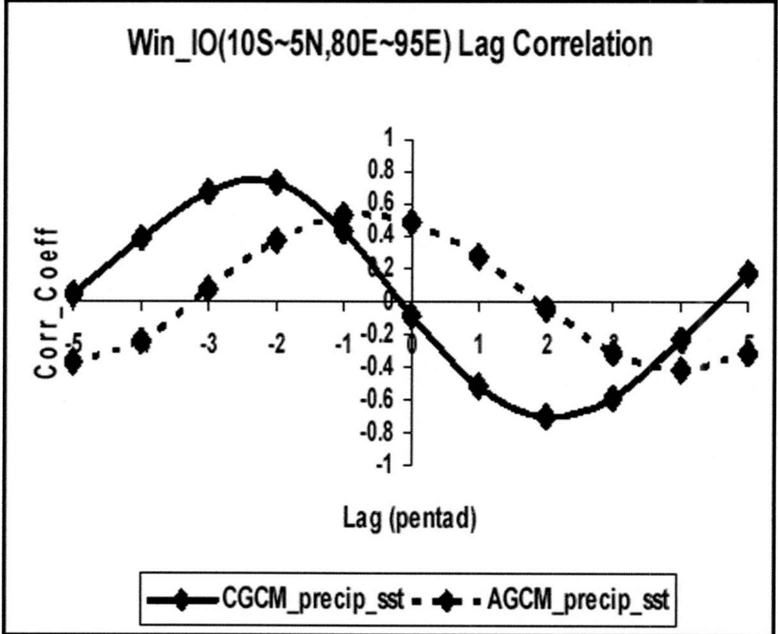

Figure 11.10 Lagged-correlation values between SST and rainfall anomalies from the NOAA Geophysical Fluid Dynamics Laboratory (GFDL) ocean-atmosphere coupled GCM (CGCM; solid) and the corresponding atmosphere-only GCM using SSTs specified from the CGCM simulation (dotted). The top plot is for the boreal summer period and is averaged over 85–105° E, 0–15° N while the bottom plot is for the boreal winter period and is averaged over 80–95° E, 10°S–5° N. From Zheng *et al.* (2004).

not the only mode of intraseasonal variability in the tropics, is the most dominant. This limitation suggests that further research is needed to examine the predictability of other intraseasonal variations in the tropics. This includes SST variability both related to and unrelated to the MJO (e.g. Kirtman *et al.*, 2001) as well as intrinsic SST modes (e.g. tropical instability waves, see Kessler, 2005), higher-frequency subseasonal modes associated with the Asian monsoon (e.g. Annamalai *et al.*, 1999; Gadgil, 2003) as well as variability over Africa (Matthews, 2004) and Pan America (Mo and Paegle, 2005).

Based on the material presented, there appears enough evidence to suggest that MJO predictions can be approached with considerable optimism as our present capabilities seem far from saturating their potential, and once exploited operationally, they will provide a unique and important bridge between the more established areas of weather and seasonal-to-interannual prediction. One of our greatest challenges remains to develop robust and realistic representations of the MJO in our weather and climate forecast models (Slingo *et al.*, 2005). Once we have such a capability, we not only have a means to improve predictions of low-frequency weather variations in the tropics that are directly impacted by the MJO, including the onsets and breaks of the Asian and Australian summer monsoons, but we will also likely improve forecasts associated with a number of processes remote to the MJO (see the Introduction).

To develop reliable prediction capabilities of intraseasonal variability and improve our understanding of its limits of predictability, there are a number of areas that warrant investigation. This includes a more complete understanding of the role that coupling to the ocean plays in maintaining, and in particular forecasting, the MJO. In addition, there has been virtually no research done on model initialisation/data assimilation issues in terms of what are the critical criteria to meet in order to adequately initialise the state of the MJO (Kalnay *et al.*, this volume; Simmons, this volume; Thorpe and Petersen, this volume). Related to this are issues regarding the importance of the basic state of the forecast model and how an incorrect basic state might negatively impact the maintenance and propagation of the MJO (e.g. Hendon, 2000; Inness *et al.*, 2003; Sperber *et al.*, 2003; Liess and Bengtsson, 2004; Allen *et al.*, this volume). Additional avenues of research include exploring the methods proposed by Krishnamurti *et al.* (1990) with other present-day forecast systems and on more MJO cases as well as exploring the possibility of assimilating empirically derived forecasts of the MJO into extended-range weather forecasts in order to improve their forecasts of the MJO as well as the remote processes and secondary circulations they interact with. Research is also needed to evaluate the best use of ensemble predictions at the intraseasonal timescale (Buizza, this volume; Kalnay *et al.*, this volume), including superensemble techniques (Krishnamurti *et al.*, 2005). It is also possible that the interactions with soil moisture and vegetation might be an influential factor that needs to be accounted for in the types of subseasonal predictions discussed here (Koster *et al.*, 2004). In addition to the above, there is clearly a need for additional dynamical predictability studies of the MJO using other GCMs as well as more sensitivity studies to test the effects of SST coupling and ENSO state, the impacts from/on midlatitude variability, and the influence of the size and type of initial condition perturbations and definition of predictability. Finally, there has been very little consideration of the economic benefits of intraseasonal predictions, namely where and when such predictions would have the greatest economic benefit, at what specific lead times, and for what sectors and industries (Richardson, this volume).

Acknowledgements

This work was supported by the Human Resources Development Fund at the Jet Propulsion Laboratory, as well as the National Science Foundation (ATM-0094416), the National Oceanographic and Atmospheric Administration (NA16GP2021), and the National Atmospheric and Aeronautics Administration (NAG5-11033). The research at the Jet Propulsion Laboratory, California Institute of Technology, was performed under contracts with the National Aeronautics and Space Administration.

References

Annamalai, H., J. M. Slingo, K. R. Sperber and K. Hodges (1999). The mean evolution and variability of the Asian summer monsoon: Comparison of ECMWF and NCEP-NCAR reanalyses. *Mon. Weather Rev.*, **127**, 1157–86.

Baldwin, M. P., D. B. Stephenson, D. W. J. Thompson, *et al.* (2003). Stratospheric memory and skill of extended-range weather forecasts. *Science*, **301**, 636–40.

Barlow, M., M. Wheeler, B. Lyon and H. Cullen (2005). Modulation of daily precipitation over Southwest Asia by the Madden-Julian Oscillation. *Mon. Weather Rev.*, **133**, in press.

Barnston, A. G., M. H. Glantz and Y. X. He (1999). Predictive skill of statistical and dynamical climate models in SST forecasts during the 1997–98 El Niño episode and the 1998 La Niña onset. *Bull. Am. Meteorol. Soc.*, **80**, 217–43.

Berbery, E. H. and J. Noguespaegle (1993). Intraseasonal interactions between the tropics and extratropics in the Southern-Hemisphere. *J. Atmos. Sci.*, **50**, 1950–65.

Bond, N. A. and G. A. Vecchi (2003). The influence of the Madden-Julian oscillation on precipitation in Oregon and Washington. *Weather Forecast.*, **18**, 600–13.

Cai, M., E. Kalnay and Z. Toth (2003). Bred vectors of the Zebiak-Cane model and their potential application to ENSO predictions. *J. Climate*, **16**, 40–56.

Cane, M. A., S. E. Zebiak and S. C. Dolan (1986). Experimental forecasts of El-Niño. *Nature*, **321**, 827–32.

Chang, C.-P. (2004). *East Asian Monsoon*. World Scientific.

Chen, S. S., R. A. Houze and B. E. Mapes (1996). Multiscale variability of deep convection in relation to large-scale circulation in TOGA COARE. *J. Atmos. Sci.*, **53**, 1380–409.

Chen, T. C. and J. C. Alpert (1990). Systematic-errors in the annual and intraseasonal variations of the planetary-scale divergent circulation in NMC medium-range forecasts. *Mon. Weather Rev.*, **118**, 2607–23.

ECMWF (2004). *ECMWF/CLIVAR Workshop on Simulation and Prediction of Intra-Seasonal Variability with Emphasis on the MJO, 3–6 November 2003*. ECMWF, Reading, UK.

Ferranti, L., T. N. Palmer, F. Molteni and K. Klinker (1990). Tropical–extratropical interaction associated with the 30–60-day oscillation and its impact on medium and extended range prediction. *J. Atmos. Sci.*, **47**, 2177–99.

Ferranti, L., J. M. Slingo, T. N. Palmer and B. J. Hoskins (1997). Relations between interannual and intraseasonal monsoon variability as diagnosed from AMIP integrations. *Quart. J. Roy. Meteor. Soc.*, **123**, 1323–57.

Fu, X. H. and B. Wang (2004). Differences of boreal summer intraseasonal oscillations simulated in an atmosphere-ocean coupled model and an atmosphere-only model. *J. Climate*, **17**, 1263–71.

Fu, X., B. Wang, T. Li and J. McCreary (2003). Coupling between northward-propagating boreal summer ISO and Indian Ocean SST: revealed in an atmosphere-ocean coupled model. *J. Atmos. Sci.*, **60**, 1733–53.

Gadgil, S. (2003). The Indian monsoon and its variability. *Annu. Rev. Earth Planet. Sci.*, **31**, 429–67.

Gill, A. E. (1980). Some simple solutions for heat-induced tropical circulation. *Quart. J. Roy. Meteor. Soc.*, **106**, 447–62.

Goswami, B. N. (2005). South Asian Summer Monsoon. In *Intraseasonal Variability of the Atmosphere-Ocean Climate System*, ed. W. K. M. Lau and D. E. Waliser. Springer.

Goswami, B. N. and P. Xavier (2003). Potential predictability and extended range prediction of Indian summer monsoon breaks. *Geophys. Res. Lett.*, 30, 1966, doi:10.1029/2003GL017,810.

Goswami, B. N., R. S. Ajayamohan, P. K. Xavier and D. Sengupta (2003). Clustering of synoptic activity by Indian summer monsoon intraseasonal oscillations. *Geophys. Res. Lett.*, **30**, art. no. 1431.

Graham, N. E. and T. P. Barnett (1995). Enso and Enso-related predictability. 2: Northern-Hemisphere 700-Mb height predictions based on a hybrid coupled Enso model. *J. Climate*, **8**, 544–9.

Gualdi, S., A. Navarra and G. Tinarelli (1999). The interannual variability of the Madden-Julian Oscillation in an ensemble of GCM simulations. *Clim. Dynam.*, **15**, 643–58.

Hendon, H. H. (2000). Impact of air-sea coupling on the Madden-Julian oscillation in a general circulation model. *J. Atmos. Sci.*, **57**, 3939–52.

Hendon, H. H. and M. L. Salby (1994). The life-cycle of the Madden-Julian Oscillation. *J. Atmos. Sci.*, **51**, 2225–37.

Hendon, H. H., C. D. Zhang and J. D. Glick (1999). Interannual variation of the Madden-Julian oscillation during austral summer. *J. Climate*, **12**, 2538–50.

Hendon, H. H., B. Liebmann, M. Newman, J. D. Glick and J. E. Schemm (2000). Medium-range forecast errors associated with active episodes of the Madden-Julian oscillation. *Mon. Weather Rev.*, **128**, 69–86.

Higgins, R. W., J. K. E. Schemm, W. Shi and A. Leetmaa (2000). Extreme precipitation events in the western United States related to tropical forcing. *J. Climate*, **13**, 793–820.

Hsu, H. H. (2005). East Asian monsoon. In *Intraseasonal Variability of the Atmosphere-Ocean Climate System*, ed. W. K. M. Lau and D. E. Waliser. Springer.

Inness, P. M., J. M. Slingo, E. Guilyardi and J. Cole (2003). Simulation of the Madden-Julian oscillation in a coupled general circulation model. II: The role of the basic state. *J. Climate*, **16**, 365–82.

Jiang, X. N., T. Li and B. Wang (2004). Structures and mechanisms of the northward propagating boreal summer intraseasonal oscillation. *J. Climate*, **17**, 1022–39.

Jones, C. (2000). Occurrence of extreme precipitation events in California and relationships with the Madden-Julian oscillation. *J. Climate*, **13**, 3576–87.

Jones, C., D. E. Waliser, J. K. E. Schemm and W. K. M. Lau (2000). Prediction skill of the Madden and Julian Oscillation in dynamical extended range forecasts. *Clim. Dynam.*, **16**, 273–89.

Jones, C., D. E. Waliser, K. M. Lau and W. Stern (2004a). The Madden-Julian oscillation and its impact on Northern Hemisphere weather predictability. *Mon. Weather Rev.*, **132**, 1462–71.

Jones, C., L. M. V. Carvalho, R. W. Higgins, D. E. Waliser and J.-K. E. Schemm (2004b). Climatology of tropical intraseasonal convective anomalies: 1979–2002. *J. Climate*, **17**, 523–39.

(2004c). A statistical forecast model of tropical intraseasonal convective anomalies. *J. Climate*, **17**, 2078–95.

Kalnay, E., R. Balgovind, W. Chao, *et al.* (1983). *Documentation of the GLAS Fourth Order General Circulation Model, Volume 1*. NASA Tech. Memo. No. 86064. NASA Goddard Space Flight Center, Greenbelt, MD.

Kalnay, E., M. Kanamitsu, R. Kistler, *et al.* (1996). The NCEP/NCAR 40-year reanalysis project. *Bull. Am. Meteorol. Soc.*, **77**, 437–71.

Kemball-Cook, S. and B. Wang (2001). Equatorial waves and air-sea interaction in the Boreal summer intraseasonal oscillation. *J. Climate*, **14**, 2923–42.

Kemball-Cook, S., B. Wang and X. H. Fu (2002). Simulation of the intraseasonal oscillation in the ECHAM-4 model: the impact of coupling with an ocean model. *J. Atmos. Sci.*, **59**, 1433–53.

Kessler, W. (2005). The oceans. In *Intraseasonal Variability of the Atmosphere-Ocean Climate System*, ed. W. K. M. Lau and D. E. Waliser. Springer.

Kessler, W. S. (2001). EOF representations of the Madden-Julian oscillation and its connection with ENSO. *J. Climate*, **14**, 3055–61.

Kessler, W. S. and R. Kleeman (2000). Rectification of the Madden-Julian oscillation into the ENSO cycle. *J. Climate*, **13**, 3560–75.

Kirtman, B. P., J. Shukla, B. H. Huang, Z. X. Zhu and E. K. Schneider (1997). Multiseasonal predictions with a coupled tropical ocean-global atmosphere system. *Mon. Weather Rev.*, **125**, 789–808.

Kirtman, B. P., D. A. Paolino, J. L. Kinter and D. M. Straus (2001). Impact of tropical subseasonal SST variability on seasonal mean climate simulations. *Mon. Weather Rev.*, **129**, 853–68.

Koster, R., M. J. Suarez, P. Liu, *et al.* (2004). Realistic initialization of land surface states: impacts on subseasonal forecast skill. *J. Hydrometeorol.*, **5**, 1049–63.

Krishnamurti, T. N., M. Subramaniam, D. K. Oosterhof and G. Daughenbaugh (1990). Predictability of low-frequency modes. *Meteorol. Atmos. Phys.*, **44**, 63–83.

Krishnamurti, T. N., M. Subramaniam, G. Daughenbaugh, D. Oosterhof and J. H. Xue (1992). One-month forecasts of wet and dry spells of the monsoon. *Mon. Weather Rev.*, **120**, 1191–223.

Krishnamurti, T. N., S. O. Han and V. Misra (1995). Prediction of the dry and wet spell of the Australian monsoon. *Int. J. Climatol.*, **15**, 753–71.

Lau, K. M. and P. H. Chan (1986). Aspects of the 40–50 day oscillation during the northern summer as inferred from outgoing longwave radiation. *Mon. Weather Rev.*, **114**, 1354–67.

Lau, K. M. and F. C. Chang (1992). Tropical intraseasonal oscillation and its prediction by the NMC operational model. *J. Climate*, **5**, 1365–78.

Lau, K. M. and T. J. Phillips (1986). Coherent fluctuations of extratropical geopotential height and tropical convection in intraseasonal timescales. *J. Atmos. Sci.*, **43**, 1164–81.

Lau, W. K. M. and D. E. Waliser (eds.) (2005). *Intraseasonal Variability of the Atmosphere-Ocean Climate System*. Springer.

Lau, K. M., T. Nakazawa and C. H. Sui (1991). Observations of cloud cluster hierarchies over the tropical Western Pacific. *J. Geophys. Res.-Oceans*, **96**, 3197–208.

Lawrence, D. M. and P. J. Webster (2001). Interannual variations of the intraseasonal oscillation in the south Asian summer monsoon region. *J. Climate*, **14**, 2910–22.

(2002). The boreal summer intraseasonal oscillation: relationship between northward and eastward movement of convection. *J. Atmos. Sci.*, **59**, 1593–606.

Li, T. M. and B. Wang (1994). The influence of sea-surface temperature on the tropical intraseasonal oscillation – a numerical study. *Mon. Wea. Rev.*, **122**, 2349–62.

Liebmann, B. and D. L. Hartmann (1984). An observational study of tropical midlatitude interaction on intraseasonal timescales during winter. *J. Atmos. Sci.*, **41**, 3333–50.

Liess, S. and L. Bengtsson (2004). The intraseasonal oscillation in ECHAM4. II: Sensitivity studies. *Clim. Dyn.*, **22**, 671–88.

Liess, S., D. E. Waliser and S. Schubert (2005). Predictability studies of the intraseasonal oscillation with the ECHAM5 GCM. *J. Atmos. Sci.*, **62**(9), 3326–36.

Lo, F. and H. H. Hendon (2000). Empirical extended-range prediction of the Madden-Julian oscillation. *Mon. Weather Rev.*, **128**, 2528–43.

Lorenz, E. N. (1965). A study of the predictability of a 28-variable atmospheric model. *Tellus*, **17**, 321–33.

(1982). Atmospheric predictability experiments with a large numerical-model. *Tellus*, **34**, 505–13.

Madden, R. and P. Julian (2005). Historical perspective. In *Intraseasonal Variability of the Atmosphere–Ocean Climate System*, ed. W. K. M. Lau and D. E. Waliser. Springer.

Madden, R. A. and P. R. Julian (1971). Detection of a 40–50 day oscillation in the zonal wind in the tropical Pacific. *J. Atmos. Sci.*, **28**, 702–8.

(1994). Observations of the 40–50-day tropical oscillation – a review. *Mon. Weather Rev.*, **122**, 814–37.

Maloney, E. D. (2002). An intraseasonal oscillation composite life cycle in the NCAR CCM3.6 with modified convection. *J. Climate*, **15**, 964–82.

Maloney, E. D. and D. L. Hartmann (2000a). Modulation of eastern North Pacific hurricanes by the Madden-Julian oscillation. *J. Climate*, **13**, 1451–60.

(2000b). Modulation of hurricane activity in the Gulf of Mexico by the Madden-Julian oscillation. *Science*, **287**, 2002–4.

Matsuno, T. (1966). Quasi-geostrophic motions in the Equatorial area. *J. Meteorol. Soc. Jpn.*, **44**, 25–43.

Matthews, A. J. (2004). Intraseasonal variability over tropical Africa during northern summer. *J. Climate*, **17**, 2427–40.

McPhaden, M. J. (1999). Climate oscillations: genesis and evolution of the 1997–98 El Niño. *Science*, **283**, 950–4.

Mo, K. and J. Paegle (2005). Pan America. In *Intraseasonal Variability of the Atmosphere–Ocean Climate System*, ed. W. K. M. Lau and D. E. Waliser. Springer.

Mo, K. C. (1999). Alternating wet and dry episodes over California and intraseasonal oscillations. *Mon. Weather Rev.*, **127**, 2759–76.

(2000). The association between intraseasonal oscillations and tropical storms in the Atlantic basin. *Mon. Weather Rev.*, **128**, 4097–107.

(2001). Adaptive filtering and prediction of intraseasonal oscillations. *Mon. Weather Rev.*, **129**, 802–17.

Moore, A. M. and R. Kleeman (1999). Stochastic forcing of ENSO by the intraseasonal oscillation. *J. Climate*, **12**, 1199–220.

Nakazawa, T. (1988). Tropical super clusters within intraseasonal variations over the western Pacific. *J. Meteorol. Soc. Jpn.*, **66**, 823–39.

Newman, M., P. D. Sardeshmukh, C. R. Winkler and J. S. Whitaker (2003). A study of subseasonal predictability. *Mon. Weather Rev.*, **131**, 1715–32.

Sardeshmukh, P. D. and B. J. Hoskins (1988). The generation of global rotational flow by steady idealized tropical divergence. *J. Atmos. Sci.*, **45**, 1228–51.

Schubert, S., R. Dole, H. v. d. Dool, M. Suarez and D. Waliser (2002). *Proceedings from a workshop on "Prospects for improved forecasts of weather and short-term climate variability on subseasonal (2 week to 2 month) timescales", 16–18 April 2002, Mitchellville, MD*. NASA/TM 2002-104606, vol. 23.

Shukla, J. (1985). Predictability. *Adv. Geophys.*, **28B**, 87–122.

Slingo, J. M., K. R. Sperber, J. S. Boyle, *et al.* (1996). Intraseasonal oscillations in 15 atmospheric general circulation models: results from an AMIP diagnostic subproject. *Clim. Dyn.*, **12**, 325–57.

Slingo, J. M., D. P. Rowell, K. R. Sperber and E. Nortley (1999). On the predictability of the interannual behaviour of the Madden-Julian Oscillation and its relationship with El Niño. *Quart. J. Roy. Meteor. Soc.*, **125**, 583–609.

Slingo, J., P. Inness and K. Sperber (2005). Modeling. In *Intraseasonal Variability of the Atmosphere-Ocean Climate System*, ed. W. K. M. Lau and D. E. Waliser. Springer.

Sperber, K. R., J. M. Slingo, P. M. Inness and W. K. M. Lau (1997). On the maintenance and initiation of the intraseasonal oscillation in the NCEP/NCAR reanalysis and in the GLA and UKMO AMIP simulations. *Clim. Dyn.*, **13**, 769–95.

Sperber, K. R., J. M. Slingo, P. M. Inness, *et al.* (2003). The Madden-Julian Oscillation in GCMs. In *Research Activities in Atmospheric and Oceanic Modelling*, Report No. 33, WMO/TD-No. 1161, p. 09–010.

Sud, Y. C. and G. K. Walker (1992). A review of recent research on improvement of physical parameterizations in the GLA GCM. In *Physical Processes in Atmospheric Models*, ed. D. R. Sikka and S. S. Singh, pp. 422–79. Wiley Eastern Ltd.

Teng, H. Y. and B. Wang (2003). Interannual variations of the boreal summer intraseasonal oscillation in the Asian-Pacific region. *J. Climate*, **16**, 3572–84.

Thompson, D. W. J. and J. M. Wallace (2001). Regional climate impacts of the Northern Hemisphere annular mode. *Science*, **293**, 85–9.

Thompson, P. D. (1957). Uncertainty of initial state as a factor in the predictability of large scale atmospheric flow patterns. *Tellus*, **9**, 275–95.

Toth, Z. and E. Kalnay (1993). Ensemble forecasting at NMC: the generation of perturbations. *Bull. Am. Meteorol. Soc.*, **74**, 2330–71.

van den Dool, H. M. and J. Qin (1996). An efficient and accurate method of continuous time interpolation of large-scale atmospheric fields. *Mon. Weather Rev.*, **124,** 964–71.

van den Dool, H. M. and S. Saha (2002). Analysis of propagating modes in the tropics in short AMIP runs. In *AMIP II workshop, Toulouse, November 12–15, 2002. WMO.*

Vecchi, G. A. and N. A. Bond (2004). The Madden-Julian Oscillation (MJO) and northern high latitude wintertime surface air temperatures. *Geophys. Res. Lett.*, **31**.

von Neumann, J. (1955). Some remarks on the problem of forecasting climate fluctuations. In *Dynamics of Climate: The Proceedings of a Conference on the Application of Numerical Integration Techniques to the Problem of the General Circulation* p. 137. Pergamon Press.

von Storch, H. and J. Xu (1990). Principal oscillation pattern analysis of the 30- to 60-day oscillation in the tropical troposphere. *Clim. Dynam.*, **4**, 175–90.

Waliser, D. E. (2005). Predictability and forecasting. In *Intraseasonal Variability of the Atmosphere-Ocean Climate System*, ed. W. K. M. Lau and D. E. Waliser. Springer.

(2006). Intraseasonal variability. In *The Asian Monsoon*, ed. B. Wang. Springer Praxis.

Waliser, D. E., K. M. Lau and J. H. Kim (1999a). The influence of coupled sea surface temperatures on the Madden-Julian oscillation: a model perturbation experiment. *J. Atmos. Sci.*, **56**, 333–58.

Waliser, D. E., C. Jones, J. K. E. Schemm and N. E. Graham (1999b). A statistical extended-range tropical forecast model based on the slow evolution of the Madden-Julian oscillation. *J. Climate*, **12**, 1918–39.

Waliser, D., Z. Zhang, K. M. Lau and J. H. Kim (2001). Interannual sea surface temperature variability and the predictability of tropical intraseasonal variability. *J. Atmos. Sci.*, **58**, 2595–14.

Waliser, D., S. Schubert, A. Kumar, K. Weickmann and R. Dole (2003a). *Proceedings from a workshop on "Modeling, Simulation and Forecasting of Subseasonal Variability", 4–5 June 2003, University of Maryland, College Park, Maryland.* NASA/TM 2003–104606, vol. 25.

Waliser, D. E., K. M. Lau, W. Stern and C. Jones (2003b). Potential Predictability of the Madden-Julian Oscillation. *Bull. Am. Meteor. Soc.*, **84**, 33–50.

Waliser, D. E., W. Stern, S. Schubert and K. M. Lau (2003c). Dynamic predictability of intraseasonal variability associated with the Asian summer monsoon. *Quart. J. Roy. Meteor. Soc.*, **129**, 2897–925.

Waliser, D. E., K. Jin, I. S. Kang, *et al.* (2003d). AGCM simulations of intraseasonal variability associated with the Asian summer monsoon. *Clim. Dynam.*, **21**, 423–46.

(2003e). AGCM simulations of intraseasonal variability associated with the Asian summer monsoon. *Clim. Dynam.*, **21**, 423–46.

Waliser, D., K. Weickmann, R. Dole, *et al.* (2006). The Experimental MJO Prediction Project. *Bull. Am. Meteor. Soc.*, in press.

Wang, B. (2005). Theories. In *Intraseasonal Variability of the Atmosphere-Ocean Climate System*, ed. W. K. M. Lau and D. E. Waliser. Springer.

Wang, B. and H. Rui (1990). Synoptic climatology of transient tropical intraseasonal convection anomalies: 1975–1985. *Meteorol. Atmos. Phys.*, **44**, 43–61.

Wang, B. and X. S. Xie (1996). Low-frequency equatorial waves in vertically sheared zonal flow. 1: Stable waves. *J. Atmos. Sci.*, **53**, 449–67.

Wang, B. and X. S. Xie (1997). A model for the boreal summer intraseasonal oscillation. *J. Atmos. Sci.*, **54**, 72–86.

Wang, B. and X. H. Xu (1997). Northern hemisphere summer monsoon singularities and climatological intraseasonal oscillation. *J. Climate*, **10**, 1071–85.

Webster, P. J. and C. Hoyos (2004). Forecasting monsoon rainfall and river discharge variability on 20–25 day timescales. *Bull. Am. Meteorol. Soc.*, **85**, 1745–65.

Weickmann, K. M. (1983). Intraseasonal circulation and outgoing longwave radiation modes during Northern Hemisphere winter. *Mon. Weather Rev.*, **111**, 1838–58.

Weickmann, K. M., G. R. Lussky and J. E. Kutzbach (1985). Intraseasonal (30–60 day) fluctuations of outgoing longwave radiation and 250-Mb stream-function during northern winter. *Mon. Weather Rev.*, **113**, 941–61.

Wheeler, M. and G. N. Kiladis (1999). Convectively coupled equatorial waves: analysis of clouds and temperature in the wavenumber-frequency domain. *J. Atmos. Sci.*, **56**, 374–99.

Wheeler, M. and K. M. Weickmann (2001). Real-time monitoring and prediction of modes of coherent synoptic to intraseasonal tropical variability. *Mon. Weather Rev.*, **129**, 2677–94.

Wheeler, M. and H. Hendon (2004). An all-season real-time multivariate MJO Index: development of an index for monitoring and prediction. *Mon. Weather Rev.*, **132**, 1917–32.

Wheeler, M. C. and J. L. McBride (2005). Australian-Indonesian monsoon region. In *Intraseasonal Variability of the Atmosphere-Ocean Climate System*, ed. W. K. M. Lau and D. E. Waliser. Springer.

Whitaker, J. S. and K. M. Weickmann (2001). Subseasonal variations of tropical convection and week-2 prediction of wintertime western North American rainfall. *J. Climate*, **14**, 3279–88.

Winkler, C. R., M. Newman and P. D. Sardeshmukh (2001). A linear model of wintertime low-frequency variability. I: Formulation and forecast skill. *J. Climate*, **14**, 4474–94.

Wu, M. L. C., S. Schubert, I. S. Kang and D. E. Waliser (2002). Forced and free intra-seasonal variability over the South Asian monsoon region simulated by 10 AGCMs. *J. Climate*, **15**, 2862–80.

Xie, P. P. and P. A. Arkin (1997). Global precipitation: a 17-year monthly analysis based on gauge observations, satellite estimates, and numerical model outputs. *Bull. Am. Meteorol. Soc.*, **78**, 2539–58.

Yasunari, T. (1979). Cloudiness fluctuations associated with the Northern Hemisphere summer monsoon. *J. Meteorol. Soc. Jpn.*, **57**, 227–42.

Zhang, C. (2005). Madden Julian Oscillation. *Rev. Geophys.*, **43**, RG2003, doi:10.1029/2004 RG000158.

Zhang, C., H. H. Hendon, W. S. Kessler and A. Rosati (2001). A workshop on the MJO and ENSO. *Bull. Am. Meteorol. Soc.*, **82**, 971–6.

Zheng, Y., D. E. Waliser, W. F. Stern and C. Jones (2004). The role of coupled sea surface temperatures in the simulation of the tropical intraseasonal oscillation. *J. Climate*, **17**, 4109–34.

Zveryaev, I. (2002). Interdecadal changes in the zonal wind and the intensity of intraseasonal oscillations during boreal summer Asian monsoon. *Tellus Ser. A-Dyn. Meteorol. Oceanol.*, **54**, 288–98.

12

Predictability of seasonal climate variations: a pedagogical review

J. Shukla, J. L. Kinter III

Center for Ocean–Land–Atmosphere Studies, Calverton, Maryland

12.1 Introduction

It is well known that the day-to-day changes in the large-scale atmospheric circulation are not predictable beyond two weeks. The small-scale rainfall patterns associated with the large-scale circulation patterns may not be predictable beyond even a few days. However, the space–time averages of certain atmospheric and oceanic variables are predictable for months to seasons. This chapter gives a pedagogical review of the ideas and the results that have led to our current understanding and the status of the predictability of seasonal climate variations.

We first review the current status of the understanding of the limits of the predictability of weather. We adopt Lorenz' classical definition of the predictability of weather as the range at which the difference between forecasts from two nearly identical initial conditions is as large in a statistical sense as the difference between two randomly chosen atmospheric states. With this definition of predictability, it is implied that the upper limit of predictability depends on the saturation value of the maximum possible error, which, in turn, is determined by the climatological variance. Lorenz provided a simple conceptual model in which the upper limit of weather prediction skill is described by three fundamental quantities: the size of the initial error, the growth rate of the error and the saturation value of the error. This simple model is able to explain the current status of the seasonal, regional and hemispheric variations of numerical weather prediction (NWP) skill. For example, winter is more

Predictability of Weather and Climate, ed. Tim Palmer and Renate Hagedorn. Published by Cambridge University Press.

predictable than summer, and the midlatitudes are more predictable than the tropics, simply because the saturation value of forecast error is much larger in winter and in the midlatitudes than in the tropics. The progress of NWP skill for the medium range over the past 25 years can be explained almost entirely by the reduction in the forecast error at day one, which, in turn, can be explained by the reduction in the initial error. The models and data assimilation techniques have steadily improved, thereby decreasing the initial error by 50%, while the growth rate of initial error has increased only modestly, resulting in steady improvement in the skill of NWP models for days 2–10.

We then will show that, in spite of the two-week upper limit of deterministic weather predictability, the effects of anomalous boundary conditions at the Earth's surface (sea surface temperature, snow, soil moisture, etc.) are sufficiently large to produce statistically significant anomalies in the seasonal mean atmospheric circulation. It is found, somewhat paradoxically, that the anomalous surface boundary conditions are much more influential in the tropics, where the deterministic limit of weather predictability is relatively short, than in midlatitudes, where the limit of weather predictability is relatively long. We review some atmospheric general circulation model (GCM) experiments which have helped advance our understanding of the boundary-forced predictability.

We then address the question of the predictability of the boundary conditions themselves. Since anomalous boundary conditions are produced by interactions between the ocean, atmosphere and land-surface processes, we review the status of the predictability of seasonal mean climate anomalies using coupled ocean–atmosphere–land models. We also address the question of seasonal predictability in a changing climate.

12.2 Predictability of weather

In a series of three papers that appeared in the late 1960s, E. Lorenz laid the theoretical groundwork for weather predictability that has been used in the subsequent decades to great advantage. In a comprehensive study of the predictability of the 28-variable model (Lorenz, 1965), in which for the first time he calculated the singular vectors, he showed that error growth is a strong function of the structure of the initial atmospheric flow and he estimated the doubling time of synoptic-scale errors to be a few days. He then employed a turbulence model (Lorenz, 1969a), assuming a $-5/3$ power law for the energy density spectrum, to compute the error saturation time, defined as the time at which the error energy at a given wave number becomes equal to the energy at that wave number prescribed in the initial conditions. He showed that scale interactions cause the error in the smallest scales to saturate the fastest, producing errors at synoptic scales within a few days after the initial time. Lorenz also devised a method for estimating the predictability of the atmosphere by searching for analogues or states that are sufficiently close in some phase space to permit using the evolution of the distance between the analogous states as a proxy for error growth in the classical predictability sense (Lorenz, 1969b). He found that the observational record available at that time was insufficient to find analogues that could be used in this way; nevertheless, he assumed a quadratic model for the error growth to estimate that the doubling time of small errors would be 2.5 days. These studies, along with several others that examined the characteristics of turbulent flows (Leith, 1971; Leith and Kraichnan, 1972) and later attempts to refine the predictability estimates with analogues (Gutzler and Shukla, 1984) and atmospheric GCMs (Smagorinsky, 1963; Charney *et al.*, 1966; Williamson and Kasahara, 1971; Lorenz, 1982; Simmons *et al.*, 1995; see Shukla, 1985 for a review) have shown that the predictability of weather is characterised completely by the growth rate and saturation value of small errors. While early atmospheric GCM estimates of error doubling time were relatively different (e.g. Charney, 1966), the diverse techniques to estimate the doubling time have converged to become remarkably consistent, all around two days. The most recent estimate available from ECMWF is 1.5 days (Simmons and Hollingsworth, 2002).

The ECMWF has shown recently that, as suggested by Lorenz (1982), the reduction in the error at day 1 has led to improvements in skill after 10 days. The expectation that the reduction in error might be overwhelmed by the increase in growth rate (smaller errors grow faster in the Lorenz model) has not occurred. A combination of improvements in the atmospheric GCM and better assimilation of available observations has led to a sustained reduction in error through at least day 7 of the forecast.

12.3 Predictability of seasonal averages – from weather prediction to seasonal prediction

One implication of the fact that the predictability of weather is a function of the error growth rate and saturation value is that predictability is quite different in the tropics and the extratropics. In the tropics, the variance of daily fluctuations, and hence the saturation value of errors, is much smaller than it is in the extratropics (Shukla, 1981a). Similarly, the error growth rate in the tropics, dominated as it is by instabilities associated with convection, is larger than in the extratropics where the error growth is primarily associated with baroclinic instability. Based on these considerations, Shukla (1981a) showed that the upper limit of deterministic predictability in the tropics is shorter than in the extratropics. By contrast, the short-term climate fluctuations in the tropics are dominated by the slowly varying boundary conditions, as will be described below, so, ironically, the seasonal means in the tropics are more predictable than the extratropics, in contrast to the situation for weather predictability.

Predictability also varies with spatial scale of motion. Several studies (Smagorinsky, 1969; Shukla, 1981b) have shown that the long waves are more predictable than the short waves in the extratropics. This is primarily due to the larger saturation value of error for the long waves, which in general have larger amplitude than the short waves. This is also consistent with the long-standing view of synoptic forecasters who have relied on systematic progressions of large-scale patterns in the atmosphere, variously called *Grosswetterlagen*, weather regimes, or centres of action (Hess and Brezowsky, 1969; Namias, 1986; Barnston and Livezey, 1987).

The main determinant of seasonal atmospheric predictability is the slowly varying boundary conditions at the Earth's surface (Charney and Shukla, 1977; Lorenz, 1979; Shukla, 1981b). It is well known that the lower boundary conditions of the Earth's atmosphere vary more slowly than the day-to-day variations of the weather and that, insofar as the boundary conditions can influence the large-scale atmospheric circulation, the time and space averages of the atmosphere are predictable at timescales beyond the upper limit of instantaneous predictability of weather. The surface boundary conditions include the sea surface temperature (SST), which governs the convergence of moisture flux as well as the sensible and latent heat fluxes between the ocean and atmosphere; the soil moisture, which alters the heat capacity of the land surface and governs the latent heat flux between continents and the atmosphere; vegetation, which regulates surface temperature as well as the latent heat flux to the atmosphere from land surfaces; snow, which participates in the surface radiative balance through its effect on surface albedo and in the latent heat flux, introducing a lag due to the storage of water in solid form in winter which is melted or evaporated in the spring and changes the soil wetness; and sea ice, which likewise participates in the energy balance and inhibits latent heat flux from the ocean. In each of these boundary conditions, anomalies can influence the surface fluxes and low-level atmospheric convergence through changes in the horizontal temperature and pressure gradients, at times leading to three-dimensional atmospheric heating anomalies, which in turn affect the entire atmospheric circulation.

12.3.1 Oceanic influences

Over the past 20 years, literally hundreds of numerical experiments have been conducted to test the hypothesis that the lower boundary conditions affect the seasonal circulation. Sensitivity studies have examined the effects of SST and sea ice anomalies in the tropical Pacific (e.g. Shukla and Wallace, 1983; Fennessy *et al.*, 1985), the tropical Atlantic (e.g. Moura and Shukla, 1981), the Arabian Sea (Shukla, 1975; Shukla and Misra, 1977), the north Pacific Ocean (Namias, 1969; Alexander and Deser, 1995), the north Atlantic Ocean (Palmer and Sun, 1985; Bhatt *et al.*, 1998), global SST anomalies (e.g. Miyakoda *et al.*, 1983; Kinter *et al.*, 1988; Shukla and Fennessy, 1988; Shukla *et al.*, 2000a, 2000b), sea ice (e.g. Randall *et al.*, 1998), mountains (Hahn and Manabe, 1975; Wallace *et al.*, 1983), deforestation (e.g. Nobre *et al.*, 1991), surface albedo anomalies associated with desertification (Charney, 1975; Xue and Shukla, 1993; Dirmeyer and Shukla, 1996), surface roughness anomalies (Sud

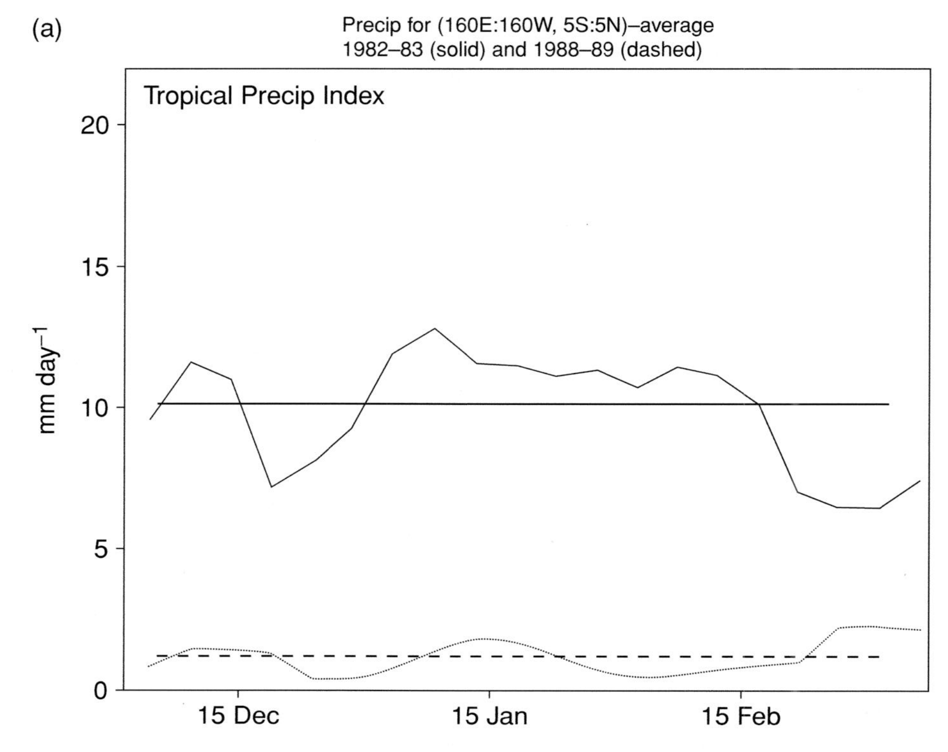

Figure 12.1 Effects of SST anomalies on tropical precipitation and extratropical atmospheric circulation. (a) and (b) show precipitation area-averaged over (160° E–160° W, 5° S–5° N), for the boreal winter season. (c) and (d) show the difference between two area-averages of the extratropical 500 hPa geopotential height: (120° W–110° W, 50° N–60° N) minus (170° W–160° W, 40° N–50° N). Anomalously warm tropical Pacific SST years are shown as a solid curve (1982–83 in a and c and 1986–87 in b and d), and anomalously cold tropical Pacific SST years are shown as dashed curves (1988–89 in a and c and 1984–85 b and d). The thick horizontal lines are the seasonal averages of the two time series.

et al., 1988), soil wetness anomalies (e.g. Dirmeyer and Shukla, 1993; Fennessy and Shukla, 1999), vegetation (Dickinson, 1984; Sellers *et al.*, 1986), and snow cover (e.g. Hahn and Shukla, 1976; Barnett *et al.*, 1988; Cohen and Rind, 1991; Bamzai and Marx, 2000). The result of all these experiments is that, under favourable conditions of the large-scale flow, and for certain structure, magnitude and location of the boundary anomalies, there is substantial evidence that the seasonal variations of the tropical climate are predictable and there is some possibility of predicting the extratropical climate as well (Palmer and Shukla, 2000).

The dependence of the predictability of the climate on location and season is illustrated in Figures 12.1–12.3. Parts (a) and (b) of Figure 12.1 show an index of tropical climate variability for the boreal winter season. Two years, shown as a solid curve, have boundary forcing, i.e. SST in the tropical Pacific, that is significantly warmer

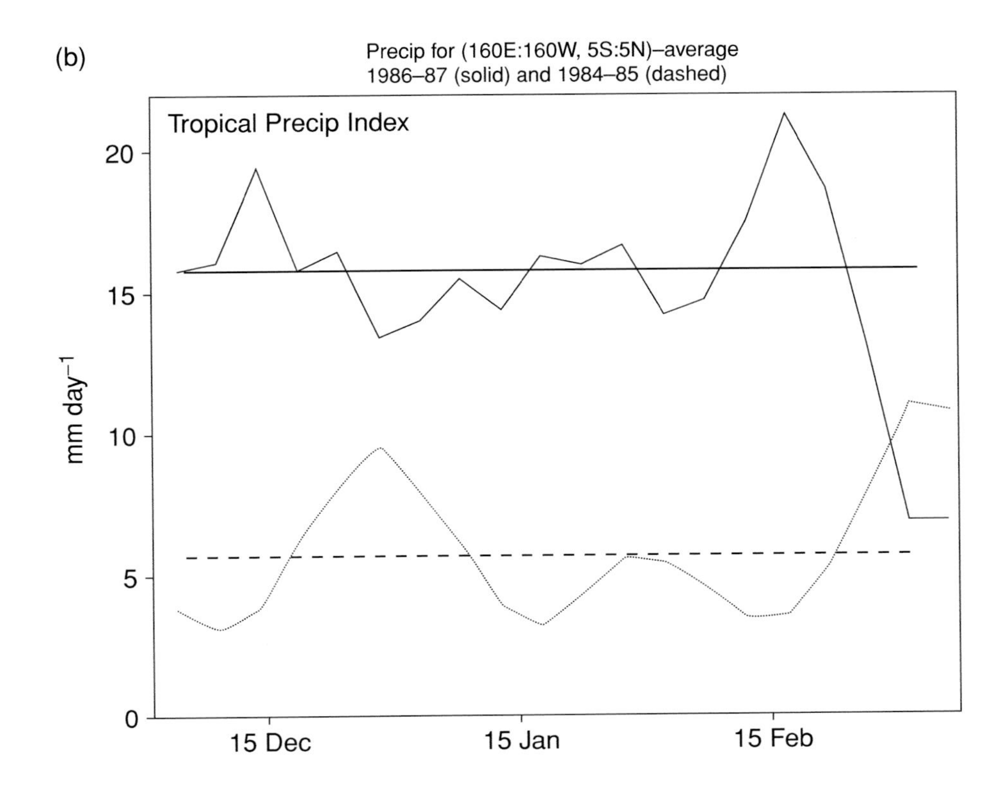

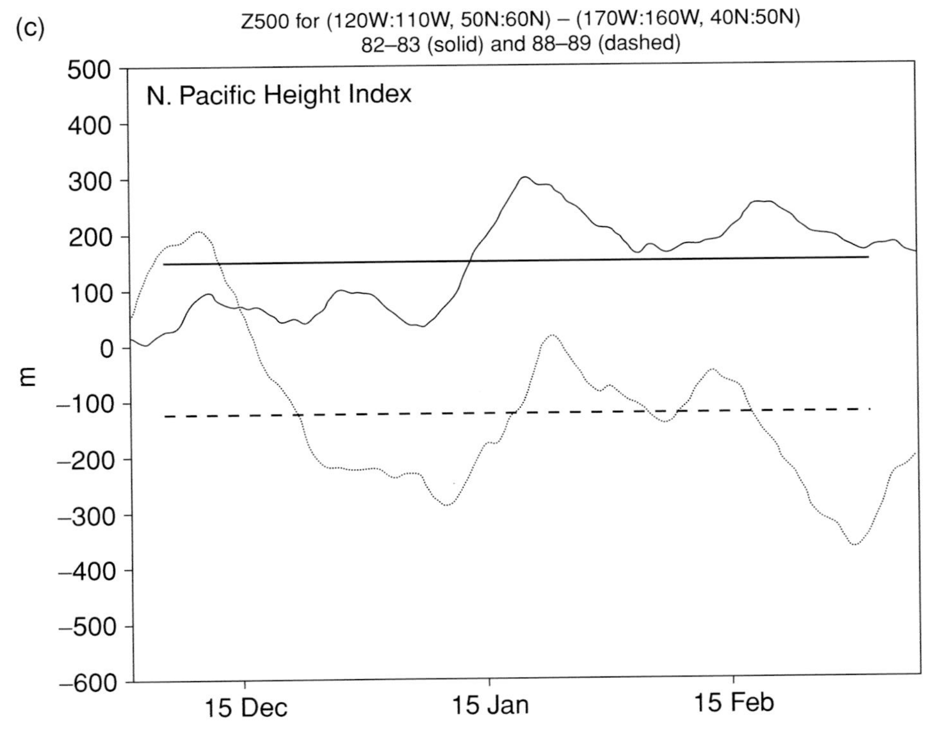

Figure 12.1 (*cont.*)

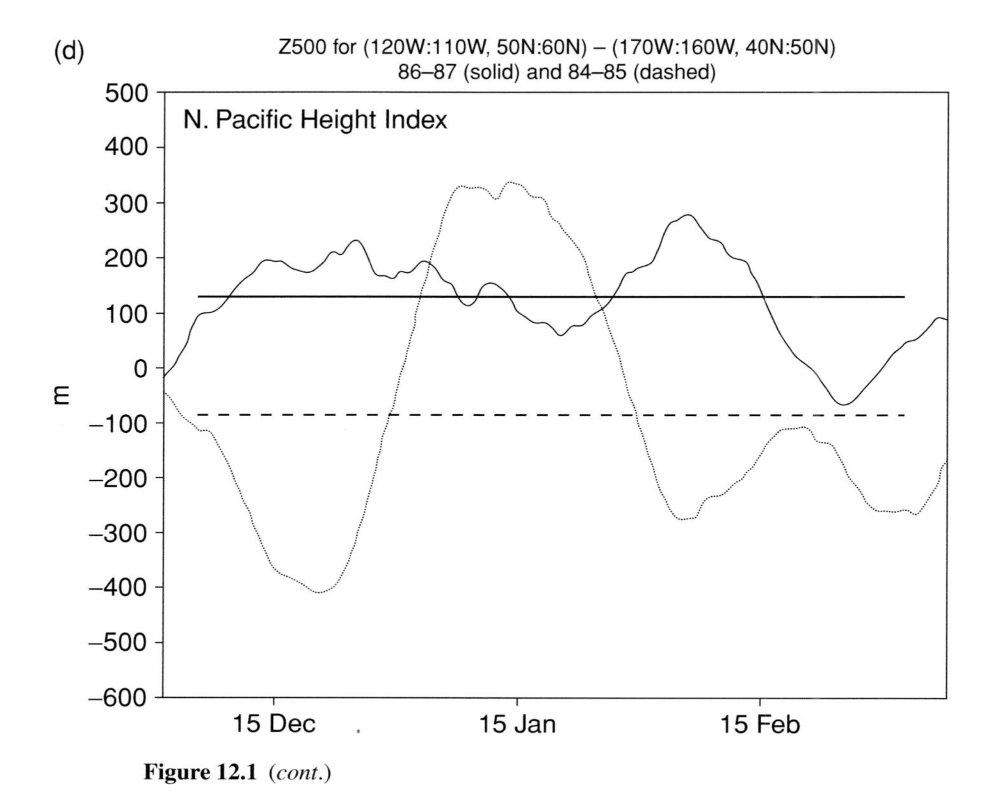

Figure 12.1 *(cont.)*

than usual. Two other years, shown as a dotted curve, have tropical Pacific SST that is significantly cooler than usual. In Figure 12.1 (c, d), an index of extratropical climate variability in boreal winter is shown for the same years. In the extratropical winter, the seasonal means of the two years are separated by a statistically significant amount that is *comparable to* the magnitude of the intraseasonal variation. In the tropics, the seasonal means are separated by a statistically significant amount that is *much larger than* the magnitude of intraseasonal variations. The variance of a typical quantity like sea-level pressure in the extratropical winter is larger than in the tropics. The boundary-forced seasonal mean differences in the tropics are relatively large. As shown schematically in Figure 12.2, the typical spread of forecasts initialised with slightly perturbed initial conditions becomes saturated at about five days or less in the tropics (dashed line) and about 10 days in the extratropics. Figure 12.3 illustrates the large difference in day-to-day variability between the tropics (typical values of 4 m/s) and the extratropics (typical values of 10 m/s).

An anomaly in the boundary conditions can affect the seasonal mean, large-scale atmospheric circulation through a relatively complex pathway, which is illustrated schematically in Figure 12.4. An SST anomaly, for example, locally alters the sensible and latent heat fluxes from ocean to atmosphere, thereby changing the temperature

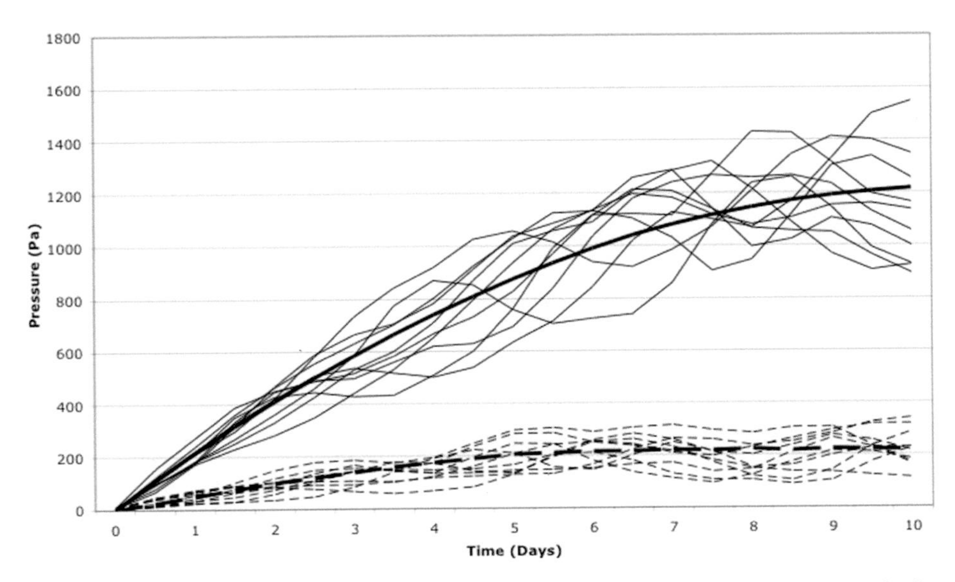

Figure 12.2 Schematic diagram illustrating the error growth in the tropics (dashed) and the extratropics (solid). Both thick lines depict the rates at which initially different states reach the boundary-forced state. The thin lines show typical spread of forecasts initialised with slightly perturbed initial conditions on day 0.

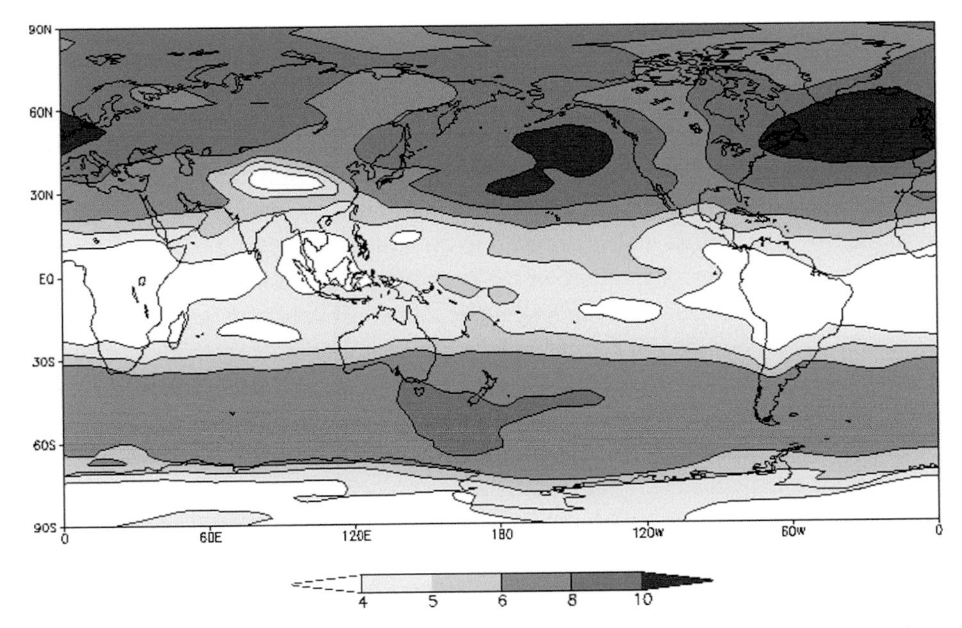

Figure 12.3 Standard deviation of daily values of 500 hPa zonal wind, computed with respect to individual monthly means for December, January and February, averaged over 1970–1999 (m/s).

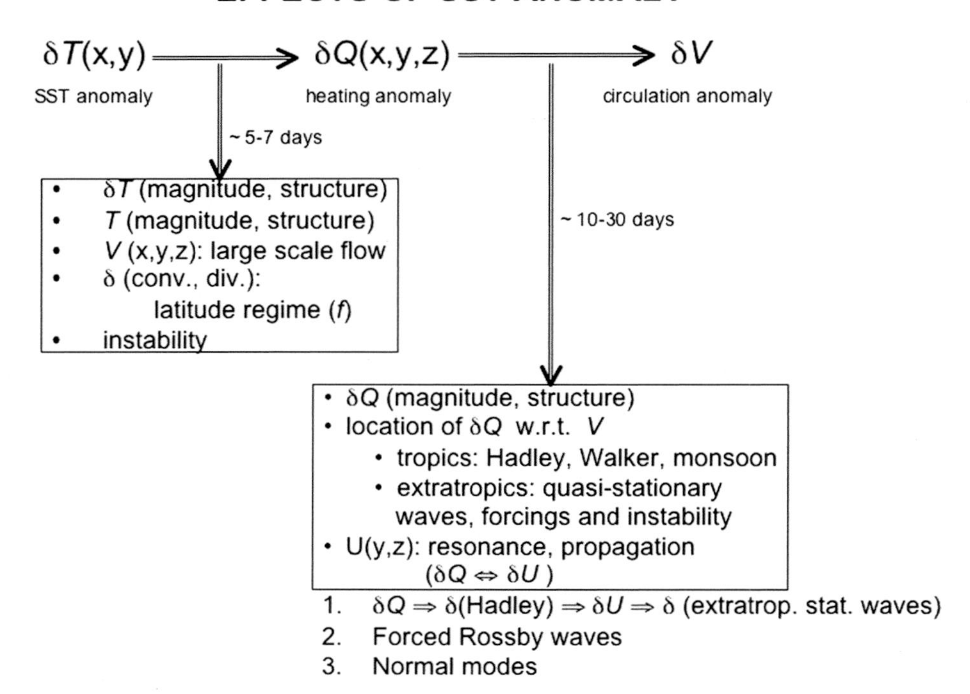

Figure 12.4 Schematic diagram of the effects of boundary condition anomalies on large-scale atmospheric heating and large-scale atmospheric circulation. See text for explanation.

and humidity in the atmosphere nearest the surface. Depending on the scale and magnitude of the SST anomaly, such changes can be relatively unimportant for the seasonal mean, or they can alter the static stability and moisture flux convergence available for convection, which can lead to an atmospheric heating anomaly over the course of 5–7 days. The nature of the heating anomaly associated with a given SST anomaly also depends on the magnitude and structure of the background SST on which it is superimposed. For example, a large warm SST anomaly that is superimposed on a very cold background ocean temperature will have little effect and likely will not lead to a large-scale atmospheric heating anomaly. The development of an atmospheric heating anomaly also depends on the structure of the large-scale atmospheric circulation overlying the SST anomaly as well as the degree to which the convergence and divergence of the atmospheric circulation are affected. A given SST anomaly may significantly alter the surface temperature gradient and, consequently, the surface pressure gradient, which then changes the low-level atmospheric convergence. The latter depends on the latitude of the anomaly, because the further the location is from the equator the less significant the divergent circulation associated with the pressure gradient anomalies is with respect to the rotational component of the flow. Changes in the divergent flow directly affect the moisture flux convergence (especially in the tropics), which is important in driving convection, thereby altering the deep heating of the atmosphere through latent heat release. Under the right conditions, a positive feedback loop involving boundary layer moisture flux convergence, precipitation and deep tropospheric heating, can significantly alter the large-scale atmospheric circulation.

Once a three-dimensional atmospheric heating anomaly has developed, it can influence the large-scale atmospheric circulation over the course of 10 to 30 days. Again, this is dependent on the size, magnitude and structure of the heating anomaly as well as several other factors. For example, if the heating anomaly is located in the tropics, its effect on the circulation depends on its position relative to the rising and descending branches of the Hadley and Walker circulations. Associated with the large land masses and the air–sea contrast in the tropics, the seasonal monsoon is the largest feature of the large-scale circulation, so the location of a heating anomaly with respect to the centres of the monsoon circulations is also an important factor. In the extratropics, matters are more complex, because there is a non-linear dynamic interaction between the zonal mean flow, the quasi-stationary eddies and the transient eddies. Similar heating anomalies located on different sides of a jet or at different phases of a stationary wave will have dramatically different effects, either through the forcing of transient waves or through an alteration of the mean flow by means of instability or wave–mean flow interaction. The effects of a given heating anomaly may be non-local as well. For example, a heating anomaly in the tropics may lead to a change in the Hadley circulation, which can thereby alter the zonal mean flow in the subtropics or mid-latitudes and change the orographically forced response. That change in the mean circulation can also, in turn, lead to an altered index of refraction for dispersive planetary waves. Through resonance, the forced waves might feed back on the tropical circulation and change the heating anomaly itself. There is also a possibility of teleconnections through propagation and dispersion of Rossby waves or through normal mode-type responses to forcing.

The role of boundary conditions in enhancing the predictability of time averages was demonstrated by Shukla (1998) through a set of numerical experiments with the COLA atmospheric GCM. Two sets of five-member ensemble simulations were conducted with very different initial conditions but identical observed SST specified as lower boundary conditions. The initial conditions were for December 1982 and December 1988, which corresponded to La Niña and El Niño years, and therefore had very different tropical atmospheric states. It was found that, in spite of very large differences in the initial conditions but the same (observed) SST for 1982–3, the simulated tropical winter mean precipitation pattern was nearly identical and had great similarity to the observed precipitation anomaly. Likewise, for the winter of 1988–9, the two simulations with very different initial states were also nearly identical, and were very similar to the observed precipitation anomaly for that year. An examination of the day-by-day simulation of tropical Pacific rainfall and circulation showed that the simulations with two very different initial conditions began to converge under the influence of the boundary conditions, and they became statistically indistinguishable within 7–10 days. This experiment confirmed what had been suspected for quite some time, namely that some parts of the tropics are so strongly forced by the underlying SST anomaly that even a very large perturbation in the initial conditions does not change the simulation of the seasonal mean rainfall.

In the same experiments, it was also found that even the extratropical seasonal mean circulation anomalies, especially over the Pacific–North American region, were largely determined by the SST anomalies. For example, the 500 hPa seasonal mean height anomalies for two very different initial conditions were nearly identical for a given SST anomaly. This suggests that the high predictability of the tropical atmosphere can also enhance the predictability of the extratropical atmosphere if the SST boundary forcing is quite strong. There has been some debate among the researchers whether boundary forcing merely changes the frequency of midlatitude modes of variability, or whether the boundary-forced variability is distinctly different from the unforced variability (Straus and Shukla, 2000).

One dramatic example of the fact, that, in spite of the high degree of variability in the extratropical atmosphere, tropical forcing can produce predictable effects in the extratropics, is shown in Figure 12.5 (colour plate). The bottom panel of the figure shows the difference in the boreal winter 500 hPa geopotential height field between large positive tropical Pacific SST anomaly (El Niño) years (1983, 1987 and 1992) and large negative (La Niña) years (1985 and 1989). The top panel shows an ensemble-mean simulation of the same field produced by an atmospheric GCM forced by the observed SST in those years. The ensemble average of several runs started from slightly different initial states was computed to filter the unpredictable component of the simulations. There is an uncanny match in both phase and amplitude of the simulated difference to the observed difference (anomaly correlation coefficient = 0.98 for the spatial domain shown in the figure).

This result has been reproduced by several atmospheric GCMs. Figure 12.6 shows the anomaly correlation coefficient for each of three different models (COLA, NSIPP and NCEP atmospheric GCMs) in each of 18 years. The years have been reordered in ascending absolute values of the corresponding NINO3 index, an indicator of the amplitude of the tropical Pacific SST anomaly, to show that the predictability of the extratropical Pacific–North American regional height anomalies depends on the magnitude of the forcing. A systematic evaluation of the possibility of dynamical seasonal prediction has been made (Shukla *et al.*, 2000a, 2000b)

It is by no means a given, however, that seasonal predictions, even in the presence of relatively strong SST anomalies in the tropical Pacific, will be highly skilful for all quantities and in all cases. As an example, a pair of seasonal hindcasts, produced using one of the models whose skill scores are shown in Figure 12.6, is shown in Figure 12.7 for two La Niña cases, 1989 and 1999. In both cases, the tropical Pacific SST forcing is fairly strong. The surface air temperature anomaly hindcast is quite good, both in terms of geographical distribution and amplitude, for the 1989 case, but only somewhat resembles the observed in the 1999 case. This is indicative of some of the difficulties

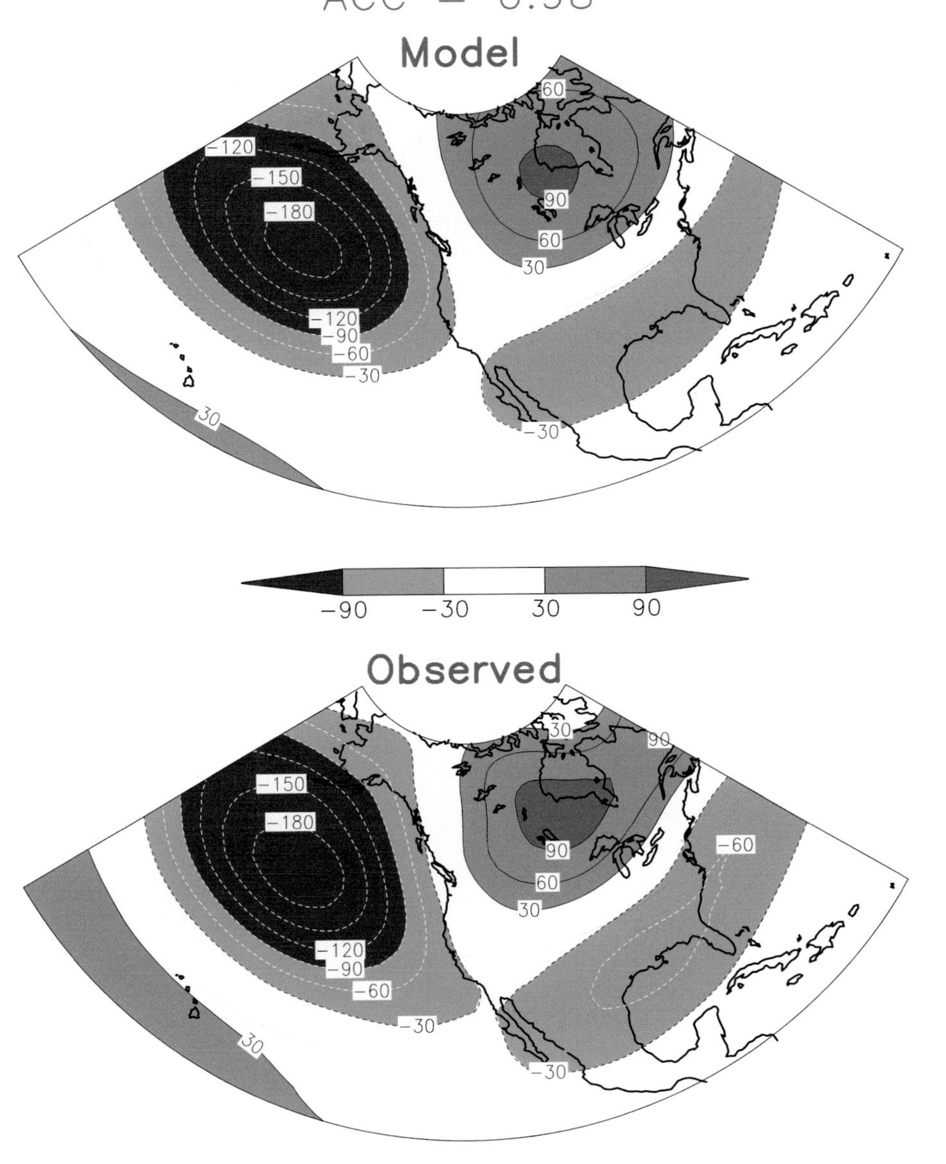

Figure 12.5 (See also colour plate section.) Composite, ensemble-mean AGCM simulation of January–February–March seasonal mean difference of 500 hPa geopotential height. Average of three years with warm SST anomalies in the eastern tropical Pacific (1983, 1987 and 1992) minus average of two years with cold SST anomalies (1985 and 1989). The ensemble of 10 model simulations were made with observed SST specified as lower boundary conditions and slightly different initial conditions in each ensemble member. The model used is the COLA atmospheric GCM.

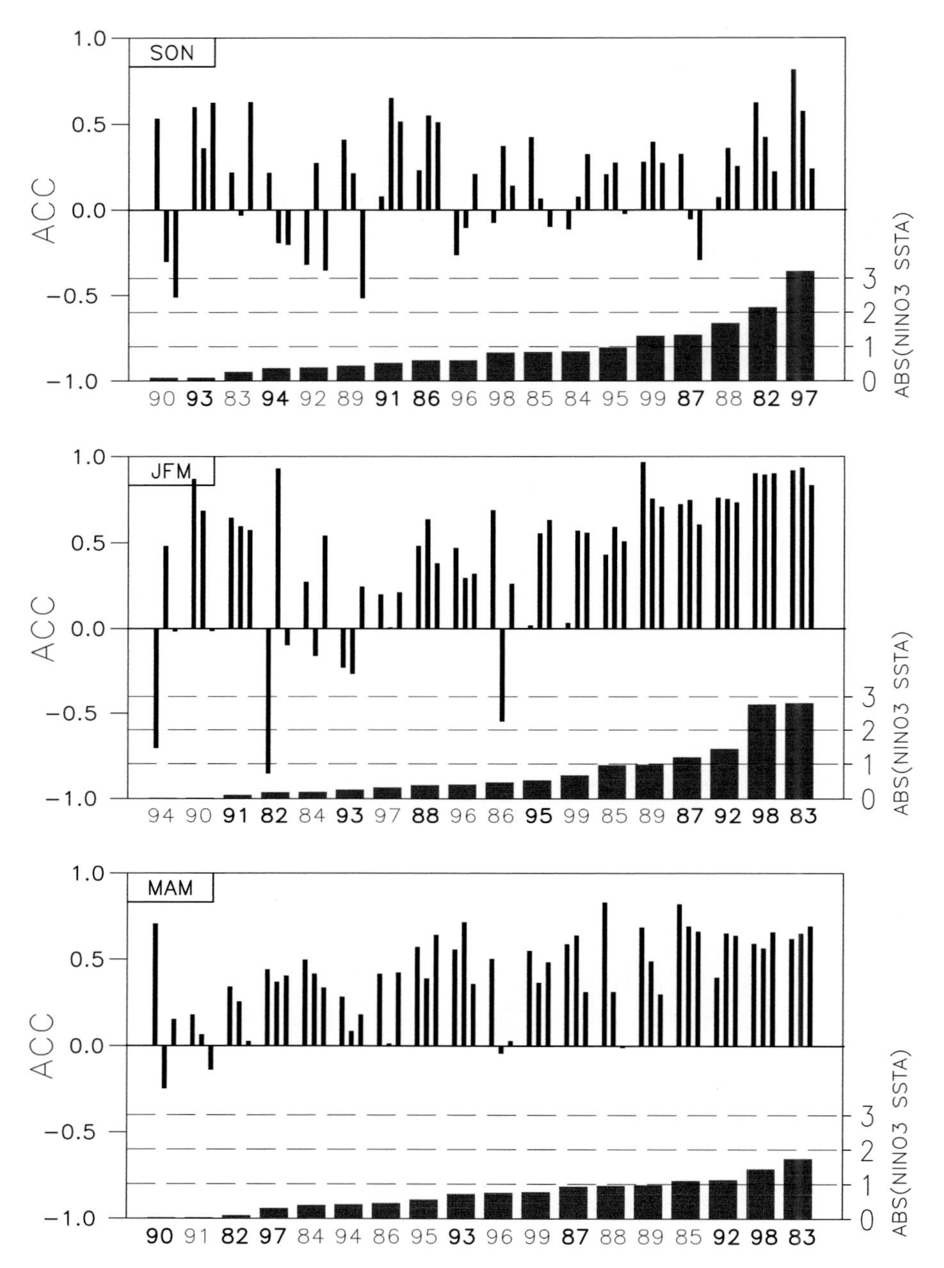

Figure 12.6 Each thin vertical bar shows the pattern correlation of the ensemble-mean seasonal mean 200-hPa height of an atmospheric GCM with the observed seasonal mean over North America 15°–70° N and 180°–60° W. The pattern correlation is computed separately for each year and for each GCM; the bars show the results for the COLA, NSIPP, and NCEP GCMs from left to right. The years are ordered by the absolute value of the NINO-3 index, increasing to the right and shown by the thick vertical bars. Bold and light numbers indicate warm and cold years respectively. Results (top) for boreal autumn (Sep–Nov), (middle) for boreal winter (Jan–Mar), and (bottom) for boreal spring (Mar–May).

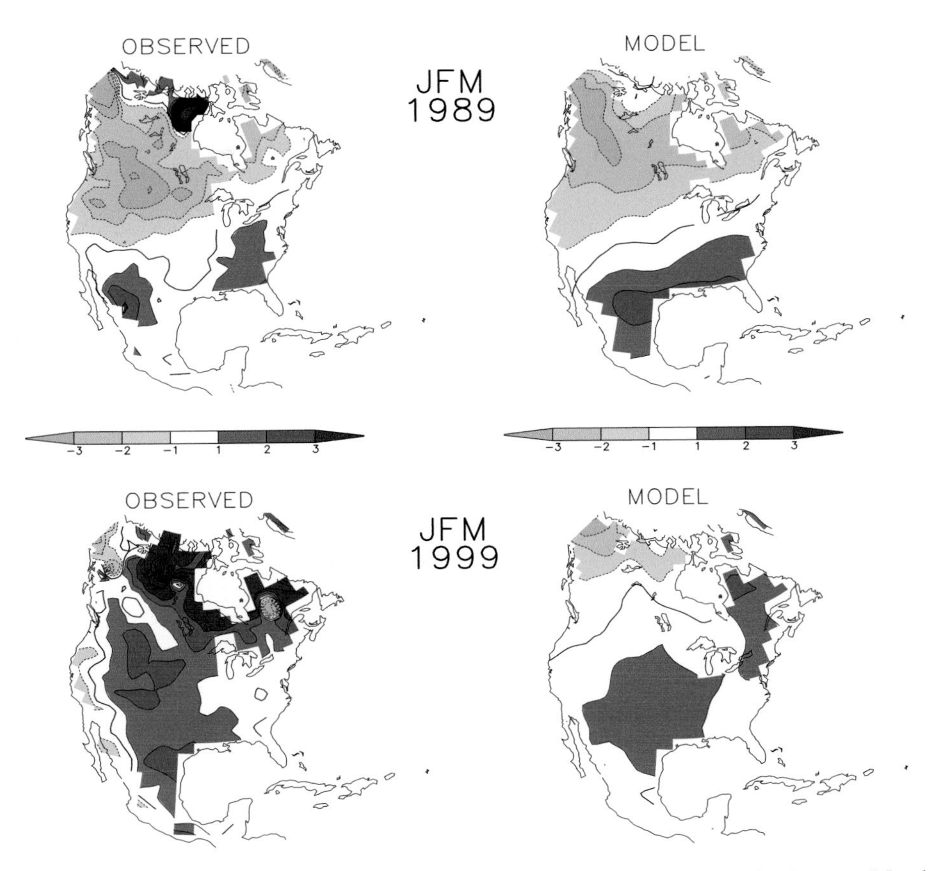

Figure 12.7 The seasonal mean surface air temperature anomaly for January–March 1989 (top row) and January–March 1999 (bottom row) from the observed (left column) and from a simulation in which the observed SST and sea ice were specified as the lower boundary conditions of an atmospheric GCM.

of translating the advances in understanding predictability into real improvements in seasonal prediction.

Likewise there is an asymmetry between warm and cold ENSO events in the extratropical variability associated with each. In particular, the warm and cold events influence the internal variability in the Pacific–North America region. Figure 12.8, comparing the ratio of interensemble variance computed for three different atmospheric GCMs, shows that all the models have more variability in cold events than in warm events, and, for some parts of the region (near the Aleutian Islands, over the west coast and south-eastern USA), the cold event variance exceeds the warm event variance by a factor of two or more.

12.3.2 Land influences

The physical processes at the land surface have effects on the climate, including the components of the hydrologic cycle and the atmospheric circulation, on a wide range

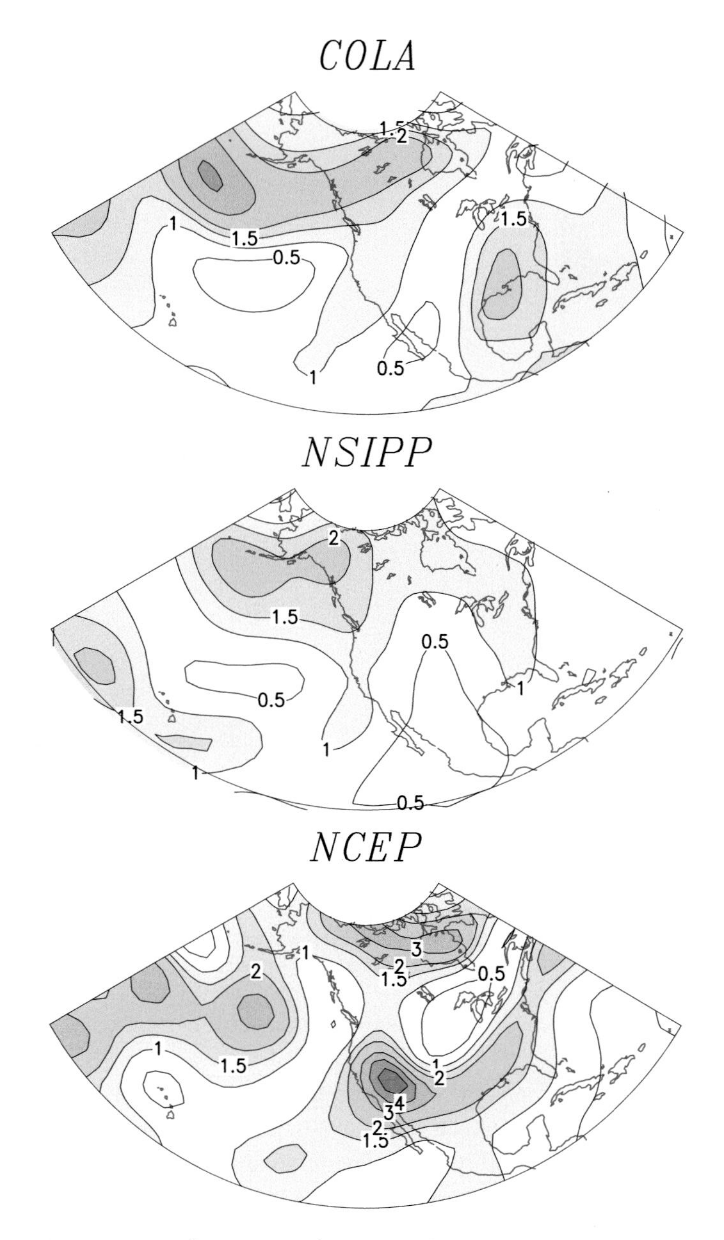

Figure 12.8 Ratio of the January–March mean intraensemble variance of Φ200 between cases with cold tropical Pacific SST (1985 and 1989) and cases with warm tropical Pacific SST (1983, 1987, 1992, and 1995).

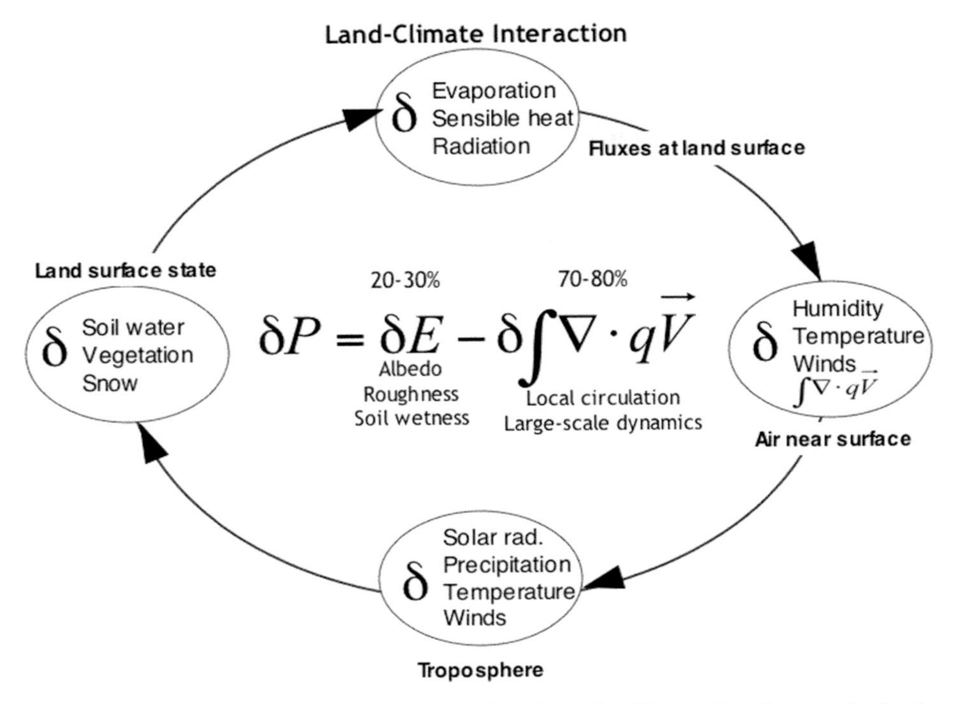

Figure 12.9 Schematic diagram showing the cycle of interactions between the land surface and the climate. Beginning with the oval at the left, a change in the land surface state (soil wetness, vegetation or snow) leads to a change in the fluxes of water, heat and radiation between the land surface and the atmosphere (oval at top). These changes can, in turn, lead to changes in the humidity and temperature of the air near the surface, as well as changes in the wind due to changes in the gradients of surface temperature or pressure (oval at right). The changes occurring near the surface can produce changes throughout the atmospheric column that occur in solar radiative flux (cloudiness), precipitation, air temperature and winds (oval at bottom). These changes can then affect the land surface state, completing the feedback loop.

of temporal and spatial scales (e.g. Dirmeyer and Shukla, 1993). The properties of the land surface, such as soil wetness, snow cover, and vegetation, affect both the evolution and predictability of climate (Dickinson, 1984). For example, soil moisture determines the rate of evapotranspiration as well as the partitioning of incoming radiation into sensible and latent heat flux. Likewise, the spatial distribution and temporal variability of vegetation, soil wetness and snow are determined by climatic conditions, so that a complete cycle of feedbacks is in operation (Figure 12.9).

The majority of modelling experiments so far have focused on the effect of land surface properties on climate predictability, beginning with the early atmospheric GCM study of Shukla and Mintz (1982) in which simulations were made with the model's land surface constrained to produce no evapotranspiration (desiccated case) or for the evapotranspiration to take place at the potential rate (saturated case), over the whole globe. They found that, in the desiccated case, there is almost no rainfall over the extratropical continents in boreal summer, and an increase in the rainfall over South-east Asia and India due to the large moisture flux convergence from the oceans. They interpreted their results to indicate that dry soil at the beginning of the summer would lead to anomalously low rainfall in the extratropical summer season.

The 'memory' effect of soil wetness is frequently invoked as the underlying mechanism for its role in predictability. Delworth and Manabe (1988, 1989), using an atmospheric GCM with a simple 'bucket' model of soil moisture, showed that soil wetness 'reddens' the spectrum of climate variability. More sophisticated models incorporating a model of the biosphere (Sellers *et al.*, 1986; Sato *et al.*, 1989) were used to show that soil wetness anomalies can persist much longer than previously thought, and that the climate is quite sensitive to variations in soil wetness.

Atlas *et al.* (1993) and Fennessy and Shukla (1999) showed that the initial soil wetness anomalies present in extreme drought (e.g. 1988) or flood (e.g. 1993) summers in North America contribute to the subsequent summer rainfall with persistent soil wetness anomalies, dry (wet) rainfall anomalies and warm (cold) surface temperature occurring with dry (wet) soil wetness anomalies. For example, Figures 12.10 and 12.11 (taken from Fennessy and Shukla, 1999) show that an initial soil wetness anomaly on 1 June (Figure 12.10) can persist for up to a season, in some regions, and have a significant impact on the evaporation (Figure 12.11) and surface air temperature (not shown). They also found that the strength and nature of the impact of initial soil wetness anomalies on precipitation and surface temperature depend on several factors, including the extent and magnitude of the initial soil wetness anomaly, the strength of the solar forcing, the proximity to moisture sources, and the strength of the regional atmospheric circulation. They interpreted their results to suggest that seasonal atmospheric predictability could be increased by realistic initial soil wetness.

It has recently been shown that the sensitivity of precipitation to soil wetness is regionally localised. Koster *et al.* (2004) used 10 different land surface models in the Global Land–Atmosphere Coupling Experiment (GLACE) to show that the coupling strength between the atmosphere and the land surface is a strong function of location. Koster *et al.* (2004) identified the areas where this coupling is strong, and therefore has a strong bearing on the variability of precipitation, during boreal summer. The places where the coupling is strong, referred to as 'hot spots', are primarily transition zones between dry and wet climatic zones where the evaporation is large and sensitive to soil moisture and where the boundary layer moisture can trigger convection in the atmosphere.

12.3.3 Ocean–atmosphere–land influences (monsoon predictability)

The monsoon circulation is a dramatic example of the combined ocean, land and atmospheric effects; therefore, an estimate of the limit of monsoon predictability

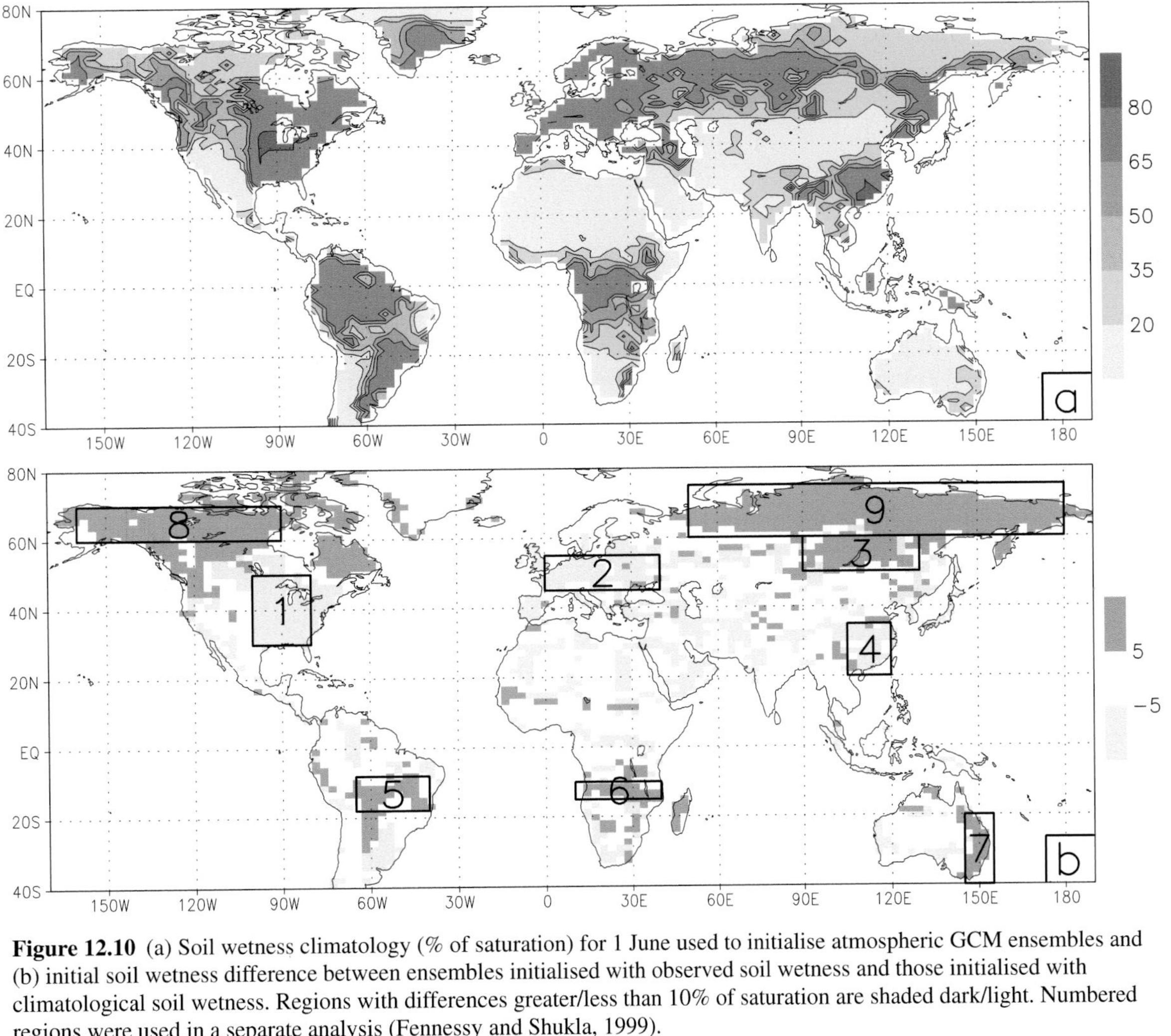

Figure 12.10 (a) Soil wetness climatology (% of saturation) for 1 June used to initialise atmospheric GCM ensembles and (b) initial soil wetness difference between ensembles initialised with observed soil wetness and those initialised with climatological soil wetness. Regions with differences greater/less than 10% of saturation are shaded dark/light. Numbered regions were used in a separate analysis (Fennessy and Shukla, 1999).

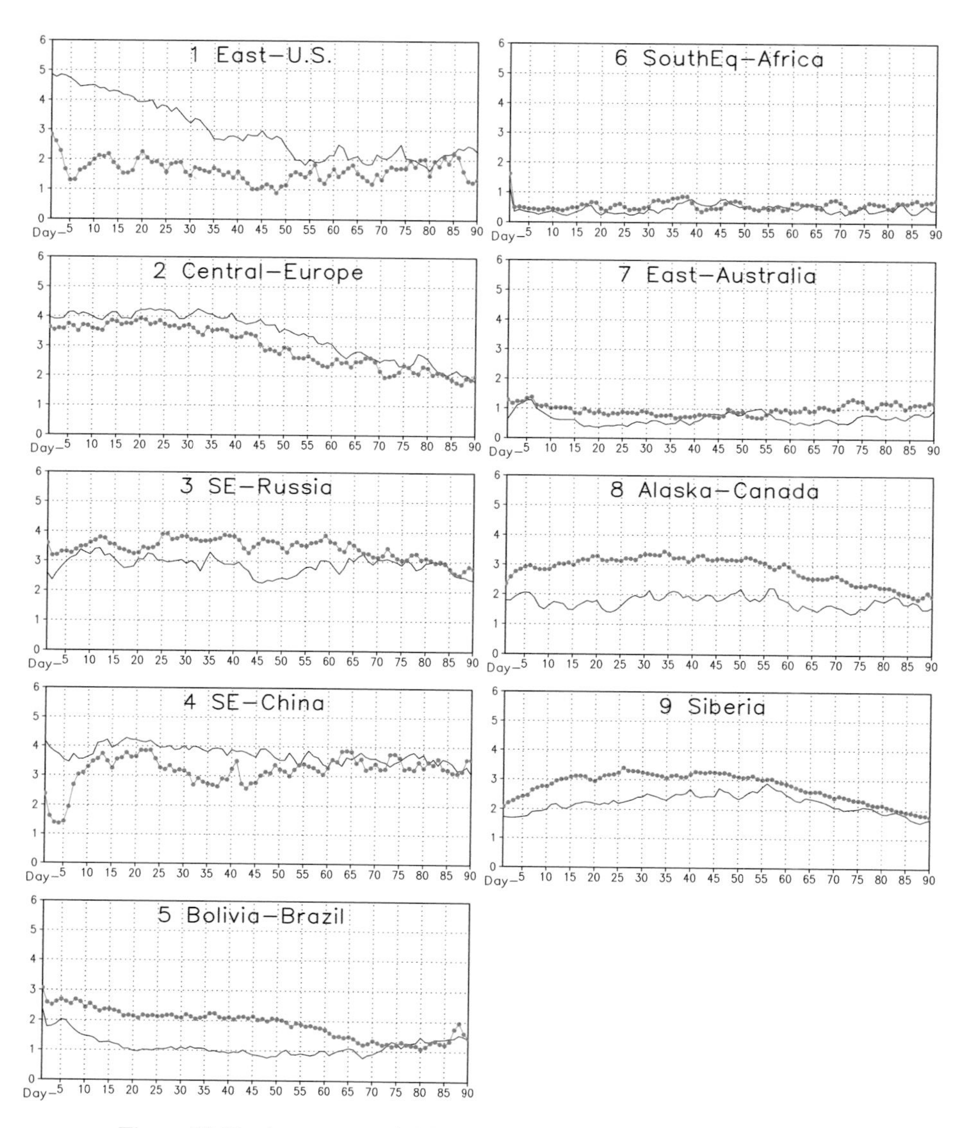

Figure 12.11 Area-averaged daily time series of evaporation for nine study regions shown in Figure 12.10(b) for control ensemble (solid) and observed initial soil wetness ensemble (dotted). Units are mm per day.

requires a clear understanding of the atmosphere–land interaction, atmosphere–ocean interaction, and the internal dynamics of the atmosphere alone. It was suggested by Charney and Shukla (1977) and later shown by Charney and Shukla (1981) that the large-scale, seasonal mean monsoon rainfall over India is largely determined by the boundary conditions over land and ocean. This was based on the observed relationship between Eurasian snow cover and monsoon rainfall (Blanford, 1884) and tropical ocean temperature and Indian monsoon rainfall (Sikka, 1980; Rasmusson and Carpenter, 1983). However, it is well known that there is a large intraseasonal variability of the regional rainfall in the monsoon region during the monsoon season. Therefore, in a simple conceptual model, the seasonal mean monsoon rainfall over any small region can be considered to consist of two components: one large-scale seasonally persistent component and one relatively small-scale intraseasonal component (Krishnamurthy and Shukla, 2001). The large-scale seasonally persistent component can be attributed to ocean and/or land surface effects and the intraseasonal component to the internal dynamics of the atmosphere. The predictability of the seasonal mean for any given season depends on the relative magnitude of the two components. Even in the presence of large boundary forcing (namely SST anomalies in 1997), if the large-scale effect is small and the intraseasonal variations are large, the seasonal mean will not be predictable. To the extent that the seasonal mean anomalies are determined by the sampling of intraseasonal variations of rainfall, and if the intraseasonal variations were independent of the boundary forcing, the seasonal mean anomalies would not be predictable. It has also not been possible either to make dynamical predictions of seasonal mean rainfall or even to make reliable estimates of the monsoon rainfall, because of various limitations in models and modelling strategies. For example, models of the monsoon circulation and rainfall have large systematic errors in the mean and variance. The current atmospheric GCM experiments in which models are forced by prescribed SST are inadequate to capture the coupled ocean–atmosphere variability, and the current models do not have adequate treatment of land surface processes. The models are also not able to simulate the intraseasonal variability over the monsoon region.

An estimate of monsoon predictability using coupled ocean–atmosphere models is even more problematic, because the coupled models have even larger systematic errors in simulating the mean climate and its variability. If the coupled model cannot capture the march of the annual cycle in SST in the Indian Ocean and the surrounding areas, it is nearly impossible to get a reasonable simulation of monsoon circulation.

12.4 Predictability of coupled system

In the previous section, we described the predictability of the tropical atmosphere with prescribed SST. In similar experiments with an ocean model (B. Huang, personal communication), it was shown that the tropical upper ocean circulation and temperature are largely determined by the overlying atmospheric forcing. When an ocean model with completely different initial conditions (for 1982 and 1988) was forced by the atmospheric fluxes of 1982–3, the resulting tropical SST anomalies were indistinguishable from each other after 3–4 months. These experiments suggest that the tropical atmosphere is highly predictable for prescribed SST and the tropical ocean is highly predictable for prescribed atmospheric fluxes. However, these results

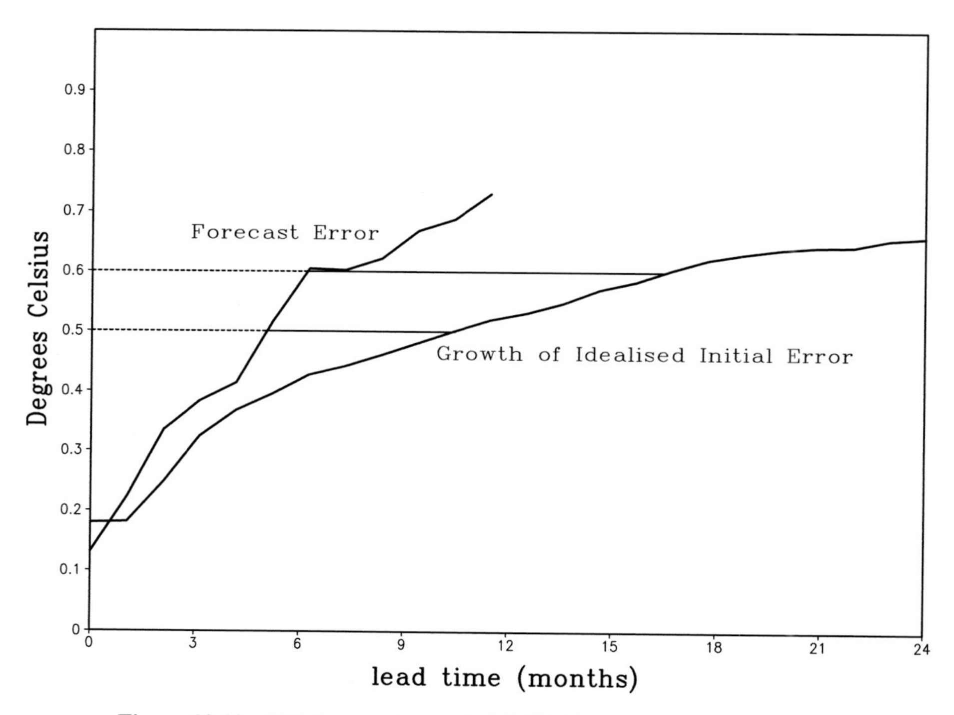

Figure 12.12 COLA anomaly coupled GCM NINO-3.4 forecast root mean squared error as a function of lead time. The forecast error is calculated based on 480 six member ensemble hindcasts initialised each January, April, July and October from 1981 to 2000 (see Kirtman, 2003, for details). The idealised NINO-3.4 root-mean-squared error growth is based on a multi-century simulation of the COLA anomaly coupled GCM (see Kirtman *et al.*, 2002) using an analogue approach to estimate the growth of initial errors.

do not necessarily imply that the coupled system is also as highly predictable as the individual components of the atmosphere and ocean. To estimate the predictability of the coupled ocean–atmosphere system is one of the major current challenges of climate research.

An estimate of the growth of initial error in the Zebiak–Cane coupled ocean–atmosphere model (Zebiak and Cane, 1987) was made by Goswami and Shukla (1991), and it was found that the coupled system is characterised by two timescales. One had an initial error doubling time of four months and the other had a doubling time of 15 months. In a more recent calculation, B. P. Kirtman (personal communication; Figure 12.12), using a global anomaly coupled ocean–atmosphere model, found that an initial error of about 0.2 K in the NINO-3.4 index of tropical Pacific SST takes about five months to double. Comparing the evolution of the forecast model's error with the growth of idealised initial errors in an 'identical twin' experiment, it was found that the forecast model reaches the same level of error in five months that is only reached after 10 months of idealised error growth. This suggests that there is substantial room for improving the prediction of tropical SST in a coupled model.

We do not yet have a clear understanding of mechanisms that determine the amplitude and the life cycle of ENSO. For example, we do not know whether ENSO should be considered to be a manifestation of an unstable oscillator, or whether it should be considered a stochastically forced damped linear system. The observations for the past 50 years can support both theories. We also do not understand the role of weather noise in the initiation and growth of ENSO events. A better understanding of ENSO dynamics, and improved methods of initialising models of the coupled ocean–atmosphere system for ENSO prediction are required before we can make reliable estimates of the predictability of ENSO.

12.4.1 Case study

A low-resolution coupled ocean–atmosphere model, a moderate-resolution atmospheric GCM and a high-resolution nested regional climate model were used to produce six-month lead forecasts for the boreal winter of 1997–8. The procedure for producing forecasts of the tropical SST anomaly, the global atmospheric circulation and precipitation and the regional climate anomalies over North America is shown schematically in Figure 12.13. It should be noted that this experimental strategy was devised only to overcome the problem of insufficient computer time to integrate high-resolution global coupled models. Briefly, the tropical Pacific SST pattern was predicted for up to 18 months in advance using the coupled ocean–atmosphere model with available input data at the time of the forecast (Kirtman *et al.*, 1997). Multiple realisations of the coupled model forecast were averaged to produce an ensemble-mean SST forecast that was then statistically extended to predict the global SST anomaly. The latter was added to an observed climatology for global SST and used as a lower boundary condition for an ensemble of integrations of the global atmospheric GCM. Each of the global model integrations was used to provide the lateral boundary conditions for a companion integration of a regional climate model. The ensembles of global and regional model predictions were made at least six months prior to the verification time of the forecasts.

12.4.2 The tropical Pacific SST forecast

The SST forecast produced by the anomaly coupled model and subsequent statistical projections called for a continuation of the unprecedented, anomalously warm surface temperature in the tropical eastern and central Pacific through boreal summer of 1998 with a peak amplitude in about December 1997 and January 1998, and diminishing thereafter. The predicted warmer than normal water in the Pacific was located east of the dateline along the equator and extended about 5–10° north and south of the equator. The peak SST value forecast by this model was about 3 °C above normal.

Up to the time that the forecast was made, during the ENSO warm event (through September 1997), the model had been quite accurate in predicting the onset and

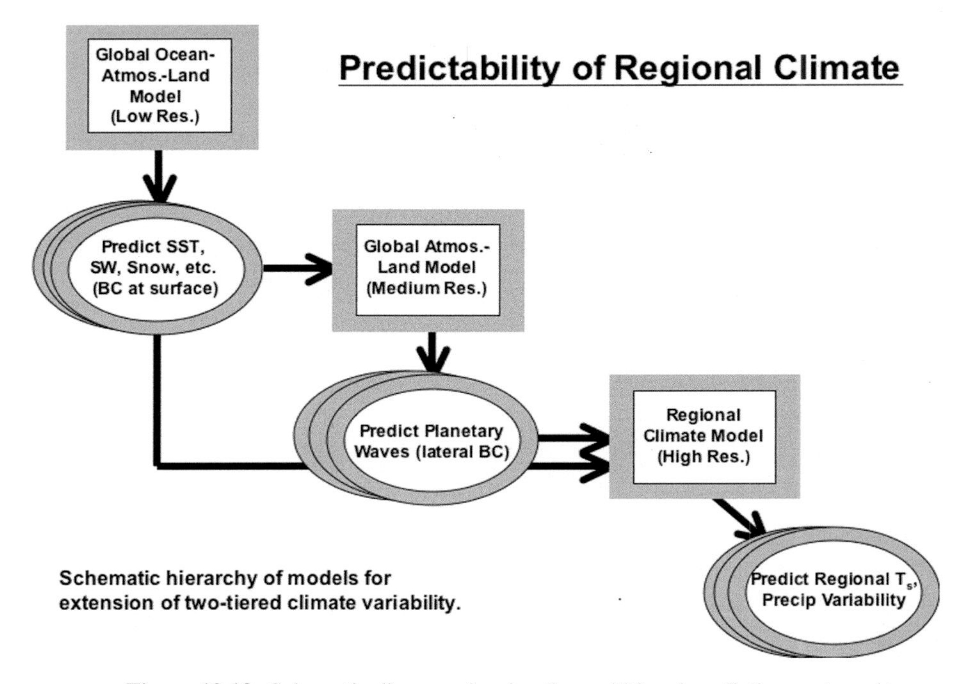

Figure 12.13 Schematic diagram showing the multitiered prediction system. At upper left is the global coupled ocean–atmosphere–land model run at low resolution to generate an ensemble of predicted SST, snow and other boundary conditions (BC) at the surface. These are then used as boundary conditions for both a global atmospheric model run at medium resolution, used to generate an ensemble of predictions of the large-scale circulation in the atmosphere, and a regional climate model run at high resolution to generate an ensemble of predictions of regional variations in surface temperature and precipitation. The large-scale circulation predicted by the medium resolution global atmospheric model is also used as a lateral boundary condition in the regional climate model.

rapid increase of the SST anomaly, although its prediction of the magnitude of the maximum anomaly was substantially smaller than observed. The JFM98 forecast made six months in advance was quite accurate in terms of amplitude, phase relative to the annual cycle and spatial pattern. The predicted maximum anomaly was west of the observed maximum and fell short of the observed amplitude by about 1 °C. Compared with similar predictions and simulations made prior to this one, the forecast was considered to be quite accurate and generally deemed successful.

12.4.3 The atmospheric circulation in the winter hemisphere

The model-based SST forecast was applied as a lower boundary condition to the COLA atmospheric GCM to produce an ensemble of nine forecasts. The ensemble members were generated by initialising the model with slightly different initial conditions to provide some measure of the uncertainty of the forecast. The COLA model ensemble mean forecast was quite skilful, comparable to the result shown in Figure 12.5.

12.4.4 Precipitation over North America

Typically, predictions of precipitation are less reliable than temperature forecasts due to the fact that precipitation is more variable, has a smaller spatial correlation scale and is non-normally distributed. Nevertheless, the large-scale characteristics of precipitation anomalies are known to be correlated with those of other fields and some information may be obtained from predictions of seasonal mean precipitation. The seasonal mean precipitation anomaly for January through March 1998 is shown in Figure 12.14: the prediction made by the global model is in the top panel, the observations (Xie and Arkin, 1996) are in the middle panel, and the prediction made using the nested Eta80 model is in the bottom panel. The main features of the predicted anomaly pattern are a swath of positive anomalies to the south and a band of negative anomalies to the north. The positive anomalies (more than 0.5 mm per day or 50 mm for the entire season above normal) extend from the Pacific north-west states of the USA, through California and the south-west USA into Mexico, and along the Gulf of Mexico into the south-east USA. The band of negative departures from normal extends from the Pacific coast of Canada through the northern plains and Great Lakes states of the USA into the maritime provinces of Canada.

This relatively limited case study shows that by employing an ensemble of models it is possible to make useful predictions of the evolution of the large SST anomalies that can have a large effect on the tropical and extratropical climate up to seasons in advance. Furthermore, these SST predictions can be used in a tier-2 system (e.g. Mason *et al.*, 1999; Goddard *et al.*, 2003) to force a reasonably good atmospheric GCM to produce predictions of the large-scale atmospheric circulation. A nested regional climate model can then be used to resolve important details of the topography and the solution itself to further constrain the forecast precipitation field and make useful six-month lead predictions of regional seasonal mean precipitation anomalies.

12.5 Predictability of seasonal variations in a changing climate

The observed current climate changes are a combination of anthropogenic influences and the natural variability. In addition to possible anthropogenic influence on climate due to changing the atmospheric composition, it is quite likely that land use in the tropics will undergo extensive changes, which will lead to significant changes in the

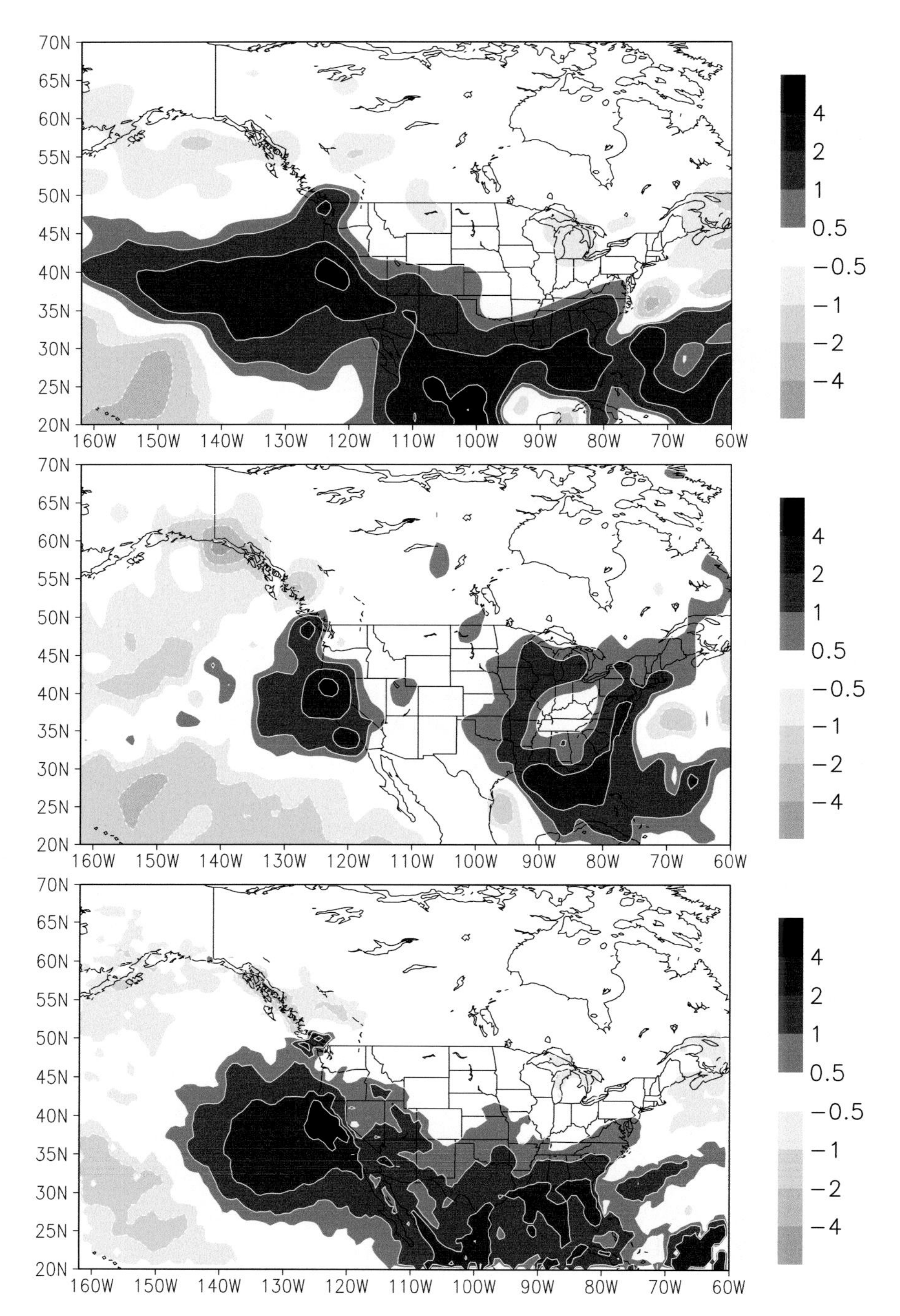

Figure 12.14 Seasonal mean precipitation anomaly over North America for January–March 1998 case. (Top) COLA atmospheric GCM ensemble mean. (Middle) Observed. (Bottom) Eta80 nested in COLA atmospheric GCM.

biophysical properties of the land surface. There is a scientific basis and a plausible mechanism for the contention that these land use changes could have significant remote influences on the climate system, and in the case of Amazon land use change, on ENSO variability in particular (Hu *et al.*, 2004). The dominant effects of changes in the land surface appear to be due to changes in the albedo and soil wetness.

The role of SST variability on the atmosphere and land in the last century, including problems such as the variability and predictability of ENSO and the ENSO/monsoon relationship have been investigated in the Climate of the 20th Century (C20C; Folland *et al.*, 2002) project using SST specified from analysis of existing data. It has been shown, for example, that the general increase in SST in the eastern tropical Indian and Pacific Oceans has led to an increase in the predictability of seasonal variations (I.-S. Kang, personal communication). This can be projected in future climate scenarios using variants of the C20C approach, in which the SST is specified from coupled GCM climate change projections. This 'two-tier' approach can illuminate the atmospheric and land surface response to the SST variability. However, it is of limited value in helping to understand the changes in the SST variability itself. A coupled approach is needed for this part of the problem.

12.6 Factors limiting seasonal predictability

Estimates of seasonal predictability are model dependent. For example, the estimates of seasonal predictability have changed considerably as the models have evolved. In Figure 12.15, the response to ENSO forcing simulated by various versions of an atmospheric GCM, as it evolved over a 20-year period, is shown. It is clear that the fidelity of the simulation, and consequently the estimate of predictability, has changed markedly over the two decades of model development. Even at this late stage of model development, model differences give rise to very large differences in estimates of predictability. Figure 12.16 shows the probability distributions of geopotential height variance explained by tropical Pacific SST in the Pacific–North America region produced by six different atmospheric GCMs, based on ensembles of simulations. Current state-of-the-art models may underestimate or overestimate the variance explained by as much as a factor of two, and there is even some variation in the spread of the models' distributions of this quantity. This is a direct consequence of the uncertainty in the models' parametrisations of subgrid-scale processes, especially convection. As shown in Figure 12.17, the six models used to produce Figure 12.16 have significantly different rainfall variance in the tropical Pacific, where the atmospheric response to SST anomalies is most sensitive. The range of rainfall variance among the models is up to a factor of 8.

Another major factor limiting progress in understanding seasonal predictability is the limited size of ensembles employed. The problem of seasonal prediction is a probabilistic one, and the simple way in which this is addressed in current prediction

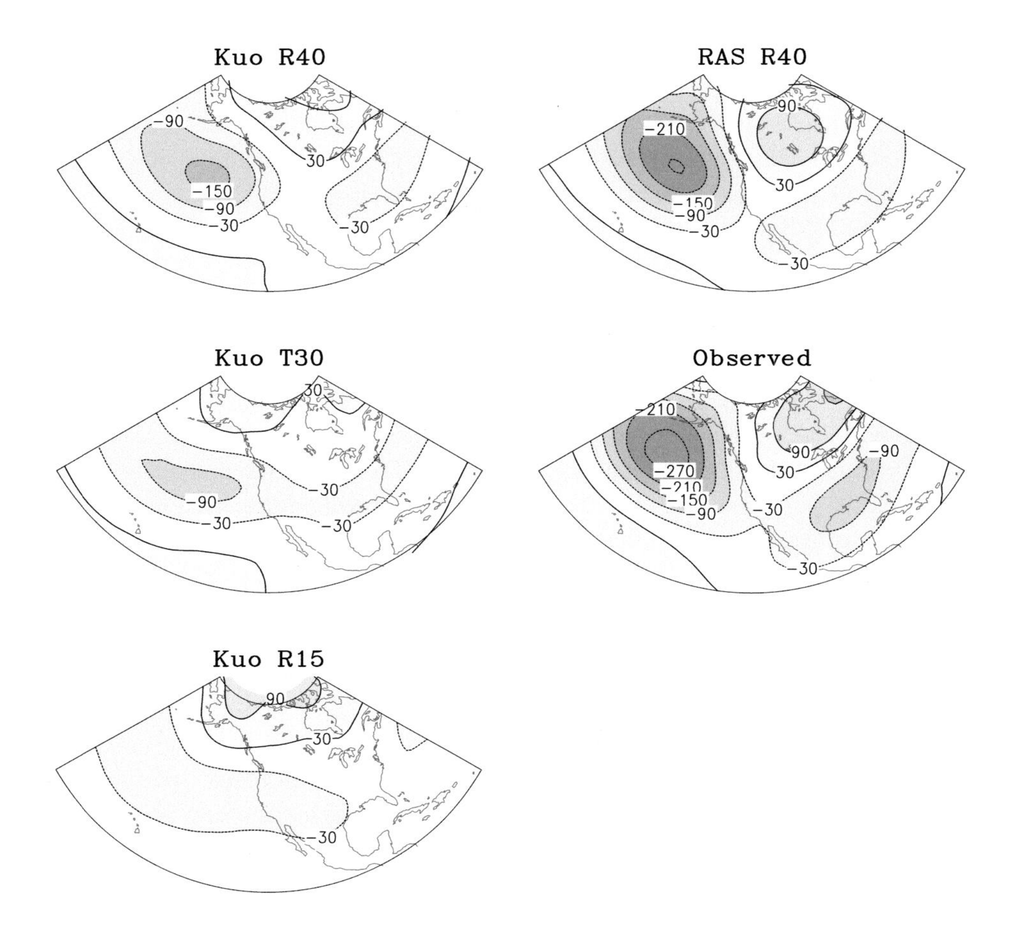

Φ_{500} Δ(JFM1983−JFM1989)

Figure 12.15 Some simulations of the difference in Jan–Feb–March mean 500 hPa geopotential height for 1983 (a year with warm SST anomalies in the eastern tropical Pacific) minus 1989 (a year with cold SST anomalies in the eastern tropical Pacific). The simulations are done with various versions of the same atmospheric GCM. (Left) Atmospheric GCM simulations with the Kuo parametrisation of cumulus convection: top, R40 horizontal resolution; middle, T30 horizontal resolution; bottom, R15 horizontal resolution. (Upper right) Atmospheric GCM simulation with RAS parametrisation of cumulus convection and R40 horizontal resolution. (Lower right) Observed difference computed using NCEP/NCAR reanalysis fields.

systems is to build ensembles of model integrations that are intended to sample the uncertainty in the initial conditions. It has been shown that fairly large ensembles are needed for cases with weak or moderate tropical forcing (Straus and Molteni, 2004). It has also been shown that the use of multiple models in an ensemble is more effective than using multiple realisations with a single model (Palmer *et al.*, 2004; T. Palmer, personal communication). The latter finding is probably because multiple

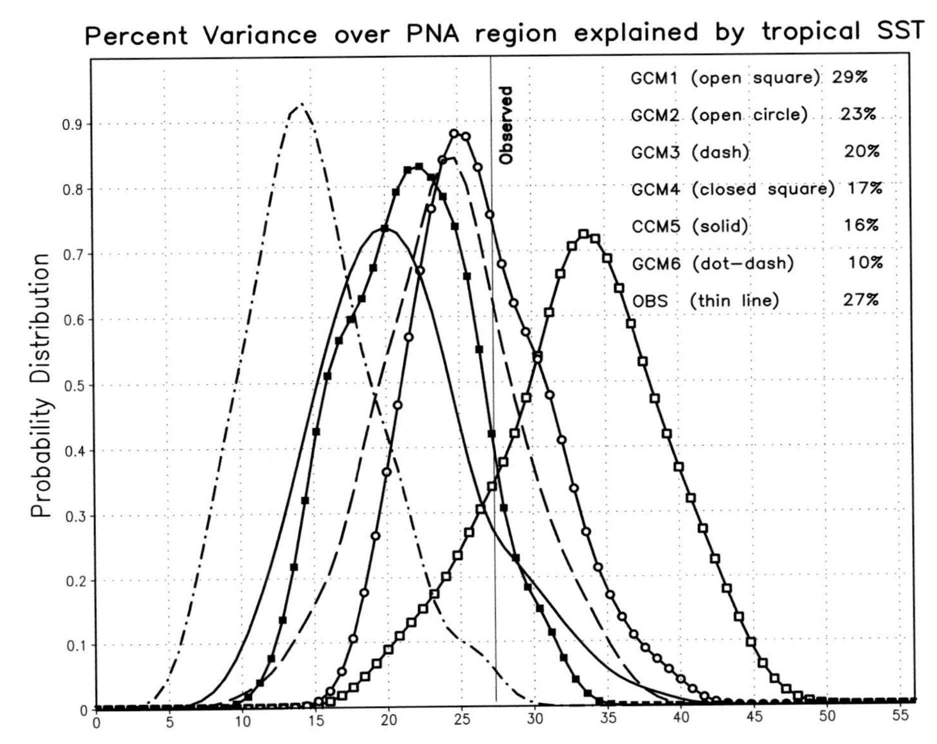

Figure 12.16 Probability distributions of (area-averaged) temporal variance of winter mean 500 hPa height explained by an observed tropical Pacific SST time series from 100 'samples' of seasonal simulations with various atmospheric GCMs. The SST time series is obtained from the leading mode of a singular value decomposition of observed tropical Pacific SST with 500 hPa height over the Pacific–North America region obtained from the NCEP reanalysis for the winters January–March 1968–97. Each GCM sample was formed from an ensemble of winter simulations for the winters of 1983–93 by picking one member of the ensemble at random for each year. The temporal variance explained by the SST time series was averaged over the Pacific–North America region (180°–60° W).

models sample the uncertainty in the physical parametrisations, and due to a non-linear interaction between the individual models' systematic error and the predictable signal. The requirement for large ensembles with multiple climate models has only been lightly explored, primarily due to limited computational resources.

A third barrier to progress is the difficulty with initialising coupled ocean–atmosphere–land models. Atmospheric data assimilation has reached a mature stage that is demonstrably extracting a large fraction of the usable information from the available observations (Simmons and Hollingsworth, 2002). Similarly, ocean data assimilation has made great strides in the recent decade, and new ocean observing systems hold the promise of providing a more complete representation of the ocean state, at least for the upper 500 m (Derber and Rosati, 1989; Ji *et al.*, 1995). The assimilation of land surface observations has only recently been attempted globally

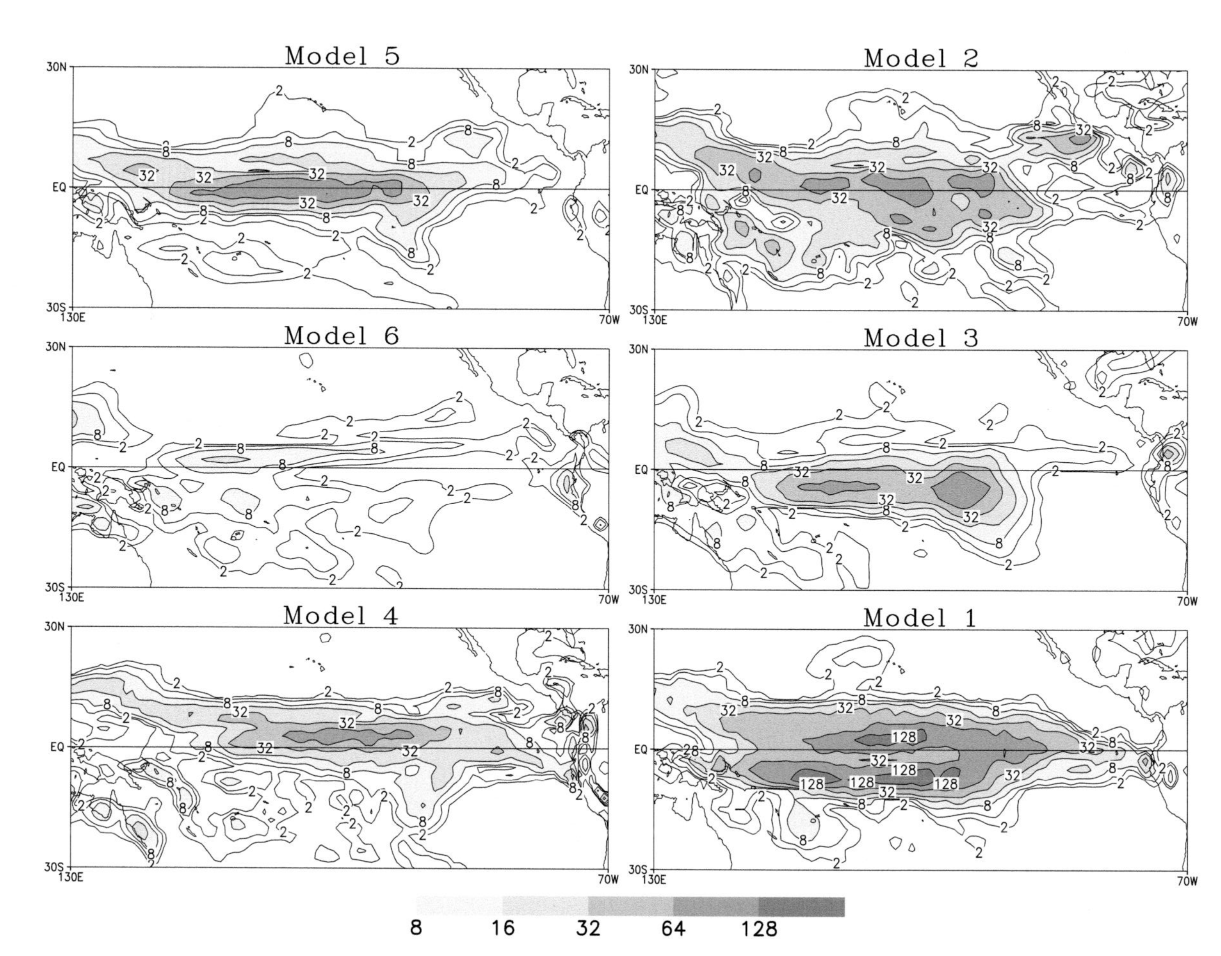

Figure 12.17 Variance of precipitation for the same six models used for Figure 12.16.

(Dirmeyer, 2000), so we have only a limited understanding of how well we are initialising land surface conditions. Furthermore, the systematic errors in the seasonal prediction models make the evolution of soil moisture from its initial state problematic. Most importantly, the climate system is known to have modes of variability that depend on the couplings between ocean and atmosphere and land and atmosphere, but these modes are not necessarily being initialised in the process of initialising the components of the coupled model.

12.7 Summary and prospects for the future

Improvements in dynamical weather prediction over the past 30 years did not occur because of any major scientific breakthroughs in our understanding of the physics or dynamics of the atmosphere. Dynamical weather prediction is challenging: progress takes place slowly and through a great deal of hard work that is not necessarily scientifically stimulating, performed in an environment that is characterised by frequent setbacks and constant criticism by a wide range of consumers and clients. Nevertheless, scientists worldwide have made tremendous progress in improving the skill of weather forecasts by advances in data assimilation, improved parametrisations, improvements in numerical techniques and increases in model resolution and computing power.

The growth rate of initial errors in NWP models is well known, and the current limits of predictability of weather are well documented. The most promising way to improve forecasts for days 2–15 is to improve the forecast at day 1. During the past 25 years, the weather forecast error at day 1 has been reduced by more than 50%. At present, forecasts for day 4 are, in general, as good as forecasts for day 2 made 25 years ago. With improved observations, better models and faster computers, it is reasonable to expect that the forecast error at day 1 will be further reduced.

There is a scientific basis for extending the successes of NWP to climate prediction. A model's ability to reliably predict the sequential evolution of the climate system depends on the model's ability to simulate the statistical properties of the observed climate system. There is sufficient evidence to indicate that, as models improve in their representation of all the statistical properties of the observed climate, they also improve in their prediction skill of the evolution of climate anomalies.

Until 25 years ago, a dynamical seasonal climate prediction was not conceivable. Over the past 25 years, steadily progressing dynamical seasonal climate prediction has achieved a level of skill that is considered useful for some societal applications. However, such successes are limited to periods of large, persistent anomalies at the Earth's surface. There is significant unrealised seasonal predictability. Progress in dynamical seasonal prediction in the future depends critically on improving coupled ocean–atmosphere–land models, improving observations, and increasing the ability to assimilate those observations. The current generation of ENSO prediction models is able to predict the average SST over the equatorial Pacific, but not the evolution and amplitude of an individual ENSO event.

Currently, about 10 centres worldwide are making dynamical weather forecasts every day with a lead time of 5–15 days with about 5–50 ensemble members, so that there are about 500 000 daily weather maps that can be verified each year. It is this process of routine verification by a large number of scientists worldwide, followed by attempts to improve the models and data assimilation systems, that has been the critical element in the improvement of dynamical weather forecasts. In contrast, if we assume that dynamical seasonal predictions, with a lead time of 1–3 seasons, could be made by 10 centres worldwide every month, each with 10–20 ensemble members, there would be fewer than 5000 seasonal mean predictions worldwide that could be verified each year. This is a factor of 100 fewer cases than are available for advancing NWP, so improvement in dynamical seasonal prediction might proceed at a pace that is much slower than that for NWP if we don't do something radically different.

It is suggested that, for accelerating progress in dynamical seasonal prediction, we reanalyse and reforecast the seasonal variations for the past 50 years, every year. This will entail annually reanalysing the observations of the atmosphere and ocean available since 1950, and making an ensemble of six-month lead forecasts, starting from initial conditions in each month of the ~50-year period. By doing so, we will exercise state-of-the-art coupled ocean–atmosphere–land models and data assimilation systems for a large number of seasonal prediction cases and verify them against observations. We should also conduct model development experiments (sensitivity to parametrisations, resolution, coupling strategy, etc.) with the specific goal of reducing seasonal prediction errors.

Acknowledgements

We would like to thank several people at the Center for Ocean–Land–Atmosphere Studies who made contributions to this paper, either through the preparation of figures or discussions. B. Kirtman and M. Fennessy provided Figures 12.12 and 12.13, respectively, and Tim DelSole, Paul Dirmeyer, Larry Marx, Dan Paolino, and David Straus helped with both the figures and the text.

References

Alexander, M. A. and C. Deser (1995). A mechanism for the recurrence of wintertime midlatitude SST anomalies. *J. Phys. Oceanogr.*, **25**, 122–37.

Atlas, R., N. Wolfson and J. Terry (1993). The effect of SST and soil moisture anomalies on GLA model simulations of the 1988 U.S. summer drought. *J. Climate*, **6**, 2034–48.

Bamzai, A. and L. Marx (2000). COLA AGCM simulation of the effect of anomalous spring snow over Eurasia on the Indian summer monsoon. *Quart. J. Roy. Meteor. Soc.*, **126**, 2575–84.

Barnett, T. P., L. Dumenil, U. Schlese and E. Roeckner (1988). The effect of Eurasian snow cover on global climate. *Science*, **239**, 504–7.

Barnston, A. G. and R. E. Livezey (1987). Classification, seasonality and persistence of low-frequency atmospheric circulation patterns. *Mon. Wea. Rev.,* **115**, 1083–126.

Bhatt, U. S., M. A. Alexander, D. S. Battisti, D. D. Houghton and L. M. Keller (1998). Atmosphere–ocean interaction in the North Atlantic: near-surface climate variability. *J. Climate*, **11**, 1615–32.

Blanford, H. F. (1884). On the connexion of the Himalayan snowfall with dry winds and seasons of droughts in India. *Proc. Roy. Soc. Lond.*, **37**, 3–22.

Charney, J. (1966). Some remaining problems in numerical weather prediction. In *Advances in Numerical Weather Prediction*, pp. 61–70. Hartford, CT: Travelers Research Center.

(1975). Dynamics of deserts and droughts in the Sahel. *Quart. J. Roy. Meteor. Soc.*, **101**, 193–202.

Charney, J. G. and J. Shukla (1977). Predictability of monsoons. *Joint IUTAM/IUGG Symposium on Monsoon Dynamics* (5–9 December, 1977), New Delhi, India.

(1981). Predictability of monsoons. In *Monsoon Dynamics*, ed. J. Lighthill and R. P. Pearce, pp. 99–109. Cambridge University Press.

Charney, J., R. G. Fleagle, H. Riehl, V. E. Lally and D. Q. Wark (1966). The feasibility of a global observation and analysis experiment. *Bull. Amer. Meteor. Soc.*, **47**, 200–20.

Cohen, J. and D. Rind (1991). The effect of snow cover on climate. *J. Climate*, **4**, 689–706.

Delworth, T. L. and S. Manabe (1988). The influence of potential evaporation on the variabilities of simulated soil wetness and climate. *J. Climate*, **1**, 523–47.

(1989). The influence of soil wetness on near-surface atmospheric variability. *J. Climate*, **2**, 1447–62.

Derber, J. and A. Rosati (1989). A global oceanic data assimilation system. *J. Phys. Oceanogr.*, **19**, 1333–47.

Dickinson, R. E. (1984). Modelling evapotranspiration for three-dimensional global climate models. In *Climate Processes and Climate Sensitivity*, pp. 58–72. Geophysical Monograph **29**, Maurice Ewing Vol. 5. American Geophysical Union.

Dirmeyer, P. A. (2000). Using a global soil wetness data set to improve seasonal climate simulation. *J. Climate*, **13**, 2900–22.

Dirmeyer, P. A. and J. Shukla (1993). Observational and modeling studies of the influence of soil moisture anomalies on atmospheric circulation (review). In *Prediction of Interannual Climate Variations*, ed. J. Shukla, pp. 1–24. NATO ASI Series I: Global Environmental Change, Vol. 6.

(1996). The effect on regional and global climate of expansion of the world's deserts. *Quart. J. Roy. Meteor. Soc.*, **122**(530), 451–82.

Fennessy, M. J. and J. Shukla (1999). Impact of initial soil wetness on seasonal atmospheric prediction. *J. Climate*, **12,** 3167–80.

Fennessy, M. J., L. Marx and J. Shukla (1985). General circulation model sensitivity to 1982–83 equatorial Pacific sea surface temperature anomalies. *Mon. Weather Rev.*, **115**, 858–64.

Folland, C. K., J. Shukla, J. L. Kinter III and M. J. Rodwell (2002). C20C: the Climate of the Twentieth Century project. *CLIVAR Exchanges*, **7**, 37–9. (Available from International CLIVAR Project Office, Southampton Oceanography Centre, Empress Dock, Southampton, SO14 3ZH, UK; (www.clivar.org/publications/exchanges/ex24/ex24.pdf)

Goddard, L., A. G. Barnston and S. J. Mason (2003). Evaluation of the IRI's "Net Assessment" seasonal climate forecasts: 1997–2001. *Bull. Am. Meteorol. Soc.*, **84**, 1761–81.

Goswami, B. N. and J. Shukla (1991). Predictability of a coupled ocean-atmosphere model. *J. Climate*, **3**, 2–22.

Gutzler, D. S. and J. Shukla (1984). Analogs in the wintertime 500 mb height field. *J. Atmos. Sci.*, **41**, 177–89.

Hahn, D. G. and S. Manabe (1975). The role of mountains in the south Asian monsoon circulation. *J. Atmos. Sci.*, **32**, 1515–41.

Hahn, D. and J. Shukla (1976). An apparent relationship between Eurasia snow cover and Indian monsoon rainfall. *J. Atmos. Sci.*, **33**, 2461–3.

Hess, P. and H. Brezowsky (1969). Katalog der Grosswetterlagen Europas. *Ber. Dt. Wetterd.*, **15**.

Hu, Z.-Z., E. K. Schneider, U. Bhatt and B. P. Kirtman (2004). Potential for influence of land surface processes on ENSO. *J. Geophys. Res.*, **109**.

Ji, M., A. Leetmaa and J. Derber (1995). An ocean analysis system for seasonal to interannual climate studies. *Mon. Weather Rev.*, **123,** 460–81.

Kinter III, J. L., J. Shukla, L. Marx and E. Schneider (1988). A simulation of the winter and summer circulation with the NMC global spectral model. *J. Atmos. Sci.*, **45**, 2486–522.

Kirtman, B. P. (2003). The COLA anomaly coupled model: ensemble ENSO prediction. *Mon. Weather Rev.*, **131**, 2324–41.

Kirtman, B. P., Y. Fan and E. K. Schneider (2002). The COLA global coupled and anomaly coupled ocean-atmosphere GCM. *J. Climate*, **15**, 2301–20.

Kirtman, B. P., J. Shukla, B. Huang, Z. Zhu and E. K. Schneider (1997). Multiseasonal predictions with a coupled tropical ocean global atmosphere system. *Mon. Weather Rev.*, **125**, 789–808.

Koster, R. D., P. A. Dirmeyer, Z. Guo, *et al.* (2004). Regions of strong coupling between soil moisture and precipitation. *Science*, **305**, 1138–40.

Krishnamurthy, V. and J. Shukla (2001). Observed and model simulated interannual variability of the Indian monsoon. *Mausam*, **52**, 133–50.

Leith, C. E. (1971). Atmospheric predictability and two-dimensional turbulence. *J. Atmos. Sci.*, **28**, 148–61.

Leith, C. E. and R. H. Kraichnan (1972). Predictability of turbulent flows. *J. Atmos. Sci.*, **29**, 1041–58.

Lorenz, E. N. (1965). A study of the predictability of a 28-variable model. *Tellus*, **17**, 321–33.

(1969a). The predictability of a flow which possesses many scales of motion. *Tellus*, **21**, 289–307.

(1969b). Atmospheric predictability as revealed by naturally occurring analogues. *J. Atmos. Sci.*, **26**, 636–46.

(1979). Forced and free variations of weather and climate. *J. Atmos. Sci.*, **8**, 1367–76.

(1982). Atmospheric predictability experiments with a large numerical model. *Tellus*, **34**, 505–13.

Mason, S. J., L. Goddard, N. E. Graham, E. Yulaeva, L. Sun and P. A. Arkin (1999). The IRI seasonal climate prediction system and the 1997/98 El Niño event. *Bull. Am. Meteorol. Soc.*, **80**, 1853–73.

Miyakoda, K., C. T. Gordon, R. Caverly, W. F. Stern, J. J. Sirutis and W. Bourke (1983). Simulation of a blocking event in January 1977. *Mon. Weather Rev.*, **111**, 846–69.

Moura, D. A. and J. Shukla (1981). On the dynamics of droughts in northeast Brazil: observations, theory and numerical experiments with a general circulation model. *J. Atmos. Sci.*, **38**, 2653–75.

Namias, J. (1969). Seasonal interactions between the North Pacific Ocean and the atmosphere during the 1960's. *Mon. Weather Rev.*, **97**, 173–92.

(1986). Persistence of flow patterns over North America and adjacent ocean sectors. *Mon. Weather Rev.*, **114**, 1368–83.

Nobre, C. A., P. J. Sellers and J. Shukla (1991). Amazonian deforestation and regional climate change. *J. Climate*, **4**, 957–88.

Palmer, T. N., A. Alessandri, U. Andersen, *et al.* (2004). Development of a European multi-model ensemble system for seasonal to inter-annual prediction (DEMETER). *Bull. Am. Meteorol. Soc.*, **85**, 853–72.

Palmer, T. N. and J. Shukla (2000). Editorial to DSP/PROVOST special issue. *Quart. J. Roy. Meteor. Soc.*, **126**, 1989–90.

Palmer, T. N. and Z. Sun (1985). A modelling and observational study of the relationship between sea surface temperature in the northwest Atlantic and the atmospheric general circulation. *Quart. J. Roy. Meteor. Soc.*, **111**, 947–75.

Randall, D., J. Curry, D. Battisti, *et al.* (1998). Status of and outlook for large-scale modeling of atmosphere–ice–ocean interactions in the Arctic. *Bull. Am. Meteorol. Soc.*, **79**, 197–219.

Rasmusson, E. M. and T. H. Carpenter (1983). The relationship between eastern equatorial Pacific sea surface temperatures and rainfall over India and Sri Lanka. *Mon. Weather Rev.*, **111**, 517–28.

Sato, N., P. J. Sellers, D. A. Randall, *et al.* (1989). Effects of implementing the Simple Biosphere model (SiB) in a general circulation model. *J. Atmos. Sci.*, **46**, 2757–82.

Sellers, P. J., Y. Mintz, Y. C. Sud and A. Dalcher (1986). A Simple Biosphere Model (SIB) for use within general circulation models. *J. Atmos. Sci.*, **43**, 505–31.

Shukla, J. (1975). Effect of Arabian sea-surface temperature anomaly on Indian summer monsoon: a numerical experiment with GFDL model. *J. Atmos. Sci.*, **32**, 503–11.

(1981a). Dynamical predictability of monthly means. *J. Atmos. Sci.*, **38**, 2547–72.

(1981b). *Predictability of the Tropical Atmosphere*. NASA Technical Memorandum 83829. Goddard Space Flight Center, Greenbelt, MD: NASA.

(1985). Predictability. In *Issues in Atmospheric and Oceanic Modeling. II: Weather Dynamics* ed. S. Manabe pp. 87–122. Advances in Geophysics 28B. Academic Press.

(1998). Predictability in the midst of chaos: a scientific basis for climate forecasting. *Science*, **282**, 728–31.

Shukla, J. and M. Fennessy (1988). Prediction of time mean atmospheric circulation and rainfall: influence of Pacific SST anomaly. *J. Atmos. Sci.*, **45**, 9–28.

Shukla, J. and Y. Mintz (1982). The influence of land-surface evapotranspiration on the earth's climate. *Science*, **214**, 1498–501.

Shukla, J. and B. N. Misra (1977). Relationships between sea surface temperature and wind speed over the Central Arabia Sea, and monsoon rainfall over India. *Mon. Weather Rev.*, **105**, 998–1002.

Shukla, J. and J. M. Wallace (1983). Numerical simulation of the atmospheric response to equatorial Pacific sea surface temperature anomalies. *J. Atmos. Sci.*, **40**, 1613–30.

Shukla, J., J. Anderson, D. Baumhefner, *et al.* (2000a). Dynamical seasonal prediction. *Bull. Am. Meteorol. Soc.*, **81**, 1–14.

Shukla, J., D. A. Paolino, D. M. Straus, *et al.* (2000b). Dynamical seasonal predictions with the COLA atmospheric model. *Quart. J. Royal. Meteor. Soc.*, **126**, 2265–91.

Sikka, D. R. (1980). Some aspects of the large-scale fluctuations of summer monsoon rainfall over India in relation to fluctuations in the planetary and regional scale circulation parameters. *Proc. Indian Acad. Sci.*, **89**, 179–95.

Simmons, A. J. and A. Hollingsworth (2002). Some aspects of the improvement in skill of numerical weather prediction. *Quart. J. Roy. Meteor. Soc.*, **128**, 647–77.

Simmons, A. J., R. Mureau and T. Petroliagis (1995). Error growth and estimates of predictability from the ECMWF forecasting system. *Quart. J. Roy. Meteor. Soc.*, **121**, 1739–71.

Smagorinsky. J. (1963). General circulation experiments with the primitive equations. I: The basic experiment. *Mon. Weather Rev.*, **91**, 99–164.

(1969). Problems and promises of deterministic extended range forecasting. *Bull. Am. Meteorol. Soc.*, **50**, 286–311.

Straus, D. M. and F. Molteni (2004). Flow regimes, SST forcing, and climate noise: results from large GCM ensembles, *J. Climate*, **17**, 1641–56.

Straus, D. M. and J. Shukla (2000). Distinguishing between the SST- forced variability and internal variability in mid-latitudes: analysis of observations and GCM simulations. *Quart. J. Royal Meteor. Soc.*, **126**, 2323–50.

Sud, Y. C., J. Shukla and Y. Mintz (1988). Influence of land-surface roughness on atmospheric circulation and rainfall: a sensitivity study with GCM. *J. Clim. Appl. Meteorol.*, **27**, 1036–54.

Wallace, J. M., S. Tibaldi and A. J. Simmon (1983). Reduction of systematic forecast errors in the ECMWF model through the introduction of an envelope orography. *Quart. J. Roy. Meteor. Soc.*, **109**, 683–717.

Williamson, D. L. and A. Kasahara (1971). Adaptation of meteorological variables forced by updating. *J. Atmos. Sci.*, **28**, 1313–24.

Xie, P.-P. and P. A. Arkin (1996). Analyses of global monthly precipitation using gauge observations, satellite estimates, and numerical model predictions. *J. Climate*, **9**, 840–58.

Xue, Y. and J. Shukla (1993). The influence of land surface properties on Sahel climate. I: Desertification. *J. Climate*, **6**, 2232–45.

Zebiak, S. E. and M. A. Cane (1987). A model El Niño–Southern Oscillation. *Mon. Weather Rev.*, **115**, 2262–78.

13

Predictability of the North Atlantic thermohaline circulation

M. Latif
Leibniz-Institut für Meereswissenschaften, Kiel

H. Pohlmann, W. Park
Max-Planck-Institut für Meteorologie, Hamburg

Sea surface temperature (SST) observations in the North Atlantic indicate the existence of strong multi-decadal variability with unique spatial structure. It is shown by means of a global climate model which does not employ flux adjustments that the multidecadal SST variability is closely related to variations in the North Atlantic thermohaline circulation (THC). The close correspondence between the North Atlantic SST and THC variabilities allows, in conjunction with the dynamical inertia of the THC, for the prediction of the slowly varying component of the North Atlantic climate system. This is shown by classical predictability experiments and greenhouse warming simulations with the global climate model.

13.1 Introduction

The North Atlantic thermohaline circulation is an important component of the global climate system. Strong and rapid changes in the THC have been reported from palaeoclimatic records (e.g. Broecker *et al.*, 1985), and a current topic for discussion is whether greenhouse warming may have a serious impact on the stability of THC (e.g. Cubasch *et al.*, 2001). The North Atlantic SST varied on a wide range of timescales during the last century (e.g. Deser and Blackmon, 1993). It has been pointed out (Bjerknes, 1964) that the short-term interannual variations are driven primarily by the atmosphere, while the long-term multidecadal changes may be forced by variations

Predictability of Weather and Climate, ed. Tim Palmer and Renate Hagedorn. Published by Cambridge University Press.

in ocean dynamics. The latter is supported by simulations with coupled ocean–atmosphere models (Delworth *et al.*, 1993; Timmermann *et al.*, 1998; Park and Latif, 2005) which show that variations in the North Atlantic THC are reflected in large-scale SST anomalies. Recently, consistency between the observed multidecadal SST variability derived from palaeo-climatic and instrumental data and that simulated by two versions of the Geophysical Fluid Dynamics Laboratory (GFDL) coupled model has been demonstrated (Delworth and Mann, 2000). The existence of multidecadal SST variability with opposite signs in the North and South Atlantic and its impact on Sahelian rainfall was described by Folland *et al.* (1984, 1986). The multidecadal SST variability in the Atlantic Ocean has also been described in many subsequent papers (e.g. Delworth and Mann, 2000, and references therein).

Changes in the THC strength may have strong implications for global and regional climates (e.g. Manabe and Stouffer, 1999). However, there currently exists no means to observe the variability of the THC. In this chapter we present a method to reconstruct past variations of the THC and to monitor the state of the North Atlantic climate system in the future by simply observing Atlantic SSTs. Additionally, we systematically explore the predictability of the North Atlantic climate system. The multidecadal variability in North Atlantic SST is described in Section 13.2 and the origin of this variability is investigated in Section 13.3. The dependence of the air–sea interactions over the North Atlantic on the ocean dynamics is analysed in Section 13.4. We present the results of the classical predictability experiments in Section 13.5 and those of the greenhouse warming simulations in Section 13.6. The chapter concludes with a summary and discussion of the major findings.

13.2 Multidecadal SST variability

We analyse first the latest Hadley Centre SST data set which covers the period 1870–1998. This data set is partly described in Folland *et al.* (1999). The monthly values were averaged to annual mean values, which is justified since we concentrate here on the multidecadal timescale. Coupled model simulations with and without ocean dynamics (Park and Latif, 2005) indicate that one of the regions of strong influence of the ocean dynamics on SST is the North Atlantic at 40–60° N. We therefore define an SST index which averages the SST over the region 40–60° N and 50–10° W. This index shows some rather strong multidecadal variability (Figure 13.1), with anomalously high SSTs around 1900 and the 1950s and increasing SSTs during the most recent years. Also shown is the low-pass filtered version of the North Atlantic SST index using a 21-year running mean filter. It is noted that the SST index does not correspond well to the North Atlantic Oscillation (NAO) index (see e.g. Hurrell, 1995), a measure of the westerlies over the North Atlantic, which suggests that the North Atlantic SST variability is not simply an in-phase response to the low-frequency variations in the NAO. As will be shown below by discussing

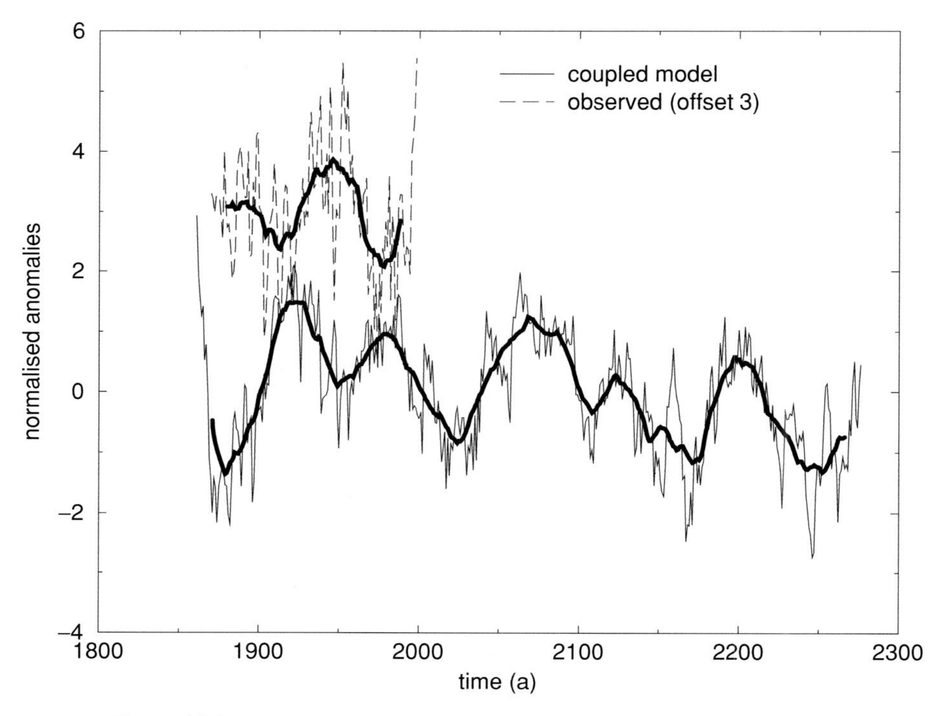

Figure 13.1 Time series of the observed (dashed curve) annual North Atlantic SST anomalies averaged over the region 40–60° N and 50–10° W and the corresponding simulation (solid line). The thick curves are the corresponding 21-year running means which highlight the multidecadal variability. The time series were normalised with their respective standard deviations. From Latif *et al.* (2004).

model results, the type of multidecadal variability considered here originates in the ocean. Furthermore, we note that the North Atlantic SST index does not exhibit any strong trend during the last several decades, but shows instead a rather oscillatory behaviour throughout the analysed period.

We computed the spatial anomaly structure of North Atlantic SST that is associated with the multidecadal variability (not shown). It is rather homogeneous and was discussed, for instance, by Delworth and Mann (2000), Folland *et al.* (1984) and Folland *et al.* (1986). Anomalies of the same sign cover basically the whole North Atlantic Ocean from the equator to the high latitudes, with strongest values near 60° N. Anomalies of opposite sign are found in the South Atlantic (not shown). This SST anomaly pattern associated with the multidecadal variability is strikingly different from the characteristic SST anomaly pattern associated with the higher-frequency interannual variability. The latter is characterised by the well-known North Atlantic tripolar SST anomaly pattern (Visbeck *et al.*, 1998) which is forced by the atmosphere through surface heat flux anomalies associated with the NAO. Thus, the multidecadal SST variability in the North Atlantic exhibits a unique spatial structure. In order to investigate the dynamics of the multidecadal SST variability further, we analyse next the results from an extended-range integration with our global climate model.

The model used in this study to explore the dynamics of the multidecadal SST variability is the Max-Planck-Institute for Meteorology global climate model (Latif *et al.*, 2004). The ocean component MPI-OM (Marsland *et al.*, 2003) is based on a C-grid version of the HOPE (Hamburg Ocean Model in Primitive Equations) ocean model and employs variable horizontal resolution, with relatively high resolution ($\sim$10–50 km) in the high latitudes and near the equator. The atmosphere model is ECHAM5 (Roeckner *et al.*, 2003), the latest cycle of the ECHAM (European Centre HAMburg) atmosphere model. It is run at T42 resolution which corresponds to a horizontal resolution of about $2.8^\circ \times 2.8^\circ$. A high vertical resolution version of ECHAM5 has been used by Giorgetta *et al.* (2002) to study the dynamics of the stratospheric quasi-biennial oscillation (QBO). A Hibler-type dynamic/thermodynamic sea ice model and a river run-off scheme are included in the climate model. Glacier calving is treated in a simple but interactive manner. The climate model does not employ flux adjustments or any other corrections. Here a 500-year control integration with the model is analysed. The model simulates the present-day climate of the North Atlantic realistically. The climate model's thermohaline circulation is consistent with observations, with a maximum overturning of about 20 Sv and a northward heat transport of about 1 PW at 30° N (Marsland *et al.*, 2003).

The model simulates the tripolar SST anomaly pattern in the North Atlantic at interannual timescales, and consistent with observations it is forced by the NAO (not shown). The model also simulates pronounced multidecadal variability in North Atlantic SST. The same North Atlantic SST index that was computed from the observations was derived from the model simulation (Figure 13.1). The figure demonstrates that, after some initial rapid adjustment, the model oscillates with a multidecadal timescale similar to that observed. However, the SST fluctuations simulated by the model appear initially to be somewhat larger than those observed. In order to highlight the multidecadal variability, the 21-year running mean is also shown in Figure 13.1. Next, we computed the spatial pattern associated with multidecadal variability of the model (not shown). The model SST anomaly pattern associated with the multidecadal variability is consistent with that derived from the observations and also characterised by a rather homogeneous pattern in the North Atlantic. Although some small regional differences exist between the observed and simulated patterns, we conclude that our climate model simulates realistically the multidecadal variability in the North Atlantic, so that it can be used to study the origin of the SST variability.

13.3 Origin of the multidecadal SST variability

The climate model offers us the possibility to investigate the physics behind the multidecadal SST changes. An investigation of the model's thermohaline circulation

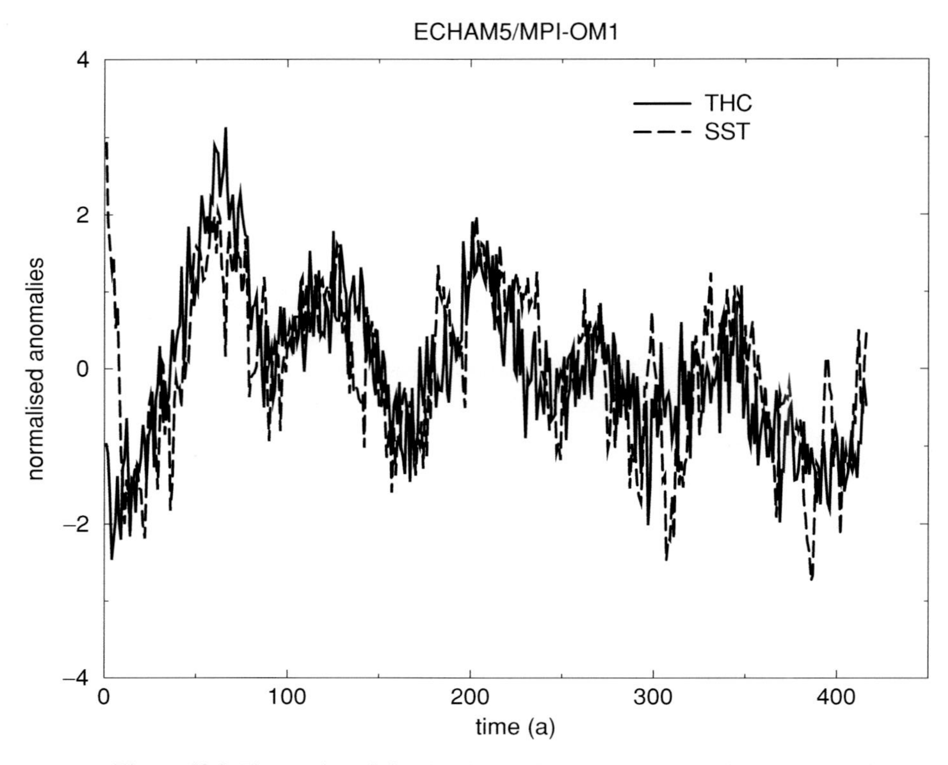

Figure 13.2 Time series of simulated annual mean North Atlantic SST anomalies (40–60° N and 50–10° W, dashed line) and annual mean anomalies of the maximum overturning at 30° N (solid line), a measure of the strength of the model's thermohaline circulation. Note that both time series are highly correlated at timescales beyond several years indicating that the low-frequency variations of the THC can be monitored by SSTs. Both time series were normalised with their respective standard deviations. From Latif *et al.* (2004).

and North Atlantic SST revealed that they are closely related to each other (Figure 13.2). Specifically, the strength of the meridional overturning at 30° N correlates almost perfectly with the North Atlantic SST index defined above at timescales beyond several years. This suggests that the multidecadal SST fluctuations are driven by ocean dynamics, which is also supported by the investigation of the surface heat flux anomalies. The latter are strongly anti-correlated with the SST anomalies, which is demonstrated by the cross-correlation function between North Atlantic SST and surface heat flux anomalies (Figure 13.3a), which exhibits the strongest negative correlations near zero lag. Thus, the surface heat flux can be regarded as a damping for the SST anomalies, a result that is also supported by observations (WOCE, 2001).

A clear connection of the SST anomalies, however, is found to the northward oceanic heat transport. The ocean heat transport leads the North Atlantic SST by several years, as shown by the cross-correlation between the meridional ocean heat

(a)

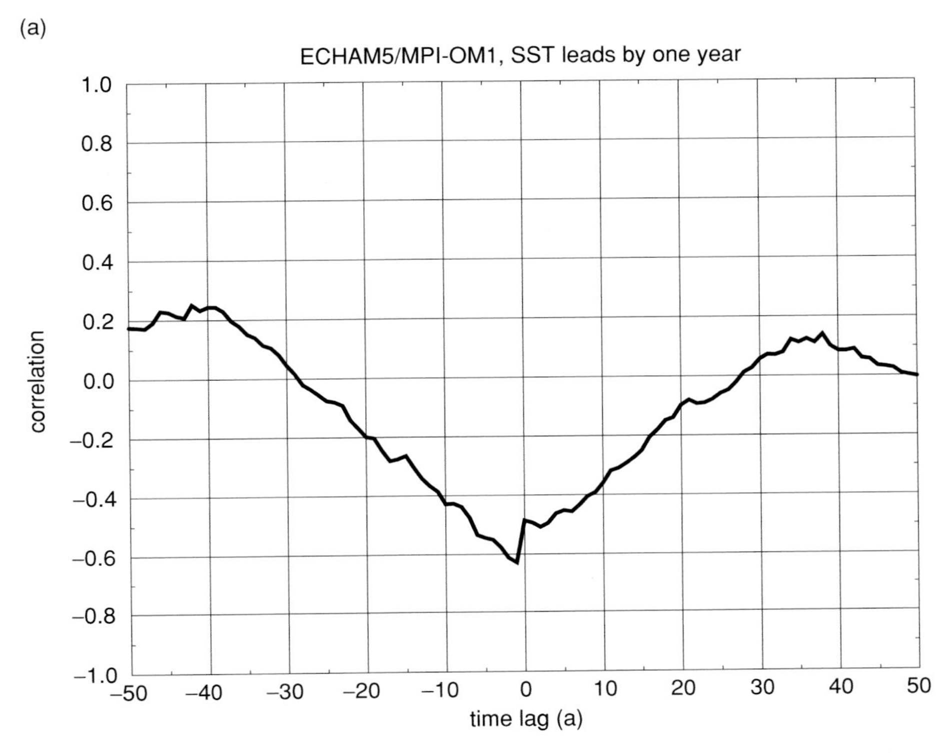

Figure 13.3 (a) Cross-correlation of the North Atlantic SST anomalies (shown in Figure 13.1) and the surface heat flux anomalies averaged over the same region as function of the time lag (a). Please note that the SST and the heat flux are negatively correlated, so that the heat flux can be regarded as a damping for the SST anomalies. (b) Cross-correlation of the North Atlantic SST index with the northward ocean heat transport at 30° N as function of the time lag (a). Maximum correlation is found at positive time lags, which indicates the ocean heat transport leads the SST. This demonstrates together with the left panel that it is the ocean dynamics that drive the SST. From Latif *et al.* (2004).

transport and the SST anomalies (Figure 13.3b). The heat transport has a thermohaline and a wind-driven part. We investigated the relative roles of the two components of the ocean heat transport for the SST variability. We found that the wind-driven part is only relevant at shorter timescales of several years and that on the multidecadal timescale it is the thermohaline part that dominates. Our results are also consistent with modelling studies investigating the stability of the THC (Manabe and Stouffer, 1988; Schiller *et al.*, 1997). In particular, the SST response to a shutdown of the THC shows large similarities to the SST anomaly pattern discussed above.

The close connection between THC strength and SST variability can be used to either reconstruct changes in the THC from SST observations or to monitor the state of the THC in the future. If our model mimics the real relationship between THC and SST correctly, the observed changes in North Atlantic SST (Figure 13.1) can be

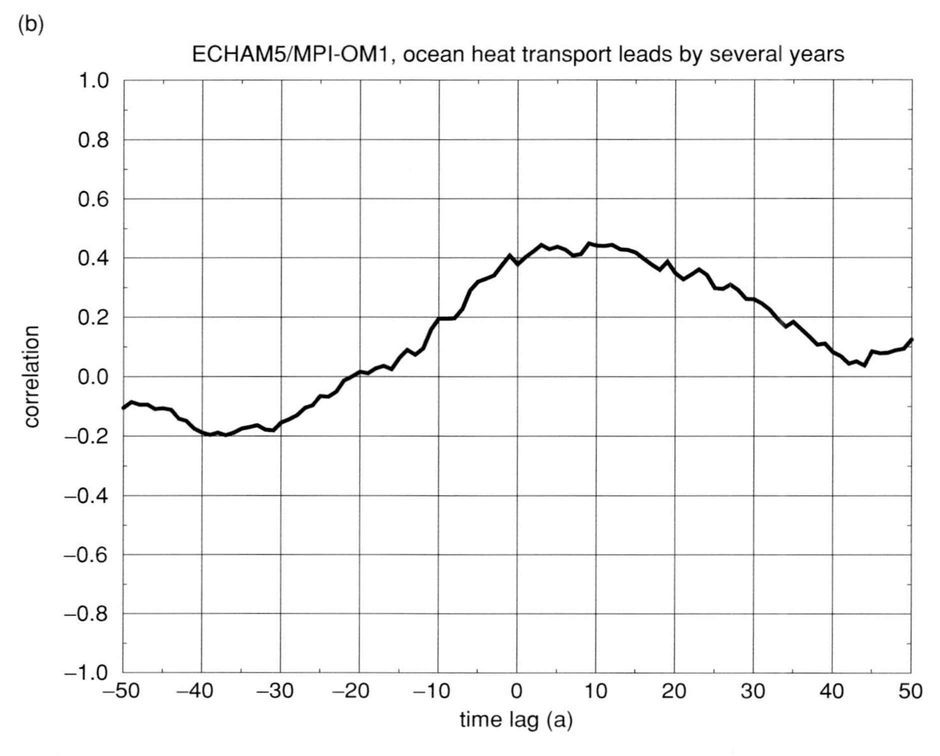

Figure 13.3 *(cont.)*

interpreted as changes in the THC strength; decade-long positive anomalies in the North Atlantic SST index can be regarded as indicators for an anomalously strong THC and vice versa. In particular, the strong cooling during the period 1960–90 may just as well be related to an anomalously weak THC as part of an internal oscillation than to anthropogenic factors, as hypothesised by some authors (e.g. Hegerl *et al.*, 1997), since the cooling is replaced by a warming during the most recent years.

13.4 Ocean dynamics and the nature of air–sea interactions

To suppress active ocean dynamics in the coupled model ECHAM5/MPI-OM the ocean model is replaced by a fixed depth (50 m) mixed layer model (MLO) and a thermodynamic sea ice module. This coupled model (ECHAM5/MLO) is a realisation of Hasselmann's (1976) simplest stochastic climate model, in which SST variability can be produced only by the integration of surface heat anomalies. There may be, however, a feedback of the ocean's SST to the atmosphere, as described below. The sea surface temperature variability in midlatitudes simulated by such a simplified model was discussed, for instance, by Dommenget and Latif (2002).

We start the analysis of the coupled runs by computing the standard deviations of 10-year mean surface temperature and sea level pressure (SLP). Since both runs have a duration of 500 years, stable estimates of the decadal standard deviations can be obtained. The ratios of the standard deviations (ECHAM5/MPI-OM and ECHAM5/MLO) are shown in Figure 13.4. Decadal surface temperature variability is enhanced by the presence of active ocean dynamics over the North Atlantic/European region, over parts of the Tropical and North Pacific and over parts of the Southern Ocean (Figure 13.4a). These were the regions highlighted by Pohlmann *et al.* (2004) to have high decadal predictability potential. The corresponding figure for SLP shows some similarity to the surface temperature pattern (Figure 13.4b). In particular, we find enhanced decadal variability over the North Atlantic. The decadal standard deviation ratios, however, are much smaller compared with those for surface temperature. Over the North Atlantic, for instance, the presence of active ocean dynamics increases the standard deviation of decadal means only by about 20%. This result, which was obtained also by other modelling groups (e.g. Manabe and Stouffer, 1996), has often been used to argue that the impact of changes in ocean currents on the atmosphere is relatively small. We show that this is not the case: although the level of the decadal variability is not strongly influenced by the presence of active ocean dynamics, the structure of the air–sea interactions and the atmospheric response are strongly affected. We concentrate below on the North Atlantic region.

The characteristic decadal-scale SST anomaly pattern simulated in the run with active ocean dynamics (ECHAM5/MPI-OM) is rather flat, with positive SST anomalies over most of the North Atlantic (Figure 13.5a). An anti-correlation exists in the South Atlantic (not shown). The SST anomaly pattern in the run with active ocean dynamics is consistent with that obtained from observations and mainly forced by variations of the THC in the coupled general circulation model (CGCM), as described by Latif *et al.* (2004). The corresponding atmospheric SLP correlation pattern is also rather flat (Figure 13.6a), with anomalously low SLP over anomalously warm SST. This SLP pattern resembling the East Atlantic Pattern (EAP) is also found by standard empirical orthogonal function (EOF) analysis. The EOF analysis was performed using the last 400 years of the integration. The leading EOF mode of low-pass filtered (applying an 11-year running mean filter) SLP anomalies over the North Atlantic accounting for about 43% of the variance is the NAO, while the second most energetic EOF mode explaining about 17% of the variance is the EAP. Only the principal component of the latter is correlated in a statistically significant manner ($r \sim 0.6$) with a North Atlantic THC index, the meridional overturning at 30° N.

Our findings in ECHAM5/MPI-OM (anomalously low SLP over anomalously warm SST) suggest that the atmospheric response is similar to the one proposed by Lindzen and Nigam (1987). The same relationship between multidecadal changes of SST and SLP was described from observations by Kushnir (1994). In principle,

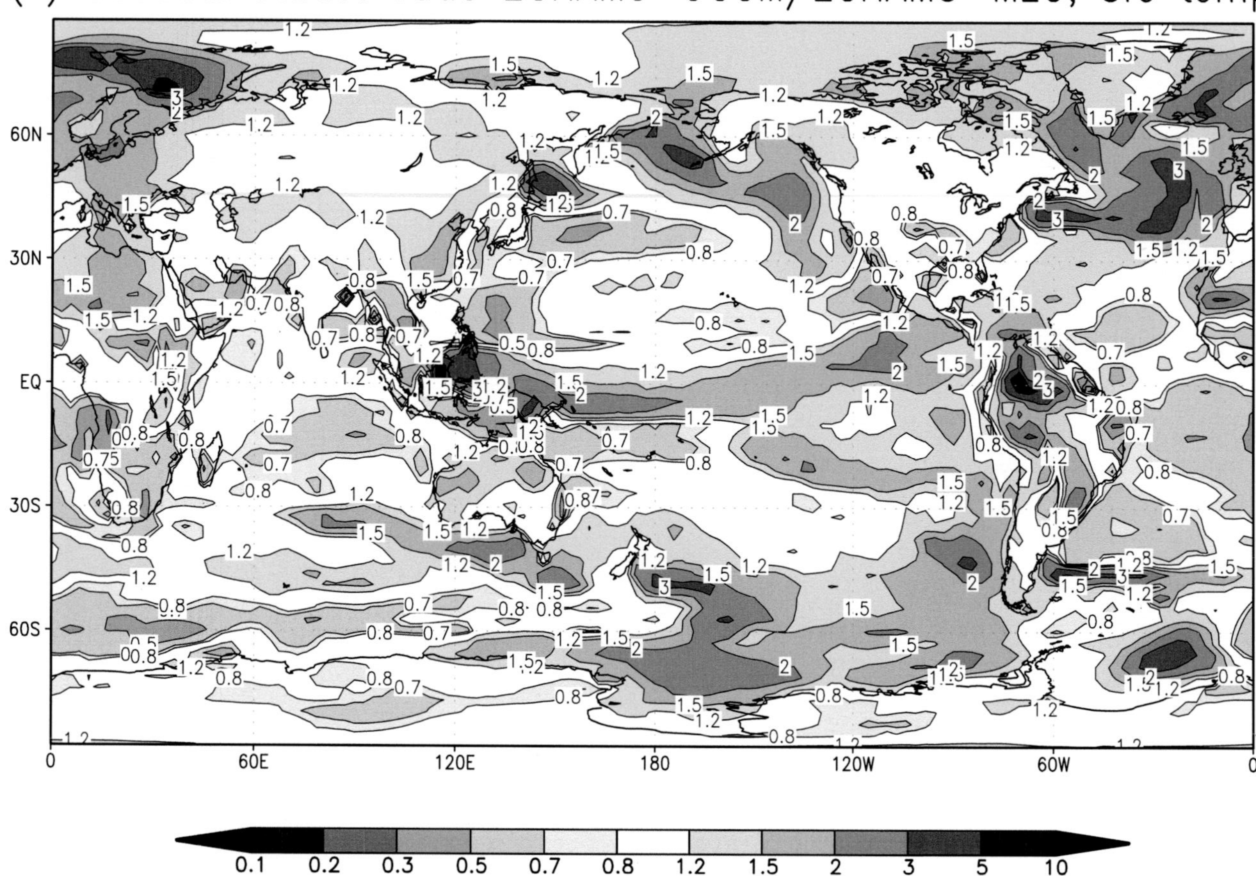

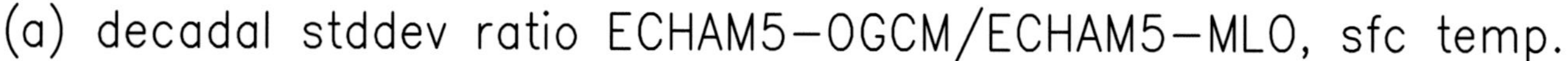

Figure 13.4 Ratio of decadal standard deviations (based on 10-year means) between the coupled runs with the dynamical ocean model (ECHAM5/MPI-OM) and those with the mixed layer ocean model (ECHAM5/MLO). (a) Surface temperature. (b) Sea level pressure. From Park and Latif (2005).

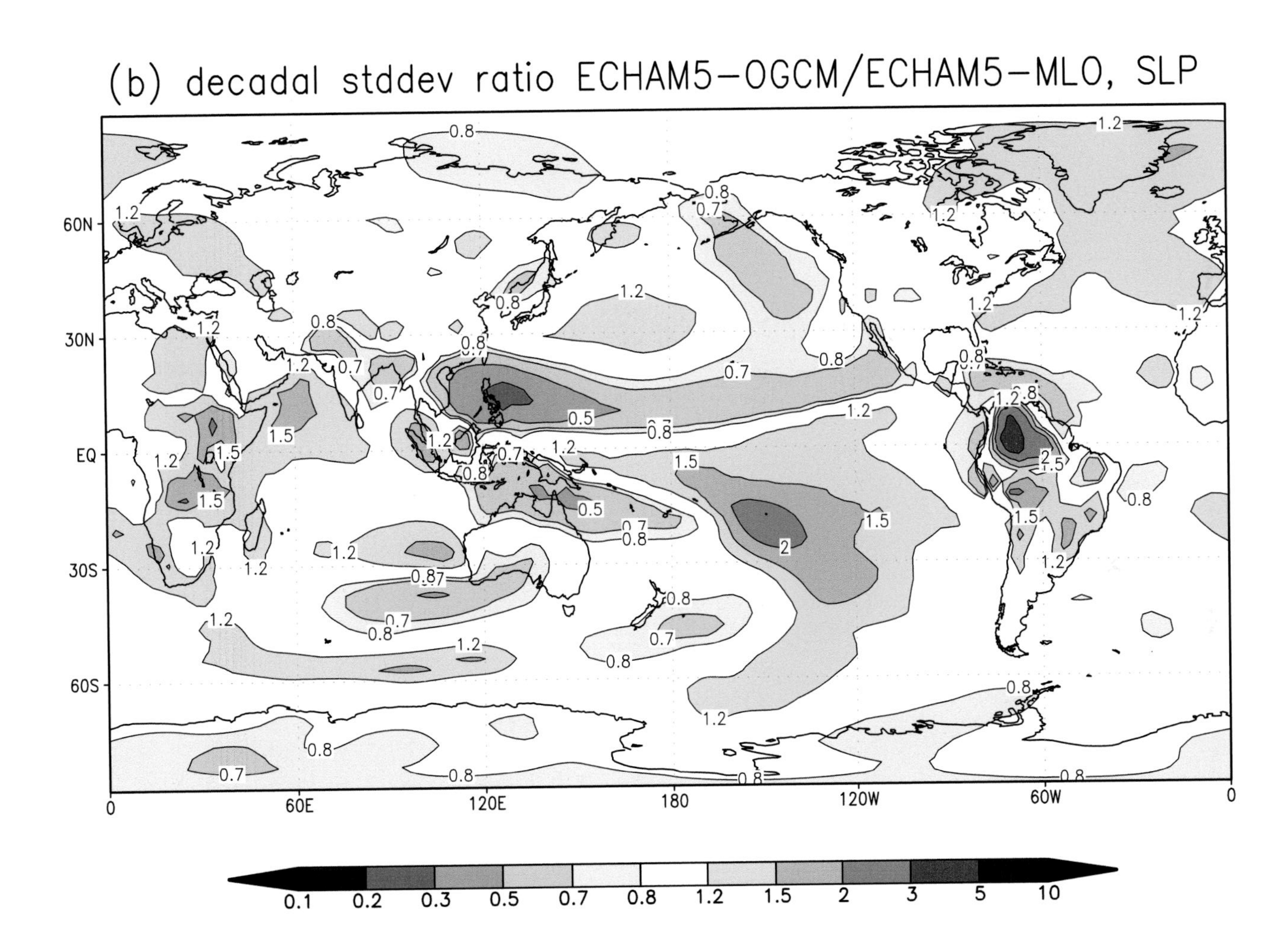

Figure 13.4 *(cont.)*

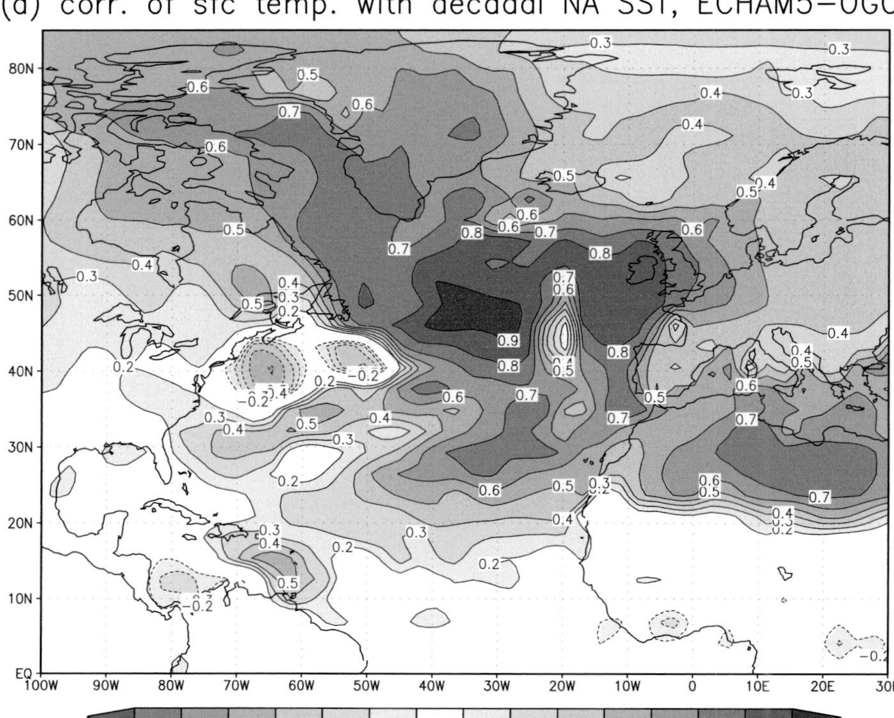

Figure 13.5 Maps of correlation coefficients of surface temperature with the North Atlantic SST index. (a) In the coupled run with the dynamical ocean model (ECHAM5/MPI-OM). (b) In the coupled run with the mixed layer model (ECHAM5/MLO). An 11-yr running mean filter was applied to both variables prior to the analysis. From Park and Latif (2005).

one would expect this type of thermal response from simple physical considerations: the SST anomaly pattern (Figure 13.5a) is rather flat without any strong large-scale horizontal gradients. Thus, the surface baroclinicity is not strongly changed by the SST anomaly pattern, so that a change in the statistics of the transient eddies and a subsequent change in the stormtrack are not to be expected.

The situation changes completely when the coupled run without active ocean dynamics (ECHAM5/MLO) is analysed, in which the dynamical ocean model was replaced by a mixed layer model. The dominant SST anomaly pattern is the well-known North Atlantic tripole (Figure 13.5b). It is forced by surface heat flux anomalies associated with the NAO, the leading mode of the atmosphere over the North Atlantic. The NAO-forced SST anomaly (tripole) pattern is characterised by strong horizontal gradients. These may well influence the surface baroclinicity and thus the transient eddy statistics, as hypothesised by Palmer and Sun (1985). This is supported by the associated SLP anomaly correlation pattern which shows a dipolar structure,

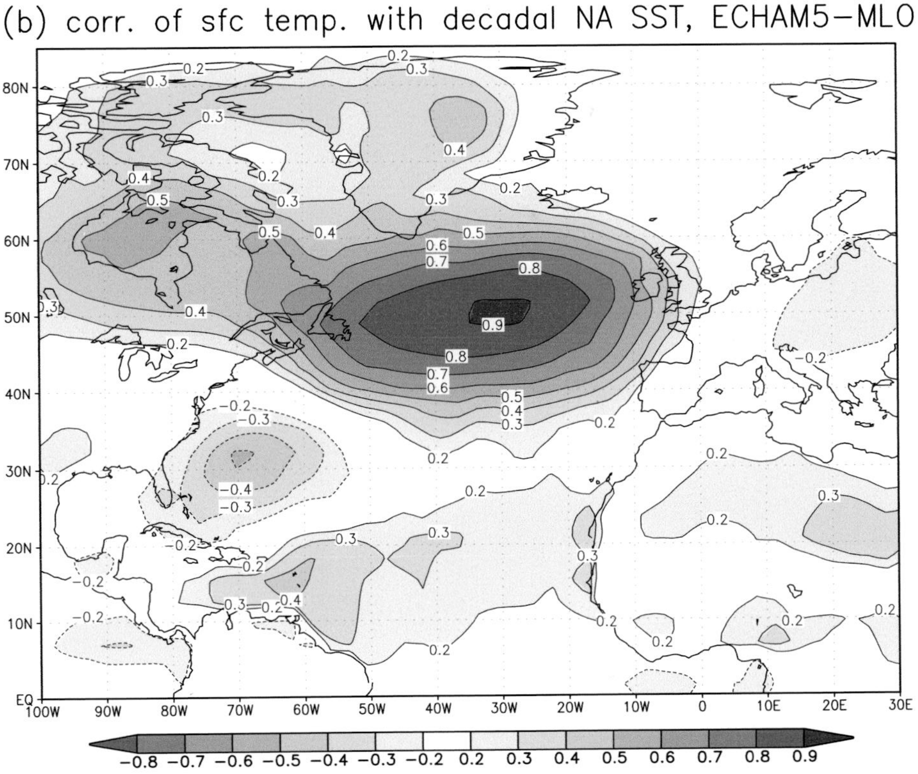

Figure 13.5 (*cont.*)

with anomalously high pressure downstream of the main positive SST anomaly (Figure 13.6b). Overall, the SLP anomaly pattern simulated in the coupled run with the mixed layer model (ECHAM5/MLO) resembles the NAO. Thus, the spatial variability characteristics at decadal timescales are completely different in the two coupled runs. In the run with active ocean dynamics, the main SLP pattern associated with decadal-scale North Atlantic SST changes resembles the East Atlantic Pattern, while, in the run without active ocean dynamics, the main pattern is the North Atlantic Oscillation. Furthermore, we find that the response in ECHAM5/MLO is barotropic in the vertical and baroclinic in ECHAM5/MPI-OM (not shown).

13.5 Classical predictability experiments

We have shown that the multidecadal variability in SST is closely related to ocean dynamics, specifically to the variability of the North Atlantic thermohaline circulation (e.g. Figures 13.2 and 13.3). This indicates that the SST variations may be predictable, even beyond interannual timescales. In order to explore the predictability

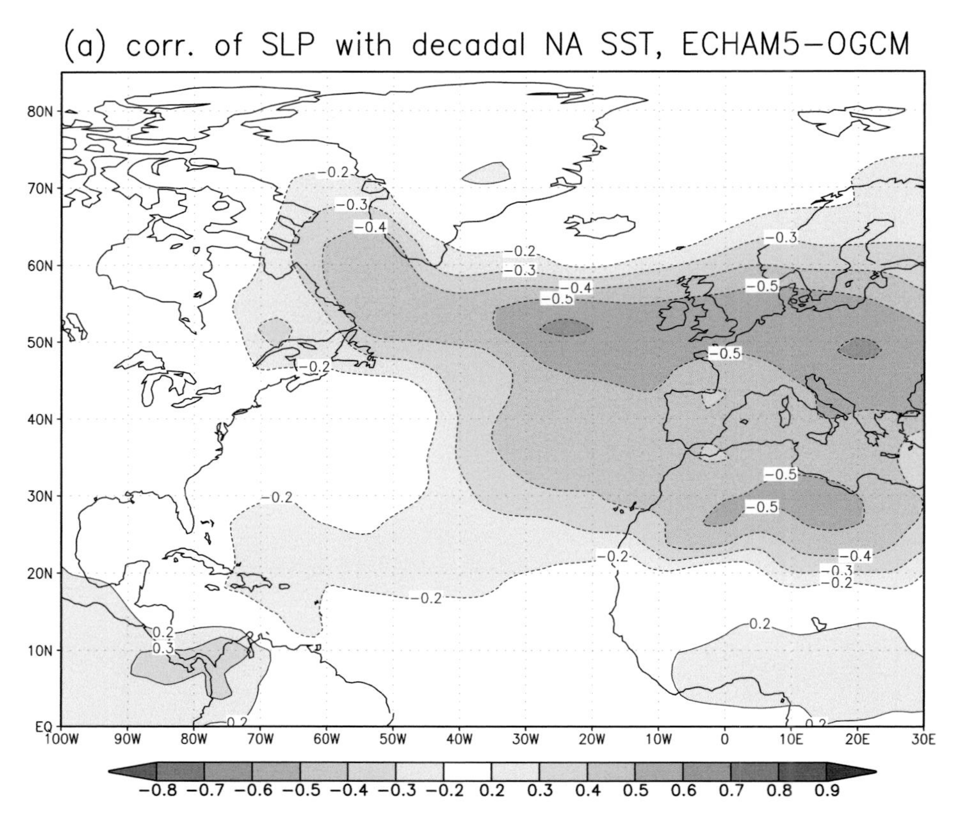

Figure 13.6 As Figure 13.3, but for sea level pressure (SLP). From Park and Latif (2005).

of the SST, we conducted an ensemble of classical predictability experiments with the climate model ECHAM5/MPI-OM. We have chosen three states from the control integration, perturbed the atmospheric initial conditions and restarted the model. We did not perturb the oceanic initial conditions, so that our predictability estimates may be regarded as upper limits of the predictability. Each perturbation experiment has a duration of 20 years, and we conducted an ensemble of six perturbation experiments for each of the three initial states. This yields a total integration time of 360 years.

The results of the predictability experiments are summarised in Figures 13.7 and 13.8. A predictability measure was defined as $P = 1 - (E/C)$. Here E is the variance between the ensemble members and C the variance of the control integration. If the spread between the individual ensemble members is small compared with the internal variability of the coupled system, the predictability measure is close to unity, indicating a high level of predictability. If, by contrast, the spread is comparable to the internal variability, the predictability measure is close to zero and predictability is lost.

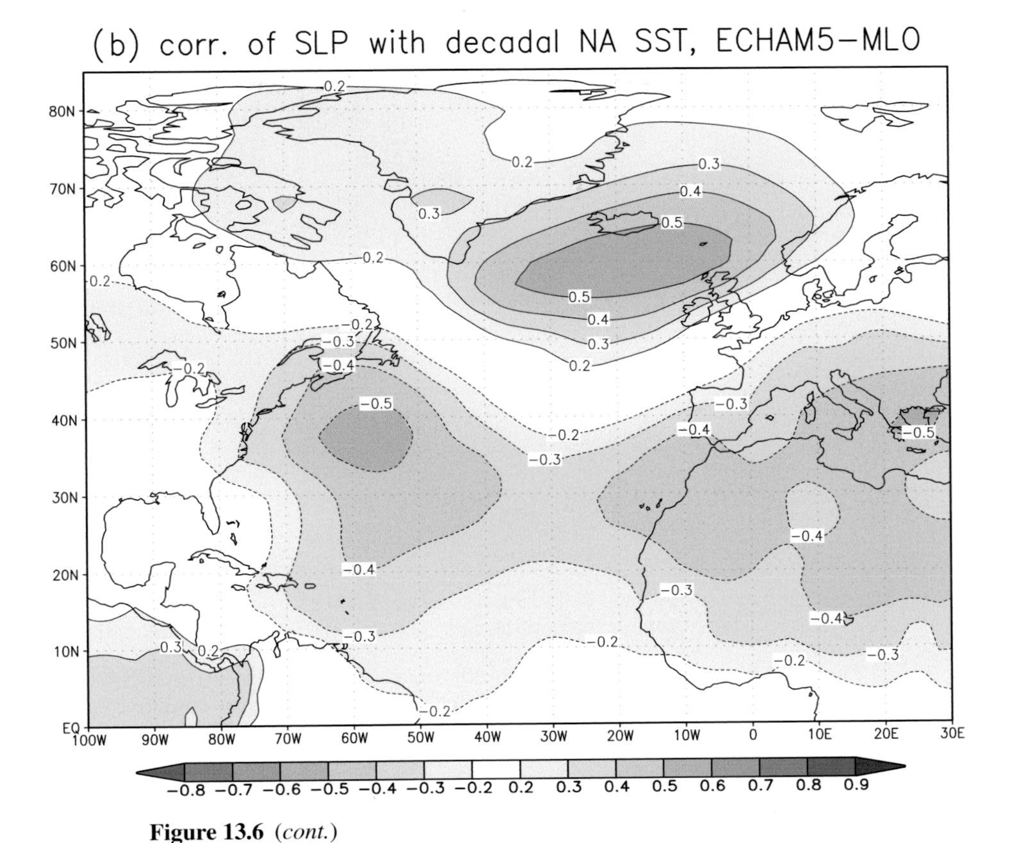

Figure 13.6 (*cont.*)

The time series of North Atlantic THC of the control integration and the predictability experiments are shown together with the predictability in Figure 13.7(a). The skill in predicting the North Atlantic THC is clearly better than that of the damped persistence forecast and exceeds the 95% significance level over the whole prediction period of 20 years. Jungclaus *et al.* (2005) analysed the mechanisms of the THC variability in the coupled model ECHAM5/MPI-OM. They found that the strength of the THC is related to the convective activity in the deep water formation regions, most notably the Labrador Sea, which is sensitive to freshwater anomalies from the Arctic. The same North Atlantic SST index as before (40–60° N and 50–10° W) is used to analyse the SST predictability (Figure 13.7b). The skill in predicting the North Atlantic SST is significant at the 95% significance level over the whole prediction period of 20 years and comparable to that of the North Atlantic THC. Our predictability experiments indicate that the North Atlantic THC and SST are predictable even at multidecadal timescales. We investigated also whether predictability exists in atmospheric quantities (not shown). The results are less impressive than those for THC and SST, but predictability in sea level pressure, for instance, exists

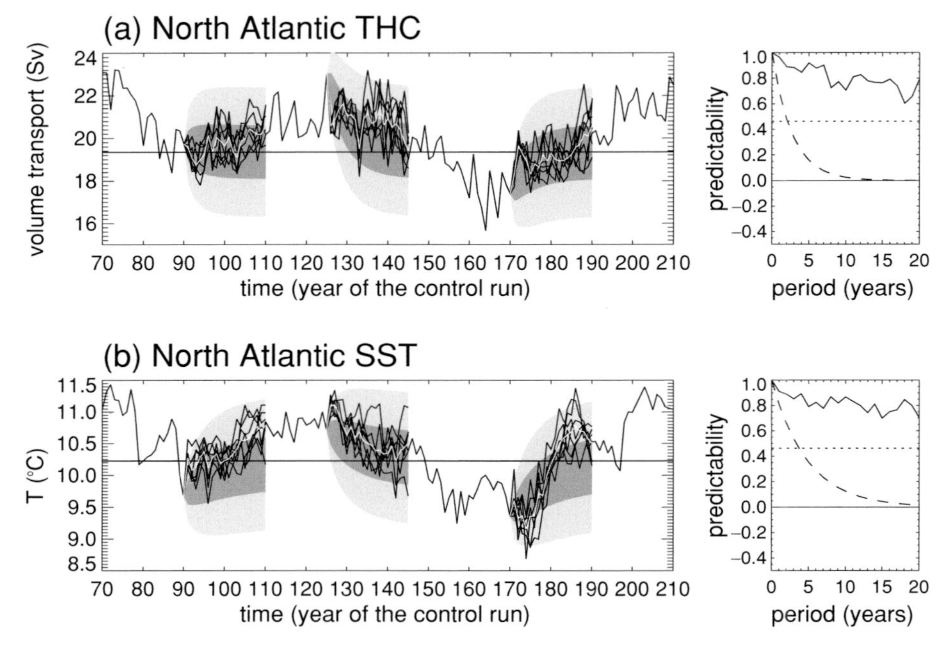

Figure 13.7 (a) (Left) Annual mean North Atlantic THC for years 70 to 210 of the control integration (thin black lines); ensemble forecast experiments initialised at the end of the years 90, 125, and 170 (thick black lines); and the ensemble means (white lines). The results of the statistical forecast method of damped persistence are shown as the range expected to contain 90% and 50% of the values from infinite size ensembles of noise driven AR-1 random processes (light and dark grey, respectively). (Right) Predictability of the North Atlantic THC averaged over the three ensemble experiments (solid curve), with the damped persistence forecast (dashed) as a function of the prediction period. Additionally, the 95% significance level according to an F test is dotted. (b) As in (a), but for North Atlantic SST. Note that the changes in the North Atlantic THC and SST indices are predictable a few decades ahead (Pohlmann *et al.*, 2004).

for at least one or two years. Interestingly, the multidecadal variability discussed here projects most strongly onto the EAP and not onto the NAO. Consequently, we find slightly higher predictability for the EAP relative to the NAO.

The predictability of surface temperature (i.e. SST over the oceans and land surface temperature elsewhere) is shown as maps averaged over the three ensemble experiments and the first and second prediction decade (Figure 13.8). Averaged over the first prediction decade the most predictable regions are in the North Atlantic, Nordic Seas, and Southern Ocean. The predictability of the second decade is everywhere less significant than that of the first decade. In this period, predictability remains significant in the North Atlantic and the Nordic Seas. In these areas, the ocean exhibits multidecadal SST predictability. Over the ocean, the predictability of surface air temperature (SAT) is very similar to that of SST. Over land, however, there is little

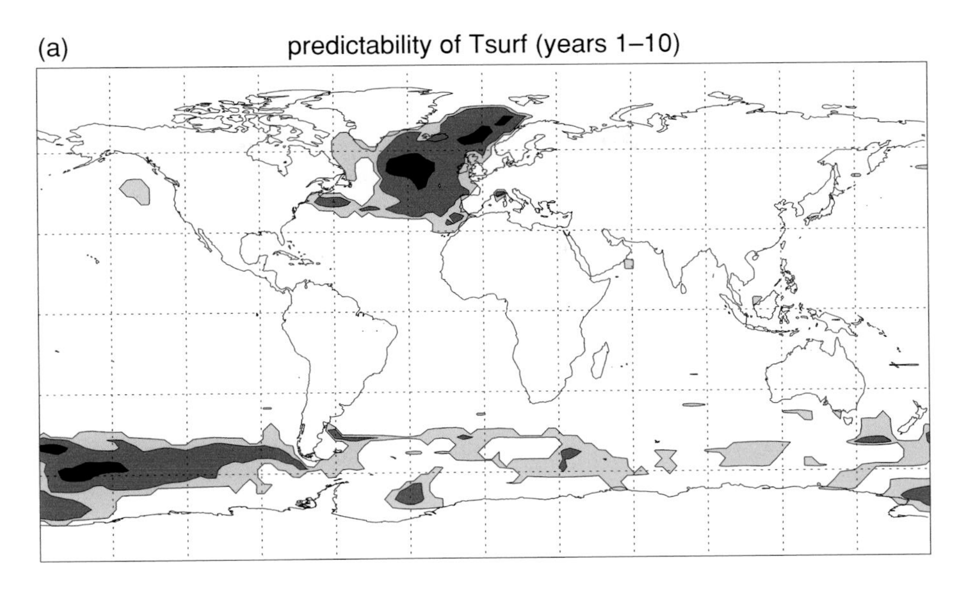

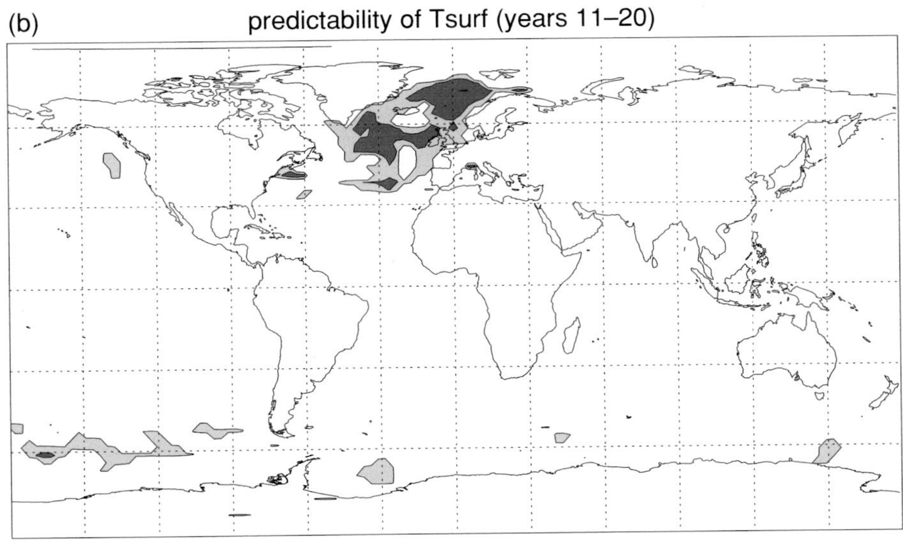

Figure 13.8 Predictability of surface temperature (Tsurf) averaged over the three ensemble experiments and the (a) first and (b) second prediction decade. The shaded values are significant on the 90% significance level according to an F-test (Pohlmann *et al.*, 2004).

evidence of decadal predictability of SAT except for some small maritime-influenced regions of Europe.

The predictability of the ECHAM5/MPI-OM is compared with that of other coupled climate models in the study of Collins *et al.* (2006). The experiments were performed as part of the European Union's project 'Mechanisms and Predictability of Decadal Fluctuations in Atlantic-European Climate' (PREDICATE). The results of the CGCM intercomparison study show that variations in the North Atlantic THC are potentially predictable on interannual to decadal timescales and also North Atlantic SAT, albeit with potential skill levels which are less than those seen for THC variations. Although the mechanisms behind the decadal to multidecadal variability in the Atlantic sector are still controversial, there is some consensus that the long-term multidecadal variability is driven by variations in the THC. In general, models with greater decadal THC variability have greater levels of potential predictability. The far more pertinent question is, of course, that of the predictability of surface climate variations over land. The predictability measure used here does not reveal robustly predictable land signals. Although the predictability of surface climate is mostly restricted to ocean regions, the probability density functions of surface temperatures over land are shown to be affected in some models by decadal variability of the large-scale oceanic circulation (Collins and Sinha, 2003; Pohlmann *et al.*, 2006). Thus, some useful decadal predictability of economic value may exist in the Atlantic sector. To exploit this decadal predictability, however, a suitable ocean observing system must be installed, since the memory of the climate system resides in the North Atlantic Ocean. In particular, the North Atlantic THC should be monitored carefully, since its variations are most interesting in the light of decadal predictability.

13.6 Greenhouse warming simulations

The SST anomaly pattern associated with the THC variability can also be used as a fingerprint to detect future changes in THC intensity. Many authors have reported a weakening of the THC in global warming simulations (see e.g. Rahmstorf, 1999; Cubasch *et al.*, 2001) which may have strong impacts on the climate of the North Atlantic/European sector. However, it is unclear how such a change in THC intensity can be observed. Our model results suggest that an easy means to monitor the THC strength can be obtained simply by observing Atlantic SSTs. However, in the presence of global warming a differential SST index which measures the contrast between the North and South Atlantic has to be used. In order to test this hypothesis, an additional ensemble of three greenhouse warming simulations was conducted (Figure 13.9a; colour plate). For this purpose the climate model was initialised from different states of the control integration that are 30 years apart from each other (years 30, 60 and 90), and the atmospheric CO_2 content was increased by 1% per year (compound). The results are analysed for the longest integration (110 years), initialised in year 60 in which the CO_2 concentration triples, and they confirm the hypothesis that changes in THC strength can be seen in the differential Atlantic SST index (Figure 13.9b).

The results also show that the THC evolution in the greenhouse warming simulations closely follows that of the control run for some decades before diverging

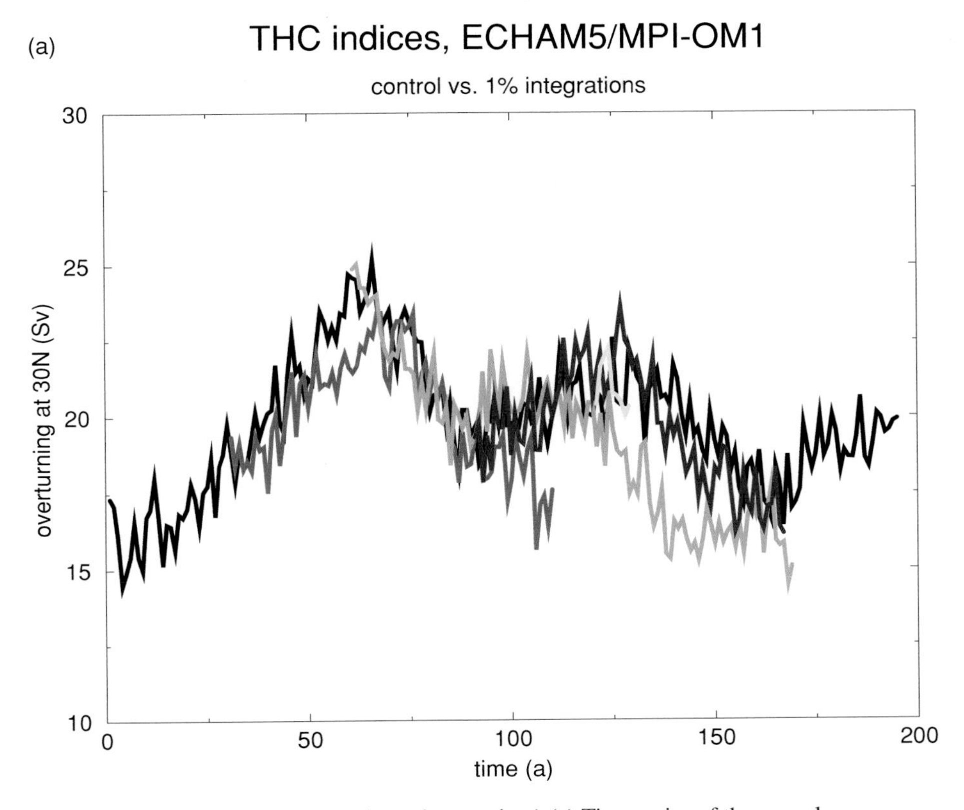

Figure 13.9 (See also colour plate section.) (a) Time series of the annual mean anomalies of the maximum overturning (Sv) at 30°N in the control integration (black line) and in the greenhouse warming simulations (coloured lines). Note that the evolutions in the greenhouse warming simulations closely follow those of the control integration for several decades, indicating a very high level of THC predictability. (b) Time series of the simulated annual mean Atlantic dipole SST index (dashed line) and annual mean anomalies of the maximum overturning at 30° N (solid line) in the longest of the greenhouse warming simulations. The dipole SST index is defined as the difference between North Atlantic (40–60° N and 50–10° W) and South Atlantic (10–40° S and 30° W–10° E) SST. Note that SST and overturning are highly correlated at timescales beyond several years in the greenhouse warming simulation. This implies that future changes in the THC can be monitored by observing SSTs. The time series were normalised with their respective standard deviations (Latif *et al.*, 2004).

from it (Figure 13.9a). This behaviour is markedly different from that of global mean surface temperature which exhibits a rather monotonic increase in all members. This implies a strong sensitivity to initial conditions but also a great deal of predictability of the multidecadal variability in the North Atlantic, provided the initial state is well known. These results are consistent with our classical predictability experiments discussed above. Furthermore, our results imply that anthropogenically forced changes in THC strength may be masked for quite a long time by the presence

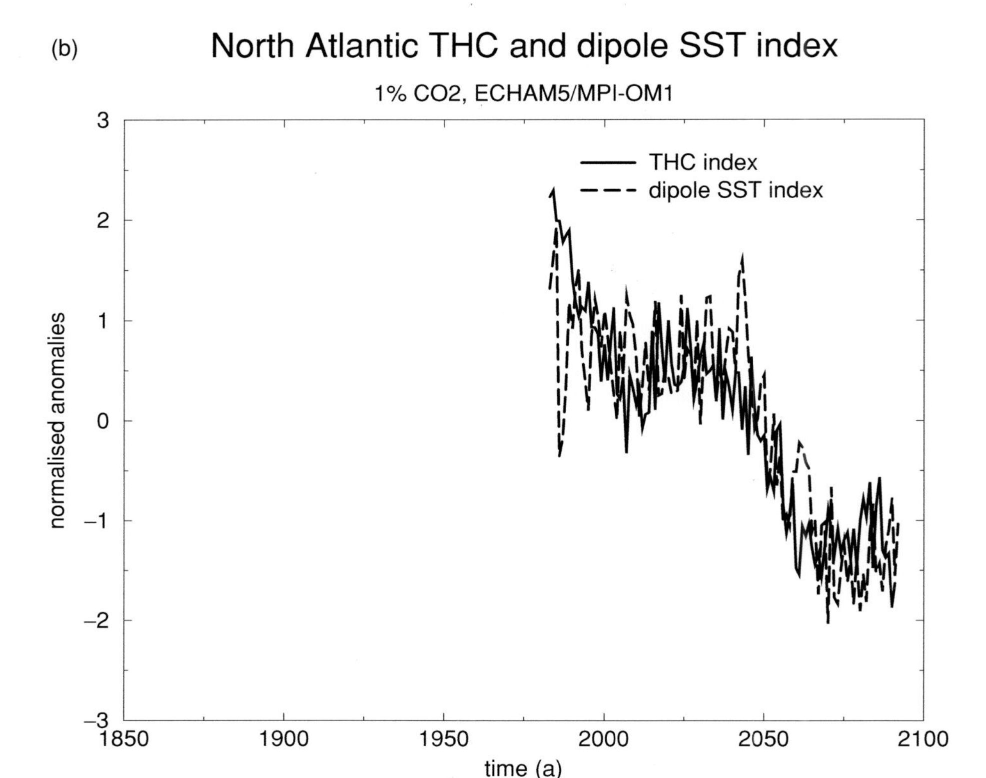

Figure 13.9 *(cont.)*

of the internal multidecadal variability. The next several decades may therefore be dominated by the internal multidecadal variability, and we have to consider a joint initial/boundary value problem when assessing how the THC will evolve during this century. Greenhouse gas simulations should therefore be properly initialised using present-day ocean conditions and they should be conducted in ensemble mode to assess the uncertainty.

13.7 Summary and discussion

A close relationship exists between multidecadal variations in the strength of the North Atlantic thermohaline circulation and Atlantic sea surface temperature. This has been shown by means of simulations with a global climate model which realistically simulates the multidecadal SST variability in the North Atlantic. The same relationship was found in another climate model (ECHAM4/OPYC, not shown) which was used by Latif *et al.* (2000), for instance, to discuss the stability of the THC in a greenhouse warming simulation. The link between THC strength and SST can be exploited to reconstruct past changes and to monitor future changes in the strength of the THC using SSTs only. Since SSTs are observable from space using passive microwave techniques, they are readily available in near real-time with good spatial and temporal coverage. Thus, the low-frequency variability of a major component of the large-scale ocean circulation, the North Atlantic thermohaline circulation, can be determined from the already existing ocean observing system.

These results are also important in view of the predictability of the North Atlantic climate system at decadal timescales. As shown by analysing the results of the classical predictability experiments and the ensemble of greenhouse warming simulations, the North Atlantic thermohaline circulation exhibits a relatively high degree of predictability at decadal timescales, which is consistent with earlier predictability studies (Griffies and Bryan, 1997; Grötzner *et al.*, 1999). Predictability, however, depends on the availability of the initial state (see chapters in this volume by Anderson, Hagedorn *et al.*, Lorenz, and Palmer). The relationship between variations in THC and SST found in our climate model can also be exploited for predictability purposes, as the initial oceanic state can be estimated from the history of SST. A simple statistical scheme, for instance, can be envisaged to reconstruct the multidecadal variations in the oceanic density structure by projecting the multidecadal SST fluctuations onto the three-dimensional oceanic density field using a model-derived statistical transfer function. These reconstructions can then easily be used in a data assimilation procedure to produce an ocean analysis from which the decadal predictions can be initialised. If skilful, such predictions of decadal changes in the THC would not only be of enormous scientific interest but also of large public interest, since they would have a substantial economic value (Richardson, this volume).

However, more model studies on this subject are necessary. Here we conclude that the potential may exist to reconstruct, monitor and predict decadal changes in the North Atlantic climate using only surface observations. The situation may be similar to that of predicting the El Niño–Southern Oscillation (ENSO) phenomenon (Philander, 1990) on interannual timescales. ENSO predictability can also be achieved by using only surface information (e.g. Oberhuber *et al.*, 1998; Segschneider *et al.*, 2000).

Acknowledgements

We thank the coupled modelling group at MPI and Professor Dr Jens Meincke for fruitful discussions. This work was supported by the European Union's PREDICATE and the German Government's DEKLIM and Ocean-CLIVAR programmes. The climate simulations were conducted at the German Climate Computer Centre (DKRZ) and the Swiss Center for Scientific Computing (CSCS). We thank Drs. Michael Botzet, Andreas Roesch, Martin Wild, and Peter Tschuck for providing the climate simulations.

References

Bjerknes, J. (1964). Atlantic air-sea interaction. *Advances in Geophysics*, **10**, Academic Press, 1–82.

Broecker, W. S., D. M. Peteet and D. Rind (1985). Does the ocean-atmosphere system have more than one stable mode of operation? *Nature*, **315**, 21–6.

Collins, M. and B. Sinha (2003). Predictability of decadal variations in the thermohaline circulation and climate. *Geophys. Res. Lett.*, **6**, doi:10.1029/2002GL016504.

Collins, M., M. Botzet, A. Carril, *et al.* (2006). Interannual to decadal climate predictability: a multi-perfect-model-ensemble study. *J. Climate*, in press.

Cubasch, U., G. A. Meehl, G. J. Boer, *et al.* (2001). Projections of future climate change. In *Climate Change: The Scientific Basis*, pp. 525–82. Cambridge University Press.

Delworth, T. L. and M. E. Mann (2000). Observed and simulated multidecadal variability in the Northern Hemisphere. *Clim. Dynam.*, **16**, 661–76.

Delworth, T. L., S. Manabe and R. J. Stouffer (1993). Interdecadal variations of the thermohaline circulation in a coupled ocean-atmosphere model. *J. Climate*, **6**, 1993–2011.

Deser, C. and M. L. Blackmon (1993). Surface climate variations over the North Atlantic during winter: 1900–1989. *J. Climate*, **6**, 1743–53.

Dommenget, D. and M. Latif (2002). Analysis of observed and simulated SST spectra in midlatitudes. *Clim. Dynam.*, **19**, 277–88.

Folland, C. K., D. E. Parker and F. E. Kates (1984). Worldwide marine temperature fluctuations 1856–1981. *Nature*, **310**, 670–73.

Folland, C. K., T. N. Palmer and D. E. Parker (1986). Sahel rainfall and world wide sea temperatures. *Nature*, **320**, 602–06.

Folland, C. K., D. E. Parker, A. W. Colman and R. Washington (1999). Large scale models of ocean surface temperature since the late nineteenth century. In *Beyond El Niño*, ed. A. Navarra, pp. 73–102. Springer.

Giorgetta, M. A., E. Manzini and E. Roeckner (2002). Forcing of the quasi-biennial oscillation from a broad spectrum of atmospheric waves. *Geophys. Res. Lett.*, **29**(8), 10.1029–32.

Griffies, S. M. and K. Bryan (1997). Ensemble predictability of simulated North Atlantic inter-decadal variability. *Science*, **275**, 181–4.

Grötzner, A., M. Latif, A. Timmermann and R. Voss (1999). Interannual to decadal predictability in a coupled ocean-atmosphere general circulation model. *J. Climate*, **12**, 2607–24.

Hasselmann, K. (1976). Stochastic climate models. I: Theory. *Tellus*, **28**, 473–85.

Hegerl, G., K. Hasselmann, U. Cubasch, *et al.* (1997). Multi-fingerprint detection and attribution analysis of greenhouse gas, greenhouse gas-plus aerosol and solar forced climate change. *Clim. Dynam.*, **13**, 613–34.

Hurrell, J. W. (1995). Decadal trends in the North Atlantic Oscillation, regional temperatures and precipitation. *Science*, **269**, 676–9.

Jungclaus, J. H., H. Haak, M. Latif and U. Mikolajewicz (2005). Arctic-North Atlantic interactions and multidecadal variability of the meridional overturning circulation. *J. Climate*, **18**, 4013–31.

Kushnir, Y. (1994). Interdecadal variations in North Atlantic sea surface temperature and atmospheric conditions. *J. Climate*, **7**, 141–57.

Latif, M., E. Roeckner, U. Mikolajewicz and R. Voss (2000). Tropical stabilisation of the thermohaline circulation in a greenhouse warming simulation. *J. Climate*, **13**, 1809–13.

Latif, M., E. Roeckner, M. Botzet, *et al.* (2004). Reconstructing, monitoring, and predicting multidecadal-scale changes in the North Atlantic thermohaline circulation with sea surface temperature. *J. Climate*, **17**, 1605–14.

Lindzen, R. S. and S. Nigam (1987). On the role of sea surface temperature gradients in forcing low level winds and convergence in the tropics. *J. Atmos. Sci.*, **44**, 2418–36.

Manabe, S. and R. J. Stouffer (1988). Two stable equilibria of a coupled ocean-atmosphere model. *J. Climate*, **1**, 841–66.

(1996). Low-frequency variability of surface temperature in a 1000-year integration of a coupled atmosphere-ocean-land surface model *J. Climate*, **9**, 376–93.

(1999). The role of thermohaline circulation in climate. *Tellus*, **51**, 91–109.

Marsland, S. J., H. Haak, J. H. Jungclaus, M. Latif and F. Röske (2003). The Max-Planck-Institute global ocean/sea ice model with orthogonal curvilinear coordinates. *Ocean Model.*, **5**, 91–127.

Oberhuber, J., E. Roeckner, M. Christoph, M. Esch and M. Latif (1998). Predicting the '97 El Niño event with a global climate model. *Geophys. Res. Lett.*, **25**, 2273–6.

Palmer, T. N. and Z. Sun (1985). A modelling and observational study of the relationship between sea surface temperature in the north-west Atlantic and the atmospheric general circulation. *Quart. J. Roy. Meteor. Soc.*, **111**, 947–75.

Park, W. and Latif, M. (2005). Ocean dynamics and the nature of air–sea interactions over the North Atlantic at decadal timescales. *J. Climate*, **18**, 982–95.

Philander, S. G. H. (1990). *El Niño, La Niña and the Southern Oscillation*. Academic Press.

Pohlmann, H., M. Botzet, M. Latif, A. Roesch, M. Wild and P. Tschuck (2004). Estimating the decadal predictability of a coupled AOGCM. *J. Climate*, **22**, 4463–72.

Pohlmann, H., F. Sienz and M. Latif (2006). Influence of the multidecadal Atlantic meridional overturning circulation variability on European climate. *J. Climate*, in press.

Rahmstorf, S. (1999). Shifting seas in the greenhouse? *Nature*, **399**, 523–4.

Roeckner, E., G. Bäuml, L. Bonaventura, *et al.* (2003). *The Atmospheric General Circulation Model ECHAM5. I: Model Description*. MPI Report 349. Max-Planck-Institut für Meteorologie, Bundesstr. 53, 20146 Hamburg, Germany.

Schiller, A., U. Mikolajewicz and R. Voss (1997). The stability of the North Atlantic thermohaline circulation in a coupled ocean-atmosphere general circulation model. *Clim. Dynam.*, **13**, 325–47.

Segschneider, J., D. L. T. Anderson and T. N. Anderson (2000). Towards the use of altimetry for operational seasonal forecasting. *J. Climate*, **13**, 3115–28.

Timmermann, A., M. Latif, R. Voss and A. Grötzner (1998). Northern Hemisphere interdecadal variability: a coupled air-sea model. *J. Climate*, **11**, 1906–31.

Visbeck, M., D. Stammer, J. Toole, *et al.* (1998). *Atlantic Climate Variability Experiment Prospectus*. Report available from LDEO, Palisades, NY, 49 pp.

WOCE (2001). Objective 8 – To determine the important processes and balances for the dynamics of the general circulation. Contr. by N. Hogg, J. McWilliams, P. Niiler, J. Price. *US WOCE Implementation Report*, **13**, 55.

(a) Control ensemble

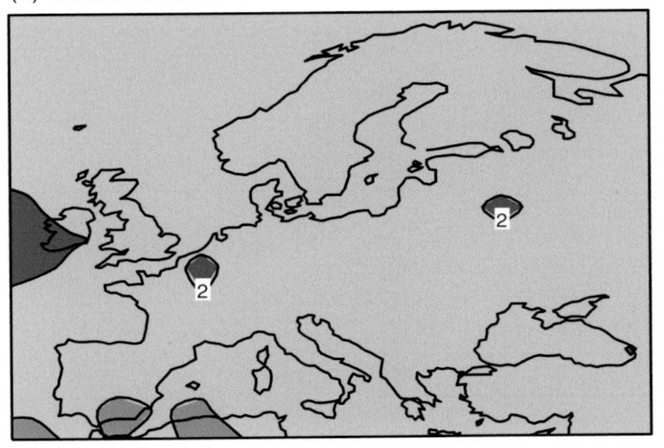

(b) Greenhouse ensemble

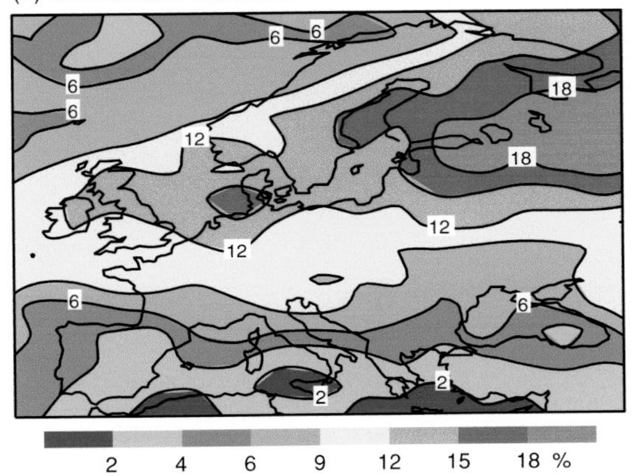

(c) Greenhouse / Control

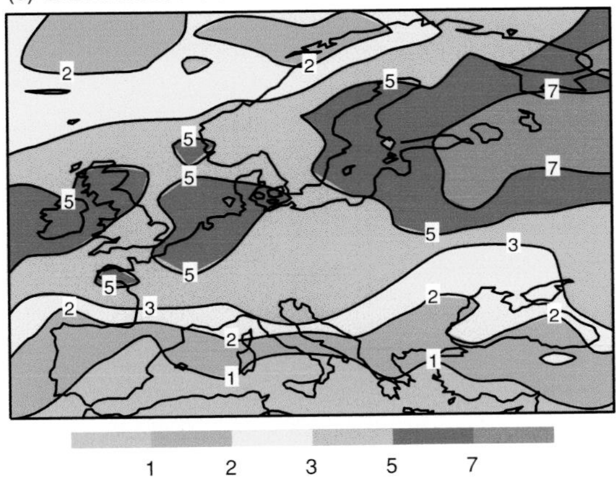

Plate 1.11 The changing probability of extreme seasonal precipitation for Europe in boreal winter. (a) The probability (in %) of a 'very wet' winter defined from the control CMIP2 multimodel ensemble with twentieth-century levels of CO_2 and based on the event E: total boreal winter precipitation greater than the mean plus two standard deviations. (b) The probability of E but using data from the CMIP multimodel ensemble with transient increase in CO_2 and calculated around the time of CO_2 doubling (years 61–80 from present). (c) The ratio of values in (b) to those in (a), giving the change in the risk of a 'very wet' winter arising from human impact on climate. From Palmer and Räisänen (2002).

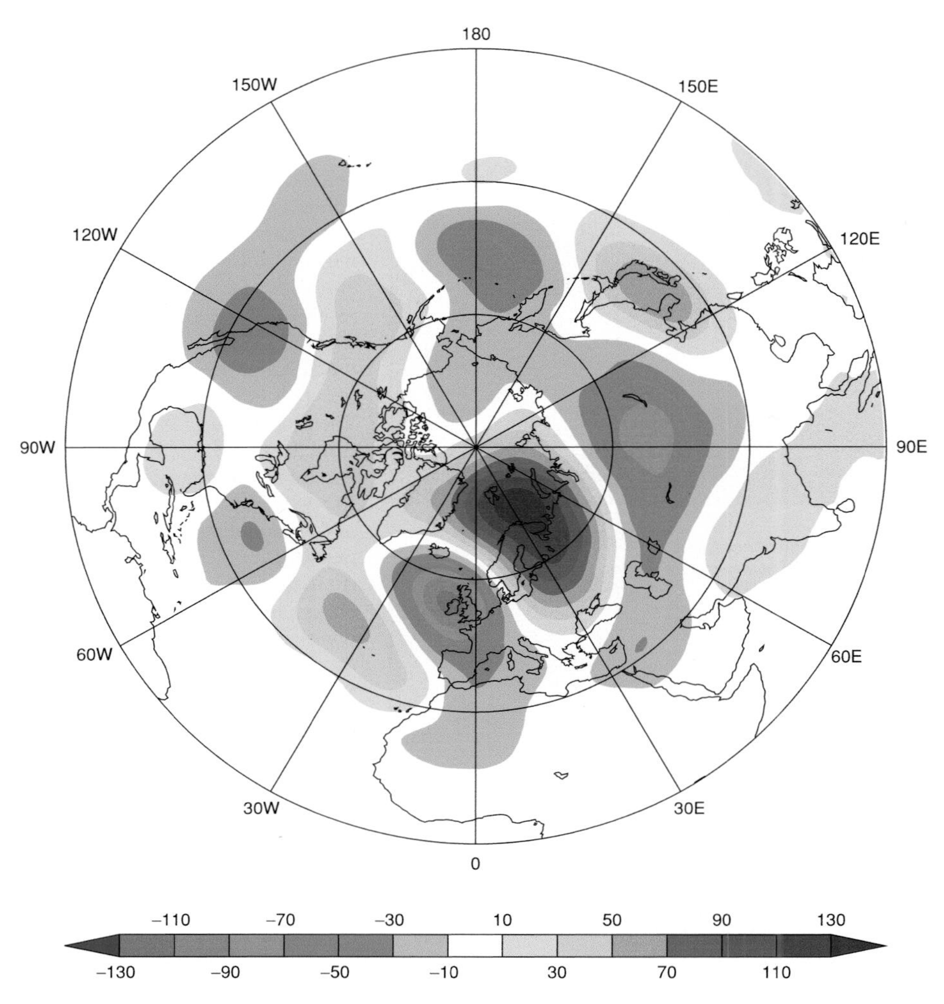

Plate 2.4 September–November 2000 300 hPa geopotential height anomalies from climate. From Blackburn and Hoskins (personal communication).

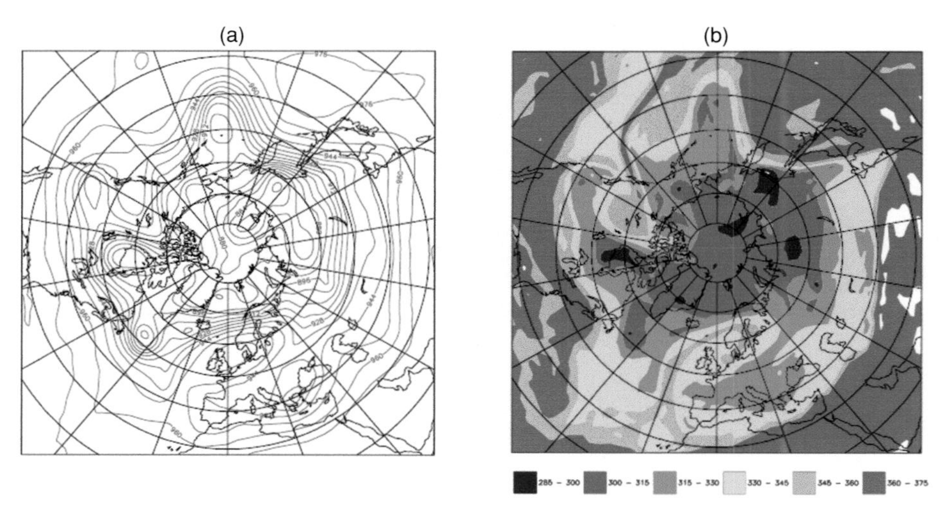

Plate 2.6 The block of 21 September 1998. Shown are the 250 hPa geopotential height field and the θ on PV2 field. From Pelly and Hoskins (2003a).

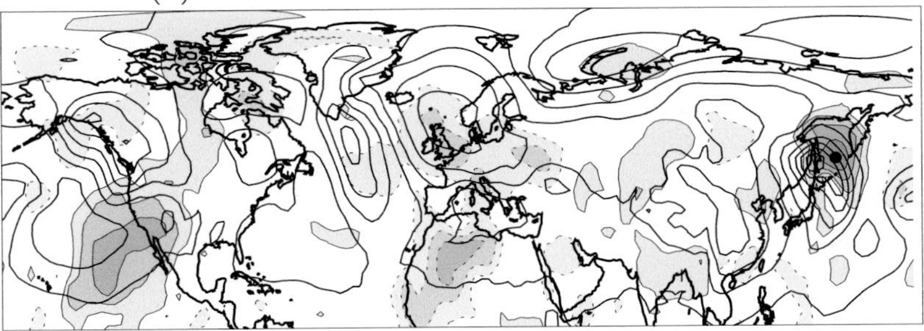

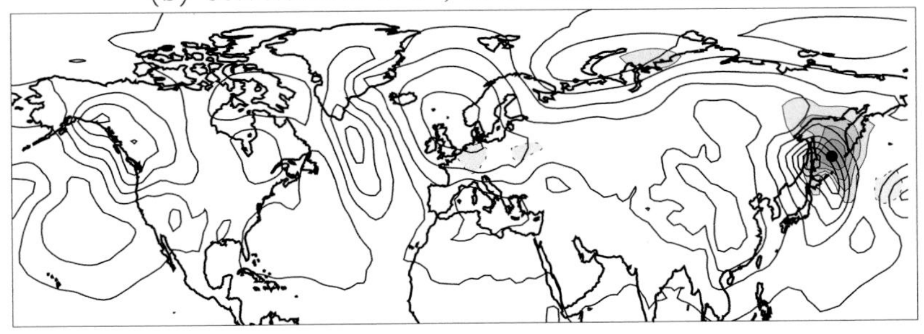

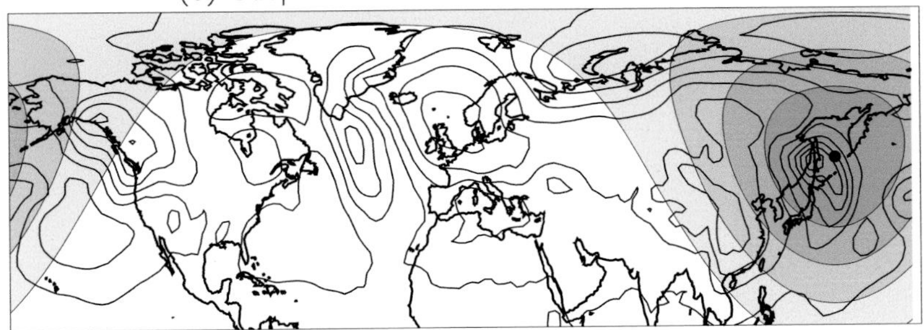

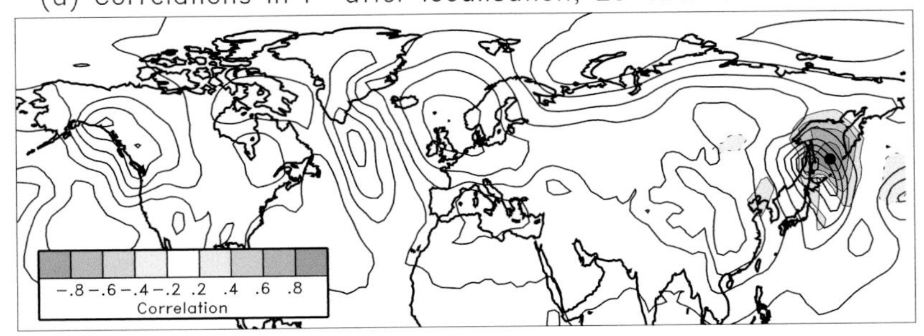

Plate 6.4 Illustration of covariance localisation. (a) Correlations of sea-level pressure directly estimated from 25-member ensemble with pressure at a point in the western Pacific (colours). Solid lines denote ensemble mean background sea-level pressure contoured every 8 hPa. (b) As in (a), but using 200-member ensemble. (c) Covariance localisation correlation function. (d) Correlation estimate from 25-member ensemble after application of covariance localisation.

Background error (shaded) and 3D- Var analysis increments

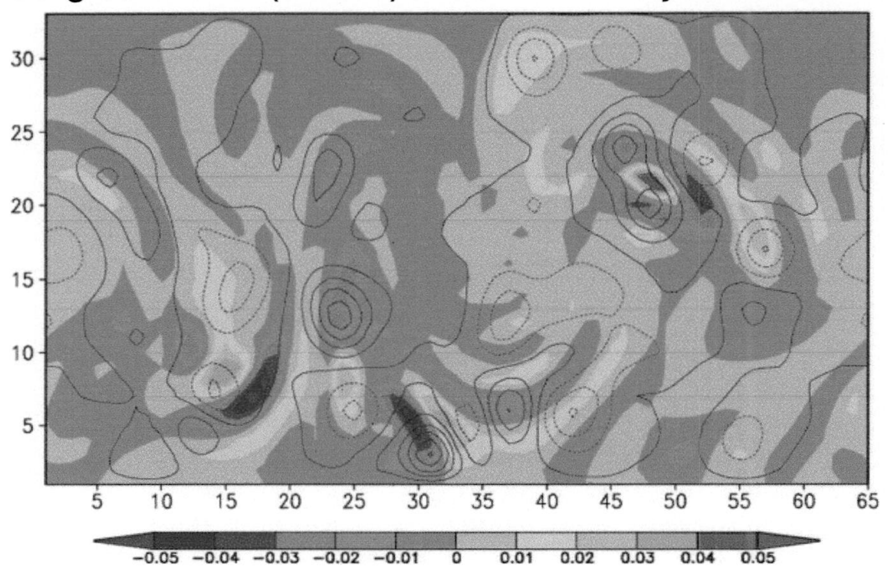

Background error (shaded) and LEKF analysis increments

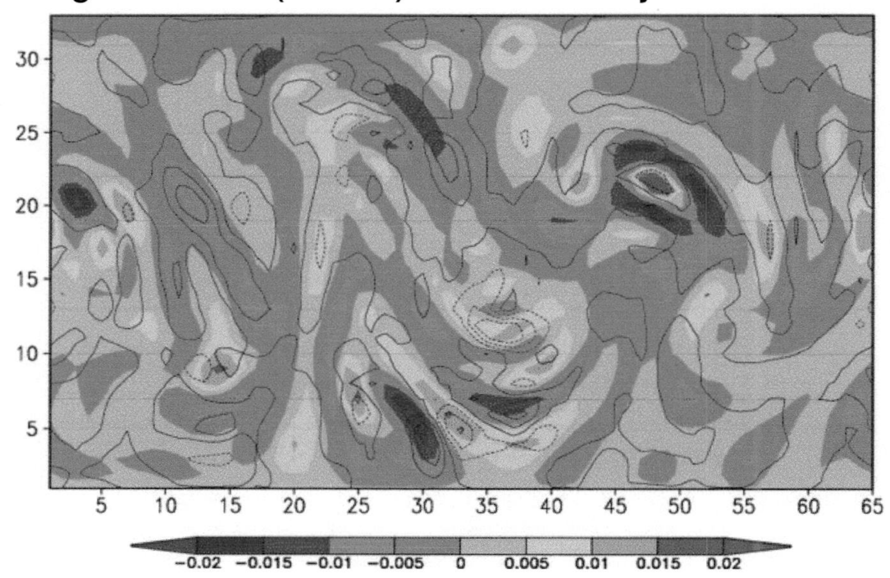

Plate 7.6 Simulation of data assimilation in a quasi-geostrophic model, assimilating potential vorticity observations at a particular day (15 June). The colours represent the 12 hr forecast (background) error and the contours the analysis corrections. (Top) 3D-Var. (Bottom) local ensemble Kalman filter. Figures courtesy of Matteo Corazza.

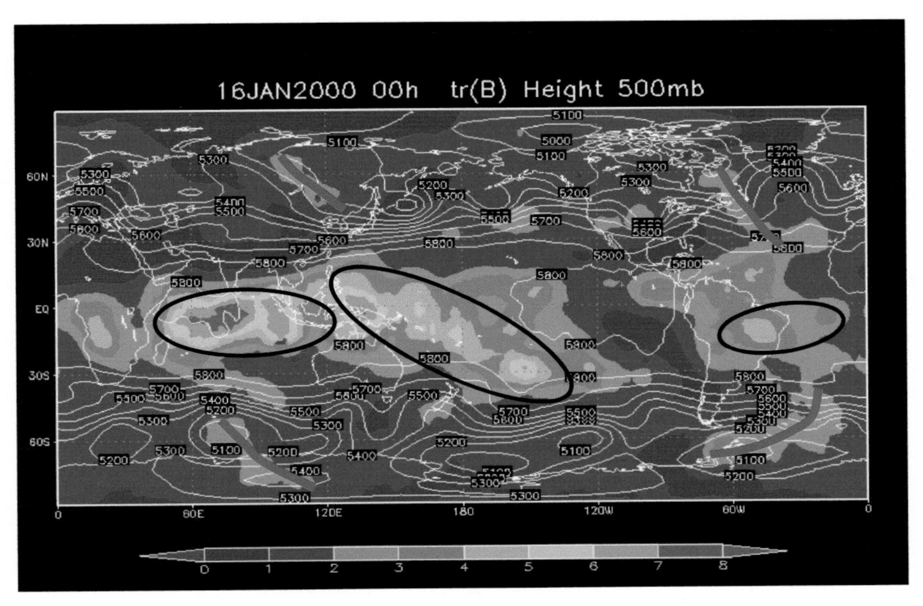

Plate 7.12 Example of a 6 hr trace of the 500 mb height forecast error covariance showing the potential use of LEKF for adaptive observations. Regions in blue and purple do not need immediate observations. Midlatitude areas marked with red have large errors but a low effective ensemble dimension, so that they are prime areas for targeting. Tropical regions with large errors (ovals), by contrast, have also large effective ensemble dimension, presumably because the error growth is dominated by convection.

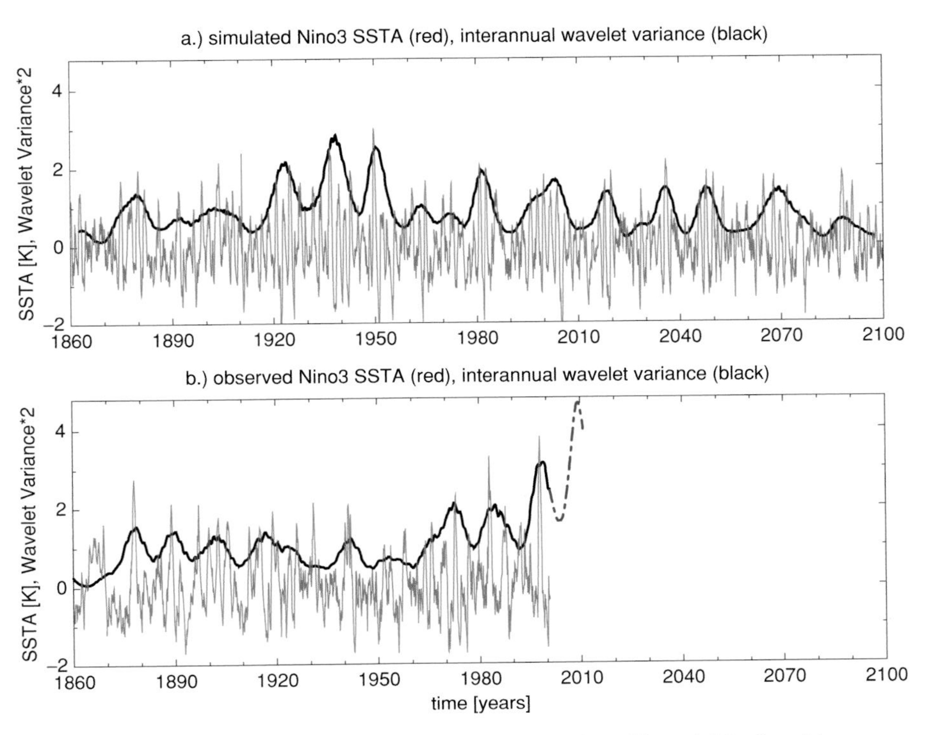

Plate 10.9 (a) NINO-3 SSTA simulated by the ECHAM4/OPYC3 model (red) and its interannual wavelet energy (black). (b) Observed NINO-3 SSTA time series (red) and its interannual wavelet energy (black), 10-year forecast of interannual wavelet energy (dashed) using non-linear prediction techniques.

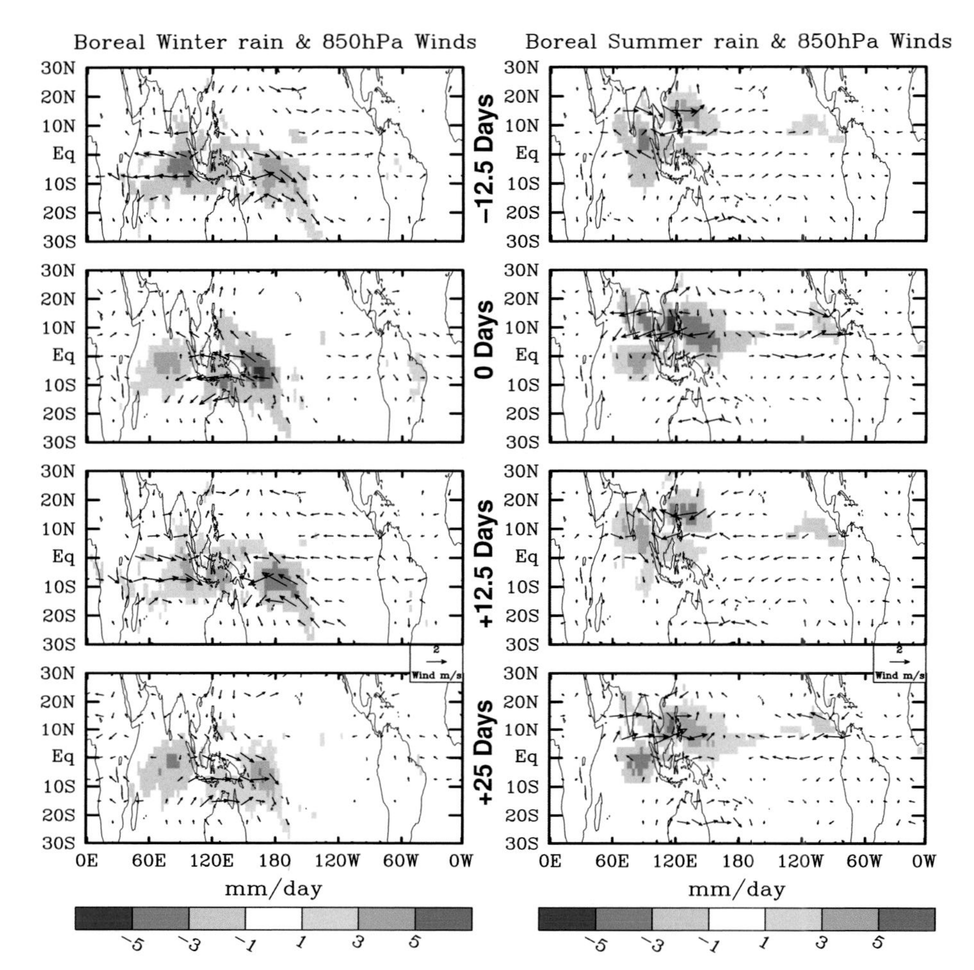

Plate 11.1 Canonical structure of an MJO event based on 5-day average (i.e. pentad) NCEP/NCAR Reanalysis (Kalnay *et al.*, 1996) and CMAP rainfall data (Xie and Arkin, 1997) from 1979 to 2000. Data were bandpassed filtered with a 30–90 day filter and then separated into boreal winter (Nov–Apr) and summer (May–Oct). Extended EOF (EEOF) analysis with +/−5 pentad lags was performed on tropical rainfall (30N to 30S, 30E to 180E) to identify the dominant 'mode' for the winter and summer separately. Composite events were constructed by selecting events if the EEOF amplitude time series exceeded 1 standard deviation ($N = 43$ (49) for winter (summer)). The resulting composites have dimensions lag (−5 to +5 pentads), latitude and longitude. In the plots above, only 4 panels of the boreal winter composite are shown, each separated by 2.5 pentads (i.e. 12.5 days). Plots show composite rainfall and 850 hPa wind vectors for (left) boreal winter and (right) boreal summer. Only values that exceed the 90% confidence limit are shown.

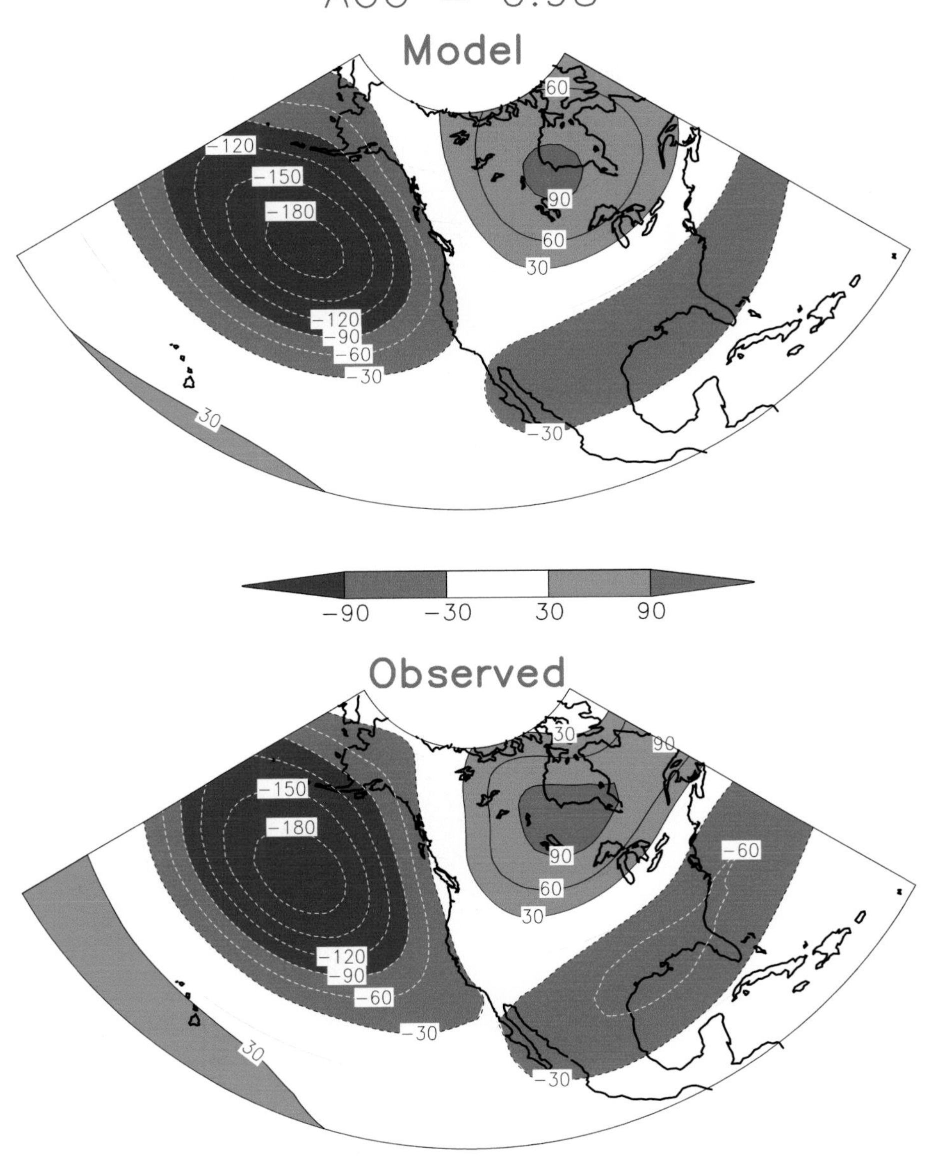

Plate 12.5 Composite, ensemble-mean AGCM simulation of January–February–March seasonal mean difference of 500 hPa geopotential height. Average of three years with warm SST anomalies in the eastern tropical Pacific (1983, 1987 and 1992) minus average of two years with cold SST anomalies (1985 and 1989). The ensemble of 10 model simulations were made with observed SST specified as lower boundary conditions and slightly different initial conditions in each ensemble member. The model used is the COLA atmospheric GCM.

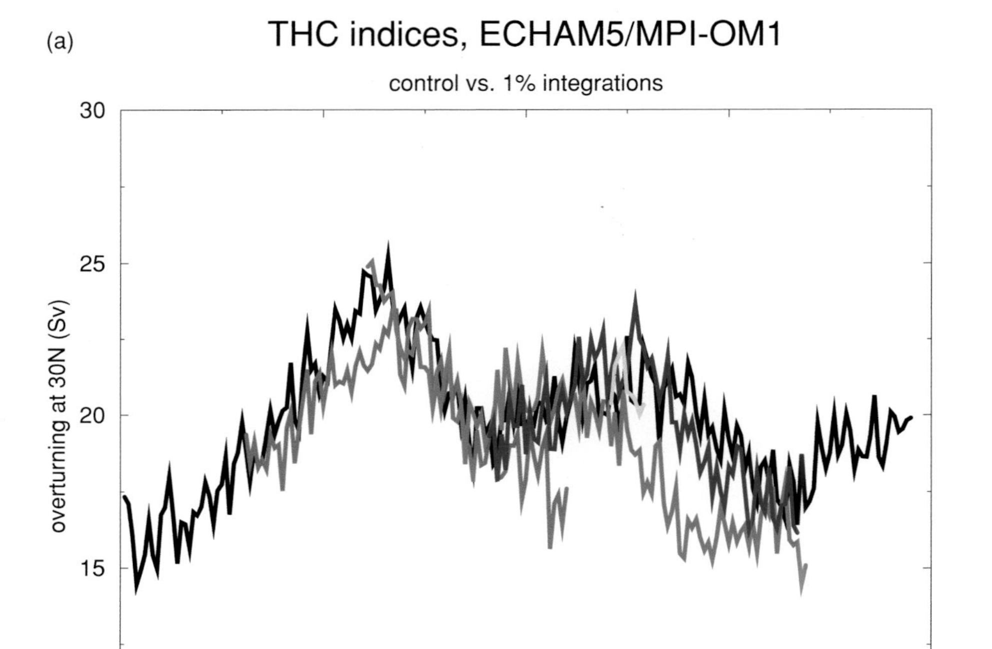
(a)
THC indices, ECHAM5/MPI-OM1
control vs. 1% integrations
overturning at 30N (Sv)
time (a)
30
25
20
15
10
0
50
100
150
200

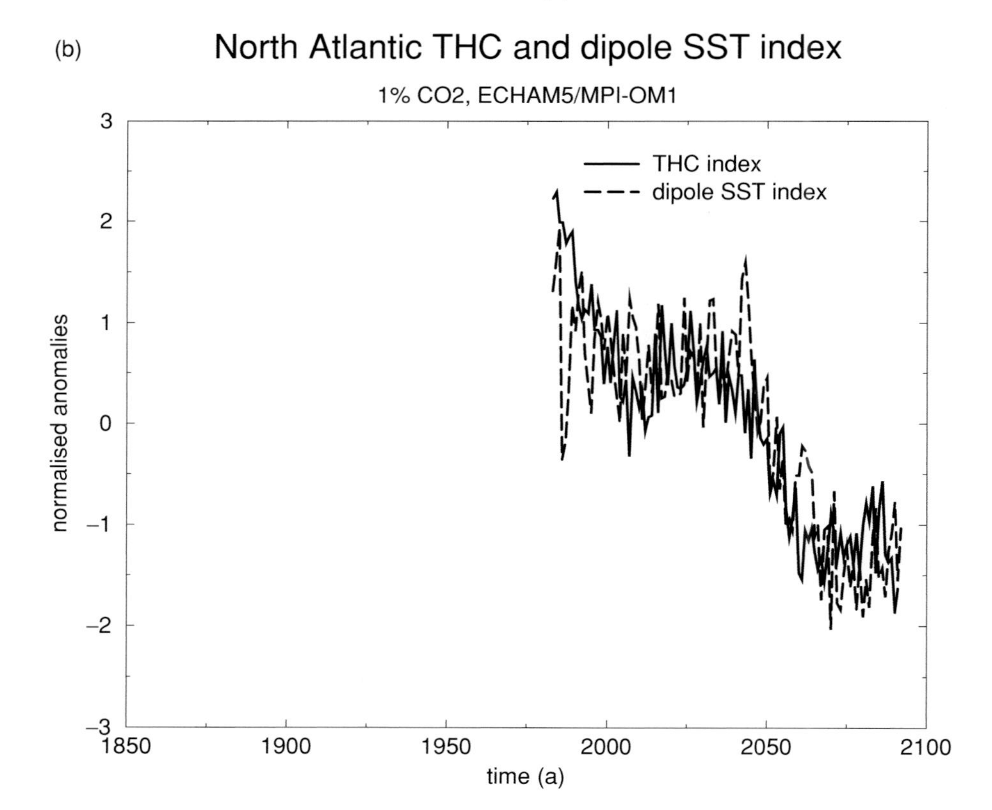
(b)
North Atlantic THC and dipole SST index
1% CO2, ECHAM5/MPI-OM1
THC index
dipole SST index
normalised anomalies
time (a)
3
2
1
0
−1
−2
−3
1850
1900
1950
2000
2050
2100

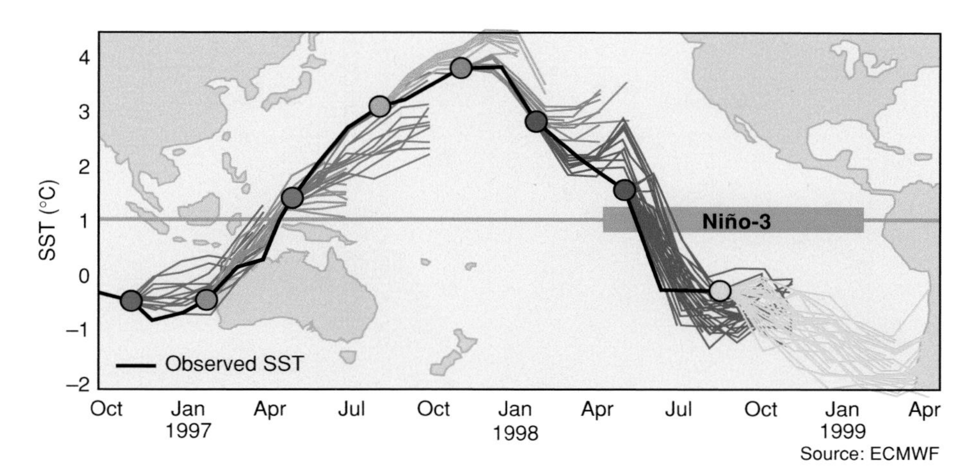

Plate 19.3 Plot of forecasts of Niño-3 for various start times throughout the large 1997–8 El Niño. Niño-3 is the region in the equatorial Pacific bounded between 5S to 5N and 150W to 90W. Different lines of the same colour indicate different ensemble members. The background indicates the location of Niño-3. This plot was produced by CLIVAR based on data from ECMWF.

←

Plate 13.9 (a) Time series of the annual mean anomalies of the maximum overturning (Sv) at 30°N in the control integration (black line) and in the greenhouse warming simulations (coloured lines). Note that the evolutions in the greenhouse warming simulations closely follow those of the control integration for several decades, indicating a very high level of THC predictability. (b) Time series of the simulated annual mean Atlantic dipole SST index (dashed line) and annual mean anomalies of the maximum overturning at 30° N (solid line) in the longest of the greenhouse warming simulations. The dipole SST index is defined as the difference between North Atlantic (40–60° N and 50–10° W) and South Atlantic (10–40° S and 30° W–10° E) SST. Note that SST and overturning are highly correlated at timescales beyond several years in the greenhouse warming simulation. This implies that future changes in the THC can be monitored by observing SSTs. The time series were normalised with their respective standard deviations (Latif *et al.*, 2004).

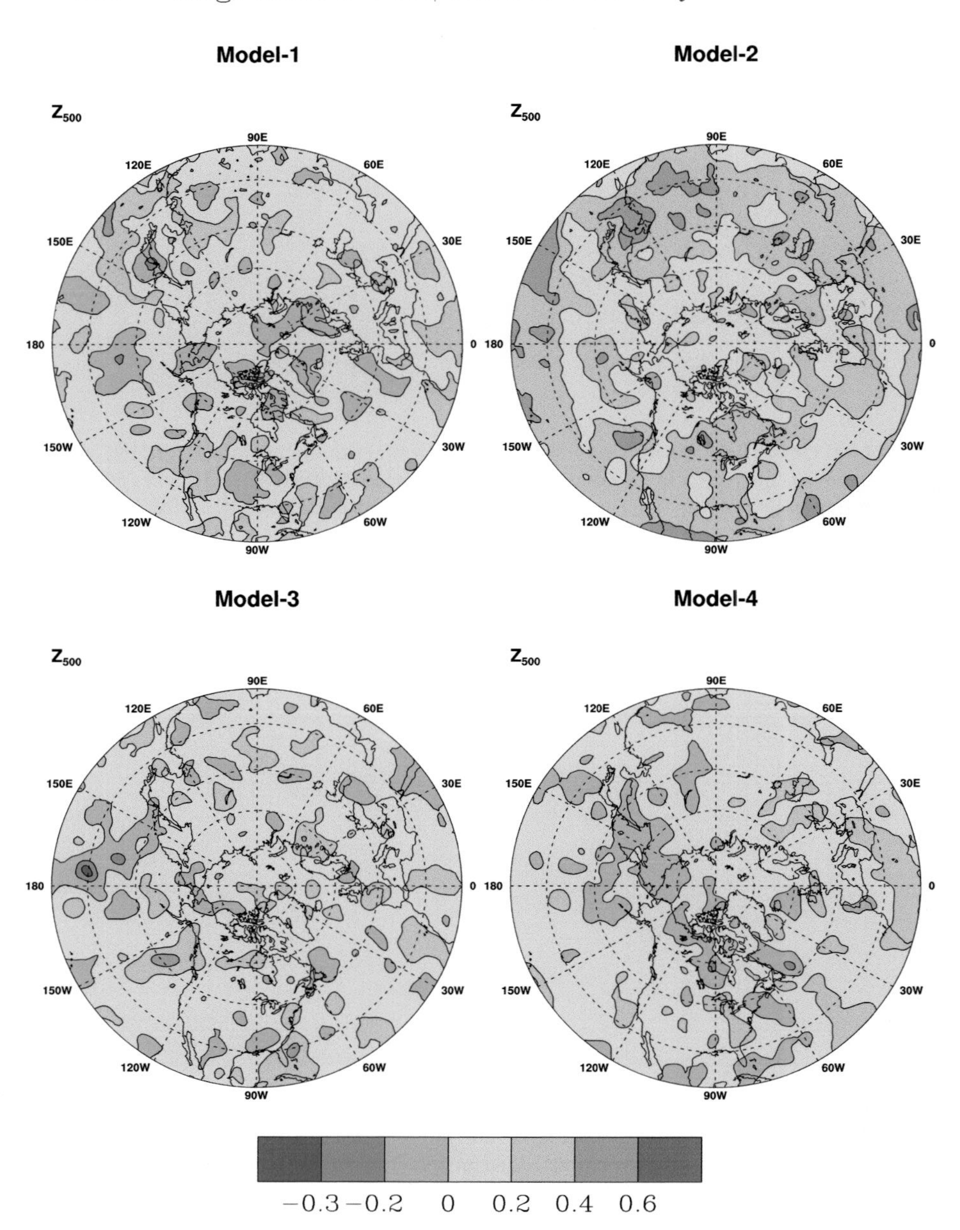

Plate 20.7 Geographical distribution of statistical weights for different member models in the northern hemisphere. Colour scale of the fractional weights is shown at the bottom.

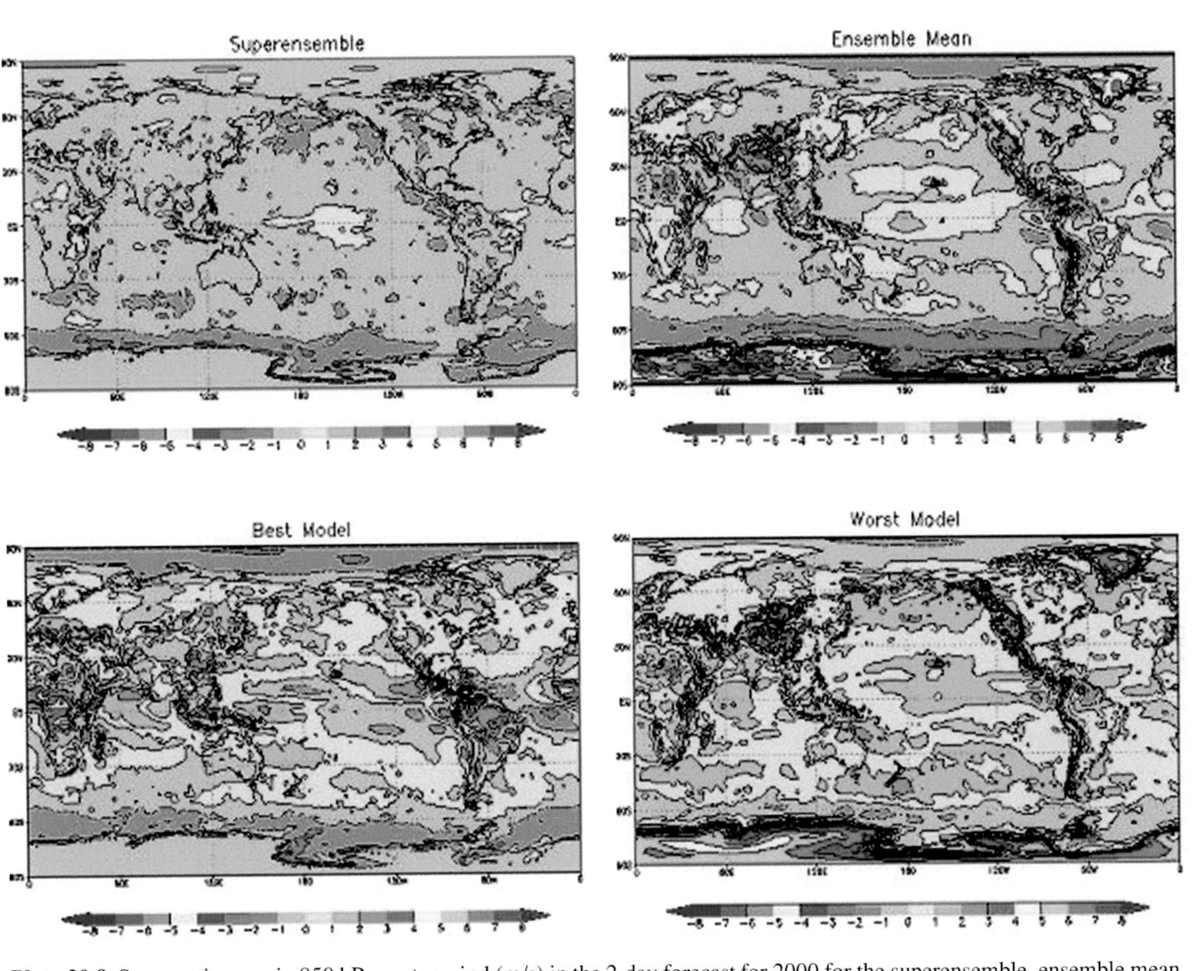

Plate 20.9 Systematic error in 850 hPa vector wind (m/s) in the 2-day forecast for 2000 for the superensemble, ensemble mean, best model, and worst model.

(a)

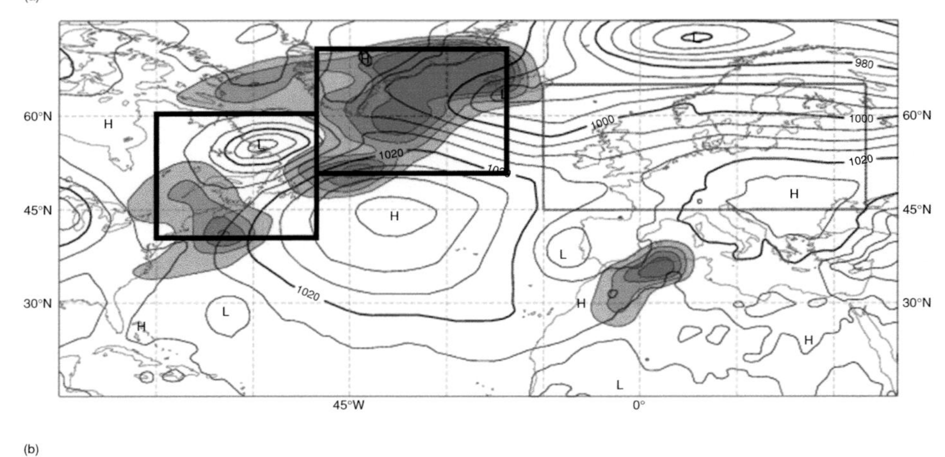

(b)

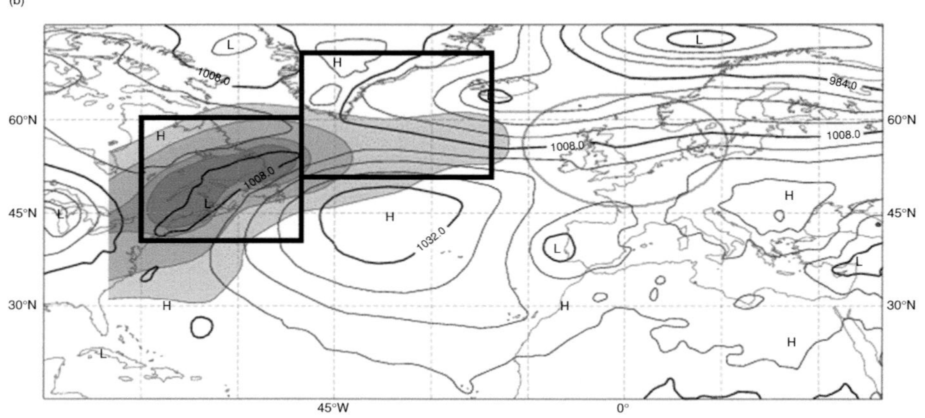

Plate 21.11 Sensitive area predictions for NA-TReC 7. (a) A prediction based on total energy singular vectors, and (b) a prediction based on the ETKF. The shaded areas are the sensitive area predictions, the darkest shade showing the most sensitive area. The sizes of the areas are 8, 4, 2 and 1×10^6 km^2. The contours show the mean sea level pressure (hPa) forecast for 66 hours in (a) and 72 hours in (b). The grey rectangle in (a) and the ellipse in (b) represent the verification area, and the bold boxes outline the region that was actually targeted.

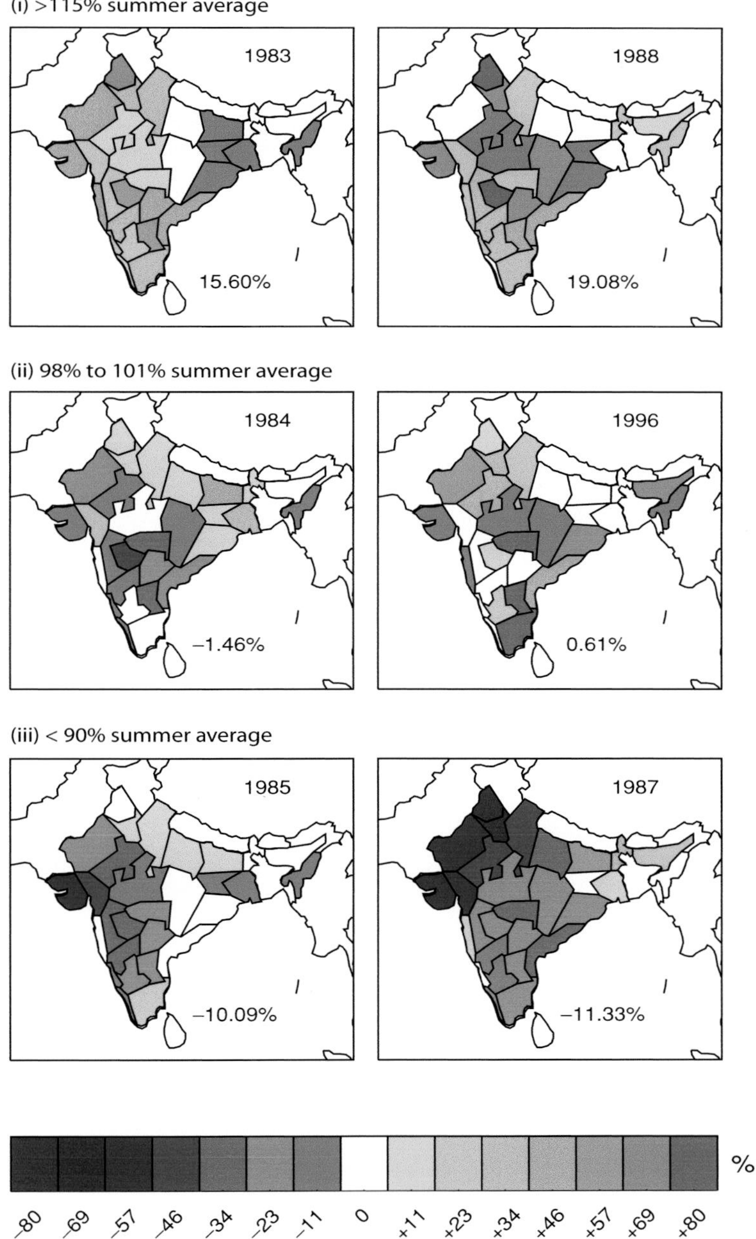

Plate 26.3 (a) Maps of departures from climatology of the total regional summer precipitation for three classifications of overall monsoon rainfall. (i) All India rainfall >115%, (ii) All India rainfall about average and (iii) All India rainfall <90%. Irrespective of the All India rainfall, there are regions of drought or flood in all classifications. Noting that most seasons fall within ±10% from normal, it is clear that even an excellent seasonal rainfall forecast will be difficult to downscale to the regional level.

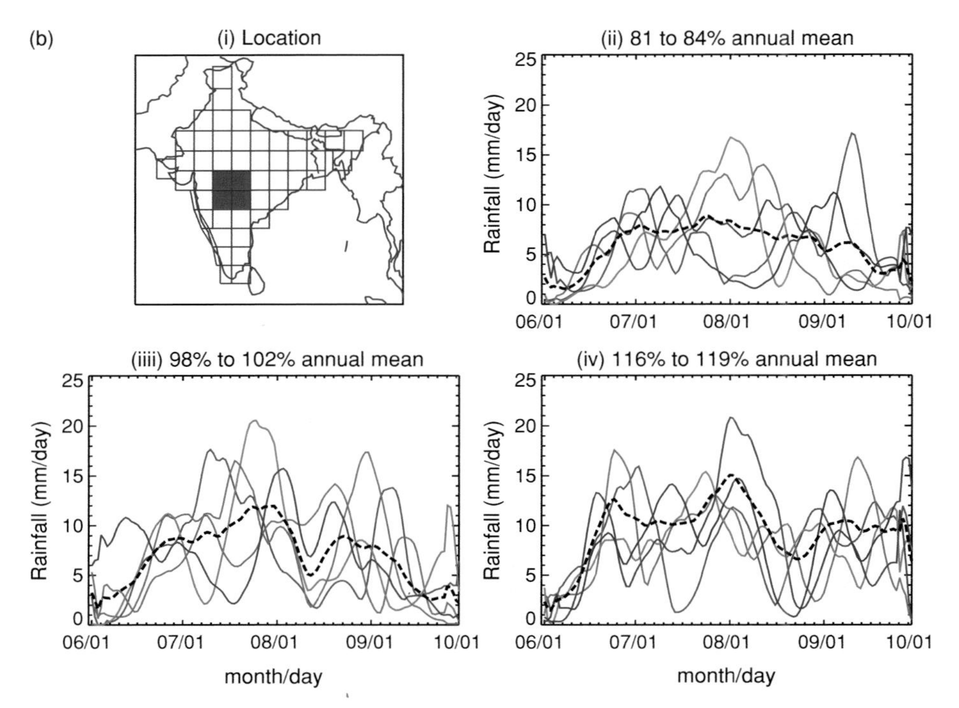

Plate 26.3 (*cont.*) (b) Schematic temporal variability of summer monsoon precipitation: a seasonal anomaly in precipitation can be made up from many temporal combinations of precipitation. Each panel shows a 20% deficit (matching the overall of 2002) distributed in different ways. Panel (i) uniformly reduced throughout the summer; (ii) a late monsoon and an early withdrawal; (iii) a normal early and late monsoon with a prolonged mid-season break (similar to 2002); and, (iv) a series of short-lived active and break periods. A perfect forecast of the overall seasonal anomaly provides no information about how the precipitation will be distributed throughout the summer season.

14

On the predictability of flow-regime properties on interannual to interdecadal timescales

Franco Molteni, Fred Kucharski
Abdus Salam International Centre for Theoretical Physics, Trieste

Susanna Corti
Institute of Atmospheric Sciences and Climate ISAC-CNR, Bologna

14.1 Introduction

Atmospheric flow regimes are usually defined as large-scale circulation patterns associated with statistical equilibria in phase space, in which the dynamical tendencies of the large-scale flow are balanced by tendencies due to non-linear interactions of high-frequency transients. The existence of states with such properties can be verified in a rigorous way in numerical simulations with simplified numerical models (as in the pioneering study of Reinhold and Pierrehumbert, 1982, or in the experiments by Vautard and Legras, 1988). By contrast, the existence of flow regimes in the real atmosphere has been strongly debated. The detection of regimes in the observational record of the upper-air field is indeed a complex task, which has been approached by a number of research groups with a variety of sophisticated statistical methods (see Section 14.3).

Although the regime classifications provided by the different observational studies were not identical, a 'core' number of regimes were consistently detected in most studies devoted to a specific spatial domain. For example, the three northern-hemisphere clusters found by Cheng and Wallace (1993) were also identified by Kimoto and Ghil (1993a), Corti *et al.* (1999) and Smyth *et al.* (1999). However, consistency does not necessarily imply statistical significance, and one may question whether the level of confidence attached to these regime classifications is sufficiently high.

Predictability of Weather and Climate, ed. Tim Palmer and Renate Hagedorn. Published by Cambridge University Press.

The search for regimes in the real atmosphere is also made complex by the fact that, unlike in simple dynamical models, the sources of energy and momentum at the lower boundary display variations on seasonal, interannual and interdecadal timescales. Therefore, one is dealing with a dynamical system subject to a continuously variable forcing, which may alter the statistical properties of flow regimes. Seasonal forcing variations are usually accounted for by analysing anomalies defined with respect to an estimated yearly cycle, and belonging to separate seasons. But how to deal with interannual and interdecadal forcing anomalies is not a trivial problem. Should one, for example, stratify the observed sample according to the El Niño–Southern Oscillation (ENSO) phase, or consider that the radiative forcing of the atmosphere has been modified by the long-term increase in greenhouse gases (GHGs)?

If regime statistics are dependent on boundary conditions and other forcing parameters, the consequent inhomogeneity of the observed record makes the detection of atmospheric flow regimes more difficult. By contrast, if we were able to determine the statistical properties of regimes as a function of the forcing anomalies, this would imply some degree of predictability of the atmospheric conditions on interannual to interdecadal timescales, which we can refer to as *regime predictability*. Given the limited size of the observed record, it is difficult to find statistically significant results on such an issue from observed data. In order to understand the dynamical meaning of interdecadal differences in regime distributions such as those shown by Corti *et al.* (1999), one has to resort to ensemble simulations made with general circulation models (GCMs), in which multiple realisations of the atmospheric flow for the same boundary (or GHG) forcing can be obtained.

Is regime predictability a sound concept? Is it feasible to obtain reliable information about the structure and frequency of regimes from observations and GCM simulations? Is the level of significance of these results acceptable? In this chapter, we will argue for positive answers to these questions, focusing on two specific issues. Firstly, we will address the issue of the statistical significance of regime estimates from the current upper-air observational record. Secondly, we will demonstrate the feasibility of regime predictions as a function of sea surface temperature (SST) conditions, investigating the effects of ENSO on extratropical regime statistics in ensembles of GCM simulations, and their impact on the predictability of interdecadal variations of the wintertime northern-hemisphere circulation.

The concept of regime predictability will be given a more formal definition in the next section. Since confidence about regimes in the real atmosphere is needed before addressing the problem of regime predictability in GCM simulations, Section 14.3 will be devoted to a brief review of relevant observational results, and will discuss what tests and what levels of statistical significance should be used in the assessment of observed statistics. In Section 14.4, results from numerical simulations performed with an intermediate-complexity atmospheric GCM will be presented, with emphasis on the model climatology during the boreal winter. Observed and modelled effects of ENSO on the northern extratropical circulation on interannual and interdecadal scales will be compared, using traditional statistics such as composite anomaly maps and their spatial correlations. In Section 14.5, the impact of ENSO will be addressed from the viewpoint of regime properties. Conclusions will be presented in Section 14.6.

14.2 A definition of regime predictability

If one regards the atmosphere as a multidimensional dynamical system, the time evolution of the atmospheric state can be described by a set of differential equations of the form

$$d_t \boldsymbol{X} = \boldsymbol{\Phi}(\boldsymbol{X}(t), \boldsymbol{F}) \tag{14.1}$$

where $\boldsymbol{X}(t)$ is the time-dependent atmospheric state vector and $\boldsymbol{F}$ a set of forcing parameters and/or boundary conditions which can be considered either as constant, or as varying on very long timescales.

In a deterministic framework, the problem of atmospheric predictability can be seen as the study of the uncertainty in our knowledge of $\boldsymbol{X}(t)$ given an 'accurate' estimate of $\boldsymbol{X}(t_0)$ at a previous time. In a probabilistic approach, the knowledge about the atmospheric state at any time t is expressed through the probability density function (pdf) in phase space, $\rho(\boldsymbol{X}, t)$.

Given the dynamical system in Eq. (14.1), the time evolution of the pdf is formally related to the time derivative of the state vector through to the Liouville equation

$$\partial_t \rho(\boldsymbol{X}, t) + div[\rho(\boldsymbol{X}, t) \cdot \boldsymbol{\Phi}(\boldsymbol{X}, \boldsymbol{F})] = 0 \tag{14.2}$$

(Gleeson, 1970; Ehrendorfer, 1994, this volume).

In weather forecasting, $\rho(\boldsymbol{X}, t)$ has to be evaluated starting from an estimate $\rho(\boldsymbol{X}, t_0)$ at the initial time t_0. Lorenz (1975) referred to such initial-value problems as 'predictions of the first kind'. After a sufficiently long time, independently from its initial value, the pdf will approach an asymptotic value $\rho^*(\boldsymbol{X})$, which is representative of the 'climate' of the system.

By contrast (again following Lorenz, 1975), a second kind of prediction can be considered, in which one wants to estimate the variations of the 'climate pdf' $\rho^*(\boldsymbol{X})$ as a function of anomalies in the forcing and boundary parameters of the system (represented by the $\boldsymbol{F}$ vector in Eq. 14.1). Estimates of future climate scenarios as a function of the concentration of atmospheric greenhouse gases (see, for example, Chapter X in IPCC, 2001) are typical examples of predictions of the second kind. As far as seasonal predictability is concerned, numerical experiments using prescribed SST to estimate the average impact of (say) El Niño events can again be regarded as predictions of the second kind, while actual seasonal forecasts with coupled ocean–atmosphere models are actually initial-value problems in a complex, multi-scale environment.

Whatever the timescale of the prediction, one is faced with the problem of estimating the properties of a pdf in a multidimensional space with many degrees of freedom. If the only required information is either the mean state or the variance of the state variables, these can be easily computed and displayed in physical (i.e. geographical) space. However, if more sophisticated probabilistic estimates are needed (typically, when the pdf is non-normal), a suitable methodology is needed to condense the large amount of information provided by a multidimensional pdf.

Let us now assume that a 'meaningful' partition of the phase space into N distinct regions could be found, and let $\{\boldsymbol{X}_j, j = 1, \ldots N\}$ be the set of the mean states (or centroids) of such partitions. Let us also assume that the variability within each of these partitions is smaller than the variability between the centroids. Then, the continuous, multidimensional pdf $\rho(\boldsymbol{X}, t)$ can be approximated by a discrete probability distribution $P(\boldsymbol{X}_j, t)$, which gives the probability of the atmospheric state vector belonging to each of the N partitions at a given time.

For probabilistic predictions of the first kind (typically provided by ensemble forecasting systems such as those described by Toth and Kalnay, 1993; Molteni *et al.*, 1996; Houtekamer *et al.*, 1996; Buizza, this volume), the phase space partitions can be defined a priori based on the properties of the climatological pdf ρ^*, or can be redefined as a function of the initial conditions (examples of both kinds can be found in Molteni *et al.*, 1996, while the former approach was adopted by Chessa and Lalaurette, 2001).

For predictions of the second kind, which by definition are concerned with the nature of the climatological pdf, the choice of the optimal partition should reflect the statistical and dynamical characteristics of the system's attractor. Specifically, if dynamical non-linearities are strong enough to generate significant anisotropic features in the climatological pdf, the system is said to display a 'regime' behaviour, where the regimes are defined by the most densely populated regions of the phase space (as in the seminal study by Reinhold and Pierrehumbert, 1982). In such a case, the regime structure provides a natural way to discretise the climatological pdf in predictability studies.

When one investigates changes in regime properties as a function of anomalies in forcing and boundary conditions, two possibilities should be considered. If the forcing variations are relatively small, then the number of distinct regimes is likely to remain the same, with minor variations of the positions of the regime centroids in phase space. In this case, one should mainly be concerned with changes in the frequency of a given set of regimes (as argued by Molteni *et al.*, 1993; Palmer 1993, 1999; Corti *et al.*, 1999). Strong forcing variations, by contrast, can take non-linear systems through bifurcation points, and the number of flow regimes can be altered (e.g. Ghil, 1987). Variations in the number of regimes caused by strong ENSO events have been reported by Molteni and Corti (1998), Straus and Molteni (2004) and Molteni *et al.* (2003) in modelling studies of the extratropical and tropical circulation, respectively. In such a case, predictions of the second kind should also provide information on the regime number and their phase space position.

14.3 Regimes in the real atmosphere: what significance?

14.3.1 The search for regimes: a brief review

The concept of weather regimes was introduced by Reinhold and Pierrehumbert (RP, 1982) to explain the preferred circulation patterns produced by the interaction of planetary-scale and synoptic-scale waves in a highly truncated two-level quasi-geostrophic model forced by large-scale topography. The RP model was an extension to synoptic scales of the model used by Charney and Straus (1980) to investigate multiple equilibria (i.e. multiple steady states) of orographically forced waves in a baroclinic framework. However, it was clear from the RP study that there was no one-to-one correspondence between steady states and weather regimes. In fact, weather regimes or flow regimes should be regarded as *statistical-dynamical* equilibria, which are defined by averaging the dynamical tendencies *on a timescale longer than the typical period of baroclinic transients*.

Although regime behaviour was detected in a number of simplified dynamical models (e.g. Mo and Ghil, 1987; Vautard and Legras, 1988; Marshall and Molteni, 1993), the relevance of such results to the real atmospheric flow was often questioned (e.g. Cehelsky and Tung, 1987). The search for observational evidence of flow regimes gave rise to even more controversy: for example, the significance of the bimodality in the distribution of the planetary-wave amplitude index found by Hansen and Sutera (1986) was strongly questioned by Nitsche *et al.* (1994) (see also the 'revisitation' by Hansen and Sutera, 1995). Further debate focused on whether regimes were best defined on a hemispheric domain (as suggested by works on planetary wave dynamics), or rather within sectors including one of the two storm tracks located over the northern oceans (e.g. Vautard, 1990; Kimoto and Ghil, 1993b; Michelangeli *et al.*, 1995; D'Andrea and Vautard, 2001).

Methodological differences in the search for regimes also contributed to the complexity of the picture. While model simulations allowed the use of methods based on the equilibration of dynamical tendencies (e.g. Vautard and Legras, 1988; Haines and Hannachi, 1995), methods to detect densely populated regions of phase space have been mostly applied to the observational record. These included univariate pdf estimation (Hansen and Sutera, 1986, 1995); multivariate pdf estimation (Kimoto and Ghil, 1993a; Corti *et al.*, 1999; Hsu and Zwiers, 2001); and different methods of cluster analysis (Mo and Ghil, 1988; Molteni *et al.*, 1990; Cheng and Wallace, 1993; Michelangeli *et al.*, 1995; Smyth *et al.*, 1999).

Although the regime classifications provided by the different methodologies were (obviously) not identical, a few 'common' regimes were detected in most studies devoted to a specific spatial domain. As mentioned in the Introduction, the three clusters identified by Cheng and Wallace (1993) for the whole northern hemisphere were also identified by Kimoto and Ghil (1993a), Corti *et al.* (1999) and Smyth *et al.* (1999). Of these clusters (shown in Figure 14.1), two show nearly opposite anomalies over the Pacific–North American region, while the third one corresponds to the negative phase of the North Atlantic Oscillation (NAO).

Among the analyses devoted to either the Atlantic or the Pacific sector of the northern hemisphere, a robust partitioning method and a detailed analysis of statistical significance were employed by Michelangeli *et al.* (1995). Their analysis yielded three and four clusters for the Pacific and the Atlantic sectors respectively; these clusters were also identified in earlier studies on regional regimes (e.g. Vautard, 1990; Kimoto and Ghil, 1993b).

14.3.2 The estimation of significance

In order to evaluate the significance of regimes found either by a pdf estimate or by cluster analysis, a common methodology has been adopted in a number of the studies cited above. Assuming that the time series of the selected phase-space coordinates are uncorrelated (as in the case of principal components), this procedure can be summarised by the following steps:

1. Define a quantity q which can be taken as a (positively oriented) measure of the likelihood of the existence of multiple modes or clusters in a given data sample.
2. Perform the pdf estimate or the cluster analysis on the selected data sample by varying the algorithm's parameters, in such a way to obtain regime partitions with an increasing number of modes or clusters, and define q_m^* as the value of q corresponding to the m-regime partition.
3. Generate a large number (N_s) of samples of pseudo-random red-noise data, with the same size and the same mean, variance and lag-1 autocorrelation of the actual data sample.
4. For each red-noise sample, repeat the pdf estimate or cluster analysis with the same parameters yielding m regimes in the actual data, and compute q_m in order to obtain a sample of N_s values, say $\{q_{mk}, k = 1, \ldots N_s\}$.
5. Since the red-noise data are assumed to have a unimodal distribution, the proportion P_m of red-noise samples for which $q_{mk} > q_m^*$ is an inverse measure of the significance of the m-regime partition of the actual data, and $1 - P_m$ is the corresponding confidence level for the existence of m regimes.

In cluster analysis, the total variance of the data sample can be divided into a fraction accounted for by the cluster means (centroids) and an intracluster part representing the mean-squared distance from the appropriate cluster centroid. The ratio of

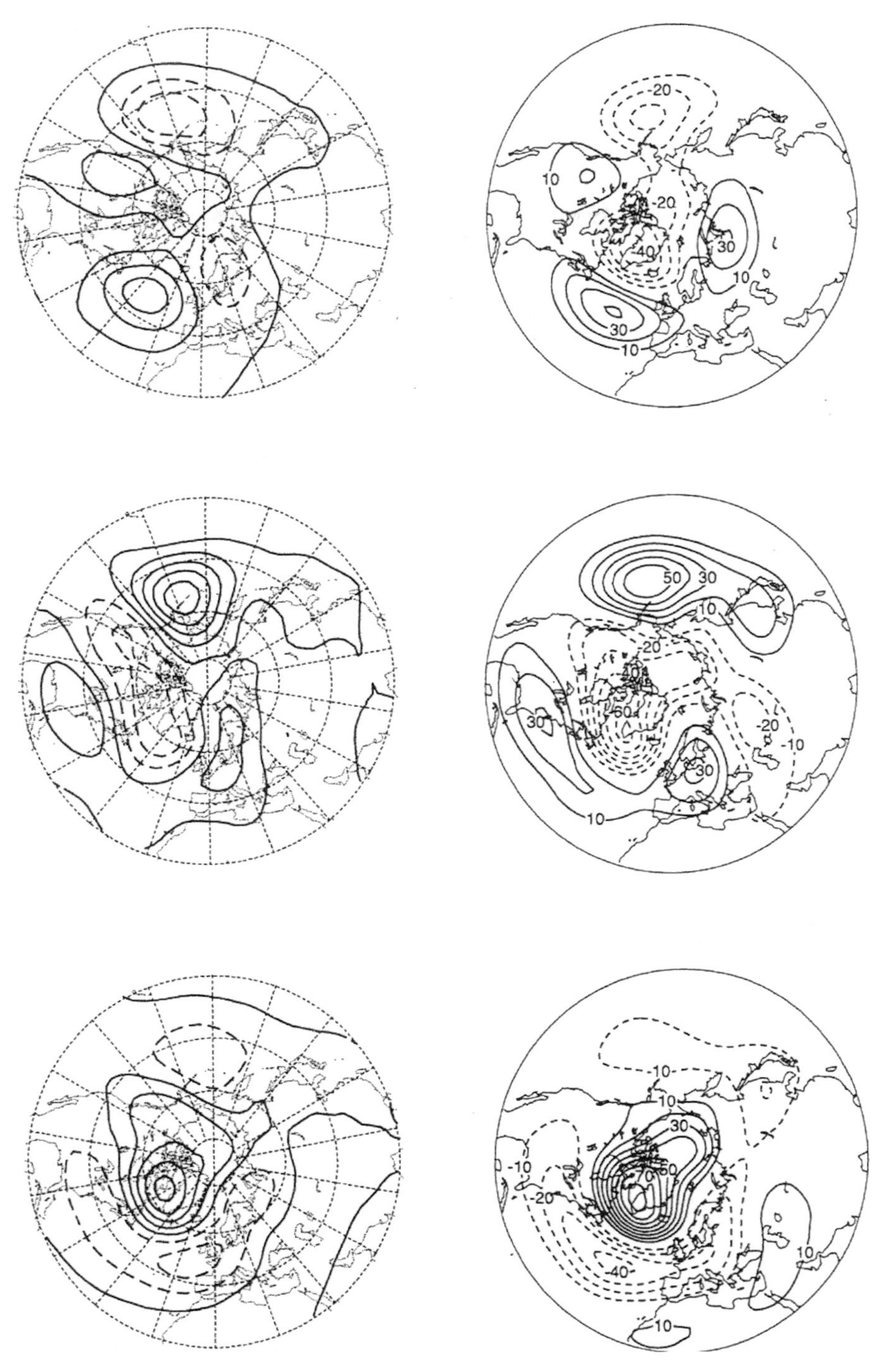

Figure 14.1 (Left column) Anomalies corresponding to the three cluster centroids of 500-hPa height found by Cheng and Wallace (1993) using a hierarchical clustering method. (Right column) Anomalies corresponding to three of the (four) density maxima in the pdf of the two leading PCs of 500-hPa monthly-mean anomalies, computed by Corti *et al.* (1999). Data are for December to February, 1946 to 1985 in Cheng and Wallace, November to April, 1949 to 1994 in Corti *et al.*

centroid variance to intracluster variance is a measure of the separation between clusters, and therefore provides a suitable definition of q (see Straus and Molteni, 2004). Other definitions used in the literature include quantities which indicate the level of reproducibility of a given cluster set using different subsets of the full data sample, or starting iterative aggregations of data from different random 'seeds' (e.g. Michelangeli *et al.*, 1995).

When regimes are studied using non-parametric pdf estimation (see Silverman, 1986), the number of regimes may be directly related to the number of local maxima (modes) in the pdf. However, such a number depends on the degree of smoothing used in the estimation, which is controlled by a disposable parameter. A widely used methodology is represented by the kernel estimation, which provides an estimate of the pdf of a (vector) variable $\boldsymbol{X}$ as the sum of elementary, unimodal functions centred around each data point:

$$\rho(\boldsymbol{X}) = N_d^{-1} \sum\nolimits_i \boldsymbol{K}[h^{-1} \| \boldsymbol{X} - \boldsymbol{X}_i \|] \qquad (14.3)$$

where $\boldsymbol{K}$ is the (normalised) kernel function, $\boldsymbol{X}_i$ are the N_d input data, $\| \cdot \|$ represents a norm in phase space and h is a parameter called the kernel width, which defines the level of smoothing in the pdf estimate.

The relationship between the kernel width h and the number of modes m in the pdf is clearly defined when a multinormal (Gaussian) form is assumed for the kernel function. In this case, it can be demonstrated that, for any data sample, m is a monotonically decreasing function of h (Silverman, 1981). The *largest* value of h for which m modes are found is called the critical kernel width for m modes (h^*_m), and can be used to estimate the significance of multimodality against red-noise data as outlined above for a generic variable q.

Note that, although many pdf estimates with increasing h are needed to determine h^*_m from the actual data sample, just one estimate is needed for each red-noise sample if a multinormal kernel is used. Because of the monotonic relationship between h and m, it is sufficient to compute the pdf using h^*_m as kernel width. If the number of modes in the pdf of k-th random sample is less than m, h_{mk} must be less than h^*_m. For example, if h^*_2 is the critical width for bimodality in the actual data, the proportion of unimodal pdfs obtained from red-noise samples setting $h = h^*_2$ gives a confidence level for bimodality.

14.3.3 On the significance of northern hemisphere regimes computed from monthly-mean anomalies

We will now apply these concepts to the estimation of the significance of multimodality in pdfs derived from a sample of monthly-mean anomalies of 500-hPa height for the northern winter (following Corti *et al.*, 1999; CMP hereafter). For the sake of comparison with the model results presented in Section 14.5, the analysis

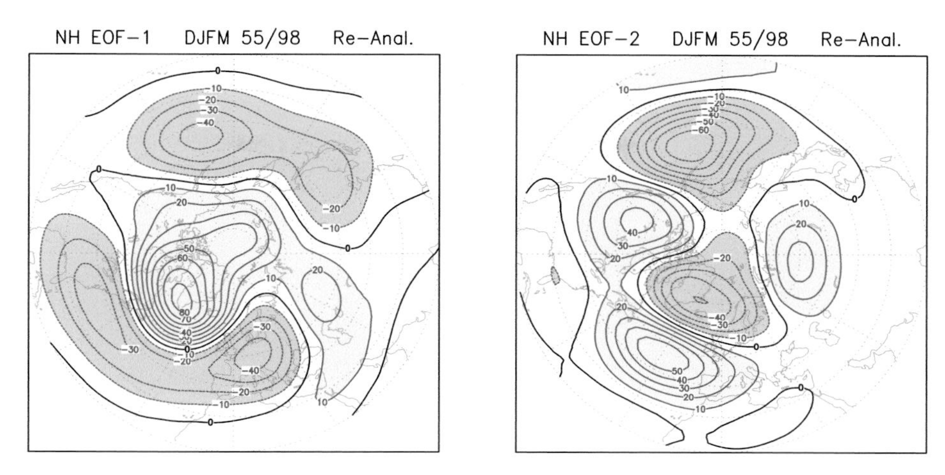

Figure 14.2 First two EOFs of 500-hPa height monthly-mean anomalies over the northern hemisphere (20–90 N), computed from the NCEP/NCAR reanalysis in the 44 winters (DJFM) 1954–5 to 1997–8. The EOFs are scaled to represent the anomaly patterns corresponding to one standard deviation of the associated PCs. Contour interval 10 m.

will cover the 44 winters from 1954/5 to 1997/8, using months from December to March (DJFM). The 500-hPa height monthly-mean data are taken from the National Centers for Environmental Predictions/National Center for Atmospheric Research (NCEP/NCAR) reanalysis (Kalnay *et al.*, 1996). Although the main results of this analysis do not differ substantially from those of CMP and other related studies (see references above), here the goal is to illustrate and discuss the significance test in greater detail.

As in many previous studies, the leading variability patterns are first identified through a principal component (PC) analysis of the 500-hPa height anomalies. The empirical orthogonal functions (EOFs) associated with the first two PCs are shown in Figure 14.2; choosing the same sign as in CMP, they are quite similar to the negative phase of the Arctic Oscillation and to the so-called Cold Ocean–Warm Land pattern respectively (Thompson and Wallace, 1998; Wallace *et al.*, 1996). The first two PCs (normalised by their respective standard deviations) are used as the coordinates of a bidimensional phase space in which pdfs will be estimated. An iterative version of the Gaussian kernel estimator (see Eq. 14.3) is used to compute the bidimensional pdfs, in order to avoid spurious local maxima in scarcely populated regions of phase space (Silverman, 1986; Kimoto and Ghil, 1993a).

The estimated pdf in the PC1–PC2 plane is shown in Figure 14.3 for three different values of the kernel width h (which becomes a non-dimensional parameter when coordinates are normalised), namely 0.3, 0.4 and 0.5. For $h = 0.3$, four local maxima are evident in the pdf. Three of them are also present in the estimate with $h = 0.4$; when the PC coordinates of these modes are multiplied by the corresponding EOFs

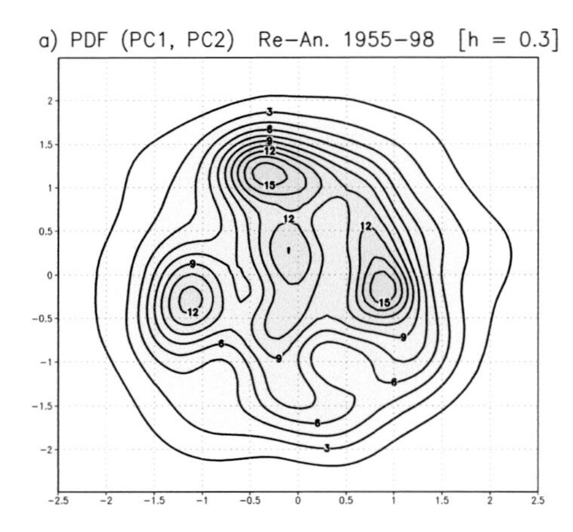

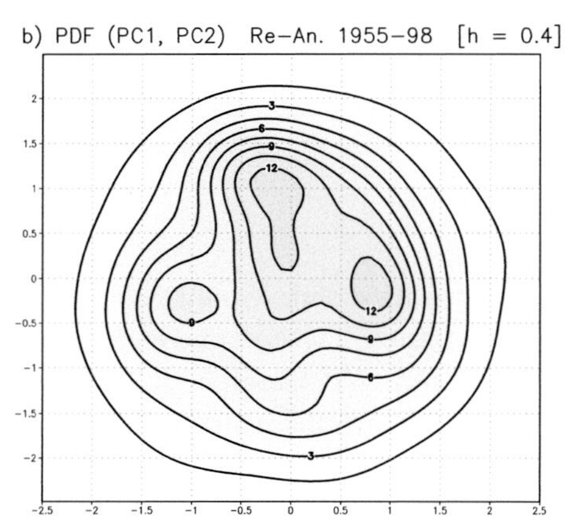

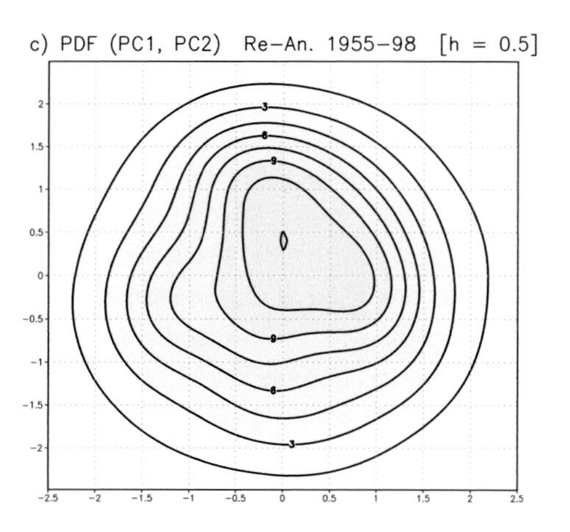

Figure 14.3 Bidimensional probability density function (pdf) of the first two PCs of 500-hPa height anomalies in winter (for the EOFs shown in Figure 14.2), computed with different values of the Gaussian kernel width h (see text). (a): $h = 0.3$; (b): $h = 0.4$; (c): $h = 0.5$.

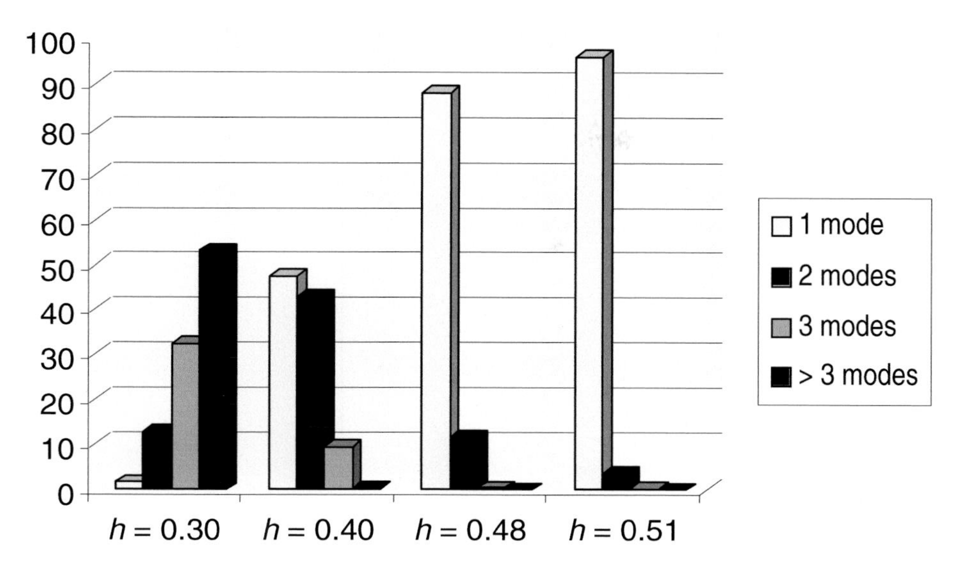

Figure 14.4 Histograms showing the proportion of pdfs with different number of modes (i.e. local maxima), computed from red-noise samples with increasing values of the Gaussian kernel width h.

to obtain a 500-hPa height anomaly, such patterns are (as expected) almost identical to the three local maxima of CMP shown on the right-hand column of Figure 14.1. By contrast, the pdf estimated with $h = 0.5$ is unimodal; further calculations show that the critical kernel width for bimodality is $h_2^* = 0.48$.

Pdfs from 500 samples of pseudo-random red-noise data with the same number of data and the same mean, variance and lag-1 autocorrelation of the actual PCs were then computed, using increasing kernel width. The histograms in Figure 14.4 show the frequency of pdfs with different number of modes for a few selected values of h. For $h = 0.3$, a very large proportion of the red-noise pdfs are multimodal, with more than half of them displaying at least four modes. Therefore, no significance statement can be made about the four modes shown in the top panel of Figure 14.3. With $h = 0.4$, most of the red-noise pdfs are multimodal; however, a three-modal pdf (as estimated from actual PCs with this kernel width) is found in just less than 10% of the cases. If we are simply concerned with the existence of more than one mode, we found that 88% of the red-noise pdfs are unimodal when the critical width for bimodality of the PC sample was used. We can therefore attach an 88% level of confidence to the multimodality of the observed PC sample.

Is 88% confidence high enough to state that multiple regimes exist in the northern hemisphere circulation? Many statisticians would not be comfortable with such a claim unless 95% confidence would be achieved (see, for example, Stephenson *et al.*, 2004). From the red-noise pdf estimates (see again Figure 14.4), one finds that a critical kernel width of 0.51 would be required in the PC sample to exceed the

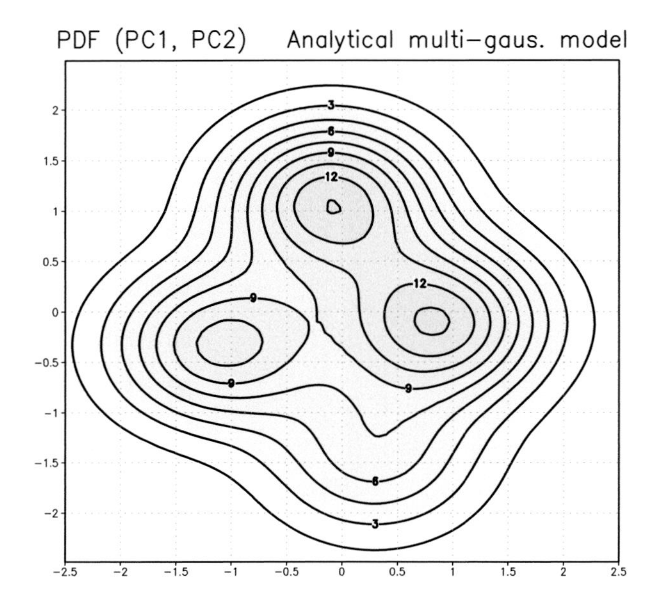

Figure 14.5 Probability density function corresponding to an analytical 3-modal distribution in the PC1–PC2 plane, used for the estimation of 'useful' significance levels for the monthly-mean record.

95% confidence level. But should one reject the multimodality hypothesis because the actual value $h_2^* = 0.48$ gives it a 12% probability to be incorrect for this specific data set?

There are different ways to answer this question. One may point out that the typical duration of extratropical flow regimes is of the order of 1–2 weeks; therefore a monthly time averaging is bound to weaken the signature of multiple regimes in the pdf. (The pdfs computed by, for example, Hansen and Sutera (1986), Kimoto and Ghil (1993a), Marshall and Molteni (1993) were derived from low-pass-filtered daily data or 5-day means.) By contrast, if one wants to stick to monthly means because of the wider data availability (for example, in comparisons with GCM simulations), one should evaluate if the required level of significance actually allows a distinction between a unimodal and a multimodal distribution, *given the size of the observed record.*

To address this issue, an *analytical* 3-modal pdf was defined by a superposition of multinormal functions centred in different points of the PC plane, close to the local maxima of the actual pdf of PC data. The analytical form of this distribution (plotted in Figure 14.5) was used to generate 500 samples of pseudo-random data with the same size as the observed monthly-mean record. From each of these samples, the pdf was recomputed using the Gaussian kernel estimator, and the critical kernel width for bimodality was recorded. It was found that 58% of the samples from the 3-modal distribution had a larger critical width than the actual PC data; however, in only 38% of the cases the critical width for multimodality was larger than the value (0.51) providing a 95% confidence level in the test against a unimodal red-noise.

These results show that, if we require a 95% confidence limit for multimodality in a sample as large as the 44-winter record of monthly means, we have a large chance (62%) that Silverman's (1981) test would *fail* to recognise a significant multimodality in a pdf derived from the analytical distribution in Figure 14.5. Therefore, the fact that the multimodality of the monthly-mean PC distribution does not pass a 5% significance test (as claimed by Stephenson *et al.*, 2004) can hardly be used to refuse the regime hypothesis. One may wish to analyse the observed record with a different time frequency, higher dimensionality or more sophisticated clustering techniques to get a larger confidence. Here, we will accept the existence of multiple flow regimes as a realistic working hypothesis, and proceed to explore the issue of regime predictability in the context of GCM simulation.

14.4 Model climatology and variability

The numerical experiments used to investigate regime predictability in the following section were run with a 7-level version of the intermediate-complexity atmospheric GCM developed at the International Centre for Theoretical Physics (Molteni, 2003). The model (nicknamed SPEEDY, for Simplified Parametrisation, primitivE-Equation DYnamics) is based on a hydrostatic, spectral dynamical core developed at the Geophysical Fluid Dynamics Laboratory (see Held and Suarez, 1994), using the vorticity-divergence form described by Bourke (1974). A set of physical parametrisation schemes has been developed from the same basic principles used in more complex GCMs, with a number of simplifying assumptions which are suited to a model with a coarse vertical resolution (details can be found in the on-line appendix to Molteni (2003) available at www.ictp.trieste.it/~moltenif/speedy-doc.html). These include short- and long-wave radiation, large-scale condensation, convection, surface fluxes of momentum, heat and moisture, and vertical diffusion. The model is currently run with a T30 spectral truncation in the horizontal; at this resolution, one year of integration takes 23 minutes on a single Xeon 2.4 GHz processor.

The experiments consist of an ensemble of eight simulations for the period 1954–99, using the EOF-reconstructed observed SST by NCEP (Smith *et al.*, 1996) as boundary conditions. The model climatology of upper-air fields and its variability on multidecadal timescales are validated here using the NCEP/NCAR reanalysis (Kalnay *et al.*, 1996). A number of verification maps for the winter and summer climatology of these experiments can be found on-line at the website above.

With respect to the 5-level version described by Molteni (2003), the wintertime climatology of the 7-level model used here shows a much improved stationary wave pattern in the Pacific sector of the northern hemisphere, while it still suffers from an underestimation of the stationary wave amplitude over the Atlantic. These features

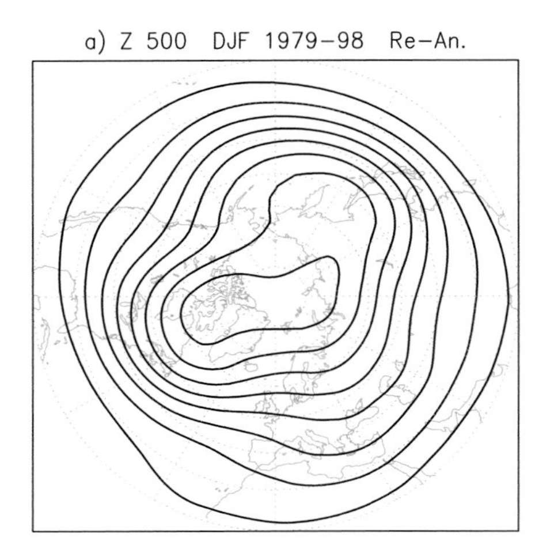

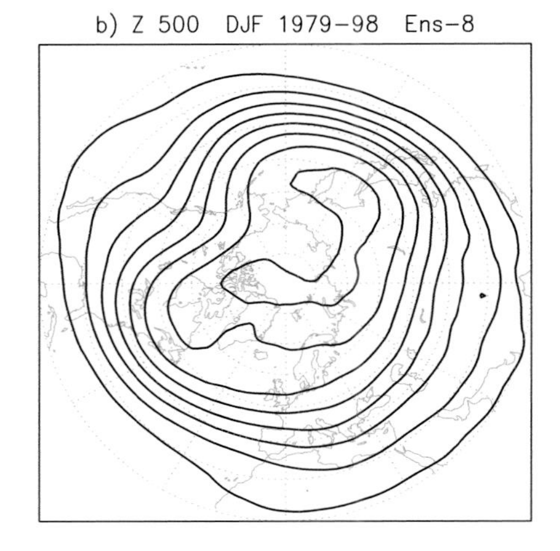

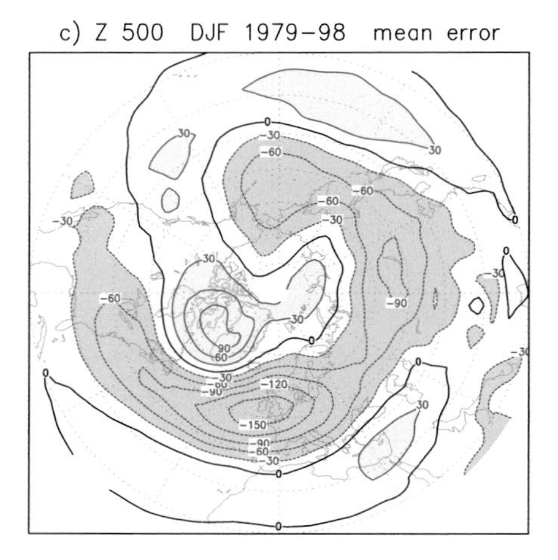

Figure 14.6 (a) Mean field of 500-hPa height for the DJF season, from the NCEP/NCAR reanalysis in winters 1978–9 to 1997–8. (b) as in (a), but from an ensemble of 8 simulations with the International Centre for Theoretical Physics (ICTP) AGCM forced by observed SST. (c) mean model error (ensemble-mean minus reanalysis). Contour interval 100 m in (a) and (b), 30 m in (c).

are illustrated by the comparison of the December-to-February (DJF) mean of 500-hPa height for the NCEP/NCAR reanalysis and the ensemble-mean model simulation in the 1979–98 period (Figure 14.6a–b). The systematic error of the model (defined by the ensemble-mean minus reanalysis difference in Figure 14.6c) reaches a large negative amplitude (over 150 m) close to the British Isles, but it just exceeds 60 m (in absolute value) in the North Pacific.

The good simulation of the Pacific stationary waves is reflected in a realistic reproduction of the extratropical response to tropical SST anomalies in opposite ENSO phases. Composites of DJF 500-hPa height in eight warm and eight cold ENSO

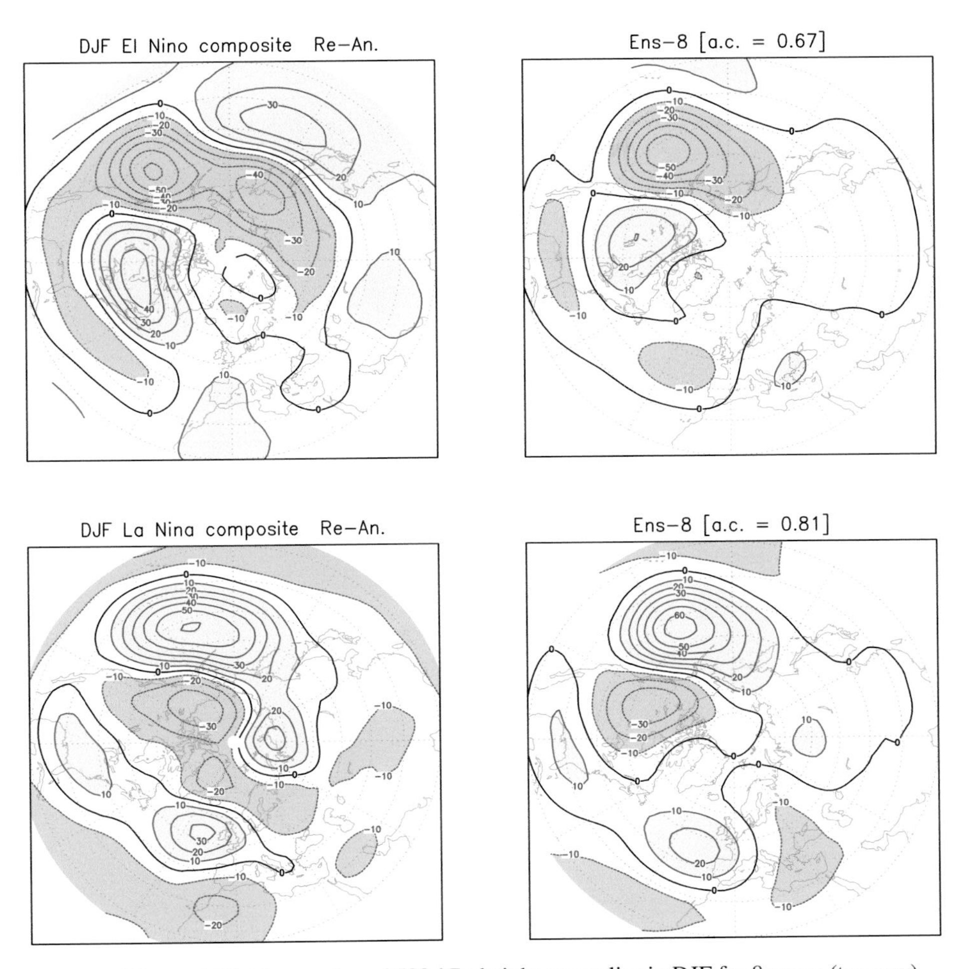

Figure 14.7 Composites of 500-hPa height anomalies in DJF for 8 warm (top row) and 8 cold (bottom row) ENSO events in the 1955-to-98 period, from the NCEP/NCAR reanalysis (left column) and from an ensemble of 8 AGCM simulations forced by observed SST (right column). Contour interval 10 m. The anomaly correlation (a.c.) between observed and modelled patterns is shown in brackets.

events during the 1954–99 period are shown in Figure 14.7 for the model ensemble mean and the NCEP/NCAR reanalysis. In order to eliminate possible effects of interdecadal variability on such composites, the anomaly for each event has been defined with respect to a 10-winter mean centred around the selected winter. Both the pattern and the amplitude of the model responses compare well to the observations in the Pacific/North American region. For the cold-event composite, even the features on the eastern Atlantic are well simulated, yielding a large spatial correlation (81%) between the model and reanalysis composites on the full hemispheric domain.

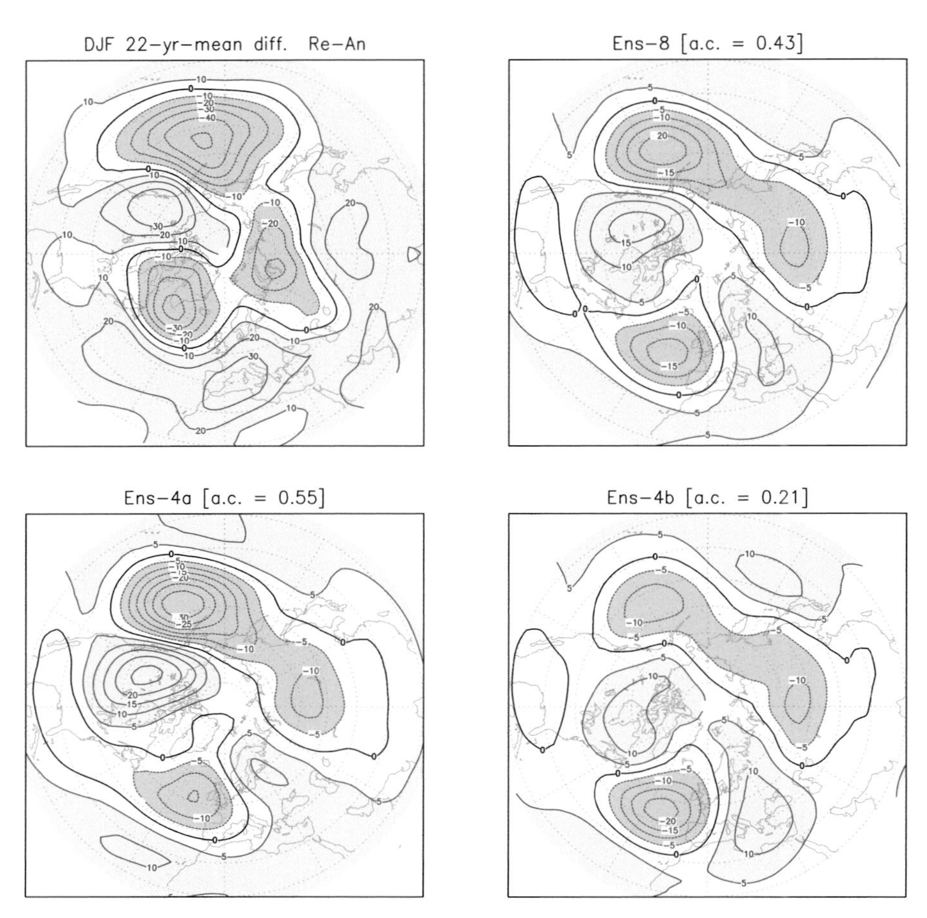

Figure 14.8 (Top left) Mean difference of 500-hPa height in DJF between two consecutive 22-year periods (1977–98 minus 1955–76), from the NCEP/NCAR reanalysis. (Top right) As in top left, but from an ensemble of 8 AGCM simulations forced by observed SST (Ens-8). (Bottom panels) As in top right but from two ensembles (Ens-4a and Ens-4b) of 4 members each, selected according to the anomaly correlation between the observed and the simulated interdecadal difference. Contour interval 10 m in top left panel, 5 m in the others. The correlation between observed and modelled patterns is shown in brackets.

Since the model reproduces ENSO-related interannual variability in a realistic way, interdecadal differences related to the well-documented SST trends in the tropical oceans should also be detectable in the simulations. Significant differences in Pacific SST before and after 1976 have been documented in many studies (e.g. Zhang *et al.*, 1997; Trenberth and Hoar, 1997); therefore, we will focus on the differences in geopotential height between the mean of 22 winters from 1976/7 to 1997/8 and the preceding 22-winter period. The 500-hPa height interdecadal differences for the eight-member ensemble-mean and the reanalysis are shown in the top panels of Figure 14.8. The spatial correlation between the modelled and observed patterns is 43%

over the northern hemisphere, while the average amplitude for the model is about half of the observed amplitude (as in Hoerling *et al.*, 2001). Looking at different regions, one finds a better correlation over the Pacific/North American sector, while the modelled negative anomalies over the North Atlantic and Siberia are located in a more southerly position than the corresponding observed features.

Computing ENSO composites and interdecadal differences from individual ensemble members, one finds that the former are quite consistent among individual members, while larger differences are found in the simulation of interdecadal variability. If the difference patterns of 500-hPa height between the post-1976 and pre-1976 periods for individual experiments are correlated with the observed difference, the correlation coefficients over the northern hemisphere range from slightly negative to greater than 60%. To illustrate the effects of internal atmospheric variability, the bottom panels of Figure 14.8 show the interdecadal differences for the ensemble means of two subensembles, one including the four members with the best interdecadal correlations, the other one including the four least-correlated members (hereafter referred to as Ens-4a and Ens-4b respectively, while Ens-8 will indicate the full ensemble). With respect to Ens-4b, the difference pattern for Ens-4a shows much stronger amplitude over the Pacific and North America, and a northward shift of the Atlantic anomalies. As a result, the correlation with the observed pattern (top left) is raised to 55% for Ens-4a, while it drops to 21% for Ens-4b. (A detailed analysis of interdecadal variability in recent ensemble simulations with the SPEEDY model can be found in Bracco *et al.*, 2004.)

14.5 ENSO-related variability in flow regime statistics

In this section, the impact of ENSO on the interannual and interdecadal variability of flow regimes will be investigated through a pdf analysis of the wintertime monthly-mean anomalies of 500-hPa height simulated by the SPEEDY model. Since the ensemble simulation gives eight times more data than the observed record, the issue of multimodality can be addressed with much stricter levels of statistical significance.

Following the same procedure used in Section 14.3.3, we first computed the two leading EOFs of monthly-mean anomalies in DJFM for the northern-hemisphere extratropics (shown in Figure 14.9). When compared with its counterpart in the reanalysis (Figure 14.2), the first EOF of the model shows a more zonally symmetric pattern, with features of comparable amplitude in the Atlantic and the Pacific sectors. Taking the same sign as in CMP, the first model EOF can be interpreted as the negative phase of the Arctic Oscillation pattern, similarly to the first reanalysis EOF. By contrast, stronger differences between model and reanalysis are found for the second EOF. The two patterns resemble each other in the Pacific sector, but the model EOF-2 has a much smaller amplitude in the Atlantic and Eurasian region. In the model, the second EOF may be described as the manifestation of the Pacific/North

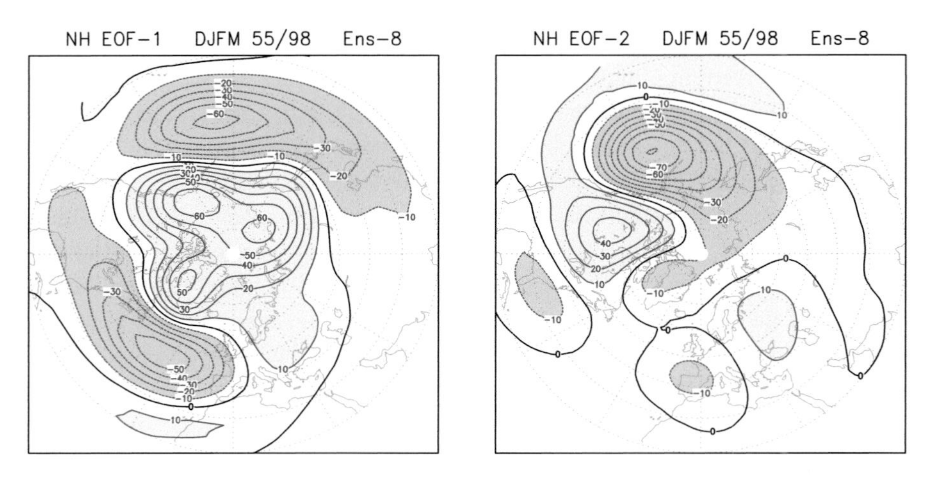

Figure 14.9 First two EOFs of 500-hPa height monthly-mean anomalies over the northern hemisphere (20–90 N) in the 44 winters (DJFM) 1954–5 to 1997–8, computed from an ensemble of 8 AGCM simulations forced by observed SST. The EOFs are scaled to represent the anomaly patterns corresponding to one standard deviation of the associated PCs. Contour interval 10 m.

American teleconnection pattern, rather than a COWL-related pattern (Wallace *et al.*, 1996).

Because of the differences in the EOF patterns, the subspace spanned by the first two PCs does not have exactly the same meaning in the model as in the observation. Still, it is interesting to see whether any variability in the model pdfs associated with different periods or ENSO phases have a counterpart in reanalysis data. Therefore, in the following we shall compare model and observed pdfs for the same subsamples, bearing in mind that only for the model simulations do we have enough data to estimate the significance of multimodality using just a fraction of the total sample. As for the observed data, bidimensional pdfs of the model PCs have been computed with an iterative Gaussian kernel estimator (Silverman, 1986).

Figure 14.10 shows pdfs in the PC1–PC2 plane computed from model anomalies in the 44-year period (DJFM 1955 to 1998, where the year refers to JFM), in its first and second halves (1955–76 and 1977–98 respectively), and in three 15-winter subsamples defined by sorting the available winters into cold, neutral and warm ENSO phases according to the bivariate index by Smith and Sardeshmukh (2000) (the cold-ENSO winter of 1998/9 was also included here). The pdf of model PCs is unimodal in the full sample and in its second half, but it is bimodal in its first half. The separation between the two modes occurs mainly along the PC2 axis, with a smaller difference along the PC1 axis. When the significance of bimodality in the 1955–76 period is tested using Silverman's (1981) procedure as described in Section 14.3.1, the confidence level exceeds 99.5%.

Looking at the pdfs for different ENSO phases, one finds that bimodality in the model phase space is primarily generated during neutral (i.e. near-normal) ENSO

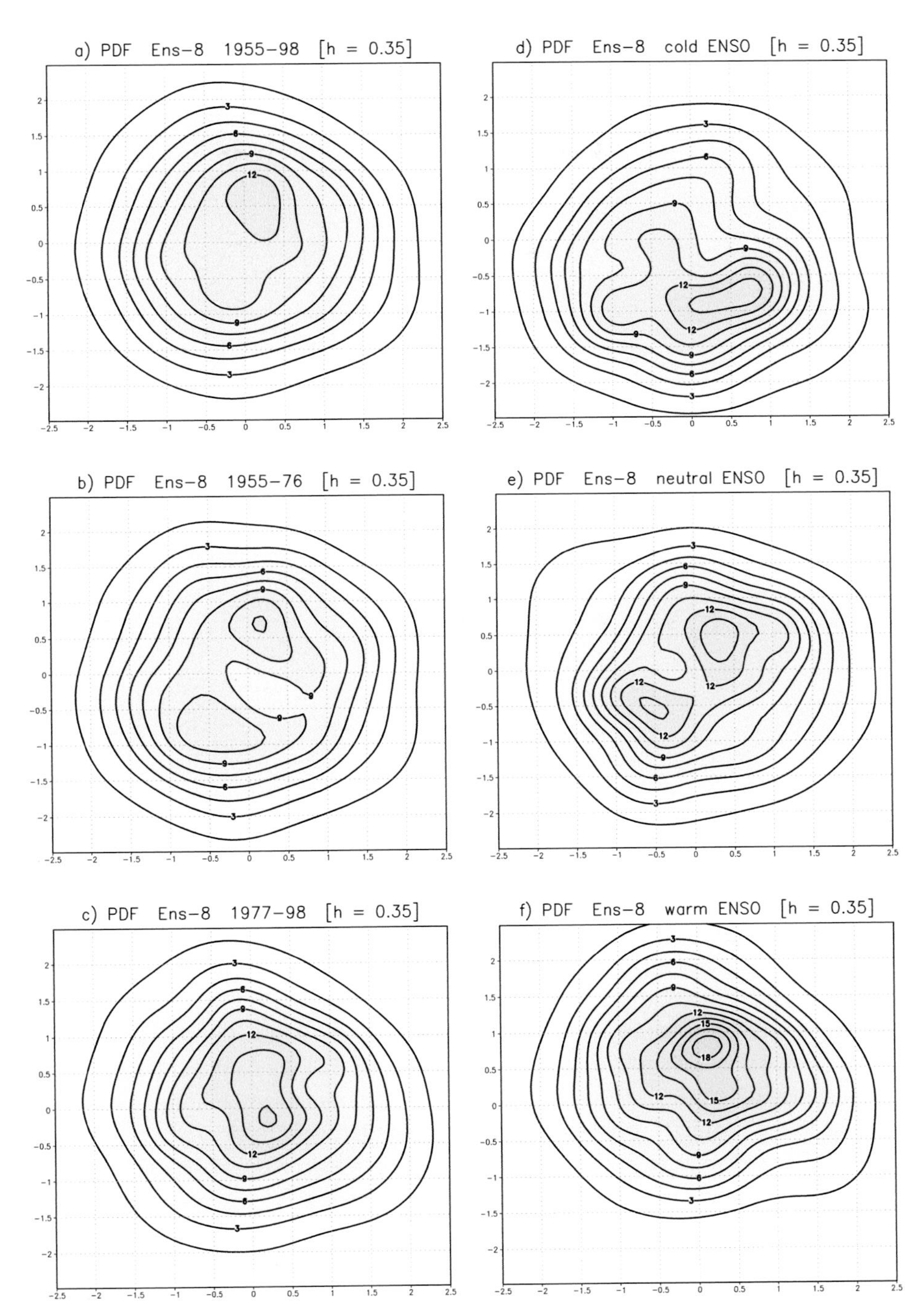

Figure 14.10 Bidimensional pdfs of the first two PCs of 500-hPa height anomalies in winter over the northern hemisphere, from an ensemble of 8 GCM simulations forced by observed SST in the period 1954 to 1999. (a) pdf for the full 44-winter period. (b) pdf for the 22-winter period 1955 to 1976. (c) pdf for the 22-winter period 1977 to 1998. (d) to (f) pdfs for samples of 15 winters with cold (d), neutral (e) and warm (f) ENSO phase.

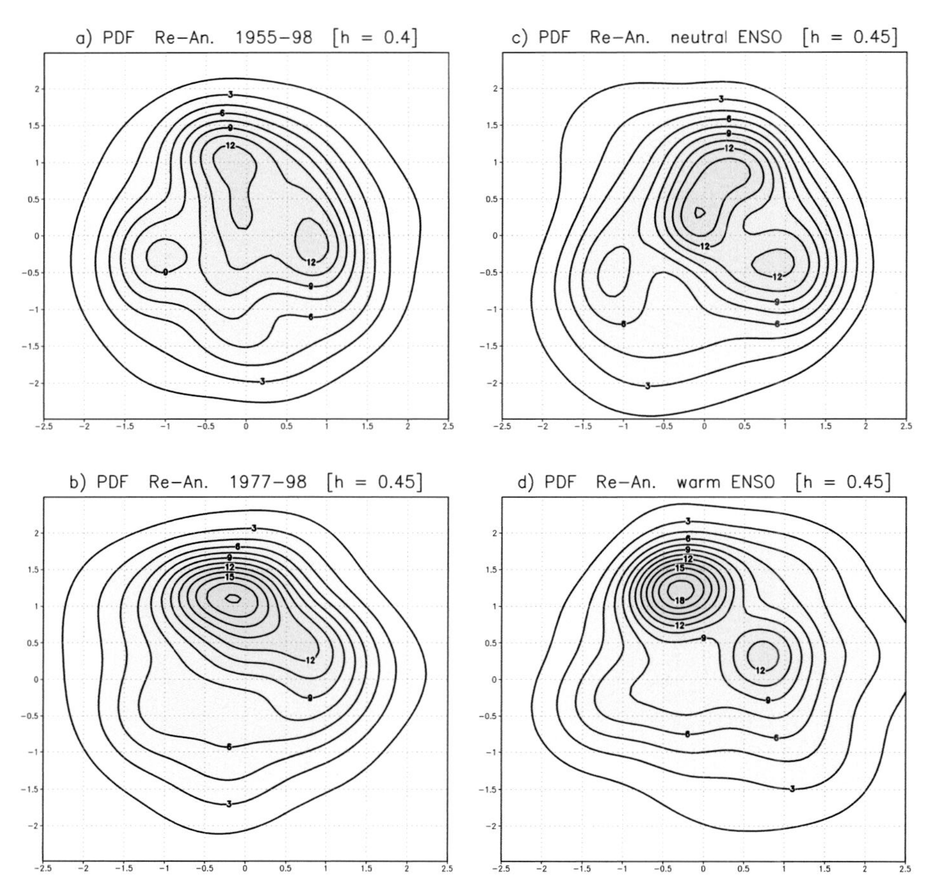

Figure 14.11 Bidimensional pdfs of the first two PCs of 500-hPa height anomalies in winter over the northern hemisphere, from the NCEP/NCAR reanalysis in the period 1954 to 1999 (a) pdf for the full 44-winter period. (b) pdf for the 22-winter period 1977 to 1998. (c) and (d) pdfs for samples of 15 winters with neutral (c) and warm (d) ENSO phase.

winters. The confidence level estimated from the neutral-ENSO sample still exceeds 96%, despite the reduction in the number of data. During the cold and warm ENSO phases, the pdf appears to be shifted along the PC2 axis (corresponding to a pattern similar to the Pacific/North American (PNA) circulation), but evidence of multiple regimes is either weak (cold phase) or totally absent (warm phase). Comparing the pdf for the second half of the record with the pdf of warm ENSO events, it appears that the unimodal nature of the distribution in the 1977–98 period is closely related to higher frequency of warm events in that period. This is consistent with the modelling results of Molteni and Corti (1998) and Straus and Molteni (2004), who found that evidence of multiple regimes in the Pacific/North American circulation was suppressed during warm ENSO events.

In order to verify whether the model behaviour has any correspondence in the observations, Figure 14.11 shows pdfs from the reanalysis record for the 1955–98

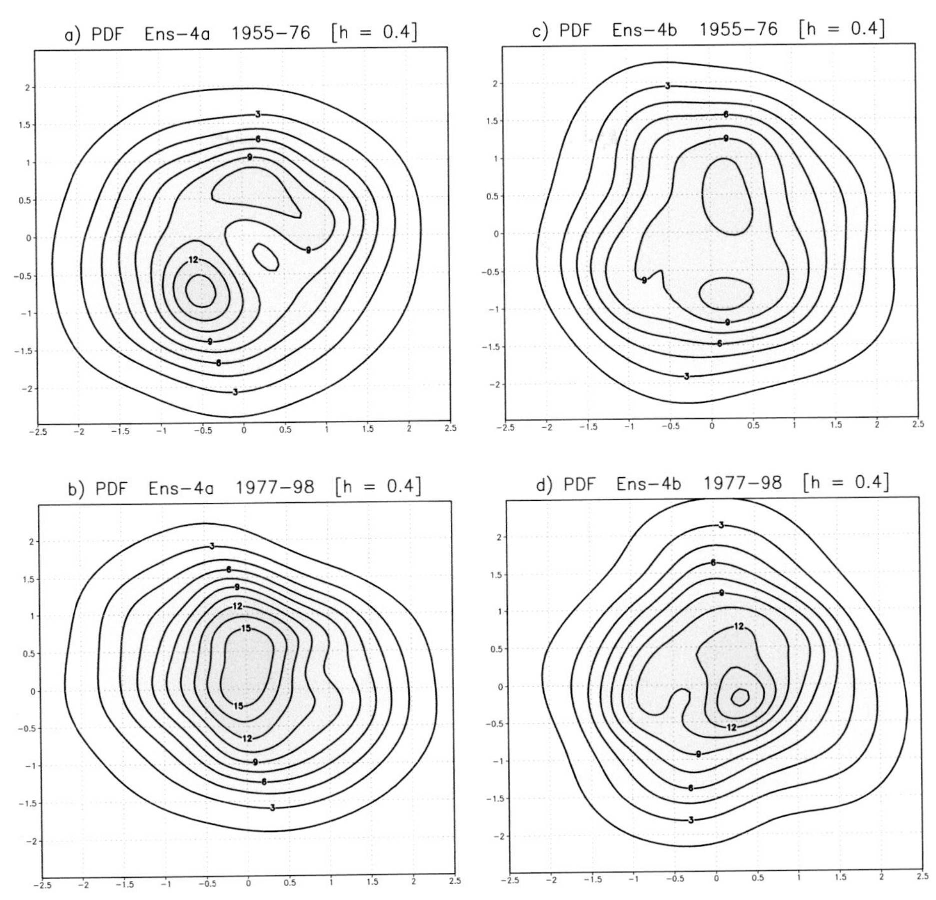

Figure 14.12 Bidimensional pdfs of the first two PCs of 500-hPa height anomalies in winter over the northern hemisphere, from two ensembles (Ens-4a and Ens-4b) of 4 members each. (Top row) pdf for the 22-winter period 1955 to 1976, from Ens-4a (a) and Ens-4b (c). (Bottom row) pdf for the 22-winter period 1977 to 1998, from Ens-4a (b) and Ens-4b (d).

and 1977–98 periods, and for the two 15-winter subsamples of neutral and warm ENSO cases. Consistently with model simulation, the multimodal nature of the 44-year pdf is made more evident by the selection of neutral ENSO years. Also, the pdfs for the second half of the record and for warm events resemble each other, and are dominated by one mode, corresponding to the positive phase of the COWL pattern (i.e. the regime in the top row of Figure 14.1).

As a further test of the robustness of the modelled interdecadal variations in flow regime properties, Figure 14.12 compares the pdfs for the 1955–76 and 1977–98 periods, evaluated from two four-member subsets of the ensemble, namely the Ens-4a and Ens-4b subsets with 'good' and 'bad' simulations of interdecadal variability (see the bottom row of Figure 14.8). It is found that the bimodal versus unimodal nature of the pdf in the two periods is reproduced in both four-member ensembles, although the difference is more evident in the ensemble (Ens-4a) that better reproduced the observed spatial pattern of interdecadal difference.

14.6 Conclusions

In this chapter, we have addressed the issue of regime predictability by focusing on two aspects:

- the significance of regimes detected in the observational record;
- the impact of interannual and interdecadal variability of the ENSO phenomenon on the statistical properties of extratropical flow regimes as simulated by a GCM.

On the first issue, we applied Silverman's (1981) methodology to test the significance of multimodality in the distribution of the two leading PCs of monthly-mean anomalies of 500-hPa height in winter, using data from a 44-year sample of NCEP/NCAR reanalyses. We estimated the confidence level for the existence of *at least* two regimes to be close to 90%. We also showed that asking for a stricter significance level has little meaning in the case of the observed monthly-mean record, since the required level of pdf smoothing would prevent the detection of multimodality even for samples generated by an analytical trimodal distribution.

With regard to the second issue, the results of our model simulations indicate that predictability 'of the second kind' (Lorenz, 1975) is indeed present in regimes statistics, as a result of their dependence on anomalous boundary forcing related to the ENSO phenomenon. In addition to a shift of the pdf along a PNA-like axis, ENSO variations change the number of regimes detected in the phase space spanned by the two leading PCs. For both model and observed data, multimodality is best revealed in years with weak ENSO forcing. By contrast, our intermediate-complexity atmospheric GCM shows no evidence of multiple regimes during the warm ENSO phase. Taking into account that the two leading PCs of model data tend to emphasise low-frequency variability in the Pacific/North American region, the latter result is consistent with the statistics of Pacific regimes from the quasi-geostrophic simulations by Molteni and Corti (1998) and the GCM ensembles analysed by Straus and Molteni (2004).

Such findings give a more complex picture of the ENSO impact on regimes than the one advocated by Molteni *et al.* (1993) and Palmer (1993), which was based on a change of frequencies in quasi-invariant regimes. While the frequency-change hypothesis is likely to be appropriate for relatively small variations of the atmospheric forcing terms (such as those related to greenhouse gas and aerosol concentrations in the twentieth century), strong ENSO events alter the forcing of stationary waves in a substantial way, possibly leading the atmospheric system through bifurcation points. Interestingly, a variation in the number of regimes as a function of the ENSO phase has also been detected in the Asian monsoon simulations described by Molteni *et al.* (2003).

The goal of this chapter was to show that the concept of regime predictability has robust statistical and dynamical foundations. We have not addressed the issue in such detail to quantify, for example, its practical relevance in the case of seasonal forecasting. However, cluster analysis has been increasingly used in recent years to condense the information provided by ensemble forecasts, so that practical examples are already available in the literature (e.g. Brankovic and Molteni, 1997). We are well aware that many aspects of atmospheric predictability may be addressed without the need for introducing flow regimes. By contrast, regimes provide a valuable framework to investigate the non-linear aspects of the relationship between anomalous forcing and atmospheric response, and may play an important role in understanding the mechanism of climate changes originated by the interactions of 'internal' climate dynamics and anthropogenic perturbations of the Earth's energy balance.

References

Bourke, W. (1974). A multilevel spectral model. I: Formulation and hemispheric integrations. *Mon. Weather Rev.*, **102**, 687–701.

Bracco, A., F. Kucharski, R. Kallummal and F. Molteni (2004). Internal variability, external forcing and climate trends in multi-decadal AGCM ensembles. *Climate Dyn.*, **23**, 659–78.

Brankovic, C. and F. Molteni (1997). Sensitivity of the ECMWF model northern winter climate to model formulation. *Clim. Dynam.*, **13**, 75–101.

Cehelsky, P. and K. K. Tung (1987). Theories of multiple equilibria and weather regimes – a critical re-examination. II: Baroclinic two-layer models. *J. Atmos. Sci*, **44**, 3282–303.

Charney, J. G., and D. M. Straus (1980). Form-drag instability, multiple equilibria and propagating planetary waves in baroclinic, orographically forced, planetary wave systems. *J. Atmos. Sci.*, **37**, 1157–76.

Cheng, X. and J. M. Wallace (1993). Cluster analysis on the Northern Hemisphere wintertime 500-hPa height field: spatial patterns. *J. Atmos. Sci.*, **50**, 2674–95.

Chessa, P. A. and F. Lalaurette (2001). Verification of the ECMWF ensemble prediction system forecasts: a study of large-scale patterns. *Weather Forecast.*, **5**, 611–19.

Corti, S., F. Molteni and T. N. Palmer (1999). Signature of recent climate change in frequencies of natural atmospheric circulation regimes. *Nature*, **398**, 799–802.

D'Andrea, F. and R. Vautard (2001). Extratropical low-frequency variability as a low dimensional problem. I: A simplified model. *Quart. J. Roy. Meteor. Soc*, **127**, 1357–74.

Ehrendorfer, M. (1994). The Liouville equation and its potential usefulness for the prediction of forecast skill. I: Theory. *Mon. Weather Rev.*, **122**, 703–13.

Ghil, M. (1987). Dynamics, statistics and predictability of planetary flow regimes. In *Irreversible Phenomena and Dynamical Systems Analysis in the Geosciences*, ed. C. Nicolis and G. Nicolis, pp. 241–283. D. Reidel.

Gleeson, T. A. (1970). Statistical-dynamical predictions. *J. Appl. Meteor.*, **9**, 333–44.

Haines, K. and A. Hannachi (1995). Weather regimes in the Pacific from a GCM. *J. Atmos. Sci.*, **52**, 2444–62.

Hansen, A. R. and A. Sutera (1986). On the probability density distribution of large-scale atmospheric wave amplitude. *J. Atmos. Sci.*, **43**, 3250–65.

(1995). The probability density distribution of planetary-scale atmospheric wave amplitude revisited. *J. Atmos. Sci.*, **52**, 2463–72.

Held, I. M. and M. J. Suarez (1994). A proposal for the intercomparison of the dynamical cores of atmospheric general circulation models. *Bull. Am. Meteorol. Soc.*, **75**, 1825–30.

Hoerling M. P., J. W. Hurrell and T. Xu (2001). Tropical origins for recent North Atlantic climatic change. *Science*, **292**, 90–2.

Houtekamer, P. L., L. Lefaivre, J. Derome, H. Ritchie and H. L. Mitchell (1996). A system simulation approach to ensemble prediction. *Mon. Weather Rev*, **124**, 1225–42.

Hsu, C. J. and F. Zwiers (2001). Climate change in recurrent regimes and modes of Northern Hemisphere atmospheric variability. *J. Geophy. Res.-Atmos*, **106**, 20145–59.

IPCC (2001). *Climate Change 2001: The Scientific Basis*, eds. J. T. Houghton, Y. Ding, D. J. Griggs, *et al.* Cambridge University Press.

Kalnay, E., M. Kanamitsu, R. Kistler, *et al.* (1996). The NCEP/NCAR 40-year reanalysis project. *Bull. Am. Meteorol. Soc.*, **77**, 437–71.

Kimoto, M. and M. Ghil (1993a). Multiple flow regimes in the Northern Hemisphere winter. I: Methodology and hemispheric regimes. *J. Atmos. Sci.*, **50**, 2625–43.

(1993b). Multiple flow regimes in the Northern Hemisphere winter. II: Sectorial regimes and preferred transitions. *J. Atmos. Sci.*, **50**, 2645–73.

Lorenz, E. N. (1975). Climate predictability. In *The Physical Basis of Climate Modeling*, pp. 132–6. GARP Publication Series, 16 WMO.

Marshall, J. and F. Molteni (1993). Toward a dynamical understanding of planetary-scale flow regimes. *J. Atmos. Sci.*, **50**, 1792–818.

Michelangeli, P.-A., R. Vautard and B. Legras (1995). Weather regimes: recurrence and quasi-stationarity. *J. Atmos. Sci.* **52**, 1237–56.

Mo, K. C. and M. Ghil (1987). Statistics and dynamics of persistent anomalies. *J. Atmos. Sci.*, **44**, 877–901.

(1988). Cluster analysis of multiple planetary flow regimes. *J. Geophys. Res.*, **93D**, 10927–52.

Molteni, F. (2003). Atmospheric simulations using a GCM with simplified physical parametrizations. I: Model climatology and variability in multi-decadal experiments. *Clim. Dynam.*, **20**, 175–91.

Molteni F. and S. Corti (1998). Long term fluctuations in the statistical properties of low-frequency variability: dynamical origin and predictability. *Quart. J. Roy. Meteor. Soc.*, **124**, 495–526.

Molteni, F., S. Tibaldi and T. N. Palmer (1990). Regimes in the wintertime circulation over northern extratropics. I: Observational evidence. *Quart. J. Roy. Meteor. Soc.*, **116**, 31–67.

Molteni, F., L. Ferranti, T. N. Palmer and P. Viterbo (1993). A dynamical interpretation of the global response to equatorial Pacific SST anomalies. *J. Climate*, **6**, 777–95.

Molteni, F., R. Buizza, T. N. Palmer and T. Petroliagis (1996). The ECMWF ensemble prediction system: methodology and validation. *Quart. J. Roy. Meteor. Soc.*, **122**, 73–119.

Molteni, F., S. Corti, L. Ferranti and J. M. Slingo (2003). Predictability experiments for the Asian summer monsoon: impact of SST anomalies on interannual and intraseasonal variability. *J. Climate*, **16**, 4001–21.

Nitsche, G., J. M. Wallace and C. Kooperberg (1994). Is there evidence of multiple equilibria in planetary wave amplitude statistics? *J. Atmos. Sci.*, **51**, 314–22.

Palmer, T. N. (1993). Extended-range atmospheric prediction and the Lorenz model. *Bull. Am. Meteorol. Soc.*, **74**, 49–66.

(1999). A nonlinear dynamical perspective on climate prediction. *J. Climate*, **12**, 575–91.

Reinhold, B. and R. T. Pierrehumbert (1982). Dynamics of weather regimes: quasi-stationary waves and blocking. *Mon. Weather Rev.*, **110**, 1105–45.

Silverman, B. W. (1981). Using kernel-density estimators to investigate multimodality. *J. Roy. Stat. Soc.*, **B43**, 97–9.

(1986). *Density Estimation for Statistics and Data Analysis*. Chapman and Hall.

Smith, C. A. and P. Sardeshmukh (2000). The effect of ENSO on the intraseasonal variance of surface temperature in winter. *Int. J. Climatol.*, **20**, 1543–57.

Smith, T. M., R. W. Reynolds, R. E. Livezey and D. C. Stokes (1996). Reconstruction of historical sea surface temperatures using empirical orthogonal functions. *J. Climate*, **9**, 1403–20.

Smyth, P., K. Die and M. Ghil (1999). Multiple regimes in Northern Hemisphere height fields via mixture model clustering. *J. Atmos. Sci.*, **56**, 3704–23.

Stephenson, D. B., A. Hannachi and A. O'Neill (2004). On the existence of multiple climate regimes. *Quart. J. Roy. Meteor. Soc.*, **130**, 583–606.

Straus, D. and F. Molteni (2004). Circulation regimes and SST forcing: results from large GCM ensembles. *J. Climate*, **17**, 1641–56.

Thompson, D. W. J. and J. M. Wallace (1998). The Arctic Oscillation signature in the wintertime geopotential height and temperature fields. *Geophys. Res. Lett.*, **25**, 1297–300.

Toth, Z. and E. Kalnay (1993). Ensemble forecasting at NMC: the generation of perturbations. *Bull. Am. Meteorol. Soc.*, **74**, 2317–30.

Trenberth, K. E. and T. J. Hoar (1997). El Niño and climate change. *Geophys. Res. Lett.*, **24**, 3057–60.

Vautard, R. (1990). Multiple weather regimes over the North Atlantic. Analysis of precursor and successors. *Mon. Weather Rev.* **118**, 2056–81.

Vautard, R. and B. Legras (1988). On the source of midlatitude low-frequency variability. II: Nonlinear equilibration of weather regimes. *J. Atmos. Sci.*, **45**, 2845–67.

Wallace, J. M., Y. Zhang and L. Bajuk (1996). Interpretation of interdecadal trends in Northern Hemisphere surface air temperature. *J. Climate*, **9**, 249–59.

Zhang, Y., J. M. Wallace and D. S. Battisti (1997). ENSO-like interdecadal variability: 1900–93. *J. Climate*, **10**, 1004–20.

15

Model error in weather and climate forecasting

Myles Allen, David Frame
Department of Physics, University of Oxford

Jamie Kettleborough
Space Science and Technology Department, Rutherford Appleton Laboratory, Didcot

David Stainforth
Department of Physics, University of Oxford

As if someone were to buy several copies of the morning newspaper to assure himself that what it said was true.

Ludwig Wittgenstein

15.1 Introduction

The phrase 'model error' means different things to different people, frequently arousing surprisingly passionate emotions. Everyone accepts that all models are wrong, but to some this is simply an annoying caveat on otherwise robust (albeit model-dependent) conclusions, while to others it means that no inference based on 'electronic storytelling' can be taken seriously at all. This chapter will focus on how to quantify and minimise the cumulative effect of model 'imperfections' (errors by any other name, but we are trying to avoid inflammatory language) that either have not been eliminated because of incomplete observations/understanding or cannot be eliminated because they are intrinsic to the model's structure. We will not provide a recipe for eliminating these imperfections, but rather some ideas on how to live with them. Live with them we must, because no matter how clever model developers, or how fast supercomputers, become, these imperfections will always be with us and represent the hardest source of uncertainty to quantify in a weather or climate forecast (Smith, this volume). This is not meant to underestimate the importance of

Predictability of Weather and Climate, ed. Tim Palmer and Renate Hagedorn. Published by Cambridge University Press.

identifying and improving representations of dynamics (see Hoskins, this volume) or parametrisations (see Palmer, this volume) or existing (and planned) ensemble-based forecast systems (Anderson, Buizza, this volume), merely to draw attention to the fact that our models will always be subject to error or inadequacy (Smith, this volume), and that this fact is especially chronic in those cases where we lack the ability to use conventional verification/falsification procedures (i.e. the climate forecasting problem).

We spend a little time reviewing how uncertainty in the current state of the atmosphere/ocean system is used in the initialisation of ensemble weather or seasonal forecasting, because it might appear to provide a reasonable starting point for the treatment of model error. This turns out not to be the case, because of the fundamental problem of the lack of an objective metric or norm for model error: we have no way of measuring the distance between two models or model versions in terms of their input parameters or structure in all but an irrelevant subset of cases. Hence there is no way of allowing for model error by sampling the space of 'all possible models' in a representative way, because distance within this space is undefinable. This is a fundamental problem that fatally undermines many 'simple' or 'intuitive' approaches to the treatment of model error which, after examination, turn out to be nothing of the kind.

If the only way of measuring model similarity is in terms of outputs, complications arise when we also wish to use these outputs to compare with observations in the initialisation of a forecast. These problems are particularly acute in seasonal or climate forecasting where the range of relevant observational datasets is extremely limited. Naïve strategies run the risk of using observations twice and hence underestimating uncertainties by a significant margin.

A possible way forward is suggested by stepping back to remind ourselves of the basic epistemological status of a probabilistic forecast. Any forecast of a single or seldom-repeated event that is couched in probabilistic terms is fundamentally unfalsifiable. The only way of verifying or falsifying a probabilistic statement is through examining a sequence of forecasts of similar situations and assessing the forecast system's 'hit rate', and we cannot compute a hit rate with a single verification point (Smith, 1995).

The objection that there are lots of different aspects of a climate forecast that we could verify against observations is a red herring, because all these variables are intimately interlinked. If we produce an ensemble forecast that purports to span a nominal 5–95% 'range of uncertainty', it is clearly absurd to expect 90% of the variables simulated by individual members of that ensemble to be consistent with observations. A forecast that gets the warming rate wrong to 2020 is likely to continue getting it wrong to 2050. A forecast that underestimates the magnitude of an extreme El Niño event in temperature is likely also to underestimate it (after correcting for model bias) in precipitation. In both cases, there is still, in effect, only a single (albeit high-dimensional) point of verification.

This verification problem applies not only to forecasts of anthropogenic climate change but also to forecasting an extreme El Niño event or even a 100-year storm. Because different aspects of model physics, and hence different model errors, may be relevant in the simulation of events of different magnitudes, a probabilistic forecasting system that performs well at forecasting 'normal' storms or 'normal' El Niños may prove incapable of even simulating, let alone forecasting, a 100-year storm or 1000-year El Niño. But these are precisely the cases in which there is most pressure on the forecaster to 'get it right' or at the very least to provide the user with some warning that conditions have reached the point where the forecast can no longer be relied upon.

This lack of any conceivable objective verification/falsification procedure has led some commentators to conclude that forecasts of anthropogenic climate change are fundamentally subjective and can never represent anything other than a formal expression of the beliefs of the forecaster(s). If you buy this view, then any probabilistic forecast of an unprecedented event goes down as subjective as well, so any El Niño forecaster with strongly held beliefs who specialises in probabilistic forecasts of unusual El Niños (which, because they are unusual, cannot be subjected to traditional out-of-sample verification procedures) could claim equality of status with the European Centre for Medium-Range Weather Forecasts. Worse still, assessment of forecasts of anthropogenic climate change degenerates all too easily into a dissection of the prior beliefs and motivations of the forecasters, 'placing climate forecasts in their sociological context'.

As die-hard old-fashioned realists, we firmly reject such New Age inclusivity. We also recognise, however, that the subjectivists have a point, in that conventional verification/falsification procedures cannot be applied to probabilistic climate forecasts, so it is up to us to clarify what we mean by a good forecast (or forecast system). The measure of goodness, we argue, ought to be whether or not a forecast is likely to have *converged*, in the sense that further developments in modelling (increasing model resolution, or including different parametrisations) are unlikely to result in a substantial revision of the estimated distribution of the forecast variable in question.

Modellers may find this counterintuitive, because they are accustomed to expecting an increase in model resolution or improved parametrisations to change the behaviour of their models in forecast mode, 'otherwise why bother?' The point is that if the next generation of models change the forecast, what guarantee do we have that the generation after that will not change it again? Only a few years ago we were told that climate forecasts 'could only be taken seriously' if performed with models that could run without surface flux adjustment. Now that this has been achieved, we are told they can only be taken seriously when they resolve oceanic eddies. But many aspects of simulated climate change have altered remarkably little, notably the overall shape (not the magnitude, which depends on the individual models' sensitivities) of the global mean response to observed and projected greenhouse gas emissions over the past and coming half-centuries.

If all the changes in resolution, parametrisations or model structure that we attempt *fail* to alter some aspect of the forecast (the ratio of warming over the past 50 years to projected warming over the next 50 years under a specific emissions scenario, in this example), then we might hope that this aspect of the forecast is being determined by a combination of the observations and the basic physical constraints which all properly formulated climate models share, such as energy conservation, and not by the arbitrary choices made by model developers.

We cannot, and will never be able to, treat model error in the formal manner in which observational error is taken into account in short-range ensemble forecasting (see, for example, Buizza, this volume), because of the problems of sampling 'all possible models' and defining a metric for changes to model parameters and structure. The aim, therefore, must be to render this kind of model error as irrelevant as possible, by ensuring that our forecasts depend on data and on the constraints that all physically conceivable models share rather than on any specific set of models, no matter how these are chosen or weighted. Such forecasts might be called STAID, or STAble Inferences from Data. STAID forecasts are unexcitable, largely immune from the whims of modelling opinion. They are also less sexy and attractive than forecasts based on a single (preferably young) super-high-resolution model. But ultimately, they are reliable: they will not change except through the painstaking acquisition and incorporation of new data.

We lay out below a methodology for STAID probabilistic forecasting by looking for convergent results across nested ensembles. Our focus will be on forecasting long-term climate change because this is the problem in which our current lack of a systematic treatment of model error is most acute, but the underlying principle could be applied to shorter-term (seasonal or even, ultimately, synoptic timescale) forecasting problems if the computing power becomes available for the necessary very large ensembles.

15.2 Model shortcomings, random errors and systematic errors

We will begin by setting out what we do *not* mean by model error. To the majority of model developers, 'model error' evokes issues such as the fact that such-and-such a climate model does not contain a dynamical representation of sea ice, or generates an unrealistic amount of low-level cloud. Their solution to these types of error is, naturally enough, to improve the model, either by increasing resolution or by introducing new physics. We will refer to this kind of research as resolving model shortcomings rather than a systematic treatment of model error, and it is not the focus of this chapter. While it is clearly desirable, it is important to recognise that the process of resolving model shortcomings is open-ended and will never represent the whole story. No matter how high the resolution or detailed the physical parametrisations of models in the future, results will always be subject to the two further types of error: random errors due to the cumulative impact of unresolved processes on the resolved flow, and systematic errors due either to parameters not being adequately constrained by available observations or to the structure of the model being incapable of representing the phenomena of interest.

Although an important area of current research, the treatment of random errors is relatively straightforward, at least in principle, and also not our focus here. Parametrisation schemes typically represent the impact of small-scale processes as a purely deterministic relationship between inputs and outputs defined in terms of the large-scale flow. Recently, experimentation has begun with explicitly stochastic parametrisation schemes as well as explicit representation of the effects of unresolved processes through stochastic perturbation of the physics tendency in the model prognostic equations (Buizza *et al.*, 1999; Puri *et al.*, 2001). The theory of 'stochastic optimals' (Moore and Kleeman, 1999; Nolan and Farrell, 1999) is being developed as a means of identifying those components of stochastic forcing that contribute most to forecast error (see also Palmer, this volume).

Whether or not stochastic forcing is included in the model, random unresolved processes will introduce unpredictable differences between model simulations and the real world, but this source of model error is relatively straightforward to treat in the context of linear estimation theory. A much more challenging problem is the treatment of systematic error, meaning those model biases that we either do not know about or have not yet had a chance to address.

Some model developers view proposals to develop a quantitative treatment of systematic error with suspicion since our objective appears to be to work out how to 'live with' model shortcomings that remain unresolved, unexplained or simply unknown. Their concern is that if usable forecasts can be made with existing models, warts and all, then the case for further model development, increasing resolution and so forth may be weakened. As it turns out, a comprehensive treatment of systematic model error demands very substantial model development resources in itself, so this is not a realistic threat. In particular, the climate*prediction*.net experiment (Stainforth *et al.*, 2004) has shown that STAID forecasting can be usefully conducted on publicly volunteered computing (following a distributed computing approach), thereby utilising an entirely new, extremely inexpensive computer resource that complements existing model development programmes.

There is, however, a genuine issue regarding the appropriate allocation of resources between quantifying the possible impact of systematic errors on the forecast system and attempting to get rid of them by improving the model. For some sources of error, it may be cheaper to eliminate them than to quantify their impact on the forecast, but this will never be true in all cases. Particularly on longer (climate) forecasting timescales, or in forecasting unprecedented events such as a record-breaking El Niño, we may simply not have the data or understanding to pin down crucial uncertainties in the model until after the event has occurred. For many of the most interesting forecasting situations, therefore, a systematic treatment of model error is essential for forecasts to be useful at all. Since no one would suggest that we should keep our probabilistic forecasts useless so that we can maintain the case for making them less useless at some unspecified time in the future, some systematic treatment of model error is essential.

15.3 Analysis error and model error: helpful analogy or cul-de-sac?

The most obvious starting point for a treatment of model error in weather and climate forecasting is as an extension of the well-established literature on the treatment of analysis error in ensemble weather forecasting. This will have been discussed extensively elsewhere in this volume, so we only provide a cursory summary, and refer the reader to, for example, Molteni *et al.* (1996) for more details and to Mylne *et al.* (2002) for the extension of these principles to the multimodel context. Suppose the analysis from which a forecast is initialised is based on a standard optimal interpolation or Kalman filter, ignoring for present purposes the many technical issues regarding how such a filter might be implemented in practice:

$$\mathbf{x}_a = \mathbf{x}_b + (\mathbf{H}^T\mathbf{R}^{-1}\mathbf{H} + \mathbf{B}^{-1})^{-1}\mathbf{H}^T\mathbf{R}^{-1}(\mathbf{y} - \mathbf{H}\mathbf{x}_b), \tag{15.1}$$

where $\mathbf{x}_a$ is the estimated state of the system at the time of the forecast initialisation, or more generally a four-dimensional description of the system over some period running up to the forecast initialisation time; $\mathbf{x}_b$ is the 'background' or a-priori estimate of the state vector $\mathbf{x}$ obtained, in a Kalman filter, by integrating the forward model from the previous analysis; $\mathbf{y}$ is a set of observations that depend on the true state $\mathbf{x}$ via $\mathbf{y} = \mathbf{H}\mathbf{x} + \mathbf{u}$; $\mathbf{R}$ is the measurement noise covariance, $\mathbf{R} = \langle \mathbf{u}\mathbf{u}^T \rangle$ (hence, for simplicity, incorporating error in the measurement operator $\mathbf{H}$ into $\mathbf{u}$); and $\mathbf{B}$ is the all-important, and typically rather obscure, 'background error covariance' into which we might hope to bury our treatment of model error.

The derivation of the Kalman filter equations is based on the assumption that the evolution of the state vector $\mathbf{x}$ and the measurement error $\mathbf{u}$ can be treated, at least at the level of analysis uncertainty, as linear stochastic processes whose properties are completely determined by the covariance matrices $\mathbf{B}$ and $\mathbf{R}$, and that these matrices are known, or at least knowable (see Judd, 2003, for a more complete discussion). For a discretisable system in which

$$\mathbf{x}_t = \mathbf{M}\mathbf{x}_{t-1} + \mathbf{N}\mathbf{z}_t, \tag{15.2}$$

where $\mathbf{z}_t$ is a vector of unit-variance, uncorrelated noise, $\langle \mathbf{z}_t\mathbf{z}_t^T \rangle = \mathbf{I}$ and the forward propagator $\mathbf{M}$ is known, then $\mathbf{B} = \mathbf{N}\mathbf{N}^T$. Note that $\mathbf{M}$ and $\mathbf{N}$ might be functions of

$\mathbf{x}$, so the theory applies to non-linear models, provided $\mathbf{M}$ and $\mathbf{N}$ can be treated as constant over the analysis increment, $\mathbf{x}_a - \mathbf{x}_b$.

If $\mathbf{x}_b$ and $\mathbf{u}$ are multivariate normal, the analysis covariance, $\mathbf{A} = \langle(\mathbf{x}_a - \mathbf{x})(\mathbf{x}_a - \mathbf{x})^T\rangle$, is given most compactly by

$$\mathbf{A}^{-1} = (\mathbf{HRH}^T)^{-1} + \mathbf{B}^{-1} \tag{15.3}$$

Since the notion that measurement errors can be treated, in some coordinate system, as multivariate normal is not too far fetched, their contribution to the total analysis error is at least conceptually straightforward, although the actual specification of $\mathbf{R}$ for a wide range of multivariate observations presents clear technical challenges. If the state vector really were generated by a linear stochastic process and the governing equations of that process are known (never the case in practice), then the contribution of the background error covariance, $\mathbf{B}$, is also straightforward. The covariance matrix defined by Eq. (15.3) defines a multidimensional ellipsoid surrounding the best-guess estimate of the state vector $\mathbf{x}_a$ within which the true state of the system might be expected to lie. If $\mathbf{z}$ is simply a vector of unit-variance, uncorrelated, Gaussian random numbers, then possible state vectors, consistent with this analysis, are given by

$$\mathbf{x}' = \mathbf{x}_a + \mathbf{A}^{\frac{1}{2}}\mathbf{z}, \tag{15.4}$$

where $\mathbf{A}^{\frac{1}{2}}$ could consist, for example, of the eigenvectors of $\mathbf{A}$, arranged columnwise, each multiplied by the square root of the corresponding eigenvalue: so each $\mathbf{x}'$ contains a random perturbation along each of the eigenvectors of $\mathbf{A}$ scaled by the standard deviation of the analysis error in that direction.

So far so good, but these may sound like sufficiently restrictive conditions such that this theory could never apply to, for example, a weather forecasting situation, since the state vector of the atmosphere is not governed by linear stochastic dynamics. In a low-order, noise-free chaotic dynamical system such as the Lorenz (1963) model, which evolves on a fractal-dimensional attractor, then neither $\mathbf{x}_a$ nor any perturbed version thereof given by Eq. (15.4) will, with probability one, lie on the attractor (Hansen and Smith, 2001). This might appear not to matter if the system is such that trajectories converge rapidly back onto the attractor, although it may cause practical difficulties if the impact of the initial conditions of the forecast being 'unbalanced' (off the system attractor) persist for a significant length of time (or worse, render the forecast numerically unstable). For this reason, considerable care is taken to 'initialise' members of an ensemble weather forecast to reduce the impact of unphysically ageostrophic terms in their initial conditions.

A more fundamental difficulty arises from the fact that the distribution of trajectories consistent with a given set of observations and the dynamics of the underlying attractor may be highly non-uniform across the ellipsoid defined by Eq. (15.4), as illustrated in Figure 15.2 of Hansen and Smith (2001). One might hope that this kind of information could be reflected in the background error covariance matrix

B, but because the space occupied by the attractor is fractal, no possible coordinate transformation could convert the pattern of accessible trajectories into a multinormal distribution.

These problems with the application of linear estimation theory to highly idealised, low-dimensional dynamical systems may, however, give a misleadingly negative impression of its applicability to much higher order systems such as weather or climate models. The reason is that small-scale processes such as cloud formation may introduce sufficient high-order 'noise' into a weather model such that the range of accessible trajectories is effectively space filling (at least in the space of 'balanced' perturbations) over regions consistent with the observations. Let us suppose for the sake of argument that the analysis is sufficiently accurate that analysis errors can be treated as linear and the notion of an ellipsoid of possible state vectors, consistent with the observations, given by Eq. (15.4) at least makes sense.

The fact that we are treating the dynamics as linear stochastic on small scales in the immediate vicinity of $\mathbf{x}_a$ does not, of course, restrict us to linear stochastic dynamics on larger scales, such as the propagation of $\mathbf{x}$ over the forecast lead time:

$$\mathbf{x}_f = \mathcal{M}(\mathbf{x}_a). \tag{15.5}$$

Indeed, exploiting the non-linearity of the system over the forecast time lies at the heart of so-called 'optimal' forecast perturbation systems such as singular vectors (Palmer *et al.*, 1994) and breeding vectors (Toth and Kalnay, 1997). Consider a two-dimensional example: in a coordinate system in which the analysis error is uniform in all directions, errors in the vertical direction grow over the forecast lead time while errors in the horizontal direction decay. Hence, for any perturbation, only its projection onto the vertical matters for the forecast spread.

In this two-dimensional system, the variance of an arbitrarily oriented vector in the vertical direction is half that of a vector of the same expected length oriented specifically in the vertical. Since this variance fraction declines with the dimension of manifold to which the state vector $\mathbf{x}$ is confined, an arbitrarily oriented perturbation on a weather or climate model might project only a tiny fraction of its variance in any given direction (for a balanced, initialised perturbation the discrepancy might be smaller, but still large). If variance in all other directions were simply to disappear over the course of the forecast without affecting what happens to the projection of the perturbation onto the directions in which errors grow, then this initial orientation would not matter: we simply have to ensure that perturbations have sufficient power in all directions to populate the n dimensional ellipsoid defined by Eq. (15.4). But since, for a weather forecasting model, n may be extremely large, the result would be that we would need very high total amplitude perturbations in order to ensure that their projection onto a small number of error-growth patterns gives a representative forecast spread, and the larger the perturbations, the more difficult it becomes to ensure they are sufficiently balanced for a stable forecast. Hence perturbations are confined to the directions on which errors might be expected to grow, on the basis of the singular vector or breeding vector analysis.

In principle, each perturbation $\mathbf{x}'$ should contain a random component consistent with the analysis error in each of the n' (mutually orthogonal) directions in which errors are expected to grow, scaled such that $\mathbf{x}'^T\mathbf{A}^{-1}\mathbf{x}' = n'$. This is complicated by the fact that n' is flow dependent and poorly determined in a complex non-linear system. In practice, perturbations are applied to rather more directions than necessary with somewhat higher amplitude (there is some ambiguity in the 'correct' amplitude in any case because of uncertainty in $\mathbf{A}$) and the dynamics of the ensemble sorts out n' (components of perturbations in directions that don't grow don't matter).

As stated in the introduction, the inclusion of random model error into this overall framework should be relatively straightforward. Uncertainty in the forward propagator of the analysis model, Eq. (15.2), arising from unknown small-scale unresolved processes, could be represented simply as an additional source of variance, augmenting $\mathbf{B}$ in the estimated analysis error. Likewise, in the generation of the ensemble forecast, individual forecast trajectories could be perturbed with the introduction of some form of 'stochastic physics' term to represent the effect of this small-scale noise on the overall ensemble spread. Stochastic physics is already used in the ECMWF ensemble prediction system (Buizza *et al.*, 1999), and has been shown to improve ensemble representativeness at the medium range (Puri *et al.*, 2001). On longer timescales, a substantial body of literature is developing around the concept of 'stochastic optimals', meaning (in a nutshell) forcing perturbations that have maximal impact on forecast spread, analogous to the optimal perturbations on initial conditions identified by singular vectors (Moore and Kleeman, 1999).

While important issues remain to be resolved in the appropriate specification of a stochastic physics term and the derivation of stochastic optimal perturbations, this is not the most challenging class of model error from a theoretical point of view. If unresolved small-scale processes exist that have an impact on the large-scale flow, then including them in the analysis or forecast model is a necessary model improvement, not a systematic treatment of model error as we interpret the term here.

15.4 Parametric and structural uncertainty and the problems of a metric for model error

Suppose the forecast model $\mathcal{M}$ in Eq. (15.5) contains a single underdetermined parameter, p, the uncertainty (prior distribution) of which is known, so $\mathcal{M} = \mathcal{M}(p)$. For the sake of simplicity, let us assume this uncertainty only affects the forecast model and not the model used to generate the analysis. Generating an ensemble forecast allowing for this uncertainty is straightforward. If p has no impact on the dynamics of error growth in the model, then we simply take initial conditions from the region defined by Eq. (15.4) and propagate these forward in time using a set of possible forecast models generated by making random perturbations to the parameter p consistent with its prior distribution. Because, *ex hypothesi*, p is independent of $\mathbf{x}_a$ there is no need to increase the size of the ensemble significantly: we simply perturb p at the same time as we perturb the initial conditions.

Generalising this to a vector of underdetermined parameters, $\mathbf{p}$, is also simple enough, provided the prior distribution of $\mathbf{p}$, $P(\mathbf{p})$, is known: we sample the distribution of possible models $\mathcal{M}(\mathbf{p})$ using random perturbations conditioned on $P(\mathbf{p})$. If perturbations to the initial conditions, $\mathbf{x}'$, have been made by perturbing a random combination of singular/breeding vectors simultaneously, then random perturbations to $\mathbf{p}$ can again be treated just like perturbations to the initial conditions.

If parametric perturbations interact with the growth of initial condition errors, then, again in principle, they can be treated with a simple extension of the singular or breeding vector technology. The goal is now to identify joint perturbations on parameters and initial conditions which maximise error growth, or for which

$$\|\mathcal{M}([\mathbf{x}', \mathbf{p}'])\|_f \gg \|[\mathbf{x}', \mathbf{p}']\|_a \tag{15.6}$$

where primes denote perturbations. The crucial element in all this is that we have a distance measure or metric for parameter perturbations analogous to the analysis error covariance for initial condition perturbations. If the distribution of $\mathbf{p}$ is multinormal and there is no interaction with $\mathbf{x}'$, then a logical distance measure for contribution of $\mathbf{p}$ to the total error is provided by the inverse covariance matrix,

$$\|\mathbf{p}'\|_a = \mathbf{p}'^T \mathbf{C}_{\mathbf{p}}^{-1} \mathbf{p}'. \tag{15.7}$$

If the distribution of $\mathbf{p}$ is not normal but known, then some form of non-Euclidean distance measure could be defined to ensure an unbiased sampling of 'possible' values of $\mathbf{p}$ where 'possible' is defined, crucially, *without* reference to the observations used in the analysis.

As soon as we begin to consider structural uncertainty, or uncertainty in parameters for which no prior distribution is available, then all this tidy formalism breaks down. Unfortunately, the most important sources of model error in weather and climate forecasting are of precisely this pathological nature. The fundamental problem is the absence of a credible prior distribution of 'possible models', defined in terms of model structure and inputs, from which a representative sample can be drawn. In order to perform such a representative sampling, we need to know how far apart two models are in terms of their structure, and how can we possibly compare the 'size' of a perturbation involving reducing the spatial resolution by a factor of ten versus introducing semi-Lagrangian dynamics without reference to model output?

Goldstein and Rougier (2004) make a useful distinction between 'measurable' inputs (like the acceleration due to gravity) and 'tuning' inputs (like the choice of numerical scheme used to integrate the equations) to a computational model. The treatment of measurable inputs is straightforward: distributions can be provided based on what is consistent with direct observations of the process in question, and provided these observations are independent of those used subsequently to initialise the ensemble forecast, everything is unproblematic. Unfortunately, many of the parameters that have the most impact on the behaviour of climate models do not correspond to directly measurable quantities (although they may share names, like 'viscosity'): defining an objective prior distribution for such tuning inputs is effectively impossible, since we have no way of comparing the relative 'likelihood' of different perturbations to model structure (how can we compare varying a parameter with introducing a new functional form for the parametrisation?). The solution (Kennedy and O'Hagen, 2001; Craig *et al.*, 2001; Goldstein and Rougier, 2004) must be to make use of the fact that models make predictions of the past as well as the future, and we discuss how this can be used to get around the problem of a lack of a defensible prior on the tuning parameters in the final sections of this chapter. We believe the practical approach we suggest here should fit nicely into the formalism proposed by Goldstein and Rougier (2004), although there is much to be done on the details.

The difficulty of defining prior distributions without reference to the observations used to initialise the ensemble is particularly acute on climate timescales where the number of independent observations of relevant phenomena are extremely limited. The point is important because a number of modelling centres are beginning to adopt the following approach to ensemble climate forecasting which we might call the 'likelihood-weighted perturbed-physics ensemble'. First a collection of models is obtained either by gathering together whatever models are available (an 'ensemble of opportunity') or by perturbing parameters in one particular model over ranges of uncertainty proposed by experts in the relevant parametrised processes. Second, members of this ensemble are weighted by some measure inversely proportional to their 'likelihood' as indicated by their distance (dissimilarity) from observed climate. Third, a 'probabilistic forecast' is generated from the weighted ensemble. Problems with this approach are discussed in the following section.

15.5 The problem with experts is that they know about things

The use of 'expert prior' information in the treatment of model error in climate forecasting is sufficiently widespread that we feel we should devote a section to the problems intrinsic to this approach. Lest it be thought that we have any problem with experts, we stress that we will end up using expert opinion to design our perturbations in the concluding sections, but we will introduce additional steps in the analysis to minimise the impact of any 'double-counting' this might introduce.

The problem with a direct implementation of a likelihood-weighted perturbed-physics ensemble is that some of the observations are almost certainly used twice, first in determining the perturbations made to the inputs and second in conditioning the ensemble. The result, of course, is that uncertainties are underestimated, perhaps by a significant margin.

Take a very simple example to begin with: suppose, no matter what parameters are perturbed in a climate model, the climate sensitivity (equilibrium response to doubling CO_2) varies in direct proportion to the net cloud forcing (CF) in the model-simulated present-day climate which, in turn, is constrained by the available observations to be zero, with standard deviation $\sigma_{CF} = 4\ \mathrm{Wm}^{-2}$ (if only things were so simple, but we simply wish to make a point here about methodology). We assemble an ensemble of models, either by collecting them up or by asking parametrisation developers to provide 'credible' ranges for parameters in a single model. Suppose we find that the model developers have done their homework well, and net cloud forcing in the resulting ensemble is also $0 \pm 4\ \mathrm{Wm}^{-2}$. We then weight the ensemble by $\exp(-CF^2/\sigma_{CF}^2)$ to mimic their 'likelihood' with respect to current observed climate and find that the variance of CF in the weighted ensemble is $\sigma_{CF}^2/2$.

If the model developers had no knowledge of the fact that the perturbations they were proposing might have an impact on cloud forcing, or no knowledge of current observations and accepted uncertainty ranges for cloud forcing, this would be the correct result: if we double the number of independent normally distributed pieces of information contributing to the same answer, then the variance in the answer is halved. But is it really credible that someone working on cloud parametrisations in a climate model could operate without knowledge of current observations? In reality, of course, the knowledge that a perturbation would be likely to affect cloud forcing and the knowledge of the likely ranges of cloud forcing consistent with the observations would have conditioned the choice of 'reasonable' perturbations to some unquantifiable extent, so uncertainties would have been underestimated by an unquantifiable margin.

In the case of 'ensembles of opportunity' obtained by assembling models from different groups, the situation is likely to be worse than this, since no one particularly wants their model to be the 2-σ outlier in something as basic as cloud forcing, which 5% of models should be if their distribution of behaviour is to be representative of current uncertainty. Hence uncertainty estimates from 'raw' unweighted ensembles of opportunity are likely to be biased low (see, for example, Palmer and Raisanen, 2002), and estimates from likelihood-weighted ensembles of opportunity would be biased even lower. To make matters even worse, it has been proposed (Georgi and Mearns, 2002; Tebaldi *et al.*, 2005) that models should be explicitly penalised for being dissimilar to each other, which would further exacerbate the low bias in uncertainty estimates resulting from the natural social tendency of modelling groups each to aspire to produce the 'best-guess' model.

The cloud forcing case is hypothetical, but there are practical examples of such problems, particularly with ensembles of opportunity. The curve and top axis in Figure 15.1, following Figure 15.1 of Allen and Ingram (2002), show an estimate

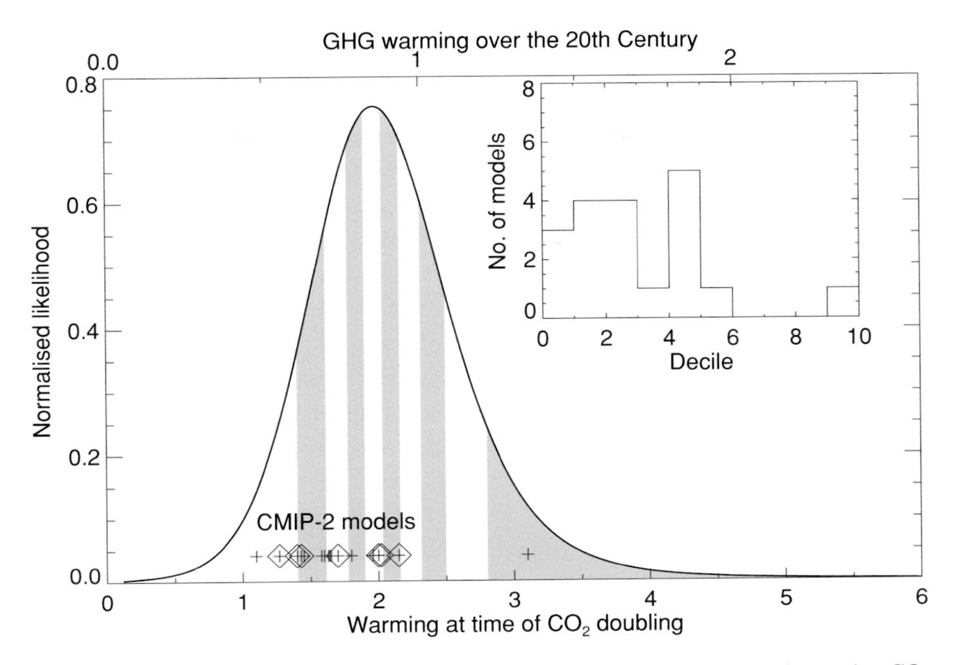

Figure 15.1 Comparison of the distribution of transient responses to increasing CO_2 expressed as attributable warming over the twentieth century (top axis) and transient climate response (TCR) (bottom axis). Crosses indicate members of the CMIP-2 intercomparison, while diamonds show models included in the summary figures of the 2001 IPCC Scientific Assessment. Inset panel shows the number of CMIP-2 models falling into each decile of the distribution.

of the distribution of warming attributable to greenhouse gases over the twentieth century based on the analysis of Stott and Kettleborough (2002). Consistent with current expert opinion, the 'best guess' greenhouse-induced warming, at 0.8 K, is slightly higher than the total warming over the twentieth century, with the net effect of other forcings estimated to be negative, but with a broad and more-or-less symmetric range of uncertainty due to internal variability and the possible confounding effects of other signals.

The bottom axis shows the same information expressed in terms of transient climate response (the expected warming at the time of CO_2 doubling under a 1% per year increasing CO_2 scenario). The advantage of transient climate response (TCR) is that it has been calculated for a wide range of models in the CMIP-2 model intercomparison experiment (Covey *et al.*, 2000), shown by the crosses on the plot. If these models were representative of behaviour consistent with observations, we would expect to see similar numbers falling in each decile (vertical stripe) of the distribution, and a more-or-less flat histogram in the inset panel. They clearly do not, with only two out of 19 models lying in the upper half of the distribution. Should we revise down our estimate of observed greenhouse-induced warming over the twentieth century because the models generally underpredict it, or should we rather suspect that our sample of models is likely to be biased?

Problems also arise in the use of explicitly defined 'expert priors', such as that used by Forest *et al.* (2002) (which we feel comfortable criticising since one of us was a co-author responsible for statistical methodology, although the author-of-intersection would like to emphasise, in self-defence, that this study was far from alone in its overall approach). In one of the cases considered, Forest *et al.* (2002) used an 'expert prior' on climate sensitivity based on a systematic survey of the opinions of a number of climate experts to generate an initial sample of possible models. These were subsequently weighted by a goodness-of-fit statistic involving, among other things, the observed warming over the twentieth century. This procedure is only valid if the opinions of the climate experts are independent of the observed warming.

There are climate experts who claim that they would still be of the opinion that the climate sensitivity (equilibrium warming on doubling of CO_2) is in the range 1.5–4.5 K even if the world had not warmed up by 0.6 K over the past century, more or less in line with a sensitivity somewhere near the middle of this range, and therefore this observation represents independent information. It is true that this range was originally proposed before much of the late-century warming took place, and also before the extent to which anthropogenic aerosol emissions might be masking CO_2-induced warming was established. Hence there might be some truth to the claim that it represented information independent of the observed warming trend back in 1979, but this is much harder to argue now: consider what the consensus view on climate sensitivity would be if the world had not warmed up at all, or had already warmed by 1.5 K. Ideally, of course, we would like the opinion of experts in the physical processes represented in climate models who happen to be unaware of any of the observations we might subsequently use to condition the ensemble, but climate model development is an empirical science, so such 'cloistered experts' (thanks to Lenny Smith for this nomenclature) are hard to find.

The problem becomes worse, not better, if the measure of model-data consistency is widened to encompass a broader range of observations. The more data are put into the ensemble-filtering step, the more likely it is that some of the data were available to the experts designing the models or parameter perturbations used to generate the initial ensemble. If models are originally designed to fit the data, and then pruned to fit better still, the resulting spread will seriously understate the likelihood that the particular cloud data set that *everyone* uses to both tune and evaluate their cirrus parametrisation might just happen to contain a fluke bias.

One response to this problem would be to exhort the modellers to try harder to produce models which are less like the observations in order to populate the tails of the distribution consistent with the data sets available. In the cloud forcing case, for example, if CF in the initial ensemble were uniformly distributed between $\pm 10\ \mathrm{Wm}^{-2}$ then the application of likelihood weighting gives almost exactly the correct variance for the weighted distribution. We have tried this approach in a seminar in one leading climate modelling centre, and would advise against it: persuading model developers to produce parametrisations that are further from uncertain observations in order that they should span the range of behaviour consistent with the observations more completely appears to be a lost cause, however theoretically desirable.

On a purely practical level, models are typically designed such that they are numerically stable only in the vicinity of observed climate, so producing a truly 'flat' prior distribution of something like *CF* would be impossible, and any convexity of the prior distribution across the range consistent with the observations would result in the variance of a likelihood-weighted ensemble being underestimated. The solution we propose in the final sections is simply to renormalise the ensemble by the density of the prior distribution in the space of the forecast variable in question, to mimic what we would have got had we been able to produce a flat prior, without actually going to the pain of producing one.

15.6 Empirical solutions and ensemble dressing: the probabilistic forecaster's flux correction

If a succession of forecasts are available, based on the same forecasting system, of the same generic situation, then workable solutions to the problem of model error may be derived from a consideration of past verification statistics. Consider for simplicity a linear prediction system in which all errors are Gaussian. If it is found that the variance between the forecast verification and ensemble mean is systematically greater than the within-ensemble variance by a factor of two, the error can be corrected simply by inflating the forecast variance above that of the ensemble spread. Systematic forecast biases can also be corrected in this empirical way.

Much more sophisticated post-processing is required to account for non-linearity, discussed in Smith (this volume), but the principle is the same: the ensemble is treated as one source of information in a probabilistic forecast, along with many others. In some respects we might compare this kind of empirical 'ensemble dressing' with the application of empirically determined flux adjustments in climate models. It works, and turns an unusable ensemble forecasting system (or climate model) into something potentially highly informative, but at the same time it could seem disconcertingly ad hoc. The question arises whether problems might be being papered over that would, if resolved, increase the utility of the forecast still further. Whatever our view on ensemble dressing, however, the fundamental problem is that it cannot be applied to what are often the most interesting forecasting situations: forecasting unprecedented or near-unprecedented events.

15.7 Probabilistic forecasts of one-off events: the verification problem

Empirical solutions may be highly effective in compensating for the impact of model error in generic forecasting situations for which a large number of verification instances can be collected. Problems arise when a forecast is to be made of an event which happens relatively seldom, or worse still is entirely unprecedented. The most obvious example of this nature is the problem of forecasting anthropogenic climate change, but similar problems arise on much shorter timescales as well. For example, in El Niño forecasting, only a couple of events have been observed in the kind of detail necessary to initialise a modern seasonal forecast, which is far too few to develop any kind of post-hoc corrections on ensemble spread. Moreover, each El Niño event evolves in a rather different way, making it likely that different sources of model error may be important in forecasting different events. Hence there is no guarantee that an empirical correction scheme, even if based on several El Niño events, will work for the next one.

The solution to this problem is intimately tied up in our basic understanding of what we mean by an 'improved' or even 'accurate' estimate of uncertainty in a forecast of the climate of 2050 or 2100. Assessing (and hence, in principle, improving) the accuracy of estimates of uncertainty in a probabilistic weather forecasting system is at least conceptually straightforward. We can examine a sequence of forecasts of a particular variable and keep track of how often the real world evolves into, say, the fifth percentile of the predicted distribution (Smith, 2000; ECMWF, 2002). If this occurs roughly 5% of the time, then the error estimates based on our forecasting system are acceptable. If it always occurs 10% of the time, then our error estimates can be recalibrated accordingly. But how can we evaluate a probabilistic forecast of anthropogenic climate change to 2050 when we will only have a single validation point, and that one available long after the forecast is of historical interest only?

This problem has led some commentators to assert (Schneider, 2002) that any estimate of uncertainty in forecast climate change will always be fundamentally subjective, ultimately representing current scientific opinion rather than an objectively verifiable or falsifiable assertion of fact. While accepting that there will always be an element of subjectivity in the choice of methodological details or the choice of climate model(s) used in the analysis, we argue that this subjective element, including the dependence of results on the climate model(s) used, can and should be second order, quantified and minimised as far as possible. To resign ourselves to any other position on an issue as contentious as climate change is to risk diverting attention from the science itself to the possible motivations of the experts or modelling communities on which current scientific opinion rests. The presence of subjective aspects in uncertainty estimates in forecasts need not preclude methodological restrictions: while we accept that there is a degree of subjective judgement in uncertainty estimates, it ought to be possible to provide limits or constraints on this subjectivity, particularly regarding specifying how and where judgements should (and should not) be made in the forecast process (see Section 15.9 for details).

Although this sounds a relatively abstract issue, there are deeply practical consequences: the headline result of Wigler and Raper (2001) was that warming rates at the upper end of the range quoted in the IPCC 2001 Scientific Assessment

(Houghton *et al.*, 2001) were extremely unlikely. This conclusion depended, to first order, on those authors' decision to assume (admirably clearly footnoted) that there was only a 1 in 20 likelihood of the climate sensitivity (equilibrium warming on doubling carbon dioxide) exceeding 4.5 K, despite the fact that the available formal surveys of expert opinion (Morgan and Keith, 1995) and estimates based on the analysis of climate observations (Andronova and Schlesinger, 2000; Forest *et al.*, 2002; Knutti *et al.*, 2002; Gregory *et al.*, 2002) suggest a substantially higher upper bound at this confidence level. Statements of the form 'this is our opinion (or, perhaps less provocative but still problematic, our model), and these are its consequences' are unlikely to engender the kind of confidence in the non-scientific community required to justify far-reaching political and economic decisions.

How can we avoid the charge of subjectivism or 'model-relativism'? The solution (Allen and Stainforth, 2002) is to focus on whether or not a probabilistic forecast for a particular climate variable has *converged*, rather than being sidetracked onto the unanswerable question of whether or not it is *correct*. In a similar vein (and, of course, not by coincidence), Smith (2002) argues that we should look for consistency of results across classes of models.

A forecast distribution for a particular variable has converged if it depends primarily on the observations used to constrain the forecast and not, to first order, on the climate model used or subjective opinions of the forecasters. Hence, in claiming that a forecast distribution (of 2100 global temperature, for example, or north-west European winter rainfall in the 2030s or, most challenging of all (Roulston and Smith, 2002), some multivariate combination thereof) has converged, we are claiming that the forecast distribution is unlikely to change substantially due to the further development of climate models or evolution of expert opinion, although uncertainties are likely to continue to be reduced as the real-world signal evolves and more observations are brought to bear on the problem (Stott and Kettleborough, 2002).

A claim of convergence is testable and falsifiable. For instance, it might be found that increasing model resolution systematically and substantially alters the forecast distribution of the variable in question, without the introduction of any new data. We can use our physical understanding of the system to assess whether this is likely to happen in any given instance, depending on the robustness of the constraints linking the forecast variable to observable quantities. In a complex non-linear system, however, any physically based arguments will need to be tested through simulation.

15.8 Stable inference from data: a pragmatic approach to STAID forecasting

Once we have agreed on what we are trying to do in the treatment of model error in a probabilistic forecast of a one-off event, there will doubtless be many different approaches to achieving this end. In these final sections, we outline one approach and discuss the implications for the design of climate forecasting systems. In essence, our approach is inspired directly by the pragmatic approach taken for many years to model tuning. All models contain some tunable parameters which must be set initially at relatively arbitrary values and subsequently adjusted in the process of model development to optimise the fit between model output and observed climate. We simply propose extending this approach to optimise the fit between the distribution of models in a perturbed-physics ensemble and known uncertainties in observable properties of the system being modelled that are directly related to the forecast quantity of interest. The key difference is that when a model is being tuned the usual practice is to adjust the parameter in question and repeat the run, whereas when an ensemble is being tuned, it may be unrealistic to rerun it, so the weights assigned to ensemble members need to be adjusted instead.

What do we mean by 'known uncertainties' and 'observable properties'? We will focus on the 'ideal' climate problem, in which the timescales of interest are much longer than the longest timescale of predictability in the system, so the role of the initial conditions can be ignored. A possible extension to problems in which initial conditions are important is discussed qualitatively in the final section. Suppose $\tilde{\mathbf{x}}_f$ is a particularly interesting quantity (externally driven global warming 2000–30, for example) derived from the full description of the system at the forecast time, $\mathbf{x}_f$, and suppose also that we find, across the members of a perturbed physics ensemble, that there is a consistent linear relationship between $\tilde{\mathbf{x}}_f$ and $\overline{\mathbf{x}}_a$, where $\overline{\mathbf{x}}_a$ is an observable property or set of observable properties of the system over the analysis period (externally driven warming 1950–2000).

We use the term 'observable property' loosely to mean a quantity whose distribution can be constrained primarily by observations with only limited use of modelling, not necessarily something that is directly observable. The requirement of a *linear* relationship between $\tilde{\mathbf{x}}_f$ and $\overline{\mathbf{x}}_a$ may seem restrictive, but it can be satisfied by construction through an appropriate redefinition of $\overline{\mathbf{x}}_a$ provided there is a monotonic and single-valued relationship between them. Aficionados of complexity in the climate system will doubtless protest that only very few forecast variables will satisfy the requirement of a monotonic, single-valued relationship to a finite set of observable quantities, but before dismissing this approach out of hand, recall that we are focusing here on climate variables, or the statistical moments of the weather attractor, which are generally much better behaved than the chaotic trajectory of the system over the attractor itself. Moreover, we do not require that this relationship be exact, simply that it should explain a significant fraction of the variance of the forecast variable of interest over the range consistent with observations: the more the variance explained, the more STAID the forecast.

Let us represent all underdetermined 'tunable inputs' (Goldstein and Rougier, 2004) to the forecasting system for which we do not have a prior distribution, including structural uncertainties, as $\mathbf{q}$. Inputs for which a prior distribution is available that we can be sure is independent of the observations used in the forecast initialisation are treated like $\mathbf{p}$ above. $P(\mathbf{q})$ is undefinable because we have no way of sampling the space of all possible models and no way of saying how far apart two models are even if we could, but we need to assume some sort of distribution for $\mathbf{q}$, $\hat{P}(\mathbf{q})$, in order to get started. The crux of the solution is that we should design the forecasting system to minimise the impact of $\hat{P}(\mathbf{q})$ on the forecast quantity of interest, and only claim forecasts are STAID to the extent that they can be shown not to depend on $\hat{P}(\mathbf{q})$ (there is a direct analogy with the analysis of the role of the a priori in satellite retrievals).

The statement that there is a consistent relationship between $\tilde{\mathbf{x}}_f$ and $\overline{\mathbf{x}}_a$ which does not depend on the choice of model is tantamount to the claim that

$$P(\tilde{\mathbf{x}}_f|\overline{\mathbf{x}}_a, \mathbf{q}) = P(\tilde{\mathbf{x}}_f|\overline{\mathbf{x}}_a) \;\; \forall(\mathbf{q}). \tag{15.8}$$

This claim remains falsifiable even if it appears to be the case for all models (values of $\mathbf{q}$) tested to date, if the incorporation of a new process either changes the relationship between $\tilde{\mathbf{x}}_f$ and $\overline{\mathbf{x}}_a$ (sometimes called a transfer function; Allen *et al.*, 2000) or renders it dependent on new tunable inputs whose values are not constrained by observations. Such claims will be on strongest ground when they can be supported by a fundamental physical understanding of *why* there must be a consistent relationship between $\tilde{\mathbf{x}}_f$ and $\overline{\mathbf{x}}_a$ which any valid model must share, like energy conservation. Many, however, will remain 'emergent constraints' on which models appear to have converged but are not necessarily dictated by the underlying physics.

Similarly, the claim that $\overline{\mathbf{x}}_a$ can be constrained primarily by observations is equivalent to the claim

$$P(\overline{\mathbf{x}}_a|\mathbf{y}, \mathbf{q}) = P(\overline{\mathbf{x}}_a|\mathbf{y}) \;\; \forall(\mathbf{q}). \tag{15.9}$$

If there are no other constraints on the observing system, so $P(\mathbf{y})$ and $P(\overline{\mathbf{x}}_a)$ are both uniform, then $P(\overline{\mathbf{x}}_a|\mathbf{y}) = P(\mathbf{y}|\overline{\mathbf{x}}_a)$, meaning the likelihood of $\overline{\mathbf{x}}_a$ taking a certain value in the light of the observations can be equated with the likelihood of us obtaining these observations given that it does take that value.

Whether or not it is reasonable to regard $P(\overline{\mathbf{x}}_\mathbf{a})$ as uniform over the range consistent with the observations will depend on the variable in question. For relatively well observed quantities like the underlying rate of global warming, this could be a reasonable assumption since, if models were to simulate warming rates consistently lower or higher than that observed, our response would be to revise the models or look for missing forcings, not revise our assessment of how fast the world was warming up. This is precisely what happened before the introduction of sulphate aerosol cooling in the early 1990s: models forced with greenhouse gas increases alone were consistently overpredicting the observed rate of warming, so the community concluded, correctly, that all models were missing some crucial aspect of the physics (aerosol-induced cooling), not that the observations were in error, which was treated as a primary observable. The 'attributable warming' shown in Figure 15.1 is a slightly more derived quantity, but could still be regarded as a primary observable that our ensemble of models should be constrained to fit, and not vice versa.

An estimate of the distribution of $\overline{\mathbf{x}}_a$ can be obtained from an ensemble of 'hindcast' simulations:

$$\tilde{P}(\overline{\mathbf{x}}_a|\mathbf{y}) = \frac{P(\overline{\mathbf{x}}_a|\mathbf{y},\mathbf{q})\hat{P}(\mathbf{q}|\mathbf{y})}{\hat{P}(\mathbf{q}|\overline{\mathbf{x}}_a,\mathbf{y})} \simeq \frac{P(\mathbf{y}|\overline{\mathbf{x}}_a)\hat{P}(\mathbf{q}|\mathbf{y})}{\hat{P}(\mathbf{q}|\overline{\mathbf{x}}_a,\mathbf{y})} \tag{15.10}$$

where the second, more tentative, equality only holds when the above statements about $\overline{\mathbf{x}}_a$ being primarily constrained by observations are true. $\hat{P}(\mathbf{q}|\mathbf{y})$ represents a sample of models (values of the tuning inputs) obtained by perturbing parametrisations by collecting models and model components from a range of different sources (an 'ensemble of opportunity'). It is not an estimate of any distribution, because the distribution of all possible models is undefined, and (no matter what the experts claim) it is conditioned on the observations: we don't know how much, but our objective is to make it the case that we don't care. The denominator $\hat{P}(\mathbf{q}|\overline{\mathbf{x}}_a,\mathbf{y})$ represents the frequency of occurrence of models in this sample in which the simulated $\overline{\mathbf{x}}_a$ lies within a unit distance from any given value, or the observed density of models in the space spanned by $\overline{\mathbf{x}}_a$, given the observations. Its role in Eq. (15.10) is rather like a histogram renormalisation in image processing: it ensures we end up with the same weight of ensemble members in each decile of the $P(\overline{\mathbf{x}}_a|\mathbf{y})$ distribution, forcing a flat inset histogram in Figure 15.1.

Suppose $\mathbf{y}$ and $\overline{\mathbf{x}}_a$ each have only a single element, being the externally forced global temperature trend over the past 50 years (Figure 15.2); $\mathbf{y}$ is observed to be 0.15 K per decade, ± 0.05 K/decade due to observational uncertainty and internal variability. For simplicity of display, we assume internal variability is independent of $\mathbf{q}$, but relaxing this assumption is straightforward. The curve in Figure 15.2 shows $P(\mathbf{y}|\overline{\mathbf{x}}_a)$ and the vertical lines show simulated $\overline{\mathbf{x}}_a$ from a hypothetical ensemble which, like the CMIP-2 ensemble shown in Figure 15.1, is biased with respect to the observations in both mean and spread. For simplicity, we have assumed a large initial-condition ensemble is available for each model version (value of $\mathbf{q}$), so the $\overline{\mathbf{x}}_a$ from the models are delta functions.

A likelihood-weighted perturbed-physics ensemble would estimate the distribution of $P(\overline{\mathbf{x}}_a|\mathbf{y})$ from

$$P'(\overline{\mathbf{x}}_a|\mathbf{y}) = P(\mathbf{y}|\overline{\mathbf{x}}_a)\hat{P}(\mathbf{q}|\mathbf{y}), \tag{15.11}$$

meaning weighting the ensemble members by some measure of their distance from observations and estimating the distribution from the result. This would only give the correct answer if $\hat{P}(\mathbf{q}|\overline{\mathbf{x}}_a,\mathbf{y})$ is uniform across the range consistent with the observations: note that this requires many more models that are almost certainly inconsistent with the observations than producing an ensemble that is consistent with the observations in the sense of providing a flat histogram in the inset panel of

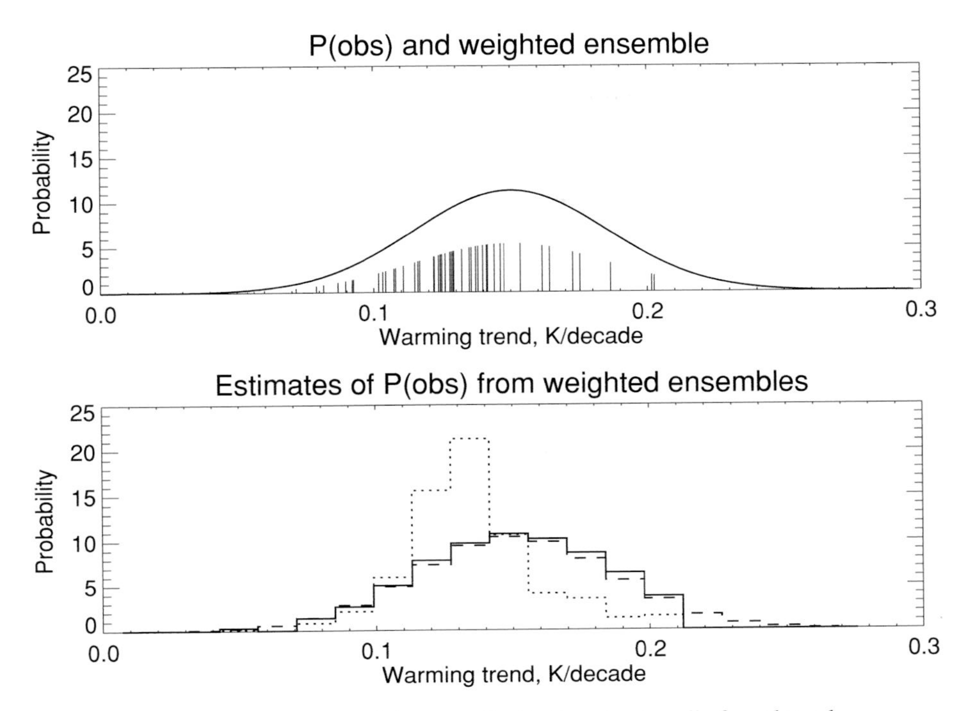

Figure 15.2 (Top panel) Schematic distribution of externally forced trends consistent with an observed trend of 0.15($\pm$0.05)K/decade, $P(\overline{\mathbf{x}}_a|\mathbf{y})$. Locations of vertical lines show model simulations from a hypothetical untargeted ensemble, $\hat{P}(\mathbf{q}, \mathbf{y})$, with heights of lines showing likelihood-based weights $P(\mathbf{y}|\overline{\mathbf{x}}_a)$. (Bottom panel) solid and dashed lines show estimates of $P(\overline{\mathbf{x}}_a|\mathbf{y})$ based on Eq. (15.10) with ensembles of 50 and 10 000 members respectively, while dotted histogram shows an estimate based on Eq. (15.11), which is biased in both mean and spread.

Figure 15.1. In this case, of course, $\hat{P}(\mathbf{q}|\overline{\mathbf{x}}_a, \mathbf{y})$ is not uniform (the vertical lines are not uniformly distributed over the 0–0.3 K/decade interval), so the likelihood-weighted perturbed-physics ensemble gives the wrong answer (dotted histogram).

A histogram-renormalised likelihood-weighted perturbed-physics ensemble (a bit of a mouthful, so let's just call it STAID) given by Eq. (15.10) gives the correct answer, but at a price. Because we need to compute the density of models in the space spanned by the observable quantity $\overline{\mathbf{x}}_a$, we need to ensure we have a large enough ensemble to populate this space. Sophisticated density-estimation methods would help (we have used a very unsophisticated one), but these would depend on assumptions about the smoothness of the response to variations in tunable inputs that could only be tested by simulation. The problem is that, for a given level of smoothness, the requisite ensemble size increases geometrically with the rank of $\overline{\mathbf{x}}_a$. In this case, $\overline{\mathbf{x}}_a$ has only a single element, and estimates $\hat{P}(\overline{\mathbf{x}}_a|\mathbf{y})$ from Eq. (15.10) are shown by the solid and dashed histograms in the figure, using 50- and 10 000-member ensembles respectively. While both are substantially closer to the true uncertainty range, there are significant distortions in the tails of the distribution (which are relevant to many policy decisions) estimated from the smaller ensemble. If no members of the ensemble happen to lie in a particular region consistent with recent climate observations, and we aren't justified in interpolating from adjacent regions, then no amount of reweighting can help establish what would happen if they did. So STAID forecasting needs large ensembles. The key advantage of the STAID ensemble, however, is that its dependence on $\hat{P}(\mathbf{q})$ is second order. Provided the ensemble is big enough to populate the space spanned by $\overline{\mathbf{x}}_a$, the impact of that troublesome prior is integrated out.

So far, this all looks like a very laborious way of recovering $P(\overline{\mathbf{x}}_a|\mathbf{y})$ which we have already said we are prepared to equate with $P(\mathbf{y}|\overline{\mathbf{x}}_a)$, so why bother? We are interested in $\overline{\mathbf{x}}_a$ because it is linearly related to some forecast quantity, $\tilde{\mathbf{x}}_f$, independent of both $\mathbf{q}$ and $\mathbf{y}$. Hence

$$\tilde{P}(\tilde{\mathbf{x}}_f|\mathbf{y}) = \frac{P(\tilde{\mathbf{x}}_f|\overline{\mathbf{x}}_a)\tilde{P}(\overline{\mathbf{x}}_a|\mathbf{y})}{P(\overline{\mathbf{x}}_a|\tilde{\mathbf{x}}_f)} \quad \forall(\mathbf{q},\mathbf{y}), \tag{15.12}$$

where the transfer function $P(\tilde{\mathbf{x}}_f|\overline{\mathbf{x}}_a)/P(\overline{\mathbf{x}}_a|\tilde{\mathbf{x}}_f)$ is provided, ideally, by fundamental physics or, in practice, by the emergent constraints indicated by the ensemble. Provided the transfer function is linear, then having determined the weights on ensemble members necessary to recover an unbiased estimate of $P(\overline{\mathbf{x}}_a|\mathbf{y})$ and established that the relationship between $\tilde{\mathbf{x}}_f$ and $\overline{\mathbf{x}}_a$ is consistent across the ensemble and not dependent on $\mathbf{q}$, we simply apply those same weights to the forecast ensemble to arrive at an estimate of $P(\tilde{\mathbf{x}}_f)$ which is not, to first order, dependent on $\hat{P}(\mathbf{q})$.

The requirement of a linear transfer function, or that $P(\tilde{\mathbf{x}}_f|\overline{\mathbf{x}}_a)/P(\overline{\mathbf{x}}_a|\tilde{\mathbf{x}}_f)$ is independent of $\mathbf{q}$, means that renormalising the ensemble to ensure a uniform predictive prior distribution in $\overline{\mathbf{x}}_a$ automatically ensures a uniform predictive prior distribution in $\tilde{\mathbf{x}}_f$. We noted above that it will always be possible in principle to redefine our observables to force a linear transfer function, but this does have counterintuitive consequences when the most obvious observable quantities are non-linear in the forecast variable of interest. A perfect example of this problem is given by the climate sensitivity, S, or long-term equilibrium warming response to an increase in greenhouse gas levels followed by indefinite stabilisation. For fundamental physical reasons, observable properties of the climate system that we have available scale with the strength of atmospheric feedbacks, λ, which is inversely proportional to the sensitivity, not with the sensitivity itself. As a result, $P(\tilde{\mathbf{x}}_f|\overline{\mathbf{x}}_a)/P(\overline{\mathbf{x}}_a|\tilde{\mathbf{x}}_f)$ is proportional to S^{-2}, so normalising the ensemble to provide a uniform prior in $\overline{\mathbf{x}}_a$ will significantly downweight the likelihood of a high climate sensitivity.

This is clearly incorrect, since the fact that directly observable properties of the climate system change less with sensitivity as sensitivity increases does not, in itself, make a high value of the sensitivity unlikely. If we had two observable quantities, one of which is linear in sensitivity and the second linear in λ, or S^{-1}, sampling uniformly in the second observable would make it appear to contain more information about high values of sensitivity, by automatically reducing the upper bound, whereas the reverse is clearly true. We need to be very careful not to rule out high values of the sensitivity by such, possibly inadvertent, sleight of hand.

Jewson and Caballero (2003) also note the importance of normalising out the impact of the prior when conditioning an ensemble with probabilistic information. They are conditioning a climatological time-series model of local weather using a probabilistic forecast whereas we are conditioning a perturbed-physics ensemble using probabilistic information about ranges of past warming rates consistent with recent observations, but the underlying objective is the same: a smooth progression of estimated distributions, making maximal use of available information, throughout the hindcast and forecast period.

Renormalising by prior density is a simple enough manoeuvre, required by a 300-year-old theorem, but it has very profound implications for experimental design. Very few forecast quantities of interest will be found to depend exclusively on only a single observable climate variable. That said, given the strong constraints linking different variables in climate models, the number of effectively independent observable dimensions might be relatively small (fewer than half a dozen) at least for large-scale forecast quantities. Sampling a four-dimensional space at decile resolution for the computation of $P(\mathbf{q}|\overline{\mathbf{x}}_a, \mathbf{y})$ requires, however, a $0(10^4)$-member ensemble of climate models, and if we also allow for initial-condition ensembles to increase the signal-to-noise and boundary-condition ensembles to allow for uncertainty in past and future forcing, the desired ensemble size runs into millions. Fortunately, such ensembles are now feasible using public-resource distributed computing (Allen, 1999; Stainforth *et al.*, 2002; Allen and Stainforth, 2002).

Interestingly, if we give up the chase for a defensible prior encompassing model error, we no longer have any universal 'best' set of weights to apply to a perturbed-physics ensemble: the weights will depend on the forecast quantity of interest. The reason is that they are determined by whatever it takes to make the ensemble consistent with current knowledge and uncertainty in observable quantities on which that forecast quantity is found to depend, and different forecast quantities will depend on different observables. There is nothing inconsistent about this, since model errors are likely to have a different impact on different variables, but it does mean that forecast applications need to be integrated much more closely into the forecasting system itself. If a climate impact, for example, depends on some combination of temperature and precipitation, we cannot simply combine results from a weighted ensemble targeting temperature with another targeting precipitation: we need to recompute the weights to identify the observable variables that specifically constrain the function of interest. The good news, however, is that provided the initial ensemble is big enough to be space filling in the relevant observables, then this is simply a post-processing exercise which does not require us to rerun the models.

15.9 The effects of prior beliefs on probabilistic climate forecasts

This section, based on Frame *et al.* (2005), provides a practical application of STAID probabilistic forecasting to the problem of constraining climate sensitivity, or the equilibrium warming response to doubling carbon dioxide levels. Climate sensitivity is generally considered a key determinant of climate change (Morgan and Keith, 1995; Houghton *et al.*, 2001), in particular of the risk that a given greenhouse gas stabilisation level might result in a 'dangerous' long-term equilibrium warming. Many recent studies (Andronova and Schlesinger, 2000; Forest *et al.*, 2002; Knutti *et al.*, 2002; Gregory *et al.*, 2002; Murphy *et al.*, 2004) have attempted to constrain climate sensitivity by comparing models with recent observations, and these studies have reported a surprisingly wide range of distributions. Here we show that much of this variation arises from different prior assumptions regarding climate sensitivity before any physical or observational constraints are applied. This apparent arbitrariness can be resolved by focusing on the intended purpose of the forecast, consistent with the approach proposed in the preceding section.

We demonstrate our point with a simple global energy balance model (EBM) and diffusive ocean (Hansen *et al.*, 1985), although the reasoning applies to any model in which atmospheric feedbacks scale linearly with surface warming and in which effective oceanic heat capacity is approximately constant under twentieth-century climate forcing. The diamonds in Figures 15.3 and 15.5 show the average warming trend caused by greenhouse gas increase over the twentieth-century (vertical axis) as a function of effective heat capacity of the troposphere–land–ocean system (horizontal) and the climate sensitivity S (symbol size: larger sizes correspond to higher sensitivities) for a range of different settings of model parameters. The black contour encloses the region consistent (at the 5% level) with observations of twentieth-century greenhouse warming and the effective heat capacity.

We isolate the greenhouse warming signal using a pattern-based attribution analysis (Stott and Kettleborough, 2002) allowing for uncertainty in both greenhouse and other forcings (Allen *et al.*, 2000). This estimate of attributable warming does not depend on climate sensitivity, although it does rely on the accuracy of patterns of temperature change and variability simulated by a climate model. Heat capacity is inferred from the observed change in global mean oceanic heat content (Levitus *et al.*, 2000, 2005) over the 1957–94 period divided by the corresponding change in decadal-mean surface temperature, allowing for the uncertainty in both quantities. Model parameters are chosen to sample heat capacity approximately uniformly, so the points are evenly spaced in the horizontal. The only difference between Figures 15.3 and 15.5 is the way we sample model parameters. In Figure 15.3, following Andronova and Schlesinger (2000), Forest *et al.* (2000) and Knutti *et al.* (2002), parameters are chosen to sample S uniformly over the range 0.17 to 20 °C.

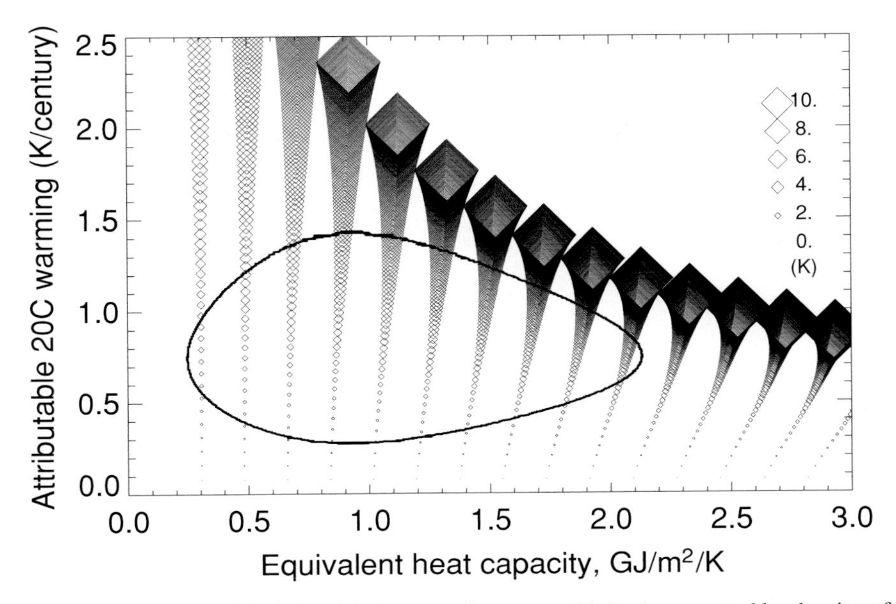

Figure 15.3 Relationship between climate sensitivity (represented by the size of the diamonds), effective ocean heat capacity and twentieth century warming attributable to changes in greenhouse gases. Diamonds show simulation results based on uniform sampling of climate sensitivity, *S*, between 0 and 20 °C. Black contour encloses the region consistent with observations at the 5% level.

Weighting each of the runs in Figure 15.3 by the likelihood of obtaining these observations (specifically, the observed level of model-data discrepancy; Forest *et al.*, 2000) if that combination of sensitivity and effective heat capacity is correct and estimating a 'posterior' distribution for *S* from the weighted ensemble (the dash-dot curve in Figure 15.4) gives a 5–95% range for climate sensitivity of 1.2–13.5 °C. Two factors contribute to this high upper bound. First, for any given ocean heat capacity, the relationship between sensitivity and transient warming to date is non-linear (Allen *et al.*, 2000; Stott and Kettleborough, 2002), which tends to concentrate the diamonds in Figure 15.3 at higher values of past warming. Second, this sampling of model parameters implies a uniform distribution for *S* before any comparison with observations, meaning a sensitivity between 2 and 3 °C is assumed to be as likely as one between 3 and 4 °C, or 9 and 10 °C.

An alternative approach is taken by Murphy *et al.* (2004): using a complex climate model they ran an ensemble of equilibrium $2\times CO_2$ experiments with expert-specified ranges of parameters, sampled parameters uniformly over these ranges and assumed that changes to parameters have an approximately linear impact on λ, the increase in energy radiated to space per degree of warming, which is proportional to $1/S$. If a single parameter dominates changes in λ, as is the case in our simple model (and as also happens to be the case in Murphy *et al.*, 2004, for high sensitivities (another

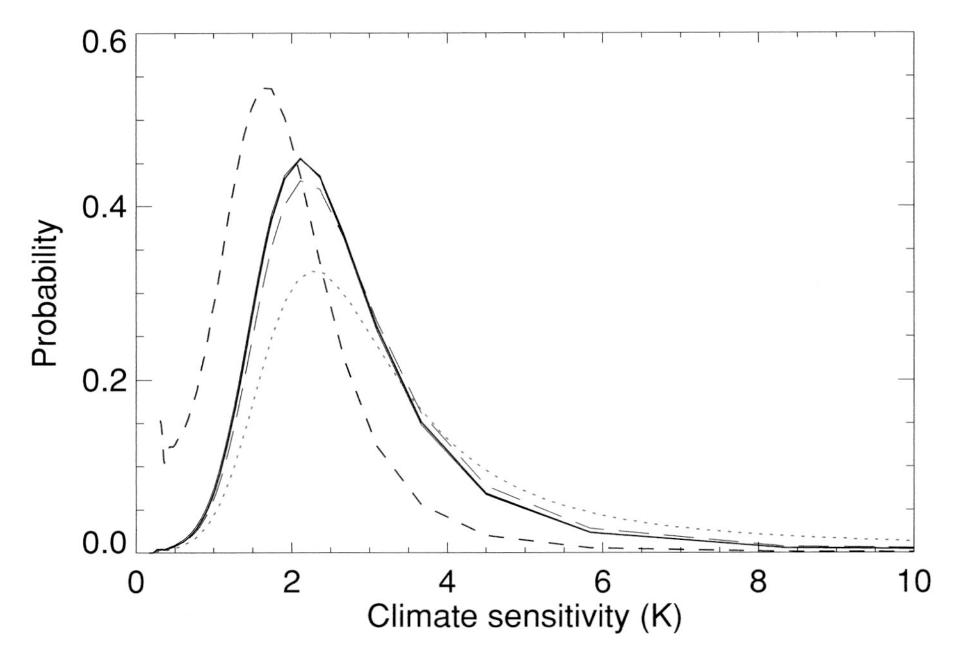

Figure 15.4 Distributions of climate sensitivity based on these observations, assuming a uniform initial distribution in sensitivity (dotted), and in feedback strength (short-dashed), as well as in forecast 1990–2100 warming under the A1FI (long dashed) and B1 (very short dashes) scenarios and in TCR (solid curve).

parameter change yields low sensitivities)), these two assumptions, in this case, imply a sensitivity between 2 and 3 °C is as likely, before any physical or observational constraints are applied, as one between 3 and 6 °C, or between 1 and 1.2 °C. The implications of such a uniform sampling of λ, for our simple model are shown in Figure 15.5: the relationship between sensitivity, warming and heat capacity is the same as in Figure 15.3, but the location of the diamonds is very different. If we weight by comparison with observations as before, we now find a low chance (<3%) of a sensitivity greater than 4.5 K. This leads to the dashed distribution in Figure 15.4 and a 5–95% range for sensitivity of 0.6–4 K.

These two approaches yield very different values for sensitivity, particularly in terms of the upper bound. Which approach is correct? We argue that it depends on what you want to forecast. This can be illustrated by considering the implications of different priors on forecasts of twentieth-century attributable warming: i.e. attempting to reconstruct the data that forms the constraint with which we began, shown in Figure 15.6. The distribution of twentieth-century warming implied by a uniform prior in sensitivity peaks above our best-guess observed warming because of the points clustering in the upper half of the region enclosed by the contour; while the distribution implied by the reciprocal of sensitivity peaks below our original best guess. As a result, neither curve adequately reconstructs the distribution of observed

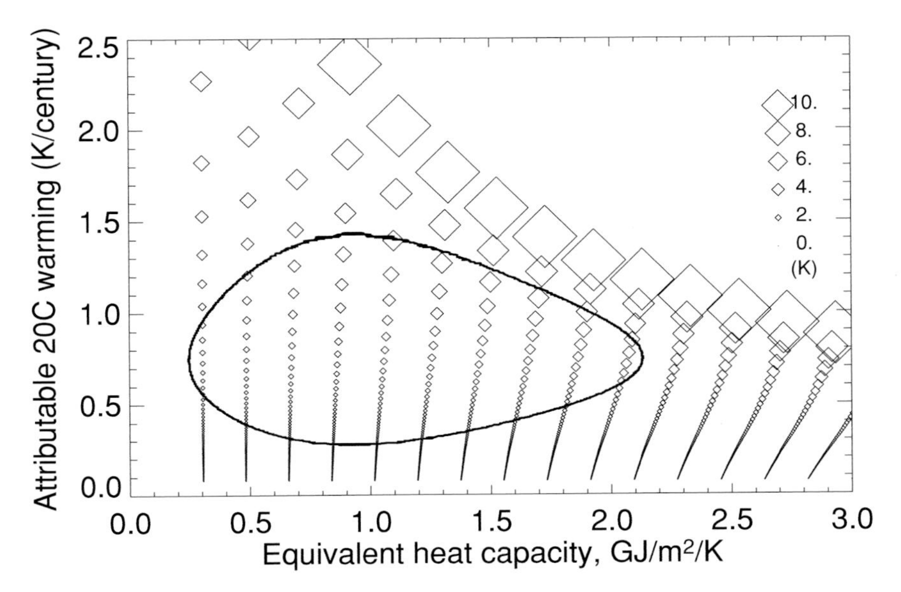

Figure 15.5 As Figure 15.3, but showing simulation results based on a uniform sampling of feedback strength, or $\lambda \propto S^{-1}$.

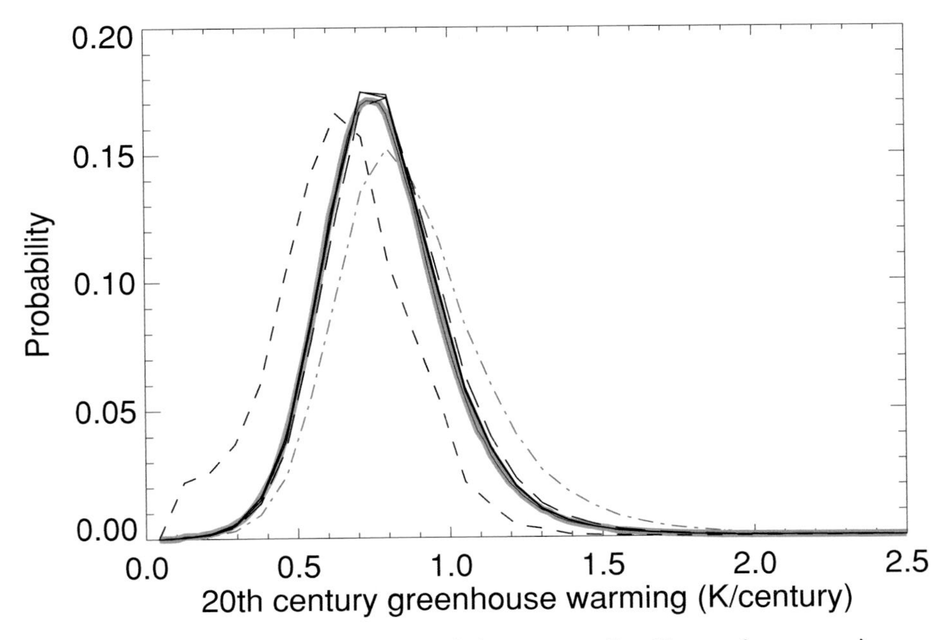

Figure 15.6 Distributions of twentieth-century attributable greenhouse warming based on original data (thick grey curve) and hindcast using a neutral prior in sensitivity (dash-dotted), feedback strength (dashed), A1FI (long dashed, near grey band) and B1 scenarios (short dashes, near grey band) and TCR (solid line)

greenhouse warming consistent with observed greenhouse warming. However, a distribution based on a uniform prior in transient climate response (solid black curve) does an excellent job of reconstructing the data.

As discussed above, many parameters in climate models do not correspond to directly observable quantities for which we can define an objective prior distribution (Kennedy and O'Hagen, 2001; Goldstein and Rougier, 2005). Equally plausible approaches using the same model and observations can yield very different estimates of the risk of high climate sensitivity. Transient and volcanic responses both fail to give an upper bound on sensitivity (Frame *et al.*, 2005; Wigley *et al.*, 2005). This perhaps intuitively surprising result is to be expected on physical grounds in any study that focuses on the transient response: a Taylor expansion of the transient temperature response to any external forcing F given a constant effective heat capacity c and feedback parameter λ is proportional to $(\lambda dt)/c$. Thus, the first sensitivity-dependent term to emerge in the limit of high sensitivity, short timescale or high heat capacity (or any combination thereof) is proportional to λ, not S. A linear relationship between λ and climatology is assumed a priori in Murphy *et al.* (2004).

If the relationship between the observational data and λ is linear, then the relationship between the data and S is non-linear, with the rate of change tending towards zero as S increases. In practical terms, this means that a change in climate system properties that takes a 5 °C to a 10 °C sensitivity has less impact on any of these observable properties of the climate system than one that takes a 0.9 °C to a 1 °C sensitivity.

Problems in which one can, equally credibly, form priors in a quantity and its reciprocal are instances of Von Kries's variant (Keynes, 1921; Rosenkrantz, 1977), of Bertrand's paradox (Bertrand, 1889). Bertrand's paradox is largely about the underdetermined nature of randomness in problems that call for random sampling (Weatherford, 1982). Though there appears to exist no canonical solution to the general problem(s), we believe that clarifying the question is a useful way out, because it presents a methodological way forward. On this view we need to clarify the purpose of the forecast distribution, and choose our prior so as to minimise our biases and maximise our ignorance in the forecast quantity, before any data are considered. This amounts to an application of the principle of indifference across the forecast variable. The principle of indifference (Weatherford, 1982; van Fraassen, 1989) states that if one has reasons for believing each element of a set of propositions is equally likely, then one's degree of belief in each of the propositions should be the same. It is a way of distributing one's ignorance evenly over a set of possibilities that, before the data are considered, are thought to be equally likely. To avoid biasing forecasts, it is desirable to use a prior that is neutral in the forecast quantity of interest, not the incidental system parameters that are used to derive it. In this case one wants to express one's indifference to warming rates before the data are available. Any other choice implies non-uniform ignorance across this forecast quantity, which implies additional information. Our choice of sampling strategy has not introduced such information about the likelihood of high or low warming rates, yet, as is apparent from Figures 15.1, 15.3 and 15.4 it skews our forecasts.

A considerable literature exists regarding possible ways of constructing non-informative priors (Howson and Urbach, 1993; Kass and Wassermann, 1996; Jaynes, 2003) and, in particular, addressing Bertrand's paradox (Heidbreder, 1996), but generally the attention has been focused on issues to do with finding invariances, exploiting symmetries, and so on in a way that will solve a canonical problem regarding the formation of prior beliefs. The area remains controversial. In the case(s) described here the focus of the problem dictates a particular strategy. We are not trying to solve Bertrand's paradox in all its forms; we are trying to avoid biasing the forecast of a particular variable, given some combination of models and data. This comparatively limited goal argues for a particular approach: one based around the principle of indifference applied to the forecast variable in question. This amounts to constructing a 'prior' which is uniform (or neutral) in the forecast quantity before any data are considered.

One of the most common alternative approaches – which does purport to offer a canonical solution to Von Kries-like problems – is to attempt to exploit symmetries in the relations between the quantities under scrutiny. This approach seeks to find invariant quantities that relate the relevant parameters. In the case where we can, equally credibly, form uniform priors in X and $1/X$, this approach invites us to find relations that are invariant under multiplication. A log-normal formulation can fulfil this requirement (Rosenkrantz, 1977), though it would not appear to be a portable solution across all variables and relationships of interest to climate scientists. In any case, the log-normal formulation still requires choices to be made about the parameters (mean and standard deviation) of the distribution. Such choices have long formed part of 'expert-elicited' priors about climate sensitivity, but there appear to be no compelling non-subjective methods for forming one's ideas about the parameters of such distributions: the distributions that have formed the basis of the scientific community's ranges for sensitivity have been admirably and transparently documented (Wigley and Raper, 1993; Houghton *et al.*, 2001) as being the result of expert subjective decisions, rather than as ranges that fall out of formal uncertainty estimates. Because this approach gives basically the same result as the suggested 'uniform prior in the forecast variable' while (a) requiring poorly constrained, subjective decisions about the parameters of the distribution and (b) holding over a more restricted range of climate-relevant relationships, we suggest that the most sound methodological approach is to condition one's ensemble forecasts on a uniform prior in the forecast variables of interest.

This is a practical methodological solution which attempts to make clear to the forecast user the relative roles of untestable prior assumptions versus observational or physical constraints. For example, users may wish to know 'what does this study tell me about X, given no knowledge of X before the study was performed?' This requires sampling non-observable parameters to simulate a uniform distribution in

X, the forecast quantity of interest, not the incidental system parameters that are used to derive it, before other constraints are applied.

For all transient climate change scenarios, future warming and hence climate impacts are related non-linearly to sensitivity, such that a distribution based on a uniform prior in either S or λ would be biased in the forecast quantity. A uniform prior in S (λ) implies a prior distribution for 1990–2100 warming that is skewed towards high (low) values. To avoid prejudging policy, we ought to be using forecasts that are neutral in the relevant warming rates, rather than interesting system properties like S or λ.

The uncomfortable conclusion is that there is no single 'correct' distribution on which to condition one's forecasts: the appropriate prior and hence uncertainty range depend on the scenario for which an estimate of sensitivity is required. Starting from a uniform prior distribution of S (Andronova and Schlesinger, 2000; Forest *et al.*, 2002; Knutti *et al.*, 2002) is appropriate only in the special case of forecasting long-term equilibrium warming under a stabilisation scenario, while starting from a uniform distribution of λ (Murphy *et al.*, 2004) may be more relevant to studies of atmospheric feedbacks or, as we discuss below, to the question of what distribution of stabilisation carbon dioxide concentrations is consistent with as target (2 K, for example) equilibrium warming (Figure 15.7). The relevant relationships here are between temperature and forcing and between forcing and concentration (Eq. 15.13). Forcing is related to concentrations of carbon dioxide by:

$$F = \frac{F_{2x}}{\ln 2} \times \ln \frac{CO_2}{CO_{2pre}} \tag{15.13}$$

in which CO_{2pre} corresponds to preindustrial concentrations of carbon dioxide (275 ppmv), and $F_{2x} = 3.74$ (corresponding to HadCM3's value for forcing in response to a doubling of CO_2).

So, since $F \propto \lambda$, we should form the relevant prior to be uniform in λ. The three forecast distributions for carbon concentration and sensitivity are shown in Figure 15.7, and in this case – because of the proportionality between λ and the CO_2 level consistent with a 2 K warming – we advocate using the dashed curve.

The implications of this point are shown by extending these weighted ensembles to 2100 under the IPCC A1FI and B1 scenarios (Houghton *et al.*, 2001). If we begin with a uniform distribution for S the weighted ensemble suggests a $>8\%$ ($>3.6\%$) chance of 1990–2100 warming $>5.0\,^{\circ}$C ($>3.0\,^{\circ}$C) under the A1FI (B1) scenario. If we begin with a uniform distribution for λ the ensemble implies only a 1.1% (0.3%) chance of the same outcome. The approach we recommend here – sampling neutrally in the forecast quantity – implies the chances of 1990–2100 warming exceeding 5.0 °C (3.0 °C) under A1FI (B1) are 3.6% and 1.5%, respectively. This agrees well with the approach taken by Allen *et al.* (2000), Stott and Kettleborough (2002) and Gregory *et al.* (2002), who sampled uniformly in the observable quantities used to constrain the

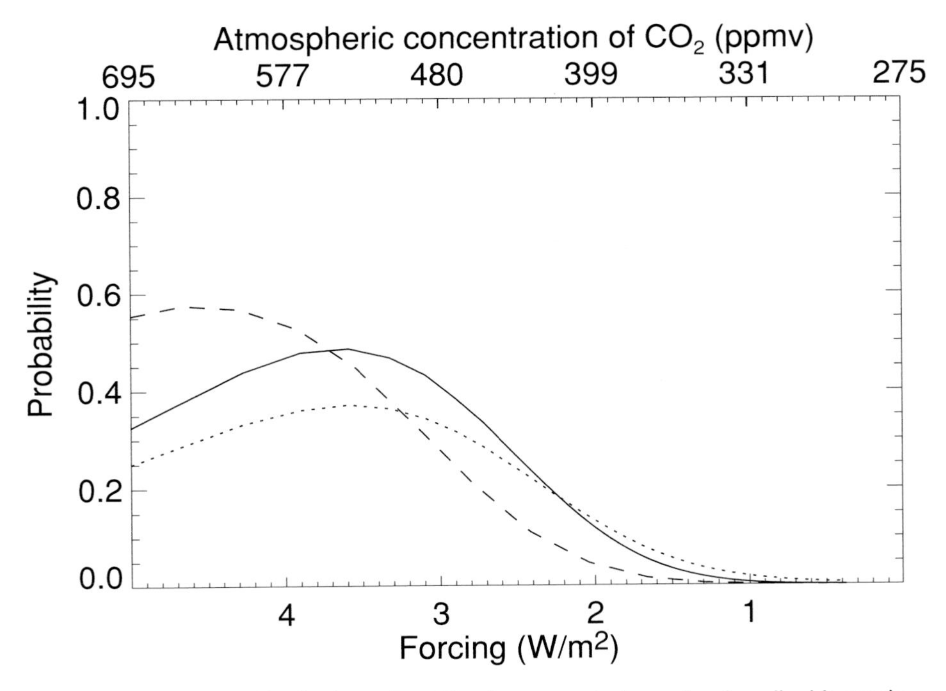

Figure 15.7 Distributions of stabilisation concentrations of carbon dioxide consistent with a 2 K target using a neutral prior in sensitivity (dotted), feedback strength (dashed), and TCR (solid curve).

forecast, and yields a distribution close to the solid line in Figure 15.4 and a chance of 1990–2100 warming exceeding 5.0 °C (3.0 °C) under A1FI (B1) of 3.6% and 1.4%, respectively. This agreement stems from the well-constrained relation between past and future warming under non-stabilisation scenarios (Allen *et al.*, 2000), so it is contingent upon the details of future emissions.

While, in principle, we need a different sampling for every scenario or forecast variable, in practice the response to almost any scenario *except* long-term stabilisation is almost linearly related to the transient climate response (TCR, or the global mean temperature change which results from a 1% per annum increase of CO_2 over 70 years) (Houghton *et al.*, 2001). Sampling sensitivity to give a uniform predictive prior distribution in TCR gives an approximately uniform distribution in most policy-relevant variables, including both past attributable warming and 1990–2100 warming under all SRES emissions scenarios. After weighting by observations as before, this approach implies a 5–95% range of uncertainty in S of 1.2–5.0 °C, with a median of 2.3 °C, suggesting that traditional ranges of uncertainty in S (Houghton *et al.*, 2001), may have greater medium-term policy relevance than recent more formal estimates using uniform priors in S or λ.

15.10 Practical implications for weather and climate forecasting

Consider the following scenario: a small number of members of a perturbed-physics or multimodel ensemble indicate a storm of unprecedented magnitude striking Paris in a couple of days' time. Examination of the ensemble statistics indicates that the magnitude of this storm is strongly correlated with the depth of a depression now forming off Cape Hatteras, independent of the model considered or perturbation made to model physics. Of course, the immediate response is to scramble our adaptive observation systems to get more data on this depression, but all available adaptive observation systems have been redeployed to some trouble spot in a remote corner of the globe. Worse still, there is a technical problem with the computer, the system manager is on a tea break (inconceivable at ECMWF, but for the sake of argument), so there is no chance of rerunning any members of the ensemble. The Mayor of Paris has heard rumours in the press and is on the line demanding an estimate of the chance that his roof is going to get blown off. Sounds tough? Welcome to climate research.

Suppose most members of the ensemble display a depression off Cape Hatteras which is consistent with the depth of the observed depression but 90% of the model depressions, $\bar{\mathbf{x}}_a$, are weaker than the observed best-guess depth $\mathbf{y}$. If we adopted a likelihood-weighted perturbed-physics approach, we would simply weight by the distance from observations (independent of sign), which would have the desirable effect of downweighting ensemble members whose depressions are much weaker than observed, but would leave a bias between the ensemble-based estimate of the depth of the depression and a model-free observation-based estimate. If we were very confident in our prior selection procedure for the inclusion of models into our ensemble, then this bias is desirable: the ensemble system is telling us the observations are, more likely than not, overestimating the depth of the depression. If, however, as this chapter has argued, we can never have any confidence in a prior selection procedure that purports to encompass model error, then we certainly shouldn't be revising the observations in the light of any model-based prior. Instead, Eq. (15.10) would imply we should renormalise the ensemble histogram to give equal weight to the 90% of members that underpredict the depression (and hence, if the relationship is monotonic, are likely to underpredict the storm) as to the 10% that overpredict: bad news for the Mayor.

Crucially, a rival forecasting centre which finds the same relationship between depression and storm but whose prior model selection is such that 90% of ensemble members overpredict the depth of the depression would, if also using Eq. (15.10) to apply a histogram renormalisation, give the Mayor the same message: good news for the scientific reputation of the forecasters. The storm is unprecedented, so neither forecast is verifiable in the conventional sense, but we can at least make sure they are STAID.

This kind of reassessment of probabilities in the light of expert judgement about the origins of biases goes on all the time already, and in time it is conceivable to imagine extended-range forecasting systems in which multimodel perturbed-physics ensembles are searched automatically for observable precursors to forecast events of interest, and then reweighted to ensure the space spanned by these observables is representatively populated. Tim Palmer asked us to discuss how the treatment of model error might apply to shorter-timescale problems than the climate timescale we normally deal with, and so we have had a go, conscious that there is a substantial literature on these shorter-term issues with which we are only vaguely familiar (apologies in advance to all the relevant authors we have failed to cite). In the shorter term, however, the ideas proposed in this chapter are likely to have much more relevance to the climate research problem, where we have the time and resources to do the problem properly, and hence no excuse not to.

The conventional climate modelling approach (tuning a maximum-feasible-resolution model to obtain an acceptably stable base climate, followed by a sequence of simulations using only a single version thereof) provides no new information unless the new model's forecasts happen to lie completely outside the forecast distribution consistent with current observations based on a lower resolution set of models. Given the large number of underdetermined parameters in even the highest-resolution models, the question of whether a single version of a high-resolution model displays a larger or smaller response to external forcing than previous models is vacuous. The interesting question, in the context of probabilistic climate forecasting, is whether the increase in resolution has systematically altered the range of behaviour accessible to the model under a comprehensive perturbation analysis, which cannot be addressed for a model being run at the maximum resolution feasible for a single integration or initial-condition ensemble.

The challenge is to use climate models to identify constraints on the response to various possible forcing scenarios: i.e. using models to establish what the climate system cannot (or is unlikely to) do, a much more demanding task than using them to describe some things it might do. Such constraints are likely to be fuzzy so, for many variables of interest, the spread of responses in currently available models may be too small to determine them effectively, as was the case for the CMIP-2 results, taken alone, in Figure 15.1. Large ensembles of models with deliberately perturbed physical parametrisations may be needed to provide enough diversity of response to identify these constraints. Adjoint-based optimal perturbation techniques used in shorter-range ensemble forecasting (Palmer, 2000) may not be extendable, even in principle, to the climate problem (Lea *et al.*, 2000), further increasing the size of ensembles required.

If constraints can be identified that, on physical grounds, we might expect any correctly formulated climate model to share, then inferences based on these constraints are likely to be robust. Some constraints will be obvious (like energy conservation) but these will typically not be enough to constrain many useful forecast quantities. Others will be more subtle and open to falsification, like the constraint that the sensitivity parameter (a measure of the net strength of atmospheric feedbacks) does not change in response to forcings up to 2–3 times preindustrial CO_2. This is a property shared by almost all climate models available to date but it could, in principle, be falsified if higher-resolution models were to display *systematically* non-linear sensitivities. Hence a probabilistic forecast of a particular variable that depends on the linearity of atmospheric feedbacks is more open to falsification than one that depended solely on energy conservation, which brings us round full-circle. We will depend on the expert judgement of modellers to assess the reliability of our probabilistic forecasts (the likelihood that they will be falsified by the next generation of models), but, in a crucial step forward, our reliance on expert judgement will be second order. We need experts to assess the reliability of our uncertainty estimates rather than to provide direct input into the uncertainty estimates themselves.

Acknowledgements

This chapter represents a synthesis of (we hope ongoing) conversations over a number of years with, among many others, Mat Collins, Jonathan Gregory, William Ingram, John Mitchell, James Murphy, Catherine Senior, David Sexton, Peter Stott, Simon Tett, Alan Thorpe and Francis Zwiers. Aficionados will recognise many of Lenny Smith's ideas in here, more than we can acknowledge specifically. We are grateful to Michael Goldstein and Jonathan Rougier for making their 2004 article available, to Tim Palmer for his invitation and good-humoured response to the talk, and to Els Kooij-Connally and Renate Hagedorn for their forbearance in the preparation of this chapter.

This work was supported by the Natural Environment Research Council and the NOAA-DoE International Detection and Attribution Group.

References

Allen, M. R. (1999). Do-it-yourself climate prediction. *Nature*, **401**, 642.

Allen, M. R. and W. J. Ingram (2002). Constraints on future climate change and the hydrological cycle. *Nature*, **419**, 224–32.

Allen, M. R. and D. A. Stainforth (2002). Towards objective probabilistic climate forecasting. *Nature*, **419**, 228.

Allen, M. R., P. A. Stott, J. F. B. Mitchell, R. Schnur and T. Delworth (2000). Quantifying the uncertainty in forecasts of anthropogenic climate change. *Nature*, **407**, 617–20.

Andronova, N. G. and M. E. Schlesinger (2000). Causes of global temperature changes during the 19th and 20th centuries. *Geophys. Res. Lett.*, **27**, 2137–3140.

Bertrand, J. (1889). *Calcul des Probabilities*. Gauthier-Villars, Paris.

Buizza, R.,M. Miller and T. N. Palmer (1999). Stochastic representation of model uncertainties in the ECMWF Ensemble Prediction System. *Quart. J. Roy. Meteor. Soc.*, **125B**, 2887–908.

Covey, C., K. M. AchutaRao, S. J. Lambert and K. E. Taylor (2000). *Intercomparison of Present and Future Climates Simulated by Coupled Ocean-Atmosphere GCMs*. Technical Report 66. Program for Climate Model Diagnosis and Intercomposition (PCMDI).

Craig, P. S., M. Goldstein, J. C. Rougier and A. H. Seheult (2001). Bayesian forecasting for complex systems using computer simulators. *J. Am. Stat. Assoc.*, **96**, 717–29.

ECMWF (2002). Talagrand diagram. Online: www.ecmwf.int/products/forecasts/guide/Talagrand_diagram.html.

Forest, C. E.,M. R. Allen, P. H. Stone and A. P. Sokolov (2000). Constraining uncertainties in climate models using climate change detection techniques. *Geophys. Res. Lett.*, **27**, 569–72.

Forest, C. E.,P. H. Stone, A. P. Sokolov, M. R. Allen and M. D. Webster (2002). Quantifying uncertainties in climate system properties with the use of recent climate observations. *Science*, **295**, 113–17.

Fraassen, B. C. van (1989). *Laws and Symmetry*. Clarendon Press.

Frame, D. J.,B. B. B. Booth, J. A. Kettleborough, *et al.* (2005). Constraining climate forecasts: the role of prior assumptions. *Geophys. Res. Lett.*, **32**, doi: 10.1029/2004GL022241.

Giorgi, F.and L. O. Mearns (2002). Calculation of average, uncertainty range, and reliability of regional climate changes from AOGCM simulations via the reliability ensemble averaging (REA) method. *J. Climate*, **15**, 1141–58.

Goldstein, M.and J. Rougier (2004). Probabilistic formulations for transferring inferences from mathematical models to physical systems. *SIAM J. Sci. Comput.* **26**, 467–87.

Goldstein, M. and J. Rougier (2005). Probabilistic formulations for transferring inferences from mathematical models to physical systems. *SIAM J. Sci. Comput.* **26**, 467–87.

Gregory, J. M.,R. Stouffer, S. Raper, N. Rayner and P. A. Stott (2002). An observationally-based estimate of the climate sensitivity. *J. Climate*, **15**, 3117–21.

Hansen, J.,G. Russell, A. Lacis, *et al.* (1985). Climate response times: dependence on climate sensitivity and ocean mixing. *Science*, **229**, 857–9.

Hansen, J. A.and L. A. Smith (2001). Probabilistic noise reduction. *Tellus*, **5**, 585–98.

Heidbreder, G. R. (1996).*Maximum Entropy and Bayesian Methods*. Kluwer Academic Publishers.

Houghton, J. T.,Y. Ding, D. J. Griggs, *et al.* (eds.) (2001). *Climate Change 2001: The Science of Climate Change*. Cambridge University Press.

Howson, C. and P. Urbach (1993). *Scientific Reasoning: The Bayesian Approach*. Open Court.

Jaynes, E. T.(2003). *Probability Theory: The Logic of Science*. Cambridge University Press.

Jewson, S. P.and R. Caballero (2003). The use of weather forecasts in the pricing of weather derivatives. *Meteorol. Appl.* **10**, 367–76.

Judd, K. (2003). Nonlinear state estimation, indistinguishable states and the extended Kalman filter. *Physica D*, **183**, 273–81.

Kass, R. and L. Wassermann (1996). The selection of prior distribution by formal rules. *J. Am. Stat. Assoc.*, **91**, 1343–70.

Kennedy, M. and A. O'Hagan (2001). Bayesian calibration of computer models. *J. Roy. Stat. Soc. B*, **63**, 425–64.

Keynes, J. M. (1921). *A Treatise on Probability. Part I.* Macmillan.

Knutti, R., T. F. Stocker, F. Joos and G. K. Plattner (2002). Constraints on radiative forcing and future climate change from observations and climate model ensembles. *Nature*, **416**, 719–23.

Lea, D. J., M. R. Allen and T. W. N. Haine (2002). Sensitivity analysis of the climate of a chaotic system. *Tellus*, **52A**, 523–32.

Levitus, S., J. I Antonov, T. P. Boyer and C. Stephens (2000). Warming of the world ocean. *Science*, **287**, 2225–9.

Levitus, S., J. I. Antonov and T. P. Boyer (2005). Warming of the world ocean: 1955–2003. *Geophys. Res. Lett.*, **32**, L02604.

Lorenz, E. N. (1963).Deterministic nonperiodic flow. *J. Atmos. Sci.*, **20**, 130–41.

Molteni, F., R. Buizza, T. N. Palmer and T. Petroliagis (1996). The ECMWF ensemble prediction system: methodology and validation. *Quart. J. Roy. Meteor. Soc.*, **122A**, 73–119.

Moore, A. M. and R. Kleeman (1999). Stochastic forcing of ENSO by the intraseasonal oscillation. *J. Climate*, **12**, 1199–220.

Morgan, M. G. and D. W. Keith (1995). Subjective judgements by climate experts. *Environ. Sci. Technol.*, **29**, 468–76.

Murphy, J., D. M. H. Sexton, D. N. Barnett, *et al.* (2004). Quantification of modelling uncertainties in a large ensemble of climate change simulations. *Nature*, **430**, 768–72.

Mylne, K. R., R. E. Evans and R. T. Clark (2002). Multi-model multi-analysis ensembles in quasi-operational medium-range forecasting. *Quart. J. Roy. Meteor. Soc.*, **128**, 361–84.

Nolan, D. S. and B. F. Farrell (1999). The intensification of two-dimensional swirling flows by stochastic asymmetric forcing. *J. Atmos. Sci.*, **56**, 3937–62.

Palmer, T. N. (2000). Predicting uncertainty in forecasts of weather and climate. *Rep. Prog. Phys.*, **63**, 71–116.

Palmer, T. N. and J. Raisanen (2002). Quantifying the risk of extreme seasonal precipitation events in a changing climate. *Nature*, **415**, 512–14.

Palmer, T. N., R. Buizza, F. Molteni, Y. Q. Chen and S. Corti (1994). Singular vectors and the predictability of weather and climate. *Philos. Tr. Roy. Soc., S-A* **348**, 459–75.

Puri, K., J. Barkmeijer and T. N. Palmer (2001). Ensemble prediction of tropical cyclones using targeted diabatic singular vectors. *Quart. J. Roy. Meteor. Soc.*, **127B**, 709–31.

Rosenkrantz, R. D. (1977). *Inference, Method and Decision: Towards a Bayesian Philosophy of Science*. D. Reidel.

Roulston, M. S. and L. A. Smith (2002). Evaluating probabilistic forecasts using information theory. *Mon. Weather Rev.*, **130**, 1653–60.

Schneider, S. H. (2002).Can we estimate the likelihood of climatic changes at 2100? *Climatic Change*, **52**, 441–51.

Smith, L. A. (1995). Accountability and error in nonlinear forecasting. In *Proceedings of the 1995 Seminar on Predictability*, pp. 351–68. ECMWF.

(2000). Disentangling uncertainty and error: on the predictability of nonlinear systems. In *Nonlinear Dynamics and Statistics*, ed. A. I. Meas, pp. 31–64. Birkhauser.

(2002). What might we learn from climate forecasts? *Proc. Nat. Acad. Sci.*, **99**, 2487–92.

Stainforth, D., J. Kettleborough, M. Allen, M. Collins, A. Heaps and J. Murphy (2002). Distributed computing for public-interest climate modeling research. *Comput. Sci. Eng.*, **4**, 82–9.

Stainforth, D. A., T. Aina, C. Christensen, *et al.* (2004). Evaluating uncertainty in the climate response to changing levels of greenhouse gases. *Nature*, **433**, 403–6.

Stott, P. A. and J. A. Kettleborough (2002). Origins and estimates of uncertainty in predictions of twenty-first century temperature rise. *Nature*, **416**, 723–6.

Tebaldi, C., R. L. Smith, D. Nychka and L. O. Mearns (2005). Quantifying uncertainty in projections of regional climate change: a Bayesian approach to the analysis of multimodel ensembles. *J. Climate*, **18**, 1524–40.

Toth, Z. and E. Kalnay (1997). Ensemble forecasting at NCEP and the breeding method. *Mon. Weather Rev.*, **125**, 3297–319.

Weatherford, R. (1982). *Philosophical Foundations of Probability Theory*. Routledge & Kegan Paul.

Wigley, T. M. L. and S. C. B. Raper (1993). Sea level changes due to thermal expansion of the oceans. In *Climate and Sea Level Change: Observations, Projections and Implications*, ed. R. A. Warrick, E. M. Barrow and T. M. L. Wigley. Cambridge University Press.

Wigley, T. M. L. and S. C. B. Raper (2001). Interpretation of high projections for global-mean warming. *Science*, **293**, 451–4.

Wigley, T. M. L., C. M. Ammann, B. D. Santer and S. C. B. Raper (2005). The effect of climate sensitivity on the response to volcanic forcing. *J. Geophys. Res.*, **110**, D09107.

16

Observations, assimilation and the improvement of global weather prediction – some results from operational forecasting and ERA-40

Adrian J. Simmons
European Centre for Medium-Range Weather Forecasts, Reading

Basic aspects of the atmospheric observing system and atmospheric data assimilation are summarised. Characteristics of the assimilation of observational data from the late 1950s onwards in the ERA-40 reanalysis project, and of medium-range forecasts run from the ERA-40 analyses, are used to illustrate improvement in the observing system and to place in context the improvement of the operational forecasts of the European Centre for Medium-Range Weather Forecasts (ECMWF) over the past 25 years. Recent advances in operational forecasting are discussed further. It is shown that the analyses of two centres, ECMWF and the Met Office, have converged substantially, but that there remain nevertheless significant differences between these analyses, and between the forecasts made from them. These differences are used to illustrate several aspects of data assimilation and predictability. Inferences from differences between successive daily forecasts and from spectra of forecast errors are also discussed for the ECMWF system.

16.1 Introduction

This chapter is based on the introductory lecture given to the Annual ECMWF Seminar for 2003, which was devoted to data assimilation. The subject had last been addressed in the Seminar Series in 1996. In the opening lecture on that occasion, the

Predictability of Weather and Climate, ed. Tim Palmer and Renate Hagedorn. Published by Cambridge University Press.

late Roger Daley discussed how, over the preceding 15 years, data assimilation had evolved from being a minor and often neglected subdiscipline of numerical weather prediction to become not only a key component of operational weather forecasting but also an approach that was important for environmental monitoring and estimation of the ocean state. The years since then have seen numerical weather prediction reap considerable benefit from the earlier and ongoing investments in the scientific and technical development of data assimilation. The application to environmental monitoring has advanced with the emergence of a broad user-base and new initiatives in reanalysis, and ocean-state estimation has become established operationally as a component of seasonal forecasting systems. Moreover, as can be seen generally in the Proceedings of the 2003 Seminar (available from www.ecmwf.int), considerable advances continue to be made in the techniques and range of application of data assimilation.

The principal aims of this chapter are to illustrate some basic aspects of atmospheric data assimilation, predictability and the improvement of forecasts. To do this, results are drawn from the recently completed ERA-40 reanalysis of the atmospheric observations made since mid 1957 (Uppala *et al.*, 2005), and from ECMWF's operational analyses and forecasts. Some results from the operational forecasting systems of the UK's Met Office and the US National Centers for Environmental Prediction (NCEP) are also used. Attention is concentrated on analyses and deterministic forecasts for the extratropical troposphere, predominantly for the 500 hPa height field. This is partly because of the continuing widespread use of such forecasts (in addition to newer products such as from ensemble prediction or related directly to weather elements), partly because it is for the 500 hPa height field that the most comprehensive records are available, and partly to enable the simplest of messages to be presented. Significant improvements have also been achieved in ensemble prediction (discussed by Buizza in this volume) and weather-element forecasts, and in general analysis and forecasting for the tropics and stratosphere, although in each case additional challenges have had to be faced.

The basics of the atmospheric observing system and data assimilation are discussed in the following section. Section 16.3 presents some results from ERA-40, discussing the evolution of the observing system since 1957 and the corresponding evolution of analysis characteristics. The changes over time in the accuracy of forecasts run from the ERA-40 analyses are also discussed, and compared with changes in the accuracy of ECMWF's operational forecasts. Further aspects of the improvement in operational forecasts are presented in Section 16.4. Section 16.5 draws some inferences relating to data assimilation and predictive skill derived from studies of the differences between analyses and forecasts from ECMWF and the Met Office, and from the differences between successive ECMWF forecasts. The scale-dependence of forecast error is discussed in Section 16.6, which is followed by some concluding remarks.

16.2 Basic aspects

Figure 16.1 presents maps of the operational coverage of atmospheric data received at ECMWF for the six-hour period beginning at 0900 UTC on 21 November 2004. Colour versions of such maps are presented daily on the ECMWF website (www.ecmwf.int). Data of each of the types illustrated are assimilated operationally at ECMWF, although in the case of ozone, data from the ERS-2 satellite (denoted by the orbits confined to the vicinity of Europe and the North Atlantic in the bottom right panel of Figure 16.1) are no longer assimilated, having been replaced by data from the SCIAMACHY instrument on ENVISAT, orbits from which are not included in the figure. Data from ENVISAT are also used in the analysis of initial conditions for the ocean-wave model that is coupled to the ECMWF atmospheric model (Janssen *et al.*, 2002)

The atmospheric observing system illustrated in Figure 16.1 comprises a range of *in-situ* and remotely sensed measurements. These different types of measurement have different accuracies and different temporal and spatial coverage. Some measurements are of quantities such as wind and temperature that are prognostic model variables, values of which are required to initiate forecasts. Others provide information that is only indirectly related to the model variables, the radiances in different parts of the electromagnetic spectrum measured from satellites in particular.

Data assimilation is a process in which a model integration is adjusted intermittently to bring it in line with the latest available observations related to the model's state variables. It is shown schematically in Figure 16.2 for a six-hourly assimilation with a three-dimensional (spatial) analysis of observations such as the three-dimensional variational (3D-Var) system used by ECMWF previously for its main operations (Andersson *et al.*, 1998), recently for ERA-40 and currently for short-cut-off analyses from which forecasts are run to provide boundary conditions for the short-range limited-area models of ECMWF Member States. 3D-Var was also used until recently by the Met Office (Lorenc *et al.*, 2000) and is used today by NCEP (Parrish and Derber, 1992), not only in operations but also in continuing the pioneering reanalysis (Kalnay *et al.*, 1996; Kistler *et al.*, 2001) carried out by NCEP in collaboration with the US National Center for Atmospheric Research (NCAR).

As illustrated in Figure 16.2, the six-hour 'background' forecast from the preceding analysis is adjusted by adding an 'analysis increment'. The increment is determined from an analysis of the 'innovations', the differences between observations and equivalents derived from the background forecast. These differences may be computed at the actual times of the observations, but are assumed to be valid at the main synoptic hour for which the analysis is carried out. This was what was done for ERA-40, for example. The analysis for a particular time in the configuration shown makes direct use of observations taken within three hours of analysis time, and is influenced by earlier observations through the information carried forward in time

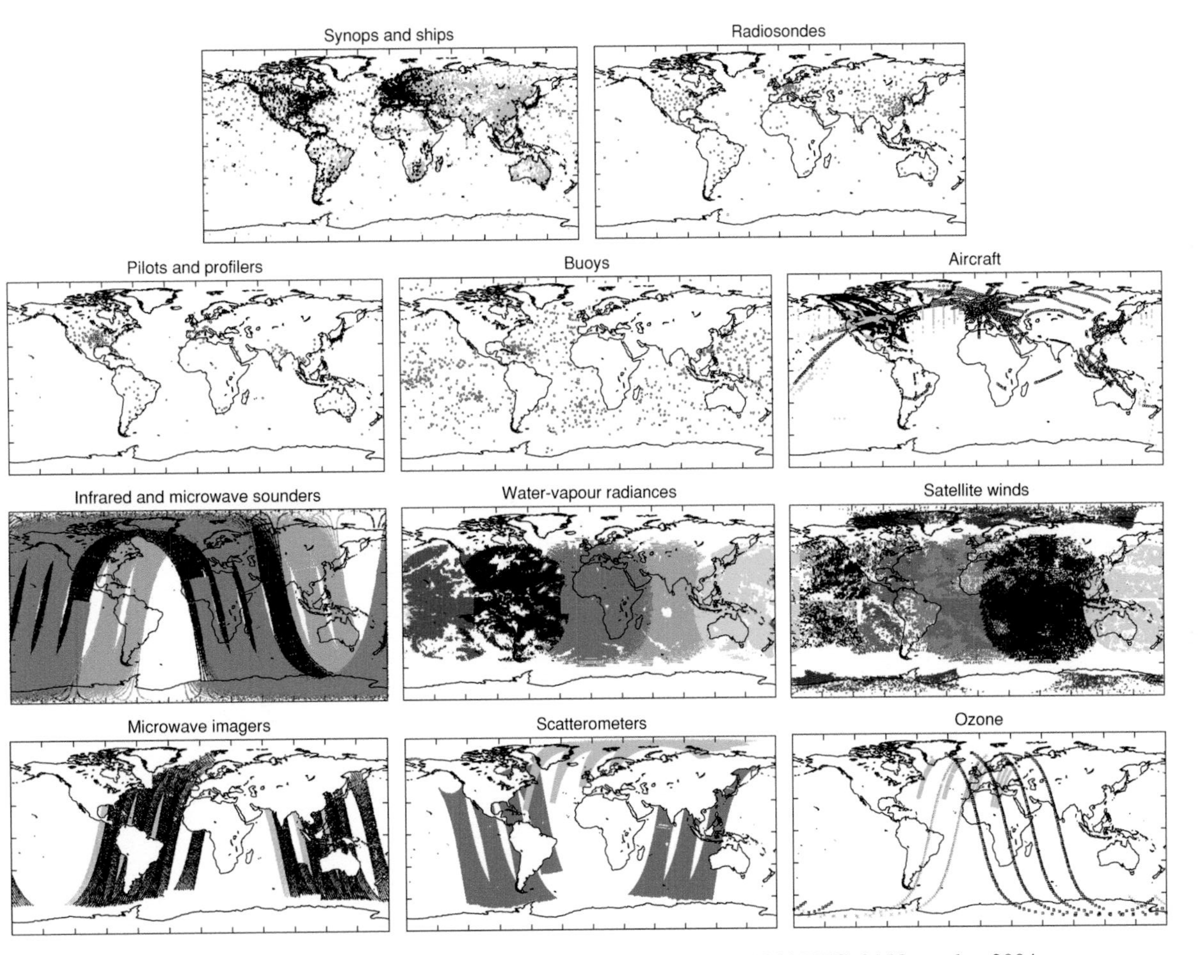

Figure 16.1 Operational data coverage plots for the six-hour period centred on 1200 UTC, 21 November 2004.

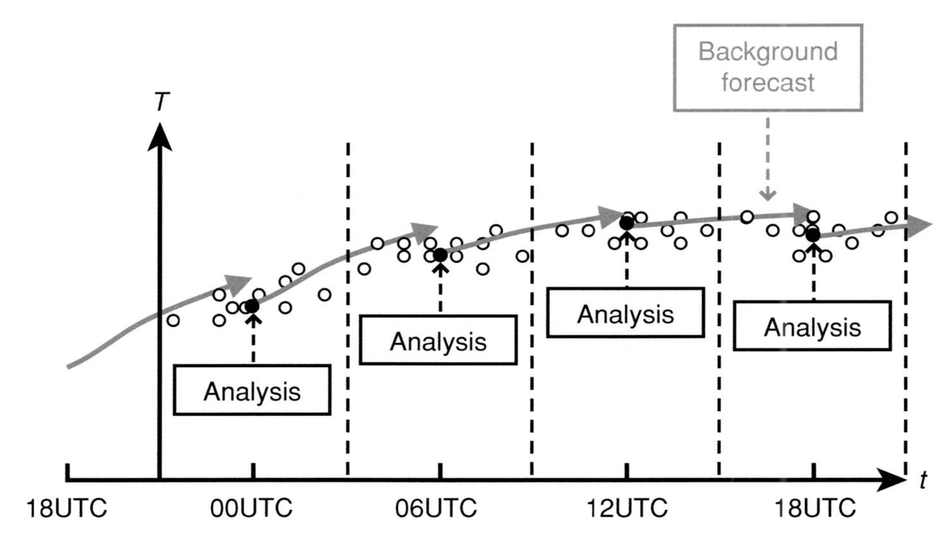

Figure 16.2 Schematic of the data assimilation process for a six-hourly three-dimensional data assimilation such as used for ERA-40. The ordinate denotes a model variable, temperature for example. Circles denote observations.

(and spread in space) by the background forecast. The weighting given to the innovations in the analysis procedure depends on the expected errors of the observations and background forecast; in variational analysis this is achieved by minimising the sum of error-weighted measures of the deviations of analysed values from the observed and background values. The variational approach in particular enables direct assimilation of raw radiances measured by satellite, as basic analysed variables can be adjusted to improve jointly the fits of simulated model-based radiances to measured radiances and the fits to measurements of other types.

The analysis increment is typically small compared with the change made by the background forecast from the preceding analysis. An example is presented in Figure 16.3, which shows maps of the mean-square changes in 500 hPa height produced by the background forecasts from 0600 UTC to 1200 UTC (left) and by the 1200 UTC analyses (centre) for the year 2001 from the 3D-Var ERA-40 data assimilation. The main Atlantic and Pacific storm tracks are evident in the northern hemisphere plot of the background changes, as is a local maximum over the Mediterranean. Larger changes occur in the southern hemisphere storm track encircling Antarctica. Mean-square analysis increments for 2001 are typically smaller than the background-forecast changes by an order of magnitude or more, as can be seen from the extratropical averages printed on each map. Also shown in the figure (right) are the corresponding increments for 1958. The differences between the increments for the two years are discussed in the following section.

The quality of the background forecast depends both on the quality of the preceding analysis and on the quality of the forecast model. The predominant role of the background model rather than the analysis increment in evolving the estimated state

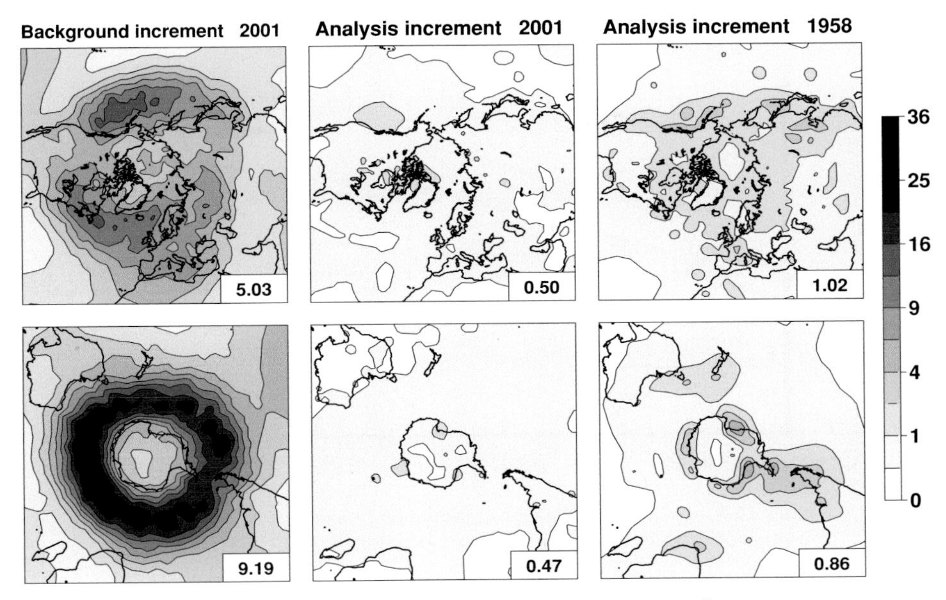

Figure 16.3 Mean-square increments in 500 hPa height (dam^2) for the ERA-40 background forecasts from 0600 UTC to 1200 UTC for the year 2001 (left) and for the 1200 UTC ERA-40 analyses for the years 2001 (centre) and 1958 (right), shown for the northern (upper) and southern (lower) hemispheres. Values averaged over each extratropical hemisphere are shown in the bottom-right corner of each map.

of the atmosphere is indicative of the need for as accurate as possible a model to enable as accurate as possible an analysis. A good model representation of synoptic- and planetary-scale dynamics and thermodynamics is needed to capture the main evolution of a field such as 500 hPa height, but a good model representation of local weather elements is needed as well, not only for direct analysis and prediction of such elements but also to enable correct extraction of the information related to larger scales that is contained in the observations of these elements.

ECMWF and Météo-France were the first to use four-dimensional variational (4D-Var) data assimilation for operational global data assimilation (Courtier *et al.*, 1994; Rabier *et al.*, 2000; Mahfouf and Rabier, 2000). They were joined in this regard by the Met Office in October 2004, and by Canada and Japan in early 2005. 4D-Var takes the time of each observation more consistently into account. The basic approach is illustrated schematically in Figure 16.4. An adjustment to the background forecast valid at the beginning of the assimilation window is determined iteratively to reduce the discrepancy between the observations within the assimilation window and the equivalent forecast values, with the adjustment limited to be consistent with the estimated errors of the observations and background forecast. Figure 16.4 shows the case of a 12-hourly window as implemented at ECMWF. Specific analysis times are marked six-hourly in the diagram because in the ECMWF system ocean-wave and land-surface conditions are adjusted at these times through separate analysis

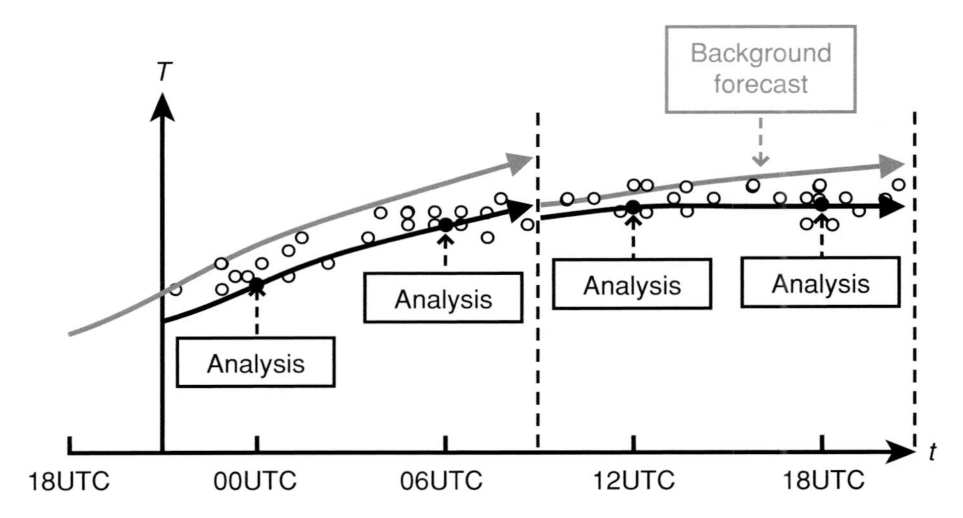

Figure 16.4 Schematic of the data assimilation process for the twelve-hourly four-dimensional data assimilation used operationally at ECMWF.

procedures. Haseler (2005) discusses the configuration recently implemented at ECMWF, which was developed to ensure as early as possible a delivery of forecast products without significant loss of accuracy. Background forecasts are run from the 0600 and 1800 UTC analyses. They are used in the cycled 12-hour 4D-Var analyses, and also in six-hour 4D-Var analyses for 1200 and 0000 UTC produced with a shorter cut-off for data receipt than used for the cycled 12-hour analyses. Ten-day deterministic and ensemble forecasts are initiated from the shorter-cut-off six-hour 4D-Var analyses.

16.3 Some results from ERA-40

ERA-40 is a project in which observations made from September 1957 to August 2002 were analysed using a six-hourly 3D-Var version of the ECMWF data-assimilation system (Uppala *et al.*, 2005). The assimilating model used the same 60-level vertical resolution as introduced in 1999 for operational forecasting at ECMWF, but the much lower horizontal resolution of T159 rather than the T511 spectral truncation used operationally since November 2000, with corresponding use of around 125 km rather than 40 km grid resolution. The data assimilation was based on a version (denoted cycle 23r4) of the forecasting system operational in the second half of 2001, modified to include a few newer features subsequently introduced operationally in cycles 25r1 and 25r4.

The global observing system changed considerably over the 45 years covered by ERA-40. The ever-present observation types were the synoptic surface observations from land stations and ships, and the soundings from radiosonde and pilot balloons.

Table 16.1 *Average daily counts of various types of observation supplied to the ERA-40 data assimilation, for five selected periods*

	1958–66	1967–72	1973–8	1979–90	1991–2001
SYNOP/SHIP	15313	26615	28187	33902	37049
Radiosondes	1821	2605	3341	2274	1456
Pilot balloons	679	164	1721	606	676
Aircraft	58	79	1544	4085	26341
Buoys	0	1	69	1462	3991
Satellite radiances	0	6	35069	131209	181214
Satellite winds	0	0	61	6598	45671
Scatterometer	0	0	0	0	7575
PAOBs	0	14	1031	297	277

The quality of radiosonde measurements improved over the period, but geographical and temporal coverage declined. In the years before availability of satellite sounding data, fixed ocean weather ships provided regular data over the North Atlantic and North Pacific, continuing a network established in 1946 with ten ships in the Atlantic and three in the Pacific. Only the 'Polarfront' (located off the coast of Norway) remained in 2002. General radiosonde coverage from land and island stations also declined in the 1990s, due both to closure of some stations and to a reduction in the frequency of soundings from others. Counts of measurements at stratospheric levels have risen nevertheless, and measurement biases have declined (e.g. Haimberger, 2005).

Other observation types have more than compensated for the decline in radiosonde coverage, at least over oceanic areas. Table 16.1 shows average daily counts of the different types of observation used in ERA-40's analysis of the primary meteorological variables. Averages are shown for five different epochs spanning the reanalysis period, each representative of a particular state of the observing system. The year 1973 was a key one that saw radiances from the first of the VTPR sounding instruments flown on the early NOAA series of operational polar-orbiting satellites. Pseudo surface-pressure observations (PAOBs) derived from satellite imagery over the southern oceans provided a further data source and many more aircraft data were assimilated from this time onwards. A major enhancement of the observing system was in place by the beginning of 1979 for FGGE, the First Global Experiment of the Global Atmospheric Research Programme. VTPR data were replaced by data from the three TOVS sounding instruments on new NOAA platforms; winds derived by tracking features observed from geostationary satellites first became available in significant numbers; and there were substantial increases in buoy and aircraft data. Observation counts declined for a while after 1979, but recovered during the 1980s. The frequency and coverage of wind and temperature measurements from aircraft increased substantially in the 1990s. There was also a substantial increase in the number and quality of wind estimates derived from geostationary satellite data. Newer satellite instruments include microwave imagers providing information on total water-vapour content and marine surface wind speed, active microwave instruments providing information on ocean winds and waves, and more advanced microwave and infrared sounders. In all, almost 30 satellite-borne instruments currently provide data that are assimilated in the operational ECMWF forecasting system, contributing to an unprecedented level of analysis and forecast accuracy (Simmons and Hollingsworth, 2002; Uppala *et al.*, 2005). Uppala *et al.* also provide more detail on the observations used in ERA-40, including qualifying comment on the numbers presented in Table 16.1.

The improvement in the observing system over the past four decades is reflected in a marked reduction in the general magnitude of analysis increments over the period of ERA-40. The maps for 1958 and 2001 presented in Figure 16.3 show the mean-square increments in 500 hPa height to be much smaller in 2001 over land and neighbouring ocean regions where there are marked increments in 1958. Isolated small-scale maxima over the oceans in 1958 indicate where radiosonde data from islands and the fixed weather ships correct the background forecast. Local impact still occurs in 2001 for isolated radiosonde stations over Antarctica and northern Russia, but is of smaller magnitude. Increments are particularly large along the west coast of North America in 1958, where background forecast error that developed over the poorly observed Pacific Ocean is corrected on its first encounter with the North American radiosonde network. Larger increments occur in 2001 than in 1958 over oceanic regions that are almost devoid of observations in 1958, due to assimilation of the satellite, buoy and aircraft data that today provide coverage of such regions. Smaller increments in 2001 over regions well covered by radiosonde data in both 1958 and 2001 can arise both from more-accurate background forecasts and from more-accurate radiosonde measurements.

Generally smaller analysis increments can also result from improvements in data assimilation systems. Uppala *et al.* (2005) compare 500 hPa height increments for 1979 from ERA-40 and from the earlier ERA-15 reanalysis, which used the optimal interpolation method superseded by 3D-Var for ECMWF operations in early 1996. Increments are generally smaller for ERA-40 than for ERA-15. This is a good result for ERA-40 if it stems from a greater accuracy of the background forecasts (in closer agreement with the observations) rather than from a failure of the analysis to draw sufficiently closely to observations. The former appears to be the case, as the 24-hour ERA-40 forecasts generally match radiosonde data better than the 24-hour ERA-15 forecasts. Increments in 500 hPa height are also generally smaller for ERA-40 than for the NCEP/NCAR reanalysis (Kistler *et al.*, 2001).

Figure 16.5 shows time series of background and analysis fits to surface pressure measurements from land stations and ships and to 500 hPa temperature measurements from radiosondes, for the extratropical northern and southern hemispheres. A general improvement in the fit to the measurements occurs over the period of ERA-40, especially for the southern hemisphere, where values approach those for the northern hemisphere for the most recent years. Changes in data coverage can affect these values, as increased coverage in the subtropics, where variance is lower, would tend

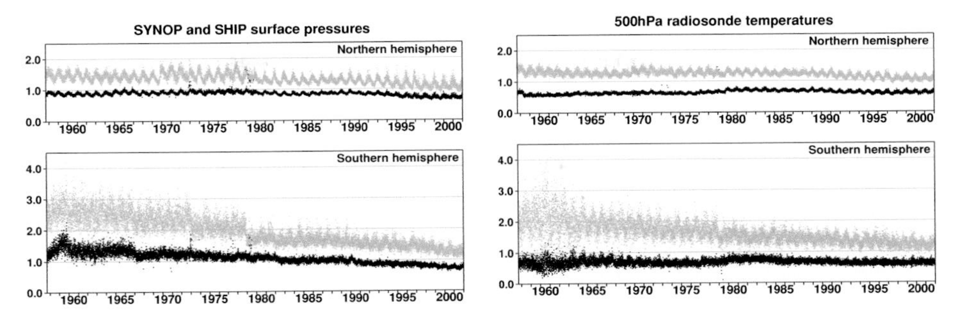

Figure 16.5 Daily values of the root-mean-square background (grey) and analysis (black) fits to 1200 UTC SYNOP and SHIP measurements of surface pressure (hPa, left) and radiosonde measurements of 500 hPa temperature (K, right) over the extratropical northern (upper) and southern (lower) hemispheres.

to reduce values in plots such as these in which no area-weighting is applied. However, the sudden improvement in fit to surface pressure data for the southern hemisphere at the beginning of 1979 is almost certainly a consequence of the major improvement of the whole observing system that took place at the time. A smaller improvement at the beginning of 1973 can also be seen. The fit of the analysis to the radiosonde data is poorer in the 1980s than earlier as the analysis has to match TOVS radiance data as well as radiosonde data in the later years. However, the fit of the background field is unchanged in the northern hemisphere and improves in the southern hemisphere, a sign of an analysis that has been improved overall by the availability of the new data.

Further evidence of the improvement of the analyses over time is provided by objective verification of the medium-range forecasts that have been run routinely from the ERA-40 analyses using the same T159 version of the model used in the ERA-40 data assimilation. Anomaly correlations of three-, five- and seven-day 500 hPa height forecasts are presented for the period of the reanalysis in Figure 16.6. Results are shown for selected regions for which the availability of radiosonde data throughout the period gives confidence in the accuracy of the verifying ERA-40 analyses. Corresponding operational results are also shown for the second half of the period. Forecast quality is relatively high throughout for the three northern hemisphere regions shown, with levels of skill for the ERA-40 forecasts that from the outset are higher than or similar to those of the operational ECMWF forecasts from the early 1980s. The quality of the ERA-40 forecasts is most uniform over time for East Asia, which is consistent with its location downstream of a land mass that is covered by conventional observations and for which the effect of any improvements in some components of the observing system is most likely to have been obscured by decline of the radiosonde coverage upstream over the former Soviet Union. Improvement is a little more evident for Europe and more evident still for North America, which, lying downstream of the broad expanse of the Pacific Ocean, benefits more than the other two regions from improvement in oceanic data coverage.

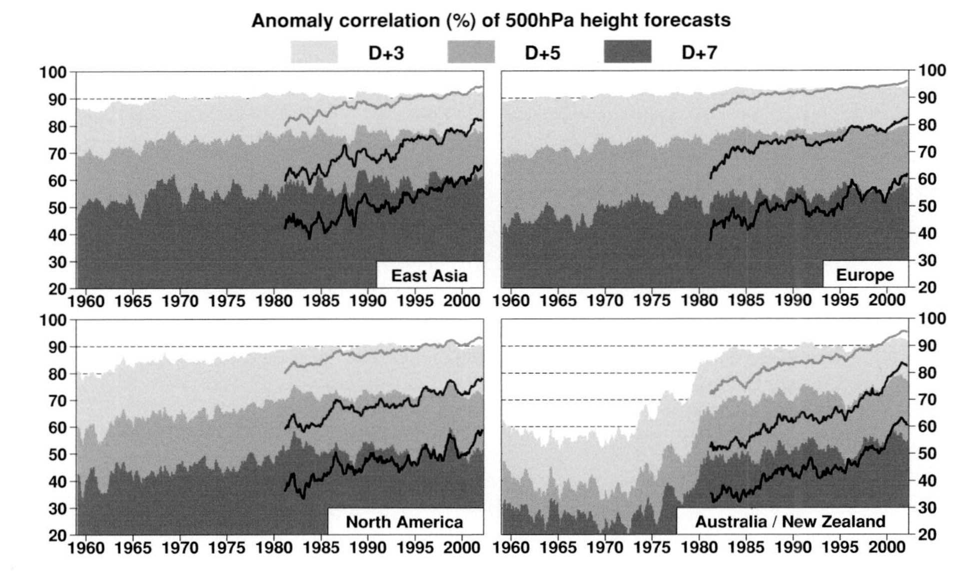

Figure 16.6 Twelve-month running mean anomaly correlations (%) of 3-, 5- and 7-day 500 hPa height forecasts for East Asia (25N–60N, 102.5E–150E), Europe (35N–75N, 12.5W–42.5E), North America (25N–60N, 120W–75W) and Australia/New Zealand (45S–12.5S, 120E–175E), from 1200 UTC ERA-40 forecasts from 1958 to 2001 (denoted by shading limits) and from operational 1200 UTC ECMWF forecasts from 1980 to 2001 (denoted by lines). Values plotted for a particular month are averages over that month and the 11 preceding months, so that the effect of an operational forecasting-system change introduced in that month is seen from then onwards.

The consequence of improved oceanic data coverage is seen most strikingly in the results shown for the Australasian region. Here, the accuracy of medium-range forecasts is very much poorer than elsewhere prior to establishment of the observing system for FGGE in 1979. Forecast scores in fact decline early in the period, which may be a consequence of degradation of observational coverage following the end of the International Geophysical Year of 1958. They begin to pick up only in the 1970s, most likely due to the assimilation of VTPR sounding data backed up by the PAOBs. They subsequently jump substantially when data from the enhanced FGGE observing system are first assimilated at the end of 1978; the improvement amounts to about a two-day gain in practical predictability in the medium range.

The ERA-40 data assimilation system was largely as used operationally at ECMWF in the second half of 2001, but with lower horizontal resolution and 3D- rather than 4D-Var analysis. The medium-range forecasts for 2001 from ERA-40 are accordingly not as accurate as the corresponding operational forecasts. However, improvement of the operational system has occurred at such a pace that the ERA-40 forecasts are similar in skill to the operational forecasts of the late 1990s. Figure 16.6 shows very similar interannual variations in skill between the ERA-40 and

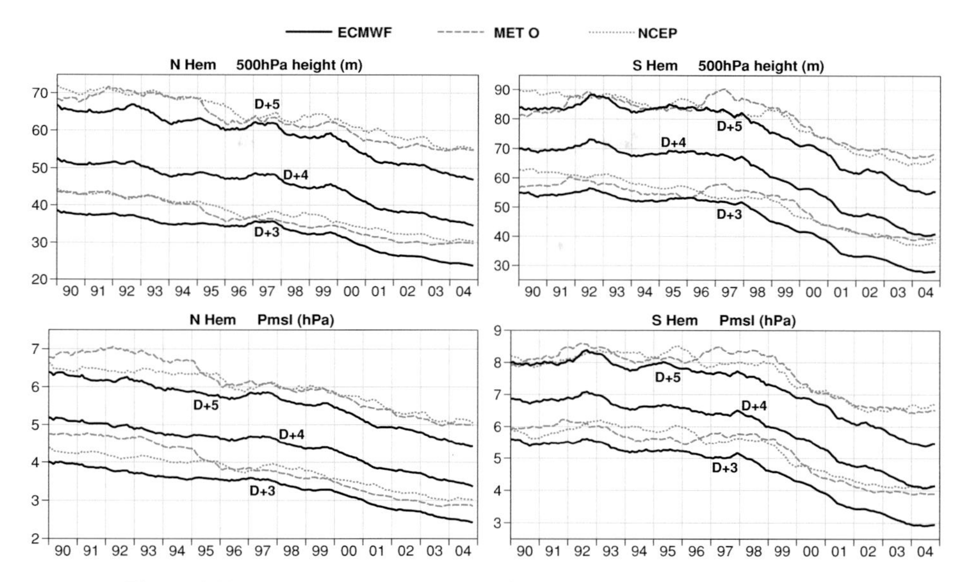

Figure 16.7 Root-mean-square errors of 3- and 5-day forecasts of 500 hPa height (upper; m) and mean-sea-level pressure (lower; hPa) for the extratropical northern (left) and southern (right) hemispheres. Results from ECMWF, the Met Office and NCEP are plotted in the form of annual running means of all monthly data exchanged by the centres from January 1989 to October 2004. ECMWF 4-day forecast errors are also shown. Values plotted for a particular month are averages over that month and the 11 preceding months.

operational forecasts at this time. This indicates that the fluctuations in operational performance were not related to short-term problems in the performance of the operational forecasting system, but rather were due to fluctuations in the predictability of the atmosphere associated with interannual variations in the atmospheric circulation or possibly with variations in the observing system. The ERA-40 analyses and forecasts provide a basis for future study of such fluctuations in predictability.

16.4 Recent improvements in operational forecasts

Figure 16.7 presents root-mean-square errors of forecasts of 500 hPa height and mean-sea-level pressure for the extratropical northern and southern hemispheres. Time series from 1990 onwards are shown for three-day and five-day forecasts from three global prediction systems, that of ECMWF and those of the Met Office and NCEP, the two national centres that come closest to matching ECMWF's performance according to these measures of forecast accuracy. ECMWF results for the four-day range are also presented. The plots show annual running means derived from the verification statistics that forecasting centres exchange monthly under the auspices of the World Meteorological Organization. Each centre's forecasts are verified by comparison with its own analyses. Results are presented for initial forecast times of 1200 UTC for ECMWF and the Met Office, and 0000 UTC for NCEP. Additional Met Office results from 0000 UTC have been available since 1995, but are not plotted as they are very similar to those presented for 1200 UTC.

Figure 16.7 shows a general trend towards lower forecast errors in both hemispheres, for both 500 hPa height and mean-sea-level pressure. The improvement since 1996 in ECMWF forecasts for the northern hemisphere amounts to around a one-day extension of the forecast range at which a given level of error is reached, today's four- and five-day forecasts being respectively about as accurate on average as the three- and four-day forecasts of six or seven years ago. The rate of improvement over this period has been especially rapid in forecasts for the southern hemisphere, amounting to a one-day gain in predictability in just three to four years. Improvement has been rather faster for ECMWF than for the other two centres shown; the difference in skill currently equates to a difference in practical predictability of a little more than 12 hours for the northern hemisphere and close to one day for the southern hemisphere.

The starting point for the rapid recent improvement in ECMWF forecasts shown in Figure 16.7 was the operational introduction of four-dimensional variational (4D-Var) data assimilation (Rabier *et al.*, 2000; Mahfouf and Rabier, 2000) in late November 1997. The improvements in short-range ECMWF forecasts since then can be linked directly to a series of subsequent forecasting-system changes. Table 16.2 shows the changes with most impact at one-day range, when they were implemented and the reductions in one-day forecast errors measured during preoperational trials. In most cases impact is larger in the southern than in the northern hemisphere. Simmons and Hollingsworth (2002) showed that the actual annual-mean root-mean-square errors of one-day 500 hPa height forecasts over the five years up to 2001 matched well the errors that would have occurred had the changes introduced between November 1997 and November 2000 given exactly the same average forecast improvements in operational use as were measured in the preoperational trials. The same result holds when the calculation is extended to include the latest changes to the forecasting system and data use. It indicates that the overall recent improvement in short-range forecasts is due overwhelmingly to changes to the forecasting system and availability of data rather than to circulation regimes that were unusually easy to predict in the last few years.

Verification of forecasts by comparison with radiosonde observations confirms the improvements shown in Figure 16.7. Figure 16.8 presents root-mean-square errors of three-, four- and five-day ECMWF forecasts of 500 hPa height and 850 hPa wind verified against radiosondes, from 1995 onwards. Values for the northern hemisphere 500 hPa height are quite similar to those from verification against analyses at these forecast ranges. The only difference of note for the southern hemisphere is that the rapid reduction in height forecast errors begins some six months earlier in 1997 in Figure 16.8 than in Figure 16.7. This earlier fall in error measured by comparison with radiosonde data appears to be associated with the introduction in May 1997 of a

Table 16.2 *The operational changes to the ECMWF forecasting system over the past seven years that had the largest impact on root-mean-square errors of one-day 500 hPa height forecasts for the extratropical northern and southern hemispheres as measured in preoperational trials*

IFS cycle	Date of change	Reduction in rms error (m) N Hem	S Hem	Principal nature of change
18r1	1997/11/25	0.47	1.77	4D-Var, with six-hourly cycling
21r1	1999/05/05	0.22	0.90	Direct assimilation of microwave (MSU and new AMSU) radiances
21r4	1999/10/12	1.49	2.31	Background error statistics from ensemble data assimilation; use of microwave imager (SSM/I) marine wind data; corrected use of humidity observations; higher resolution in the planetary boundary layer; improved parametrisations of cloud, convection and orography
22r3	2000/06/27	0.57	1.94	New background and observation error variances; additional ATOVS data; improved parametrisations of land-surface, sea-ice and longwave radiation
23r1	2000/09/12	0.42	0.65	Twelve-hourly 4D-Var and related data assimilation refinements
23r3	2000/11/21	0.80	0.66	Increased (T511/T159) model/analysis resolution; increased angular resolution in wave model
26r3	2003/10/07	0.58	0.42	Assimilation of high-resolution infrared (AIRS) radiances and additional geostationary and AMSU radiances; revised use of low-resolution infrared (HIRS) radiances; use of additional wind-profiler data; new humidity analysis; improvements to radiation parametrisation and ocean-wave data assimilation

new formulation for the background error constraint in the then-operational 3D-Var system (Derber and Bouttier, 1999). This change brought a marked improvement in tropical forecasts, and may thus have given a greater improvement in verification against those radiosondes located in the southern subtropics than in verification against analyses over the whole southern extratropics. A major element of recent changes to the forecasting system has been wider utilisation and new methods of assimilation of satellite data, so it is particularly reassuring to see a large reduction in forecast errors as measured against the southern hemisphere radiosonde network.

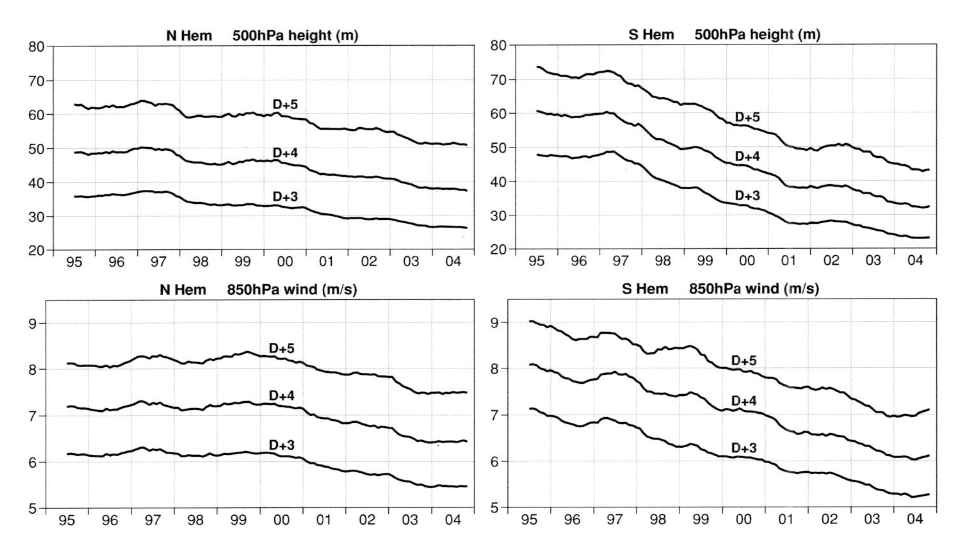

Figure 16.8 Root-mean-square errors of 3-, 4- and 5-day ECMWF 500 hPa height (upper; m) and 850 hPa vector wind (lower; m/s) forecasts for the extratropical northern (left) and southern (right) hemispheres, plotted in the form of annual running means of monthly data for verification against radiosondes from July 1994 to October 2004. Values plotted for a particular month are averages over that month and the 11 preceding months.

The levels of skill of northern and southern hemisphere forecasts cannot be compared simply in terms of root-mean-square errors because of interhemispheric differences in natural levels of variance. Comparison can, however, be made directly in terms of anomaly correlation coefficients, which are closely related to mean-square errors normalised by corresponding variances (e.g. Simmons *et al.*, 1995). The left-hand panel of Figure 16.9 presents anomaly correlations of 500 hPa height based on ECMWF's operational three-, five- and seven-day forecasts from January 1980 to October 2004. Running 12-month means of the monthly-mean skill scores archived routinely over the years are plotted for the two hemispheres. Corresponding results for the forecasts run from the ERA-40 analyses are presented in the right-hand panel.

Figure 16.9 shows a higher overall rate of improvement in the operational forecasts for the southern hemisphere. In the early 1980s, the skill levels of the three- and five-day forecasts for this hemisphere were only a little better than those of the five- and seven-day northern hemisphere forecasts. At the time this was not surprising in view of the sparsity of conventional ground-based and aircraft observations in the southern hemisphere (Bengtsson and Simmons, 1983). Today, however, the skill at a particular forecast range in the southern hemisphere is only a little lower on average than that at the same range in the northern hemisphere. The ERA-40 results show some increase in time in forecast accuracy for the southern hemisphere, which indicates improvement of the observing system over time if there is no significant trend in the underlying predictability of the atmosphere. It is nevertheless clear from Figure 16.9

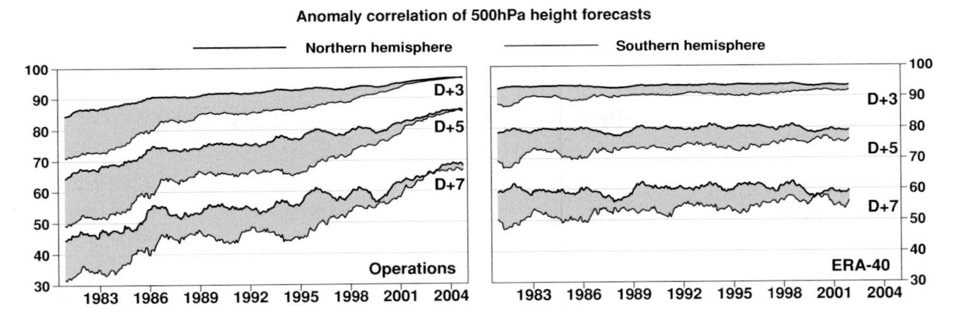

Figure 16.9 Anomaly correlation coefficients of 3-, 5- and 7-day operational ECMWF 500 hPa height forecasts for the extratropical northern and southern hemispheres, plotted in the form of annual running means of archived monthly-mean scores for the period from January 1980 to October 2004 (left) and corresponding forecasts to December 2001 from ERA-40 (right). Values plotted for a particular month are averages over that month and the 11 preceding months. The shading shows the differences in scores between the two hemispheres.

that most of the improvement in operational forecasts from 1980 to 2001 has stemmed from better data assimilation and modelling techniques and higher resolution (supported by substantial increases in computer power) rather than from net improvement of the observing system. The data assimilation improvements include a considerable refinement in the use of satellite data, which provide a more important component of the observing system in the southern than the northern hemisphere, as shown by observing-system experiments such as reported by Bouttier and Kelly (2001). It should be noted, however, that some of the most recent improvement in operational forecasts has come from using types of satellite data that were not assimilated in ERA-40: frequent water-vapour radiances from geostationary satellites (Köpken *et al.*, 2004), high-latitude winds derived by tracking features in images from polar-orbiting satellites (Bormann and Thépaut, 2004) and radiances from the first in a new generation of high-resolution infrared sounders (McNally *et al.*, 2006).

Interannual variations in skill are also evident in Figure 16.9, especially for the northern hemisphere at the five- and seven-day time ranges. In particular, there is a pronounced minimum in the northern hemisphere scores arising from relatively poor performance over the year to August 1999. A corresponding maximum can be seen in the time series of root-mean-square errors shown in the left panels of Figure 16.7. This is evident for the Met Office forecasts as well as for those of ECMWF.

16.5 Analysis and forecast differences

Several aspects of the performance of data assimilation and forecast systems, and of predictability in general, can be illustrated by comparing results from the systems of different centres, or by comparing successive forecasts from one particular centre.

Table 16.3 *Root-mean-square differences between ECMWF and Met Office 500 hPa height analyses (m) over the extratropical northern and southern hemispheres for December to February (DJF) from 1997/8 to 2003/4 (labelled 1998 to 2004) and for June to August (JJA) from 1998 to 2004*

	N Hem DJF	S Hem DJF	N Hem JJA	S Hem JJA
1998	14.1	21.2	10.6	29.7
1999	15.6	21.4	10.3	27.4
2000	12.2	16.3	8.9	17.0
2001	9.8	11.6	8.2	14.2
2002	10.1	11.5	8.3	12.3
2003	7.8	10.2	6.7	12.1
2004	7.5	10.3	8.4	11.2

16.5.1 Differences between ECMWF and Met Office analyses and forecasts

Table 16.3 presents root-mean-square differences between ECMWF and Met Office analyses of 500 hPa height evaluated over the extratropical northern and southern hemispheres. Results are shown for the periods from December to February (DJF) and from June to August (JJA) for each of the past seven years. Differences between the analyses have been reduced substantially over these years, particularly for the southern hemisphere. The fact that the analyses from the two centres have become much closer to each other does not necessarily imply that both sets have become much more accurate, but given the substantial improvements in forecast accuracy achieved by both centres it may be inferred that both centres' analyses have indeed become significantly closer to the truth.

Maps of the mean-square differences between the two sets of analyses show reductions over time at virtually all locations. Difference patterns remain largely the same, however. Figure 16.10 shows them for DJF 2003/4 and JJA 2004 over the two hemispheres. They are relatively small over substantial areas of the northern continental land masses, but are larger over much of central Asia where radiosonde coverage is poorer than over other northern land areas, and also over far northern land regions and the western boundaries of Europe and America, where the background fields of the data assimilation suffer from the lower accuracy of upstream analyses over the Arctic, Pacific and Atlantic oceans. Differences are larger over the mid- and high-latitude oceans, particularly in the southern hemisphere, and largest generally at polar latitudes. Local minima occur nevertheless at the South Pole and around parts of the coastline of Antarctica, reflecting the availability of radiosonde observations that both sets of analyses fit quite closely.

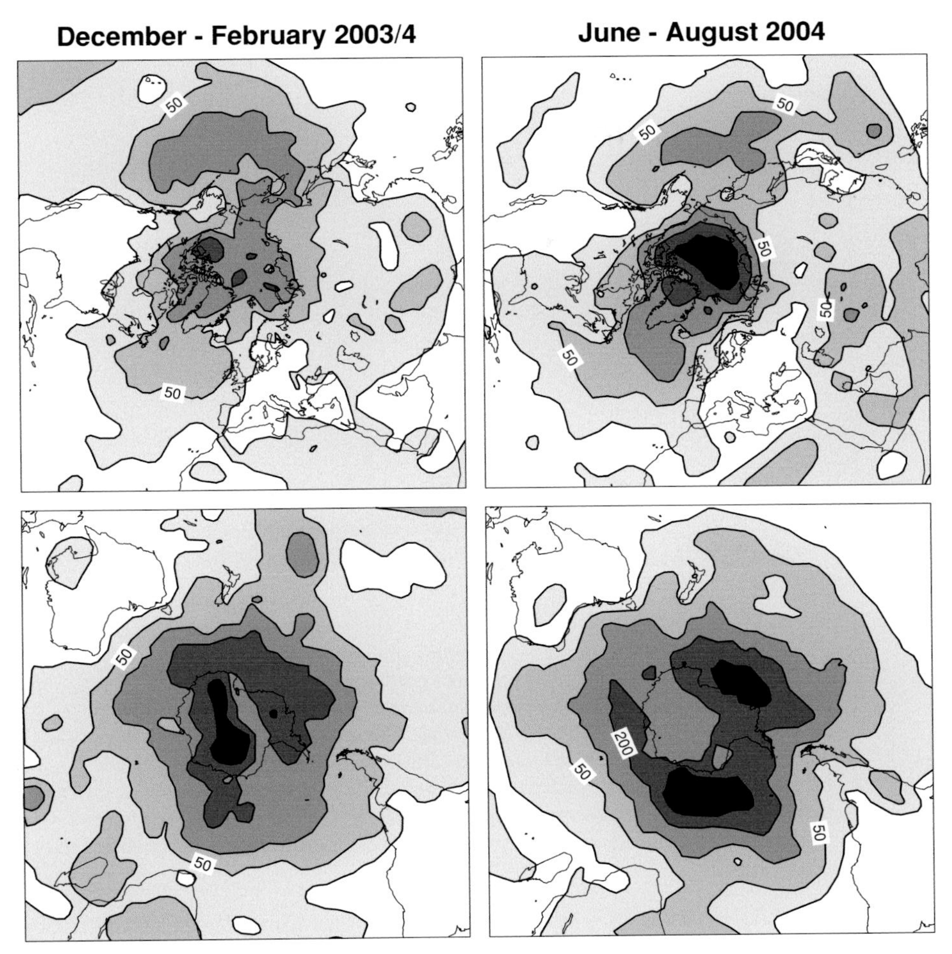

Figure 16.10 Mean-square differences (m^2) between the 500 hPa height analyses of ECMWF and the Met Office for the northern (upper) and southern (lower) hemispheres computed over the periods December–February 2003–4 (left) and June–August 2004 (right). Contours are drawn at intervals of 25, 50, 100, 200, 400 and 800 m^2.

The largest northern hemisphere differences (and by inference relatively large analysis errors) occur over the Arctic in summer, where *in-situ* measurements are sparse and satellite data are difficult to use in the lower troposphere due to the nature of the underlying surface. The geographically and seasonally fixed background-error correlations used in the ECMWF analysis system may also be less representative for this region than elsewhere. Propagation and amplification of forecast error that originates from analysis error over the Arctic is a known source of medium-range error over Europe in summer. It was, for example, particularly prevalent in the summer of 1999 (Klinker and Ferranti, 2000; Simmons *et al.*, 2000). Another region that has

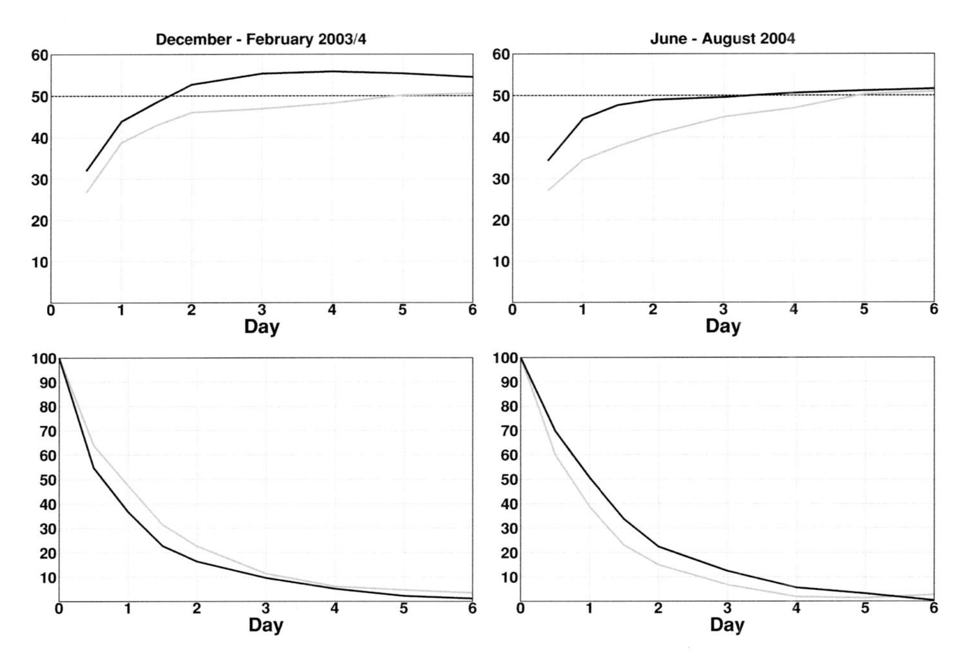

Figure 16.11 Correlations (%) as functions of forecast range computed for the 500 hPa height field over the extratropical northern (black) and southern (grey) hemispheres for 1200 UTC verifying dates in the periods December–February 2003–4 (left) and June–August 2004 (right). The upper panels show the correlations between the errors of ECMWF and Met Office operational forecasts (with the forecast from a particular system verified against subsequent analyses from the same system). The lower panels show the correlations between the differences in the forecasts and the differences in the verifying analyses of ECMWF and the Met Office.

proved to be a source of medium-range prediction error for Europe is the convectively active eastern USA in late spring and summer; here the analysis differences can be seen in Figure 16.10 to be larger in JJA than DJF.

The fact that differences between the analyses of ECMWF and the Met Office (and between ECMWF and NCEP for that matter) have become substantially smaller in recent years does not imply for certain that analysis errors have declined, as the data assimilation systems of the different centres could have converged to such an extent that each share common pronounced errors. In this case, however, one would expect that the short-range forecast errors of the centres would be highly correlated, which is not the case.

The upper panels of Figure 16.11 show correlations between ECMWF and Met Office forecast errors (with verification against each centre's own analyses) computed for DJF 2003/4 and JJA 2004. They are plotted as functions of forecast range, for the extratropical northern and southern hemispheres. Under perfect-model assumptions and neglecting error in the verifying analyses, such correlations would asymptote to the value 50% for large forecast ranges (when the two sets of forecasts and the verifying analyses can be regarded as three sets of random deviations

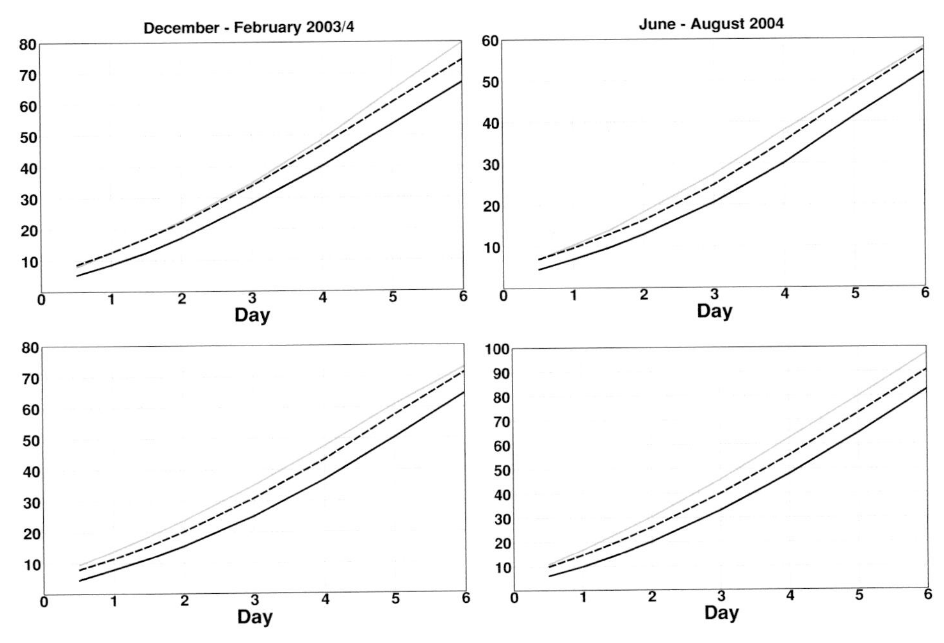

Figure 16.12 Root-mean-square errors of 500 hPa height forecasts (m) for the extratropical northern (upper) and southern (lower) hemispheres verifying at 1200 UTC in the periods December–February 2003–4 (left) and June–August 2004 (right). Results are shown for the operational forecasts from ECMWF (black, solid) and the Met Office (grey, solid) at the forecast range indicated on the ordinate. Also shown (black, dashed) are the errors of operational ECMWF forecasts verifying at the same times but initiated 12 hours earlier (plotted with 12-hour shift in the forecast range).

from climatology). The panels show that correlations close to 50% are in fact reached quite early in the forecast range, after about day 2. Correlations are weaker at shorter time ranges. Extrapolating correlations and the measured forecast errors themselves (shown in Figure 16.12) enables estimates of analysis errors to be made (Simmons and Hollingsworth, 2002). This gives 500 hPa height analysis errors of about 4.5 m for the northern hemisphere in both seasons and for the southern hemisphere in summer, and 5.5 m for the southern hemisphere winter, in the case of ECMWF. These values are close to half the typical observation error of radiosonde measurements.

The lower panels of Figure 16.11 show corresponding correlations between differences in the forecasts and differences in the verifying analyses of ECMWF and the Met Office. The correlations are 100% at the start of the forecast range, when the forecast difference is equal to the analysis difference. The correlation drops towards zero with increasing forecast range, as the assimilation of observations drives successive (verifying) analyses and subsequent background forecasts further away from the free-running forecast. The correlation falls slightly more quickly for the winter than the summer hemisphere. It takes of the order of five days for sufficient observations to have been assimilated for 500 hPa height analysis differences to lose virtually all memory of earlier differences in background forecasts. A similar estimate for the memory of the ECMWF system has been derived from the initial growth of differences in ensemble data assimilation and from the initial degradation of forecast quality in data-denial experiments (Fisher *et al.*, 2005).

If the correlation between differences in forecasts and differences in verifying analyses illustrated in Figure 16.11 is representative of the correlation between true forecast error and error in the verifying analyses, then estimates can be made of the effect of error in the verifying analysis on the measured forecast errors. Simmons and Hollingsworth (2002) discuss this, and show that the effect appears to be rather small. Measured hemispheric root-mean-square errors of recent 500 hPa height forecasts are estimated to be within a metre or so of the true errors at the one- and two-day ranges. In contrast, root-mean-square errors of short-range forecasts verified against radiosonde data tend to be dominated by the observation error of the radiosonde measurements, as shown for example by Simmons and Hollingsworth (2002) for the 500 hPa height field and by Simmons *et al.* (2005) in a study of stratospheric temperature and wind forecasts.

Figure 16.12 compares root-mean-square errors of ECMWF and Met Office 500 hPa height forecasts as functions of forecast range, evaluated over the extratropical northern and southern hemispheres. Results are shown for all forecasts from 1200 UTC start times verifying in DJF 2003/4 and JJA 2004. ECMWF forecasts verifying at 1200 UTC but starting from 0000 UTC are also shown, with the forecast range shifted by 12 hours in the plots so that, for example, a 36-hour forecast error is plotted as if the range was 24 hours.

Figure 16.12 shows that the forecasts from ECMWF are considerably more accurate on average than those from the Met Office made with the same starting time, by around 12 hours or a little more in the northern hemisphere and by more still in the southern hemisphere, as judged by the forecast range at which a particular level of error occurs. The advantage is at least 12 hours throughout the forecast range. The implication is that the initial analysis error is considerably lower in the ECMWF system, leading to lower subsequent forecast errors at all ranges.

A further indication of the link between reducing forecast error in the very short range (and by implication reducing analysis error) and reducing error in the medium range is provided by smoothed time series of the ranges at which ECMWF's operational 500 hPa height forecasts have reached certain levels of anomaly correlation over the past two decades. In recent years the improvement of forecasts as measured by the increases in these forecast ranges has been by an amount that varies little beyond a day or so ahead. Improvements in medium-range 500 hPa height forecasts thus appear to have stemmed directly from model, analysis and observing-system improvements that have reduced analysis and short-range forecast error.

Figure 16.12 can be interpreted as showing that the value of the difference in performance between the ECMWF and Met Office assimilation systems is more than the value of the latest 12 hours of observations, at least as regards the hemispheric

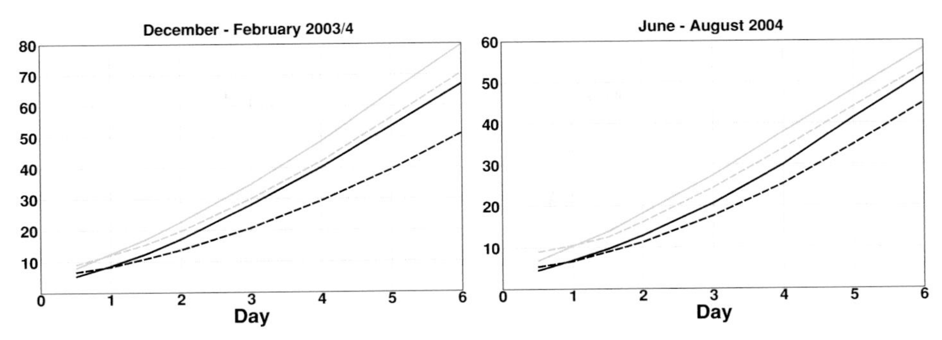

Figure 16.13 Root-mean-square errors and differences of 500 hPa height forecasts (m) for the extratropical northern hemisphere verifying at 1200 UTC in the periods December–February 2003–4 (left) and June–August 2004 (right). Errors are shown for the operational 1200 UTC forecasts from ECMWF (black, solid) and the Met Office (grey, solid) at the forecast range indicated on the ordinate. Also shown are the differences between these ECMWF and Met Office forecasts (grey, dashed), and the differences between these ECMWF forecasts and ECMWF forecasts verifying at the same times but initiated 12 hours earlier (black, dashed).

accuracy of analyses and forecasts of 500 hPa height. Forecasts from an analysis that uses the most recent observations are not necessarily more accurate than forecasts from an analysis for an earlier synoptic hour that has been produced by a better data assimilation system. In designing an operational forecasting system, compromises have to be made regarding the data cut-off (the wall-clock time at which the data assimilation is started), the times by which forecast products have to be delivered, and the complexity (and hence computational cost) of the data assimilation and forecast model integration. Traditionally, the data cut-off for short-range prediction has been determined first and foremost by the need to receive the bulk of the latest radiosonde ascents, and a quite stringent limit is placed on the time that can be taken by the forecast-production process. The increasing importance of asynoptic observations, particular from satellites, and the development of more accurate but more computationally demanding data assimilation systems and forecast models are making appropriate choices less clear. Whatever choices are made, analysis and forecast accuracy remains dependent on the amount of data available for assimilation, and effective telecommunication systems to ensure rapid transmission of information from the point of measurement (satellite or otherwise) to forecasting centres remain a key requirement for optimal operational forecasting.

Figure 16.13 casts a different light on the performance of the 1200 UTC ECMWF and Met Office forecasts. It repeats (for the northern hemisphere) the plots of root-mean-square errors of the 500 hPa height forecasts, and adds plots of the root-mean-square differences between the 1200 UTC ECMWF and Met Office forecasts and between the 1200 UTC ECMWF forecasts and the 12-hour older ECMWF forecasts made from 0000 UTC. Results for the southern hemisphere are essentially similar. The main point of note is that the 1200 UTC ECMWF forecasts differ from the corresponding Met Office forecasts by a little more than they differ from the verifying analyses (consistent with the forecast error correlations of about 50% noted earlier). The ECMWF and Met Office forecasts do not predominantly share common errors that have resulted from each analysis system having drawn the same inaccurate observations. Instead, differences in data analysis (quality control decisions, basic analysis methods and specified error statistics) and the growth of differences through the background forecasts of the data assimilation cycles result in two analyses which, were it not for the overall superiority of the ECMWF system, would essentially be two randomly drawn and equally likely estimates of the true state of the atmosphere. As such the two analyses would provide two sets of initial conditions for an ideal ensemble prediction system. Reproducing such variability among analyses is a challenge for ensemble data assimilation systems aimed at providing either optimal perturbations for ensemble prediction or flow-dependent background-error estimates for improved data assimilation.

The 1200 UTC ECMWF forecasts can be seen in Figure 16.13 to be much closer to the earlier 0000 UTC ECMWF forecasts than they are either to the verifying analyses or to the 1200 UTC Met Office forecasts. The observations that are assimilated over one 12-hour assimilation window are clearly insufficient to remove dependence of the 1200 UTC analysis on the background forecast initiated 12 hours earlier. Thus, whilst the 1200 UTC Met Office 500 hPa height forecast is on average similar in accuracy to the earlier 0000 UTC ECMWF forecast, it provides a more distinct possible alternative to the (probably more correct) 1200 UTC ECMWF forecast.

16.5.2 Differences between successive ECMWF forecasts

Lorenz (1982) discussed the comparison of root-mean-square forecast errors with root-mean-square differences between ECMWF forecasts started a day apart and verifying at the same time. He argued that if the forecast model in operational use at the time was realistic enough for small differences in initial conditions to cause forecasts to diverge at a rate close to that at which separate but similar atmospheric states diverge, then the rate of growth of the forecast differences would provide a limit to the potential accuracy of the forecast that could not be surpassed without analysis or model changes which reduced the one-day forecast error. The evolution of the forecast differences (or 'perfect-model' errors) would in particular provide a basis for estimating the intrinsic rate of growth of initially small forecast errors.

Measures of the differences between successive numerical forecasts valid for the same time are indicators of forecasting-system performance that are also of direct relevance to bench forecasters. These measures provide indications of the consistency of the forecasting system, the extent to which the latest forecast is consistent with the forecast provided 12[1] or 24 hours earlier. High consistency (low values for measures of the differences between successive forecasts) is clearly a desirable feature if it stems from a basic high accuracy of the forecasts. For forecasts of limited accuracy the

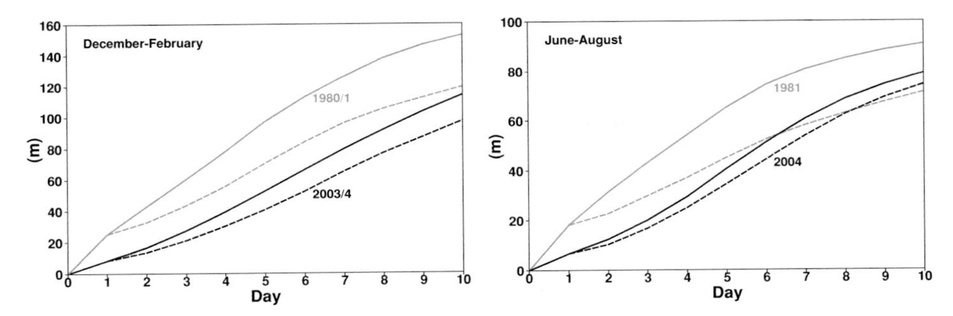

Figure 16.14 Root-mean-square forecast errors (solid) and root-mean-square differences between successive forecasts (dashed), based on 1200 UTC ECMWF 500h Pa height (m) forecasts for the extratropical northern hemisphere verifying in the periods December–February 1980–1 and 2003–4 (left) and June–August 1981 and 2004 (right).

situation is less clear. Inconsistent forecasts in such cases provide the forecaster with indications of a less predictable situation and of possible alternative developments, but if issued to end users may unduly undermine confidence in the forecasts.

Figure 16.14 compares root-mean-square errors and root-mean-square differences between successive daily 1200 UTC operational forecasts of 500 hPa height for the extratropical northern and southern hemispheres for DJF 1980/1 (the period studied by Lorenz, 1982) and 2003/4, and for JJA 1981 and 2004. The 2003/4 errors are substantially lower than the 1980/1 errors throughout the forecast range. The relative difference is largest at short range; 2003/4 values are about a third of 1980/1 values at day 1. The gap between the forecast-error and forecast-difference curves is much smaller in 2003/4 than in 1980/1, indicative of model improvement since 1981. This is known to include a significant reduction in the systematic component of forecast error (Simmons *et al.*, 1995; Ferranti *et al.*, 2002).

The model improvements over the past two decades have brought with them, as an unavoidable by-product, a faster rate of amplification of forecast errors, and a faster rate of growth of inconsistency between successive forecasts (Simmons and Hollingsworth, 2002). Lower absolute values for forecast errors and the differences between successive forecasts have occurred nevertheless, due to the very much lower starting point for the growth of errors and inconsistency that has been provided by the substantial reduction in analysis and short-range forecast error. Figure 16.15 shows, for both extratropical hemispheres, the difference (or inconsistency) curves plotted for each DJF from 1980/1 to the present. Evident in this figure is the general reduction in one-day forecast errors, with a distinct fall in particular between 1999/2000 and 2000/1, consistent with the operational changes listed in Table 16.2. Also evident is a much more rapid growth rate of differences following model changes in the late 1980s and early 1990s. Both are more pronounced for the southern than the northern hemisphere. The increased consistency of forecasts in recent years stems mainly from the reduction in short-range forecast error, as doubling times for small forecast differences have not increased (Simmons and Hollingsworth, 2002).

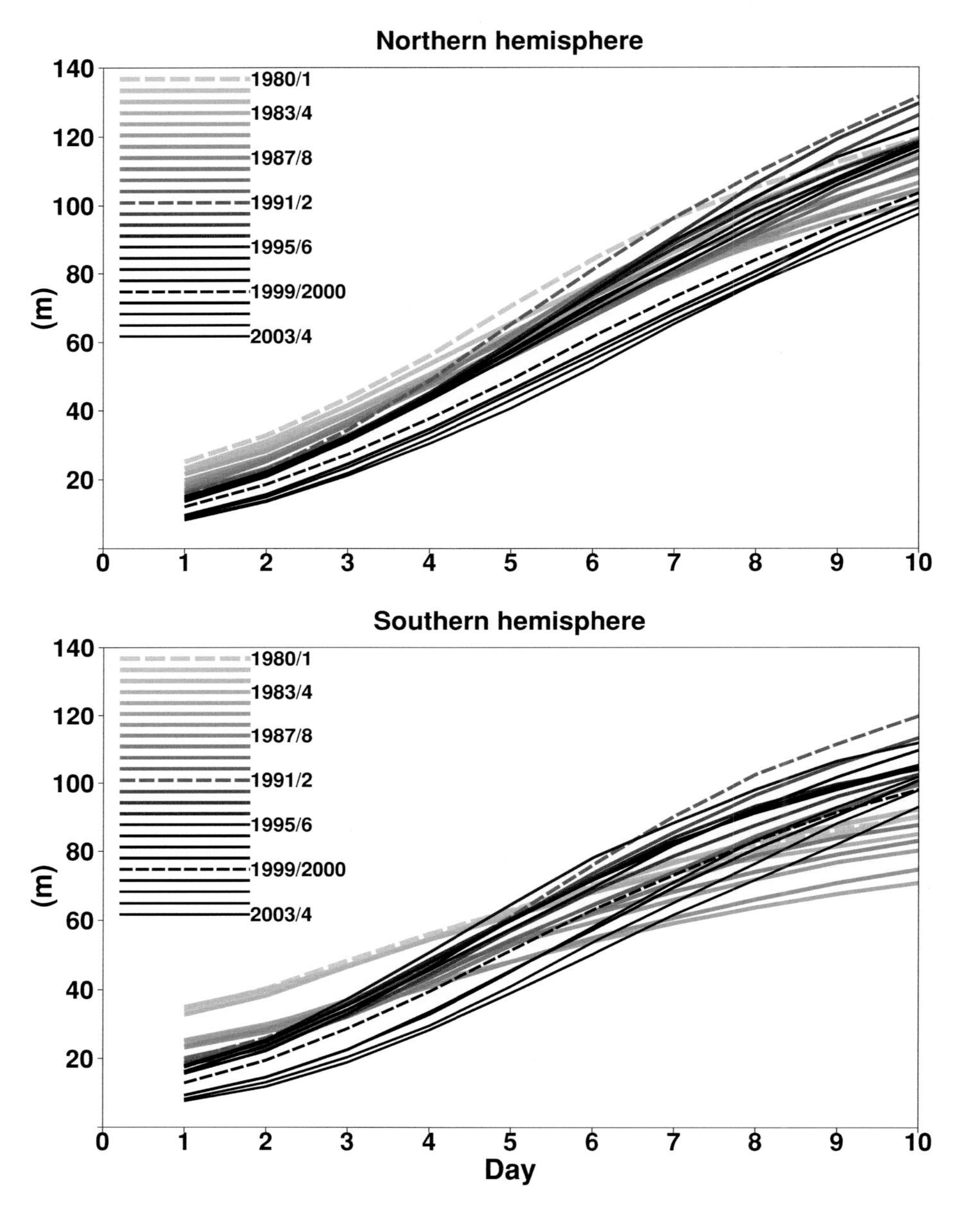

Figure 16.15 Root-mean-square differences between successive 1200 UTC forecasts of 500 hPa height (m) as functions of forecast range. Results are shown for each December–February period from 1980–1 to 2003–4, for the extratropical northern (upper) and southern (lower) hemispheres.

There is a relatively slow growth of differences for 1999/2000, especially in the southern hemisphere. This is due to a deficiency in the operational humidity analysis between October 1999 and April 2000. The lower stratosphere was much too moist over this period, leading through the radiation parametrisation in the medium-range forecasts to a changed structure of potential vorticity near the tropopause and a consequent damping of tropospheric transient-eddy activity. What looked at first sight to be a systematic model error in fact was due to a bias in the analysis of a field that adjusted only slowly towards reality as the forecast range increased. The relatively fast growth of differences for 1991/2 was due to teething problems with a new semi-Lagrangian advection scheme and a new 31-level vertical resolution, both of which were introduced in September 1991 and caused excessive transient eddy activity.

16.6 Dependence of forecast error on horizontal scale

The dependence of forecast error on horizontal scale is of interest as regards predictability and forecasting-system performance in general, and more specifically as regards the potential for improved analysis of smaller scales.

Figure 16.16 presents log-linear plots of spectra of the global error (measured by verification against analyses) of operational one-day ECMWF 500 hPa height forecasts for each DJF since 1980/1. Results are shown for wave numbers up to 40, the highest wave number for which operational forecast results were saved in the early years. Spectra of the temporal variance of error rather than the mean-square error are presented to avoid the plots being complicated by the relatively large time-mean component of error that occurred at low wave numbers early in the period.

Figure 16.16 shows that one-day 500 hPa height forecast errors have been substantially reduced over time at the large synoptic scales represented by a band of wave numbers centred around about wave number nine. In the early years there was a quite pronounced spectral peak of error at these wave numbers. Error is more uniformly distributed across the spectrum for more recent years, with the flattest spectrum for the latest years. The sharp decrease in small-scale error with increasing wave number seen for the first few years presumably reflects the characteristics of the forecast (and assimilating) model of the time rather than a low level of actual small-scale error. Subsequent forecasting-system changes resulted in an increase in small-scale error, and only following the system changes made in 1999 and 2000, not least changes to model and analysis resolution (Table 16.2), does the one-day error as measured against analyses drop across the whole spectral range for which results are presented.

Spectra of mean-square errors for DJF 2003/4 are presented in Figure 16.17 for forecast ranges from 12 hours to 10 days. Log-log plots are shown for wave numbers up to the limit of the model's T511 truncation. Results are presented for 850 hPa temperature and vorticity as there is relatively more amplitude in small scales for these fields (especially vorticity) than for 500 hPa height.

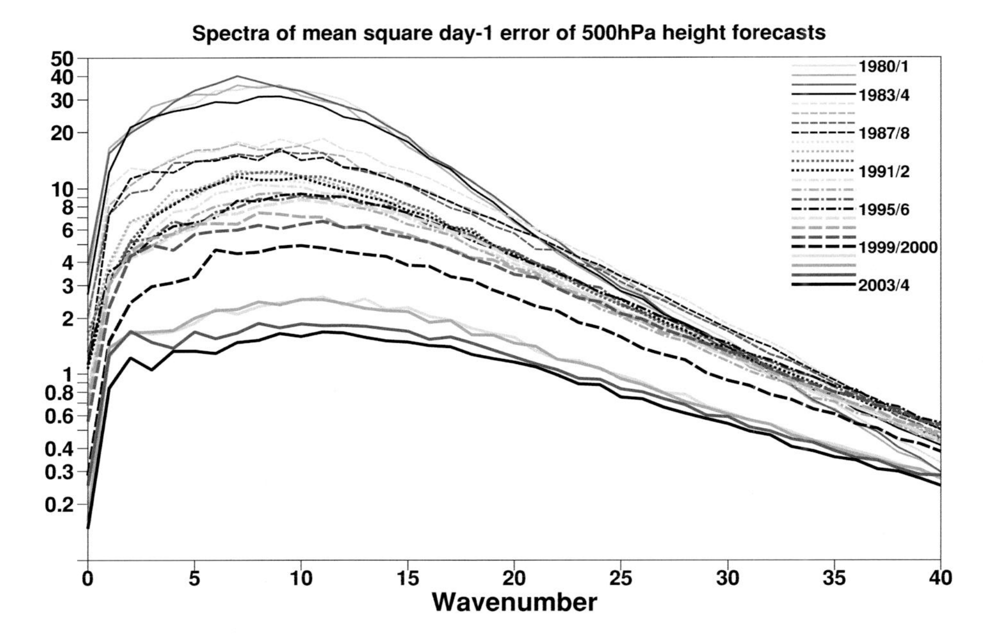

Figure 16.16 Log-linear plot of the spectra of the variance of the global 500 hPa height forecast error (m^2) at one-day range for forecasts verifying in each December–February period from 1980–1 to 2003–4.

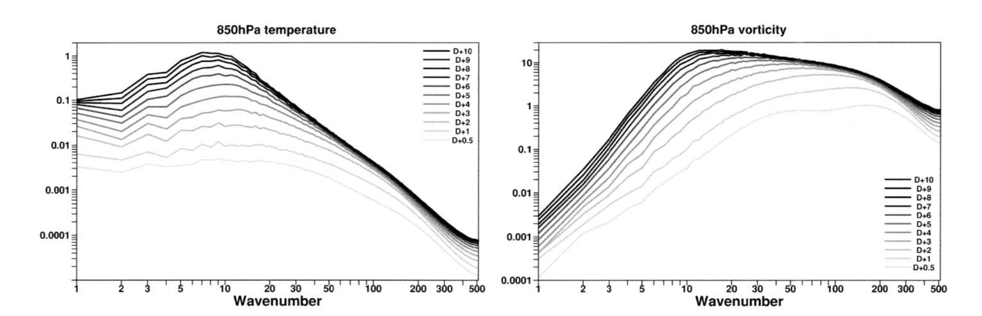

Figure 16.17 Log-log plot of the spectra of the mean-square global 850 hPa forecast errors for temperature (K^2) and relative vorticity (10^{-12} s^2) at forecast ranges from 12 hours to 10 days, for forecasts verifying at 1200 UTC in the period from 1 December 2003 to 29 February 2004.

The spectrum of 850 hPa temperature error for 12-hour forecasts is rather flat over a quite wide range of wave numbers. A more peaked error spectrum evolves with increasing forecast range, with error continuing to grow throughout the 10 days near the spectral peak and at longer wavelengths. The spectral peak in vorticity error shifts from short towards synoptic scales as the forecast range increases. At smaller scales error largely saturates within the forecast range, but a slow component of error growth is evident near the truncation limit. Boer (1994) reported similar small-scale

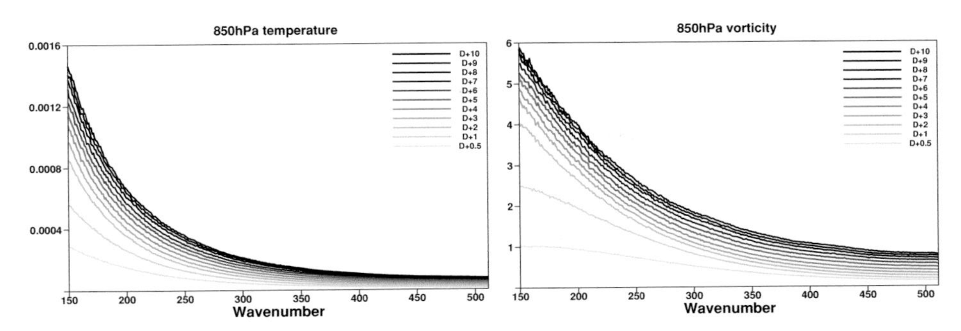

Figure 16.18 Spectra of 850 hPa forecast errors in temperature and relative vorticity as in Figure 16.17, but shown in a linear-linear plot for wave numbers 150 and above.

behaviour for an earlier version of the ECMWF forecasting system, although he has subsequently shown such behaviour to be much weaker in Canadian Meteorological Centre forecasts (Boer, 2003).

A clearer picture of the small-scale behaviour emerges when graphs are redrawn using linear-linear axes, as shown in Figure 16.18 for wave numbers 150 and higher. The figure shows more clearly that there are essentially two timescales for the growth of error in small spatial scales. There is rapid growth and apparent saturation early in the forecast range of a component of error that is likely to represent intrinsic dynamics with small temporal as well as spatial scales. This component decreases markedly in amplitude with decreasing scale, as would be expected given the model's high damping rates for the smallest scales. There is also a second component of error (with a less marked decrease in amplitude with decreasing scale) that evolves much more slowly throughout the forecast range. It is likely that this arises from small-scale components of the temperature and vorticity fields that are directly forced by large-scale dynamical and thermodynamical processes that have much slower intrinsic error growth rates.

16.7 Concluding remarks

A general introduction to the atmospheric observing system and to the way that data from this system are assimilated into numerical models to provide the starting point for numerical weather forecasts has been given. The improvement in synoptic-scale weather prediction that has resulted from improvements made to the observing and forecasting systems has been illustrated, and some general aspects of predictability have been discussed. Use has been made of results from the ERA-40 reanalysis project and from the operational forecasting activities of ECMWF and other centres.

The results from ERA-40 show that there has been a clear long-term improvement in the observing system. This is especially so for the southern hemisphere, for which ERA-40 provides striking evidence of the substantial benefit of the changes made to the observing system in the build-up to FGGE in the late 1970s. ERA-40 also shows nevertheless that a modern data assimilation system is capable of providing analyses of the state of the northern hemisphere more than 40 years ago that are of sufficient accuracy to yield forecasts that remain skilful well into the medium range. Lack of today's analysis and modelling techniques and computer power appear to have been more of a hindrance to general northern hemisphere forecast quality 40 years ago than lack of today's observing system. It must, however, be kept in mind that this remark applies to routine synoptic-scale prediction beyond a day or two ahead, and not necessarily to the short-range prediction of severe weather events where satellite imagery (and observing systems such as ground-based radar) can play such an important role. Moreover, the newer observing systems have had to compensate for a decline in radiosonde coverage over key oceanic and high-latitude regions.

There has been a particularly large improvement in operational forecasts over the past seven or so years. Evidence discussed here and elsewhere indicates that improvements have stemmed in particular from improved data assimilation (improved assimilating models as well as improved analysis techniques), helped by the availability of new or improved types of observation. It should be noted that attention has been concentrated here on the accuracy of forecasts of 500 hPa height, and to a lesser extent wind and temperature in the free troposphere, rather than on forecasts of weather elements such as near-surface temperature, cloud and precipitation. The latter benefit directly from model improvements as well as from the improved definition of the synoptic environment that is documented by the type of verification presented here. General forecast improvements have been achieved by many centres, and it has been demonstrated how the initial analyses of two centres, ECMWF and the Met Office, have converged substantially. Nevertheless, significant (and informative) differences in analyses and forecasts remain.

The spectral breakdown of error shows that there has been a distinct recent improvement in the handling of smaller scales of motion in the ECMWF system. In the system's incremental 4D-Var data assimilation, the spectral truncation of the highest-resolution minimisation is at wave number 159 and the assimilation period is 12 hours. For this system, forecast error is still some way from saturation after 12 or even 24 hours for a range of wave numbers higher than 159, suggesting that there may well be scope for extending the resolution of the minimisation. This might be a rather optimistic view given that the small-scale short-range forecast 'errors' considered here are determined by verifying against an analysis that at small scales is very heavily influenced by the model background forecast. A clear benefit accrued, however, from the resolution changes made in November 2000.

Acknowledgements

Production of the ERA-40 analyses was possible only because of the considerable external support the project received, as documented by Uppala *et al.* (2005), who also presented the originals of several of the figures that have been reproduced or adapted here.

The study in Section 16.5 utilized 500 hPa height analyses and forecasts made available by the Met Office to ECMWF on a routine basis.

Note

1. Forecasts at 12-hourly frequency using comparable data cut-off times have been produced regularly by ECMWF only since 28 March 2001.

References

Andersson, E., J. Haseler, P. Undén, *et al.* (1998). The ECMWF implementation of three-dimensional variational assimilation (3D-Var). III: Experimental results. *Quart. J. Roy. Meteor. Soc.*, **124**, 1831–60.

Bengtsson, L. and A. J. Simmons (1983). Medium-range weather prediction – operational experience at ECMWF. In *Large-scale Dynamical Processes in the Atmosphere*, ed. B. J. Hoskins and R. P. Pearce, pp. 337–63 Academic Press.

Boer, G. J. (1994). Predictability regimes in atmospheric flow. *Mon. Weather Rev.*, **122**, 2285–95.

(2003). Predictability as a function of scale. *Atmos. Ocean*, **41**, 203–15.

Bormann, N. and J.-N. Thépaut (2004). Impact of MODIS polar winds in ECMWF's 4D-Var data assimilation system. *Mon. Weather Rev.*, **132**, 929–40.

Bouttier, F. and G. A. Kelly (2001). Observing system experiments in the ECMWF 4D-Var data assimilation system. *Quart. J. Roy. Meteor. Soc.*, **127**, 1469–88.

Courtier, P., J.-N. Thépaut and A. Hollingsworth (1994). A strategy for operational implementation of 4D-Var, using an incremental approach. *Quart. J. Roy. Meteor. Soc.*, **120**, 1367–88.

Derber, J. C. and F. Bouttier (1999). A reformulation of the background error covariance in the ECMWF global data assimilation system. *Tellus*, **51A**, 195–222.

Ferranti, L., E. Klinker, A. Hollingsworth and B. J. Hoskins (2002). Diagnosis of systematic forecast errors dependent on flow pattern. *Quart. J. Roy. Meteor. Soc.*, **128**, 1623–40.

Fisher, M., M. Leutbecher and G. A. Kelly (2005). On the equivalence between Kalman smoothing and weak-constraint four-dimensional variational data assimilation. *Quart. J. Roy. Meteor. Soc.*, **131**, 2581–3.

Haimberger, L. (2005). *Homogenization of Radiosonde Temperature Time Series Using ERA-40 Analysis Feedback Information*. ERA-40 Project Report Series 22. ECMWF (available from www.ecmwf.int).

Haseler, J. (2005). *The Early Delivery System. Technical Memorandum* 454. ECMWF (available from www.ecmwf.int).

Janssen, P. A. E. M., J. D. Doyle, J. Bidlot, *et al.* (2002). Impact and feedback of ocean waves on the atmosphere. In *Atmosphere-Ocean Interactions*, Vol. I, ed. W. Perrie, pp. 155–97. Advances in Fluid Mechanics 33. WIT Press.

Kalnay, E., M. Kanamitsu, R. Kistler, *et al.* (1996). The NCEP/NCAR 40-year reanalysis project. *Bull. Am. Meteorol. Soc.*, **77**, 437–71.

Kistler, R., E. Kalnay, W. Collins, *et al.* (2001). The NCEP-NCAR 50-year reanalysis: monthly means CD-Rom and documentation. *Bull. Am. Meteorol. Soc.*, **82**, 247–67.

Klinker, E. and L. Ferranti (2000). Forecasting system performance in summer 1999. 1: Diagnostics related to the forecast performance during spring and summer 1999. Technical Memorandum 321. ECMWF.

Köpken, C., G. Kelly and J.-N. Thépaut (2004). Assimilation of Meteosat radiance data within the 4D-Var system at ECMWF: assimilation experiments and forecast impact. *Quart. J. Roy. Meteor. Soc.*, **130**, 2277–92.

Lorenc, A. C., S. P. Ballard, R. S. Bell, *et al.* (2000). The Met. Office global three-dimensional variational data assimilation scheme. *Quart. J. Roy. Meteor. Soc.*, **126**, 2991–3012.

Lorenz, E. N. (1982). Atmospheric predictability experiments with a large numerical model. *Tellus*, **34**, 505–13.

Mahfouf, J.-F. and F. Rabier (2000). The ECMWF operational implementation of four-dimensional variational assimilation. II: Experimental results with improved physics. *Quart. J. Roy. Meteor. Soc.*, **126**, 1171–90.

McNally, A. P., P. D. Watts, J. A. Smith, *et al.* (2006). The assimilation of AIRS radiance data at ECMWF. *Quart. J. Roy. Meteor. Soc.*, in press.

Parrish, D. F. and J. C. Derber (1992). The National Meteorological Center's Spectral Statistical-Interpolation analysis system. *Mon. Weather Rev.*, **120**, 1747–63.

Rabier, F., H. Järvinen, E. Klinker, J. F. Mahfouf and A. J. Simmons (2000). The ECMWF operational implementation of four dimensional variational assimilation. 1: Experimental results with simplified physics. *Quart. J. Roy. Meteor. Soc.*, **126**, 1143–70.

Simmons, A. J. and A. Hollingsworth (2002). Some aspects of the improvement in skill of numerical weather prediction. *Quart. J. Roy. Meteor. Soc.*, **128**, 647–77.

Simmons, A. J., R. Mureau and T. Petroliagis (1995). Error growth and predictability estimates for the ECMWF forecasting system. *Quart. J. Roy. Meteor. Soc.*, **121**, 1739–71.

Simmons, A. J., E. Andersson, M. Fisher, *et al.* (2000). Forecasting system performance in summer 1999. 2: Impact of changes to the ECMWF forecasting system. Technical Memorandum 322. ECMWF.

Simmons, A., M. Hortal, G. Kelly, A. McNally, A. Untch and S. Uppala (2005). Analyses and forecasts of stratospheric winter polar vortex break-up: September 2002 in the southern hemisphere and related events. *J. Atmos. Sci.*, **62**, (3), 668–89.

Uppala, S. M., P. W. Kållberg, A. J. Simmons, *et al.* (2005). The ERA-40 Reanalysis. *Quart. J. Roy. Meteor. Soc.*, **131**, 2961–3012.

17

The ECMWF Ensemble Prediction System

Roberto Buizza
European Centre for Medium-Range Weather Forecasts, Reading

There are two key sources of forecast error: the presence of uncertainties in the initial conditions and the approximate simulation of atmospheric processes achieved in the state-of-the-art numerical models. These two sources of uncertainties limit the skill of single, deterministic forecasts in an unpredictable way, with days of high/poor quality forecasts randomly followed by days of high/poor quality forecasts. One way to overcome this problem is to move from a deterministic to a probabilistic approach to numerical weather prediction, and try to estimate the time evolution of an appropriate probability density function in the atmosphere's phase space. Ensemble prediction is a feasible method to estimate the probability distribution function of forecast states. The European Centre for Medium-Range Weather Forecasts (ECMWF) Ensemble Prediction System (EPS) is one of the most successful global ensemble prediction systems run on a daily basis. In this chapter the ECMWF EPS is described, its forecast skill documented, and potential areas of future development are discussed.

17.1 The rationale behind ensemble prediction

The time evolution of the atmospheric flow, which is described by the spatial distribution of wind, temperature, and other weather variables such as specific humidity and surface pressure, can be estimated by numerically integrating the mathematical differential equations that describe the system time evolution. These equations include

Predictability of Weather and Climate, ed. Tim Palmer and Renate Hagedorn. Published by Cambridge University Press.

Newton's laws of motion used in the form 'acceleration equals force divided by mass' and the laws of thermodynamics. Numerical time-integration is performed by replacing time-derivatives with finite differences, and spatial-integration either by finite difference schemes or spectral methods. In the ECMWF model, dynamical quantities such as pressure and velocity gradients are evaluated in spectral space, while computations involving processes such as radiation, moisture conversion, turbulence, are calculated on a mesh of grid points, each representing a geographical point in space.

Numerical integration starts from the system's initial conditions constructed using meteorological observations made all over the world. At ECMWF, observations are assimilated in a dynamically consistent way using a four-dimensional scheme to estimate the actual state of the atmosphere (Talagrand and Courtier, 1987; Courtier and Talagrand, 1987; Courtier *et al.*, 1994). The fact that only a limited number of observations are available (limited compared with the degrees of freedom of the system) and that part of the globe is characterised by a very poor coverage introduces some uncertainties in the system's initial conditions. The presence of these *initial uncertainties* is the first source of forecast errors.

Data assimilation and forward time integration is performed using models designed to simulate physical processes associated with radiative transfer, turbulent mixing and moist processes. These processes are active at scales smaller than the grid size used in the numerical integration, and thus remain unresolved in terms of model-resolved variables and can only be approximated. The problem of the simulation of these unresolved processes using model-resolved variables is referred to as parametrisation, and is one of the most difficult and controversial areas of weather modelling (Holton, 1992). At ECMWF, the assimilation and forecasting system (Simmons *et al.*, 1989; Courtier *et al.*, 1991; Simmons *et al.*, 1995) includes a parametrisation of many physical processes such as surface and boundary layer processes (Viterbo and Beljaars, 1995), radiation (Mocrette, 1990) and moist processes (Tiedtke, 1993; Jacob, 1994). The approximate nature of the parametrisation schemes included in the numerical weather prediction models is one of the key reasons of *model uncertainties*, and is the second source of forecast errors.

These two sources of forecast errors cause weather forecasts to deteriorate with forecast time. It is worth stressing that initial conditions will always be only approximately known, since all data are characterised by an error that depends on the instrumental accuracy. Observational errors, usually in the smaller scales, amplify and through non-linear interactions spread to longer scales, eventually affecting the skill of the longer scales (Somerville, 1979). Similarly, numerical models will never be able to resolve all physical processes: thus, numerical weather prediction has to find a way to deal with these two sources of forecast error, and take them into account when generating weather predictions.

One way to take these sources of forecast errors into consideration is to move from a single deterministic to a probabilistic approach, whereby the weather prediction problem is stated in terms of the time evolution of an appropriate probability density function in the atmosphere's phase space. Although this problem can be formulated exactly through the continuity equation for probability (Liouville equation; see Ehrendorfer, 1994 and this volume), ensemble prediction based on a finite number of deterministic integrations appears to be the only feasible method to predict the probability density function beyond the range of linear error growth. An ensemble system can take into account the two sources of model errors, for example by using different initial conditions and different parametrisation schemes in each ensemble member, and then using the ensemble of forecasts to estimate the probability of occurrence of any event of interest.

Since December 1992, the US National Center for Environmental Predictions (NCEP, previously NMC; Tracton and Kalnay, 1993) and ECMWF (Palmer *et al.*, 1993) have been producing operationally global ensemble forecasts. These operational implementations followed the theoretical and experimental work of, among others, Epstein (1969), Gleeson (1970), Fleming (1971a, 1971b), Leith (1974) and Lorenz (this volume). In 1998, ECMWF and NCEP were followed by the Meteorological Service of Canada (MSC; Houtekamer *et al.*, 1996). Today, these three centres can be considered the leaders in the field of operational ensemble prediction, which includes nine centres running a global system on a daily basis (ECMWF in the UK, MSC in Canada, NCEP and the US Navy in the USA, BMRC in Australia, CPTEC in Brazil, NMC in China, JMA in Japan, and KMA in Korea).

In this chapter, the ECMWF EPS is described, its performance is documented, and plans for future developments are discussed. Two appendices present some methodological details: the singular vector technique used to generate the EPS initial perturbations is discussed in Appendix A, and the accuracy measures most commonly used to assess the accuracy of probabilistic forecasts are briefly revised in Appendix B.

17.2 The ECMWF Ensemble Prediction System (EPS)

17.2.1 The development and implementation of the EPS

The fact that ensemble prediction systems should simulate the effect both of initial uncertainties and of model uncertainties was shown, for example, by the works of Downton and Bell (1988), Richardson (1998) and Harrison *et al.* (1999), who studied the relative importance of the initial and model uncertainties on the forecast error. Downton and Bell and Richardson compared forecasts given by the UK Met Office and the ECMWF forecasting systems, and indicated that forecast differences between the ECMWF and the UK Met Office operational forecasts could be linked mainly to differences between the two operational analyses, rather than between the two forecast models. Harrison *et al.* (1999) compared multimodel, multianalysis ensemble systems constructed using the ECMWF and the UK Met Office analysis and forecast systems, and concluded that the impact of model uncertainties on the forecast error should not be ignored, especially in the medium range (say after forecast day 3).

The first version of the ECMWF EPS (Palmer *et al.*, 1993; Molteni *et al.*, 1996), implemented operationally in December 1992, and its subsequent upgrades reflected the key conclusion of these studies that initial uncertainties are the most important sources of forecast error in the short forecast range, but then model errors also play a role. The first version of the EPS included only a simulation of initial uncertainties, but since October 1998 the EPS also included a stochastic scheme designed to simulate the random model errors due to parametrised physical processes (Buizza *et al.*, 1999). Table 17.1 lists the key upgrades of the ECMWF from the implementation of a 33-member T63L19 system (spectral truncation T63 with 19 vertical levels) in December 1992. This list includes:

- In December 1996, an upgrade to a 51-member T_L159L31 system, with spectral triangular truncation T159 with linear grid (Buizza *et al.*, 1998).
- In March 1998, the introduction of the evolved singular vectors (Barkmeijer *et al.*, 1999).
- In October 1998, the introduction of the stochastic physics scheme (Buizza *et al.*, 1999).
- In November 2000, a second major resolution upgrade to a T_L255L31 system (Buizza *et al.*, 2003).
- In January 2002, the introduction of tropical initial perturbations (Barkmeijer *et al.*, 2001).

17.2.2 The EPS configuration operational in 2004

Formally, each ensemble member e_j can be written as the time integration

$$e_j(d,t) = e_j(d,0) + \int_{t=0}^{t} [A(e_j,t) + P'_j(e_j,t)]dt \tag{17.1}$$

of the perturbed model equations

$$\frac{\partial e_j}{\partial t} = A(e_j,t) + P'_j(e_j,t), \tag{17.2}$$

where A and P' identify the contribution to the full equation tendency of the non-parametrised and parametrised physical processes. For each grid point $\mathbf{x} = (\lambda, \phi, \sigma)$ (identified by its latitude, longitude and vertical hybrid coordinate), the perturbed

Table 17.1 *List of key changes introduced in the ECMWF EPS configuration*

		Singular vector's characteristics						Forecast characteristics				
Date	Description	HRES	VRES	OTI	Target area	EVO SVs	sampl	HRES	VRES	Tend	#	Mod Imp
Dec 1992	Oper Impl	T21	L19	36h	globe	NO	simm	T63	L19	10d	33	NO
Feb 1993	SV LPO	"	"	"	NHx	"	"	"	"	"	"	"
Aug 1994	SV OTI	"	"	48h	"	"	"	"	"	"	"	"
Mar 1995	SV hor resol	T42	"	"	"	"	"	"	"	"	"	"
Mar 1996	NH+SH SV	"	"	"	(NH+SH)x	"	"	"	"	"	"	"
Dec 1996	resol/mem	"	L31	"	"	"	"	TL159	L31	"	51	"
Mar 1998	EVO SV	"	"	"	"	YES	"	"	L31	"	"	"
Oct 1998	Stoch Ph	"	"	"	"	"	"	"	"	"	"	YES
Oct 1999	ver resol	"	L40	"	"	"	"	"	L40	"	"	"
Nov 2000	FC hor resol	"	"	"	"	"	"	TL255	"	"	"	"
Jan 2002	TC SVs	"	"	"	(NH+SH)x+TC	"	"	"	"	"	"	"
Sep 2004	sampling	T42	L40	48h	(NH+SH)x+TC	YES	Gauss	TL255	L40	10d	51	YES

parametrised tendency (of each state vector component) is defined as

$$P'_j(e_j, t) = [1 + \langle r_j(\lambda, \phi, t)\rangle_{D,T}]P(e_j, t), \tag{17.3}$$

where P is the unperturbed diabatic tendency, and $<\ldots>_{D,T}$ indicates that the same random number r_j has been used for all grid points inside a $D \times D$ degree box and over T time steps (Buizza *et al.*, 1999).

The introduction of space and time coherence in the stochastic perturbations was based on the assumption that organised systems have some intrinsic space and time scales that may span more than one model time step and more than one model grid point. Making the stochastic uncertainty proportional to the tendency was based on the concept that organisation (away from the notion of a quasi-equilibrium ensemble of subgrid processes) is likely to be stronger, as the parametrised contribution becomes stronger. In the EPS, the *time-t* forecast from the initial day d is computed integrating Eq. (17.1) starting from perturbed initial conditions

$$e_j(d, 0) = e_0(d, 0) + \delta e_j(d, 0), \tag{17.4}$$

where each initial perturbation $\delta e_j(d,0)$ is generated using *singular vectors* computed to maximise the total energy norm over a 48-hour time interval (Buizza and Palmer, 1995), and scaled to have an amplitude comparable to analysis error estimates. Appendix A gives a brief summary of the singular vector definition, while for a more detailed description of the singular vector characteristics and a discussion of the nature of singular vector growth the reader is referred to Hartmann *et al.* (1995) and Hoskins *et al.* (2000).

The use of singular vectors (Buizza and Palmer, 1995) is a key feature of the ECMWF EPS that distinguishes it from the other two leading global ensemble systems implemented at MSC and NCEP (Buizza *et al.*, 2005). Singular vectors identify perturbations of maximum growth during a finite time interval, named the optimisation time interval: small errors in the initial conditions along these directions would amplify most rapidly and affect the forecast accuracy. Singular vectors are usually located in regions of strong barotropic and baroclinic activity: at initial time, they have most of their energy confined in the small scale and are confined vertically in the lower troposphere. During the optimization time interval, they change shape and grow in scale, and vertically propagate upward. As an example, Figure 17.1 shows the amplification rate (i.e. the singular value) of the leading 25 northern hemisphere singular vectors used in the EPS started at 12 UTC of 1 December 2003, and Figure 17.2 shows the average vertical distribution of total energy and the total energy spectra for these singular vectors. These two figures summarise two of the key characteristics of the singular vectors: the decreasing spectra of singular values, the upward energy propagation during the optimisation time interval coupled with the conversion of initial-time potential energy into final-time kinetic energy, and the upscale energy propagation from the small to the large (synoptic) scales.

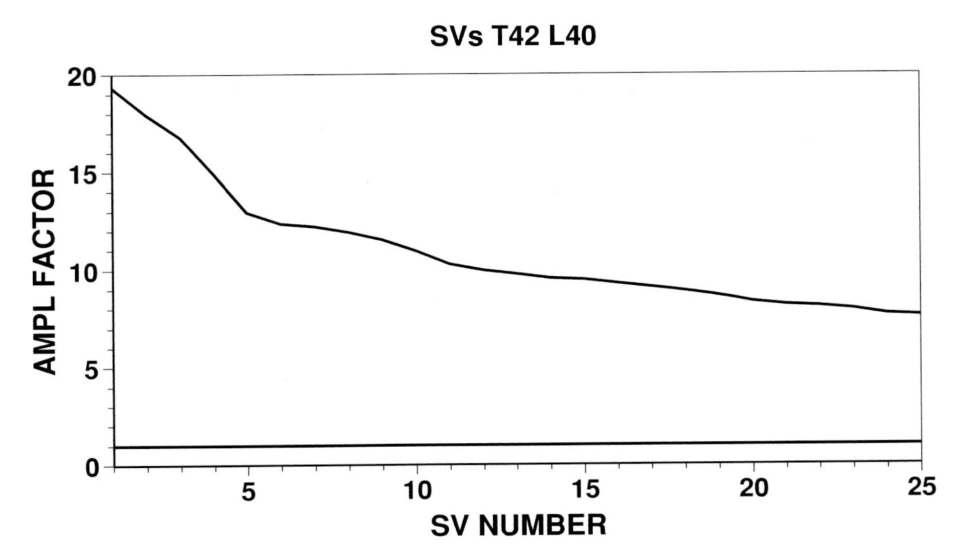

Figure 17.1 Amplification rates (i.e. singular values) of the leading 25 singular vectors used in the operational EPS started at 12 UTC of 1 December 2003. These singular vectors were computed at T42L40 resolution, with simplified dry physics, a 48-h optimisation time interval, and final time total energy norm maximised over the northern hemisphere extratropics (latitude $\lambda \geq 30°$ N).

The EPS configuration operational in 2004 (Table 17.1) includes 50 perturbed members and one unperturbed member (the control forecast) run at T_L255L40 resolution: the control starts from the unperturbed analysis, while the 50 perturbed members start from perturbed initial conditions generated by adding to the unperturbed analysis a linear combination of the leading singular vectors growing to have maximum energy, at optimisation time, inside three sets of area covering the whole globe (Appendix A). During this linear combination, the leading singular vectors are rescaled to have amplitude comparable to analysis error estimates (Molteni *et al.*, 1996; Buizza *et al.*, 2005). Formally, each initial perturbation is defined as

$$\delta e_j(d,0) = \sum_{A=1}^{N_A} \left\{ \sum_{i=1}^{N_{SV}} \left[\alpha_{i,j}^A v_i^A(d,0) + \beta_{i,j}^A v_i^A(d-t,t) \right] \right\}, \tag{17.5}$$

where $\boldsymbol{v}_i^A$ *(d,0)* is the i-th initial-time singular vector growing between d and $d+t$ inside, the number of different regions N_A varies between 2 and 8 (Appendix A), and the number of selected singular vectors N_{SV} is set to 25. Analogously, $\boldsymbol{v}_i^A$ $(d-t,t)$ is the i-th final-time t singular vector growing between $d-t$ and d inside area A. Following the latest changes introduced in 2004, the coefficients $\alpha_{i,j}$ and $\beta_{\mathrm{i,j}}$ are sampled from a Gaussian distribution (Ehrendorfer and Beck, 2003), still with the constraint that, on average, the ensemble standard deviation (which is a measure of the average distance of a single member from the ensemble mean) is comparable to

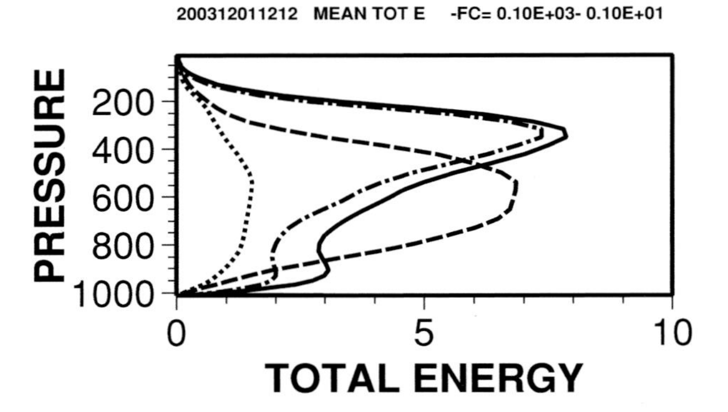

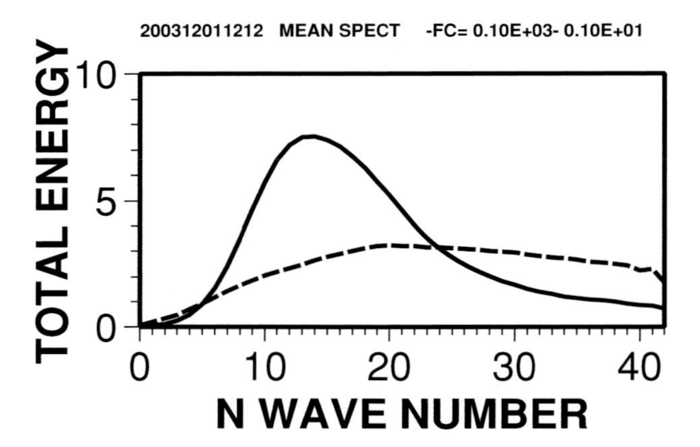

Figure 17.2 (Top panel) Average initial-time total (dashed line) and kinetic (dotted line) energy, final-time total (solid line) and kinetic (chain-dashed line) energy vertical cross-section. (Bottom panel) Average initial-time (dashed line) and final-time (solid line) total energy spectra. The averages have been computed considering the leading 25 singular vectors used in the operational EPS started at 12 UTC of 1 December 2003.

the error of the ensemble mean (which is a measure of the distance of the analysis from the ensemble mean). This guarantees that, on average, the analysis has the same distance from the ensemble mean as a perturbed member.

Figure 17.3 shows the EPS ensemble-mean forecast and the ensemble standard deviation, which is a measure of the ensemble spread, at initial and at three forecast steps, for the EPS started on 1 December 2004. The ensemble standard deviation at initial time shows the areas where the EPS initial perturbations were located, and their average amplitude. Figure 17.3 shows that the initial perturbations were located in regions of strong gradient (e.g. the exit of the North Atlantic jet stream) and intense baroclinic activity (e.g. the area of cyclonic depression over Spain). Figure 17.3 also indicates how the ensemble standard deviation can be used to

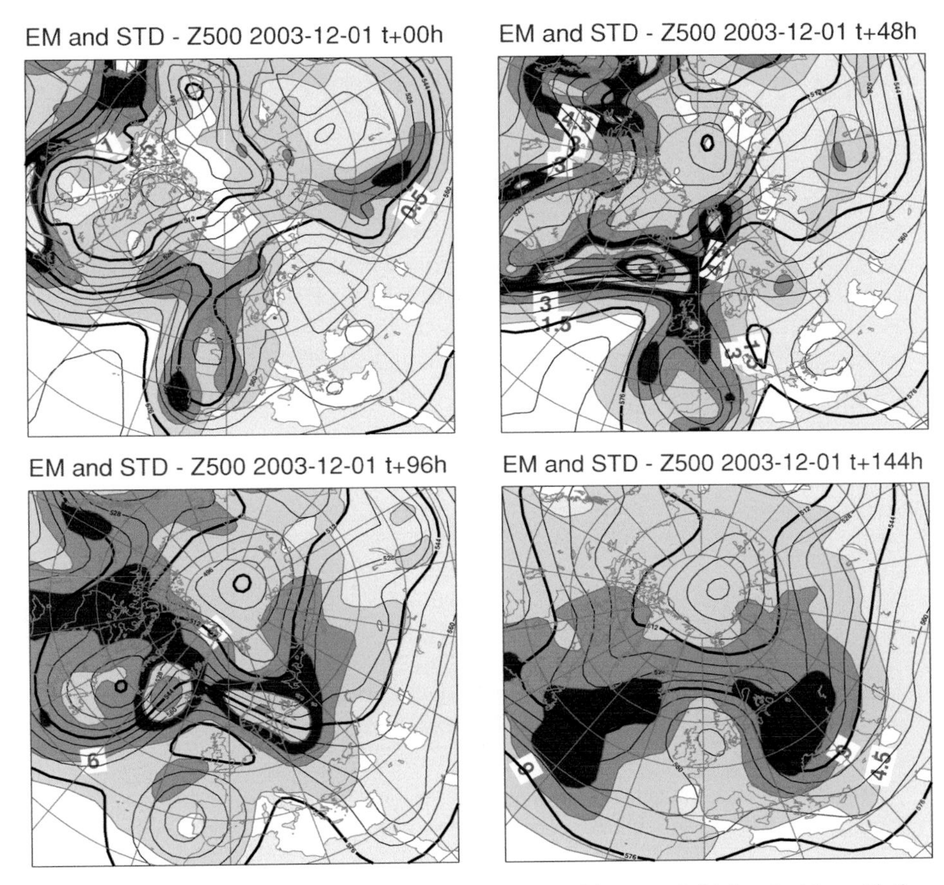

Figure 17.3 (Top left panel) Initial time ensemble mean (which coincides with the unperturbed analysis) and standard deviation. (Top right panel) Ensemble mean and standard deviation at forecast day 2. (Bottom left panel) Ensemble mean and standard deviation at forecast day 4. (Bottom right panel) Ensemble mean and standard deviation at forecast day 6. Fields shown refer to the 500 hPa geopotential height field of the EPS started at 12 UTC of 1 December 2003. Contour interval for ensemble mean is 80 m; contour shading for the ensemble standard deviation is 5 m at initial time, 15 m at day 2, 30 m at day 4 and 45 m at day 6.

estimate predictability: regions with small standard deviation (i.e. with small ensemble spread) should be more predictable than regions with large values, since in these regions the verifying analysis should be closer to the forecast states. If we consider Europe, for example, the ensemble standard deviation is small compared with the other regions during the early forecast range, but starts being relatively large at forecast day 4.

The average distance between the perturbed forecasts and the control is another measure of the ensemble spread, which can be compared with distance of the control forecast from the analysis. In a perfect ensemble system, the average (computed for

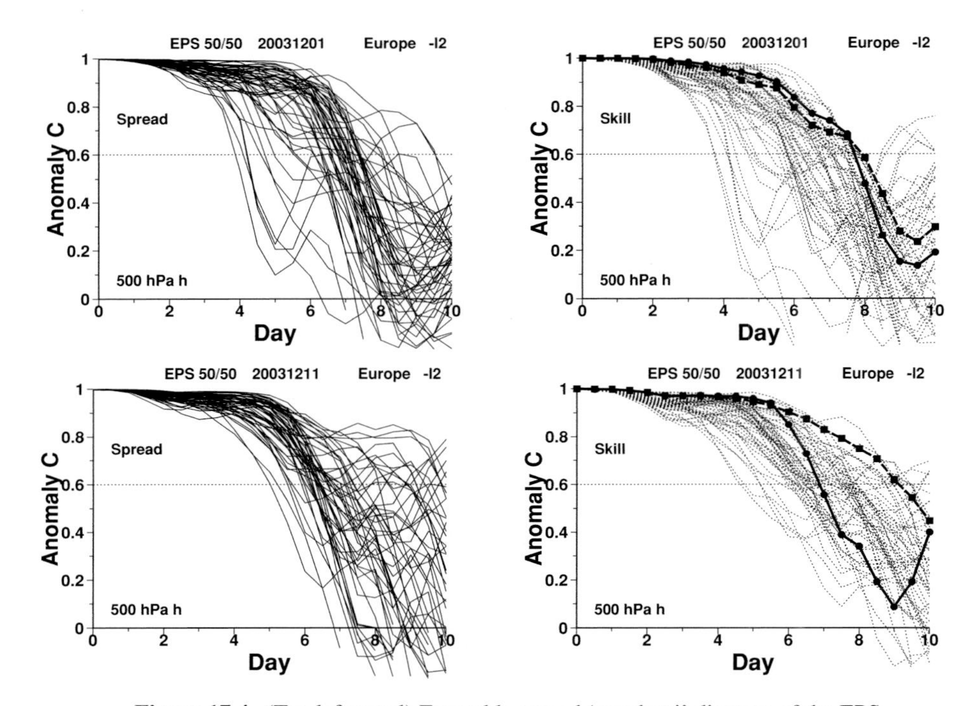

Figure 17.4 (Top left panel) Ensemble spread 'spaghetti' diagram of the EPS started at 12 UTC of 1 December 2003 over Europe, with spread measured by the anomaly correlation coefficient between each perturbed member and the control. (Top right panel) As top left panel but for the skill of the perturbed members (dotted line), the control (solid line with full circles) and the ensemble mean (dashed line with squares), with skill measured by the anomaly correlation coefficient between each forecast and the verifying analysis. (Bottom panels) As top panels but for the EPS started at 12 UTC of 11 December 2003.

a large sample of cases) ensemble spread should be comparable to the average error of the control forecast. Furthermore, as mentioned above, a small ensemble spread should indicate a predictable case, or, in other words, a small control error. These two characteristics can be checked, for example, using so-called spaghetti diagrams of ensemble spread and forecast error, with spread and error measured by the anomaly correlation coefficient.

Figure 17.4 shows the 'spaghetti diagrams' of the spread of each ensemble perturbed-member from the control forecast and the forecast error measured using the anomaly correlation coefficient between the forecast and the verifying analysis, for two cases (1 December and 11 December 2003) over Europe. Considering the EPS started on 1 December, the growth of ensemble spread of each individual member with the forecast time reflects the growth of the ensemble standard deviation shown in Figure 17.3. It is worth pointing out that the skill of the individual ensemble member starts crosses the 0.6 value at around forecast day 4, the time when the ensemble spread over Europe starts being relatively large (0.6 is usually considered the lower limit for a skilful forecast). Note also that for this case, the skill of the control, the ensemble perturbed members and the ensemble mean fall below the 0.6 threshold after forecast day 7.5.

Figure 17.4 also shows the spread and skill spaghetti diagram for the EPS started on 11 December. Note that between forecast day 4 and 6 the ensemble spread is smaller in this second case, and thus one would expect this second situation to be more predictable in this forecast period. Indeed, the skill of the control and the perturbed members is shown to be higher in this second case. Finally, it is worth pointing out that in both cases the skill of the ensemble-mean forecast is higher than the skill of the control forecast: this reflects the fact that by averaging all ensemble members, small-scale unpredictable features have been filtered out, thus providing a smoother but more skilful forecast.

17.2.3 Comparison of the methodologies used at MSC, NCEP and ECMWF

At MSC, the ensemble initial perturbations are generated by randomly perturbing each observation and by generating a set of initial analyses by running the MSC data assimilation system with a set of different models. Because the analysis and forecast system is repeated several times with different random input, the perturbed-observation method can be considered as a classical example of a Monte Carlo approach (see, for example, Hamill, this volume). The rationale behind the MSC choice of using non-selective, purely random ensemble perturbations is presented in Houtekamer *et al.* (1996) and by Anderson (1997). At NCEP, the ensemble initial perturbations are generated in a similar way as at ECMWF, but bred vectors (Toth and Kalnay, 1993) are used instead of singular vectors. The bred vector method is based on the argument that fast-growing perturbations develop naturally in a data-assimilation cycle and continue to grow as short- and medium-range forecast errors. Bred vectors can be considered as a non-linear extension of the Lyapunov vectors (Boffetta *et al.*, 1998). In the NCEP ensemble, the bred vectors are rescaled to follow the geographically varying level of estimated analysis uncertainty (Iyengar *et al.*, 1996). Table 17.2 lists the key characteristics of the ECMWF, MSC and NCEP ensemble prediction systems.

Buizza *et al.* (2005) compared three subsets of the ECMWF, MSC and NCEP EPS defined by only 10 perturbed members starting at 00 UTC, for a three month period (May, June and July 2002). It is worth quoting directly from this work the consensus position that emerged on some key aspects of the three ensemble systems:

- Overall, the ECMWF EPS exhibits the most skilful performance when measured using root-mean-square error, anomaly correlation, Brier skill score and the area under a relative operating characteristic curve.

Table 17.2 *Key characteristics of the MSC, ECMWF and NCEP ensemble systems*

	MSC	ECMWF	NCEP
Simulation model uncertainty	Yes (different models)	Yes (1 model + stoch phys)	No
Simulation initial uncertainty	Yes (analysis cycles)	Yes (singular vectors)	Yes (bred vectors)
Hor-res pert members	TL149	TL255 (d0–10)	T126(d0–3.5) > T62(d3.5–16)
Vertical levels (c&pf)	23 and 41, 28	40	28
Top of the model	10 hPa	10 hPa	3 hPa
Perturbed members per cycle	16	50	10
Cycles per day	1 (0UTC)	2 (0, 12UTC)	4 (0, 6, 12, 18UTC)
Forecast length	10 days	10 days	16 days
Operational impl.	February 1998	December 1992	December 1992

From Buizza *et al.* (2005).

- When a measure called perturbation versus error correlation analysis (Wei and Toth, 2003) is used to assess the correlation between perturbation and forecast-error patterns, the ECMWF EPS does not show any superior performance. At short lead times, the error patterns are best described by the NCEP EPS if one considers the small scales and by the MSC EPS if one considers the large scales.
- The superior skill of the ECMWF EPS may be mostly due to its superior model and data-assimilation systems, and should not be considered as proof of a superior performance of singular vector-based initial perturbations. In other words, at MSC and NCEP ensemble performance is negatively affected in the short range by the relatively low quality of the ensemble of data-assimilation systems, and in the long range by the relatively low model resolution.
- As for statistical reliability, the superior outlier statistics of the MSC EPS may be due to the use of multiple model versions. This technique may capture large-scale model-related errors in longer lead times.
- The spread in the (single-model) ECMWF EPS grows faster than that in the other two systems due to a combined effect of sustained singular vector-based perturbations' growth and of the stochastic simulation of random model errors.
- There are indications that the stochastic simulation of random model error scheme implemented in the ECMWF EPS improves the forecast statistical reliability.

Considering these 10-member ensembles, Buizza *et al.* (2005) showed that the ECMWF EPS probabilistic forecasts of geopotential height anomalies had about one day more skill than the other two ensembles at around forecast day 5. The reader is referred to Buizza *et al.* (2005) for a detailed discussion of these results.

17.3 EPS performance from May 1994 to September 2004

The ECMWF EPS forecasts are used to generate deterministic products (e.g. the forecasts given by the EPS control or the ensemble mean) and probabilistic products, such as the probability of occurrence of some selected events (e.g. the probability of occurrence of positive anomalies, or of positive/negative anomalies greater than or smaller than one standard deviation of monthly variability). The accuracy of deterministic and probabilistic forecasts has been assessed using a range of accuracy measures, but for reason of space attention will be focused hereafter to 500 hPa geopotential height fields over the northern hemisphere and Europe, from 1 May 1994 to 1 September 2004. Appendix B briefly summarises the definition of the three most common measures used hereafter to assess the accuracy of probabilistic forecasts: the area under the relative operating characteristic curve, the Brier score and skill score, the ranked probability score and skill score. The reader is referred to Wilks (1995) for a comprehensive discussion of accuracy measures that can be used to assess different attributes of forecast accuracy (see also Toth *et al.*, this volume).

Skill values have been translated into predictability gains, measured in days-per-decade (d/de), using the following formula:

$$gain(t)_{1994-2004} = \frac{sk(2004, t) - sk(1994, t)}{\langle sk(t + 12h) - sk(t - 12h)\rangle_{1994-2004}}, \tag{17.6}$$

where $sk(Y, t)$ is a measure of the skill of the t-day forecast issued in the Yth year, and $\langle \ldots \rangle_{1994\text{–}2004}$ denotes the average between 1994 and 2004.

Considering deterministic forecasts, Figure 17.5 shows the time evolution of the skill of the control and the ensemble-mean 5- and 7-day forecasts over the northern hemisphere. Results indicate a continuous improvement, equivalent for the control 5-day (7-day) forecast to a predictability gain of 0.95 (0.98) d/de, and for the ensemble mean to a gain of 1.40 (1.66) d/de. Considering probabilistic forecasts, Figure 17.6 shows the time evolution of the area under the relative operating characteristic curve, the Brier score and skill score, the ranked probability score and skill score of 5- and 7-day probabilistic forecasts over the northern hemisphere. These results indicate predictability gains ranging between 2 and 3.3 d/de.

Figure 17.7 summarises the predictability gains of the deterministic and probabilistic forecasts achieved between 1994 and 2004, for both the northern hemisphere and Europe. It is worth pointing out that the accuracy of the probabilistic forecasts

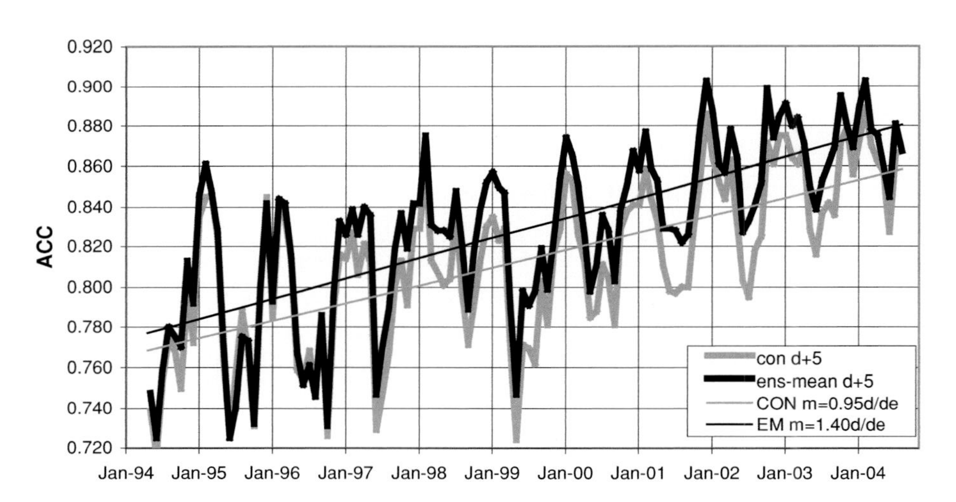

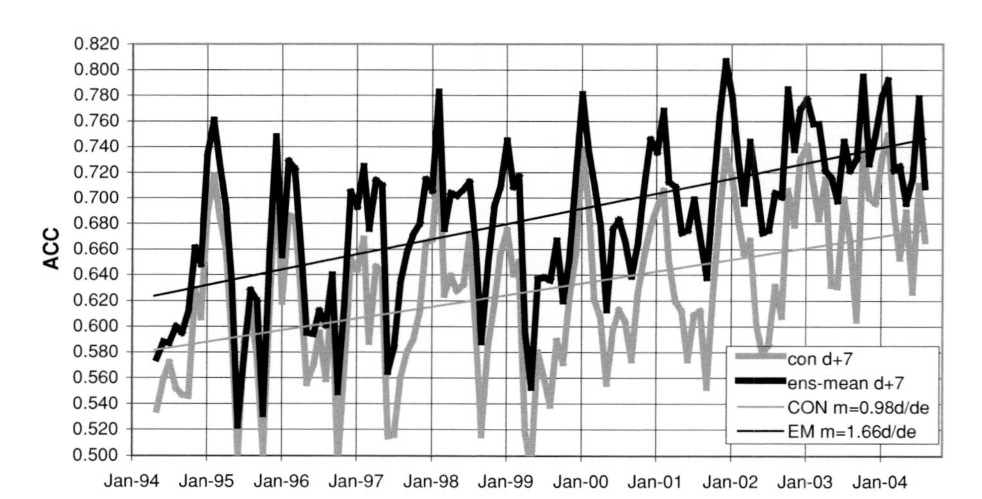

Figure 17.5 (Top panel) Monthly average anomaly correlation coefficient of the control (bold solid black line) and the ensemble mean (bold solid grey line) 5-day forecasts of 500 hPa geopotential height fields over the northern hemisphere. Thin lines show linear regression curves. (Bottom panel) As top panel but for the 7-day corresponding forecasts.

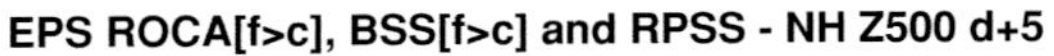

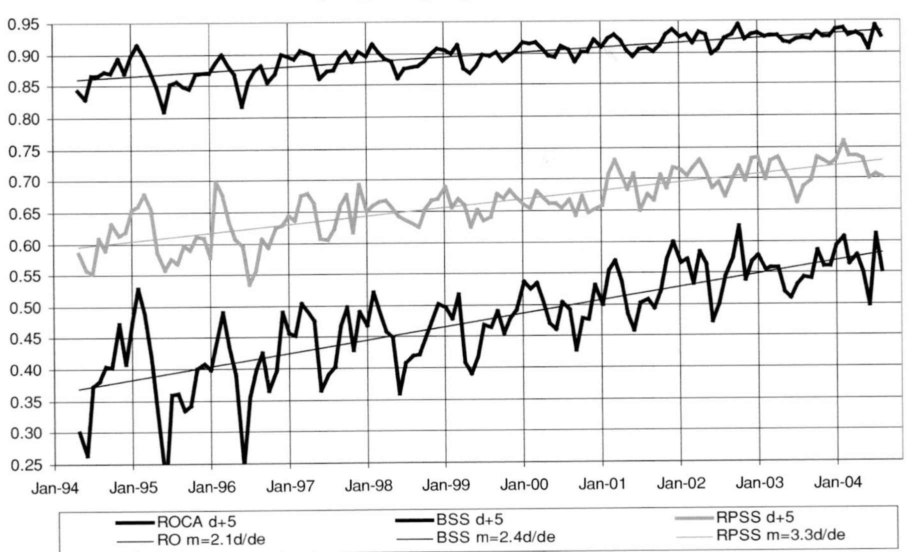

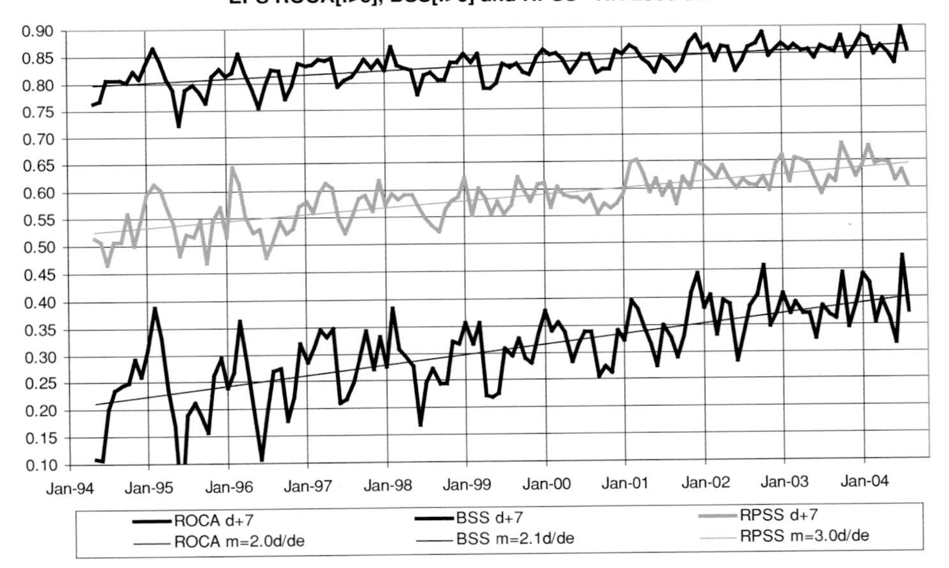

Figure 17.6 (Top panel) Monthly average area under the relative operating characteristic curve (higher bold solid black line) and Brier skill score (lower bold solid black line) of the 5-day probabilistic forecasts of positive anomalies, and ranked probability skill score (grey bold solid line) of the 5-day probabilistic forecasts of 500 hPa geopotential height fields over the northern hemisphere. Thin lines show linear regression curves. (Bottom panel) As top panel but for the 7-day corresponding forecasts.

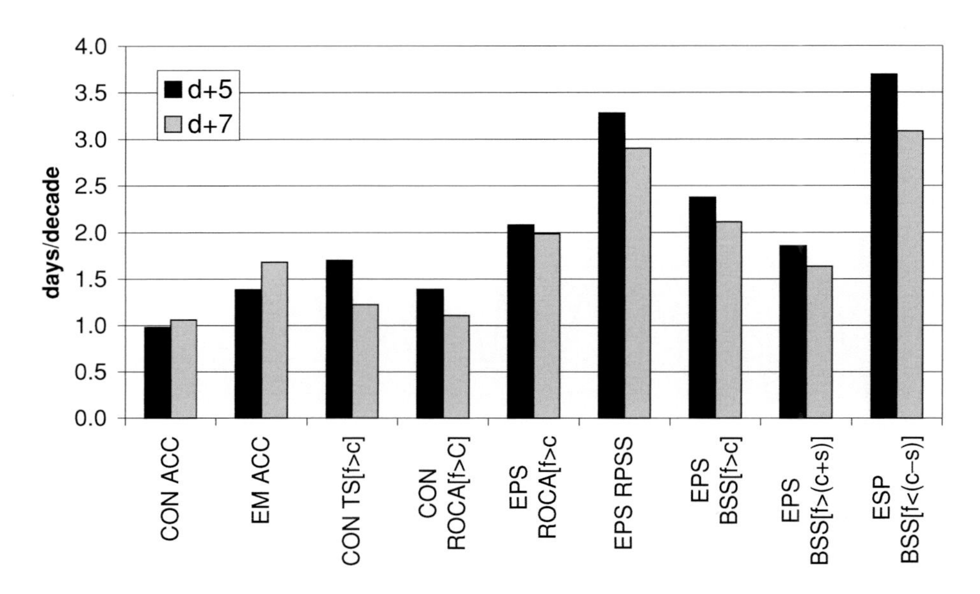

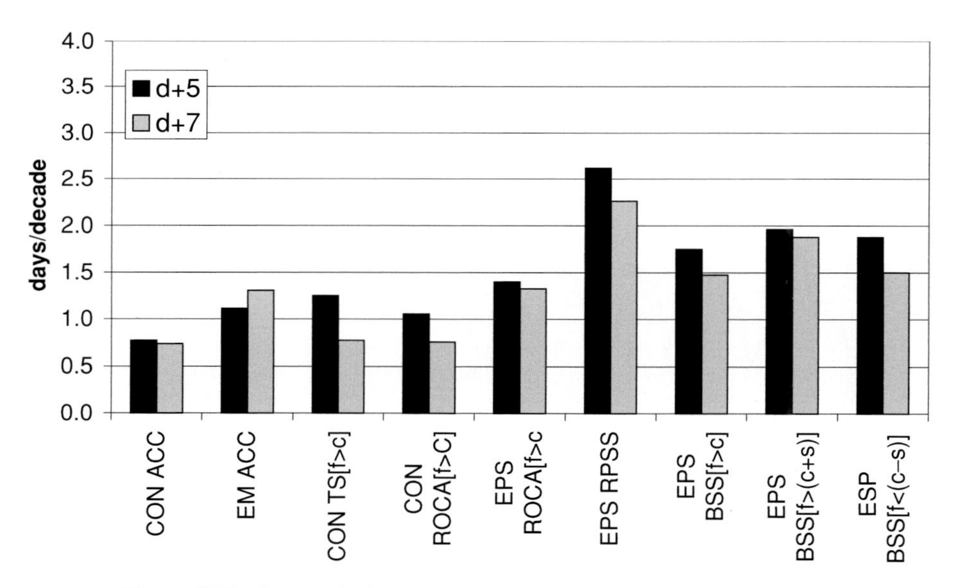

Figure 17.7 (Top panel) Gains in predictability, computed using Eq. (17.6), of 5-day (black bars) and 7-day (grey bars) forecasts of 500 hPa geopotential height fields over the northern hemisphere, computed from different forecasts:

has improved by about twice the value shown by the deterministic forecasts. This improvement first of all confirms the fact that the EPS benefits from ameliorations of the ECMWF data assimilation and forecast model, but also indicates that the changes introduced in the EPS configurations listed in Table 17.1 played an important role in improving the value of the EPS. Although it is difficult to identify clearly which of the changes of the EPS configurations (Table 17.1) had the largest impact on the EPS scores, published works suggest that the improvements shown in Figure 17.7 are mainly due to the increases of the EPS resolution in 1996 and 2000 (Buizza *et al.*, 1998, 2003), the increase in the ensemble size in 1996 (Buizza *et al.*, 1998; Buizza and Palmer, 1998; Mullen and Buizza, 2002), the introduction of evolved singular vectors (Barkmeijer *et al.*, 2001) and of stochastic physics in 1998 (Buizza *et al.*, 1999; Mullen and Buizza, 2001).

17.4 Future developments of the ECMWF EPS

One of the most important advances in numerical weather prediction of the last 15 years has been the operational implementation of ensemble prediction systems. Ensemble systems provide a possible way to estimate the probability distribution function of forecast states, and have been shown to be particularly useful, if not necessary, to provide early warnings of extreme weather events (Buizza and Hollingsworth, 2001). The economic value of information generated using ensemble forecasts has been recognised not only in meteorology (Richardson, 2000, this volume; Lalaurette and van der Grijn, 2005; Mylne, 2005) but also in business applications (Smith *et al.*, 2001), more specifically in real business applications in the energy sector (Taylor and Buizza, 2003, 2004) and in ship routeing (Saetra, 2004).

At ECMWF, work is in progress to improve the current system in three key areas:

(a) the simulation of initial uncertainties
(b) the simulation of model imperfection
(c) the system design.

Figure 17.7 (*cont.*)

- Bars 1–2: the control anomaly correlation coefficient
- Bars 3–4: the ensemble-mean anomaly correlation coefficient
- Bars 5–6 and 7–8: the area under the relative operating characteristics of the probabilistic forecast of positive anomalies given by the control (5–6) and the EPS (7–8)
- Bars 9–10: the ranked probability skill score of the EPS
- Bars 11–12: the Brier skill score of the probabilistic prediction of positive anomalies
- Bars 13–14 and 15–16: the Brier skill score of the probabilistic prediction of positive anomalies greater-than and smaller-than 1 standard deviation.

(Bottom panel) As top panel but for Europe.

In the area of the simulation of initial uncertainties, developments are expected in the definition of the norm used to compute the singular vectors, in the use of moist, higher-resolution singular vectors, and in the combination of ensemble data assimilation and singular vectors. In the area of the simulation of model uncertainties, developments are expected in the simulation of the effect of near-grid-scale and subgrid-scale processes. Finally, in the area of system design, changes in the ensemble size and resolution are under investigation.

17.4.1 Simulation of initial uncertainties

Definition of the initial-time norm used in the singular vectors' computation

The current EPS uses singular vectors computed with initial- and final-time total energy norm, i.e. without using any information of analysis error statistics. One way to use this type of information would be to use a norm defined by the Hessian of the cost function of the three-dimensional (or four-dimensional) variational assimilation system (3D/4D-Var) to define singular vectors (Barkmeijer *et al.*, 1998). These so-called Hessian singular vectors would be constrained at initial time by analysis error statistics and still produce fast perturbation growth during the first few days of the forecast.

Use of higher-resolution, 24-hour, moist singular vector

Considering the second point, the current EPS uses extratropical singular vectors computed at T42L40 resolution and simplified dry physics to grow over a 48-hour time period. Coutinho *et al.* (2004) have shown that the inclusion of large-scale latent heat release results in a shift to smaller horizontal scales and enhanced growth, and to a change in their location that depends on the availability of moisture. They have also shown that while for the dry singular vectors a T42 resolution is sufficient, the moist singular vectors require a T63 resolution to resolve their structure and growth. Furthermore, they have suggested that a 24-hour optimisation time appears to be more appropriate for the moist singular vectors because of the larger growth. Work is in progress to test the use of these singular vectors in the EPS.

Combined use of ensemble data assimilation and singular vectors

An ensemble approach to data assimilation was tested a few years ago (Buizza and Palmer, 1999), and work is now resuming to assess whether a combined use of different perturbation methods could be beneficial for the EPS. Following Houtekamer *et al.* (1996), but with the ECMWF approach to represent model uncertainties, the plan is to generate an ensemble of initial perturbations using the ECMWF 3D- and/or

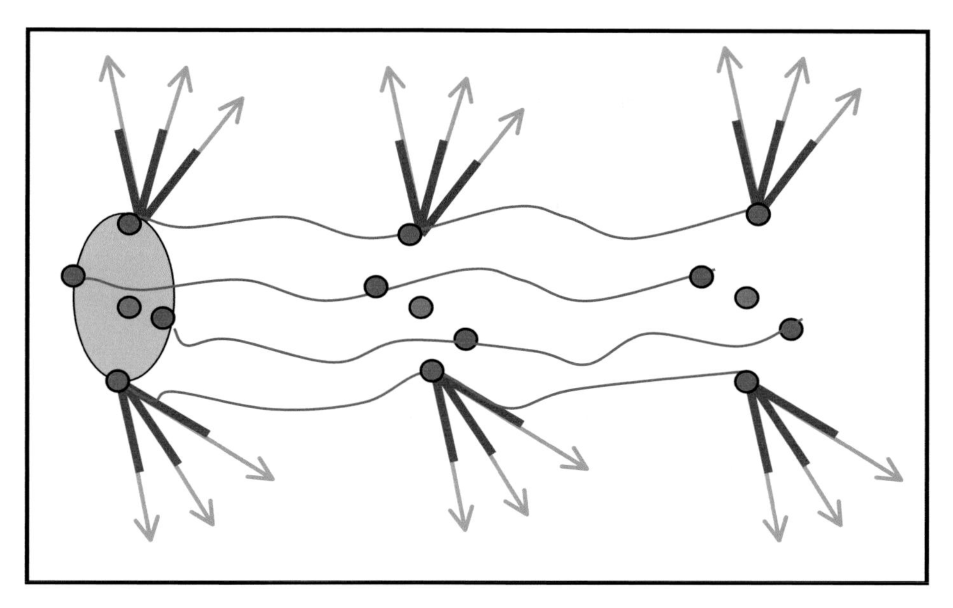

Figure 17.8 Schematic of a possible future configuration of the ECMWF EPS, the '*caterpillar*' system. The caterpillar is a probabilistic assimilation and prediction system, whereby the EPS initial conditions (full black circles) are defined using a combination of ensemble data assimilation and singular vectors, and the ensemble forecasts (arrows) are run with a higher resolution in the short and medium forecast range (say during week 1) and a lower resolution in the long forecast range (say week 2).

4D-Var data assimilation schemes. This, combined with the use of a variable resolution during the forecast integration (see Section 17.4.3), could lead to the development and implementation of the ECMWF probabilistic data-assimilation and prediction system (the 'caterpillar' system), schematic illustrated in Figure 17.8.

17.4.2 Simulation of model imperfection

Simulation of near-grid and subgrid scale processes

Work is in progress to revise the current scheme used to simulate the effect of random model errors linked to the parametrisation of physical processes. A new scheme, designated the Cellular Automaton Stochastic Backscatter Scheme (Shutts, 2004), has been designed to introduce kinetic energy sources during the numerical integration to counteract for energy dissipation processes at the near-grid scale. Results indicate that this new scheme may be more effective in generating effective spread in the EPS, especially in the medium- and long-forecast range. Substitution of the old with the new scheme, or a combined use of the two schemes, is under investigation.

17.4.3 System design

Ensemble size

Earlier sensitivity results (Buizza and Palmer, 1998; Mullen and Buizza, 2002) have indicated that an increase in ensemble size between 10 and about 30 induces sizeable positive impacts, but the impact decreases if size is further increased. Later work has suggested on the one hand that a larger ensemble size is required to predict rare events, but on the other hand that if ensemble forecasts are first postprocessed then the impact of an ensemble size increase is smaller (Wilks, 2002). The plan is to reassess whether the current membership should be revised or not.

Variable Resolution EPS (VAREPS)

The rationale behind VAREPS is that the predictability of small scales is mainly lost relatively early in the forecast range. Therefore, while forecasts benefit from a resolution increase in the early forecast range, they do not suffer so much from a resolution reduction in the long range. Thus, it is more cost efficient to use any extra computer resources to increase the EPS's resolution as much as possible in the early forecast range rather than implement a more modest resolution increase over the whole forecast range. Furthermore, resources saved by running at higher resolution only during the early forecast range give the possibility to extend the EPS' forecasts from day 10 to day 14 with a limited extra cost. Work to implement VAREPS, with a $T_L399L40$ resolution and a 1200 s time step between day 0 and day 7 and a $T_L255L40$ resolution with a 2700 s time step from day 7 to day 14, is progressing. Developments in these areas, sustained by improvements in data assimilation and numerical modelling should further increase the value of the ECMWF EPS and promote its use in other sectors. The sector where the operational use of ensemble systems is expected to start in the near future is hydrology (Gouweleeuw *et al.*, 2004). The use of ensemble methods in hydrology will further convince users that ensemble predictions are extremely valuable because they not only offer an estimate of the most probable future state of a system but they also provide an estimate of the range of possible future outcomes.

Acknowledgements
The ECMWF Ensemble Prediction System is based on the Integrated Forecasting System/Arpege software, developed in collaboration by ECMWF and Météo-France, and is the result of the work of many ECMWF staff members and consultants. Without their contributions the ECMWF EPS would not have reached the mature development stage and accuracy that it has reached now: their work is fully acknowledged.

Appendix A Singular vectors' definition and methodology

Lorenz (1965) was the first author to indicate that perturbation growth in realistic models is related to the eigenvalues and eigenvectors of the operator product of the tangent forward linear and adjoint operators, i.e. to the *singular values* and *singular vectors* of the tangent forward operator. Subsequently, Farrell (1982), studying the growth of perturbations in baroclinic flows, showed that, although the long time asymptotic behaviour is dominated by discrete exponentially growing normal modes when they exist, physically realistic perturbations could present for finite time intervals amplification rates greater than the most unstable normal mode amplification rate. In later works, Farrell (1988, 1989) confirmed Lorenz's (1965) indications that the singular vectors could be computed solving the eigenvalue problem of the product of the tangent forward and adjoint model propagators. Considering the more complex system of the atmospheric flow, singular vectors were computed, for example, by Borges and Hartmann (1992) using a barotropic model, and by Molteni and Palmer (1993) using a barotropic and a three-level quasi-geostrophic model at spectral triangular truncation T21. Buizza *et al.* (1993) first identified singular vectors in a primitive equation model with a large number of degrees of freedom. In this appendix, the methodology followed at ECMWF to compute the singular vectors is briefly reviewed.

A.1 Linearised model equations

Let χ be the state vector of a generic autonomous system, whose evolution equations can be formally written as

$$\frac{\partial \chi}{\partial t} = A(\chi). \tag{17.7}$$

Denote by $\chi(t)$ an integration of Eq. (17.7) from t_0 to t. The time evolution of a small perturbation x around the time evolving trajectory $\chi(t)$ can be described, in a first approximation, by the linearised model equations

$$\frac{\partial x}{\partial t} = A_l x, \tag{17.8}$$

where

$$A_l = \left.\frac{\partial A(x)}{\partial x}\right|_{\chi(t)}$$

is the tangent operator computed at the trajectory point $\chi(t)$. Denote by $L(t,t_0)$ the integral forward propagator of the dynamical Eq. (17.8) linearised about a non-linear trajectory $\chi(t)$:

$$x(t) = L(t, t_0)x(t_0), \tag{17.9}$$

that maps a perturbation x at initial time t_0 to the optimisation time t.

A.2 Inner product

Consider two perturbations x and y, and a positive definite Hermitian matrix E. Denote by $(\ldots;\ldots)_E$ the inner product between the two vectors

$$(x; y)_E = \langle x; Ey \rangle, \tag{17.10}$$

where $\langle \ldots; \ldots \rangle$ identifies the canonical Euclidean scalar product,

$$\langle x; y \rangle = \sum_{i=1}^{N} x_i y_i. \tag{17.11}$$

Denote by $\|\ldots\|_E$ the norm associated with the inner product $(\ldots;\ldots)_E$,

$$\|x\|_E^2 = (x; x)_E = \langle x; Ex \rangle. \tag{17.12}$$

A.3 Adjoint operator

Denote by L^{*E} the adjoint of L with respect to the inner product $(\ldots;\ldots)_E$,

$$(L^{*E}x; y)_E = (x; Ly)_E. \tag{17.13}$$

The adjoint of L with respect to the inner product defined by E can be written in terms of the adjoint L^* defined with respect to the canonical Euclidean scalar product,

$$L^{*\mathrm{E}} = E^{-1}L^*E. \tag{17.14}$$

A.4 Singular values and singular vectors

From the definitions above, it follows that the squared norm of a perturbation x at time t can be computed as

$$\|x(t)\|_E^2 = (x(t_0); L^{*E}Lx(t_0))_E. \tag{17.15}$$

From Eq. (17.15) it follows that the problem of finding the phase space directions x for which $\|x(t)\|_E^2 / \|x(t_0)\|_E^2$ is maximum can be reduced to the computation of the eigenvectors $v_I(t_0)$ with the largest eigenvalues σ_i^2, i.e. to the solution of the eigenvalue problem

$$L^{*E} L v_i(t_0) = \sigma_i^2 v_i(t_0). \tag{17.16}$$

The square roots of the eigenvalues, σ_i, are called the *singular values* and the eigenvectors $v_i(t_0)$ the (right) *singular vectors* of L with respect to the inner product E (see, for example, Noble and Daniel, 1977). The singular vectors with largest singular values identify the directions characterised by maximum growth. The time interval $t - t_0$ is called the optimisation time interval.

At optimisation time t, the i-th singular vector evolves into

$$v_i(t) = L(t, t_0) v_i(t_0), \tag{17.17}$$

a vector with total norm equal to

$$\|v_i(t)\|_E^2 = \sigma_i^2. \tag{17.18}$$

Since any perturbation $x(t)/\|x(t_0)\|_E$ can be written as a linear combination of the singular vectors $v_i(t)$, it follows that

$$\max_{\|x(t_0)\|_E \neq 0} \left(\frac{\|x(t)\|_E}{\|x(t_0)\|_E} \right) = \sigma_1, \tag{17.19}$$

which implies that maximum growth as measured by the norm $\|...\|_E$ is associated with the dominant singular vector v_1.

At ECMWF, the singular vectors are computed by solving the eigenvalue problem defined in Eq. (17.4) using a Lanczos iterative method (Golub and Van Loan, 1983).

A.5 Local projector operator (in physical space)

Denote by x_g the grid point representation of the state vector x, by S the spectral-to-grid point transformation operator, $x_g = Sx$, and by Gx_g the multiplication of the vector x_g, defined in grid point space, by the function $g(s)$:

$$\begin{aligned} g(s) &= 1 \forall s \in \Sigma \\ g(s) &= 0 \forall s \notin \Sigma \end{aligned}, \tag{17.20}$$

where s defines the coordinate of a grid point, and Σ is a geographical region. Consider a generic vector x. The application of the local projection operator T defined as

$$T = S^{-1} G S, \tag{17.21}$$

to the vector x sets the vector x to zero for all grid points outside the geographical region Σ.

The projection operator T can be used either at initial or at final time, or at both times. As an example, these operators can be used to solve the following problem: find the perturbations with (a) the fastest growth during the time interval $t - t_0$, (b) unitary E_0-norm at initial time and (c) maximum E-norm inside the geographical region Σ at optimisation time. The solution to this problem is given by the singular values of the operator

$$K = E^{-1/2} T L E_0^{-1/2}. \tag{17.22}$$

A.6 Singular vectors in the ECMWF EPS

At the time of writing (2004), the EPS initial perturbations are generated using localised singular vectors computed by solving the eigenvalue problem defined in Eq. (17.6) for up to eight different regions in the following configuration:

- T42 spectral truncation and 40 vertical levels
- 48 hour optimisation time interval
- total energy norm, computed inside a localised region
- simplified dry physical processes in the extratropics (Buizza, 1994)
- moist tangent forward and adjoint physics (Mahfouf, 1999) in the tropics
- no moisture perturbations
- regions:
  - the northern hemisphere extratropics (latitude north of 30° N)
  - the southern hemisphere extratropics (latitude south of 30° S)
  - up to six tropical regions, selected to cover areas of tropical cyclone development.

Appendix B Accuracy measures of probabilistic forecasts

The three measures most commonly used to assess the accuracy of probabilistic forecasts are briefly reviewed hereafter. The reader is referred to Stanski *et al.* (1989) and Wilks (1995) for more details.

B.1 The Brier score and skill score

The most common measure of accuracy of a probabilistic forecast is the Brier score (BS; Brier, 1950):

$$BS = \frac{1}{N_g} \sum_{g=1}^{N_G} (p_{f,g} - o_g)^2, \tag{17.23}$$

which is the mean-squared error of the forecast probability $p_{f,g} = p_f(\lambda, \phi)$, where the index $g = 1, N_G$ denotes the forecast/event pairs of all considered grid points. The observed probability function is defined to be $o_g = 1$ if the event occurs, and $o_g = 0$ if the event does not occur. The Brier skill score (BSS) is defined with respect to a reference forecast (e.g. a climatological forecast, a forecast based on the sample climate, or consistency)

$$BSS = \frac{BS - BS_{ref}}{0 - BS_{ref}}. \tag{17.24}$$

In this work, the reference forecast has been defined by the sample climatology. One of the properties of the BS is that it can be considered a strictly proper score, in the sense that the BS cannot be improved by forecasting something other than one's true beliefs about future weather events (i.e. hedging, see Wilks, 1995).

B.2 The area under the relative operating characteristic curve

A second measure of probabilistic forecast accuracy is the area under a relative operating characteristic curve (ROC) defined in signal detection theory (Mason, 1982; Stanski *et al.*, 1989). Consider a forecast probability distribution $p_f(\lambda, \phi)$, stratified according to observation into 51 categories. For any given probability threshold j, the entries of this table can be summed to produce the four entries of a 2×2 contingency table:

$$\begin{aligned} a_j &= \sum_{k=j+1}^{51} x_k \quad & b_j &= \sum_{k=j+1}^{51} y_k \\ c_j &= \sum_{k=1}^{j} x_k \quad & d_j &= \sum_{k=1}^{j} y_k \end{aligned} \tag{17.25}$$

From each of the j-th contingency tables, the probability of detection POD_j and the probability of false detection PFD_j can be computed. The 51 pairs (PFD_j,POD_j) can be plotted one against the other on a graph. The result is a smooth curve called the relative operating characteristic curve.

B.3 The ranked probability score and skill score

The ranked probability score (RPS; Epstein, 1969; Wilks, 1995) is another measure of probabilistic forecast accuracy. Given a set of events related to the same variable and characterised by a different amount (e.g. consider precipitation events characterised by different rainfall amounts), the RPS can be considered an extension of the Brier score to multicategory events (Wilks, 1995). Let J_{ev} be the number of (ranked) forecast events, $p^j_{f,g}$ the forecast probability for the j-th event and o^j_g the observed probability function (with $o^k_g = 1$ if the k-th event is observed, and $o^j_g = 0$ for $j \neq k$), where $g = 1,N_g$ denotes the g-th grid point. The grid point ranked probability score

RPS_g is computed from the squared error of the cumulative forecast and observed probabilities

$$RPS_g = \sum_{m=1}^{J} \left[\left(\sum_{j=1}^{m} p_{f,g}^{j} \right) - \left(\sum_{j=1}^{m} o_g^j \right) \right]^2 . \tag{17.26}$$

The area-average ranked probability score RPS is defined as

$$RPS = \frac{1}{N_g} \sum_{g=1}^{N_g} RPS_g. \tag{17.27}$$

As for the Brier skill score, the ranked probability skill score RPSS is defined with respect to a reference forecast

$$RPSS = \frac{RPS_{ref} - RPS}{RPS_{ref}}. \tag{17.28}$$

In this work, the reference forecast has been defined by the sample climatology.

References

Anderson, J. L. (1997). The impact of dynamical constraints on the selection of initial conditions for ensemble predictions: low-order perfect model results. *Mon. Weather Rev.*, **125**, 2969–83.

Barkmeijer, J., M. van Gijzen and F. Bouttier (1998). Singular vectors and estimates of the analysis error covariance metric. *Quart. J. Roy. Meteor. Soc.*, **124**, 1695–713.

Barkmeijer, J., R. Buizza and T. N. Palmer (1999). 3D-Var Hessian singular vectors and their potential use in the ECMWF Ensemble Prediction System. *Quart. J. Roy. Meteor. Soc.*, **125**, 2333–51.

Barkmeijer, J., R. Buizza, T. N. Palmer, K. Puri and J.-F. Mahfouf (2001). Tropical singular vectors computed with linearized diabatic physics. *Quart. J. Roy. Meteor. Soc.*, **127**, 685–708.

Boffetta, G., P. Guliani, G. Paladin and A. Vulpiani (1998). An extension of the Lyapunov analysis for the predictability problem. *J. Atmos. Sci.*, **55**, 3409–16.

Borges, M. and D. L. Hartmann (1992). Barotropic instability and optimal perturbations of observed non-zonal flows. *J. Atmos. Sci.*, **49**, 335–54.

Brier, G. W. (1950). Verification of forecasts expressed in terms of probability. *Mon. Weather Rev.*, **78**, 1–3.

Buizza, R. (1994). Sensitivity of optimal unstable structures. *Quart. J. Roy. Meteor. Soc.*, **120**, 429–51.

Buizza, R. and T. N. Palmer (1995). The singular-vector structure of the atmospheric general circulation. *J. Atmos. Sci.*, **52**(9), 1434–56.

(1998). Impact of ensemble size on the skill and the potential skill of an ensemble prediction system. *Mon. Weather Rev.*, **126**, 2503–18.

(1999). Ensemble data assimilation. In *Proceedings of the 17th Conference on Weather Analysis and Forecasting*, 13–17 September 1999, Denver, Colorado. ECMWF.

Buizza, R. and A. Hollingsworth (2001). Storm prediction over Europe using the ECMWF Ensemble Prediction System. *Meteor. Appl.*, **9**, 289–306.

Buizza, R., J. Tribbia, F. Molteni and T. N. Palmer (1993). Computation of optimal unstable structures for a numerical weather prediction model. *Tellus*, **45A**, 388–407.

Buizza, R., T. Petroliagis, T. N. Palmer, *et al.* (1998). Impact of model resolution and ensemble size on the performance of an ensemble prediction system. *Quart. J. Roy. Meteor. Soc.*, **124**, 1935–60.

Buizza, R., M. Miller and T. N. Palmer (1999). Stochastic simulation of model uncertainties. *Quart. J. Roy. Meteor. Soc.*, **125**, 2887–908.

Buizza, R., D. S. Richardson and T. N. Palmer (2003). Benefits of increased resolution in the ECMWF ensemble system and comparison with poor-man's ensembles. *Quart. J. Roy. Meteor. Soc.*, **129**, 1269–88.

Buizza, R., P. L. Houtekamer, Z. Toth, *et al.* (2005). A comparison of the ECMWF, MSC and NCEP Global Ensemble Prediction Systems. *Mon. Weather Rev.*, **133**, 1076–97.

Courtier, P. and O. Talagrand (1987). Variational assimilation of meteorological observations with the adjoint vorticity equation. 2: Numerical results. *Quart. J. Roy. Meteor. Soc.*, **113**, 1329–47.

Courtier, P., C. Freydier, J.-F. Geleyn, F. Rabier and M. Rochas (1991). The Arpege project at Météo-France. In *Proceedings of the ECMWF Seminar on Numerical Methods in Atmospheric Models*, Vol. II, pp. 193–231. ECMWF, Shinfield Park, Reading RG2 9AX, UK.

Courtier, P., J.-N. Thepaut and A. Hollingsworth (1994). A strategy for operational implementation of 4D-Var, using an incremental approach. *Quart. J. Roy. Meteorol. Soc.*, **120**, 1367–88.

Coutinho, M. M., B. J. Hoskins and R. Buizza (2004). The influence of physical processes on extra tropical singular vectors. *J. Atmos. Sci.*, **61**, 195–209.

Downton, R. A. and R. S. Bell (1988). The impact of analysis differences on a medium-range forecast. *Meteor. Mag.*, **117**, 279–85.

Ehrendorfer, M. (1994). The Liouville equation and its potential usefulness for the prediction of forecast skill. I: Theory. *Mon. Weather Rev.*, **122**, 703–13.

Ehrendorfer, M. and A. Beck (2003). *Singular Vector-based Multivariate Normal Sampling in Ensemble Prediction*. ECMWF Technical Report 416. Available at ECMWF, Shinfield Park, Reading RG2 9AX, UK (www.ecmwf.int/publications/library/).

Epstein, E. S. (1969). Stochastic dynamic prediction. *Tellus*, **21**, 739–59.

Farrell, B. F. (1982). The initial growth of disturbances in a baroclinic flow. *J. Atmos. Sci.*, **39**(8), 1663–86.

(1988). Optimal excitation of neutral Rossby waves. *J. Atmos. Sci.*, **45**(2), 163–72.

(1989). Optimal excitation of baroclinic waves. *J. Atmos. Sci.*, **46**(9), 1193–206.

Fleming, R. J. (1971a). On stochastic dynamic prediction. I: The energetics of uncertainty and the question of closure. *Mon. Weather Rev.*, **99**, 851–72.

(1971b). On stochastic dynamic prediction. II: Predictability and utility. *Mon. Weather Rev.*, **99**, 927–38.

Gleeson, T. A. (1970). Statistical-dynamical predictions. *J. Appl. Meteor.*, **9**, 333–44.

Golub, G. H. and C. F. Van Loan (1983). *Matrix Computation*. North Oxford Academic Publ. Co. Ltd.

Gouweleeuw, B., P. Reggiani and A. de Roo (2004). *A European Flood Forecasting System (EFFS)*. Full report of the EFFS project of the European Commission. Available from Ad de Roo, Institute for Environment and Sustainability, JRC, Via E. Fermi, 21020 Ispra (Va), Italy.

Harrison, M. S. J., T. N. Palmer, D. S. Richardson and R. Buizza (1999). Analysis and model dependencies in medium-range ensembles: two transplant case studies. *Quart. J. Roy. Meteor. Soc.*, **125**, 2487–515.

Hartmann, D. L., R. Buizza and T. N. Palmer (1995). Singular vectors: the effect of spatial scale on linear growth of disturbances. *J. Atmos. Sci.*, **52**, 3885–94.

Holton, J. R. (1992). *An Introduction to Dynamic Meteorology*. Academic Press.

Hoskins, B. J., R. Buizza and J. Badger (2000). The nature of singular vector growth and structure. *Quart. J. Roy. Meteor. Soc.*, **126**, 1565–80.

Houtekamer, P. L., L. Lefaivre, J. Derome, H. Ritchie and H. Mitchell (1996). A system simulation approach to ensemble prediction. *Mon. Weather Rev.*, **124**, 1225–42.

Iyengar, G., Z. Toth, E. Kalnay and J. Woollen (1996). Are the bred vectors representative of analysis errors? In *Preprints of the 11th AMS Conference on Numerical Weather Prediction*, 19–23 August 1996, Norfolk, Virginia, pp. J64–J66. American Meteorological Society.

Jacob, C. (1994). The impact of the new cloud scheme on ECMWF's Integrated Forecasting System (IFS). In *Proceedings of the ECMWF/GEWEX workshop on Modelling, Validation and Assimilation of Clouds*, 31 October–4 November 1994. ECMWF, Shinfield Park, Reading RG2 9AX.

Leith, C. E. (1974). Theoretical skill of Monte Carlo forecasts. *Mon. Weather Rev.*, **102**, 409–18.

Lorenz, C. E. (1965). A study of the predictability of a 28-variable atmospheric model. *Tellus*, **17**, 321–33.

Mason, I. (1982). A model for assessment of weather forecasts. *Aust. Meteorol. Mag.*, **30**, 291–303.

Mahfouf, J.-F. (1999). Influence of physical processes on the tangent-linear approximation. *Tellus*, **51A**, 147–66.

Mocrette, J.-J. (1990). Impact of changes to the radiation transfer parametrisation plus cloud optical properties in the ECMWF model. *Mon. Weather Rev.*, **118**, 847–73.

Molteni, F. and T. N. Palmer (1993). Predictability and finite-time instability of the northern winter circulation. *Quart. J. Roy. Meteor. Soc.*, **119**, 1088–97.

Molteni, F., R. Buizza, T. N. Palmer and T. Petroliagis (1996). The new ECMWF ensemble prediction system: methodology and validation. *Quart. J. Roy. Meteor. Soc.*, **122**, 73–119.

Mullen, S. and R. Buizza (2001). Quantitative precipitation forecasts over the United States by the ECMWF Ensemble Prediction System. *Mon. Weather Rev.*, **129**, 638–63.

Mullen, S. and R. Buizza (2002). The impact of horizontal resolution and ensemble size on probabilistic forecasts of precipitation by the ECMWF Ensemble Prediction System. *Weather Forecast.*, **17**, 173–91.

Noble, B. and J. W. Daniel (1977). *Applied Linear Algebra*. Prentice-Hall.

Palmer, T. N., F. Molteni, R. Mureau, R. Buizza, P. Chapelet and J. Tribbia (1993). Ensemble prediction. In *Proceedings of the ECMWF Seminar on Validation of Models over Europe,* Vol. I. ECMWF, Shinfield Park, Reading, RG2 9AX, UK.

Richardson, D. S. (1998). The relative effect of model and analysis differences on ECMWF and UKMO operational forecasts. In *Proceedings of the ECMWF Workshop on Predictability*. ECMWF, Shinfield Park, Reading RG2 9AX, UK.

Richardson, D. S. (2000). Skill and economic value of the ECMWF Ensemble Prediction System. *Quart. J. Roy. Meteor. Soc.*, **126**, 649–68.

Saetra, O. (2004). *Ensemble Ship-routing*. ECMWF Research Department Technical Memorandum 435. Available from ECMWF, Shinfield Park, Reading RG2 9AX, UK.

Shutts, G. (2004). *A Stochastic Kinetic Energy Backscatter Algorithm for Use in Ensemble Prediction Systems*. ECMWF Research Department Technical Memorandum 449. Available from ECMWF, Shinfield Park, Reading RG2 9AX, UK.

Simmons, A. J., D. M. Burridge, M. Jarraud, C. Girard and W. Wergen (1989). The ECMWF medium-range prediction models development of the numerical formulations and the impact of increased resolution. *Meteor. Atmos. Phys.*, **40**, 28–60.

Simmons, A. J., R. Mureau and T. Petroliagis (1995). Error growth and predictability estimates for the ECMWF forecasting system. *Quart. J. Roy. Meteor. Soc.*, **121**, 1739–71.

Smith, L. A., M. S. Roulston and J. von Hardenbergm (2001). *End to End Ensemble Forecasting: Towards Evaluating the Economic Value of the Ensemble Prediction System.* ECMWF Research Department Technical Memorandum 336. Available from ECMWF, Shinfield Park, Reading RG2 9AX, UK.

Somerville, R. C. J. (1979). Predictability and prediction of ultra-long planetary waves. In *Preprints of the AMS Fourth Conference on Numerical Weather Prediction, Silver Spring, MD*, pp. 182–5. American Meteorological Society.

Stanski, H. R., L. J. Wilson and W. R. Burrows (1989). Survey of common verification methods in meteorology. World Weather Watch Technical Report 8, WMO/TD. 358. World Meteorological Organization.

Talagrand, O. and P. Courtier (1987). Variational assimilation of meteorological observations with the adjoint vorticity equation. 1: Theory. *Quart. J. Roy. Meteor. Soc.*, **113**, 1311–28.

Taylor, J. and R. Buizza (2003). Using weather ensemble prediction in energy demand forecasting. *Int. J. Forecasting*, **19**, 57–70.

(2004). A comparison of temperature density forecasts from GARCH and atmospheric models. *J. Forecasting*, **23**, 337–55.

Tiedtke, M. (1993). Representation of clouds in large-scale models. *Mon. Weather Rev.*, **121**, 3040–60.

Toth, Z. and E. Kalnay (1993). Ensemble forecasting at NMC: the generation of perturbations. *Bull. Am. Meteorol. Soc.*, **74**, 2317–30.

Tracton, M. S. and E. Kalnay (1993). Operational ensemble prediction at the National Meteorological Center: practical aspects. *Weather Forecast.*, **8**, 379–98.

Viterbo, P. and C. M. Beljaars (1995). An improved land surface parametrisation scheme in the ECMWF model and its validation. *J. Climate*, **8**, 2716–48.

Wei, M. and Z. Toth (2003). A new measure of ensemble performance: perturbation versus error correlation analysis (PECA). *Mon. Weather Rev.*, **131**, 1549–65.

Wilks, D. S. (1995). *Statistical Methods in Atmospheric Sciences*. Academic Press.

Wilks, D. S. (2002). Smoothing ensembles with fitted probability distributions. *Quart. J. Roy. Meteorol. Soc.*, **128**, 2821–36.

18

Limited-area ensemble forecasting: the COSMO-LEPS system

Stefano Tibaldi, Tiziana Paccagnella, Chiara Marsigli, Andrea Montani, Fabrizio Nerozzi

ARPA-SIM, Bologna

18.1 Introduction

The improvement of quantitative precipitation forecasting (QPF) is still one of the major challenges in numerical weather prediction (NWP). Despite the constant increase of computer power resources, which has allowed the development of more and more sophisticated and resolved NWP models, accurate forecasts of extreme weather conditions, especially when related to intense and localised precipitation structures, are still difficult beyond day 2 (Mullen and Buizza, 2001) and, in rare and selected cases, even at 24 hours. This limitation is due, among other reasons, to the inherently low degree of predictability typical of the relevant physical phenomena. The probabilistic approach has been recently increasingly explored to try to come to terms with the chaotic behaviour of the atmosphere and to help forecasting phenomena with low deterministic predictability.

In addition to this, almost twenty years ago Henk Tennekes, at the time member of the ECMWF (European Centre for Medium-Range Weather Forecasts) Scientific Advisory Committee, raised the question of the opportunity of producing a-priori estimates of forecast skill stating that 'no forecast is complete without a forecast of the forecast skill'. It is not an overstatement to say that his bold assertion contributed greatly to the development, at least at ECMWF, of forecast skill studies, estimates and prediction techniques (e.g. Palmer and Tibaldi, 1988) and to the related development of statistical-dynamical prediction methods like ensemble forecasting.

Predictability of Weather and Climate, ed. Tim Palmer and Renate Hagedorn. Published by Cambridge University Press.

In fact, global ensemble prediction systems, implemented operationally for several years now by some of the major meteorological centres (Tracton and Kalnay, 1993; Molteni *et al.*, 1996; Houtekamer *et al.*, 1996), have become extremely important tools to tackle the problem of predictions beyond day 3-to-4 and are becoming more and more the bread-and-butter of operational forecasters all around the world. In order, however, to fully exploit the potential of the stochastic-dynamic approach – that is, to predict within a certain detail the evolution of the probability density function (pdf) of the meteorological system (or at least of some important observable variable) in the atmosphere's phase space – the population of the ensemble should be of the same order of magnitude as the dimensionality of the unstable subspace of the phase space itself (Montani, 1998). This implies a number of integrations probably much higher than what can be achieved by present-day operational ensemble populations (order 50 at most). Furthermore, probabilistic global ensemble systems are usually run at a coarser resolution with respect to (single) deterministic global predictions. Hence, ensemble skill in forecasting intense and localised events in the short and medium range is currently still limited. In order to enhance the present-day prediction capabilities of operational ensemble systems, several approaches have been attempted.

As far as global models are concerned, the horizontal resolution of the ECMWF Ensemble Prediction System (EPS) was increased at the end of 2000 (Buizza *et al.*, 1999) and is now based on a TL255L40 model (spectral model with triangular truncation at wave number 255 and 40 vertical levels), approximately equivalent to a grid spacing of 80 km (Buizza, this volume). During 1999, ECMWF also developed a Targeted Ensemble Prediction System (TEPS; Hersbach *et al.*, 2000), where the perturbations applied to the analysis to obtain the different initial conditions were selected in order to maximise the 48-hour total energy perturbation growth over the European area (about 35N–75N, 40W–30E), instead of over the whole extratropical northern and southern hemisphere as is the case in the operational EPS. This TEPS system was based on a T159L40 model (the same EPS configuration in operation at that time) and was developed within an ECMWF special project as a collaboration amongst KNMI (Royal Meteorological Institute of the Netherlands), ARPA-SIM (the Hydro-Meteorological Service of Emilia-Romagna Region of Italy, formerly ARPA-SMR) and the Norwegian Meteorological Institute (DNMI). The project aim was to increase the ensemble spread over the European area in the short range and early medium range. Results concluded that TEPS performs marginally better than EPS in the 72-to-96 hour forecast range in terms of probabilistic prediction of severe events over Europe while, for shorter ranges, the two systems have comparable skill (Hersbach *et al.*, 2000).

As an alternative approach (trading ensemble size for model horizontal resolution), a smaller-size global ensemble, but at a higher resolution, was tested by Molteni *et al.* (2001). Although the system turned out to be very expensive from a computational point of view, it also gave promising results in the probabilistic prediction of heavy precipitation.

In recent years, much attention has been devoted to the development of a multi-model multi-analysis ensemble system (MMAE) based on the UK Met Office and ECMWF ensembles. It was found that, in a large number of cases, MMAE almost invariably performs at least as well as the best individual ensemble and, sometimes, better than either of the two single-model ensemble systems (Mylne *et al.*, 2002).

Regarding short-range global ensembles, Météo-France has recently developed an 11-member global ensemble system referred to as PEACE (Prévision d'Ensemble à Courte Échéance), where the initial perturbations are generated using the targeted singular vector technique. This system, operational since June 2004, is run up to 60 hours. A systematic evaluation of the system is currently being performed. In addition to this, the UK Met Office is experimenting with a global ensemble system for short-range predictions based on perturbations of initial conditions, model parameters and physics. The main purpose of this ensemble is to provide boundary conditions to a regional ensemble. First results show positive impact in the prediction of surface parameters (Arribas, 2004).

Turning to limited-area model (LAM) applications of the ensemble technique, different approaches are being explored, taking also into account the constraints imposed by lateral boundary conditions on inner domain growing perturbations. One of the possible approaches is the use of the LAM to perform a dynamical downscaling of the different global ensemble forecasts produced by the driving global EPS. Other methodologies rely on inducing perturbations during the limited-area model run employing different techniques based on changing physical schemes, perturbing model parameters or applying stochastic perturbations on physical tendencies.

As for operational implementations, NCEP (the US National Centers for Environmental Predictions) developed the first operational multi-initial-condition multi-model Short-Range Ensemble Forecasting system (SREF; Tracton *et al.*, 1998; Du *et al.*, 2003). This system is composed of 15 elements, five members from the Regional Spectral Model and 10 members from the Eta model; members from the same model are differentiated by changing the convection scheme. Perturbations on initial conditions are based on 'regional' breeding cycles. SREF has been operationally running since May 2001; the system is now run for 63 hours twice a day (at 9 and 21 UTC) at the horizontal resolution of 32 km.

At the DNMI, a limited-area ensemble (LAMEPS) was generated by nesting the operational limited-area model (HIRLAM) in each element of a 21-member TEPS set (Frogner and Iversen, 2001, 2002). This approach (sometimes referred to as the 'brute-force' approach, BFA) appears to provide better results than global TEPS for the prediction of heavy rainfall events, but the computational burden makes it difficult to afford on an operational basis. The semi-operational version of the system is called NORLAMEPS, where TEPS and LAMEPS are combined to obtain a 41-member ensemble (Haakenstad and Frogner, 2004).

The UK Met Office is currently planning to implement a limited-area ensemble system of about 16 members at 24 km horizontal resolution over a European domain, with initial and boundary conditions provided by the global ensemble, while Météo

France is testing a LAMEPS system based on coupling PEACE with the regional model ALADIN (10 km of horizontal resolution, 11 members, 48 hours forecast range).

In the framework of the SRNWP (Short Range Numerical Weather Prediction project of the EUMETNET network) cooperation, a 'poor-man' ensemble (PEPS) has been implemented at Deutscher Wetterdienst by collecting and combining many (19 at present) operational deterministic limited-area models (www.dwd.de/PEPS). The system has been operationally running since December 2004.

A few years ago, ARPA-SIM proposed the LEPS methodology (Limited-area Ensemble Prediction System; Molteni *et al.*, 2001; Marsigli *et al.*, 2001; Montani *et al.*, 2001, 2003a), which will be described in more detail in Section 18.2. The LEPS methodology attempts to combine the benefits of the probabilistic approach with the high-resolution capabilities of the LAM integrations, limiting the computational investment. The method is based on an algorithm to select a number of members out of a global ensemble system. The selected ensemble members (called representative members, RMs) provide initial and boundary conditions to integrate a limited-area model. The ensemble size reduction is necessary to keep the overall computational load operationally affordable. The transfer of information from the large scale to the mesoscale can be viewed as a dynamical downscaling of the forecast provided by the global-model probabilistic system. The good results of the early experimentation phase of the LEPS system led the limited-area modelling consortium COSMO (COnsortium for Small-scale Modelling; www.cosmo-model.org) to the decision of implementing the LEPS technique within the COSMO framework on a regular basis, giving rise to COSMO-LEPS (Montani *et al.*, 2003b) as further described in Section 18.2.

In Section 18.3, some case studies produced using LEPS are analysed in order to investigate in some more detail the behaviour of the system, while Section 18.4 is devoted to an objective and more systematic evaluation of the COSMO-LEPS system performance. Some conclusions and future plans are outlined in Section 18.5.

18.2 The COSMO-LEPS system

As already mentioned in the introduction, the LEPS methodology is based on the idea of reducing the number of LAM integrations needed by an order of magnitude by retaining a hopefully large amount of the global ensemble information while decreasing the number of ensemble elements subjected to LAM runs. This is achieved by, first, grouping the global ensemble members into a number of clusters and then choosing a 'representative member' (RM) within each cluster. Each RM is considered to be representative of the possible evolution scenario associated with each particular cluster and provides both initial and boundary conditions for a high-resolution LAM integration.

During the entire preliminary phase of the experimentation of LEPS, the LAM used has been LAMBO (Limited-Area Model BOlogna), the limited-area model running operationally at the time at ARPA-SMR, a hydrostatic LAM based on an early version of the NCEP Eta model (Mesinger *et al.*, 1988; Janjic, 1990). The system was then constituted by five LAMBO integrations driven by five RMs selected from five clusters derived from the operational ECMWF EPS set of global forecasts. LAMBO was integrated at a horizontal resolution of approximately 20 km and with 32 vertical levels (domain: 1–25E, 36–50N). Montani *et al.* (2001, 2003a) and Marsigli *et al.* (2001, 2004) have shown that, over a number of test cases (including episodes in or close to mountain areas) and for all forecast ranges (from 48 to 120 hours), the LEPS system performs better than the ECMWF EPS (or the Targeted EPS) in probabilistic quantitative prediction of heavy precipitation events, both in terms of a better geographical localisation of the regions most likely to be affected by the events and of a more realistic intensity of the associated rainfall patterns.

Following the encouraging results of the early experimental phase, the regular (daily) generation of a pre-operational limited-area ensemble prediction system, the COSMO-LEPS system, was started in November 2002 on the ECMWF computing system (Montani *et al.*, 2003b) under the auspices of the COSMO Consortium, which involves the operational meteorological institutions of Germany, Italy, Switzerland, Greece and Poland cooperating on the development of the limited-area non-hydrostatic model Lokal Modell (LM). COSMO-LEPS has been designed from the outset for the 'short-to-medium range' timescale (48–120 hours).

The first step of the procedure is the application of a cluster analysis procedure to the merge of the two most recent ECMWF operational global ensembles. The ECMWF Ensemble Prediction System (EPS) is now based on a TL255L40 model (global spectral model with triangular truncation at wave number 255 and 40 vertical levels), corresponding to a horizontal resolution on a linear latitude-longitude grid of about 80 km, and is formed by 51 members (Buizza, this volume).

Two successive 12-hour-lagged EPS ensembles (started at 00 and 12 UTC) are therefore grouped together so as to generate a 102-member so-called superensemble. The introduction of this superensemble technique was aimed at increasing the spread of the starting global ensemble, thereby improving the spanning of the phase space (Montani *et al.*, 2003a).

A multivariate hierarchical cluster analysis is then performed on the resulting 102 members, so as to group all elements into ten clusters of different populations; the clustering algorithm being based on the Complete Linkage method (Wilks, 1995) with the number of clusters fixed to ten. The clustering variables are the geopotential height, the two components of the horizontal wind and the specific humidity at three pressure levels (500, 700, 850 hPa) and at two forecast times (fc+96h and fc+120h for the 'youngest' EPS, started at 12 UTC). The cluster domain covers the region 30N–60N, 10W–30E (see Figure 18.1).

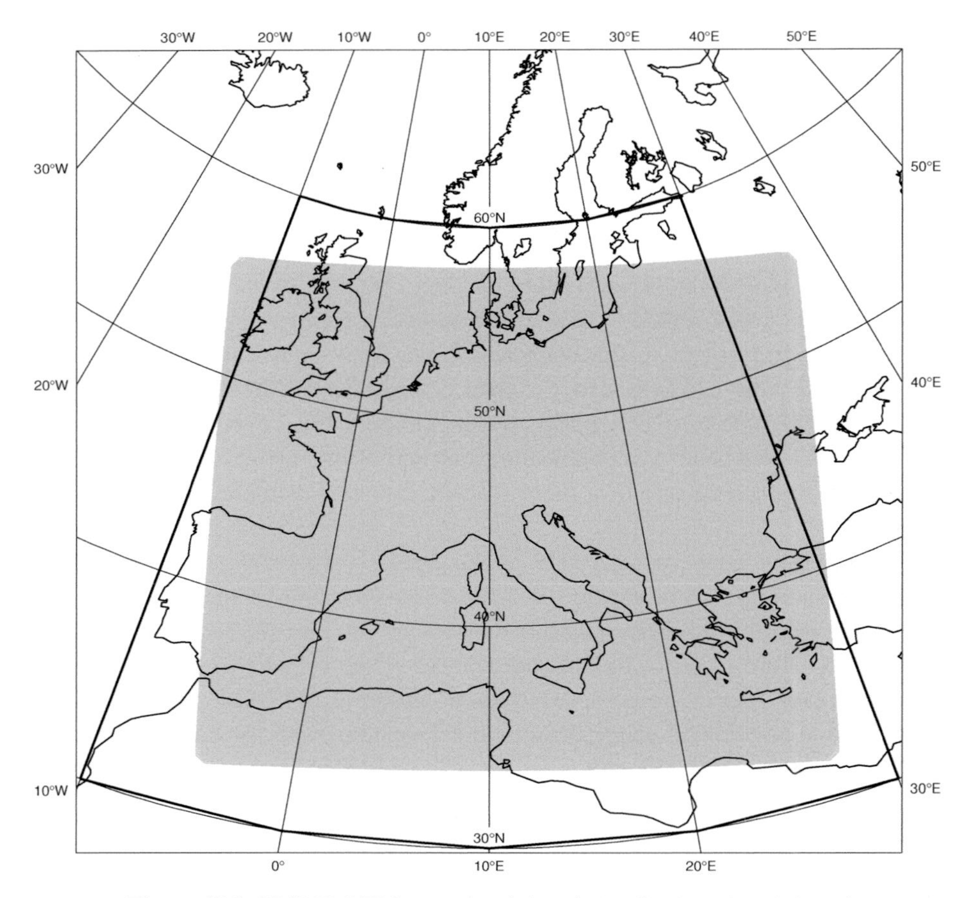

Figure 18.1 COSMO-LEPS operational domain (outlined area) and clustering area (shaded rectangle).

The configuration just described has been running since June 2004. From November 2002 to June 2004, COSMO-LEPS was based on five (instead of ten) RMs selected clustering three (instead of two) 12-hour lagged consecutive EPS (also the EPS starting at 12 UTC of the previous day was used). The reasons for such changes were based on the results of an objective verification exercise, reported in some detail in Section 18.4.

Within each cluster, one representative member (RM) is selected according to the following criteria: the RM is that element closest to the members of its own clusters and most distant from the members of the other clusters; distances are calculated using the same variables and the same metric as in the cluster analysis; in this way, ten RMs are selected, one for each cluster. Each RM provides initial and boundary conditions for the integrations of LM, which is run ten times for 120 hours, always starting at 12 UTC. LM is run with a horizontal resolution of about 10 km and with 32 levels in the vertical.

Assuming a relationship between cluster population and the probability of occurrence of its associated RM, probability maps are generated by assigning to each LM integration a weight proportional to the population of the cluster from which the RM was selected.

A number of deterministic and probabilistic products are routinely produced and disseminated to the COSMO community for regular pre-operational evaluation. The deterministic products for each of the ten LM runs are: precipitation, mean sea level pressure, 700 hPa geopotential, 850 hPa temperature. The main probabilistic products include: probability of 24-h cumulated rainfall exceeding 20, 50, 100, 150 mm, probability of daily Tmax exceeding 20, 30, 35, 40 degrees C, probability of daily Tmin below −10, −5, 0, +5 degrees C, probability of daily Vmax exceeding 10, 15, 20, 25 m/s, probability of 24-h cumulated snowfall exceeding 1, 5, 10, 20 mm of equivalent water.

COSMO-LEPS product dissemination started in November 2002 and at the time of writing (early 2005) the system is being tested to assess its usefulness in met operations rooms, particularly in terms of the assistance given to forecasters in cases of extreme events. In addition to COSMO partners, COSMO-LEPS products are made available also to other interested ECMWF member states.

18.3 Results from selected case studies

As examples of the capability of COSMO-LEPS to predict, already at the short and early-medium range, the occurrence of local severe weather, the behaviour of the system is presented for two heavy precipitation events that recently occurred over the alpine area. For each of the two case studies, the experimentation included:

- rerunning of the global ECMWF-EPS (hor. res.: 80 km, 40 vertical levels; forecast range: 120 hours), archiving model output every 3 hours (1 EPS per case study; no superensemble technique applied);
- nesting LM (hor. res. 10 km; 35 vertical levels; forecast range: 120 hours) in each EPS member (brute-force approach), in a configuration identical to the operational set-up;
- evaluating different ensemble-size reduction techniques, the clustering variables being exactly the same as in the operational suite (Z, U, V, Q at 500, 700, 850 hPa, at the forecast ranges fc+96h and fc +120h);
- assessing the quality of 5-member, 10-member, . . . , 51-member COSMO-LEPS.

18.3.1 Description of the case studies

In this section, we report a short description of the two case studies (more details can be found in Montani *et al.*, 2001).

1. Friuli event: a mesoscale convective system developed due to the intrusion of cold air masses travelling from west to east. Precipitation maxima exceeding 300 mm/day were recorded over Friuli (north-east Italy) between 28 and 29 August 2003, secondary rainfall peaks being observed also over the Ticino area.
2. Piedmont event: an upper-level trough over France caused moist south-westerly flow to blow over the Alps for several days, this leading to widespread heavy precipitation over north-western Italy as well as over Switzerland; rainfall peaks above 150 mm/day were observed between 15 and 16 November 2002.

The two panels of Figure 18.2 show the observed precipitation cumulated over a 24-hour period starting at 12 UTC, for each case study. It is worth pointing out that, while the former case refers to a localised weather event (mesoscale convective system), the latter case deals with a heavy precipitation event for which the large-scale forcing plays a major role (mid-tropospheric trough). Therefore, the analysed case studies span two different types of flood events, on which the performance of COSMO-LEPS system can be assessed.

In the following two subsections the behaviour of COSMO-LEPS at different prediction ranges is analysed, as well as the impact of ensemble size on the quality of the probabilistic forecast.

18.3.2 Behaviour of the system at the different forecast ranges

The behaviour of the COSMO-LEPS system for the Piedmont case (right panel of Figure 18.2) is investigated considering the probabilistic forecast of the events '24-hour rainfall exceeding 20 mm' and '24-hour rainfall exceeding 50 mm' for three different forecast ranges (2, 3 and 4 days). Figure 18.3 shows the consistency of the rainfall probability maps for these events (top-row panels for the 20 mm event; bottom-row panels for the 50 mm event). All maps verify at 12 UTC of 16 November 2002. It can be noticed that, already at the 96-hour range (top-left and bottom-left panels, respectively), a high probability of occurrence is predicted over northern Italy, in the areas where heavy precipitation was actually observed. This kind of information would enable a hypothetical bench forecaster to issue a flood warning, to be either confirmed or dismissed on the basis of forecasts at ranges closer to the predicted event. Thanks to the relatively long lead time (four days), preventive actions could have been taken so as to limit damage as much as possible.

The possibility of a heavy rainfall scenario is confirmed and reinforced by the 72- and 48-hour forecasts (middle-column and right-column panels, respectively), the probability of occurrence exceeding 90% in the correct locations.

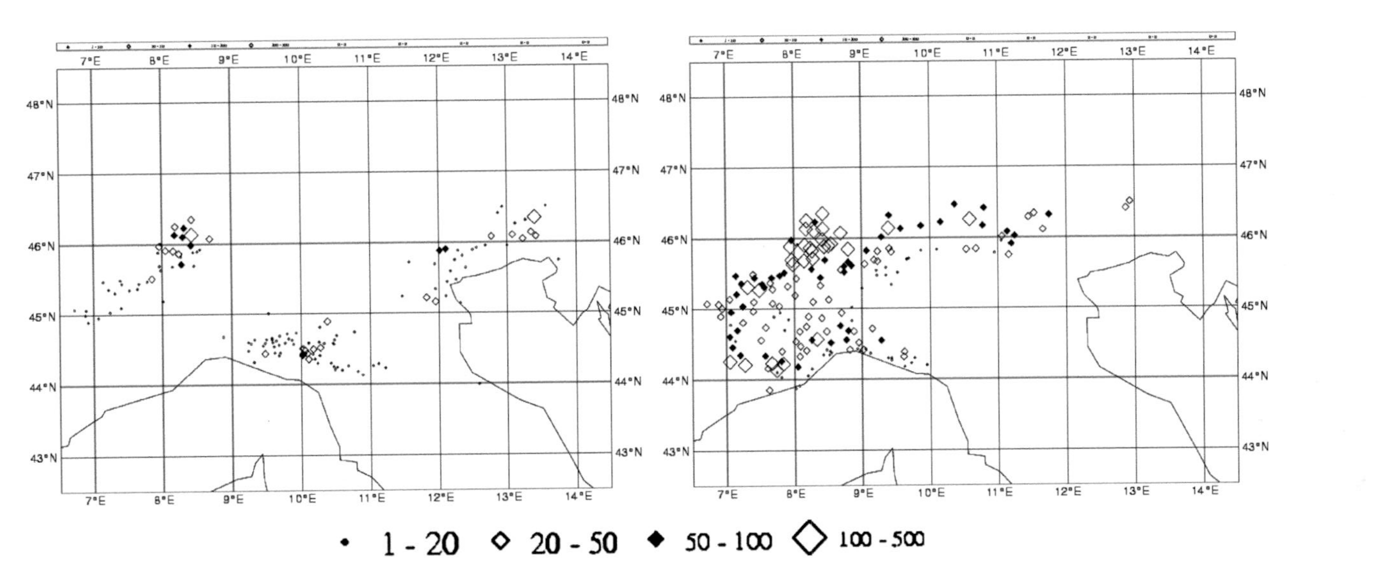

Figure 18.2 Twenty-four-hour observed precipitation for the Friuli (left; cumulation period starting at 12 UTC of 28 August 2003) and Piedmont (right; cumulation period starting at 12 UTC of 15 November 2002); thresholds: 1 mm (black dots), 20 mm (empty diamonds), 50 mm (full diamonds), 100 mm (large empty diamonds).

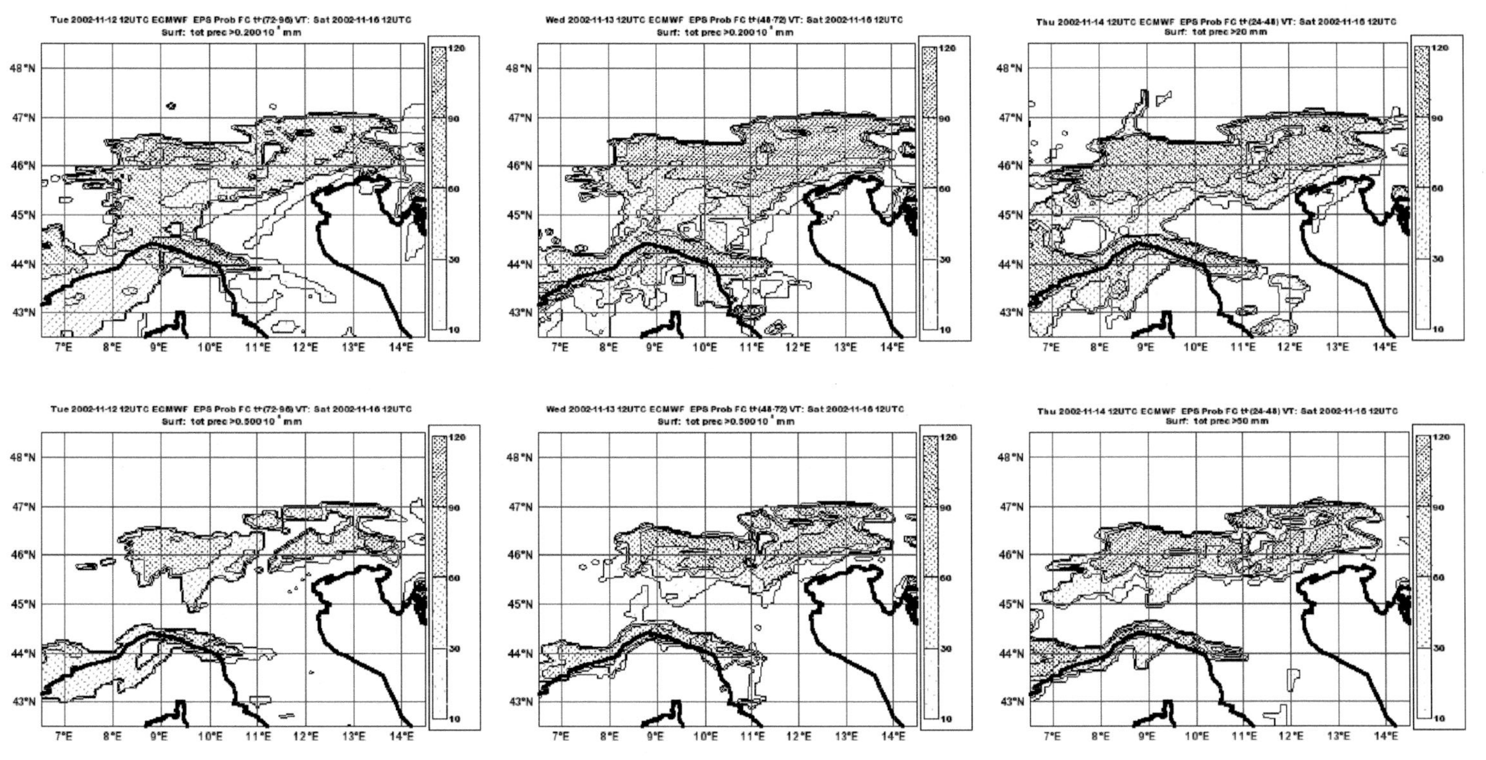

Figure 18.3 Piedmont case (all maps verify at 12 UTC of 16 November 2002); COSMO-LEPS probability maps for 24-h rainfall exceeding 20 mm (top-row panels) and 50 mm (bottom-row panels) at the 96-h range (left-column panels), 72-h range (middle-column panels) and 48-h range (right-column panels). Contour intervals: 10%, 30%, 60%, 90%.

Similar considerations apply also for the Friuli case, where the performance of COSMO-LEPS is again accurate and consistent (Figure 18.4). Despite this case study being characterised by a local-forcing and the predictability being more limited than in the previous one, the probability maps indicate, as much as 96 hours before the event, the possibility of a heavy rainfall scenario (24-hour cumulated rainfall above the 50 mm threshold) over north-eastern Italy, as evident in the bottom-left panel of Figure 18.4. This critical situation is confirmed by the 3-day and 2-day probabilistic forecasts, where also the Ticino (central Alps) area is correctly highlighted as a region possibly affected by heavy precipitation.

18.3.3 Impact of the ensemble reduction and comparison with the ECMWF EPS

As described in Marsigli *et al.* (2001) and Molteni *et al.* (2001), the main motivations behind the ensemble-size reduction of LEPS methodology are dictated by computing power limitations as well as by the (supposed) possibility to capture, thanks to the clustering-selection technique, alternative weather patterns, with higher or lower probability of occurrence. This latter aspect is critical and deserves further investigation. In fact, it is not fully clear the extent to which the ensemble-size reduction may induce a loss of information about the spectrum of possible atmospheric flows and thus affect the overall performance of LEPS.

In order to assess the impact of the ensemble-size reduction on the rainfall probabilistic prediction, the cluster analysis has been performed several times with the number of clusters increasing from 5 to 51 generating limited-area ensembles, with a corresponding increasing population from 5 to 51. Attention is focused on the 4-day predictability of the event '24-hour precipitation exceeding 50 mm'; the performances of the 5-member, 10-member and 51-member ensembles are assessed both for the global EPS and for the limited-area LEPS (the system configurations will be referred to as EPS5, EPS10, EPS51, LEPS5, LEPS10 and LEPS51, respectively).

Figure 18.5 illustrates the results obtained for the Piedmont case: it is clear that EPS forecasts (top-row panels) underestimate the rainfall intensity, since the probability of occurrence is, in all configurations, well below 30% and, additionally, poorly localised. It can also be noticed that only the EPS51 configuration (top-right panel) shows three different areas over northern Italy that are highlighted as possible locations of heavy precipitation (as actually happened); nevertheless, the probability values are very low and poor guidance would have been given to a forecaster involved in alert procedures.

The situation is completely different when the better description of orographic and mesoscale-related processes comes into play in the high-resolution forecasts of the LEPS system. The bottom-left panel of Figure 18.5 indicates that the LEPS5 system is already able to highlight the possibility of occurrence of a heavy precipitation event

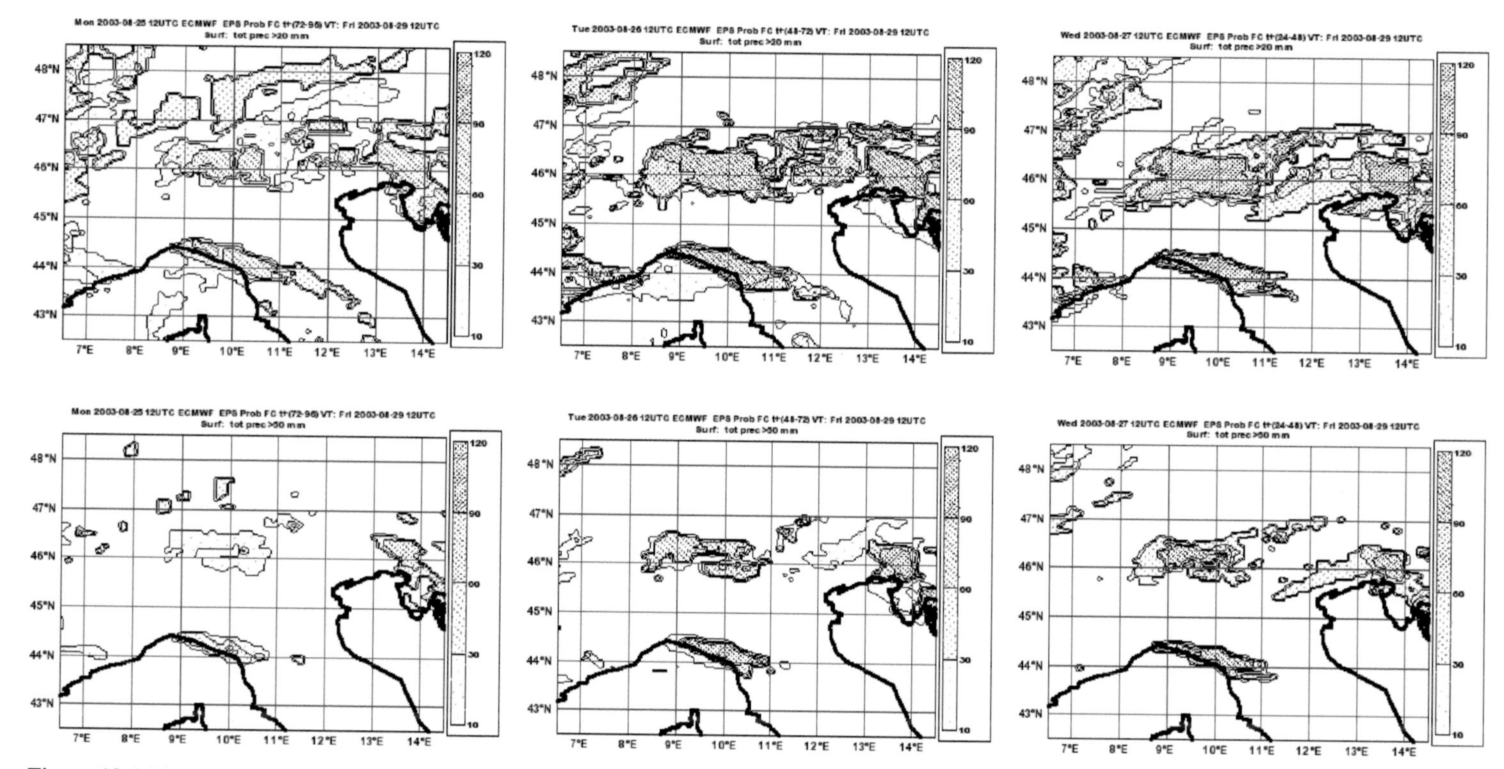

Figure 18.4 The same as Figure 18.3 but for the Friuli case (all maps verify at 12 UTC of 28 August 2003).

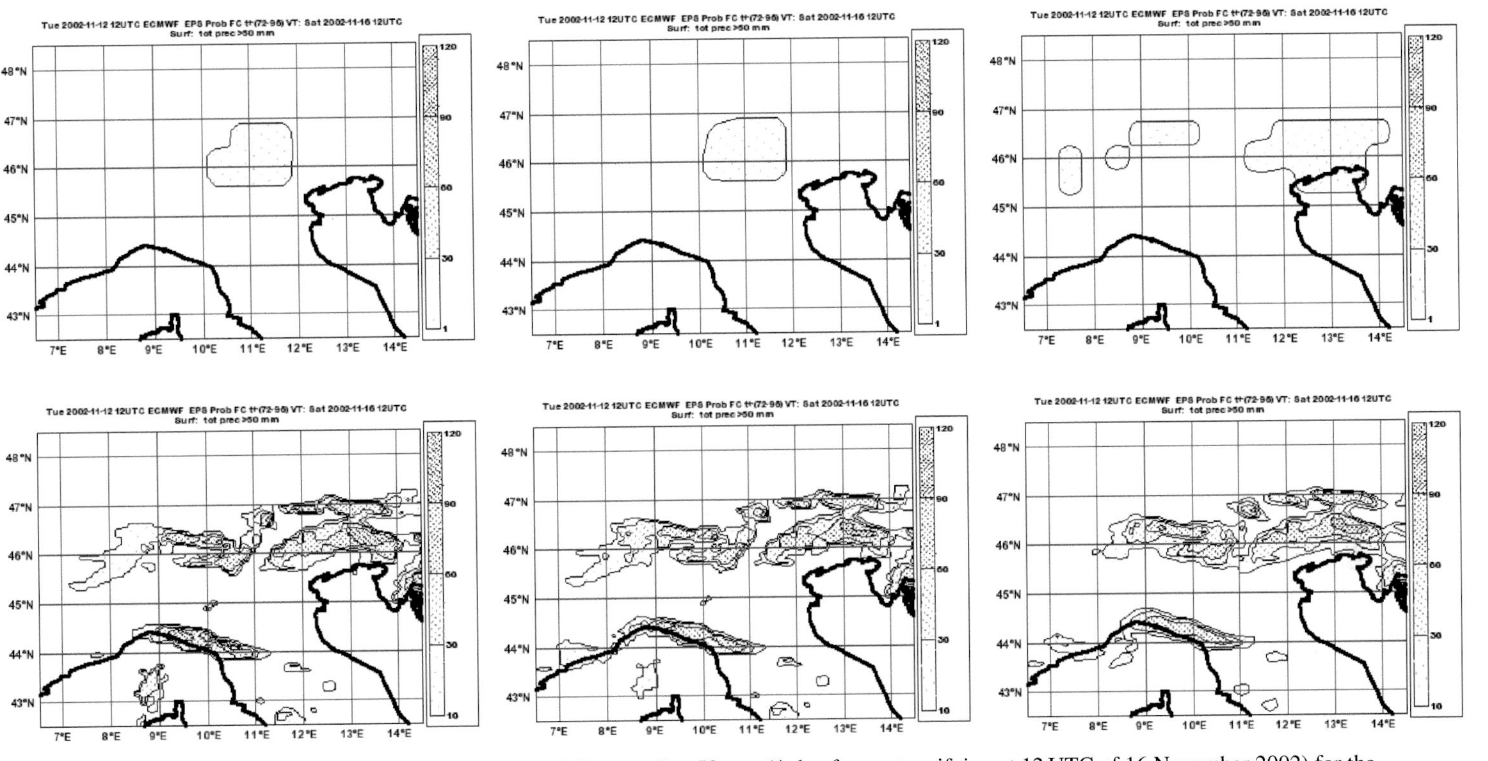

Figure 18.5 Piedmont case; probability maps for 24-h rainfall exceeding 50 mm (4-day forecast verifying at 12 UTC of 16 November 2002) for the 5-member EPS (top-left panel), 10-member EPS (top-middle), 51-member EPS (top-right), 5-member LEPS (bottom-left), 10-member LEPS (bottom-middle) and 51-member LEPS (bottom-right). Contour intervals: 1%, 30%, 60%, 90% for top-row panels; 10%, 30%, 60%, 90% for bottom-row panels.

in the correct locations. As the population of the LEPS system increases (bottom-middle and bottom-right panels), the results obtained are very similar, the probability of occurrence being only slightly modified.

As concerns the Friuli case, the top-row panels of Figure 18.6 show that EPS rainfall probability maps (for precipitation exceeding 50 mm/day) fail completely to indicate north-eastern Italy as an area possibly affected by heavy rainfall. In the EPS51 configuration (top right) a weak signal is evident over the Ticino region, where heavy precipitation was actually observed; but also in this case the probability is below 30%. By contrast, the information provided by LEPS forecasts is much more accurate than that to be found in the global runs. The bottom-row panels show that both the Ticino and the Friuli area are highlighted as regions most likely to be hit by the severe weather event and the pattern of the probability maps is roughly unchanged as the LEPS population increases from 5 to 51. Regarding high probability values over the Liguria region, it is worth noting that this false alarm is not due in particular to the LEPS ensemble system but rather to a generalised systematic error of the model in this region.

From the above results, the impact of the ensemble-size reduction on the forecast accuracy does not seem to be crucial, since, for both cases, the heavy rainfall scenario is properly captured in the 'LEPS5' as well as in the 'LEPS51''configuration. This result is, however, very encouraging: at least for these two case studies, the clustering-selection technique highlights the most important evolution scenarios and enables the generation of reliable and accurate probability maps. It is clear that the results obtained by these studies may not have any statistical significance, since they are based on only two cases, both characterised by heavy precipitation events. Nevertheless, the outcome of these experiments is important, since it can indicate the potential of COSMO-LEPS methodology, so justifying future investigation and system development.

18.4 Statistical evaluation

In this section, a more systematic and objective evaluation of the COSMO-LEPS forecast performance is presented, so as to assess overall abilities and shortcomings of this prediction system. The attention remains focused on the probabilistic prediction of heavy precipitation, one of the main causes of damage and loss of life over Europe.

The local effects of precipitation are related to two main characteristics: the cumulative volume of water deployed over a specific region and the rainfall peaks which occur within this region. The relative importance of these two features of the intense event is related to the geomorphological features of the area affected. Both these aspects need to be taken into account when the verification of precipitation is concerned. This is accomplished by considering respectively the mean precipitation over an area as well as the precipitation maxima in the same area.

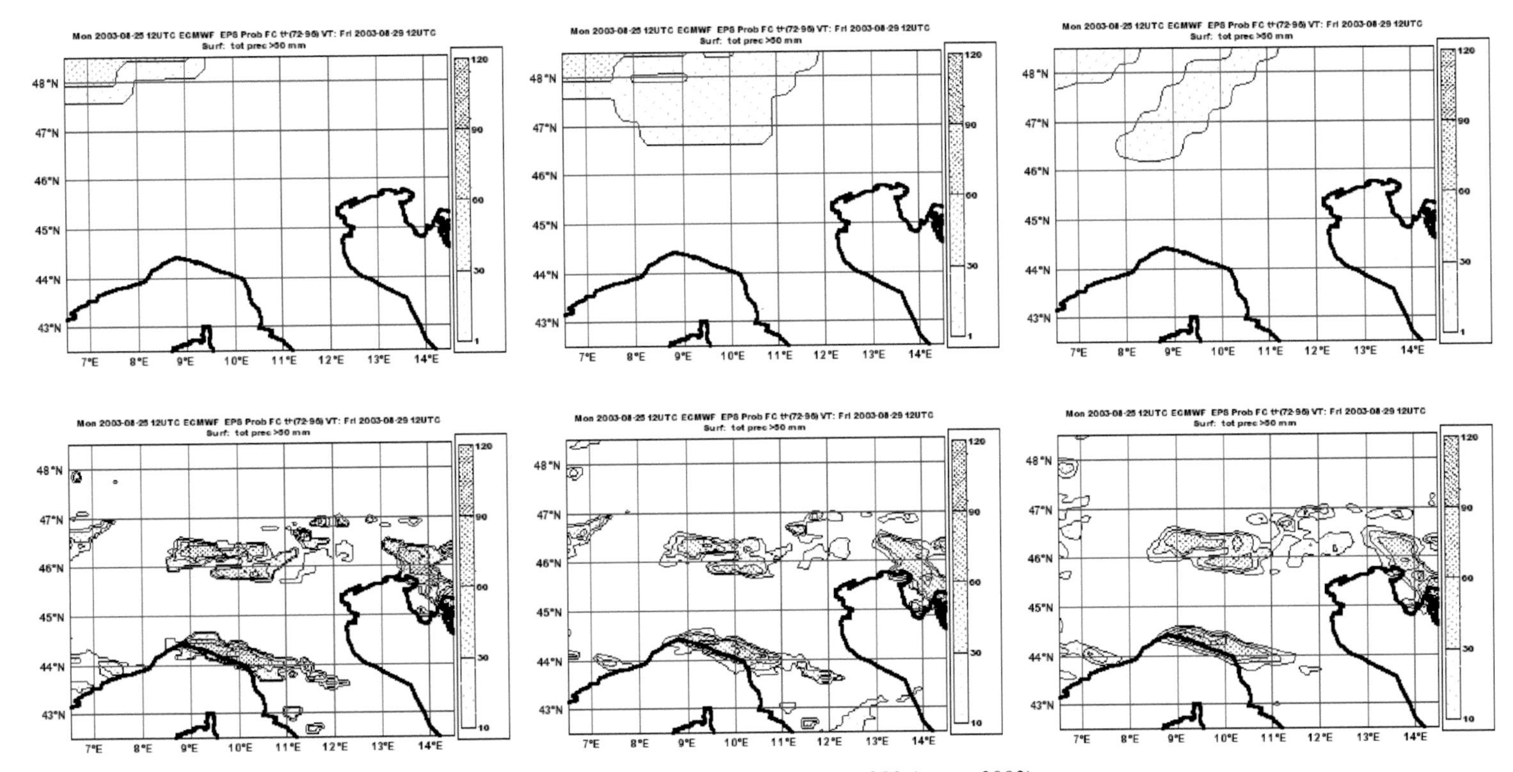

Figure 18.6 The same as Figure 18.5 but for the Friuli case (all maps verify at 12 UTC of 28 August 2003).

For this new probabilistic system to be shown as potentially valuable, it has to be compared with the state-of-the-art ensemble system, namely ECMWF EPS, which is already available to European national weather services at no extra cost. Therefore, we investigate the extent to which the high-resolution forecast details and intensities provided by the COSMO-LEPS system produce real added value with respect to the global EPS (Marsigli *et al.*, 2005).

COSMO-LEPS forecasts are here verified on the basis of the traditional probabilistic scores: Brier Skill Score (Wilks, 1995), ROC area (Mason and Graham, 1999), Cost-loss Curve (Richardson, 2000), Percentage of Outliers (Buizza, 1997). Results are shown for the autumn 2003 period (September, October and November).

From the results shown in Tibaldi *et al.* (2003), no positive impact of the weighting procedure was detected regarding probabilistic forecast of precipitation; therefore, the results presented in this section refer to probabilities computed giving the same weight to each of the reduced-ensemble members.

This analysis is relative to the COSMO-LEPS configuration which has been running pre-operationally from November 2002 to June 2004: five RMs were selected clustering three 12-hour lagged consecutive EPS (FORMER SUITE). Since June 2004 the suite was changed (10 RMs selected from a superensemble of two EPSs, CURRENT SUITE) according to the results shown in the following subsections.

Verification is subdivided into two main streams: a verification of the methodology on which COSMO-LEPS is based (Section 18.4.1) and a verification of the system performance against other available probabilistic systems (Section 18.4.2).

18.4.1 Analysis of the methodology

The superensemble approach, grouping together three consecutive EPSs, has been adopted to enlarge the size of the ensemble on which the RM selection algorithm is applied. This allows us to increase the ensemble spread and to explore better the phase space, even if a price in terms of skill is paid: the older the EPS, the less skilful its members are likely to be. In order to evaluate the comparative effects of the increased spread and of the decreased skill, the representative members chosen with the FORMER-SUITE methodology are compared with those chosen using only one or two EPS. Furthermore, the impact of the reduced-ensemble size is also evaluated by comparing each 5-RMs configuration with the correspondent 10-RMs configuration. The six examined configurations are listed in Table 18.1.

This analysis is performed in terms of 24-hour precipitation. The forecast values at each grid point are compared with a proxy for the true precipitation that occurred, chosen as the +24 hour forecast by the ECMWF deterministic model (horizontal resolution 40 km). The extent to which this proxy is a good approximation for the

Table 18.1 *Characteristics of the analysed COSMO-LEPS configurations*

Configuration name	EPS on which the cluster analysis is based	Number of RMs selected	Suite name
3eps-5rm	3 12-hour lagged EPS	5 RMs	FORMER SUITE
2eps-5rm	2 12-hour lagged EPS	5 RMs	
1eps-5rm	1 12-hour lagged EPS	5 RMs	
3eps-10rm	3 12-hour lagged EPS	10 RMs	
2eps-10rm	2 12-hour lagged EPS	10 RMs	CURRENT SUITE
1eps-10rm	1 12-hour lagged EPS	10 RMs	

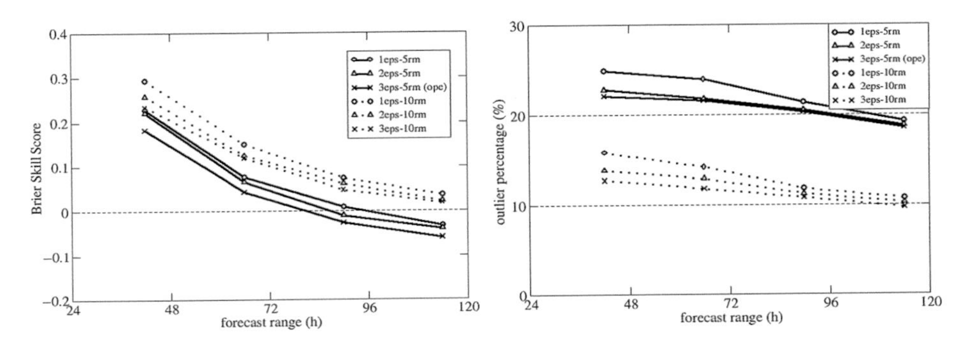

Figure 18.7 Brier skill score and percentage of outliers for different configurations of representative members.

truth is not crucial, because this is a comparison between different configurations of the same model. The period chosen for this test is September–November 2003 and the verification area is the entire clustering area (the shaded rectangle in Figure 18.1).

Regarding the 5-RM ensembles, results show that the Brier Skill Score (Figure 18.7, left panel) is highest when the clustering is based on the most recent EPS only (1eps-5rm, solid line with circles), while it is lowest for the 3eps-5rm (solid line with crosses). The difference between the two is not so large, but it is clearly detectable at every forecast range. The 2eps-5rm (solid line with triangles) has an intermediate skill, closer to the 1eps-5rm.

The percentage of outliers of the systems is also shown. This is the percentage of times the 'truth' falls out of the range spanned by the forecast values. The percentage of outliers (Figure 18.7, right panel) of the 1eps-5rm (solid line with circles) is rather higher than the other two, for every forecast range, while there is almost no difference in terms of outliers between the 2eps-5rm (solid line with triangles) and the 3eps-5rm (solid line with crosses).

These results would seem to suggest that the use of just two EPS in the superensemble can be a good compromise, permitting us to decrease the percentage of outliers significantly but leading only to a small decrease of the skill.

Regarding the impact of the ensemble size, the difference between each 5-member ensemble and the corresponding 10-member ensemble is remarkable, being about 0.1 in terms of Brier Skill Score for every configuration. This is shown in the left panel of Figure 18.7, comparing each solid line with the dotted line carrying the same symbols. The impact of doubling the ensemble size is almost the same for every configuration and is larger than the impact of changing the number of EPSs on which the cluster analysis is performed (two or three).

These results led to two major modifications of the COSMO-LEPS methodology, as applied to the pre-operational suite, at the beginning of June 2004: since this date the superensemble has been constructed by using only the two most recent EPSs and the number of clusters has been fixed to ten, nesting Lokal Modell on each of the ten selected RMs.

18.4.2 Comparison of the limited-area ensemble with global ensembles

In order to quantify the added value brought about by the mesoscale probabilistic system, COSMO-LEPS is compared with ECMWF EPS. The comparison has to take into account two important issues: the difference in the number of ensemble members and the difference in terms of model resolution.

As far as the population of the ensembles is concerned, it is worth pointing out that the verification is carried out during the period September–November 2003, when the pre-operational COSMO-LEPS population was fixed to five members. Therefore, COSMO-LEPS is compared not only with the operational EPS (51 members), but also with the reduced EPS ensemble made up by the five representative members. This permits us to quantify the impact of the increased resolution alone.

The problem of the different resolutions of the two systems (10 km for COSMO-LEPS and 80 km for EPS) is tackled by upscaling both systems to an even lower-resolution common interface: the grid-point forecasts of both models are averaged over boxes of 1.5×1.5 degrees.

The comparison is performed in terms of 24-hour precipitation against observed data. In order to compare properly forecast values over boxes and observed values on station points, the observations within a box are averaged and the obtained values are compared with the averaged forecast values. The comparison is carried out over a large fraction of the COSMO-LEPS domain. In fact, a very dense network of stations (about 4000) recording daily precipitation (cumulated from 06 to 06 UTC) are made available by Germany, Switzerland and Italy; see Figure 18.8.

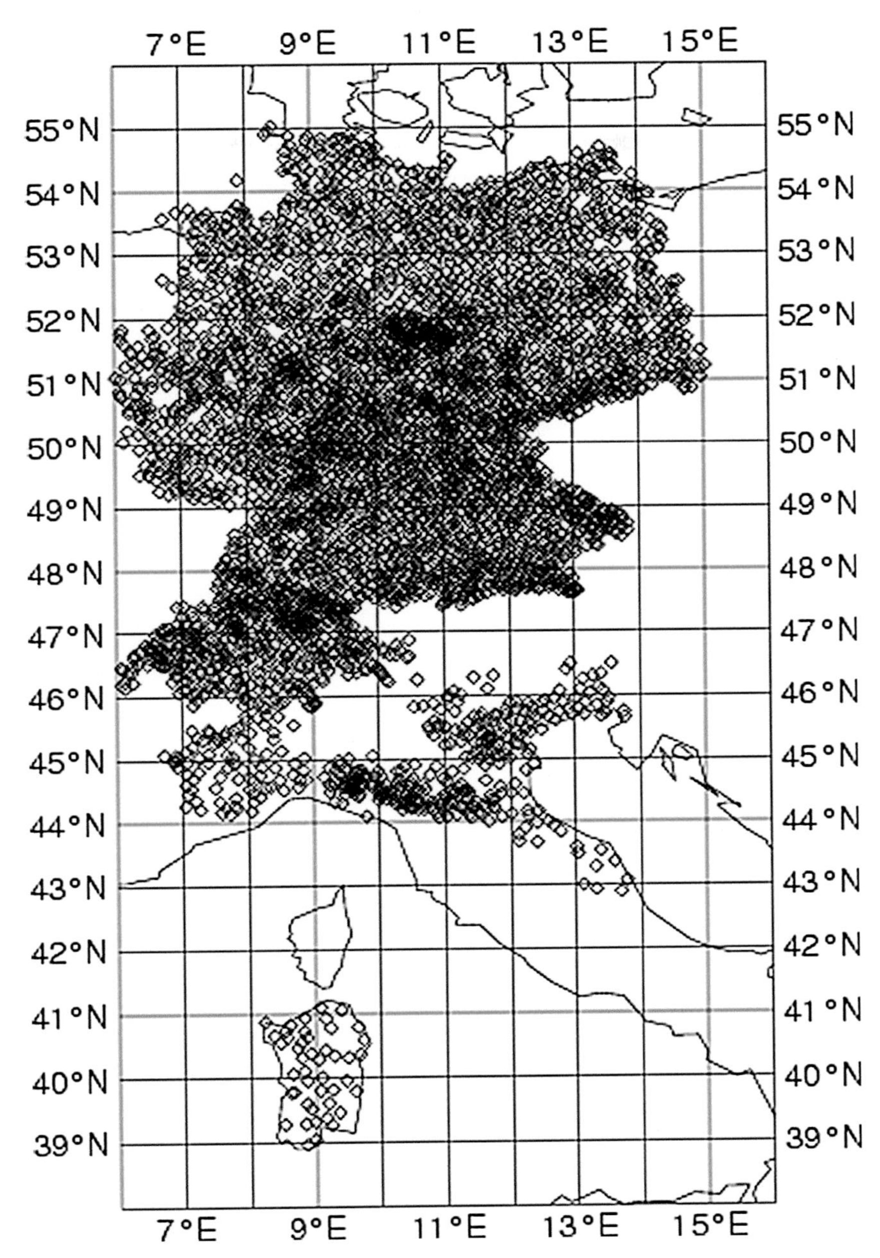

Figure 18.8 Network of about 4000 stations in Germany, Switzerland and Italy recording daily precipitation (cumulated from 06 to 06 UTC the following day). The network was used to compute daily mean and maximum precipitation in 1.5 × 1.5 degree boxes used to verify forecasts in Section 18.4.2.

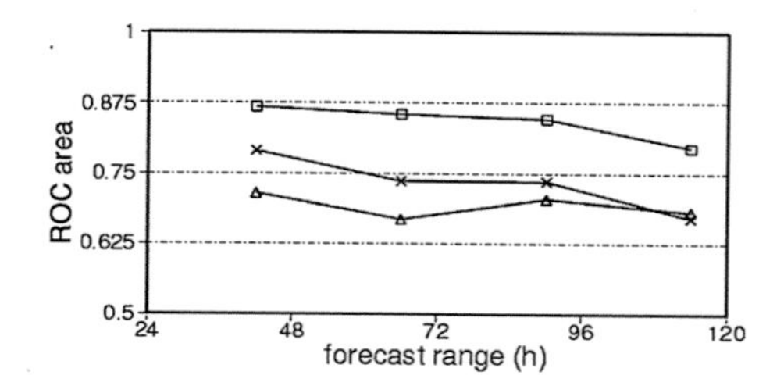

Figure 18.9 Average forecasts against average observations: ROC area values for precipitation over 1.5 × 1.5 boxes exceeding 20 mm/24 h. The crosses are relative to the COSMO-LEPS system, the squares to the 51-member EPS and the triangles to the 5-RM EPS.

The three ensemble systems compared are:

- the COSMO-LEPS system, made up of five members, 10 km of horizontal resolution, referred to as 'cleps';
- the EPS reduced-ensemble made up by the five representative members chosen from the superensemble, 80 km of horizontal resolution, referred to as 'epsrm';
- the operational 51-member ECMWF EPS starting at the same initial time as COSMO-LEPS (the 'youngest' EPS constituting the superensemble), 80 km of horizontal resolution, referred to as 'eps51';

In Figure 18.9, the average observed value of each box is compared with the average forecast value relative to the same box, for each of the three forecasting systems. The event considered is 'precipitation exceeding 20 mm/24 h over 1.5 × 1.5 degree boxes'. Since the observed and forecast values are averaged over an area of 1.5 × 1.5 degrees, this threshold has been chosen as representative of an intense precipitation event.

In terms of relative operating characteristic (ROC) area, the eps51 configuration (Figure 18.9, solid line with squares) shows the best scores at this threshold for every time range. The ensemble size, therefore, plays a major role in the computation of the probabilistic indices, making it difficult to carry out a proper comparison between cleps and eps51. When the two systems with the same ensemble size are compared, cleps (solid line with crosses) has higher scores than those of epsrm (solid line with triangles).

A comparison in terms of precipitation maxima has also been performed: the maximum forecast value within a 1.5 × 1.5 box is compared with the maximum observed value in the same box.

The ROC area values for cleps (Figure 18.10, line with crosses) are higher (and therefore better) than both the epsrm (line with triangles) and eps51 (line with squares), indicating that COSMO-LEPS is more skilful in forecasting correctly high precipitation values over a rather large area. It is worth pointing out that these results

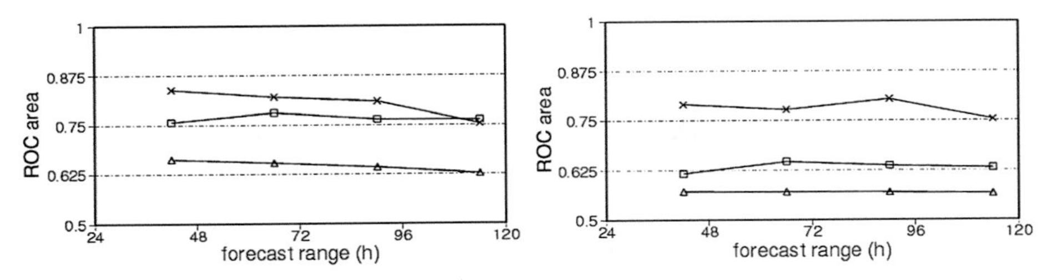

Figure 18.10 Maximum forecasts against maximum observations: ROC area values for precipitation over 1.5 × 1.5 boxes exceeding 20 (left panel) and 50 (right panel) mm/24 h. The crosses are relative to the COSMO-LEPS system, the squares to the 51-member EPS and the triangles to the 5-RM EPS.

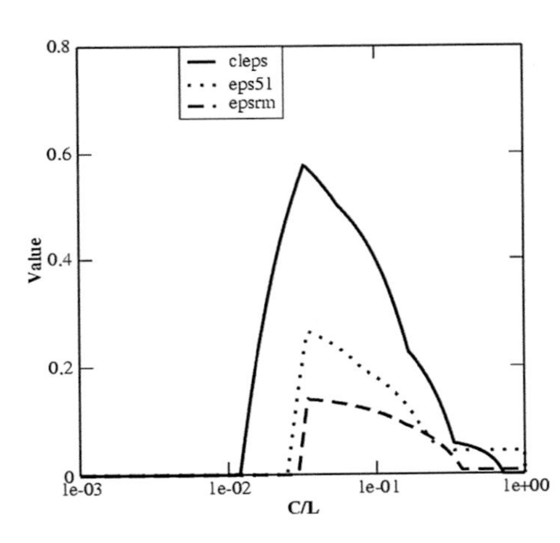

Figure 18.11 Maximum forecasts against maximum observations: cost–loss curves for precipitation over 1.5 × 1.5 boxes exceeding 50 mm/24 h (90 hour forecast range). The solid line is relative to the COSMO-LEPS system, the dotted line to the 51-member EPS and the dashed line to the 5-RM EPS.

are based on a relatively large number of occurrences (about 600 and 150 for the 20 and 50 mm thresholds, respectively).

The same conclusions apply when the cost–loss curves relative to the three systems are considered for the event 'maximum precipitation exceeding 50 mm/24 h' at the 90-hour forecast range (Figure 18.11). To keep the diagram readable, only the envelope curve is plotted for each system.

The curve relative to cleps (solid line) is well above the curves of both eps51 and epsrm. This is especially true for low cost–loss ratios (left portion of the x-axis), where the global model ensembles have almost no value.

18.5 Summary and concluding remarks

Since the day, almost 20 years ago, in which Henk Tennekes stated, during a meeting of the Scientific Advisory Committee of the European Centre for Medium-Range Weather Forecasts, that 'no forecast is complete without a forecast of the forecast skill', the demand has been ever increasing for numerical forecasting tools which should have the capability at the same time of providing quantitative estimates of forecast reliability and of casting quantitative forecasts of various meteorological parameters in terms of probabilities. Consistently with this, among ECMWF's operational users, output from the now twice-daily EPS system is eroding more and more the historical primacy detained by the single, daily, purely deterministic, highest possible resolution, 'main', 12:00 GMT, 10-day forecast run, which for a long time has represented by far the main source of reliable, good-quality, numerical short- and medium-range forecast information.

EPS products, however, suffer from a number of drawbacks, most of them related to the comparatively low resolution of the global model used to produce them. The size of the ensemble, i.e. the number of the forecasts which compose it, is a very important factor in producing a sufficiently adequate 'exploration' of the phase space of all possible future atmospheric states which are compatible with our imperfect knowledge of the atmospheric initial conditions. This imposes a compromise between model resolution and the total number of elements needed in the ensemble, i.e. the model resolution has to be kept low enough to make the total computational requirements affordable. This has as a consequence that in the EPS model a number of local (mainly orography-related) atmospheric features/phenomena/processes are still misrepresented or underestimated. Orographically related local and/or intense precipitation is one of the best possible such examples, but not the only one. This carries with it a reduced capability of the EPS system to provide useful guidance in cases of intense meteorological events (often catastrophical in terms of consequences on life and property), notably those characterised by large rainfall amounts, possibly leading to floods.

In the attempt, therefore, to push the point of compromise between resolution and number-of-elements further toward a better representation of local effects (such as the one typical of high-resolution limited-area models), without having to sacrifice too much to completeness of the sampling of the atmospheric phase space, the LEPS (Limited-area Ensemble Prediction System) was developed at ARPA-SIM. The basic goal is to obtain most of the advantages of a brute-force approach, which can hardly be afforded on an operational basis (run a high-resolution LAM integration from each and every element of an ECMWF EPS, for example), without having to pay the full computational cost.

The LEPS idea is in fact very simple: once all EPS elements have been grouped in clusters, from all the elements of each cluster a most 'representative member' (RM) is defined, which carries with it a weight proportional to the population of the cluster it represents and from which a high-resolution LAM is integrated. The LAM integrations can be combined, using the cluster-population weights, to produce all statistically based predictions typical of a complete EPS system, such as for example the probability for a meteorological variable to overcome given thresholds as a function of space and time.

Preliminary case-study-based results were considered promising enough to launch a COSMO-LEPS pre-operational, daily experimentation exercise, carrying out nesting, in LEPS mode, a limited-area model (Lokal Modell) within the ECMWF EPS model output. At the time of writing (January 2005) the experimentation is carried out constructing a 102-element superensemble from two 12-hour lagged operational EPS ensembles and integrating ten times a 10-km mesh version of LM once a day for 120 hours. The limited-area model integrations using Lokal Modell are carried out on a domain covering the geographical territory of all countries participating in COSMO. The experiment is carried out on the ECMWF computer system, to minimise file transfer requirements. This extended pre-operational test started during the first week of November 2002 on computer resources coming from the ECMWF member states which are also COSMO partners (that is, Germany, Greece, Italy and Switzerland).

As examples of the high quality of COSMO-LEPS products it was shown that, for cases of heavy precipitation events, the system is able to predict the possibility of a flood event four days in advance and to confirm (and reinforce) the scenario as the forecast range gets shorter.

The impact of the large ensemble-size reduction on the forecast accuracy has been studied for a number of case studies. It is found that the heavy rainfall scenarios are already captured when the limited-area ensemble size is reduced from 51 to either 5 or 10 members. This result is very encouraging: for these particular case studies, the clustering-selection technique highlights the most important evolution scenarios and enables the generation of reliable and accurate probability maps.

A statistical analysis of the methodology on which COSMO-LEPS was based for the first part of experimentation (November 2002 to June 2004) was also undertaken, focusing the attention on the prediction of precipitation. The results indicate that the use of two EPS in the superensemble is a reasonable compromise between the decrease in the percentage of outliers (the cases for which no LEPS member comes close to the observed rainfall intensity) and a modest decrease of the forecast skill. In addition to this, it was found that doubling the ensemble size permitted us to halve the percentage of outliers. These results led us to modify the COSMO-LEPS operational suite at the beginning of June 2004 as follows: use of two EPS (instead of three) to construct the superensemble and increase the limited-area ensemble size from 5 to 10 members.

As a final step, an objective verification of the COSMO-LEPS system was shown for autumn 2003 for forecasts up to 120 hours. It was found that, as far as precipitation averaged over 1.5×1.5 degree boxes is concerned, the operational full-size EPS exhibits a higher skill than COSMO-LEPS. This appears to be due to the higher population of the global model ensemble. In fact, when COSMO-LEPS is compared with a reduced size (comparable population) version of the EPS, the limited-area ensemble has noticeably greater skill. By contrast, COSMO-LEPS outperforms both full-size and reduced-size EPS in terms of prediction of precipitation maxima over the boxes.

These results could be conducive to a truly operational implementation of the system to assist forecasters and agencies involved in civil protection tasks for the issue of warnings about localised severe weather and are strengthening the value of ensemble forecasting systems at ranges shorter than the full medium range.

References

Arribas, A. (2004). Results of an initial stochastic physics scheme for the Met Office Unified Model. Forecasting Research Technical Report No 452. Available at Met Office, Exeter, UK.

Buizza, R. (1997). Potential forecast skill of ensemble prediction and spread and skill distributions of the ECMWF ensemble prediction system. *Mon. Weather Rev.*, **125**, 99–119.

Buizza, R., J. Barkmeijer, T. N. Palmer and D. Richardson (1999). Current status and future developments of the ECMWF Ensemble Prediction System. *Meteorol. Appl.*, **6**, 1–14.

Du, J., G. DiMego, M. S. Tracton and B. Zhou (2003). *NCEP short-range ensemble forecasting (SREF) system: multi-IC, multi-model and multi-physics approach.* Research Activities in Atmospheric and Oceanic Modelling, Report 33, CAS/JSC Working Group Numerical Experimentation (WGNE), WMO/TD-No. 1161.

Frogner, I. and T. Iversen (2001). Targeted ensemble prediction for northern Europe and parts of the North-Atlantic Ocean. *Tellus*, **53A**, 35–55.

(2002). High-resolution limited-area ensemble predictions based on low-resolution targeted singular vectors. *Quart. J. Roy. Meteor. Soc.*, **128**, 1321–41.

Haakenstad, H. and I-L. Frogner (2004). Extended tests with a limited area ensemble prediction system at the Norwegian Meteorological Institute – LAMEPS. Met.no report 7. Meteorologist Institutt.

Hersbach, H., R. Mureau, J. D. Opsteegh and J. Barkmeijer (2000). A short-range to early-medium-range Ensemble Prediction System for the European Area. *Mon. Weather Rev.*, **128**, 3501–19.

Houtekamer, P. L., J. Derome, H. Ritchie and H. L. Mitchell (1996). A system simulation approach to ensemble prediction. *Mon. Weather Rev.*, **124**, 1225–42.

Janjic, Z. I. (1990). The step-mountain co-ordinate: physical package. *Mon. Weather Rev.*, **118**, 1429–43.

Marsigli, C., A. Montani, F. Nerozzi, *et al.* (2001). A strategy for high-resolution ensemble prediction. II: Limited-area experiments in four Alpine flood events. *Quart. J. Roy. Meteor. Soc.*, **127**, 2095–115.

Marsigli, C., A. Montani, F. Nerozzi and T. Paccagnella (2004). Probabilistic high-resolution forecast of heavy precipitation over Central Europe. *Nat. Hazard Earth Sys. Sci.*, **4**, 315–22.

Marsigli, C., F. Boccanera, A. Montani and T. Paccagnella (2005). The COSMO-LEPS ensemble system: validation of the methodology and verification. *Nonlinear Proc. Geoph.*, **12**, 527–36.

Mason, S. J. and N. E. Graham (1999). Conditional probabilities, relative operating characteristics and relative operating levels. *Weather Forecast.*, **14**, 713–25.

Mesinger, F., Z. I. Janjic, S. Nickovic, D. Gavrilov and D. G. Deaven (1988). The step-mountain co-ordinate: model description and performance for cases of Alpine lee cyclogenesis and for a case of Appalachian redevelopment. *Mon. Weather Rev.*, **116**, 1493–518.

Molteni, F., R. Buizza, T. N. Palmer and T. Petroliagis (1996). The ECMWF Ensemble Prediction System: methodology and validation. *Quart. J. Roy. Meteor. Soc.*, **127**, 2069–94.

Molteni, F., R. Buizza, C. Marsigli, A. Montani, F. Nerozzi and T. Paccagnella (2001). A strategy for high-resolution ensemble prediction. I: Definition of representative members and global-model experiments. *Quart. J. Roy. Meteor. Soc.*, **127**, 2069–94.

Montani, A. (1998). Targeting of observations to improve forecasts of cyclogenesis. PhD thesis. University of Reading, Reading, UK.

Montani, A., C. Marsigli, F. Nerozzi, T. Paccagnella and R. Buizza (2001). Performance of ARPA-SMR Limited-area Ensemble Prediction System: two flood cases. *Nonlinear Proc. Geoph.*, **127**, 2095–115.

Montani, A., C. Marsigli, F. Nerozzi, T. Paccagnella, S. Tibaldi and R. Buizza (2003a). The Soverato flood in Southern Italy: performance of global and limited-area ensemble forecasts. *Nonlinear Proc. Geoph.*, **10**, 261–74.

Montani, A., M. Capaldo, D. Cesari, *et al.* (2003b). Operational limited-area ensemble forecasts based on the Lokal Model. *ECMWF Newsletter*, **98**, 2–7.

Mullen, S. L. and R. Buizza (2001). Quantitative precipitation forecast over the United States by the ECMWF Ensemble Prediction System. *Mon. Weather Rev.*, **129**, 638–63.

(2002). The impact of horizontal resolution and ensemble size on probabilistic forecasts of precipitation by the ECMWF Ensemble Prediction System. *Weather Forecast.*, **17**, 173–91.

Mylne, K. R., R. E. Evans and R. T. Clark (2002). Multi-model multi-analysis ensembles in a quasi-operational medium-range forecasting. *Quart. J. Roy. Meteor. Soc.*, **128**, 361–84.

Palmer, T. N. and S. Tibaldi (1988). On the prediction of forecast skill. *Mon. Weather Rev.*, **116**, 2453–80.

Richardson, D. S. (2000). Skill and relative economic value of the ECMWF ensemble prediction system. *Quart. J. Roy. Meteor. Soc.*, **126**, 649–67.

Tibaldi, S., T. Paccagnella, C. Marsigli, A. Montani and F. Nerozzi (2003). Short-to-medium-range Limited-area Ensemble Prediction: the LEPS system. *Proceedings of ECMWF Seminars on Predictability of Weather and Climate*, 9–13 September 2002, Reading, UK.

Tracton, M. S. and E. Kalnay (1993). Operational ensemble prediction at the National Meteorological Centre: practical aspects. *Weather Forecast.*, **8**, 379–98.

Tracton M. S., Du, J. and Juang, H. (1998). Short-range ensemble forecasting (SREF) at NCEP/EMC. *Proceedings of the 12th AMS/NWP Conference, Phoenix AZ*, pp. 269–72.

Wilks, D. S. (1995). *Statistical Methods in Atmospheric Sciences*. Academic Press.

19

Operational seasonal prediction

David L. T. Anderson

European Centre for Medium-Range Weather Forecasts, Reading

Representing the ECMWF Seasonal Forecasting Section, Magdalena Balmaseda, Laura Ferranti, Tim Stockdale, Alberto Troccoli, Kristian Mogensen, Arthur Vidard, Frederic Vitart

At ECMWF, a seasonal forecast system has been operating for several years. This system is described and some results presented. The forecasts are made by fully coupled atmosphere ocean models covering the globe. A multimodel forecast system is also well advanced. This approach avoids to some degree the tendency for individual models to be too confident in their predictions. Model error is still a major issue and considerable effort is needed to improve the models.

19.1 Introduction

For several years now ECMWF has been running, operationally, a seasonal forecast suite. This consists of an ocean data assimilation system to provide initial conditions for the forecast, a fully coupled ocean–atmosphere model to create the forecast ensemble and a post-processing procedure to generate forecast products. This system is being generalised to include other coupled models and to produce multimodel products. In this chapter we will consider the various components of the forecasting system.

Weather forecasts have a limited forecast range on account of the chaotic nature of the atmosphere (see Lorenz, this volume); depending on what variable one seeks to predict and on what scale, the predictability horizon might be roughly ten days.[1] Why then do we think we can predict climate months or even years ahead? The

Predictability of Weather and Climate, ed. Tim Palmer and Renate Hagedorn. Published by Cambridge University Press.

information on which the predictability of such long timescale processes is based cannot be simply atmospheric (Palmer and Anderson, 1994). The longer timescales come mainly from the ocean, which has a much larger heat capacity and slower dynamics than the atmosphere. Including ocean variability can give rise to enhanced atmospheric predictability if we are dealing with processes that depend on both media interacting. By contrast, it is quite possible to have some memory in the ocean and some predictability of ocean variability but with little or no associated atmospheric predictability if the ocean is not driving the atmosphere by the predictable part of its variability (Latif *et al.*, 2002, and Latif *et al.*, this volume). So including the ocean in the forecast system does not of necessity lead to enhanced atmospheric predictability. The ocean may have greater predictability than the atmosphere but unfortunately, in general, we are more interested in atmospheric prediction than ocean prediction. See Timmermann and Jin (this volume) for a discussion of coupled processes involving fast and slow systems relevant to ENSO.

The largest source of climate variability on interannual timescales is that associated with El Niño, or ENSO (El Niño–Southern Oscillation) as it is frequently now referred to. Originally the ocean variability was associated with El Niño and the atmospheric variability was associated with the Southern Oscillation. These are now known to relate to the same process though it took several decades to appreciate that this was the case and that they are manifestations of a coupled atmosphere–ocean process. El Niño is mainly located in the tropical Pacific but its reach extends to almost all parts of the globe. Of course in distant regions other processes may also be affecting climate variability and ENSO may not be the dominant process. ENSO involves a positive feedback between the sea surface temperature (SST) gradient along the equator and the winds blowing along the equator. It also involves ocean dynamics whereby information in the west equatorial Pacific can influence events in the east equatorial Pacific months later. To the extent that we know enough about the processes by which this information is propagated via equatorial Kelvin waves and how these come to later influence the atmosphere, one has a basis for prediction. The tropical Atlantic and Indian oceans may have zonal modes of this type too but they are less dominant than in the Pacific, less clearly identifiable against a relatively noisier background and of shorter duration. There is some indication that there may also be meridional modes of climate variability but again these have not been clearly identified. See McCreary and Anderson (1991), Neelin *et al.* (1998), Wang and Picaut (2004) and Philander (2004) for some review articles on ENSO, and the chapters by Shukla and Kinter, Krishnamurti *et al.*, Molteni *et al.*, and Palmer, this volume.

In middle latitudes the ocean seems to act much less favourably for seasonal forecasts. SST variability in middle latitudes does not influence the atmosphere as strongly as at the equator. In low latitudes, the atmosphere exhibits convection throughout the depth of the troposphere, the location and intensity of which is influenced by the SST. In most of what follows we will discuss primarily tropical processes associated with ENSO. The influence of El Niño and La Niña (when the east equatorial Pacific is cold) on the atmosphere can extend to higher latitudes and indeed to other oceans through the so-called atmospheric bridge.

The ECMWF seasonal forecasting strategy was to develop a single coupled atmosphere–ocean suite to predict simultaneously the SSTs and the atmospheric variability. This is the so-called one-tier approach. Others have tried a two-tier system, to first predict SST and then calculate the atmospheric response to these SSTs (Behringer *et al.*, 1998). While the two-tier approach may have some short-term advantages in that it avoids, to some degree, climate drift, it was felt unlikely to be a good long-term strategy. The strategy of two-tier forecasting is potentially flawed as it does not properly represent ocean–atmosphere interaction. In addition, the climate drift in a one-tier approach is not small compared with the signal one is trying to predict and this is likely to reduce the skill achievable. However, seasonal forecasting is a relatively new endeavour and experience will tell which approach is best.

The first coupled model forecast system at ECMWF was assembled in 1996 and real-time forecasts produced in 1997. This system was called System-1 (S1). For reasons that will become clear later, related to the handling of model error, these forecast systems are not changed frequently. In fact S1 was running until March 2003. Its successor, S2, has been running operationally since January 2002 (Anderson *et al.*, 2003). S3 is in development for implementation in 2006.

In Section 19.2, we will consider the ECMWF coupled system, concentrating on the ocean component, and in Section 19.3 we will present some results. The use of dynamical models for seasonal forecasting is now becoming more widespread, but before these were available, forecasts were made by statistical methods. In Section 19.4 dynamical model results are compared with operational statistical systems.

19.2 The ECMWF system

19.2.1 Introduction

In this section we will describe the current ECMWF seasonal forecast system. Other systems based on coupled models will differ in detail but in principle the ECMWF system is representative of the components needed. Since the information on the climate system on which a forecast is based lies in the upper ocean, it is necessary to know the state of the ocean at the start of the forecast quite well. For this purpose we have an ocean analysis system to use the available ocean observations.

19.2.2 The ocean observing system

Ocean observations are mainly of thermal data (T) and mainly in the upper 500 m. Some salinity (S) measurements are now available but only in usable numbers in the last few years. There are few measurements of velocity, except in the surface layer. Figure 19.1 shows the thermal data coverage in a typical 10-day window. Most

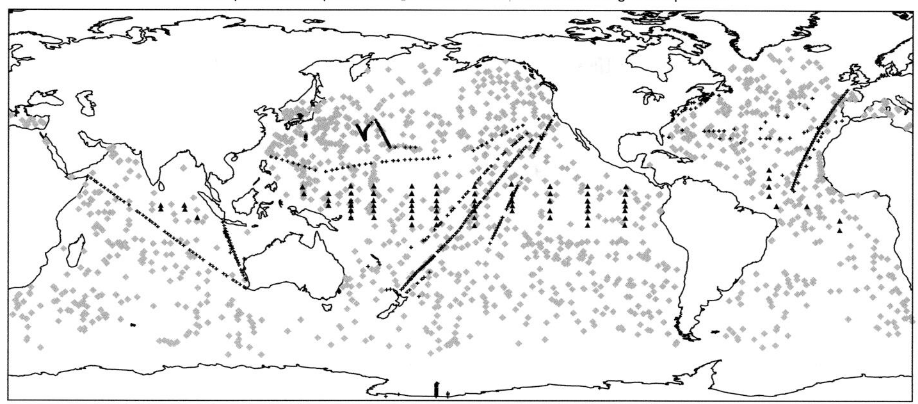

Figure 19.1 Data coverage for a 10-day period centred on 1 February 2005 showing the TRITON/TAO/PIRATA mooring array in the Pacific/Atlantic oceans (black triangles), the voluntary ship network (black crosses) and the ARGO floats (grey diamonds). The moorings report daily in the Pacific to the east of the Dateline and in the Atlantic. Hourly reports are received from moorings west of the Dateline.

observations are now received at operational centres such as ECMWF within a day or so of being taken. Although the coverage in Figure 19.1 looks reasonable, it is barely adequate. Ten or twenty years ago the coverage was much worse. It is thanks to the Tropical Ocean Global Atmosphere (TOGA) programme that the coverage is as good as it now is. Starting in 1985, TOGA steered oceanography to a free and open exchange of data, most of it in real time, and so usable in current operational forecast systems. See Anderson *et al.* (1998) for a comprehensive analysis of the TOGA experiment.

The array of triangles in the tropical Pacific and Atlantic oceans in Figure 19.1 is the TRITON/TAO/PIRATA. The straight or slightly curved lines of data are from merchant ships (Voluntary Observing Ships) making measurements of temperature to a depth of $\sim$500 m; a few measure salinity. The diamonds are ARGO float measurements from buoys which drift at $\sim$1000 m, but every 10 days pop up to the surface, measuring temperature and often salinity, and relay the information via satellite to a ground station for onward distribution.

When data are received at ECMWF, each individual observation is checked. It is compared against the model first guess and also with an analysis performed without the datum being checked. The actual quality control is quite complex and will not be covered in detail here. See Smith *et al.* (1991) for further discussion.

19.2.3 The ocean analysis system

The scheme currently in operational use is OI (Optimal Interpolation) using a time window of 10 days: all observations in the 10-day window are taken to apply at the

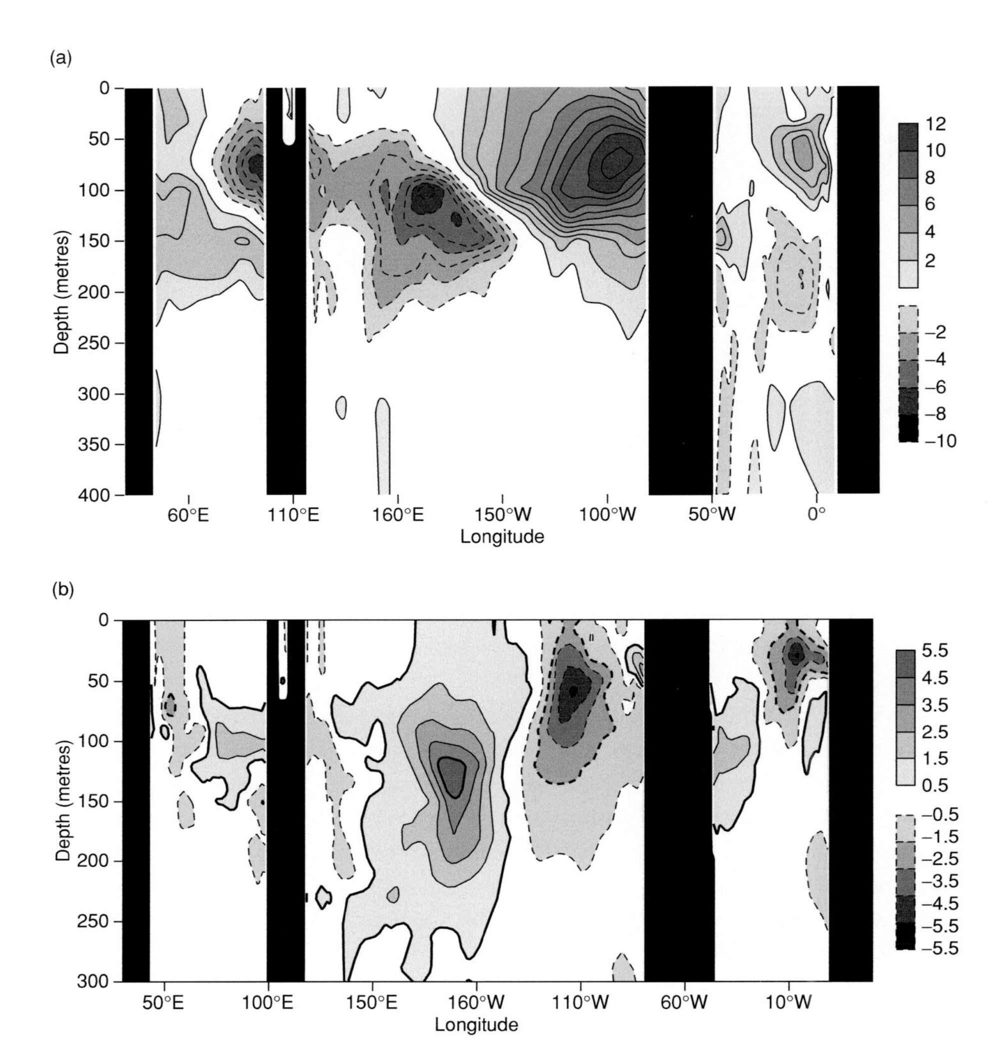

Figure 19.2 Vertical section along the equator of the temperature anomaly. The section on the left corresponds to the Indian, the middle section to the Pacific and the right section to the Atlantic. (a) The end of 1997. (b) The end of June 2004. Solid contours indicate that the water is anomalously warm, dashed contours that it is cold. The contour interval is 1 K. From the ECMWF operational ocean analysis system.

centre of the window. Although only T is analysed directly, velocity and salinity are corrected. Burgers *et al.* (2002) and Balmaseda (2003) describe the velocity correction. Salinity is corrected by applying an S correction such that the T–S relationship is preserved below the surface mixed layer. In the ocean when T changes, so generally does S in such a way as to preserve the T–S relationship. Further details are given in Troccoli *et al.* (2002).

A novel feature of the ocean analysis system is that not just a single analysis but multiple analyses are performed. The purpose of the analysis is to provide initial conditions for monthly and seasonal forecasts. Such forecasts must be probabilistic. This implies that an ensemble of forecasts must be made. In our case the ensemble size is 40. The ensemble is there to sample uncertainty arising from the chaotic nature of the atmosphere. However, it should also take into account uncertainty in the ocean initial conditions. One method of representing this uncertainty is through running an ensemble of ocean analyses. In our case the size of the ocean analysis ensemble is five. This ensemble of analyses is not an ensemble Kalman filter (EnKF); the ensemble size is too small to estimate analysis error covariances. Hamill (this volume) discusses EnKF methods. See also Palmer (2000) for an excellent discussion of many aspects of ensemble generation.

Almost all the information on which an atmospheric analysis is based comes from observations of the atmosphere. In the case of the ocean, a substantial amount of information on the ocean state can be obtained not through ocean observations but through atmospheric observations – in fact all the observations that are involved directly or indirectly in defining the surface wind, heat and freshwater fluxes. For seasonal forecasting the most important of the surface forcings is the wind, but it has uncertainty. We estimate this uncertainty and then force an ensemble of five ocean analyses with wind fields that are perturbed commensurate with the estimated uncertainty. In addition, the sea surface temperature (SST) field is not known sufficiently accurately. So perturbations to SST are also applied. This is discussed more fully in Vialard *et al.* (2005), who show the spread generated by different ensemble generation strategies.

Figure 19.2 shows sections along the equator from the surface down to 300 m, the part of the ocean most active in El Niño. Two times are shown: the end of December 1997 at the peak of El Niño and the end of June 2004 preceding the weak El Niño of 2004. Panel (a) shows the huge mass of anomalously warm water in the central east Pacific during the height of the 1997 El Niño. It also shows anomalously cold water in the west Pacific. This cold water had no surface expression but over the next few months it travelled eastwards, eventually surfacing and bringing an end to the El Niño. Figure 19.3 (colour plate) shows the forecasts made at the time, which captured the peak and decline of El Niño reasonably well.

Figure 19.2(b) shows quite a lot of anomalously warm water in the central Pacific in June 2004. The central east Pacific is cool. The big question at that time was, ‘Is there another big El Niño on its way?’ The forecasts are shown in Figure 19.4(b). The subsequently observed SST is also plotted, showing that indeed these forecasts verified very well: there would be warming over the next six months, but not to very large amplitude. This was even more surprising since there were several intraseasonal oscillations. These impose a signal in the ocean, which travels eastwards along the equator and sometimes leads to a rapid growth in El Niño – though not on this occasion. The ocean signal of the intraseasonal oscillations can be clearly seen in Figure 19.4(a), as well as the eastward propagation. The presence of ISOs can really confuse the prediction of El Niño: the temptation is to think that because there is a significant anomaly in the ocean which will travel eastwards, it will lead to warming

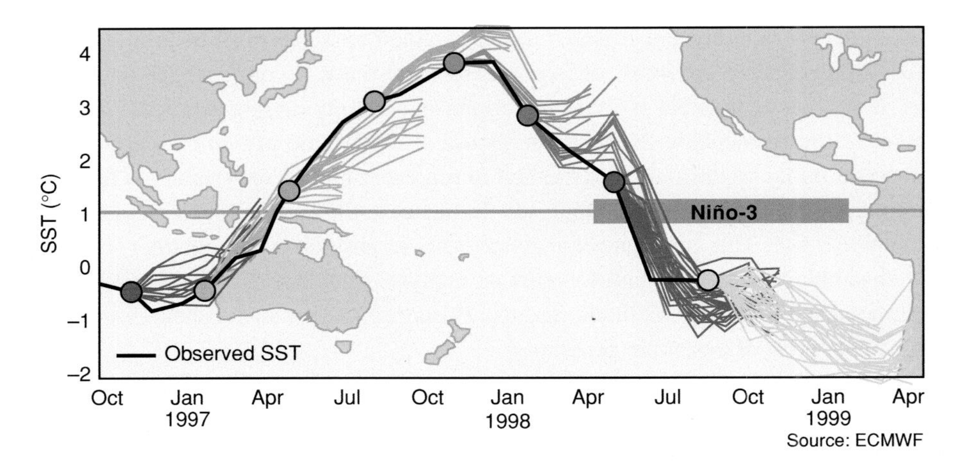

Figure 19.3 (See also colour plate section.) Plot of forecasts of Niño-3 for various start times throughout the large 1997–8 El Niño. Niño-3 is the region in the equatorial Pacific bounded between 5S to 5N and 150W to 90W. Different lines of the same colour indicate different ensemble members. The background indicates the location of Niño-3. This plot was produced by CLIVAR based on data from ECMWF.

SSTs in the central east equatorial Pacific. Frequently the signal propagates eastwards but dies as it approaches the eastern Pacific, with little impact on SSTs. See Lengaigne *et al.* (2004) for further discussion.

A set of experiments to show the impact of assimilating ocean data on forecast quality was performed by Alves *et al.* (2004). Four sets of ocean analyses were created and four ensembles of forecasts were run from these analyses. Two different wind products were used and for each, experiments with and without data assimilation were performed. The forecasts based on analyses with data assimilation were clearly better than those without: the root-mean-square (rms) errors were smaller and the anomaly correlations were higher at all forecast lead times. Any results probably apply only to the system being tested, as improvements in either the ocean model or the forcing fields through improved atmospheric analyses could change the results. As wind fields get better, then analyses without data will also improve, narrowing the gap between the forecasts initialised from analyses performed with and without ocean data assimilation.

19.3 Some results

19.3.1 Introductory remarks

As mentioned in the introduction, in a fully coupled atmosphere–ocean model, as used at ECMWF, there is nothing to constrain the model climate to that of nature. Climate drift can and does occur. In fact drift occurs in the atmospheric model and leads to erroneous land temperatures. In the coupled model the SST can also drift.

(a)

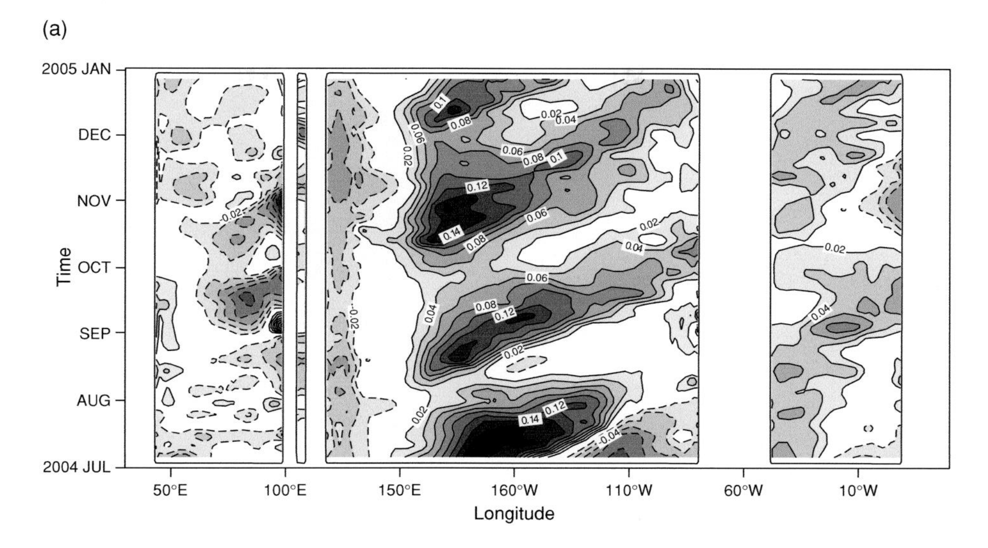

(b)

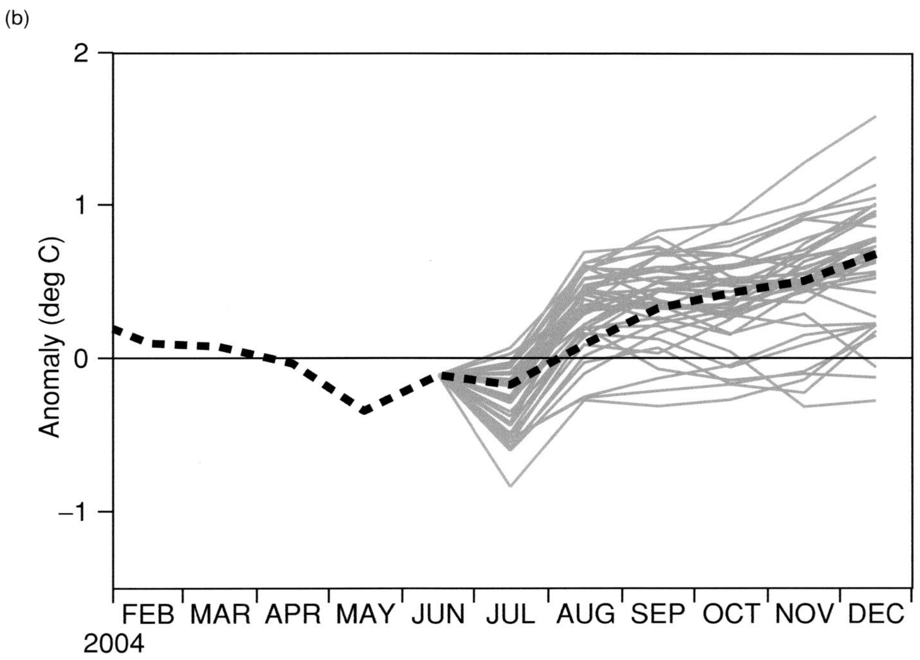

Figure 19.4 (a) A plot of the sea level along the equator as a function of time showing the signals imposed on the ocean by the intraseasonal oscillation (MJO) and the eastward propagation of these signals. The contour interval is 2 cm, with the zero contour not plotted. (b) Plot of forecasts of SST in the Niño-3 region started from 1 July 2004. The thin curves give the individual ensemble members. The dashed curve shows the subsequently observed SST.

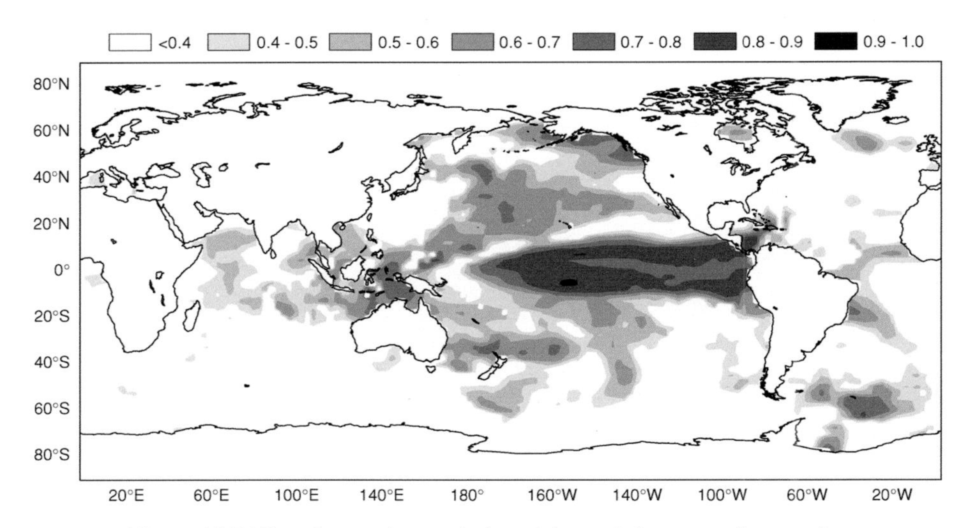

Figure 19.5 Plot of anomaly correlation of 6-month forecasts of sea surface temperature from S2. Dark shading, mainly in the equatorial Pacific, indicates high correlations (high skill). Skill in the Atlantic is considerably lower at this time range (the latter half of a 6-month forecast).

As the drift is not small compared with the signal one is trying to predict, a strategy to deal with it is essential. That used at ECMWF is to calculate the model climate as a function of lead time from an extensive set of integrations of the model spanning many past years. The difference between the actual forecast and the model climate is then a measure of the predicted anomaly. This is discussed more fully in Stockdale (1997), Stockdale *et al.* (1998) and van Oldenborgh *et al.* (2005).

The El Niño of 1997/8 was the most intense in the last 100 years, apart possibly from 1982/3. Predictions of El Niño developing and of its magnitude were generally not good (Landsea and Knaff, 2000). The forecasts by the ECMWF S1 were, by contrast, quite accurate, though by no means perfect. In particular the forecasts from around spring 1997 underpredicted the amplitude (Vitart *et al.*, 2003; Anderson *et al.*, 2003). The predictability of the 1997 El Niño and the role which the February MJO (Madden Julian Oscillation or intraseasonal oscillation) played is still under debate (Kessler and Kleeman, 2000).

The overall skill of a forecast system can be indicated from the anomaly correlation. Figure 19.5 shows the anomaly correlation of sea surface temperature from S2 illustrating that the skill in the tropical Pacific is much higher than in the Atlantic and Indian oceans and that skill falls off rapidly with latitude. The overall skill is very similar to that of S1 (not shown here but see Anderson *et al.*, 2003).

The rms error of forecasts for the Niño-3 region is shown in Figure 19.6. The skill of the system beats persistence at all lead times. This is true for all start months although you cannot see this from the figure. The correlation of predicted SST with observed SST is also higher than that from persistence. So, in that sense coupled

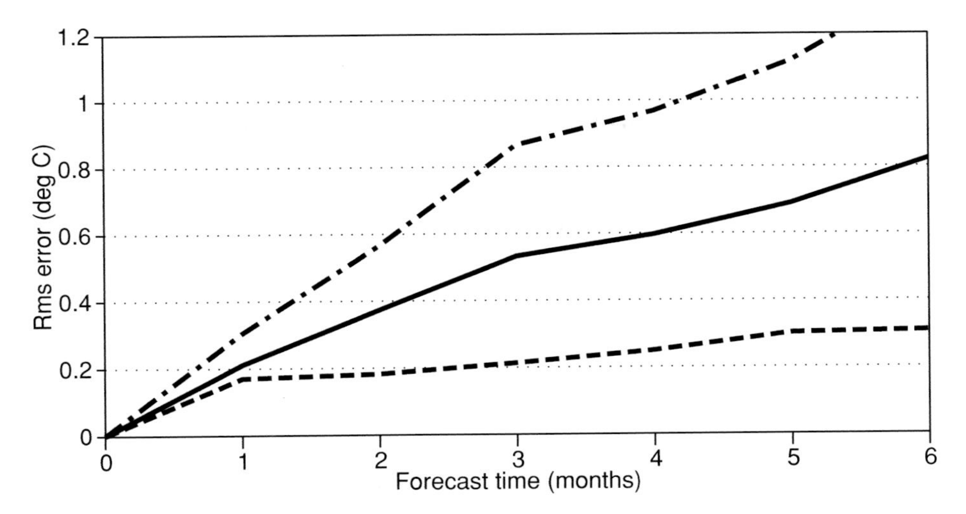

Figure 19.6 The growth of error in the forecasts of NINO-3 (solid curve) together with the ensemble spread (dashed). The fact that the latter is less than the former indicates that the model is too confident. The two curves should be close in a well-balanced system. The dot-dashed curve indicates the skill of persistence and shows that the model easily beats persistence at all lead times.

model forecasts are good. Figure 19.6 also shows the growth of the spread in the ensemble (dashed curve). This grows less fast than the error. One can interpret this result in two ways. The negative interpretation says that the spread is smaller than the error and therefore the forecast system is poorly calibrated: the model forecasts are too confident when in fact the observed SST frequently lies outside the range spanned by the forecast ensemble.

An alternative, more optimistic, interpretation is to take the model estimate of spread as a measure of potential predictability by interpreting one ensemble member as truth and measuring the differences of other members from that. This then gives the potential limit of predictability in the absence of model error. Our system is far from that limit. So by working harder and reducing model error we should (hopefully) be able to improve the forecasts. Of course the current model might underestimate the limits to predictability since the model does not do a good job of reproducing intraseasonal variability (Madden Julian Oscillation) which, it is thought, might play a role in limiting predictability of ENSO (Waliser, this volume). However, even if the optimistic interpretation of the limit of predictability were correct, the reality is that we are not there. We have to work with the practical reality that for now our model is not well calibrated. This limitation is not specific to ECMWF but applies to other models as well.

One way of improving the forecast reliability is to sample model error in the ensemble probability distribution function (pdf) and one way to do that is to develop a multimodel approach. This has already been done in the context of DEMETER in a non-real-time mode (see Hagedorn *et al.*, this volume). At ECMWF we are in the process of developing a real-time operational multimodel forecast system. This currently consists of forecasts from the Met Office and Météo France as well as ECMWF. The amplitude of the wind perturbations used in the Met Office system is half that of ECMWF as they draw less strongly to the data in the ocean analysis and their coupled model is more active – perhaps too active – whereas the ECMWF coupled model is not active enough (Anderson *et al.*, 2003).

In order to get some feel for the potential improvement in forecast skill as a result of the multimodel approach, we plot in Figure 19.7 the rms error for the Niño-3.4 region for the three models that will participate in real-time multimodel predictions at ECMWF. The models are actually from DEMETER and as such are earlier versions than will be used in real-time operational applications but they should give a fairly good assessment of what to expect. The error growth is shown for two regions, one in the Pacific and one in the Atlantic. Both are equatorial. Consistent with Figure 19.5, the skill in the Atlantic is lower than in the Pacific: actually the error growth is similar but the size of the interannual signal is smaller in the Atlantic, so the error is more serious. This can also be seen in the anomaly correlation (not shown) which drops more rapidly in the Atlantic than the Pacific.

A selection of graphical products from the seasonal forecast system is displayed on the ECMWF website. Spatial maps of 2-metre temperature, precipitation and mean sea level pressure are shown, in the form of probabilities for tercile and 15 percentile categories as well as the ensemble mean anomaly and the probability of exceeding the climate median. The Niño SST indices include the Niño-3.4 and Niño-4 regions as well as Niño-3, and the ocean analysis plots include several meridional sections, as well as zonal and horizontal maps. Website addresses are given at the end of this chapter.

19.3.2 Further verification

For a correct interpretation of seasonal predictions the user needs to complement the forecast products with knowledge of the forecast skill. It is not possible in this chapter to show all the verification that has been done but an extensive assessment is available on the ECMWF website[2]. Estimates of model bias for a wide range of variables, including zonal averages, time series of a set of indices of SST and large-scale patterns of variability such as the Southern Oscillation Index (SOI), the Pacific North American Pattern (PNA) and the North Atlantic Oscillation (NAO) are available. A suite of verification scores for deterministic (e.g. spatial anomaly correlation and mean square skill score error (MSSE)) and probabilistic forecasts can be viewed for the operational system (S2).

The robustness of verification statistics is always a function of the sample size. For the operational seasonal forecast system, the sample size of 15 years is considered barely sufficient. Verification is performed in cross-validation mode (Michaelson,

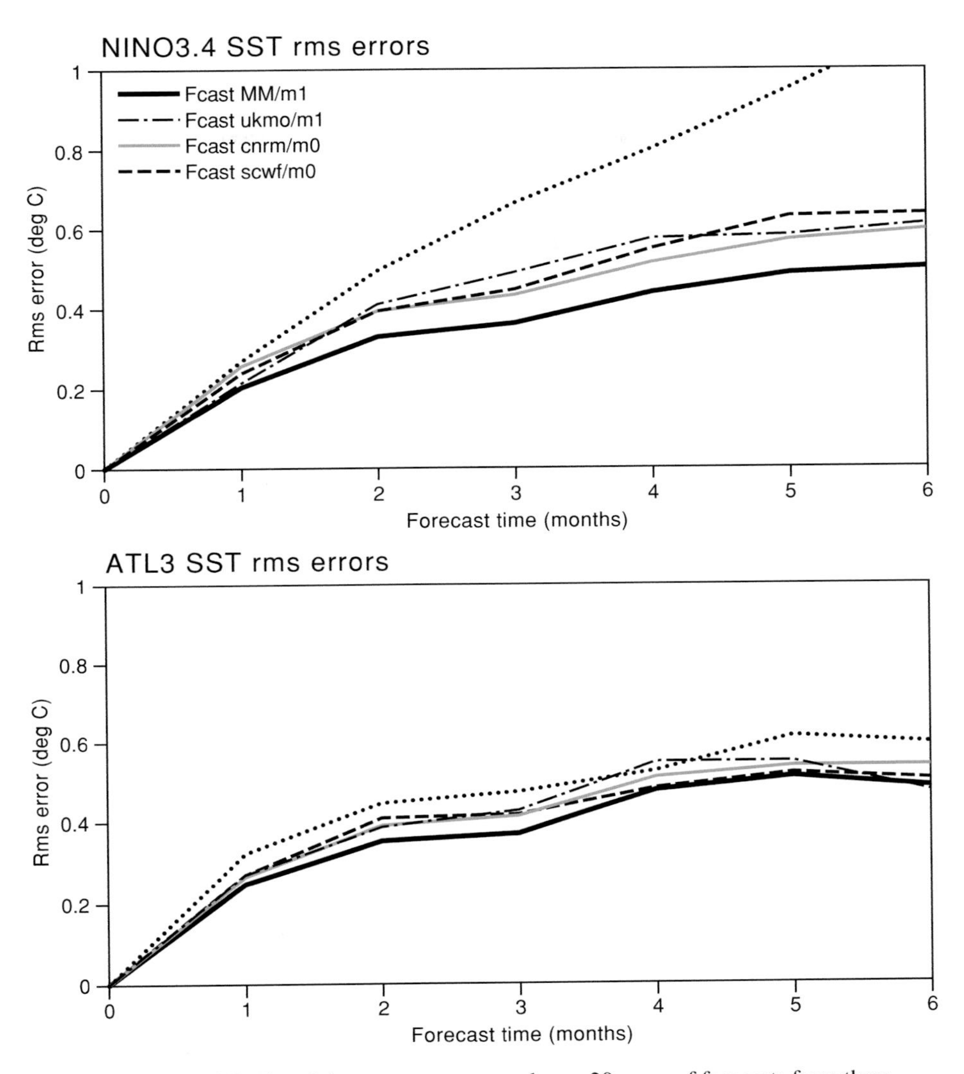

Figure 19.7 Plot of the rms error averaged over 20 years of forecasts from three models (ECMWF, Met Office and Météo France) as a function of lead time. The heavy solid line indicates the multimodel and the dotted line indicates persistence. Two regions are shown: Niño-3.4 in the central east equatorial Pacific where all forecasts significantly beat persistence, and ATL3 in the central-east equatorial Atlantic where the skill of the forecasts relative to persistence is modest. In both cases, however, the multimodel seems the best.

1987) using the whole set of forecast data available, i.e. both hindcasts and real-time forecasts. The seasonal forecast skill depends strongly on the season; so forecasts are evaluated separately for different starting months. Rainfall is a very important variable, but it is difficult to verify as observations are difficult to make. Global Precipitation Climatology Project (GPCP) data are used to verify precipitation. Other atmospheric parameters are verified against the ERA-40 reanalysis (Uppala *et al.*, 2005).

19.3.3 Single-value forecast scores

The most usual way to summarise the information from an ensemble of forecasts into one value is to use the grand mean of all members. Although such values are often referred to as deterministic forecasts, the grand mean acts to remove random, unpredictable errors from the forecast. As a consequence, ensemble-mean charts are unrealistic, if compared with analyses, as many small-scale features are smoothed out, but they are usually the best single-value estimates, if evaluated using root mean square error measures. For such forecasts, verification of anomaly correlation based on all the years available and mean square skill score (MSSS) are provided both as global maps and averaged over predefined areas (Murphy, 1988).

19.3.4 Probabilistic scores

The full content of the information provided by the seasonal ensemble forecasts is only accessible in multivalued, probabilistic mode. Basic methods for verifying probabilistic forecasts have been in use for several years at ECMWF for medium-range EPS products and the methodology is now being naturally extended to seasonal forecasts. The relative operating characteristics (ROC) curve in Figure 19.8 shows, for a range of different probability thresholds, hit rates versus false-alarm rates of forecasts of a particular event in different regions. Values lying above the diagonal indicate the forecasts have skill; one would like the area above the diagonal to be as large as possible. (See Buizza, this volume, including appendices.) The event thresholds are defined with respect to terciles from model and verifying climatologies. Figure 19.8(a) shows the ROC diagrams for 2-metre temperature summer forecasts for the northern hemisphere. This figure indicates that on average there is skill in predictions several months ahead for the extratropics. For some regions such as Europe forecast skill can be limited, however. Figure 19.8(b) is a ROC curve for the prediction of tropical rainfall anomalies for June to August, indicating that there is skill in the forecast of rainfall anomalies in the tropics, even though rainfall is a very chaotic variable. Grid point values of ROC scores and ranked probability skill scores are available on the Web as global maps.

19.4 Statistical prediction

In principle, numerical models that represent the dynamics of the atmosphere, ocean and land should be able to give better seasonal forecasts than purely statistical approaches, because of their ability to handle a wide range of linear and non-linear

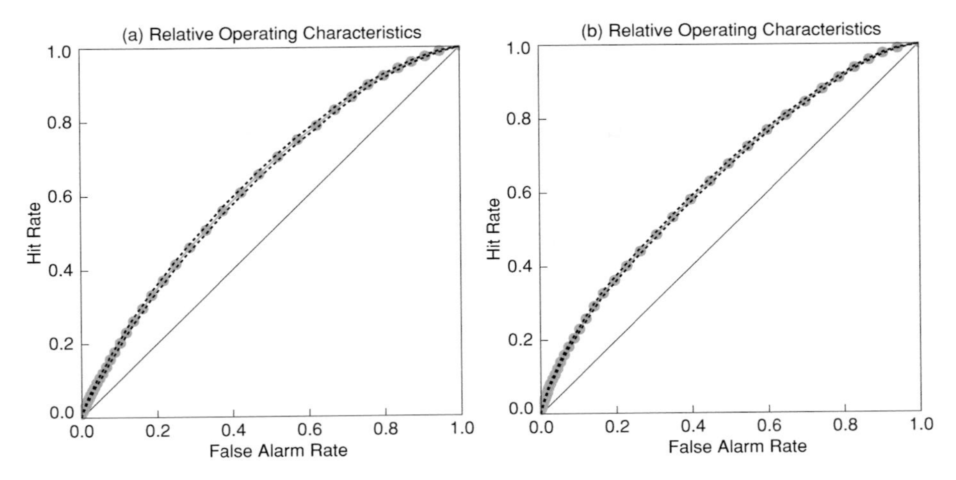

Figure 19.8 (a) ROC diagrams for 2 meter temperature anomaly being in the upper tercile. The forecasts are for a lead time 2–4 month forecasts (i.e. JJA for forecasts starting in May) for the period 1987–2002. Results for the northern extratropics. (b) As for (a) but for tropical precipitation.

interactions and their potential ability to do so in a changing climate. In practice, model deficiencies are still a substantial source of error (Latif *et al.*, 2001; Palmer *et al.*, 2004). If the climate record were sufficiently long, then statistical methods should be able to do a reasonable job. In practice, the climate record on which to train statistical models is short and this also limits their applicability.

Van Oldenborgh *et al.* (2005) sought to determine to what extent the present generation of numerical forecast models is able to challenge existing operational empirical methods for seasonal forecasting. They compared the seasonal forecasting performance of S1 and S2 with the Markov model by Xue *et al.* (2000) and the Constructed Analogue (CA) model by van den Dool (1994) and van den Dool and Barnston (1994), which are in operational use at the National Centers for Environmental Prediction, USA.

Both dynamical models are better at forecasting the Niño-3 index than the statistical models, including some other non-operational statistical models they included in their evaluation. Figure 19.9 intercompares the seasonal dependence of the skill of the different models. Plotted is the correlation coefficient of the monthly indices at lead time +3 as a function of the *target* season. The anomaly correlation coefficient is used as the skill measure, since it is not affected by the biases in the mean state and amplitude of the variability. The first-order forecast, the ensemble mean, is used. Figure 19.9 shows that the statistical models have a marked seasonality in skill, which is much less evident in the dynamical models. The statistical models show the spring persistence barrier as the low predictability of June–August Niño-3 (Webster, 1995; Balmaseda *et al.*, 1995). Overall, the coupled models are better able to predict El Niño than statistical models.

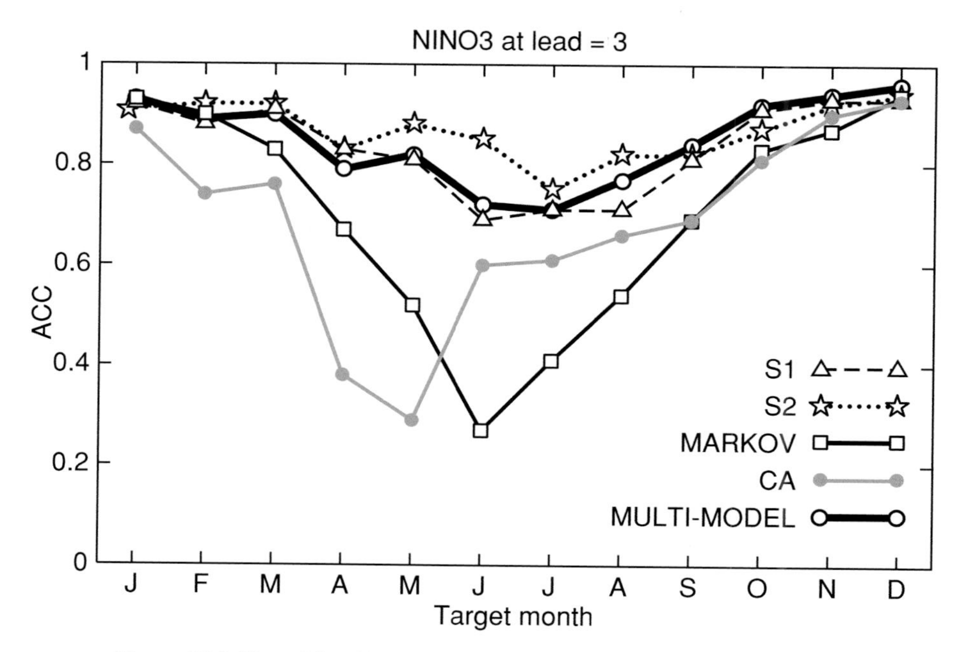

Figure 19.9 Plot of the skill in predicting NINO-3 SSTs in the ECMWF coupled models S1 and S2, the Markov model, and the Constructed Analogue model (CA). The multimodel consisting of S1, S2, Markov and CA is also plotted.

The relatively small seasonality in the skill of the dynamical models is probably in part due to the assimilation and propagation of subsurface oceanic information: a Kelvin wave takes about two months to cross the Pacific Ocean, and slower oceanic processes give some skill beyond that time. A statistical model using subsurface information also has better skill in crossing the spring barrier (Balmaseda *et al.*, 1995; Xue *et al.*, 2000; McPhaden, 2003; Balmaseda, 2003).

A multimodel average could combine the strong points of the dynamical and statistical models. Figure 19.9 shows that the multimodel is slightly better than the dynamical models in predicting winter values of Niño-3 but gives worse forecasts in summer. This is related to the poor skill of the statistical models in crossing the predictability barrier.

19.5 Conclusions

The seasonal forecasting system at ECMWF has been described briefly. Its ability to forecast El Niño-type SST variability is useful and well established internationally, although the forecasts are not yet completely reliable. Based on a limited sample of ~15 years, the statistics suggest that there are many areas and parameters for which the atmospheric forecasts have skill, but the results are geographically variable and subject to sampling error. Model error is a serious source of forecast error but this can be partly addressed by the use of several models. Work is well advanced to include the Met Office and Météo France models as part of the seasonal forecasting system and hopefully to include other models later. There is much work still to be done to improve model forecasts, and to improve the representation of forecast uncertainties.

The skill of the ECMWF seasonal forecast models has been compared with that of a set of statistical forecast models used operationally for predicting ENSO; based on anomaly correlation, the physically based models are better than operational statistical models. The move towards multimodel forecasting provides a context for producing more accurate and reliable forecasts than individual models. No multimodel product combining all the existing models yet exists but such an idea is being discussed.

Acknowledgements

The assistance of the whole of the Seasonal Forecasting Section at ECMWF is much appreciated. Thanks also to G.-J. van Oldenborgh for assistance in preparing Figure 19.9.

Notes

1. There are atmospheric phenomena with a longer timescale such as the intraseasonal oscillation (~40–60) days and some aspects of these may have predictability beyond ten days. (See Waliser, this volume.)
2. The website at www.ecmwf.int/products/forecasts/d/charts/seasonal/ verification provides a comprehensive documentation of skill levels, using methods that have been agreed at the international (WMO) level for the evaluation of long-range forecast systems.

   Spatial distribution of the mean errors (biases) are provided at www.ecmwf.int/products/forecasts/d/charts/seasonal/verification/bias/.

References

Alves, O., M. Balmaseda, D. L. T. Anderson and T. Stockdale (2004). Sensitivity of dynamical seasonal forecasts to ocean initial conditions. *Quart. J. Roy. Meteor. Soc.*, **130**, 647–68.

Anderson, D. L. T., E. Sarachik, P. Webster and L. Rothstein (eds) (1998). The TOGA DECADE: reviewing the progress of El Niño research and Prediction. *J. Geophys. Res.*, **103**, 14167–510.

Anderson, D., T. Stockdale, M. Balmaseda, *et al.* (2003). *Comparison of the ECMWF seasonal forecast Systems 1 and 2, including the relative performance for the 1997/8 El Niño*. ECMWF Technical Memorandum 404. Also available online at www.ecmwf.int.

Balmaseda, M. A. (2003). Ocean data assimilation for seasonal forecasts. In '*Recent developments in data assimilation for atmosphere and ocean*' *Seminar Proceedings*, September 2003, pp. 301–25. ECMWF. Also available online at www.ecmwf.int.

Balmaseda, M. A., M. K. Davey and D. L. T. Anderson (1995). Seasonal dependence of ENSO prediction skill. *J. Climate*, **8**, 2705–15.

Behringer, D., Ming Ji and A. Leetmaa (1998). An improved coupled model for ENSO prediction and implications for ocean initialisation. 1: The ocean data assimilation. *Mon. Weather Rev.*, **126**, 1013–21.

Burgers, G., M. Balmaseda, F. Vossepoel, G. J. van Oldenborgh and P. J. van Leeuwen (2002). Balanced ocean data assimilation near the equator. *J. Phys. Oceanogr.*, **32**, 2509–19.

Kessler, W. S. and R. Kleeman (2000). Rectification of the Madden-Julian oscillation into the ENSO cycle. *Mon. Weather Rev.*, **13**, 3560–75.

Landsea, C. W. and J. A. Knaff (2000). How much skill was there in predicting the very strong 1997–8 El Niño? *Bull. Am. Meteorol. Soc.*, **81**, 2107–19.

Latif, M. *et al.* (2001). ENSIP: the El Niño simulation intercomparison project. *Clim. Dynam.*, **18**, 255–76.

Latif, M., A. Timmermann, A. Grötzner, C. Eckert and R. Voss (2002). On North Atlantic interdecadal variability: a stochastic view. In '*Ocean Forecasting*', ed. N. Pinardi and J. Woods, pp. 149–78. Springer Verlag.

Lengaigne, M., J.-P. Boulanger, C. Menkes, E. Guilyardi, P. Delecluse and J. Slingo (2004). Westerly wind events and their influence on the coupled atmosphere ocean system: a review. In *Earth's Climate*, ed. C. Wang, S.-P. Xie and J. A. Carton, pp. 49–70. Geophysical Monograph Series 147. American Geophysical Union.

McCreary, J. P. and D. L. T. Anderson (1991). An overview of coupled ocean–atmosphere models of El Niño and the Southern Oscillation. *J. Geophys. Res.*, **96**, 3125–50.

McPhaden, M., 2003. Tropical Pacific Ocean heat content variations and ENSO persistence barriers. *Geophys. Res. Lett.*, **30**, 1480–3.

Michaelson, J. (1987). Cross-validation in statistical climate forecast models. *J. Clim. Appl. Meteorol.*, **26**, 1589–1600.

Murphy, A. H. (1988). Skill scores based on the mean square error and their relationship to the correlation coefficient. *Mon. Weather Rev.*, **116**, 2417–24.

Neelin, D. *et al.* (1998). ENSO theory. *J. Geophys. Res.*, **103**, C7, 14261–90.

Palmer, T. N. (2000). Predicting uncertainty in forecasts of weather and climate. *Rep. Prog. Phys.*, **63**, 71–116.

Palmer, T. N. and D. L. T. Anderson (1994). Prospects for seasonal forecasting. *Quart. J. Roy. Meteor. Soc.*, **120**, 755–94.

Palmer, T. N. *et al.* (2004). Development of a European multi-model ensemble system for seasonal to interannual prediction. *Bull. Am. Meteorol. Soc.*, **85**, 853–72.

Philander, S. G. (2004). *Our Affair with El Niño*. Princeton University Press.

Smith, N. R., J. E. Blomley and G. Meyers (1991). A univariate statistical interpolation scheme for subsurface thermal analyses in the tropical oceans. *Prog. Oceanogr.*, **28**, 219–56.

Stockdale, T. N. (1997). Coupled atmospehere-ocean forecasts in the presence of climate drift. *Mon. Weather Rev.*, **125**, 809–18.

Stockdale, T. N., D. L. T. Anderson, J. O. S. Alves and M. A. Balmaseda (1998). Global seasonal rainfall forecasts using a coupled ocean-atmosphere model. *Nature*, **392**, 370–3.

Troccoli, A., M. Balmaseda, J. Segschneider, *et al.* (2002). Salinity adjustments in the presence of temperature data assimilation. *Mon. Wea. Rev.*, **130**, 89–102. Also available online as ECMWF Technical Memorandum 305, at www.ecmwf.int.

Uppala S. M., P. Kallberg, A. Simmons and 43 others (2005). The ERA-40 reanalysis. *Quart. J. Roy. Meteor. Soc.*, Submitted. See also ECMWF Newsletter 101 (online) at www.ecmwf.int/publications/newsletters/

van Oldenborgh, G. J., M. Balmaseda, L. Ferranti, T. N. Stockdale and D. L. T. Anderson (2005). Did the ECMWF Seasonal forecast model outperform statistical ENSO forecast models over the last 15 years? *J. Climate*, **18**, 3240–9.

van den Dool, H. M. (1994). Searching for analogues: how long must one wait? *Tellus*, **46A**, 314–24.

van den Dool, H. M. and A. G. Barnston (1994). Forecasts of global sea surface temperature out to a year using the constructed analogue method. In *Proceedings of the 19th Climate Diagnostics Workshop*, College Park, MD, pp. 416–19.

Vialard J., F. Vitart, M. Balmaseda, T. Stockdale and D. Anderson (2005). An ensemble generation method for seasonal forecasting with an ocean-atmosphere coupled model. *Mon. Weather Rev.*, **133**, 441–53.

Vitart F., M. Balmaseda, L. Ferranti and D. Anderson (2003). Westerly wind events and the 1997/98 El Niño in the ECMWF Seasonal forecasting system. *J. Climate*, **16**, 3153–70.

Wang C. and J. Picaut (2004). Understanding ENSO physics: a review. In *Earth's Climate*, pp. 21–48. Geophysical Monograph Series 147. American Geophysical Union.

Webster, P. J. (1995). The annual cycle and the predictability of the tropical coupled ocean-atmosphere system. *Meteorol. Atmos. Phys.*, **56**, 33–55.

Xue, Y., A. Leetmaa and M. Ji (2000). ENSO prediction with Markov models: the impact of sea-level. *J. Climate*, **13**, 849–71.

20

Weather and seasonal climate forecasts using the superensemble approach

T. N. Krishnamurti, T. S. V. Vijaya Kumar, Won-Tae Yun, Arun Chakraborty, Lydia Stefanova

Department of Meteorology, Florida State University, Tallahassee

In this chapter we present a short overview of the Florida State University (FSU) superensemble methodology for weather and seasonal climate forecasts and cite some examples on application for hurricanes, numerical weather prediction (NWP) and seasonal climate forecasts. This is a very powerful method for producing a consensus forecast from a suite of multimodels and the use of statistical algorithms. The message conveyed here is that the superensemble reduces the errors considerably compared with those of the member models and of the ensemble mean. This is based on results from several recent publications, where varieties of skill scores such as anomaly correlation, root-mean-square (rms) errors and threat scores have been examined. The improvements in several categories such as seasonal climate prediction from coupled atmosphere–ocean multimodels and NWP forecasts for precipitation exceed those of the best models in a consistent manner and are more accurate compared with the ensemble mean. It is difficult to state, soon after a forecast is made, as to which among the member models would have the highest skill. The superensemble is very consistent in this regard and is thus more reliable. In this study, we show walk-through tables that illustrate the workings of the superensemble for a hurricane track and heavy rain forecast for a flooding event. A number of features of the superensemble – number of training days, behaviour as the number of models increased, reduction of systematic errors and use of a synthetic superensemble – illustrate the strength of this new forecast experience.

Predictability of Weather and Climate, ed. Tim Palmer and Renate Hagedorn. Published by Cambridge University Press.

20.1 Introduction

Ensemble averaging from either multimodel or even from several runs of the same model has been an active area of research and operation in global modelling (Baumhefner, 1968; Palmer *et al.*, 1992; Mullen and Baumhefner, 1994; Kalnay *et al.*, 1996; Buizza, this volume). Invariably the clear message emerges from these studies that an ensemble mean has somewhat higher skill than the member models. The notion of a multimodel superensemble was first proposed some five years ago (Krishnamurti *et al.*, 1999). Here the skills of recent past performances of a number of models were assessed in order to define a training phase of a multimodel superensemble. This was carried out via a vast array of statistics, which was later carried over to a forecast phase. Several major refinements of this simple methodology have been brought out by forecasts of hurricanes/typhoons (track, intensity, and timing), precipitation and floods, global NWP and seasonal climate. It was necessary to collect the ongoing forecasts from a diverse group of operational and research modellers in order to evaluate the skill of superensemble methodology in the above areas of interest. A series of papers dwell on the performance of the FSU superensemble in each of the above areas: Krishnamurti *et al.* (1999, 2000a, 2000b, 2001, 2002, 2003a, 2003b), Stefanova and Krishnamurti (2002), Williford *et al.* (2003), Kumar *et al.* (2003), Yun *et al.* (2003) and Ross and Krishnamurti (2005). The most recent advancement has come from the design of what we call a synthetic multimodel superensemble for seasonal climate forecasts (Yun *et al.*, 2005). This chapter is intended to provide an introduction to this powerful methodology that carries a higher skill than participating models and their ensemble mean.

20.2 Superensemble methodology

Figure 20.1 illustrates the conventional procedure that is being used for the construction of a multimodel superensemble for global NWP. This includes the training phase to the left of time 0, where the forecasts from multimodels are regressed against the observed (assimilated) estimates. This being done for $\boldsymbol{m}$ multimodels at $\boldsymbol{n}$ grid points (along the horizontal and vertical) for $\boldsymbol{p}$ variables and $\boldsymbol{q}$ time intervals constituted as many as $\boldsymbol{m*n*p*q}$ statistical coefficients (which came to around 10^7 for a reasonably high resolution global model). This degree of detail for the construction of the superensemble was found necessary. The superensemble grid was at a common denominator resolution (T126 – triangular truncation at 126 waves, corresponding to approximately 110 km latitude/longitude transform grid resolution) interpolated from the participating member model forecasts. For NWP we used as many as 11 global models (Krishnamurti *et al.*, 2001). The methodology for this conventional

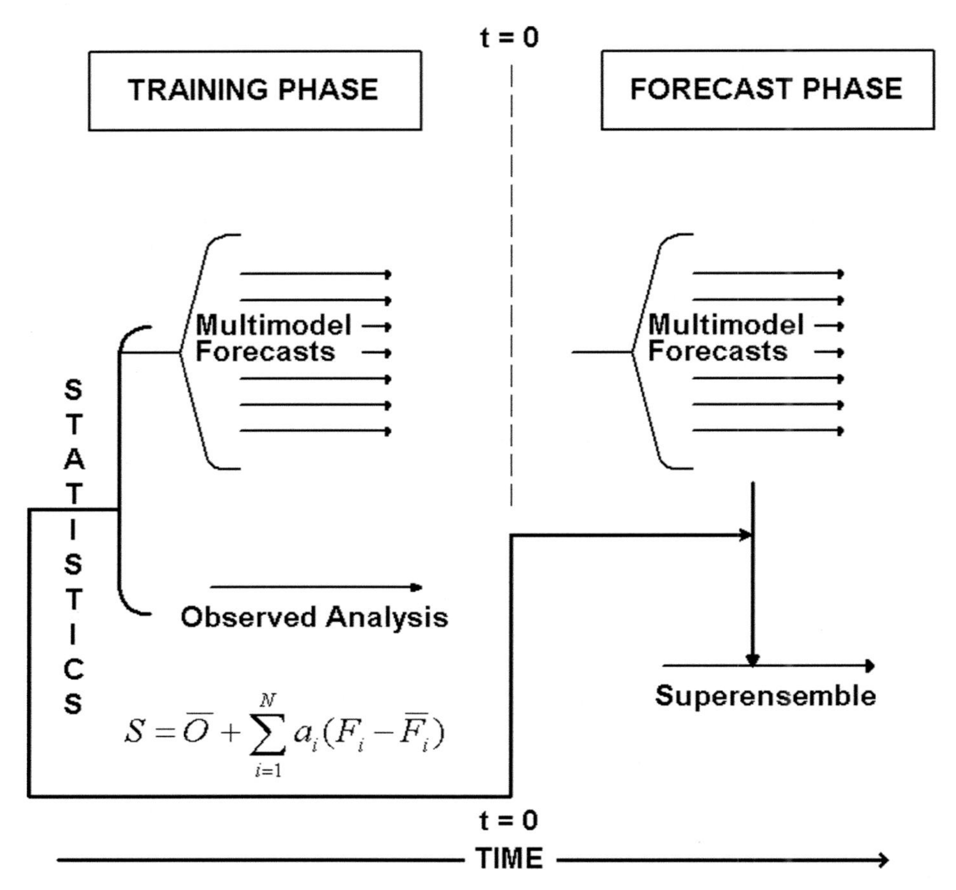

Figure 20.1 A schematic outline of the FSU superensemble that includes training and forecast phases. It also requires the observed analysis fields during the training phase. The training phase statistics are passed to the forecast phase.

procedure consists of a definition of the superensemble forecast:

$$S = \overline{O} + \sum_{i=1}^{N} a_i (F_i - \overline{F}_i), \tag{20.1}$$

where S is the superensemble prediction, $\overline{O}$ is the observed time mean, a_i are the weights for individual models i, F_i is the predicted value from model i, $\overline{F_i}$ is the time mean of prediction by model i for the training period, and N is number of models.

The weights a_i are computed at each of the transformed grid points by minimising the objective function $\boldsymbol{G}$ for the mean square error of the forecasts:

$$G = \sum_{t=0}^{t=train} (S_t - O_t)^2, \tag{20.2}$$

where t denotes the length of a training period.

In the conventional superensemble methodology, a collection of a sequence of individual forecasts from several models is subjected to a multiple regression against the observed (or assimilated) counterpart fields. These multiregression coefficients are collected during the training phase of the superensemble. The length of this training data phase varies for each type of forecast addressed in this chapter. This statistic, collected during the training phase, is simply passed on to a forecast phase of the superensemble. In this forecast phase, we again have forecasts from the same member models that are corrected based on their past collective behaviour. This type of weighted bias removal is more effective than a bias removed ensemble mean. The latter assigns a weight of 1.0 to all models after bias removal. The superensemble includes fractional and even negative weights depending on past behaviour of member models.

A variant of the above formulation was necessary for improved skills for seasonal climate forecasts. From the member model forecast data sets we generate additional data sets named the 'Synthetic Datasets'. We found a significant improvement in the skills of the seasonal climate forecasts using this variant of the superensemble. The synthetic data set is created from a combination of the past observations and past forecasts. We determine a consistent spatial pattern among the observations and forecasts. This is simply a linear regression problem in the empirical orthogonal function (EOF) space. Sets of such synthetic data are then obtained, one for each available forecast, for the creation of superensemble forecasts. A method of creating the synthetic data and the associated statistical procedure is given in Appendix A.

20.3 Hurricane/typhoon forecasts

Many modelling groups from the USA, UK and Japan provide data sets from their daily real-time hurricane/typhoon forecasts. These data sets are used for consensus forecasts. In our studies, the hurricane and typhoon forecasts are carried out using a one-dimensional superensemble. The formulation of the one-dimensional forecasts is centred along the track of the minimum pressure of the storm. The data sets employed for the one-dimensional problem are the latitude and longitude positions of the centre of the storm, the timing along the track including landfall, the intensity along the track and the heading angle of the storm.

We started hurricane/typhoon forecasts with the FSU superensemble in 1998 (Krishnamurti *et al.*, 1999; Williford *et al.*, 2003; Kumar *et al.*, 2003). The track forecast skills of the FSU superensemble show significant improvement over those of the member models. Over the Atlantic basin, during the 1998 season, the improvements over the best model from the superensemble were 20, 70, and 120 km for days 1, 2 and 3 of forecasts. It is worth noting that these skills are higher than those of the ensemble mean as well. The performance of the superensemble was somewhat similar during the subsequent years. The position and intensity errors for year 2002

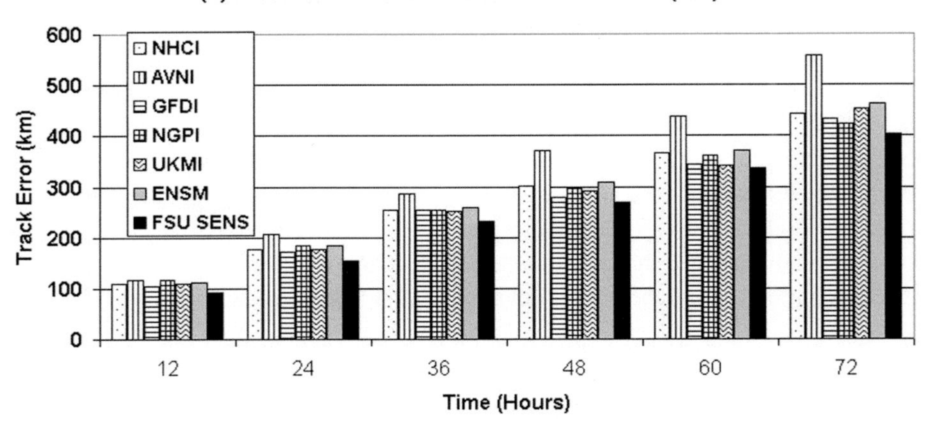

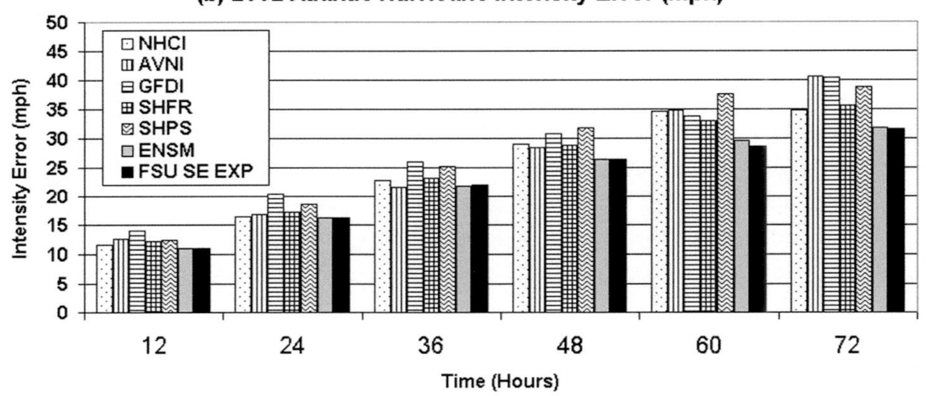

Figure 20.2 Atlantic hurricane forecast skills of multimodels and the superensemble for 2002. (a) Track errors in km and (b) intensity error in mph. Various models in this list include: NHC (National Hurricane Center Consensus forecast), AVN (NCEP Aviation Model), GFD (NOAA/Princeton Geophysical Fluid Dynamics Institute Hurricane Model), NGP (Navy NOGAPS Model), UKM (UK Met Office Model), ENSM (ensemble mean), FSU SENS and FSU SE EXP(FSU superensemble experiment). The intensity models include SHIFOR (SHFR) and SHIPS (SHPS); these are statistical in-house intensity models of NHC. Suffix 'I' stands for interpolated versions of the model forecasts.

for all of the hurricanes that traversed the Atlantic, Caribbean, and the Gulf of Mexico regions are presented in Figure 20.2. During the first 72 hours of forecasts, the position errors of the superensemble are less than those of the member models and the ensemble mean. The errors increase from roughly 90 km at hour 12 to around 400 km at hour 72. The intensity forecast errors from the superensemble relative to the ensemble mean and the member models are quite similar to those above. The superensemble errors are clearly lower than those of the member models. The intensity errors increase from roughly 10 mph (at hour 12) to around 32 mph (at hour 120). Thus, at the larger time range of forecasts, the intensity error can be as large as a category of wind intensity.

In order to illustrate the workings of the superensemble, it is possible to walk through the computation of superensemble forecasts. An example of such tabulation for Hurricane Lenny of November 1999 is presented in Table 20.1. This shows the walk-through computation for a forecast at hour 72 of forecasts. First we shall address the latitudinal positions of forecasts. Column one lists four models $\boldsymbol{i}$. The second column provides the weights a_i for these four models; these entries were obtained from the training phase behaviour of these four models. The next column provides the normalised forecast increments for each model (F_i') with respect to the forecast mean ($\overline{F}_i$), where prime denotes increments. The superensemble function $a_i(F_i' - \overline{F_i'})$ for each member model and the superensemble forecast, i.e. $\overline{O} + a_i(F_i' - F_i)$, are shown in the next two columns. The last two columns show the forecast position increments and the forecast errors. Below that are the entries for the observed position (OBS; latitude) of the storm at hour 72, the superensemble based forecast (SUP) and the ensemble mean (ENS). We note here that the mean latitudinal position error for Hurricane Lenny was 0.5 degrees latitude for the superensemble, which was lower than that of all member models. The lower part of the same table shows entries for the longitudinal errors. The number of models at hour 72 were somewhat reduced since the longitudinal positions were too far off (outliers) and were not used for the construction of the superensemble. The overall longitudinal error of the superensemble is −2.53 degrees and was still the least among the different model members.

The superensemble forecast of Hurricane Lenny was unique in another important manner. Generally, if we have a suite of forecasts from multimodels, they will show some spread in the distribution of these tracks. The ensemble mean generally resides within the envelope of this spread. The superensemble-based track forecasts can even lie outside such an envelope. Lenny was clearly one such storm; Figure 20.3 illustrates such a spread of tracks. This is a special strength of the superensemble since it is able to portray such bias corrections from the past performance of these models. The intensity forecasts have generally been quite poor from the numerical prediction modelling. The issues of adequate model resolution, treatment of physics, lack of microphysics, hydrostatic versus non-hydrostatic motions are some of the important factors that preclude a proper modelling of the cloud's life cycle and the hurricane intensity. These do seem to impact the growth and maintenance of the hurricane scale motions. The multimodel superensemble does improve intensity skills slightly over those of the member models and the ensemble mean.

The timing within track forecasts, especially for the landfall, is an important practical and scientific issue. In principle, it is possible to tailor forecasts of landfall parameters such as position, intensity, timing and the heading angle. This calls for the construction of a single parameter superensemble from multimodel forecast data sets. This landfall forecast is the lowest dimension superensemble. Powell and Aberson

Table 20.1 *Walking-through forecast of Hurricane Lenny from superensemble*

72-hr multimodel superensemble forecast of Hurricane Lenny. Valid 19991118112

Latitude position						
Model	Coefficient a_i	Normalised difference of forecast increment $(F_i' - \overline{F}_i')$	$a_i(F_i' - \overline{F}_i')$	Superensemble forecast increment $\overline{O}' + \sum_1^n a_i(F_i' - \overline{F}_i')$	Forecast position (60 hr pos. + increment)	Forecast error (latitude)
1	−0.460971	−0.372857	0.171876	1.022222–0.340774 =	23.3	5.5
2	0.092239	−0.210714	−0.01944	0.681448	20.5	2.7
3	0.551240	−0.1	−0.05512		16.4	−1.4
4	0.578033	−0.757895	−0.43809		18.7	0.9
		$\sum_1^n a_i(F_i' - \overline{F}_i')$	−0.340774		17.8	
OBS	60 hr position = 17.7					
SUP	60 hr forecast = 17.62				18.30	0.50
ENS					19.2	1.4

Longitude position						
Model	a_i	$(F_i' - \overline{F}_i')$	$a_i(F_i' - \overline{F}_i')$	$\overline{O}' + \sum_1^n a_i(F_i' - \overline{F}_i')$	Forecast	Error (longitude)
1	0.046219	−1.707143	−0.0789	−0.7–2.24686 =	77.5	13.9
2	1.069466	−2.027143	−2.16796	−2.94686	71.7	8.1
		$\sum_1^n a_i(F_i' - \overline{F}_i')$	−2.24686			
OBS	60 hr position = 64.1				63.6	
SUP	60 hr forecast = 64.02				61.07	−2.53
ENS					68.65	5.05

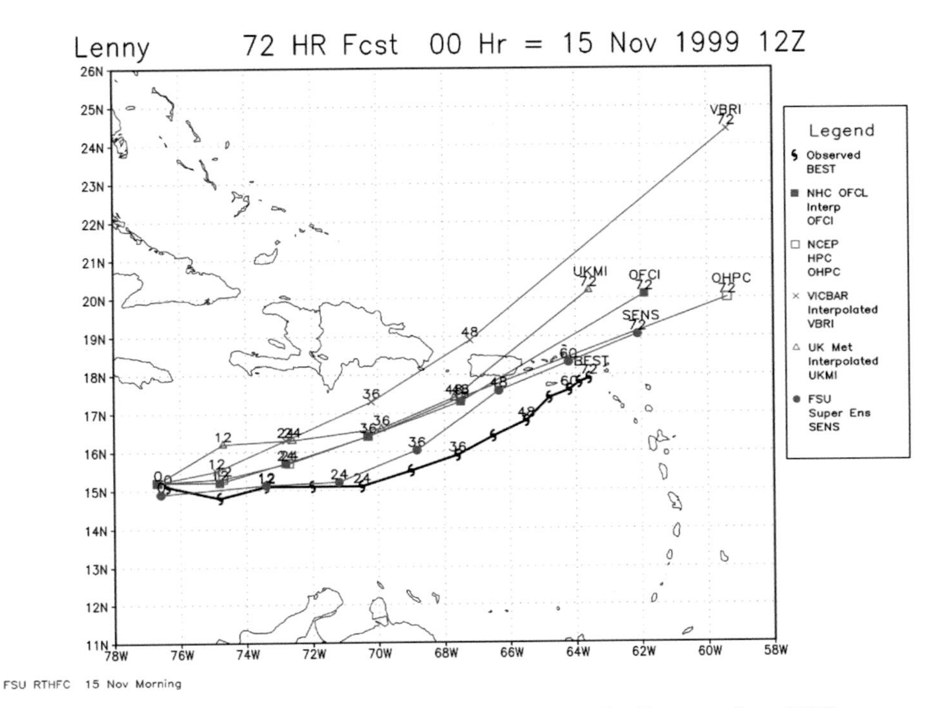

Figure 20.3 Hurricane Lenny: 72 hours forecast tracks. Forecasts from NHC in-house model VICBAR (VBRi), UK Met Office forecast (UKMI), official forecast of the NHC (OFCL), the NCEP aviation model (OHPC), the FSU superensemble SENS and the observed best track (BEST) are shown in this diagram.

(2001) have addressed the landfall issues from large samples of past forecasts over the Atlantic. Using those data sets and recent multimodel inventories, we have also addressed these landfall issues. Figure 20.4 provides some of these research findings for the forecasts of landfall time 24 hours prior to the event. It is possible to improve the timing of landfall of hurricanes within hours 0 to 24, 24 to 48 and 48 to 72 of forecasts by 1.2, 2.8 and 8.5 hours compared with the best models. Although this gain in timing is not substantial, nevertheless it suggests the possibilities of further improvements. Those improvements can come about from member model forecasts and superensemble methodologies that can contribute towards improved landfall-based forecasts.

Forecasts of tropical cyclones covering the entire Pacific Ocean were also studied using this same methodology (Kumar *et al.*, 2003). The statistical summary for the 1998, 1999 and 2000 seasons for track and intensity forecasts of typhoons up to 120 hours of forecast is presented in Figure 20.5. These results essentially reconfirm similar findings on the skill score of the hurricane forecasts for the Atlantic Ocean. The forecasts and error reduction by the superensemble over the Pacific Ocean were in general somewhat larger compared with those of the Atlantic basin; the reasons for that are not quite apparent at this stage.

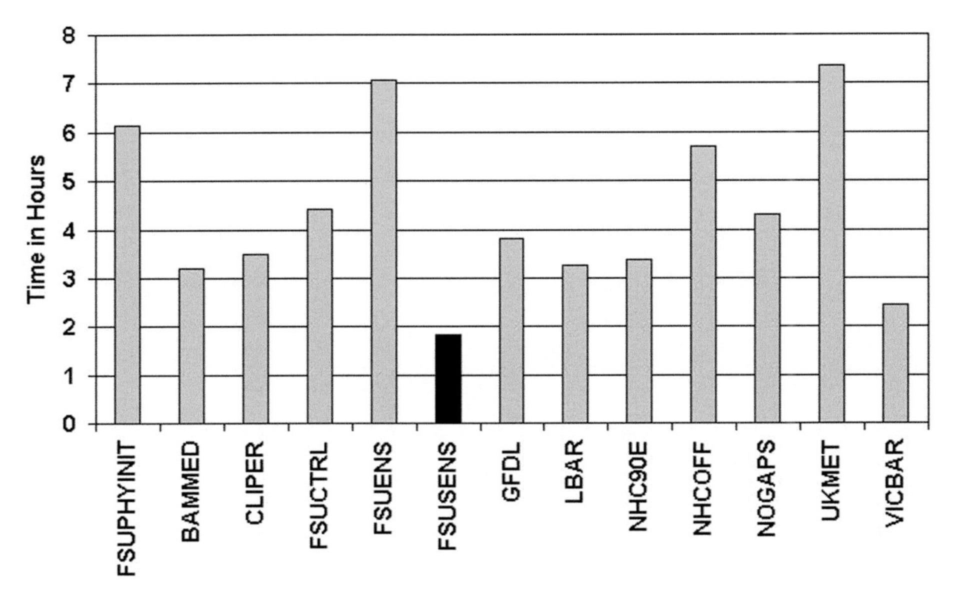

Figure 20.4 Timing errors at the landfall of hurricanes from the Atlantic during 1999 and 2000. These errors are in the last one-day of forecast prior to landfall. Models include FSU control and FSU with physical initialisation, several versions of NHC's in-house models, operational dynamical models and the FSU ensemble and superensemble (dark bar).

Further improvements in hurricane and typhoon forecasts are possible from the generation and use of synthetic model components. A variant for the superensemble procedure, called the quadratic synthetic approach, is tested for this purpose (Szymczak, 2004). This method uses a quadratic regression technique to generate space-smoothed track for each model forecast and these space-smoothed tracks act as additional proxy members. In most instances we do see some improvements in the latitude and longitude locations of the storm centre from the use of the synthetic superensemble. We have noted that by having $2n$ models in place of n original models, it is possible to reduce the forecast errors generally. Several such innovations are possible for further small but significant improvements on hurricane/typhoon forecasts.

20.4 Medium-range numerical weather prediction

A total of eleven global multimodels are used for the construction of the multimodel superensemble at the Florida State University. This is an ongoing real-time effort. The data sets include six global operational models' daily forecasts from NCEP/EMC in the USA, JMA in Japan, NOGAPS the US Navy's global model, BMRC the Australian model, RPN the Canadian model and the UKMO model from England.

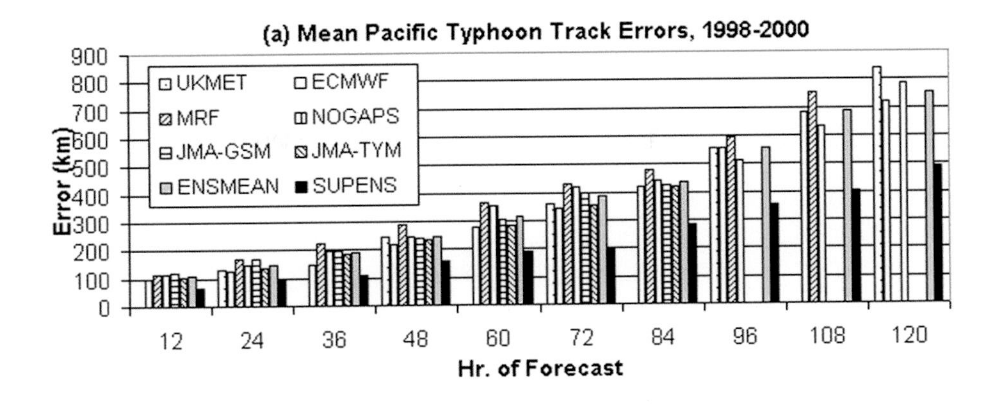

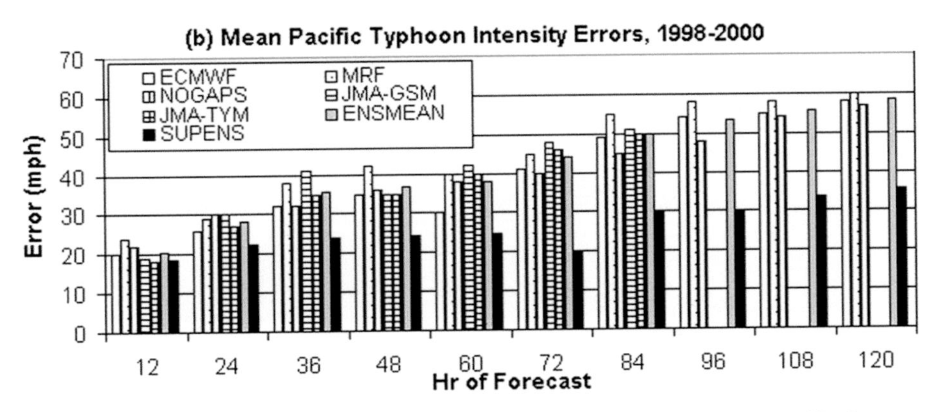

Figure 20.5 Track and intensity forecast errors for typhoons over the Pacific Ocean during 1998–2000. The models included here are from ECMWF, NCEP's MRF, US Navy's NOGAPS, Japan JMA-GSM, Japan JMA-TYM, the ensemble mean and the superensemble. Track errors are in km and the intensity errors are in mph. JMA forecasts are available through 84 hrs only.

Besides those, we use a suite of five versions of the FSU model (Krishnamurti *et al.*, 2001). For the training phase of the NWP superensemble for heavy rain events, we use satellite-based rainfall estimates from the NASA satellite TRMM and the US DMSP satellites. The rainfall estimates are derived from several rain rate algorithms developed by Kummerow *et al.* (1998, 2000), Huffman *et al.* (1995), Turk *et al.* (2001), Olson *et al.* (1990, 1996), Ferraro (1997) and Ferraro *et al.* (1998). These algorithms are based on the microwave scattering temperatures regressed to ground truth estimates of rainfall rates. Several versions of the FSU global spectral model are subjected to physical initialisation (which is a rain rate initialisation procedure) using these different rain rate estimates. A detailed account of this procedure is described in Krishnamurti *et al.* (2001). The different initialisations provide additional forecasts for the construction of the NWP superensemble.

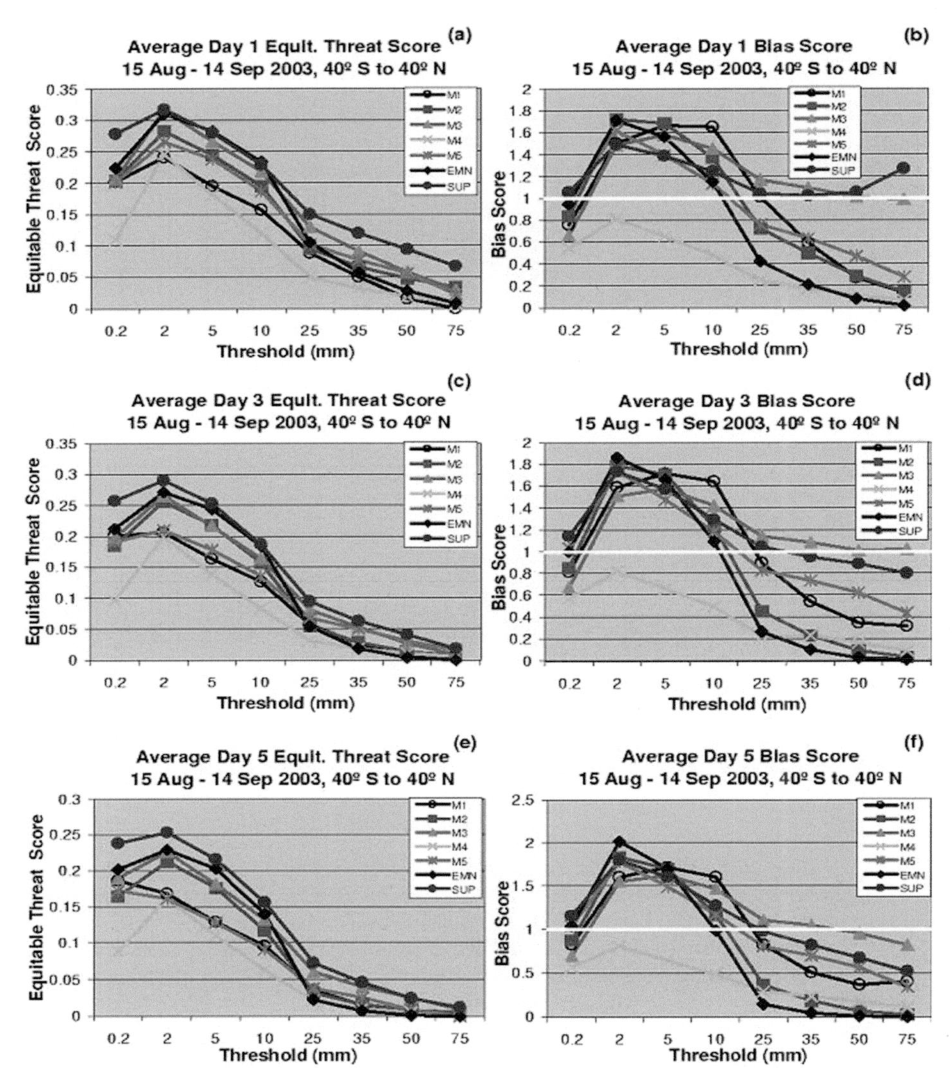

Figure 20.6 Recent skill (equitable threat score and bias score) of precipitation forecasts from real-time FSU superensemble for day 1 (a,b), day 3 (c,d) and day 5 (e,f) for the period from 15 August to 14 September 2003. Skills for different member models, ensemble mean and superensemble are shown here for the global tropics (40S–40N).

The skills of precipitation forecasts from the superensemble have been quite high compared with those of the member models. The metrics for the measure of skill include anomaly correlation, rms error and probabilistic equitable threat scores. Recent skills of equitable threat score and bias score for global tropics (40° S to 40° N) are illustrated in Figure 20.6. These skills are computed for the month from 15 August to 14 September 2003 for day 1 (Figure 20.6 a and b), day 3 (Figure 20.6

Table 20.2 *Walking through a day 3 superensemble precipitation forecast*
Multimodel superensemble precipitation forecast from h48 to h72. Valid from 20020709/12 UTC thru 20020710/12 UTC. Valid at 15.43 deg N, 120.00 deg E (western shore of Luzon, Philippines)

		Precipitation (mm)				
Model	Coef. a_i	Pcp. F_i	Mean Pcp $\overline{F}_i$	$a_i(F_i - \overline{F}_i)$	$\overline{O} = 26.14$ $\overline{O} + (F_i - \overline{F}_i)$	Error (mm) $F_i -$ OBS
BMRC	0.60191	88.38	31.52	34.23	83.00	−22.06
FSUFER	0.01805	19.38	11.40	0.14	34.12	−91.06
JMA	0.08609	60.46	12.73	4.11	73.87	−49.98
NCEP	0.22313	91.68	15.89	16.91	101.92	−18.76
NRL	0.22343	57.33	10.64	10.43	72.82	−53.11
RPN	0.08697	78.36	20.21	5.06	84.29	−32.08
FSUCTL	0.43546	24.98	18.00	3.04	33.12	−85.46
FSUOLS	0.08800	23.94	11.46	1.10	38.61	−86.50
FSUTRM	0.06130	20.32	11.23	0.56	35.22	−90.12
ENSMEAN		51.65				−58.79
OBS		**110.44**				
SUPENS		101.71				−8.73
BIAS-REM ENSMEAN		61.89				−48.55

c and d) and day 5 (Figure 20.6 e and f) from the real-time FSU superensemble. The illustrations include the skills of member models, the ensemble mean and those of the superensemble. Operational models skills of precipitation threat scores are generally around 0.3 or lower for day 1 to day 5 of forecasts. In this regard some higher skills, as much as 0.5 to even 0.6, have been possible from the use of the superensemble.

Many instances of flooding arise from landfall of hurricanes and other severe weather phenomena. The precipitation superensemble provides some useful guidance for floods. Many such examples are presented in Krishnamurti *et al.* (2003b). We shall next present a walk-through illustration (Table 20.2) on heavy rains that resulted in flooding over the western shore of northern Philippines (near Luzon) from the landfall of Typhoon Halong of July 2002. Table 20.2 is analogous to Table 20.1. The day 3 of forecasts (between hours 48 and 72) over a grid location 15.43 N and 120 E are illustrated. Here the various vertical columns show the models, the training coefficients a_i, the predicted model precipitation F_i, the mean of the predicted precipitation $\overline{F_i}$ during the training phase, the function $a_i(\mathrm{F}_i - \overline{F_i})$, the observed mean precipitation during the training phase $\overline{O}$, the superensemble function $\overline{O} + a_i(\mathrm{F}_i - \overline{F_i})$ and the forecast error for each model. Also shown is the observed rain during

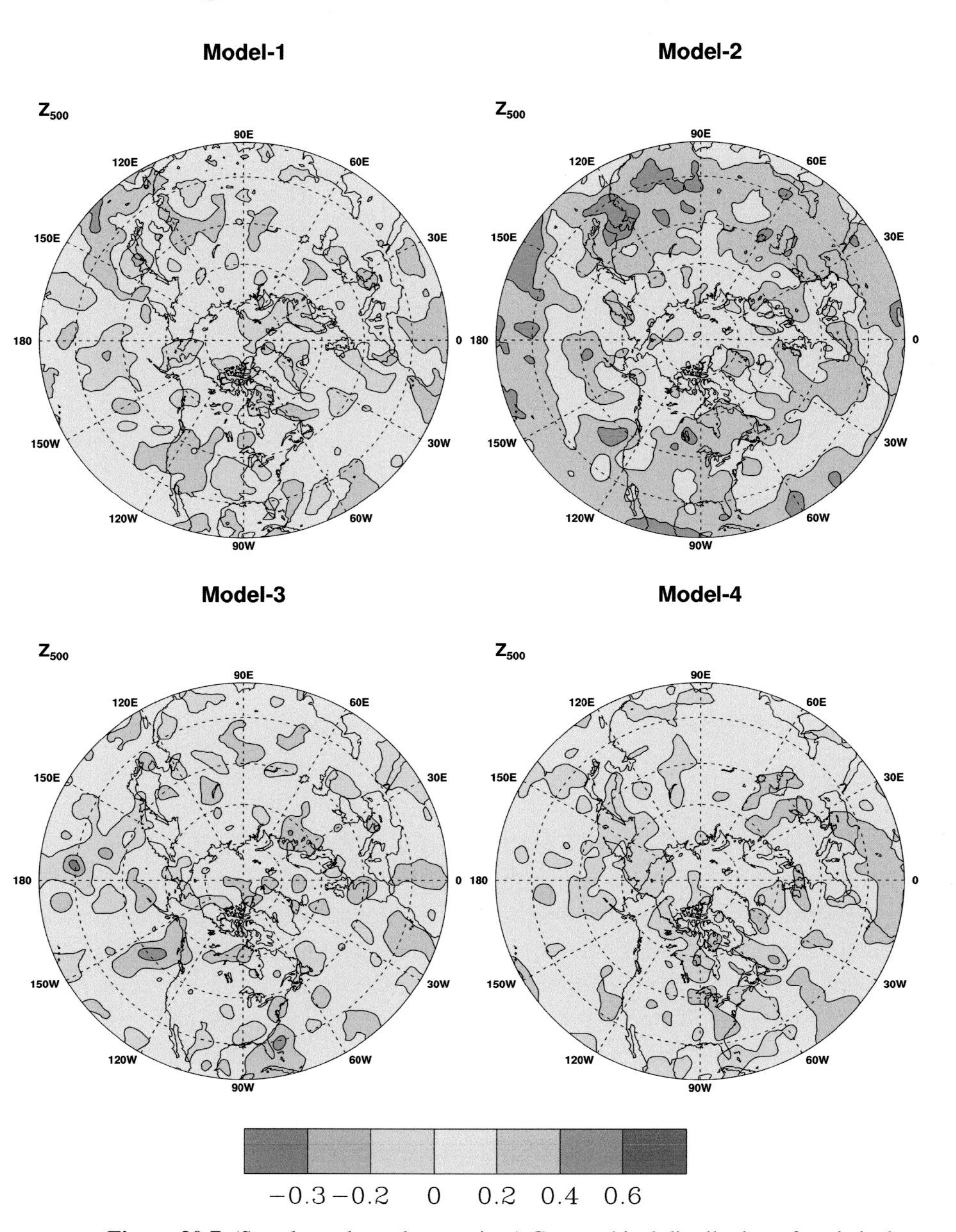

Figure 20.7 (See also colour plate section.) Geographical distribution of statistical weights for different member models in the northern hemisphere. Colour scale of the fractional weights is shown at the bottom.

this event. The superensemble forecasts, the bias removed ensemble mean and their errors are shown in the lower part of the same table. Overall, it is clear that the best rainfall forecast of 101.7 mm/day (observed 110.4 mm/day) was achievable from this exercise. It is the accumulated effort for such point-by-point improvements which result in providing higher regional and global skills for the superensemble.

The multimodel superensemble for global NWP (i.e. for all other variables) is also being constructed using the data sets from the same 11 models. Here the training database for all variables came from the real-time data assimilation from ECMWF (at 0.5 degrees latitude/longitude resolution) that is used as a benchmark for forecast validation as well. Detailed assessments of the NWP superensemble are published in Ross and Krishnamurti (2005) and Krishnamurti *et al.* (2002, 2003a). Some of the main findings thus far are:

(a) Roughly 120 days of data sets for the training phase seem to provide the best results for the superensemble.
(b) Typically weights from the participating models show a great degree of regional variability from one model to another. The superensemble weights can be positive or negative; Figure 20.7 (colour plate) illustrates the spread of weights for day 6 of forecasts for the 500 hPa heights for the northern hemisphere (Krishnamurti *et al.*, 2003a). It was our experience that the best models were not necessarily the best in all regions. Some overall improvements of skill to the superensemble did come from the poorer models over certain regions.
(c) Major model changes always lead to problems for the applicability of training phase statistics into the forecast phase. This requires a monitoring of the modelling change activities of the participating models of the superensemble.
(d) As one increases more and more member models for the construction of the superensemble we note that the superensemble errors keep decreasing and levelling off somewhat. The ensemble mean, however, seems to have increasing errors as the number of models are increased (Figure 20.8). The latter assigns a weight of 1.0 for all models, even after bias correction, whereas the fractionally weighted combination of member models of the superensemble appears clearly superior. Poorer models upon bias correction do not meet the same levels of performance as the top models.
(e) The largest reduction of rms errors from the superensemble was noted for the meridional wind component v. The meridional wind carries more of the divergent wind (zonally averaged v is entirely divergent) compared with the zonal wind.
(f) The improvement as well as the overall skill of the southern hemisphere anomaly correlation of the 500 hPa heights exceeded those of the northern hemisphere (Table 20.3; Krishnamurti *et al.*, 2003a). The seasonal skills for

Table 20.3 *Six -day global 500 hPa geopotential height anomaly correlation for the period 20 August–17 September 2000*

	Day-1	Day-2	Day-3	Day-4	Day-5	Day-6
Superensemble	0.992	0.979	0.958	0.928	0.881	0.799
Ensemble mean	0.983	0.962	0.935	0.891	0.827	0.756
Model-1	0.984	0.967	0.936	0.889	0.824	0.713
Model-2	0.981	0.957	0.932	0.880	0.796	0.623
Model-3	0.963	0.930	0.885	0.815	0.706	0.579
Model-4	0.962	0.925	0.871	0.786	0.697	0.578
Model-5	0.956	0.918	0.858	0.767	0.665	0.549
Model-6	0.941	0.889	0.846	0.739	0.632	

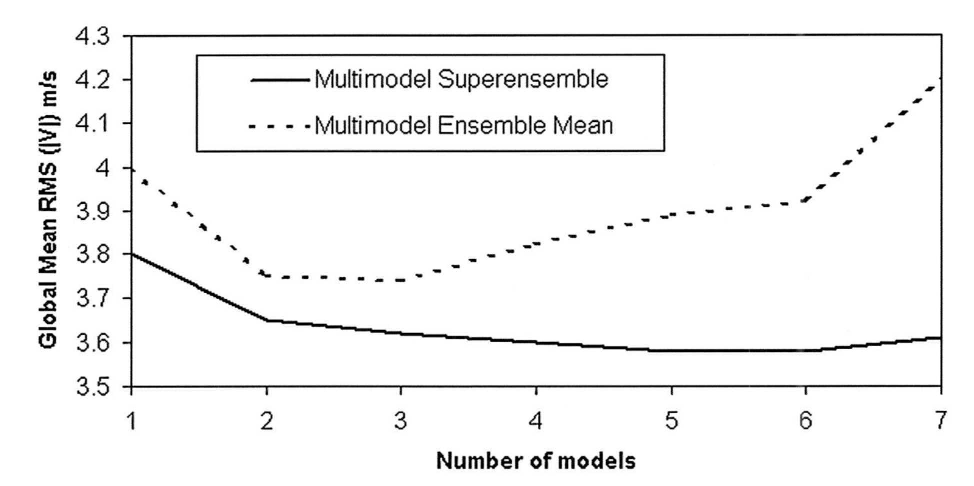

Figure 20.8 Global mean rms error of total wind from superensemble and ensemble mean as a function of number of models.

the best model and the superensemble were: NH: 0.69 and 0.75; SH: 0.71 and 0.81. The southern hemisphere forecast skills have been greatly improving in recent years.

(g) The reductions of the systematic errors by the superensemble were very robust (Ross and Krishnamurti, 2005). Shown in Figure 20.9 (colour plate) are typical magnitudes of the systematic errors for the best model, for a model with the lowest skill, for the ensemble mean and for the superensemble. These are results for the total wind field at the 850 hPa level covering daily 48-h forecasts for an entire year 2000. The results for other variables demonstrate similar improvements for the reduction of the systematic errors.

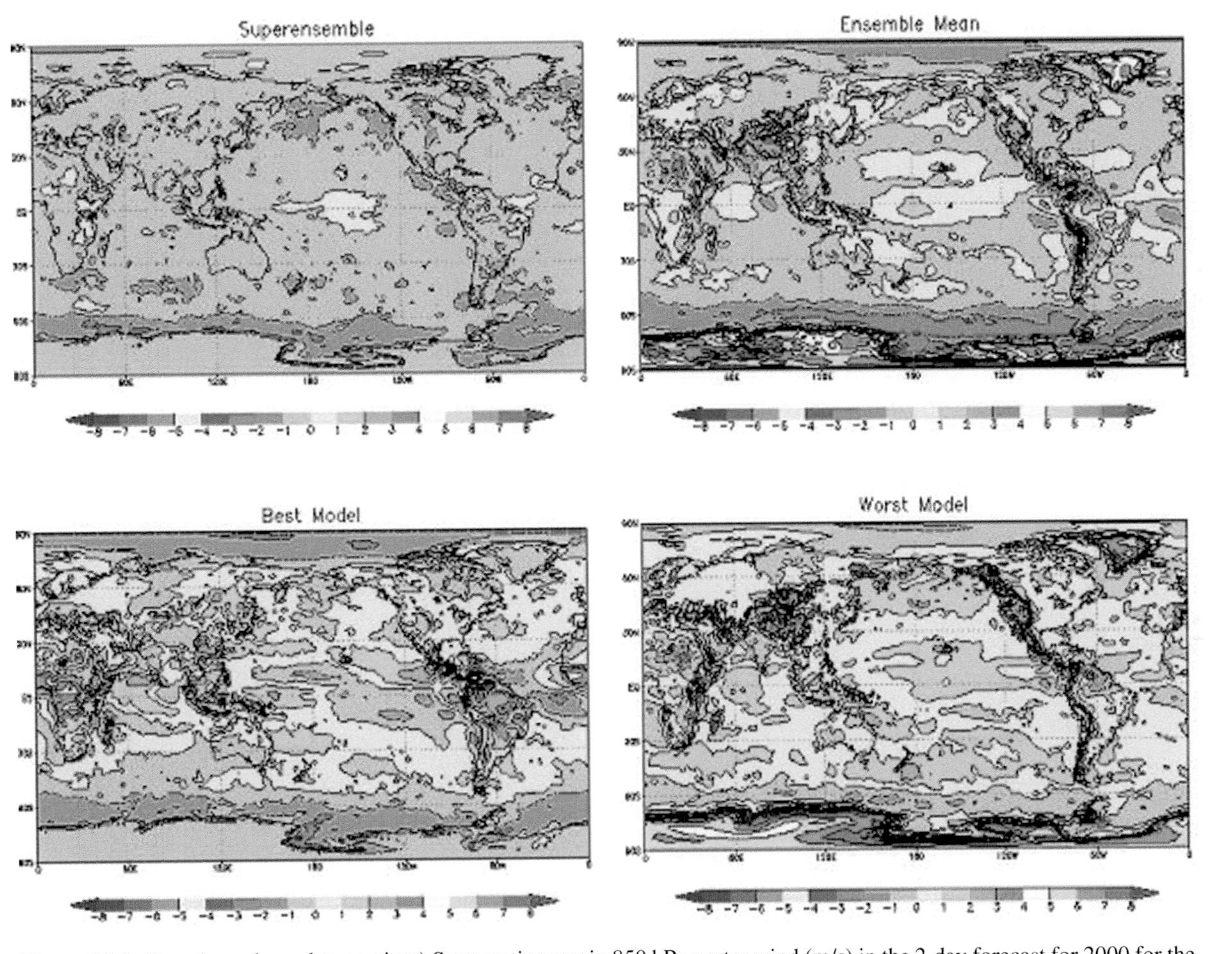

Figure 20.9 (See also colour plate section.) Systematic error in 850 hPa vector wind (m/s) in the 2-day forecast for 2000 for the superensemble, ensemble mean, best model, and worst model.

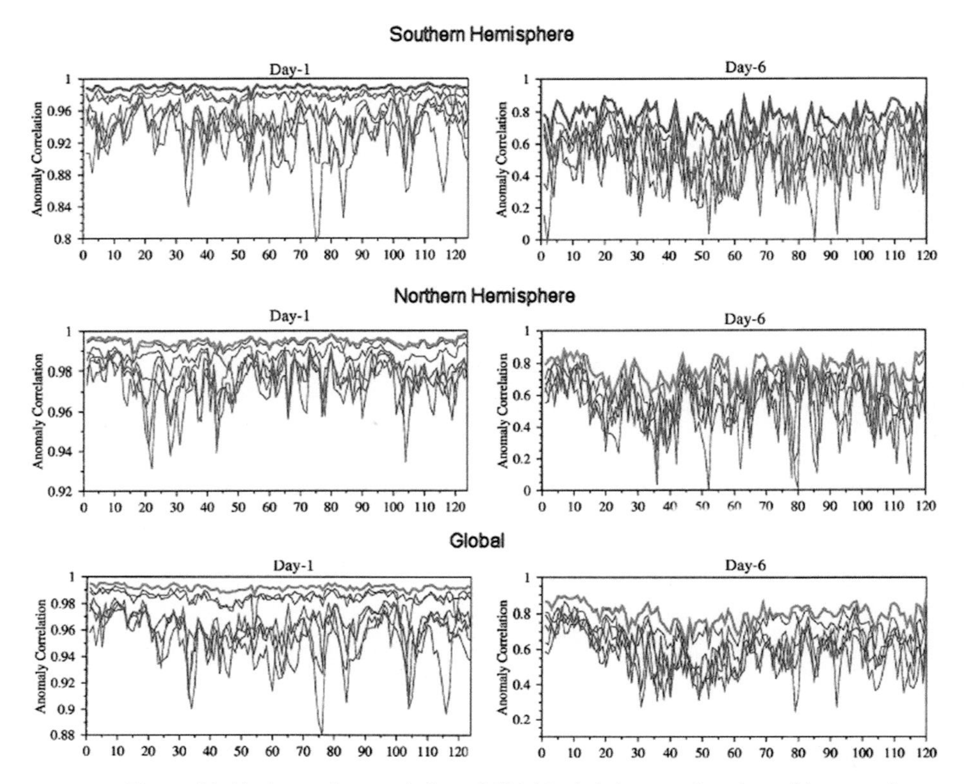

Figure 20.10 Anomaly correlation of 500 hPa heights as a function of forecast days: thin lines show results for member models, thick dark lines in the top of each panel show results for the superensemble. Top two panels: day 1 and day 6 forecasts for southern hemisphere. Middle two panels: day 1 and day 6 forecasts for northern hemisphere. Bottom two panels: day 1 and day 6 forecasts for the entire globe.

(h) The day-to-day superensemble forecast is generally very consistent. The 500 hPa anomaly correlation of geopotential height for northern hemisphere, southern hemisphere and entire globe in Figure 20.10 (Krishnamurti *et al.* 2003a) shows that regardless of the daily up and down swings in the performance of the member model's skills, the superensemble stays consistently as the one with the highest skill at both day-1 and day-6 forecasts. Also shown in this string of forecasts are the anomaly correlations for the ensemble mean that demonstrates less of a consistency. As multimodels provide forecasts for the next week, one does not know which of the member models would provide the best forecast; however, we can anticipate that the forecast provided by the superensemble can be most reliable in that regard. This is clearly apparent in Figure 20.10.

A post-processing algorithm based on multiple regressions of synthetic multi-model solutions towards observed fields during a training period is one of the best solutions for extended prediction. Our study shows that the proposed technique further reduces the forecast errors below those of the conventional superensemble technique and increases the predictive skill of forecasts. The usefulness of these improved two-week forecasts is still open to question since an anomaly correlation of around 0.5 still has a marginal value. Small further improvements of the member models and of the superensemble strategies can move these skills to 0.6 (or higher) for two-week forecasts.

20.5 Seasonal climate forecasts

This is clearly one of the most difficult forecasting areas. For seasonal timescale, the perennial question as to whether it will be cold or warm or wet or dry over a given region has been a major challenge. Clearly some progress has emerged in seasonal model forecasts when strong SST anomaly signals are present, especially in the El-Niño scenarios. The seasonal climate forecasts for these situations have improved by many modelling groups: Frederiksen *et al.* (2001), Goddard *et al.* (2001), Roads *et al.* (2001), Kanamitsu *et al.* (2002, 2003), Wang *et al.* (2002), Voldoire *et al.* (2002), Palmer *et al.* (2004) and Anderson (this volume). However, climate anomalies abound in all scenarios and many difficulties in forecasts stem from weak bottom boundary-forcing situations where a more important role may even come from the internal variabilities. In our earlier study on seasonal climate forecasting (Krishnamurti *et al.*, 2002), it was noticed that the superensemble-based anomaly forecasts have higher skill compared with the bias-removed ensemble mean of member models, individual bias removed ensemble mean of the member models and the climatology. Our applications of the proposed synthetic superensemble (Yun *et al.* (2005) show that many improvements in seasonal forecast skills are possible above those of the participating member models when a suite of models is used. A brief description of the method is given in Appendix A. These results are outlined here.

In a recent BAMS paper Palmer *et al.* (2004) reported on the seasonal forecast capabilities from the use of several coupled atmosphere–ocean multimodels. This is a well-known DEMETER data set (Hagedorn *et al.*, this volume) and is available online at the ECMWF website www.ecmwf.int/research/demeter/verification. The DEMETER consists of coupled atmosphere–ocean models that include three coupled models from France (CERFACS, LODYC, METEO FRANCE), the ECMWF coupled model, the UK Met Office coupled model, an Italian model (INGV), and a model from Germany (MPI). Each of these models had carried out several years of seasonal forecast experiments. In our study, we used 15 years of data covering the period from 1987 to 2001, and each of these models provided an ensemble of nine experiments based on separate initial conditions for respective start date. In our data compilations we include as many as 4000 individual seasonal forecasts from the DEMETER and FSU coupled model data sets. The FSU suite of four coupled atmosphere–ocean models uses permutation of physical parametrisation algorithms: Arakawa Schubert and Kuo cumulus parametrisations and a band and a simple emissivity/absorptivity model for the radiative transfers. Same initial states are provided to each of these models for their respective seasonal runs (Krishnamurti *et al.*, 2002). We have compared the monthly and seasonal forecast skills of these member models, of the ensemble mean, and those from our conventional and the synthetic superensemble. In all, as many as 67 forecast experiments per season were available from each of these models. Our preliminary studies were in three parts, results based on European coupled model data sets (the DEMETER data set), those based on FSU coupled models and those based from combining all of these models.

The metrics for forecast evaluation include standard deterministic ensemble mean scores such as anomaly correlation coefficient (ACC), root mean square error (RMSE) and mean square skill score (MSSS). We have also examined probabilistic skill measures such as reliability diagrams, relative operating characteristic (ROC), Brier skill scores, equitable threat scores, potential economic value curves and significance tests for these skill scores, scatter diagrams of area-averaged skill measures and probability density functions of grid point skill scores. Some of these follow the studies of Palmer *et al.* (2004). The performance of the FSU synthetic superensemble method is assessed by comparing its cross-validated skill in terms of RMSE and ACC with the corresponding simple ensemble mean and the conventional superensemble. We also looked at the geographical distribution for the skills of precipitation, surface temperature and the low-level winds over some regions of interest. For the synthetic superensemble data set, we note a very marked decrease of the RMS scores and some increase of the ACC for both precipitation and surface temperature forecast compared with those of the member models and their ensemble mean. The spread in the RMS is reduced compared with those of the member models. However, the spread in ACC is seen enhanced compared with member models. This shows the effects of the intermodel spreads.

The cross-validated RMS and ACC for precipitation and the surface temperature from the DEMETER models using the synthetic superensemble are shown in Figure 20.11. The skill scores are for the global belt between 0° to 60° N latitudes. Each of the DEMETER coupled models were run for six months of duration. We considered forecasts for the month 2 through 4, since a 1-month lead time for the forecast was inherent in the DEMETER database. The results presented here are averages for the months 2 through 4 of forecasts. These illustrations show the skill scores for the seven DEMETER models – each one of these is in fact an ensemble mean from nine individual model runs. Synthetic ensemble mean and the superensemble from synthetic data are also presented here. We note a drastic reduction of errors in terms of RMSE for the precipitation and surface temperature for the years 1987 through 2001. The 14-year summary of the mean scores is also presented on the right side

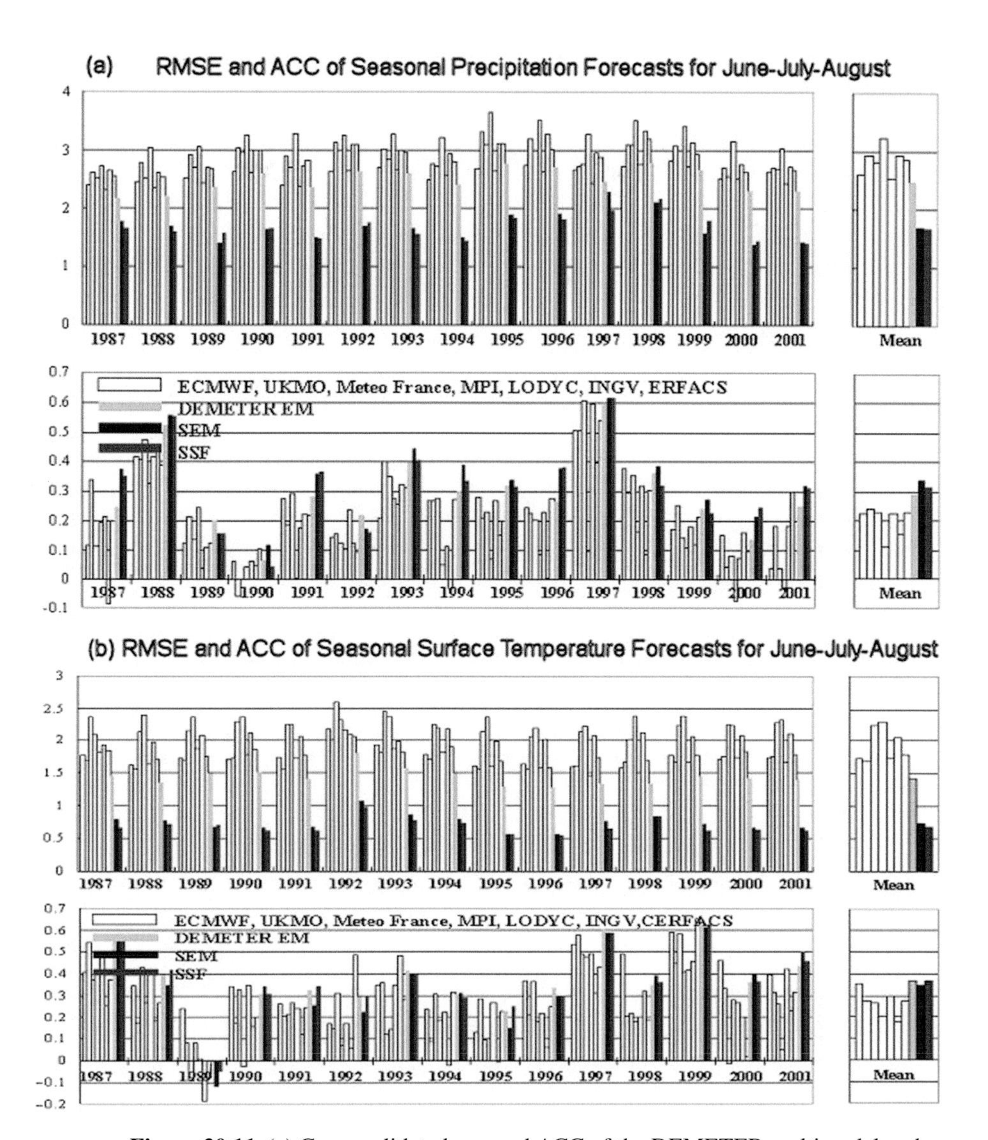

Figure 20.11 (a) Cross-validated rms and ACC of the DEMETER multimodel and synthetic superensemble (0°–60°N northern hemispheric summer, June-July-August, average of precipitation forecasts for 2 through 4 months). (b) Same as (a) but for surface temperature.

of Figure 20.11. Compared with the member models, this time mean demonstrates a 45% improvement for the overall RMSE. The improvements are quite similar for the ACC of precipitation and temperature. The improvements for the ACC are around 50% of the original values of the individual models. We had noted similar improvements for the lower level winds (850 hPa), both for the North American and the Indian monsoon regions.

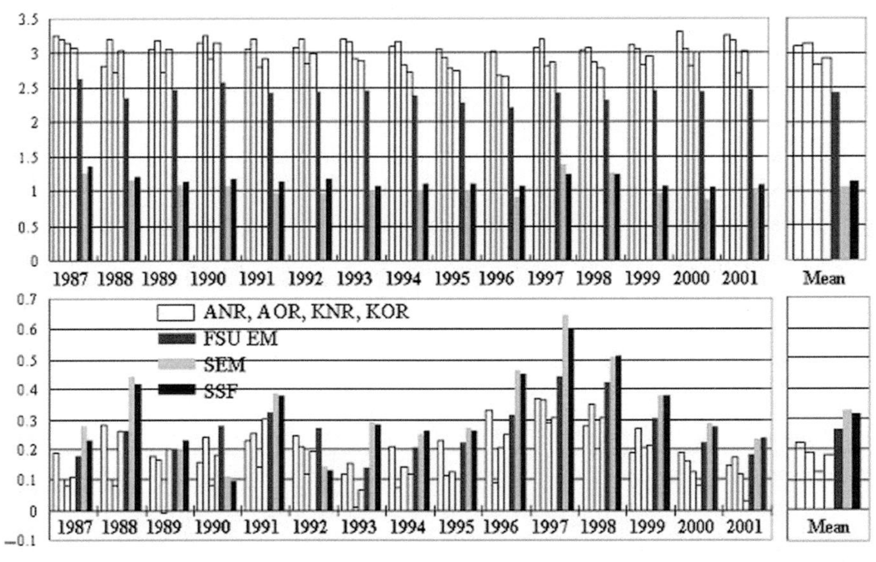

Figure 20.12 Seasonal precipitation forecast skills (rms error and anomaly correlation) from four FSU models over the northern hemisphere belt 0° to 60°N. Results from 15 seasons are shown here. Also shown are the results from the ensemble mean, the synthetic ensemble mean and the synthetic superensemble. The small box on the right shows 15 season average skills for the rms and the anomaly correlation for seasonal precipitation forecasts.

Apart from the statistical skill scores, we have also examined the geographical distributions of the precipitation, surface temperature and lower level winds for the summer season (not shown here). The simple ensemble mean was unable to capture the intensity of precipitation, fine structure and intensity of the surface thermal gradients and strength of the wind at 850 hPa, compared with the respective observations, whereas the synthetic ensemble mean and the synthetic superensemble forecast were able to capture both the gradients and intensity seen for the precipitation, surface temperature and winds reasonably well. For all these three fields over the North American region, the agreement of synthetic superensemble forecasts with the observations was excellent compared with the simple ensemble mean. Over the Asian monsoon region also we noted that the synthetic ensemble mean and the synthetic superensemble forecasts are able to capture the intensity and the contrasts of the precipitation distributions for the Asian monsoon system. All these results confirm that for seasonal climate forecasts it is necessary to go beyond a simple ensemble mean.

Histograms in Figure 20.12 show the cross-validated RMS and ACC of northern hemisphere (0° to 60° N) summer season precipitation from a suite of four FSU member coupled atmosphere–ocean models, their ensemble mean, ensemble mean of the synthetic forecasts and the synthetic superensemble forecasts respectively.

These are averages for one to three months of forecasts. We find that the RMSE for the member models are quite similar for each of the years from 1987 through 2002. For the simple ensemble mean the skill scores are slightly less compared with the results for the member models. However, for the synthetic forecasts we note a drastic reduction in these errors. The magnitude of the reduction of error is around 65% compared with the individual member models. The ACC of precipitation forecasts indicates a significant reduction in error for synthetic superensemble compared with those of the individual member models and their ensemble mean. Overall, we note a 50% improvement for the ACC score for the tropical belt (30° N to 30° S). The errors in tropics are slightly higher compared with those of the northern hemisphere. The new synthetic scheme is able to reduce the model errors over the tropical belt by significant amounts (more than 50% in RMSE and around 40% in ACC). This is a major improvement on the performance of current day single models for seasonal climate forecasting using statistical-dynamical methods.

Next, we have taken all of the DEMETER plus the FSU models to construct an overall synthetic superensemble. Figure 20.13 shows the precipitation forecasts for the Asian summer monsoon for the year 2000 from this synthetic superensemble technique. Forecasts from different members of the DEMETER, their ensemble mean, ensemble mean of FSU member models and the FSU superensemble are shown in this figure. The Xie–Arkin precipitation data set is used as the observational benchmark for training and for validating these forecasts. We note that a straightforward ensemble mean is unable to capture the intensity (amplitude) of the monsoon rains, while the member models have many difficulties in representing the geographical distribution of precipitation. Much superior performance of the synthetic superensemble is clearly apparent from this illustration, where the forecasts from the superensemble closely match to the observed precipitation over the Asian monsoon region. The anomaly forecasts for 2000 from the FSU superensemble, shown in Figure 20.14, are further evidence of the superior performance of this approach. The observed anomaly for 2000 (a wet year over central/peninsular India and southern Indochina) carries an alternating (west to east) anomaly pattern. Below normal values over the central Arabian Sea, a wet area over central/peninsular India, a below normal anomaly over the northern Bay of Bengal and a wet anomaly over southern Indochina are seen in these observed fields. These features are reasonably reproduced by the FSU synthetic superensemble. Among the DEMETER models, the alternating (west to east) pattern was only seen for the LODYC model of France and the UKMO coupled model. These two did capture the general features quite well although the northern India rainfall had a somewhat excessive spread of heavy rains, which was also reflected in the UKMO model forecasts. The FSU ensemble mean (bottom left panel of Figure 20.14) does not capture the below normal rain over the central Arabian Sea. Overall, the synthetic superensemble seems to carry the seasonal forecasts of these wet and dry spells somewhat better than the member models and the ensemble mean.

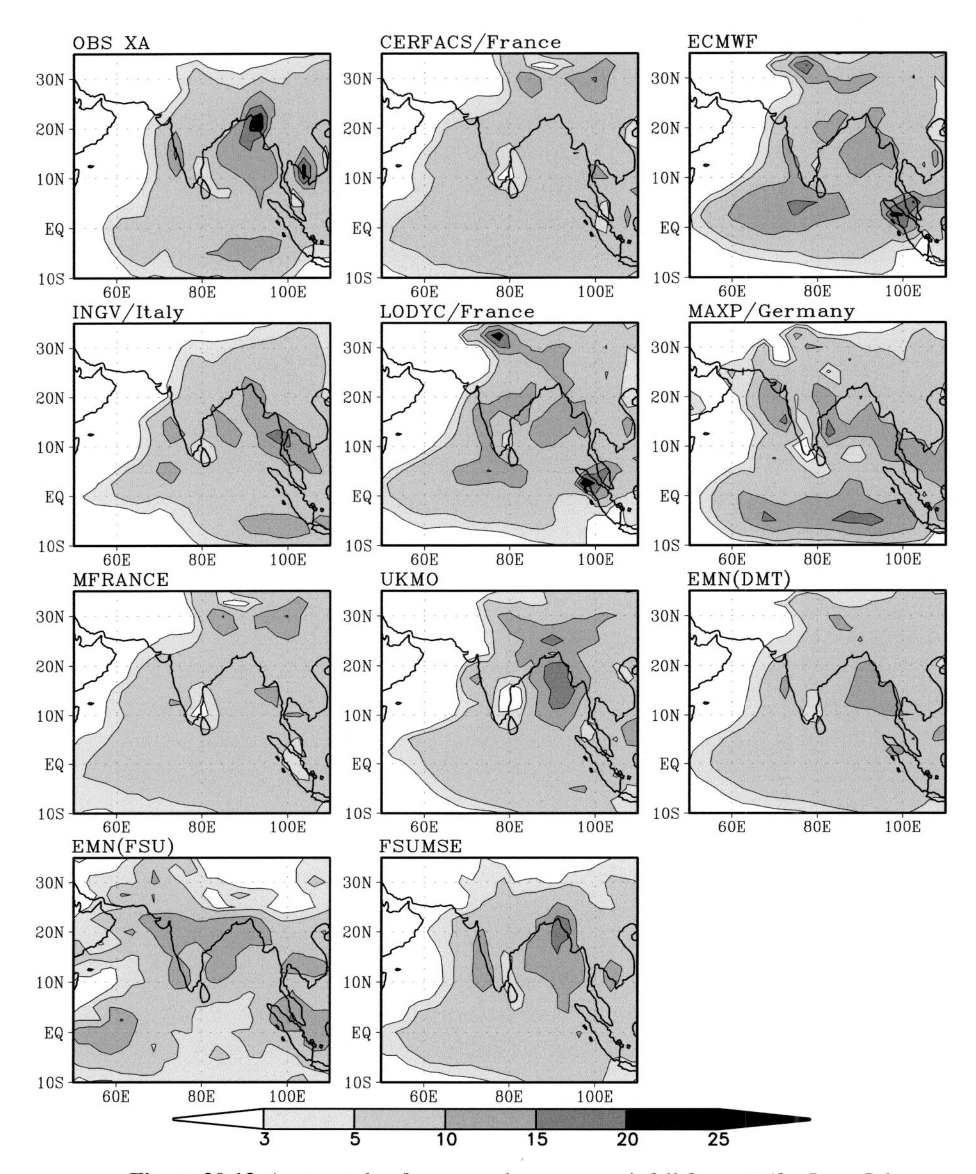

Figure 20.13 An example of a seasonal monsoon rainfall forecast (for June, July, August 2000) in mm/day. Panels illustrate, from top left: observed estimates, DEMETER coupled models (7 panels), ensemble mean of DEMETER coupled models, ensemble mean of FSU models and the synthetic superensemble.

20.6 Summary and future work

Watching the performance of several multimodels, it became evident that large errors abound in the numerical prediction of weather and seasonal climate. Single models carry large systematic errors. A suite of multimodels does provide a means for

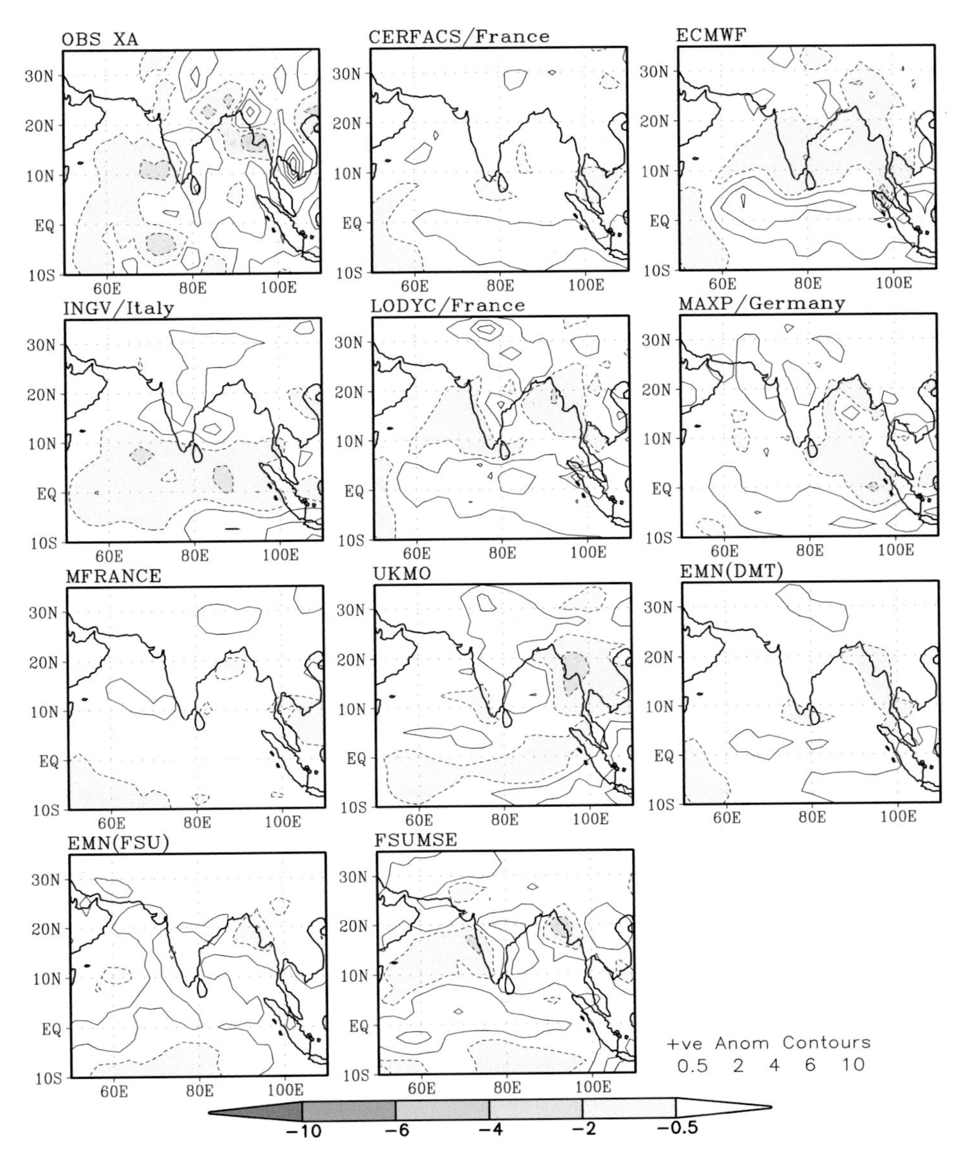

Figure 20.14 Same as Figure 20.13 but for precipitation anomaly forecasts.

the improvement of the collective skills via various ways of ensemble averaging. Improvement of the skills of single models was faster during the decades of the 1970s and 1980s, when faster supercomputers, improved data sets (space and surface based), data assimilation methodologies, improved resolutions, improved physics and the massively parallel computer software contributed to a rapid improvement. These improvements in skill have slowed down somewhat in recent years. An example worth citing here is that of skill improvements from a switch to improved cloud-radiation transfer algorithms for the medium-range NWP models after some four to five years of dedicated work (Lee, 2000). A single model was improved by this procedure to yield an increase of the anomaly correlation of geopotential height by 0.02 at the 500 hPa level. The use of a superensemble using multimodels yields an improvement of nearly 0.2 for the same variable. This conveys the importance of multimodels – however, the slow progress in the skill improvements of single models must continue since any kind of ensemble averaging still relies critically on the performance of the member models.

We have summarised here the results of multimodel forecast experiments with the ensemble/superensemble approach carried out at Florida State University. During the past three years we have experimented on the issue of improved skills from multimodel forecasts of hurricanes (track, intensity and timing), global NWP, heavy rains associated with flood and the seasonal climate forecasts. In all of these areas, we noted an ascending order of skill from the member models, the ensemble mean and the superensemble. We learned that performing an ensemble averaging (after bias removal of individual member models) increased the skill above those of member models, but not to the level of the multimodel superensemble; the latter distinguishes between poorer and better models regionally. Consistently the multimodel superensemble provided superior skills compared with these other ways of looking at forecasts for the hurricane, NWP and heavy rainfall issues. We developed an array of deterministic and probabilistic skills (the METRICS) for forecast evaluation. The thorny issue of seasonal climate anomaly forecast remained only marginally better than that of the climatological values, when the conventional superensemble was used.

An extension of the multimodel superensemble was next carried out using an additional suite of proxy models called the synthetic superensemble components. This shows the most promise for seasonal climate anomalies. In several tests, with a large array of coupled atmosphere–ocean multimodel forecasts (using the European DEMETER model plus a suite of FSU models) we noted somewhat increasing skills for the seasonal climate forecast issues. Our preliminary examinations of these algorithms suggest that the rms error can be reduced by almost a factor of 3 compared with those of member models. This suggests that the proposed method somewhat improves the current state of the art for the climate anomaly forecasts. Thus, further research is expected to be most rewarding. The issue of usefulness of a forecast may not be necessarily met over all regions even from this prescription. Further work is clearly needed to address the usefulness issue. Variants of the synthetic superensemble are also being explored in the areas of hurricane, NWP and heavy rain forecasts. Preliminary tests of these seem quite promising and are summarised here. We recognise the difficulties in assessing the useful skills for improving regional high-resolution mesoscale forecasts. That will be another major area of thrust using the multimodel superensemble strategies.

Acknowledgements
We gratefully acknowledge the ECMWF for providing observed analysis and seven DEMETER coupled model data sets. The research reported here was supported by NSF grant number ATM-0108741, NOAA grant number NA06GPO512, FSURF grant number 1338–831–45, and NASA grants NAG5–13563.

Appendix A: Method for synthetic data generation

The time series of observation can be written as a linear combination of EOFs such as

$$O(x,t) = \sum_n P_n(t) \,.\, \phi_n(x), \tag{20.3}$$

where n is the number of modes selected. The two terms on the right-hand side of the above equation represents the time (principal component PC) and space (EOF) decomposition respectively. PC time series $P(t)$ represents how EOFs (spatial patterns) evolve in time. PCs are independent of each other. Similarly the forecast data can be projected into the PCs and EOFs as below for i member models,

$$F_i(x,T) = \sum_n F_{i,n}(T) \,.\, \varphi_{i,n}(x). \tag{20.4}$$

Here index i represents a particular member model; i can vary from 1 to m. We are interested in knowing the spatial patterns of forecast data, which evolve in a consistent way with the EOFs of the observation for the time series considered. Here we use a regression relationship between the observation PC time series and a number of PC time series of forecast data:

$$P(t) = \sum_n \alpha_{i,n} F_{i,n}(t) + \varepsilon(t). \tag{20.5}$$

In the above equation the observation time series $P(t)$ is expressed in terms of a linear combination of forecast time series $F(t)$ in EOF space. The regression coefficients α_n are found such that the residual error variance $E\ (\varepsilon^2)$ is a minimum. Once the regression coefficients are determined, the PC time series of synthetic data can be written as:

$$F_i^{reg}(T) = \sum_n \alpha_{i,n} F_{i,n}(T). \tag{20.6}$$

Then the synthetic data set is reconstructed with EOFs and PCs as:

$$F_i^{syn}(x,T) = \sum_n F_{i,n}^{reg}(T) \,.\, \phi_n(x). \tag{20.7}$$

These synthetic data (m sets) generated from m member models' forecasts are now subjected to the conventional FSU superensemble technique (Krishnamurti *et al.*, 2002) described in Section 20.2 of this chapter.

References

Baumhefner, D. P. (1968). Application of a diagnostic numerical model to the tropical atmosphere. *Mon. Weather Rev.*, **96**, 218–28.

Ferraro, R. R. (1997). Special sensor microwave imager derived global rainfall estimates for climatological applications. *J. Geophys. Res.*, **102**, 16715–35.

Ferraro, R. R., E. A. Smith, W. Berg and G. J. Huffman (1998). A screening methodology for passive microwave precipitation retrieval algorithms. *J. Atmos. Sci.*, **55**, 1583–600.

Frederiksen, C. S., H. Zhang, R. C. Balgovind, N. Nicholls, W. Drosdowsky and L. Chambers (2001). Dynamical seasonal forecasts during the 1997/98 ENSO using persisted SST anomalies. *J. Climate*, **14**, 2675–95.

Goddard, L., S. J. Mason, S. J. Zebiak, C. F. Ropelewski, R. Basher and M. A. Cane (2001). Current approaches to seasonal-to-interannual climate predictions. *Int. J. Climatol.*, **21**, 1111–52.

Huffman, G. J., R. F. Adler, B. Rudolph, U. Schneider and P. Keehn (1995). Global precipitation estimates based on a technique for combining satellite-based estimates, rain gauge analysis, and NWP model precipitation information. *J. Climate*, **8**, 1284–95.

Kalnay, E., M. Kanamitsu, R. Kistler, *et al.* (1996). The NCEP/NCAR 40-year reanalysis project. *Bull. Am. Meteorol. Soc.*, **77**, 437–71.

Kanamitsu, M., A. Kumar, H. M. H. Juang., *et al.* (2002). NCEP dynamical seasonal forecast system 2000. *Bull. Am. Meteorol. Soc.*, **83**, 1019–37.

Kanamitsu, M., L. Cheng-Hsuan, J. Schemm and W. Ebisuzaki (2003). The predictability of soil moisture and near-surface temperature in hindcasts of the NCEP Seasonal Forecast Model. *J. Climate.*, **16**, 510–21.

Krishnamurti, T. N., C. M. Kishtawal, T. E. LaRow, *et al.* (1999). Improved weather and seasonal climate forecasts from multimodel superensemble. *Science*, **285**, 1548–50.

Krishnamurti, T. N., C. M. Kishtawal, Z. Zhang, *et al.* (2000a). Multi-model super ensemble forecasts for weather and seasonal climate, *J. Climate*, **13**, 4196–216.

Krishnamurti, T. N., C. M. Kishtawal, D. W. Shin and C. E. Williford (2000b). Improving tropical precipitation forecasts from a multianalysis superensemble. *J. Climate*, **13**, 4217–4227.

Krishnamurti, T. N., S. Surendran, D. W. Shin, *et al.* (2001). Real-time multianalysis-multimodel superensemble forecasts of precipitation using TRMM and SSM/I products, *Mon. Weather Rev.*, **129**, 2861–83.

Krishnamurti, T. N., L. Stefanova, A. Chakraborty, *et al.* (2002). Seasonal forecasts of precipitation anomalies for North American and Asian monsoons. *J. Meteorol. Soc. Jpn.*, **80**, 1415–26.

Krishnamurti, T. N., K. Rajendran, T. S. V. Vijaya Kumar, *et al.* (2003a). Improved skill for the anomaly correlation of geopotential heights at 500 hPa. *Mon. Weather Rev.*, **131**, 1082–102.

Krishnamurti, T. N., T. S. V. Vijaya Kumar, B. P. Mackey, N. Cubukcu, R. S. Ross and R. Adler. (2003b). Current thrusts on TRMM and SSM/I based modeling studies on heavy rains and flooding episodes. *Mausam*, **54**, 121–40.

Kumar, T. S. V. V., T. N. Krishnamurti, M. Fiorino and M. Nagata (2003). Multimodel superensemble forecasting of tropical cyclones in the Pacific. *Mon. Weather Rev.*, **131**, 574–83.

Kummerow, C., W. Barnes, T. Kozu, J. Shiue and J. Simpson (1998). The tropical rainfall measuring mission (TRMM) sensor package. *J. Atmos. Ocean., Tech.*, **15**, 809–17.

Kummerow, C., J. Simpson, O. Thiele, *et al.* (2000). The status of the Tropical Rainfall Measuring Mission (TRMM) after two years in orbit. *J. Appl. Meteorol.*, **39**, 1965–82.

Lee, H. S. (2000). *Cloud Specification and Forecasts in the Florida State University Global Spectral Model*. FSU Report No. 2000–1, January 2000. Available from Dept. of Meteorology, Florida State University, Tallahassee, Fl 32306–4520, USA.

Mullen, S. L. and D. P. Baumhefner (1994). Monte Carlo simulations of explosive cyclogenesis. *Mon. Weather Rev.*, **122**, 1548–67.

Olson, W. S., F. J. LaFontaine, W. L. Smith and T. H. Achtor (1990). Recommended algorithms for the retrieval of rainfall rates in the tropics using the SSM/I (DMSP-8). Manuscript, University of Wisconsin, Madison, WI.

Olson, W. S., C. Kummerow, G. M. Heymsfiled and L. Giglio (1996). A method for combined passive-active microwave retrievals of cloud and precipitation profiles. *J. Appl. Meteorol.*, **35**, 1763–89.

Palmer, T. N., F. Molteni, R. Mureau, R. Buizza, P. Chapelet and J. Tribbia (1992). *Ensemble prediction*. ECMWF Technical Memorandum, 188.

Palmer, T. N., A. Alessandri, U. Andersen, *et al.* (2004). Development of a European multi-model ensemble system for seasonal to inter-annual prediction (DEMETER). *Bull. Am. Meteorol. Soc.*, **85**, 853–4.

Powell, M. D. and S. D. Aberson (2001). Accuracy of United States tropical cyclone landfall forecasts in the Atlantic Basin (1976–2000). *Bull. Am. Meteorol. Soc.*, **82**, 2749–68.

Roads, J. O., S-C. Chen, and F. Fujioka (2001). ECPC's weekly to seasonal global forecasts. *Bull. Am. Meteorol. Soc.*, **82**, 639–58.

Ross, R. S. and T. N. Krishnamurti (2005). Reduction of forecast error for global numerical weather prediction by the Florida State University (FSU) Superensemble. *Meteorol. Atmos. Phys.*, **88**, 215–35.

Stefanova, L. and T. N. Krishnamurti. (2002). Interpretation of seasonal climate forecast using Brier skill score, FSU superensemble and the AMIP-I dataset. *J. Climate*, **15**, 537–44.

Szymczak, H. (2004). Skill of synthetic superensemble hurricane forecasts for the Canadian maritime provinces. Masters Dissertation. Available from Department of Meteorology, Florida State University, Tallahassee, FL 32306, USA.

Turk, J., C.-S. Liou, S. Qiu, R. Scofield, M. Ba and A. Gruber (2001). Capabilities and characteristics of rainfall estimates from geostationary and geostationary+microwave-based satellite techniques. In *Symposium On Precipitation Extremes: Prediction, Impacts and Responses, Albuquerque, NM.* Preprints, pp. 191–4. American Meteorological Society.

Voldoire, A., B. Timbal and S. Power (2002). *Statistical-dynamical seasonal forecasting*. BMRC Research Report No. 87. BMRC Australia (Bureau of Met.).

Wang, G., R. Kleeman, N. Smith and F. Tseitkin (2002). The BMRC coupled GCM ENSO forecast system. *Mon. Weather Rev.*, **130**, 975–91.

Williford, C. E., T. N. Krishnamurti, R. J. C. Torres, S. Cocke, Z. Christidis and T. S. V. V. Kumar (2003). Real-time multimodel superensemble forecasts of Atlantic tropical systems of 1999. *Mon. Weather Rev*, **131**, 1878–94.

Yun, W. T., L. Stefanova and T. N. Krishnamurti (2003). Improvement of the multimodel superensemble technique for seasonal forecasts. *J. Climate*, **16**, 3834–40.

Yun, W. T., L. Stefanova, A. K. Mitra, T. S. V. V. Kumar, W. Dewar and T. N. Krishnamurti (2005). Multi-model synthetic superensemble algorithm for seasonal climate prediction using DEMETER forecasts. *Tellus*, **57**, 280–9.

21

Predictability and targeted observations

Alan J. Thorpe, Guðrún Nína Petersen
University of Reading

The aim of this chapter is to provide a summary of the development of the ideas behind, and experiments undertaking, so-called targeted observations of the atmosphere. The scientific issue is the assessment of the role of such targeted observations in improving the skill of numerical weather predictions for time periods up to two weeks ahead. Particular reference will be made to the problem of forecasting extra-tropical cyclones. Within the context of the international programme THORPEX, a vision of the numerical weather prediction (NWP) system of the future will be given involving a two-way interaction between the observing system and the NWP system.

21.1 Introduction

Severe windstorms and precipitation cause substantial societal and economic impact. It is therefore important to consider how we can accelerate improvements in predictive skill. There have been tremendous strides forward taken in numerically predicting the weather, and the three day forecasts of surface pressure are now about as accurate as the one day forecasts were 20 years ago. This is one of the greatest scientific achievements of the twentieth century, with huge societal and economic benefits. These advances in numerical weather prediction arise from developments in modelling as well as in making and utilising observations. Ensemble predictions enable us to do probability estimations, the observational capability of satellites have increased tremendously and there have been great advances in variational data assimilation.

Predictability of Weather and Climate, ed. Tim Palmer and Renate Hagedorn. Published by Cambridge University Press.

However, inaccuracy in initial conditions as well as uncertainties in model formulations still remain a problem. They still cause significant failures in high impact weather forecasts and an inability to significantly extend the range of skilful predictions into 'week 2'. The influence of the tropics on extratropical forecasts is poorly described and there is inadequate skill in predicting mesoscale weather such as precipitation.

Storms are the principal natural hazard in north-western Europe. In fact, it has been estimated that an individual has a higher chance of being killed in a UK windstorm than in a hurricane in Florida. The societal and economic impact of such windstorms can be large: for example, two storms in December 1999 caused 100 fatalities as well as blowing down 400 million trees, 3.5 million electricity users were affected for up to 20 days, 12% of mobile phone transmitters were inoperative and 3 million people were left without water. The annual average insurance loss due to European windstorms is around €1.2 to 1.7 billion. In fact the total insurance loss due to windstorms can be compared to the loss in the USA due to hurricanes.

To improve predictive skill of such storms a range of scientific developments are taking place. Model improvements are being made using process studies from research field experiments, and observations from the new generation of satellite instruments can improve initial conditions. Nevertheless significant data-poor regions will remain, such as in cloud layers in both the extratropics and the tropics.

The concept of dynamically determined adaptive observing was first discussed during the meeting in October 1994 of the First Prospectus Development Team of the US Weather Research Program (Emanuel *et al.*, 1995). The idea was developed further during planning for the 'Fronts and Atlantic Storm-Track Experiment' (Joly *et al.*, 1997), during the meeting in May 1995 of the Second Prospectus Development Team of the US Weather Research Program (Dabberdt and Schlatter, 1996) and at a workshop at National Center for Atmospheric Research (NCAR) in May 1995 (Snyder, 1996). In Emanuel *et al.* (1995), it is noted that 'Another intriguing technique that should be explored is to use ensemble forecasting methods to make a priori estimates of the distribution of sensitivity to observational error, so that programmable observation platforms, such as unmanned aerial vehicles or programmed deployment of dropsondes from commercial aircraft, can be directed to focus on sensitive regions. Adaptive observational strategies may serve to help optimize observations in aid of numerical weather prediction.' That report concludes by saying that 'Advanced applications of adjoint techniques to numerical weather prediction may reveal, in near real time, those parts of the atmosphere that are particularly susceptible to initial error, allowing us to target such regions for observational scrutiny and thereby greatly reduce numerical forecast errors.'

Dabberdt and Schlatter, 1996 used the term adaptive to mean the same as on-demand. To avoid any confusion here we shall use the term targeting in the following way. Targeting is *the process of locating regions in which observations would maximally improve the skill of a weather forecast, using knowledge of the 'flow-of-the day' or more generally dynamically determined information obtained from the forecast model.* Such observations will be called targeted observations. The regions that targeting locates are usually referred to as sensitive regions. Sensitive regions are the localised *zones of the atmosphere from which analysis errors grow significantly and thereby degrade forecast skill.* The sensitive regions depend strongly on many factors including the forecast time-range and the verification region. Sensitive regions may exist in geographical zones, e.g. in areas where flare-up of tropical convection leads to downstream wave-train propagation, they may be associated with certain weather types, such as an extratropical storm track or tropical cyclone genesis regions or they might be flow dependent in zones from where analysis errors grow rapidly.

The use of adjoint products to predict the location of sensitive regions was first suggested in 1995. It is now possible, in principle, to make such predictions operationally, that is 'target', and thereby also decide how optimally to make observation in the sensitive area and design perturbations for regionally targeted ensembles.

21.2 Theory of predicting the location of sensitive regions

Since the idea of adaptive observing using dynamical means was first raised in 1995, a number of distinct, but related, mathematical methods to identify the location of sensitive regions have been proposed and tested. This class of methods builds upon the use of the perturbations to initial conditions that are used in operational ensemble prediction systems. It uses a full non-linear forecast trajectory, which here we call the preliminary forecast, and estimates of the (linear) growth of small perturbations to the trajectory.

Before summarising the different variants within this class of methods, it is useful to note that another class of more empirical methods of targeting has also been proposed. An example of this second class is the so-called water vapour-potential vorticity method described by Demirtas and Thorpe (1999). Localised mismatches between an analysis and a timely water vapour satellite image can be interpreted as a tropopause-level potential vorticity analysis error. Inversion of the potential vorticity error allows (balanced) targeted wind and temperature 'observations' to be available to the forecast model. Other examples include the practice of bogusing hypothetical observations creating a tropical cyclone vortex in analyses to enable such cyclones to be more effectively initiated. Also weather forecasters can apply their experience to suggest subjectively key, or sensitive, regions where synoptic development may be most strongly affected by analysis errors. Browning *et al.* (2000) provide a perspective on the forecasting of ex-Hurricane Lili and compare the empirical class with the class based on linear growth of perturbations. The conclusion is that the sensitive regions highlighted by the two classes are in similar locations at least for that case.

We now return to the linear perturbation class of methods. One might wish that the targeting method was in some senses statistically optimal in that over a number of cases the predicted locations of the sensitive regions would lead, if targeted observations were obtained therein, to a significant reduction in forecast error (Berliner *et al.*, 1999). The required estimates of the change in the forecast error covariance due to the addition of targeted observations could be found using an extended Kalman filter. However, this is not currently feasible due to the large dimension of the state space of the numerical weather prediction model.

In principle the problem of finding sensitive regions is one of estimating where analysis errors may be large and of estimating the rate at which such errors will grow during the forecast. Targeting is, in simplified terms, aimed at finding regions with potentially large analysis errors that will also grow rapidly. Lorenz and Emanuel (1998) and Hamill and Snyder (2002) focus on locations where initial condition uncertainty is large or where targeted observations would reduce the analysis error the most. By contrast, Buizza and Montani (1999) and Montani *et al.* (1999) focus on estimating the growth rate of singular vectors during the forecast. The leading singular vectors (SV) are the structures that grow the most rapidly, in a linear sense, over a fixed forecast period. As well as the SV method, adjoint sensitivities (Langland and Rohaly, 1996) and quasi-inverse linear integrations (Pu *et al.*, 1997) use the adjoint or tangent-linear of the full forecast model trajectory.

It is important to also take into account the characteristics of the data assimilation system, and various methods for targeting have been proposed that aim to do so. Baker and Daley (2000) and Doerenbecher and Bergot (2001) examine sensitivity with respect to the observations. By contrast, Bishop and Toth (1999) and Bishop *et al.* (2001) use the already computed ensemble members. These are manipulated by linear combination to evaluate the likely forecast error reduction resulting from a localised analysis error reduction presumed to have arisen from targeted observations located there. These are referred to as the ensemble transform (Kalman filter) methods. These methods include a number of approximations, e.g. the size of the ensemble is usually too small to give a good estimate of the background error covariance, the description of the observation network is crude, and observations up to and at the verification time are excluded. Differences between covariances described by the methods and in the operational data assimilation, on which the ensemble is based, impact the results and, as for the other methods in this class, a linear assumption has to be made (Bishop and Toth, 1999). However, despite these assumptions the ensemble transform Kalman filter (ETKF) is a valuable tool and targeting in ETKF sensitive areas has decreased forecast errors substantially (see Section 21.4).

Another way to predict the likely forecast error reduction resulting from deployment of targeted observations is by a reduced-rank state estimation using the leading order Hessian singular vectors (Leutbecher, 2003). In this way the dependence of the likely forecast error reduction on the particular deployment of observations within the sensitive region can be evaluated.

For further details of these various methods, and other related ones, the reader is referred to the published papers quoted herein. To provide one example of what is possible we now present some results from Leutbecher *et al.* (2002) on the Lothar storm that caused significant damage and loss of life in France and neighbouring countries on 26 December 1999. Observing system simulation experiments over a 48-hour forecast period were carried out using the European Centre for Medium-Range Weather Forecasts (ECMWF) forecast model and its 4D-Var assimilation system. Both total energy and Hessian singular vectors were computed for comparison purposes. A truth trajectory (using the Météo-France analysis and ECMWF model) and a poor forecast (using the ECMWF analysis and model) were selected from an ensemble of forecasts. It was then possible to locate sensitive regions with the poor forecast trajectory and transplant soundings taken from the analysis leading to the truth trajectory as the theoretical targeted observations.

The sensitive region was formed from a weighted average of the first five singular vectors and is shown in Figure 21.1 for both total energy (TESV) and Hessian (HSV) singular vectors. There is clearly significant overlap between the two predictions of the sensitive region. However, there is a distinct difference: while the TESV sensitive region covers the east coast of North America, the HSV region doesn't. The reason for this difference is that the HSV take into account an observational network. The east coast of North America is an area of dense observations, where the initial errors are more constrained than over the ocean and other remote areas.

In order to assess whether these linear estimates of the most sensitive region are accurate, a set of 14 test sensitive regions was defined. The impact of adding 40 targeted soundings into each of these was found by integration of the full nonlinear forecast model. In this way the optimal zone for observations (OZO) could be found and compared with the sensitive regions highlighted in Figure 21.1. The OZO was well estimated by the HSV providing evidence for the accuracy of this targeting method. In Figure 21.2 the evolution of the forecast error for Lothar is shown using targeted observations in a variety of locations. It can be seen that putting these observations in the HSV region is almost as good as in the OZO, showing the potential power of targeting in reducing forecast error.

In Figure 21.3 the implications of the forecast error reduction in total energy for the sea-level pressure field is shown. The verification is the truth forecast and it is clear that the ECMWF forecast exhibited a significant error, in this case, at 48-hour range. The final frame shows the excellent improvement possible by including 40 targeted soundings in the sensitive region.

21.3 Experimental testing of targeting in FASTEX

The first opportunity to test the real-time targeting of observations came in the Fronts and Atlantic Storm-Track Experiment, FASTEX, which took place during January and February 1997. FASTEX involved research aircraft, ship and additional routine

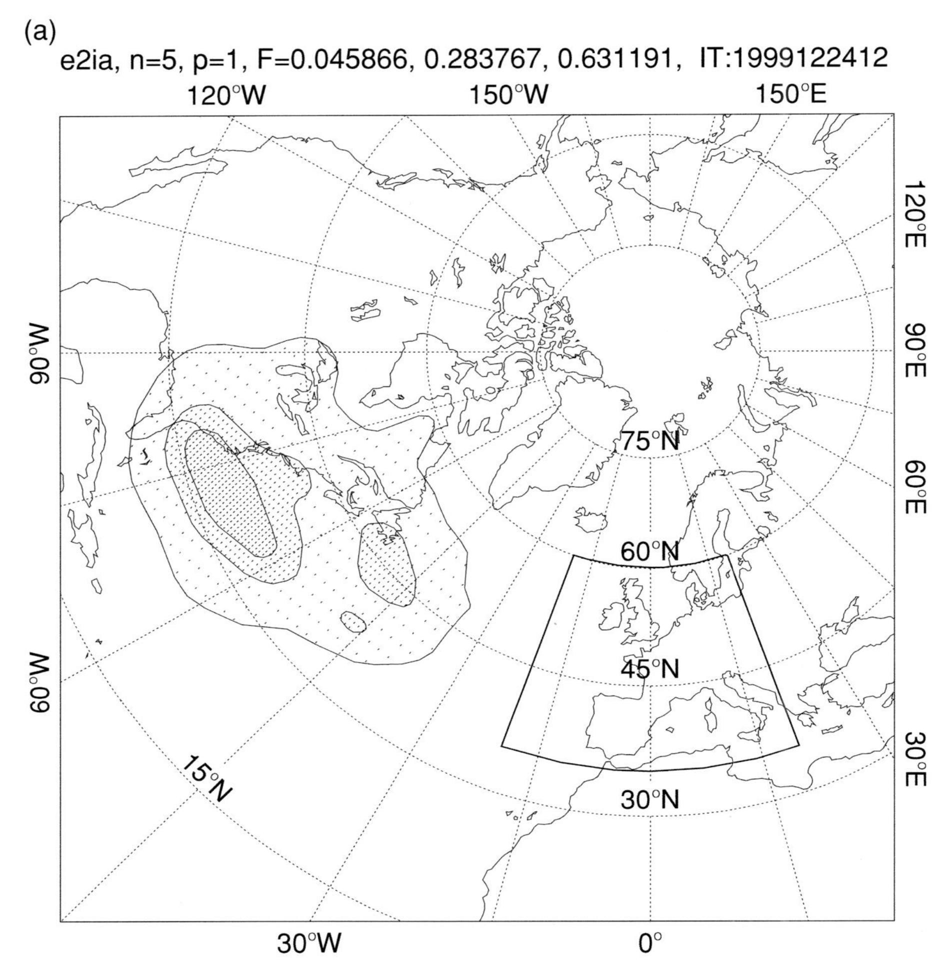

Figure 21.1 Sensitive regions for Lothar predicted based on (a) total energy and (b) Hessian singular vectors. The box indicates the verification area. From Leutbecher *et al.* (2002).

radiosondes being utilised to provide an enhanced observational description of the evolution of extratropical cyclones across the North Atlantic storm track. Real-time targeting calculations were made by a variety of centres: ECMWF, Météo France, US National Centers of Environmental Prediction (NCEP) and US Naval Research Laboratory (NRL). These allowed the location of sensitive zones to be predicted and various research aircraft to be directed into these regions to release dropsondes. The resulting profiles thereby produced targeted observations for later inclusion in the operational forecast suites. A number of sample hindcasts with and without the targeted observations were then carried out by the various participating groups. Results of these various studies were published in the *Quarterly Journal of the Royal Meteorological Society* Special Issue on FASTEX (October C 1999). The overall

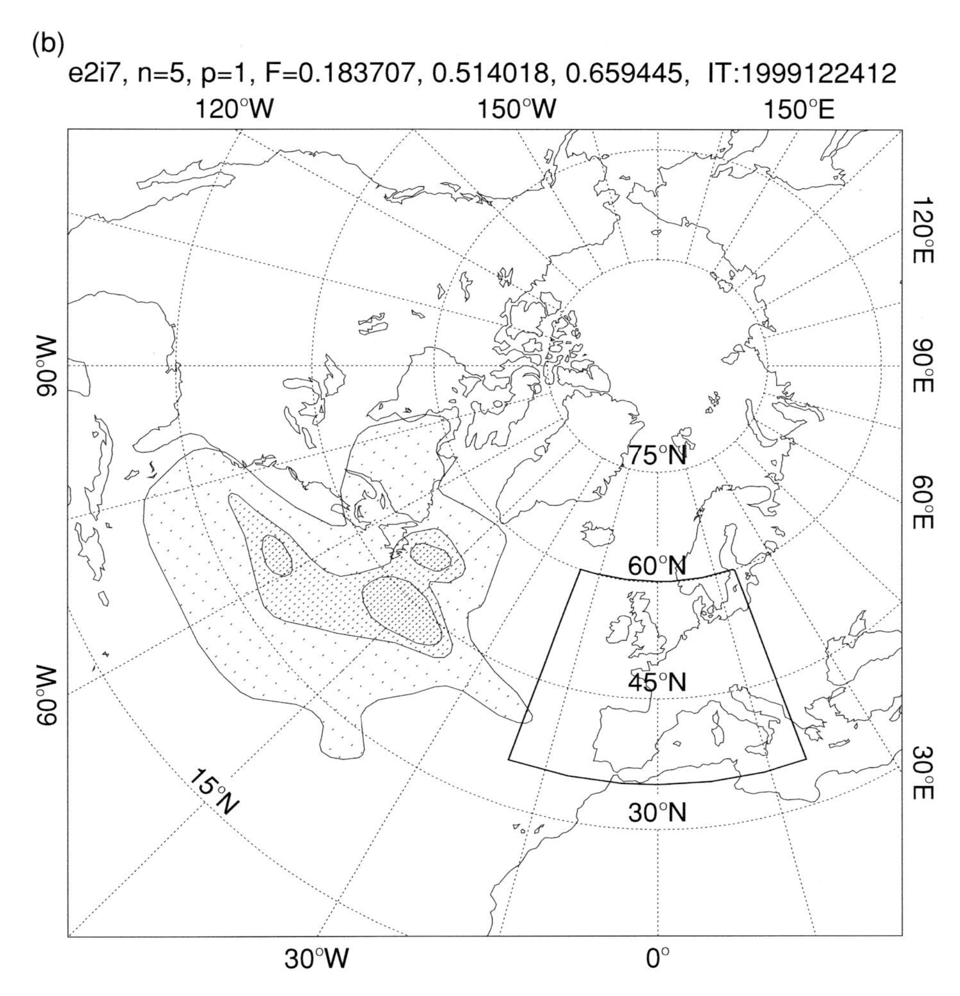

Figure 21.1 (*cont.*)

conclusion was that there were cases of significantly improved skill when using the targeted observations but also there was a significant number of small and even negative impact cases. One important point to bear in mind was that many of the cyclones in FASTEX were relatively well forecast using the routine observing network and so the potential for improvements by including additional targeted observations was somewhat limited.

The study that used the ECMWF forecast system gave some of the more encouraging results, which may reflect the relative sophistication of the data assimilation system and the quality of the model in terms of both resolution and parametrisations. In Figure 21.4 we show results from Montani *et al.* (1999). This study used the ECMWF total energy norm singular vectors to provide the estimates of the locations of the sensitive regions in the western Atlantic sector. Dropsonde data obtained in

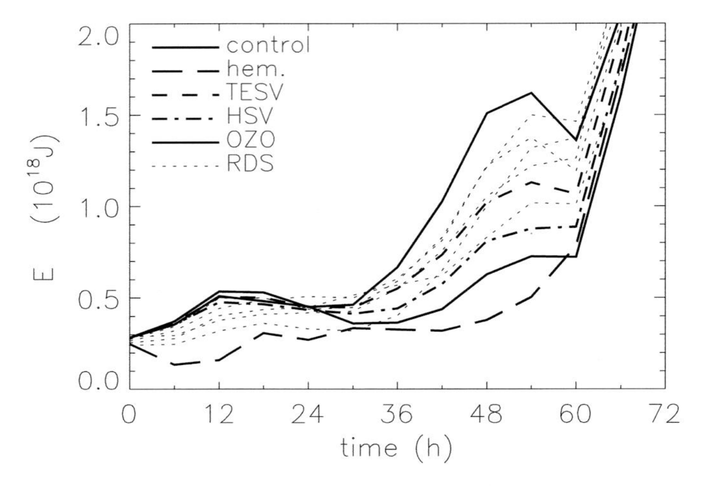

Figure 21.2 Evolution of the forecast error for Lothar including 40 targeted profiles in various locations. TESV: total energy singular vectors; HSV: Hessian singular vectors, OZO: optimal zone for observing; RDS: random distribution scheme. From Leutbecher *et al.* (2002).

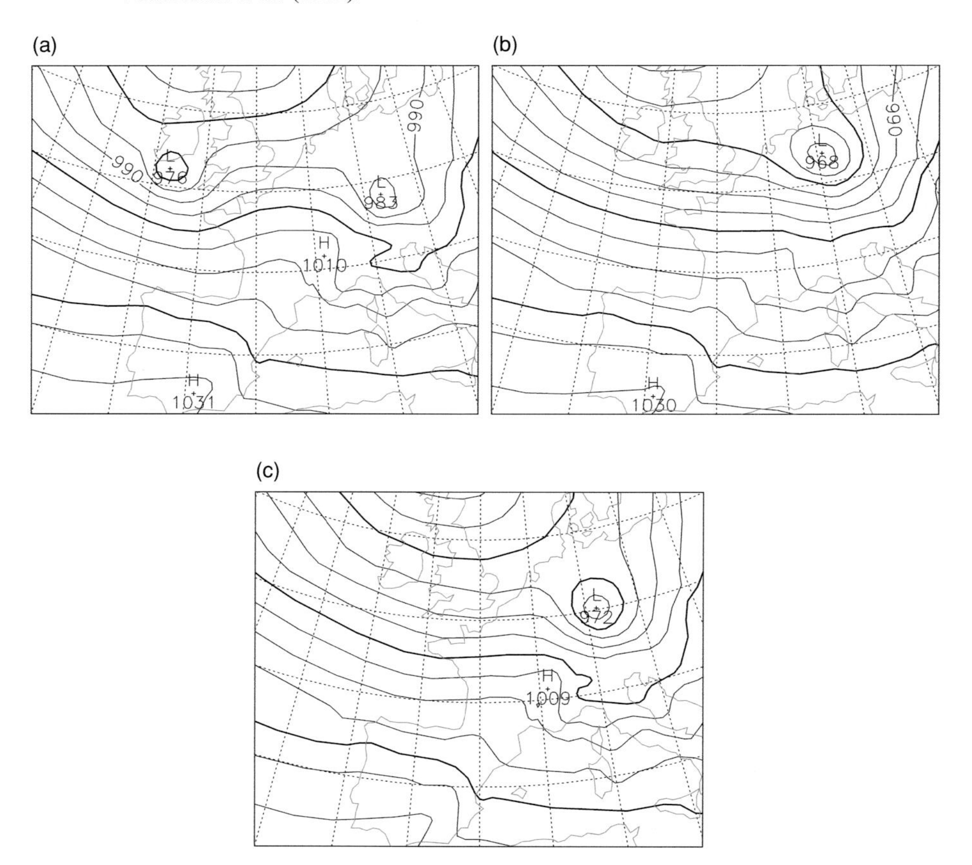

Figure 21.3 Observation system simulation experiments for Lothar. Mean sea level pressure on 26 December 1999, 12 UTC. (a) 48-hour forecast from the ECMWF analysis, (b) the truth forecast and (c) ECMWF forecast with 40 targeted soundings. Contour interval is 5 hPa. The figure is adapted from Leutbecher *et al.* (2002).

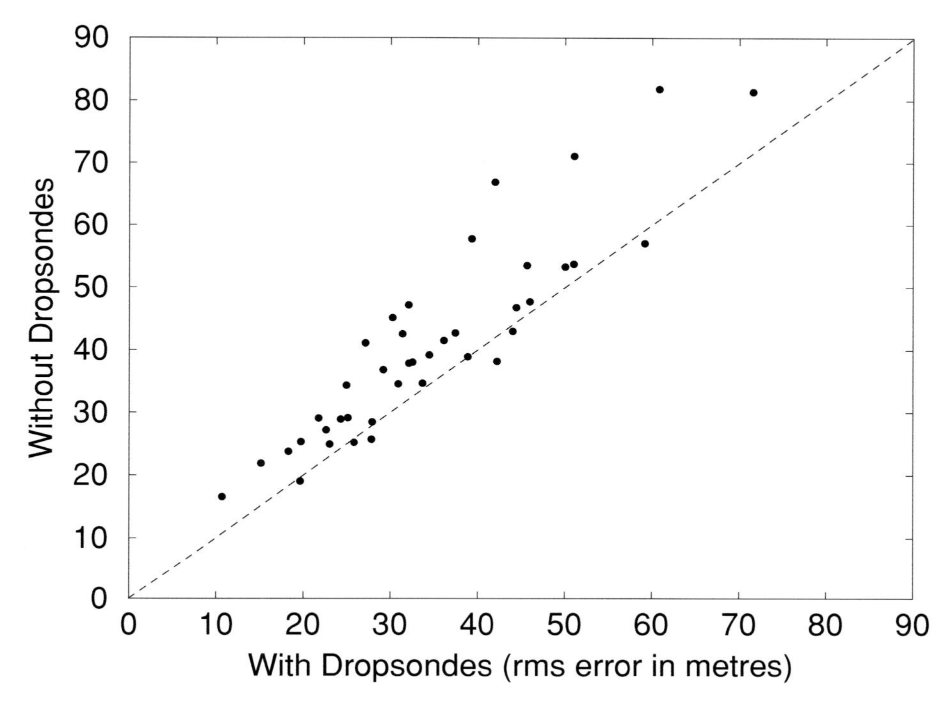

Figure 21.4 Scatter plot over SVs' verification regions for FASTEX cases, comparing the rms forecast error (m) of 500 hPa and 1000 hPa geopotential height for forecasts including the FASTEX dropsondes with those without the targeted supplementary data. The forecast error is averaged over a verification region over north-west Europe. From Montani *et al.* (1999).

the predicted sensitive area were used as the targeted observations. Figure 21.4 shows a scatter plot of the forecast error (root-mean-square) of 500 hPa and 1000 hPa height fields averaged over a verification region in north-west Europe encompassing the UK. Forecast periods, every 6 hours, between 30 and 48 hours are included and the error is plotted for forecast reruns with the targeted observations included against those where they were not included. Very few points are below the dashed line (representing zero error reduction) indicating that in most cases the targeted observations reduced the forecast error. The error reduction is particularly large when the control forecast, without the targeted observations, exhibited large prediction errors. On average the gain in forecast accuracy is about 15% with the largest error reduction being about 37%.

21.4 An operational system for the Pacific: WSR programme

The FASTEX field campaign in the north Atlantic sector in early 1997 was followed by a targeting trial of a similar type in the north Pacific in early 1998 (NORPEX;

Langland *et al.*, 1999). NOAA, via activities in NCEP, built on the success of these field campaigns by establishing the Winter Storm Reconnaissance (WSR) field programme in early 1999 (Toth *et al.*, 1999). Toth *et al.* (2002) note that 'The aim of the WSR program is to reduce forecast errors for significant winter weather events over the contiguous US and Alaska in the 24–96 hour lead time range through the use of adaptive observations over the data sparse northeast Pacific'. NOAA and USAF manned aircraft flew from bases in Hawaii and Alaska to launch dropsondes into predicted sensitive regions.

The WSR 1999, and the follow-up WSR 2000, exercises were successes with the majority of the forecasts with targeted observations having significantly improved skill. The National Weather Service made the WSR programme operational in January 2001 and has run it for 60 days every winter, covering the 24–60 hours forecast time range over the Pacific coast. The programme is now run utilising the Ensemble Transform Kalman Filter (ETKF) to locate sensitive regions.

The WSR programme is operational in the sense that the flights are triggered by forecasters. The process involves case selection followed by automated sensitivity calculations being triggered and choice of predesigned flight tracks in the light of the location of the sensitive regions being made. This chain of events is under the complete control of the forecasters, with no input from the research scientists. For further information about the WSR programme see, for example, Majumdar *et al.* (2002), Szunyogh *et al.* (2002) and Holland *et al.* (2004).

During WSR 2001 there were a total of 27 flights/cases during a one-month period. Forecasts were improved in 60–70% of all WSR cases and the average rms error reduction in the 24–96 hour forecasts in the preselected verification regions was about 10%, with some cases in excess of 25%. This implies a lead-time gain for a 2-day forecast of about 12 hours. Similar results have been found for the other WSR programmes. In Figure 21.5, adapted from Szunyogh *et al.* (2002), we see the propagation in WSR 2000 of the regions of positive forecast improvement from the location of targeted observations in the Pacific eastwards until at 72 hours the improvement covers many parts of the USA and Canada.

21.5 Cloudy sensitive regions

The WSR programme utilises the availability of manned meteorological research aircraft to make targeted observations. There is a wide potential for many other observing methods to be used for targeting. These methods, which could ultimately be made operational, include targeting of satellite data, ASAP ship soundings, AMDAR aircraft reports, supplementary mobile and fixed-location soundings as well as other supplementary in-situ instruments. New satellite instruments will provide huge data sets, but *targeting* is essential for selective dynamic utilisation of the huge data flows from satellites and determining locations for higher scan rates.

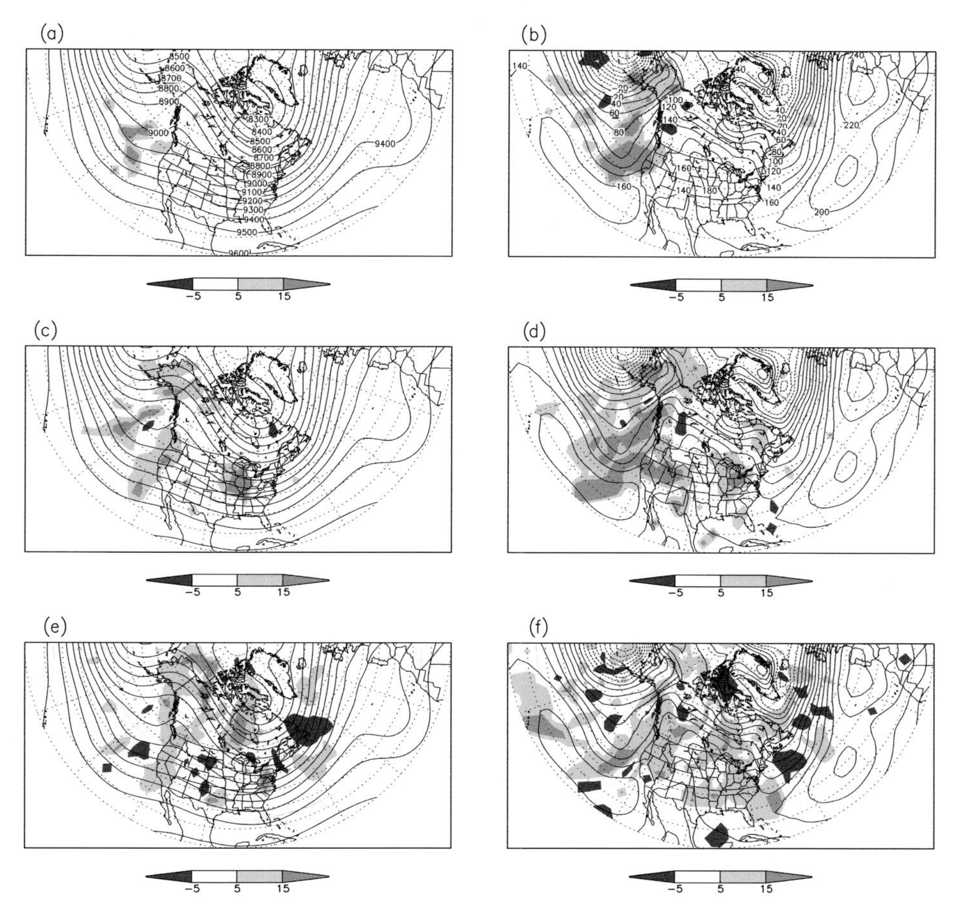

Figure 21.5 Forecast error reduction in geopotential height (%, shaded) due to soundings over the mid Pacific during WSR 2000, at (a), (c), (e) the 300 hPa pressure level and (b), (d), (f) for surface pressure at 24, 48 and 72 h lead times. Contours show the time mean of geopotential height at (a), (c), (e) the 300 and (b), (d), (f) 1000 hPa pressure level. From Szunyogh *et al.* (2002).

However, in cloud layers and below cloud-base the new satellite instruments will have poor resolution. Figure 21.6 shows the huge volume of GOES satellite derived winds in three layers at a single analysis time in mid Pacific. Given such huge data volumes there is a need to thin such data sets for real-time data assimilation. Furthermore, due to the satellite observation error estimate being uncertain, it is possible that a huge volume of information from satellites might deteriorate the analysis that forecasts are based on. One way to carry out such thinning is via dynamical targeting. If we retain only the data in the layer 400–699 hPa, which is known from the singular vector calculations to be a key sensitive region, then far fewer data are available (see Figure 21.7). This dramatic reduction in data availability in the sensitive region is because of the difficulty of sensing from satellites in significant cloud layers.

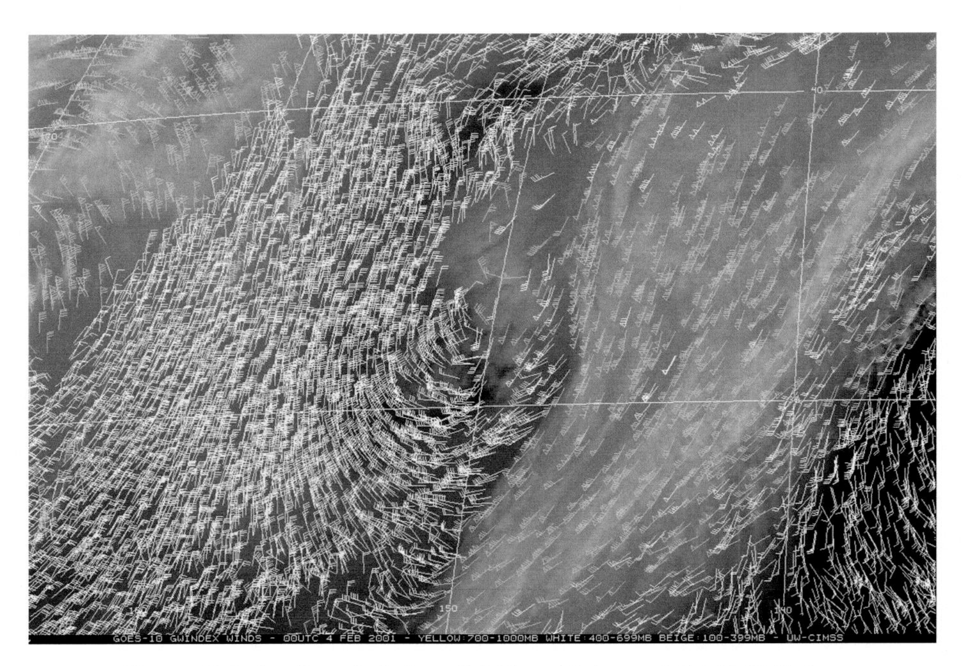

Figure 21.6 Satellite winds from GOES on 4 February 2001 00 UTC. Image courtesy of Chris Velden/CIMSS.

Figure 21.7 Satellite winds from GOES on 4 February 2001 00 UTC. Only winds in the sensitive layer 400–699 hPa are shown. Image courtesy of Chris Velden/ CIMSS.

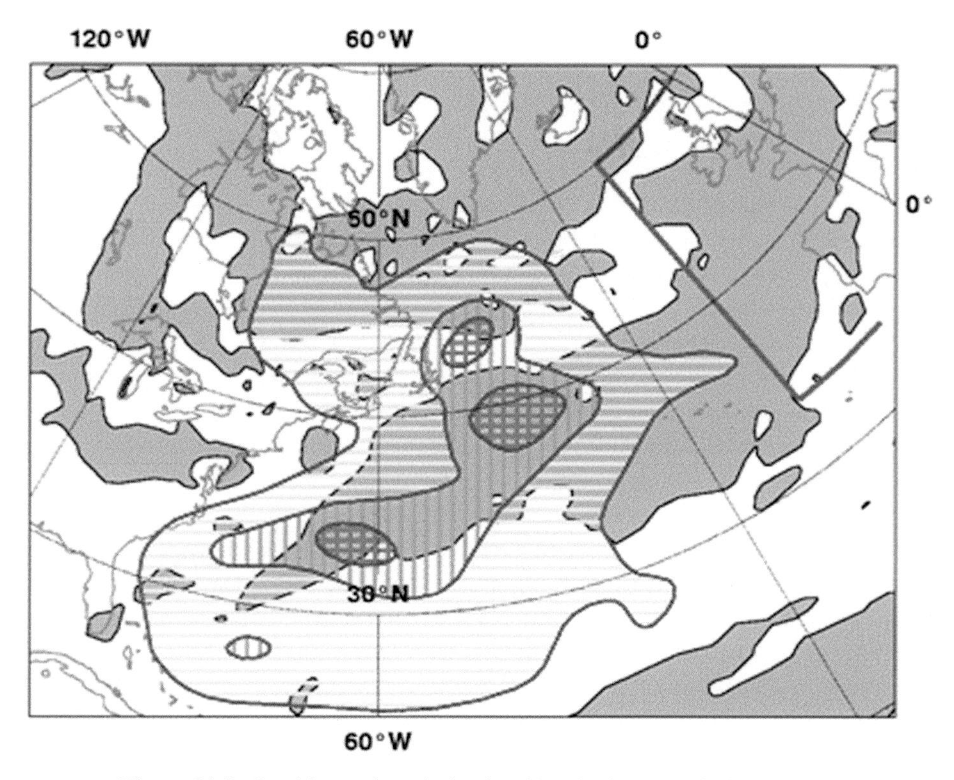

Figure 21.8 Sensitive regions (striped and hatched) occur within cloud layers (shaded). Adapted from Leutbecher *et al.* 2002.

Calculations made by Leutbecher *et al.* (2002) of the location of the sensitive region for the Lothar storm are shown as striped and hatched regions in Figure 21.8. Also plotted are the model cloudy regions, given with continuous grey shading. It can be seen that there is a large overlap between cloud layers and sensitive regions.

Given that the currently planned generation of satellite instruments will have poor resolution in cloud layers, it is clear that other sources of targeted observations will have to be investigated.

Fortunately there is substantial potential to enhance the observing system with new low-cost *in-situ* instruments. This opens up the exciting prospect of targeting additional mobile *in-situ* instruments into these zones.

21.6 THORPEX: A Global Atmospheric Research Programme

An *internationally coordinated* programme of research and field experimentation is needed to determine the optimal utilisation of satellite and *in-situ* observing systems, leading to an improved global observing system and advances in data assimilation methodologies. The aim of such a programme would be *to enable forecast skill to make a 'leap forward' necessary to solve the problems described earlier.*

The Global Atmospheric Research Programme, GARP, led to the Global Weather Experiment called the First GARP Global Experiment, FGGE, in 1979. This led to a significant acceleration in NWP skill improvements.

Recent advances in, for example, data assimilation, ensemble prediction and targeting suggest that it is now possible to make another leap forward. It is time for a second GARP, which is called THORPEX: A Global Atmospheric Research Programme. For more detailed information on THORPEX see www.wmo.int/thorpex. THORPEX was established in 2003 and is one of the most ambitious, fundamental, complex and promising international efforts in the field of atmospheric and related sciences. It is a decade-long international research programme aiming to accelerate improvements in the accuracy of one-day to two-week high-impact weather forecast for the benefit of the society and the environment. It is clear that one of the major societal challenges of the twenty-first century is to reduce the impact of weather-related hazards. Skilful weather forecasts can decrease this impact. THORPEX is based on collaboration between the operational and research community as well as users of forecast products.

In order to achieve its goals THORPEX needs to facilitate advances in predictability theory, data assimilation and understanding of dynamical processes. By using probabilistic forecasts that quantify the likelihood of an event we can now provide more specific information on the likely outcome of an event. Probabilistic forecasts are especially valuable tools for predicting events of low probability but high risk, i.e. extreme events. Data assimilation is the process leading to an estimate of the state of the atmosphere and the ocean at a particular time – an analysis – as well as measuring the uncertainty associated with the analysis. The uncertainties in the analysis arise from uncertainties in the observations, the first-guess forecasts and approximations in the assimilation schemes. There have been recent advances in many aspects of assimilation and observing systems, e.g. improvement in assimilation algorithms, targeting techniques and an increased volume and quality of atmospheric observations. These advances provide an opportunity to make substantial improvements in the forecast skill. The limits of predictability depend on the properties of the atmosphere that are forecast. Increased knowledge of the global-to-regional influences of the evolution and predictability of high-impact weather can increase the spatial and temporal range of predictability.

THORPEX will explore the idea of a reactive two-way flow of information between the observational and NWP systems and develop a design of an interactive forecast system. Until now this flow has been one way from observations into the NWP system. An interactive system allows information to flow between forecast users, NWPs, data assimilation systems and observations to maximise forecast skill.

We can, for example, dynamically utilise satellite and other data, so we can determine the optimal mix of diverse observational types and design ensembles so as to target supplementary observations.

THORPEX is structured with four subprogrammes of research:

1. Predictability and Dynamical Processes
2. Observing Systems
3. Data Assimilation and Observing Strategies
4. Societal and Economic Applications.

THORPEX must also facilitate development of a diverse range of *in-situ* observing technologies to supplement satellite data in cloudy sensitive regions. As mentioned in Section 21.5 there is a range of new and innovative *in-situ* low cost instruments under current development capable in principle of obtaining targeted observations. THORPEX will act as a spur to the development and testing of these instruments, which include:

(a) driftsonde
(b) robotic aircraft
(c) aircraft-deployable ocean surface data buoys
(d) rocketsonde
(e) bi-directional radiosondes.

In Figure 21.9 we show a schematic of the NCAR driftsonde system involving a zero-pressure balloon flying in the stratosphere with a gondola holding 24 dropsondes. The launch times of the driftsonde and the deployment time of the individual dropsondes will allow targeting to be implemented. The design specification of the driftsonde is that the total cost of each sounding of the atmosphere is similar to that of a routine radiosonde profile.

An example of the type of coverage possible over the north Pacific sector from four launch sites in Japan is given in Figure 21.10.

A number of observing system tests and regional campaigns are planned during THORPEX. One such campaign, The North Atlantic THORPEX Regional Campaign (NA-TReC), was carried out in 2003 (see Section 21.7).

As a culmination of THORPEX research it is envisaged that there will be a 'THORPEX Global Experiment' by analogy with FGGE. This will last for a year, say in 2009 or 2010, and include all candidate *in-situ* systems and available remote sensing systems. The experiment will consider all predictable spatial and temporal scales out to 14 days and the data from the experiment will be available in real time.

THORPEX aims to assist global coordination to develop the observing system for weather forecasting; for example, it has strong links with the Commission for Basic

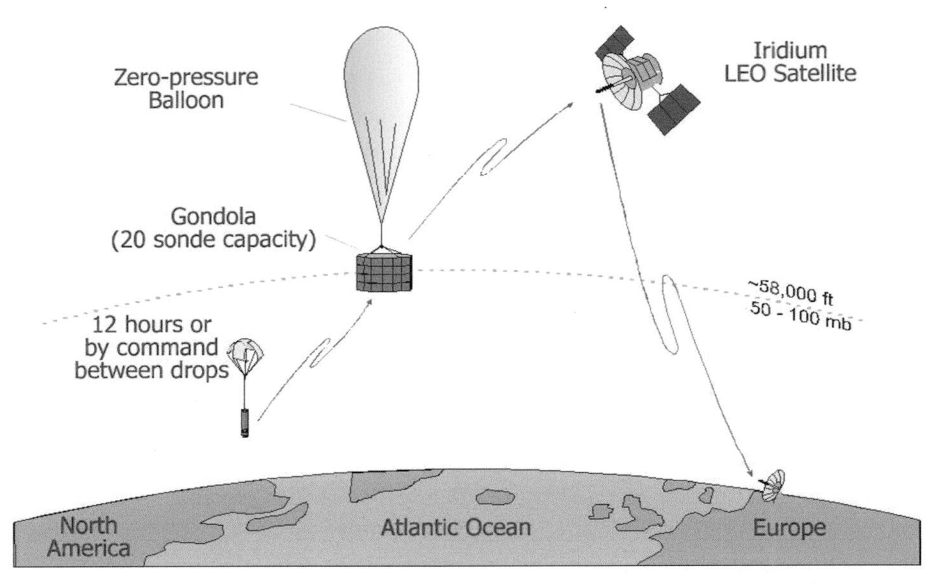

Figure 21.9 A schematic figure of a driftsonde system. Illustration courtesy of Hal Cole, UCAR.

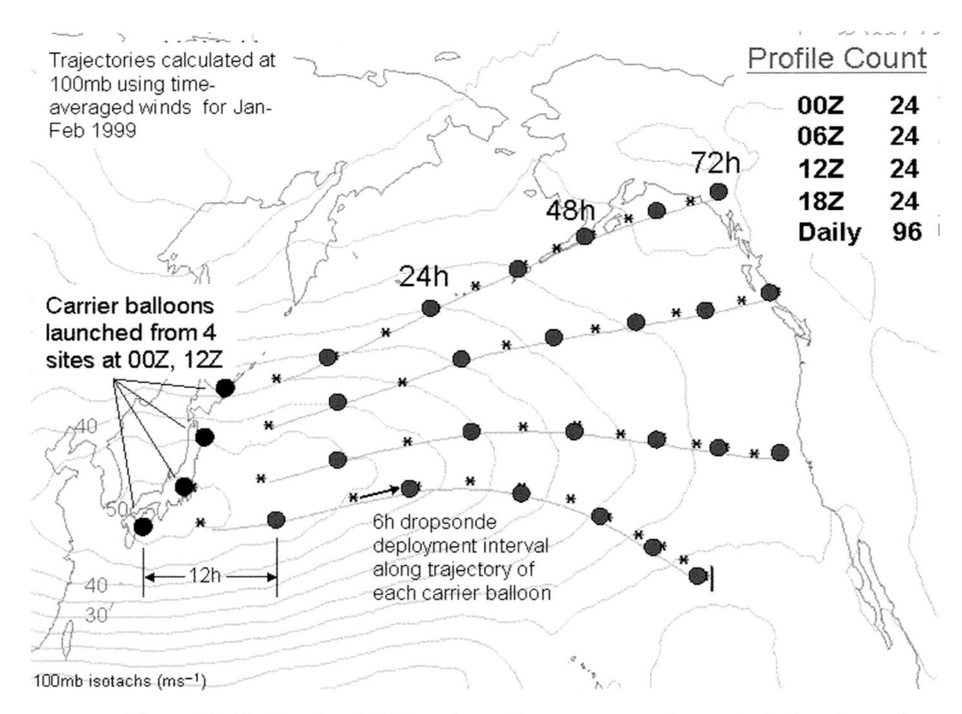

Figure 21.10 Simulated driftsonde profile coverage on data assimilation time, after three days of deployment from 100 hPa. The carrier balloons were launched from four sites in Japan. Each dot represents a separate carrier balloon/gondola and GPS dropsonde profile location at 0000 or 1200 UTC. Stars are profile locations at 0600 or 1800. Illustration courtesy of Rolf Langland, NRL.

Systems and the EUropean Composite Observing System (EUCOS). It also needs to coordinate with the design of the Global Climate Observing System so that advances will also benefit climate. THORPEX is a programme of the World Weather Research Programme and is supported by the Working Group on Numerical Experimentation.

21.7 The North Atlantic THORPEX Regional Campaign

The North Atlantic THORPEX Regional Campaign (NA-TReC) took place from 17 October to 15 December 2003. The NA-TReC was organised and run by EUCOS but other organisations played a vital role including the 18 European National Met Services that are members of EUMETNET, National Oceanic and Atmospheric Administration (NOAA), National Aeronautics and Space Administration (NASA), Environment Canada, EUMETSAT, National Center for Atmospheric Research (NCAR), German Aerospace Center (DLR) and NRL.

The campaign's main objectives were to test the real-time quasi-operational targeting of observations using a variety of platforms, such as AMDAR, ASAP ships, extra radiosonde ascents, research aircraft and meteorological satellites. The focus of the campaign was on short range, 24 to 72 hour forecasts, regional scale numerical weather prediction over Europe and the eastern coast of North America. The campaign utilised different targeting techniques: SV methods, the ETKF and adjoint models. The sensitive area calculations were carried out at ECMWF, UK Met Office, Météo France, NCEP and NRL. The targeted region in each case was selected using all available sensitive area predictions. During the field campaign 21 cases were targeted and 23 000 additional AMDAR observation reports generated, as well as 65 extra ascents from ASAP ships, 214 additional radiosonde ascents and 277 aircraft dropsonde profiles. In addition, satellites were applied to make targeted observations of rapid scan winds (Mansfield *et al.*, 2005). Further information regarding the planning of targeted observations can also be found in Leutbecher *et al.* (2004).

Figure 21.11 (colour plate) shows an example of sensitive area predictions by the TESV and the ETKF methods. The contours show the forecast mean sea level pressure while the shaded areas represent the regions predicted sensitive by each method. The area that was targeted in this case is shown by the bold boxes. Even though there are similarities in the sensitive area predictions, in this case there are obvious differences between the two predictions. A possible explanation is that the verification areas are not of the same size, but that doesn't explain all differences, for example that the centre of the surface low is predicted highly sensitive by the ETKF method but is outside the 1×10^6 km contour for the TESV method. One of the scientific objectives of the NA-TReC is to contribute to the understanding

(a)

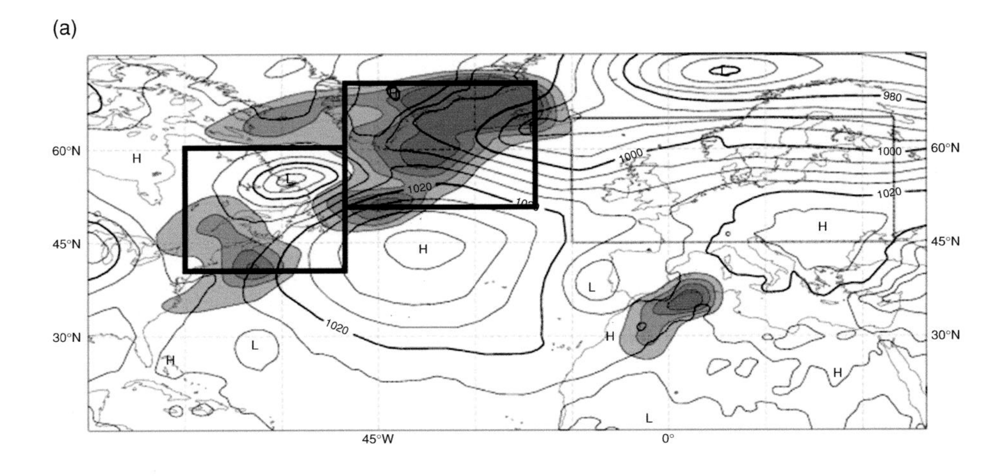

(b)

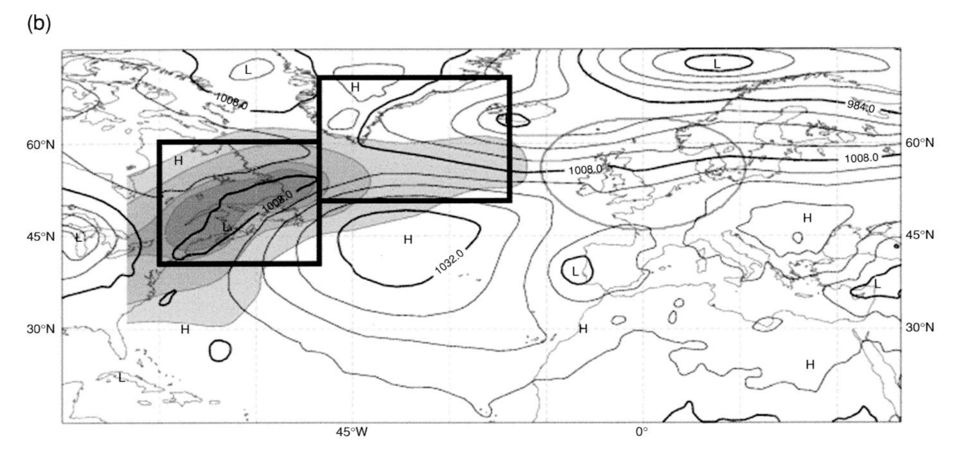

Figure 21.11 Sensitive area predictions for NA-TReC 7. (a) A prediction based on total energy singular vectors, and (b) a prediction based on the ETKF. The shaded areas are the sensitive area predictions, the darkest shade showing the most sensitive area. The sizes of the areas are 8, 4, 2 and 1×10^6 km^2. The contours show the mean sea level pressure (hPa) forecast for 66 hours in (a) and 72 hours in (b). The grey rectangle in (a) and the ellipse in (b) represent the verification area, and the bold boxes outline the region that was actually targeted.

of differences between the targeting methods as well as the impact of the targeted observations.

Figure 21.12 shows the locations of the targeted observations associated with this NA-TReC 7. The AMDAR observations cover the targeted areas quite well, while extra radiosonde ascents are only at the eastern coast of Canada. In addition to the observations shown in the figure, there were satellite observations. The observations in the Mediterranean region are in relation to another case.

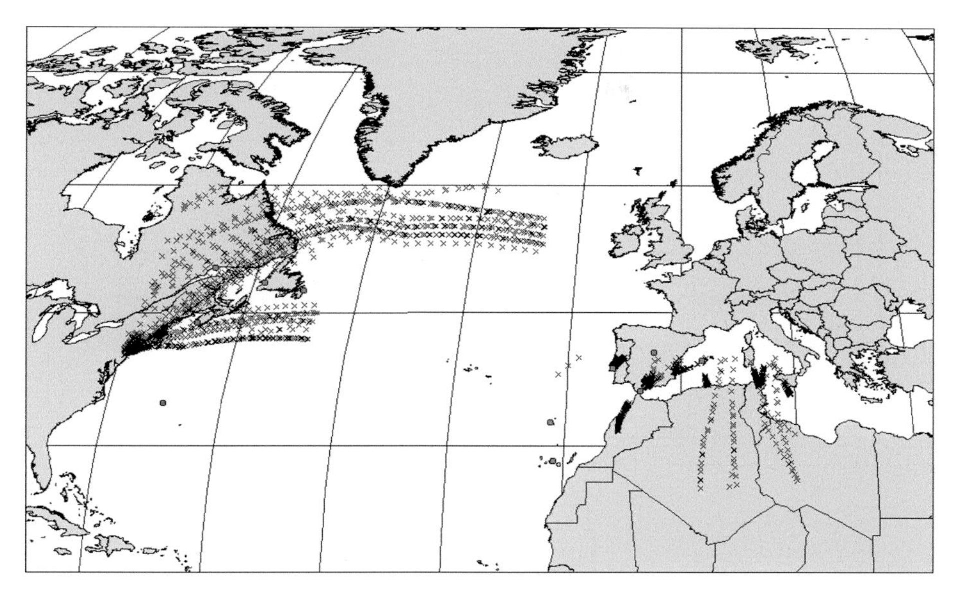

Figure 21.12 An overview of the targeted observations 28 October 2003. The crosses represent AMDAR observations, the dots radiosonde ascents and an ASAP ship radiosonde ascent.

This is the first time that the real-time targeting control of such a complex set of observing platforms has been attempted and the campaign might be considered an essential preparation for future targeting field campaigns. However, it is possible that the number of cases in the campaign is not sufficiently large to give significant results regarding the different targeting techniques.

21.8 Concluding remarks

Building on the ideas of targeting observations we can propose a vision of the way numerical weather prediction could be done in the future. This vision involves a preliminary forecast being made using observations from the routine observing system and the NWP model. Ensembles of the future will involve an ensemble design that is optimised for particular customer needs. As computing resources become more readily available, suites of ensembles each separately designed for different customers can be contemplated. Part of the output from the ensemble design step is the identification of sensitive regions in which to make additional targeted observations to improve the skill of each particular ensemble. The additional targeted observations are then ingested into the data assimilation system and the production of the optimised ensemble occurs. This represents a two-way flow between observations

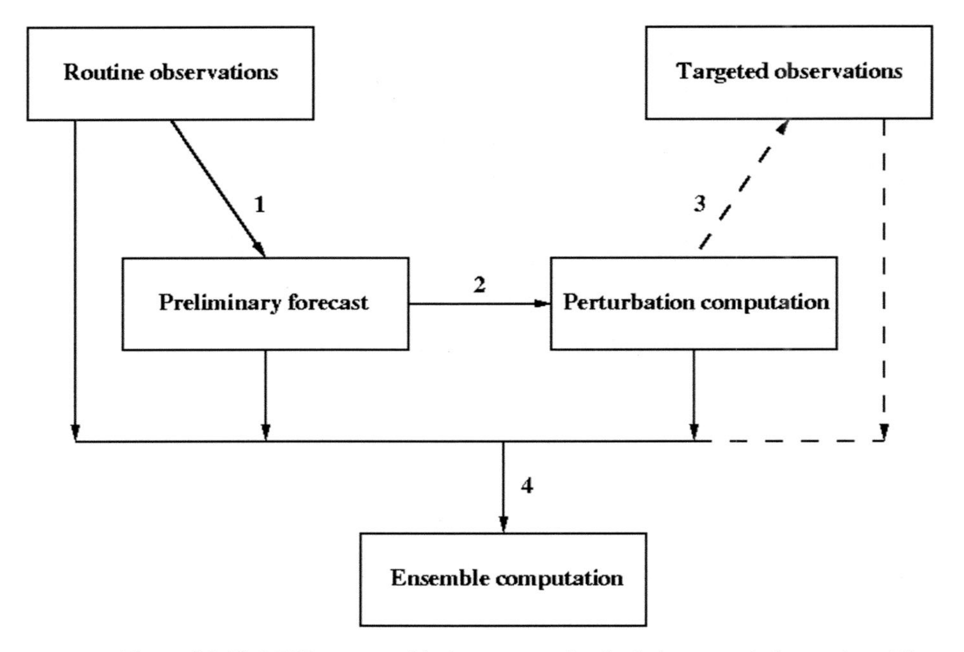

Figure 21.13 NWP system with the new step 3 to include targeted observations. The advent of dynamical targeting allows the new step 3 to be introduced into the NWP system. It consequently allows, in principle, information to flow to and fro between the observing system and the NWP system in an interactive way to optimise forecast skill. The international programme THORPEX will carry out global testing and refinement of these ideas to assess the potential impact on operational forecast skill.

and the model, with the data assimilation system acting as the flow controller. It can be described as a four-step process:

Step 1: Preliminary forecast

Observations from the routine observing system with the highest resolution model version used for that problem, generating what we call here the routine analysis and the preliminary numerical forecast. The information flows from the observing system to the model.

Step 2: Perturbation computation

Computation of a set of initial condition perturbations, from a simpler (probably lower resolution and simpler physics) version of the model, using the preliminary numerical forecast trajectory and customer-targeted metrics such as maximum error growth over north-west Europe. There will be a suite of these sets of perturbations, depending on the metrics chosen, feeding, in principle, a suite of ensembles.

Step 3: Targeted observations

Definition of sensitive regions for additional targeted observations to be obtained for that particular member of the ensemble suite, serving particular customers' needs. Either additional targeted observations can be made, or otherwise discarded observations utilised, in the sensitive regions, if resources permit. Now the information flows from the model back to the observing system. These targeted observations are used along with the routine observations to generate an improved analysis.

Step 4: Ensemble computation

Addition of the perturbations, from step 2, to the improved analysis to initialise the individual ensemble members within that particular member of the suite of ensembles. The ensemble forecast is produced and disseminated to the relevant customers.

This structure of the NWP system is shown schematically in Figure 21.13.

Acknowledgements

This chapter incorporates material collected for the THORPEX presentation to the WMO's Commission for Atmospheric Science meeting in February 2002. That material was assembled by Mel Shapiro (NCAR), Rolf Langland (NRL) and Alan Thorpe and includes figures provided by several members of the THORPEX community. The authors would like to thank everyone who has contributed in this way. Original source material, including references and background papers, can be found on the THORPEX website, the address for which is given in this chapter.

References

Baker, N. L. and R. Daley (2000). Observation and background adjoint sensitivity in the adaptive observation-targeting problem. *Quart. J. Roy. Meteor. Soc.*, **126**, 1431–54.

Berliner, L. M., Z.-Q. Lu and C. Snyder (1999). Statistical design for adaptive weather observations. *J. Atmos. Sci.*, **56**, 2536–52.

Bishop, C. H., B. J. Etherton and S. J. Majumdar (2001). Adaptive sampling with the ensemble transform Kalman filter. I: Theoretical aspects. *Mon. Weather Rev.*, **129**, 420–36.

Bishop, C. H. and Z. Toth (1999). Ensemble transformation and adaptive observations. *J. Atmos. Sci.*, **56**, 1748–65.

Browning, K. A., A. J. Thorpe, A. Montani, D. Parsons, M. Griffiths, P. Panagi and E. M. Dicks (2000). Interactions of tropopause depressions with an ex-tropical cyclone and sensitivity of forecasts to analysis errors. *Mon. Weather Rev.*, **128**, 2734–55.

Buizza, R. and A. Montani (1999). Targeting observations using singular vectors. *J. Atmos. Sci.*, **56**, 2965–85.

Dabberdt, W. F. and T. W. Schlatter (1996). Research opportunities from emerging atmospheric observing and modeling capabilities. *Bull. Am. Meteorol. Soc.*, **77**, 305–23.

Demirtas, M. and A. J. Thorpe (1999). Sensitivity of short-range weather forecasts to local potential vorticity modifications. *Mon. Weather Rev.*, **127**, 922–39.

Doerenbecher, A. and T. Bergot (2001). Sensitivity to observations applied to FASTEX cases. *Nonlinear Proc. Geoph.*, **8**, 467–81.

Emanuel, K., D. Raymond, A. Betts, *et al.* (1995). Report of the First Prospectus Development Team of the US Weather Research Program to NOAA and the NSF. *Bull. Am. Meteorol. Soc.*, **76**, 1194–208.

Hamill, T. M. and C. Snyder (2002). Using improved background-error covariances from an ensemble Kalman filter for adaptive observations. *Mon. Weather Rev.*, **130**, 1552–72.

Holland, L., Z. Toth, J. Moskaitis, S. Majumdar, C. Bishop and R. Smith (2004). Results from the Winter Storm Reconnaissance Program 2002–2003. In *Proceedings of the 84th Annual Meeting, 11–15 January 2004, Seattle, WA*. American Meteorology Society.

Joly, A., D. Jorgensen, M. A. Shapiro, *et al.* (1997). Definition of the Fronts and Atlantic Storm-Track Experiment (FASTEX). *Bull. Am. Meteorol. Soc.*, **78**, 1917–40.

Langland, R. H. and G. D. Rohaly (1996). Adjoint-based targeting of observations for FASTEX cyclones. In *Preprints, Seventh Conference on Mesoscale Processes*. pp. 369–71. American Meteorology Society.

Langland, R. H., Z. Toth, R. Gelaro, *et al.* (1999). The North Pacific experiment (NORPEX-98): targeted observations for improved North American weather forecasts. *Bull. Am. Meteorol. Soc.*, **80**, 1363–84.

Leutbecher, M. (2003). Reduced rank estimate of forecast error variance changes due to intermittent modifications of the observing network. *J. Atmos. Sci.*, **60**, 729–42.

Leutbecher, M., J. Barkmeijer, T. N. Palmer and A. J. Thorpe (2002). Potential improvement of forecasts of two severe storms using targeted observations. *Quart. J. Roy. Meteor. Soc.*, **128**, 1641–70.

Leutbecher, M., A. Doerenbecher, F. Grazzini and C. Cardinali (2004). Planning of adaptive observations during the Atlantic THORPEX Regional Campaign 2003. *ECMWF Newsletter, Winter Issue*, **102**, 16–25.

Lorenz, E. N. and K. A. Emanuel (1998). Optimal sites for supplementary weather observations: simulation with a small model. *J. Atmos. Sci.*, **55**, 399–414.

Majumdar, S. J., B. J. Etherton and Z. Toth (2002). Adaptive sampling with the ensemble transform Kalman filter. II: Field program implementation. *Mon. Weather Rev.*, **130**, 1356–69.

Mansfield, D., D. Richardson and B. Truscott (2005). An overview of the Atlantic THORPEX Regional Campaign A-TReC. In *Proceedings of the 1st THORPEX International Science Symposium, 6–10 December 2004, Montreal.*

Montani, A., A. J. Thorpe, R. Buizza and P. Unden (1999). Forecast skill of the ECMWF model using targeted observations during FASTEX. *Quart. J. Roy. Meteor. Soc.*, **125**, 3219–40.

Pu, Z.-X., E. Kalnay, J. Sela and I. Szunyogh (1997). Sensitivity of forecast error to initial conditions with a quasi-inverse linear method. *Mon. Weather Rev.*, **125**, 2479–503.

Snyder, C. (1996). Summary of an informal workshop on adaptive observations and FASTEX. *Bull. Am. Meteorol. Soc.*, **77**, 953–61.

Szunyogh, I., Z. Toth, A. V. Zimin, S. J. Majumdar and A. Persson (2002). Propagation of the effect of targeted observations: the 2000 Winter Storm Reconnaissance Program. *Mon. Weather Rev.*, **130**, 1144–65.

Toth, Z., I. Szunyogh, S. Majumdar, *et al.* (1999). The 1999 Winter Storm Reconnaissance Program. In *Preprints for the 13th Conference on Numerical Weather Prediction, 13–17 September 1999, Denver CO*, pp. 27–32. American Meteorology Society.

Toth, Z., I. Szunyogh, C. Bishop, *et al.* (2002). Adaptive observations at NCEP: past, present and future. In *Preprints for the 82nd Annual Meeting, Symposium on Observations, Data Assimilation and Probabilistic Prediction, 13–17 January 2002, Orlando FL*, pp. 185–190. American Meteorology Society.

22

The attributes of forecast systems: a general framework for the evaluation and calibration of weather forecasts

Zoltan Toth
National Centers for Environmental Prediction, Washington DC

Olivier Talagrand
Laboratoire de Météorologie Dynamique, Paris

Yuejian Zhu
National Centers for Environmental Prediction, Washington DC

Reliability and resolution are the two main attributes of forecast systems. These attributes statistically relate the performance of a forecast system to verifying data in an abstract sense. Forecast attributes have been separately defined in the literature for systems that generate forecasts of particular formats or types. In this chapter, statistical reliability and resolution are defined in a general sense, irrespective of the type or format of a forecast. Statistical reliability is concerned only with the form of forecasts, whereas statistical resolution is concerned only with the predictive capability of a forecast system, related to the time evolution of the system that is being forecast.

The two main attributes are independent characteristics of a forecast system and can be quantitatively assessed by a host of different verification measures. The general definition of forecast attributes allows a systematic discussion of the relationship between the verification and calibration of forecasts. Calibration as defined here is an adjustment of the form of the forecasts, to match the distribution of verifying observations that follow the issuance of forecasts of a particular form.

Resolution, as the inherent predictive value of forecast systems, is the attribute most sought after by developers of forecast systems. Reliability, however, is equally important in real world applications. That calls for the generation of a long enough

Predictability of Weather and Climate, ed. Tim Palmer and Renate Hagedorn. Published by Cambridge University Press.

record of hindcasts to allow for a good calibration of forecasts, or, preferably, for improvements in forecast systems that directly lead to better reliability.

22.1 Introduction

There exists a vast array of statistics for the description of various aspects of forecast systems, such as those discussed for weather and climate in this volume by Allen *et al.*, Anderson, Buizza, Hagedorn *et al.*, Kalnay *et al.*, Krishnamurti *et al.*, Lalaurette and der Grijn, Mylne, Tibaldi *et al.*, Waliser, and Webster *et al.* Some of these statistics are based solely on the forecast system investigated, while others, called verification statistics, depend both on the forecast values and the corresponding observations from the system that is being forecast (the atmosphere in the case of weather forecasts). The specifics of these statistics or forecast verification measures are not the subject of the present study. Interested readers can find a review of many of these statistics, with additional references, for example, in a recent handbook edited by Joliffe and Stephenson (2003).

Instead, this study focuses on the underlying *statistical verification attributes* of forecast systems. The main statistical forecast verification attributes, *statistical reliability* and *statistical resolution* (from here on, reliability and resolution), have long been discussed in the literature (see, for example, Murphy and Daan, 1985, and references therein). Yet these attributes have been discussed only with respect to particular forecast formats (single value, categorical, or one or another of the probabilistic forecast format types; see, e.g., Stansky *et al.*, 1989; Wilks, 1995; Joliffe and Stephenson, 2003) and not for weather forecasts of any type in general.

Sections 22.2 and 22.3 will introduce a general definition and discuss some characteristics of statistical forecast verification attributes (in short, forecast attributes), respectively. Section 22.4 will explore the statistical limits of measuring forecast attributes. Based on the general definition of the forecast attributes, and on an analysis of the statistical limitations in assessing them, an examination of the relationship between forecast verification and the calibration of weather forecasts (that is, the enhancement of certain statistical properties of the forecasts) follows in Section 22.5. Section 22.6 will explore the significance of the two main forecast attributes to developers and users of forecast systems, while Section 22.7 offers a summary of the main findings of this study.

22.2 Definition of forecast attributes

Forecast attributes, as their name suggests, are abstract concepts that the various verification statistics, using different metrics, quantify. Taking an example from physics, length is an attribute that can be measured by a number of different metrics. As mentioned in the Introduction, forecast attributes have been discussed so far in the context of specific types of forecasts (see, e.g., Murphy and Daan, 1885; Stansky *et al.*, 1989; Wilks, 1995; Toth *et al.*, 2003). Forecast attributes are defined below in a general sense, allowing for a comprehensive discussion of weather forecasts and their statistical calibration.

The verification attributes discussed below are defined in a statistical sense, which is related to forecast systems, and not to individual forecasts generated by them. Forecasts can be of any format but are assumed to belong to a finite number of different 'classes', called F_i. The set of verifying observations corresponding to a large number of forecasts of the same class are characterised by an empirical frequency distribution, called observed frequency distribution (ofd), and marked by o_i.

22.2.1 Reliability

When defining the first forecast attribute, statistical reliability, consider a particular forecast class, F_i. Consider further the frequency distribution of observed outcomes that follow forecasts from class F_i, that is o_i. If forecast F_i has the exact form of o_i for all forecast classes (i), the forecasts are statistically consistent with the observations and the forecast system is called (perfectly) reliable. Different measures of reliability are based on various methods for comparing forecast F_i and the corresponding observed frequency distribution o_i for all forecast classes (i), and measuring their difference.

22.2.2 Resolution

The second forecast attribute, statistical resolution, is defined as a forecast system's ability to distinguish, ahead of time, between different outcomes of the natural system (in case of weather forecasts, the future state of the real atmosphere).

For a more formal definition of resolution, let us assume that the observed events are classified into a finite number of classes, marked by O_i. If each observed class O_i is preceded by a distinctly different forecast class F_i, the forecast system is said to have perfect resolution. Conversely, if the forecast is the same prior to each observed class O_i (i.e., the forecasts do not vary, $F_i = F$ for all i), or if the forecasts vary but the observed frequency distribution o_i following the issuance of different forecasts F_i is the same (i.e., $o_i = c$, the climatological distribution, for all i), the forecast system has no resolution at all.

Resolution in a forecast system can be measured by the degree of separation among the frequency distributions of observed events (o_i), conditioned on different forecast classes (F_i). In practice, this can be achieved by comparing the observed frequency distributions (o_i), constructed from observed events that follow different forecast classes, with the overall climatological distribution of observations (c, that is the reference for a forecast system with no resolution). Different measures of resolution are based on various methods for carrying out this comparison.

22.3 Some characteristics of forecast attributes

(a) *Reliability and resolution are two independent attributes.* Reliability is concerned only with the statistical consistency between each class of forecasts F_i and the corresponding distribution of observations o_i that follow such forecasts, whereas resolution is not affected at all by this consistency. By contrast, resolution reflects how well different forecast classes can separate cases with different subsequent observed events, whereas reliability is unaffected by this property of forecast systems.

While the format and the actual values used by a forecast system are irrelevant to its resolution, they are critical for its reliability. By contrast, a forecast system with perfect reliability does not necessarily have good resolution. Two examples are interesting to note here. A forecast system always issuing the observed climatological distribution has perfect reliability and no resolution by definition, while a system using forecast anomalies that are systematically reversed compared with observed anomalies would have perfect resolution but no reliability.

(b) *In principle, reliability can always be statistically 'enforced' or corrected.* This is true as long as both the forecast and observed systems are stationary in time, and there is a long enough record of forecast-observed data pairs. This is because reliability reflects only the statistical consistency between forecast and observed distributions. All one has to do to achieve the desired consistency is to replace the forecasts in a given forecast class with the frequency distribution of observations that follow such forecasts.

(c) *Unlike reliability, resolution cannot be improved by statistically correcting the forecasts so they follow the distribution of ensuing verifying observations.* This is because resolution does not depend on statistical consistency. Resolution reflects the inherent value of forecast systems, and can be improved only through the modification of the forecast scheme based on additional knowledge about the temporal evolution of the observed system.

(d) *Reliability and resolution, as defined above, are general attributes of forecast systems.* They can be interpreted for systems generating forecasts of any type, such as single value, categorical, or probabilistic.

It is interesting to note that single value (out of a continuum) forecasts can be perfectly reliable only if they have perfect resolution as well. This is the only way the observed frequency distribution would exactly match the Dirac function form of the forecasts.

As mentioned earlier, forecast attributes have been interpreted in the past for forecasts issued in specific formats (i.e. not necessarily in the general form of a probability distribution). While this can be useful for special purposes, it must be noted that such narrow definitions of forecast attributes are not fully consistent with the general definition introduced in this study.

Consider, for example, the case of a forecast system with less than perfect resolution that issues single value forecasts. In this case, it could be possible to define statistical reliability (or statistical consistency, as it is also often referred to; see, e.g., Wilks, 1995) as a lack of conditional systematic bias. According to this narrow definition, a forecast system is considered reliable if for all forecast values the frequency distribution of corresponding observations has the same mean as the forecast value. It is easy to see that this feature is a necessary but not sufficient condition for reliability as defined in the present study. In fact, the no-spread single value forecasts, even if they have no systematic bias, will have less than perfect reliability for any system with less than perfect resolution. Such a narrow definition of reliability will have an implication for statistical calibration as well, as will be discussed in Section 22.5.

22.4 The limits of assessing reliability and resolution

22.4.1 Measures of forecast attributes

As discussed by Toth *et al.* (2003) for forecasts in probabilistic format, some existing verification measures assess reliability, some resolution, while still others provide a combined measure of both. Note that some measures can be calculated for selected subsets of all forecast cases – like the reliability and resolution components of the Brier score verifying for only one of a set of categorical events. These measures can be related to reliability and resolution as defined in the present study only if the measure is aggregated over all observed categories.

22.4.2 Factors limiting the statistical accuracy of verification statistics

While forecast attributes can theoretically be defined assuming that the number of forecast cases goes to infinity, in practice verification measures are always computed based on finite samples. Therefore, verification results can be considered estimates whose accuracy will depend on the sample size. Knowledge about the uncertainty in verification results is important (see, e.g., Hamill, 1997), especially when one compares two or more competing forecast systems. In such cases it is especially important to assess the statistical significance of the comparative verification results (see, e.g., Candille and Talagrand, 2005). The associated uncertainty in the verification results can be reduced only through increasing the sample size, which is often impossible when evaluating real life forecast systems.

Another factor limiting the accuracy of verification estimates is the uncertainty in the verifying data (Candille, 2003). Observations used to verify forecasts are generally associated with measurement and other errors. For properly assessing reliability and resolution of a forecast system, such errors in the observations need to be carefully accounted for, otherwise the results will either be biased and/or will look statistically more certain than they are. Observational errors can be considered in forecast verification by replacing an observed value (Delta function) with a probability density function (pdf) that reflects the observational uncertainty. The use of incorrect observational error estimates (such as assuming perfect observations in the presence of errors, as in the case of most verification studies) will introduce errors in the verification (and pursuant calibration) results.

A third factor influencing the accuracy of forecast verification statistics is the choice of the level of granularity introduced in the calculations, which is a function of the level of detail sought in the results. The granularity of verification studies can be controlled through a number of choices.

First, forecasts can theoretically take an infinite number of forms. Yet, when in practice a finite sample of forecasts are evaluated statistically, forecasts of a similar form must be grouped into a finite number of classes. For more detailed verification statistics one might possibly wish to establish a large number of forecast classes. The number of different classes is limited, however, by the requirement that there be enough forecast cases in each of the classes established.

Second, forecast probability distributions can theoretically be defined and manipulated as continuous functions. In practice, however, calculations are always carried out over finite intervals. And because the sample size is limited, the width of the intervals cannot be reduced arbitrarily, otherwise most intervals would contain no data points.

Finally, if the overall sample size is small, one may need to group together forecast–observed pairs from similar geographical regions and/or similar parts of the annual cycle.

In practice, when choosing the level of granularity in verification calculations, one seeks a compromise between having a large enough sample for all forecast classes and verification intervals, while retaining as many classes, intervals, and geographical, seasonal distinctions as possible, given the total number of forecast–observation pairs (Atger, 2003). Obviously, the larger the overall sample of forecast–observation pairs is for verification, the more questions about the performance of the forecast system can be answered. As we will see in the next section, the same holds true for the number of adjustment types that can be made as part of a statistical calibration algorithm.

22.5 Calibration

The goal of calibration is to make the form for each class of forecasts statistically more consistent with the distribution of the corresponding verifying observations. Calibration, as defined here, is the replacement of the forecast, whatever form it may have (i.e. single value, categorical, or probabilistic), with an estimate of the corresponding odf (which describes the distribution of observations that in the past followed the issuance of forecasts from the same forecast class). The success of calibration can be measured by comparing the reliability of the calibrated forecasts with that of the raw, uncalibrated forecasts.

Note that calibration is directly related to the verification of statistical reliability, since both are based on estimating the distribution of observations following different forecast classes. While verification assesses the statistical reliability of a forecast system over a period in the past, calibration adjusts the forecasts with the intention to make them more consistent with observed statistics in the future. Calibration is based on the assumption that the statistical behaviour of the forecast and observed systems, as analysed over a period in the past, will not change in the future. Calibration, therefore, is subject to an additional limitation beyond those discussed with respect to verification, namely that the quality of calibration will suffer if either the natural or the forecast system is non-stationary in time. As with verification, the use of a small sample size, errors in describing uncertainty in the verifying observations, and an inappropriate choice for the level of granularity in the calculations will also adversely affect calibration results.

There are a number of ways that forecasts from different classes, geographical regions or different parts of the annual cycle can be grouped together for computing verification statistics that are also needed for calibration. The resulting formation of larger subsamples allows a more robust statistical estimate of the underlying distribution of the observations corresponding to a broader group of forecasts – at the expense of reducing the level of details in the verification, and consequently in the pursuant calibration results. Therefore, careful compromises are needed when the level of granularity is chosen for the computation of statistics for calibration. Allow too many details in the verification (i.e. use too many different forecast classes), and the calibration will suffer from sampling noise. Conversely, the lack of enough detail in verification (i.e. grouping forecasts from areas with distinctly different verification statistics together; see Atger, 2003) can also adversely affect the calibration by leaving the biases present in the smaller subsamples uncorrected.

It should be noted that calibration, as discussed earlier with respect to verification, can be introduced in a narrower sense than that defined above. Forecasts, for example, can be corrected only to reduce their systematic bias in the first moment. An application based on such a narrow definition of calibration will necessarily be limited since other, higher moment aspects of the forecasts will not be statistically corrected. By contrast, calibration, if applied in a general sense as defined above on single value, categorical, or any other type of forecasts, will naturally change the format of the forecasts to the more general probabilistic format.

22.6 Significance of attributes to forecast developers and users

Neither the reliability nor the resolution of real life weather forecast systems is perfect. What is the significance of either attribute to the developers or users of weather forecasts? Is one or the other attribute more important?

22.6.1 Developers' perspective

We recall that the inherent value of forecast systems lies in their ability to predict future events, as reflected in the statistical resolution of forecast systems. This is equivalent to a forecast system issuing uniquely different signals prior to different observed events. For example, if a system systematically gives a prediction of 'heavy snow' (or 'red') and 'light snow' (or 'blue') prior to observed rain and no rain events respectively, it has a high resolution.

Since the forecast signals issued by this forecast system are significantly different from the subsequent observed verification events, however, the forecasts have poor reliability. If such behaviour is systematic, the forecasts can be calibrated and the developers of the forecast system may be content with the good resolution and may not be overly concerned with the apparent lack of reliability.

22.6.2 Users' perspective

It must be noted that when forecasts from the system described above are taken by the users at their 'face value', they can be worthless or even harmful. A user who believes what the forecast says and acts on that information can be seriously hurt (e.g. Zhu *et al.*, 2002). Even forecast systems with high predictive skill (high resolution) have no value to users unless they also have good reliability. This explains why users often emphasise reliability in their evaluation of forecast systems, based on the principle of 'do no harm'.

22.6.3 Need for calibration

Generally, a long enough record of observed–forecast pairs will allow an adjustment or calibration of the forecast signal to match the distribution of observations that follow a particular forecast class. Incidentally, a similarly long record of observed-forecast pairs may be needed for the precise assessment of resolution in a forecast system (see Section 22.4). In the case of a forecast system with high resolution, calibration can significantly enhance the utility of forecast systems. This underlies the need for the provision of a large enough set of hindcasts (forecasts generated on past events). Exactly how large a hindcast dataset is needed for calibration is the subject of ongoing research. A large enough hindcast dataset will allow a proper assessment of both the resolution and reliability of the forecast system, and will facilitate a subsequent calibration of the forecasts in case the forecast system lacks statistical reliability. In such a case, statistical reliability can be achieved through a statistical adjustment via calibration.

22.6.4 Value of forecasts

As discussed above, beyond resolution, the users also critically depend on the reliability of the forecasts. It is therefore important that when (typically after they are calibrated) the value of forecast systems is assessed for the users, both resolution and reliability are considered.[1] One can argue that for a forecast system to show genuine improvement, its resolution must be measurably enhanced. An experimental forecast system with enhanced resolution, but an insufficient hindcast data set for calibration, however, may degrade utility. One may argue that enhanced resolution forecast systems be operationally implemented only if their reliability is not affected negatively, or if at least a sufficient hindcast dataset is generated to ameliorate any problem with reliability through calibration.

22.6.5 Future directions

As forecast systems mature, there is a natural tendency to use more detail from the forecasts. For that to happen, one needs to include more detail in the calibration of the forecasts as well. That, as discussed earlier, calls in turn for longer periods of past observed–forecast pairs. Unfortunately, the number of such pairs is usually severely limited due to the lack of long periods of detailed observations. This is of particular concern when extreme events are considered. Such events, by definition, occur rarely (Zhu and Toth, 2001). Therefore, their statistical calibration is especially problematic (Legg and Mylne, 2005). Yet these rare events are often of the greatest interest to users.

It follows that as forecast and application methods improve and more details are demanded from a system, the potential value added by statistical calibration will likely diminish. Since under such conditions statistical corrections are of little or no help, directly improving the reliability of a forecast system itself will become more important and sometimes will offer the only tractable solution. When the realism of models representing weather systems (that is directly related to reliability) is improved, the changes may also lead to improvements in predictive skill (i.e. resolution). Prediction of tropical storms is a prime example of a situation where the role of statistical calibration is limited due to the highly non-linear nature of these systems.

If a storm, due to model deficiencies (e.g. too low spatial resolution), is not predicted (well) by a forecast system, the insertion (modification) of a storm into the forecast via statistical inference/calibration may require an impractically large training data set. In such cases the reliability (and utility) of the forecasts can be improved only by enhancing the realism of the numerical weather prediction model itself.

22.7 Conclusions

This study introduced a distinction between the abstract notion of forecast system attributes and the statistical measures used to assess them. Unlike earlier studies, a general definition of the forecast attributes was proposed, irrespective of the format of the forecasts. Both of the two main attributes, reliability and resolution, were interpreted in a statistical sense. Reliability was defined as a perfect match between the form of a forecast and the distribution of verifying observations that follow the issuance of that particular forecast form. A forecast system is said to have perfect resolution, by contrast, if it consistently gives different signals prior to the occurrence of different observations.

Reliability and resolution were shown to be independent of each other. Of the two attributes, forecast system developers are more concerned about resolution since that is related to the intrinsic predictive capability of forecast systems. For the users who take the weather forecasts at face value, reliability is equally or even more important. This is because users who use the raw (uncalibrated) forecasts act upon the actual form of the forecasts, and it is reliability that is used to assess how this form compares statistically with what is being observed.

A number of verification measures exist for the assessment of reliability and resolution. These measures, like any other statistics based on finite samples, are subject to sampling and other types of errors. These same errors were also shown to affect calibration, where the reliability of forecast systems is enhanced. Calibration was defined in general terms as the replacement of the form of the forecasts by the distribution of observations that follow the issuance of any particular forecast form, based on a set of observed–forecast data pairs.

It follows from the general definition of the main forecast attributes and calibration that the general format of forecasts is that of a probability density function (pdf) since that is the only format that can, in general, be consistent with the distribution of ensuing observations. A pdf format allows the forecast system to reflect case-by-case variations not only in the expected first moment of future weather parameters but also in the higher moments, such as error variance. For example, forecasts in pdf format can distinguish, given a certain expected value, between cases with higher and lower uncertainty (Toth *et al.*, 2001). Such information is known to have potentially great economic value for the users (Zhu *et al.*, 2002), yet cannot be provided by a forecast system using a single value format. To what extent ensemble forecast systems can provide useful information beyond the first moment of the distribution is still an open question (see, e.g., Atger, 1999).

Acknowledgements
This chapter is an outgrowth of work on a chapter (Toth *et al.*, 2003) written by the authors and a collaborator (G. Candille) for a textbook edited by Joliffe and Stephenson (2003). The authors acknowledge the stimulating discussions with the editors of that volume. The first author is indebted to Professor Eugenia Kalnay, who asked him to contribute to a lecture series on Statistics in Meteorology at the Department of Meteorology, University of Maryland. This chapter is also intended as a draft contribution to the lecture notes accompanying that course.

Note

1. As discussed in Zhu *et al.* (2002), some measures of forecast performance, such as the potential economic value, assume that the forecasts can be perfectly calibrated (i.e. forecasts are automatically calibrated as part of the computation of potential economic value, using the *dependent* and not an independent set of data for calibration). These results will overestimate the actual utility of forecasts that in practice will necessarily be lowered by the limits of calibration discussed in Section 22.5.

References

Atger, F. (1999). The skill of ensemble prediction systems. *Mon. Weather Rev.*, **127**, 1941–53.

Atger, F. (2003). Spatial and interannual variability of the reliability of ensemble-based probabilistic forecasts: consequences for calibration. *Mon. Weather Rev.*, **131**, 1509–23.

Candille, G. (2003). *Validation of probabilistic meteorological forecast (in French).* Doctoral Dissertation, Universite Pierre-et-Marie-Curie, Paris, France.

Candille, G. and O. Talagrand (2005). Evaluation of probabilistic prediction systems for a scalar variable. *Quart. J. Roy. Meteor. Soc.*, **131**, 2131–50.

Hamill, T. (1997). Reliability diagrams for multicategory probabilistic forecasts. *Weather Forecast.*, **12**, 736–41.

Joliffe, I. T. and D. B. Stephenson (eds.). (2003). *Forecast Verification: A Practitioner's Guide in Atmospheric Science*. Wiley.

Legg, T. P. and K. Mylne (2005). Early warnings of severe weather from ensemble forecast information. *Weather Forecast.*, **19**, 891–906.

Murphy, A. and H. Daan (1985). Forecast evaluation. In *Probability, Statistics, and Decision Making in the Atmospheric Sciences*, ed. A. H. Murphy and R. W. Katz, pp. 379–437. Westview Press.

Stanski, H. R., L. J. Wilson and W. R. Burrows (1989). *Survey of Common Verification Methods in Meteorology*. World Weather Watch Technical Report 8. World Meteorological Organization.

Toth, Z., O. Talagrand, G. Candille and Y. Zhu (2003). Probability and ensemble forecasts. In *Forecast verification: A Practitioner's Guide in Atmospheric Science*, ed. I. T. Jolliffe and D. B. Stephenson, pp. 137–63. Wiley.

Toth, Z., Y. Zhu and T. Marchok (2001). On the ability of ensembles to distinguish between forecasts with small and large uncertainty. *Weather Forecast.*, **16**, 436–77.

Wilks, D. S. (1995). *Statistical Methods in the Atmospheric Sciences*. Academic Press.

Zhu, Y. and Z. Toth (2001). Extreme weather events and their probabilistic prediction by the NCEP Ensemble Forecast System. In *Preprints of the AMS Symposium on Precipitation Extremes: Prediction, Impact, and Responses, 14–19 January 2001, Albuquerque, NM*, pp. 82–5. American Meteorological Society.

Zhu, Y., Z. Toth, R. Wobus, D. Richardson and K. Mylne (2002). The economic value of ensemble based weather forecasts. *Bull. Am. Meteorol. Soc.*, **83**, 73–83.

23

Predictability from a forecast provider's perspective

Ken Mylne
Met Office, Exeter

For a forecast provider, predictability means striking a balance between the needs of an end user to make decisions and the limitations of what is scientifically possible. Scientifically the best information is probabilistic, normally generated from ensemble forecasts, but to make effective use of this information we need to understand the decision-making process of the user. This chapter will discuss some of the issues related to the calculation of relevant probabilities, and how to transmit that information to users and help them with decision-making.

23.1 Introduction

Predictability is not a new issue for forecast providers, such as the UK Met Office. Forecasters have always dealt with uncertainty, usually describing it subjectively with terms such as 'mainly in the north-west', or 'a risk of patchy fog affecting the airfield, but you should get in OK'. The second example here immediately shows an understanding by the forecaster of the decision which the pilot has to make, and many forecasters' daily jobs involve providing bespoke services to individual customers. By understanding those customers' businesses, forecasters are able to provide them with information on some of the risks and uncertainties which impinge on their activities and affect their decision-making, and tune their forecasts accordingly. Nevertheless, two major changes in recent years are altering the way we deal with

Predictability of Weather and Climate, ed. Tim Palmer and Renate Hagedorn. Published by Cambridge University Press.

forecast uncertainty. First, new methods, such as ensemble prediction, are improving the ability of forecast providers to assess uncertainty quantitatively, in an objective and verifiable fashion. Second, forecast services are increasingly provided automatically in order to minimise costs and delays, and allow flexible production of forecasts for many sites. Use of objective methods allows us to include consistent measures of uncertainty in these automatically generated forecasts. The challenge is therefore to find ways of expressing uncertainty which are meaningful and useful to customers. We also need to work with end users to help them understand what the numbers mean and how to make use of them in risk management and decision-making.

This chapter will discuss the use of probabilities in providing forecast services to customers, and describe some of the ways that ensembles are used in the Met Office to support and improve our services.

23.2 Predictability – refining climatology

For a forecast provider like the Met Office, predictability is about balancing customer desires for certainty, with what is actually predictable. Customers would like certainty to ease decision-making but this is normally impossible due to the effects of chaos and processes we cannot resolve. So what can we predict with any certainty? A good starting point is climatology. Past statistics can tell us the climatological frequency of an event: for example, if snow falls on 17 out of every 100 January days, the daily probability of snow in January is 17%. Assuming the climatology is static and representative, this provides a perfectly reliable probability forecast, although it has no resolution beyond the seasonal variation of the climate. (Following Murphy, 1973, *reliability* measures how well forecast probabilities match the frequency of observation, while *resolution* measures the ability of the forecasts to discern subsamples with different relative frequencies of the event; see Wilks, 1995.) A good benchmark for any forecast system is therefore that it should improve on climatology. Where we have no predictive capability, climatology provides the best available guidance to a customer. For example, an insurance company providing cover against possible weather disruption many months ahead will assess risks and set premiums based on climatology.

As well as setting a benchmark for the skill of probability forecasts, climatology is also useful in interpreting probabilities. A common criticism of probability forecasts is that forecasters are simply 'covering themselves' or 'don't know', particularly when a mid-range probability such as 50% is issued. However, a forecast of 50% can be extremely informative. Consider the following forecast issued in November: 'There's a 50% probability of snow in London tomorrow.' While not impossible, climatology tells us that snow in London is rare in November, so a 50% probability for the next day is indicating a very high risk compared with normal. This forecast therefore contains a strong signal, even though the forecaster could quite honestly

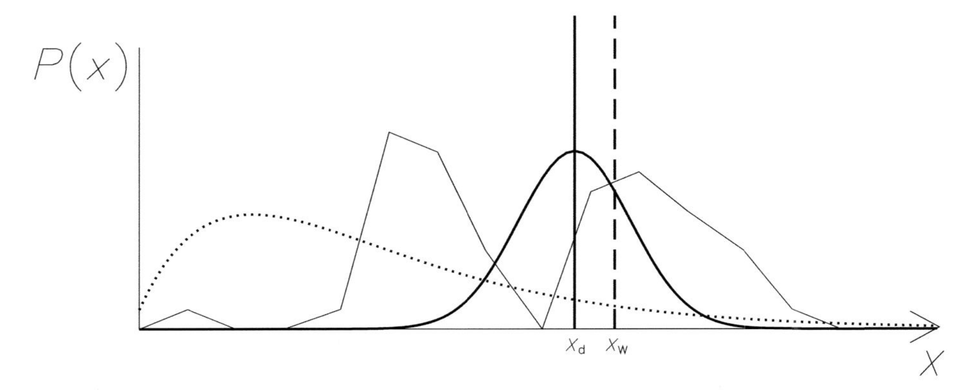

Figure 23.1 A schematic illustration of methods of generating probability forecasts for a variable x which has a climatological distribution shown by the dotted line. x_d represents a deterministic forecast value, and the thick solid curve represents a distribution of forecast errors. The thin solid line represents an ensemble forecast. x_w is a warning threshold for which the probability of exceedence is required.

say 'I do not know if it will snow in London tomorrow'. By presenting the forecast relative to climatology, the signal is made much clearer. More generally, there is nothing wrong with an individual forecast of 50% provided it is part of a large set of forecasts issued with a range of probabilities. Even where forecasts have good resolution and can discriminate well between events and non-events, there will be some occasions when the occurrence or otherwise of an event is marginal. To issue a more categorical forecast on such an occasion would be misleading.

Climatology thus provides a baseline for predictability, and any forecast should be an attempt to refine the probability to give more information. Figure 23.1 illustrates forecasts of a parameter x which has a climatological distribution shown by the dotted line. There are several standard ways to generate forecasts of x based on numerical weather prediction (NWP). A deterministic forecast is the outcome of a single run of an NWP model, and gives a single solution x_d. Over a number of previous forecasts it is possible to generate statistics of the errors of x_d, and from these the deterministic forecast may be supplemented by an error distribution function as shown by the solid curve, providing a simple estimate of the forecast probability density function (pdf). In Figure 23.1 this is illustrated with a Gaussian distribution but the method may also be applied using different functional forms. A gamma distribution may be more appropriate for parameters such as rainfall which typically have a skewed climatological distribution (Wilks, 1995). Whatever distribution is used, its mathematical form does not generally vary with the meteorological situation or the value of the deterministic forecast. On occasions when the deterministic forecast is extreme, and therefore of particular interest, the pdf from the error distribution is least likely to be representative of the true forecast probabilities, and may even predict significant probabilities for values of x well outside the climatological range.

By contrast, an ensemble forecast attempts to sample the forecast pdf in a meteorologically consistent way, taking account of the synoptic situation and thus the variable predictability of the atmosphere. Ensemble pdfs may take on more complex forms, including multimodal distributions as illustrated by the thin solid line in Figure 23.1. Assuming the climatology of the ensemble model provides a reasonable approximation to that of the real atmosphere, the ensemble pdf should always be consistent with climatology, even in extreme meteorological situations.

Each of these forecast methods provides increasingly sophisticated refinements of the climatological distribution, from which forecasts of x may be expressed as probabilities. The distributions in Figure 23.1 could be used to generate probabilities of exceeding the warning threshold x_W, given by the area under the pdf to the right of x_W. In this schematic example, climatology gives a low probability (around 7%) in the tail of the distribution. Used simplistically, the deterministic forecast would give a probability of zero, but when enhanced by the Gaussian it gives a probability of 24%; the ensemble suggests a probability around 38%. While the ensemble method is clearly the most expensive way to generate probability forecasts, its flow-dependent nature means that for most applications it is likely to give better results than other methods. An important area of research for operational centres is to assess for which types of forecast the additional information in ensemble forecasts is of sufficient value to justify the cost of generating them.

23.3 Interpretation and verification of probability forecasts

Describing uncertainty quantitatively is only beneficial if the numbers can be shown to be meaningful and of value to the end user. The event to which a numerical probability is attached must be unambiguous and relevant to the user's application – both provider and customer must be clear exactly what the probability refers to. For example, if a forecast states there is 'a 30% probability of rain in England', does this mean 30% at any one location, or 'somewhere in England'? Is this a 30% risk of a trace being recorded, or of a downpour? It must be clearly stated exactly what is being predicted. The following example is much more meaningful: '30% probability of more than 5 mm of rain at Heathrow Airport between 1200 and 1800'.

The user must then understand how to interpret and verify probabilities. If the forecast probability of exceeding x_W is 10%, and x_W is indeed exceeded, this neither makes the forecast right nor wrong. But out of 100 independent forecasts of 10%, x_W should be exceeded 10 times. If this is the case the forecasts are said to be perfectly reliable. A single probability forecast cannot be right or wrong, and verification must be done over many forecasts. A reliability diagram plots the frequency of occurrence of an event against the forecast probability, and is readily understood by most users.

Reliability is not sufficient on its own for probability forecasts to be useful. Climatology provides perfectly reliable probabilities, but contains no forecast information specific to the occasion. Useful forecasts also need to have discrimination. Discrimination indicates the ability of the forecast system to distinguish between occasions when an event does occur from ones when it does not and is related to the slope on a reliability diagram – if the graph is horizontal, the probability of the event occurring is independent of the forecast probability, and the forecasts are useless. Discrimination may be measured quantitatively by the relative operating characteristic (ROC) taken from signal detection theory (Stanski *et al.*, 1989), but ROC is difficult to interpret for end users. A much more user-oriented measure of discrimination is whether forecasts have economic value in decision-making, as discussed by Richardson elsewhere in this volume. A full discussion of verification methods for probability forecasts is provided by Jolliffe and Stephenson (2003), and some examples will be shown in the remainder of this chapter.

23.4 Use of ensemble forecasts at the Met Office

23.4.1 Long-range forecasting

In monthly and seasonal-range forecasts, predictability is inherently low, and forecast systems aim to skew the climatological distribution slightly in the right direction. Long-range predictability comes from slowly varying changes in sea-surface temperature (SST), and is greatest in the tropics where the atmosphere is more directly forced by SST. Forecasts are based on multimodel ensembles of coupled atmosphere and ocean models, including both ECMWF and Met Office models, to estimate time-averaged behaviour of the atmosphere compared with climatology. When run over extended periods the climates of the models may differ significantly from the real climate, so model climatologies are determined by running the models over many past seasons. Forecasts are then expressed as anomalies relative to model climatology, and may be interpreted or calibrated by reference to real climatology. Probabilities issued are frequently close to climatology, due to low predictability, but verification shows that there is nevertheless some discrimination. Forecasts can therefore have value for users who can adjust their actions in response to small changes in probability and benefit in the long term, averaged over a number of occasions. Skill is greatest in the tropics, but some extratropical areas also have useful skill. The Met Office issues seasonal forecasts on its website (http://www.metoffice.gov.uk/weather/seasonal/index.html), and monthly forecasts are provided to a number of customers.

23.4.2 Medium-range forecasting

Medium-range forecasting (3–10 days) has been transformed since the mid 1990s by the availability of the ECMWF EPS (Ensemble Prediction System). Prior to that forecasts were essentially deterministic, based heavily on the Met Office's global model, with statements about confidence and uncertainty based on the agreement or otherwise of a few other models. Today the main deterministic products are still produced, including isobaric charts with frontal systems, but they are now based on what is perceived to be the most probable solution taking account of both the EPS and the other models available. Forecasters have access to a wide range of tools for visualisation of the ensemble, and use field modification software (Carroll, 1997) to produce meteorologically consistent fields representing the most probable outcome. Figure 23.2 presents results from monthly verification, assessing day 3 (T+72) and day 5 (T+120) mean sea level pressure (MSLP) forecasts before and after modification, and shows that in most months the forecasters are able to improve on the unmodified products (Hewson, 2004).

In addition to generating the most-probable chart solution, medium-range guidance forecasters are also able to generate alternative solutions where the EPS suggests a different solution with a probability of occurrence of more than about 20%. Such alternatives are normally derived from clustering of ensemble members, and probabilities based on numbers of members in different clusters. Chart products are supplemented with detailed discussion of confidence and the risks indicated by the ensemble. Guidance includes tables giving confidence ranges for temperature and probabilities for rainfall accumulations at a number of locations.

While the use of ensembles has helped forecasters to improve the quality of conventional deterministic products, the greatest potential in ensemble forecasts is in the generation of probabilistic forecasts. The remainder of this section will focus on development of services which take full account of probability to aid end users in risk management.

23.4.3 Forecasting severe weather

Much emphasis is now being put on improving predictions of severe weather. Since the development of severe weather is frequently highly non-linear, this is an appropriate application of ensembles; at the same time it is a particularly demanding application. It is also difficult to verify since severe weather occurs relatively rarely, so data samples are small. For long-range forecasting we noted that model climatology is often significantly different from real climatology – the same is true in the medium or short range when considering severe weather, since many severe developments depend on quite small-scale processes which are not fully resolved. It is therefore often necessary to calibrate forecasts rather than interpreting model output directly.

Over recent years the Met Office has started to use the EPS to generate early warnings of severe weather in support of the UK National Severe Weather Warning Service (NSWWS). Early warnings can be issued up to five days in advance when the probability of an event occurring 'somewhere in the UK' is 60% or more, this

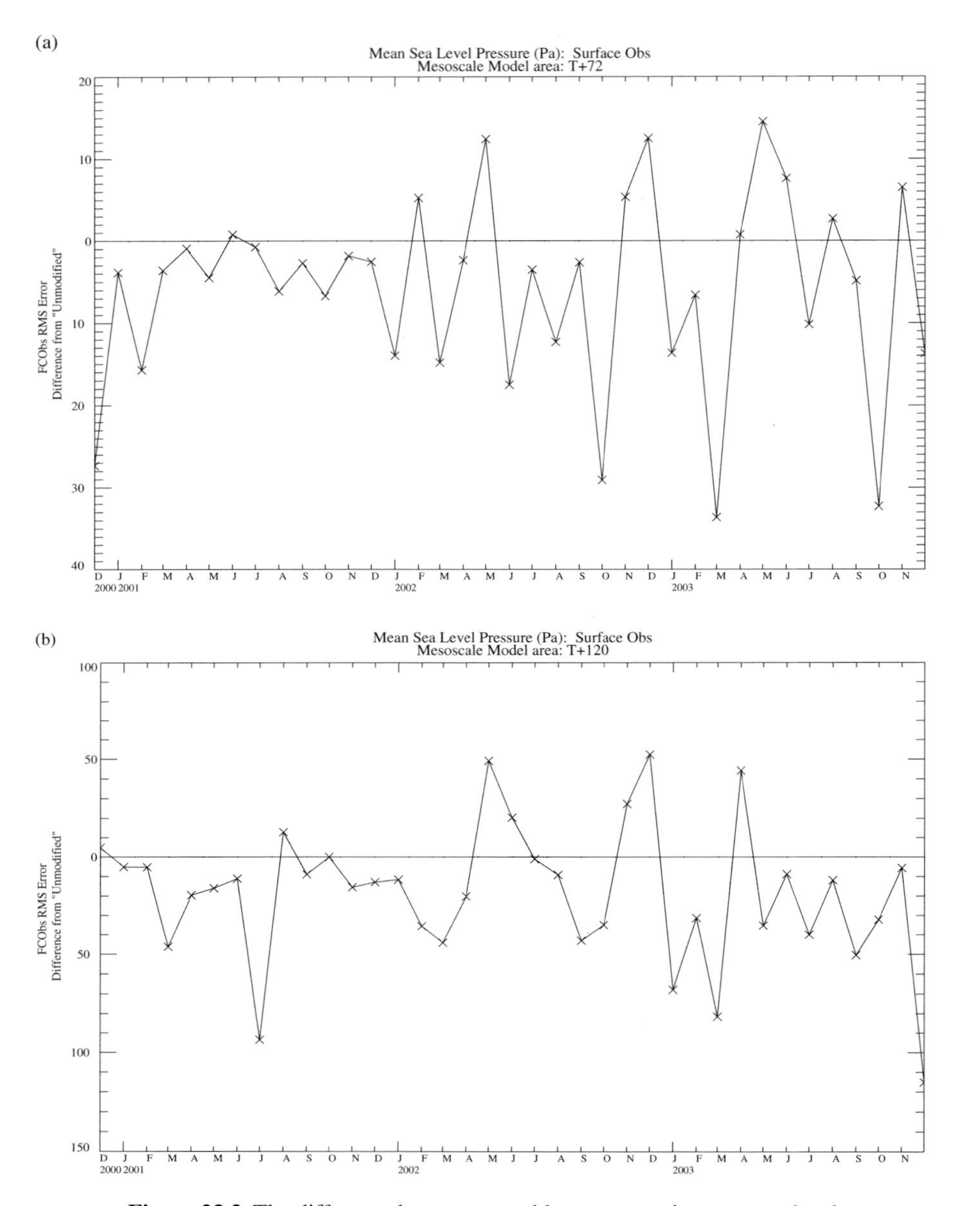

Figure 23.2 The difference between monthly rms errors in mean sea-level pressure with and without forecaster modification at (a) day 3 and (b) day 5 (Hewson, 2004).

threshold being specified by the NSWWS customers. In addition to an overall UK probability, probabilities are also given for 12 local regions. In the past probabilities were assessed subjectively, and forecasters rarely had confidence to issue warnings more than 36 hours in advance. In order to encourage earlier issue and reduce the overall miss rate, the First-Guess Early Warning (FGEW) system was developed to provide forecasters with consistent and verified objective probabilities from the EPS (Legg and Mylne, 2004).

FGEW warnings provide a good example of the need to calculate the relevant probability for the application, in this case that severe weather will occur somewhere within the UK (or a subregion). In calculating probabilities an ensemble member is therefore counted if it generates severe weather at any grid point in the region. Also, at 3–5 days ahead the precise timing of severe weather is not critical so the probability calculation looks inside a time window of ± 12 hours. As a result, probabilities are much higher than those seen at fixed times at individual grid points. As well as calculating probabilities relevant to users, verification of a system such as FGEW also needs to be user oriented. Standard verification methods normally compare forecasts with observations (or analyses) at fixed times, whereas FGEW verification has to make allowance for the fact that warnings are of variable length and may be verified by an observation which only overlaps with part of the warning period. From a user perspective it would be misleading to count such a warning as a false alarm during the non-overlapping period (assuming the warning is not excessively long). In the case of FGEW, scores were calculated for each calendar day, and warnings spanning more than one day were allocated to the first day to avoid double counting, but each system must be designed according to the needs of the end users.

The 60% probability threshold defined for the issue of early warnings reflects customer desire for high confidence, but in practice this is rarely attained. The development of severe weather normally involves the non-linear interaction of quite small-scale flow anomalies in the atmosphere. Small differences in the position, intensity, or timing of such anomalies in the model can lead to large differences in forecast evolution. Most members of an ensemble will therefore produce different interactions from what happens in the atmosphere, and although the ensemble can be expected to include members with severe events, it would be unusual for it to predict high probabilities of severe weather. This indeed turns out to be the case in practice.

Figures 23.3 and 23.4 show examples of verification of 4-day forecasts from the FGEW system which illustrate that most of the forecast information is contained in low-probability forecasts. Figure 23.3 shows a reliability diagram for early warnings of heavy rain, with a corresponding histogram of the number of times each forecast probability was issued (sharpness diagram). The histogram shows that forecasts were rarely issued with high probabilities. (Note that because the events are rare the probability bins used have been concentrated towards the low probability end, and that most forecasts give probabilities below 10%.) Forecasts issued at lower probabilities, below about 30%, provide excellent reliability with the curve lying close to the ideal diagonal. At higher probabilities the reliability diagram is very noisy due to small data samples, but there is some indication of the right general trend, with severe weather increasingly more likely to occur when higher probabilities are issued. It was noted above that for severe weather it is important to calibrate the forecasts relative to model climatology. This was done for the FGEW system by adjusting

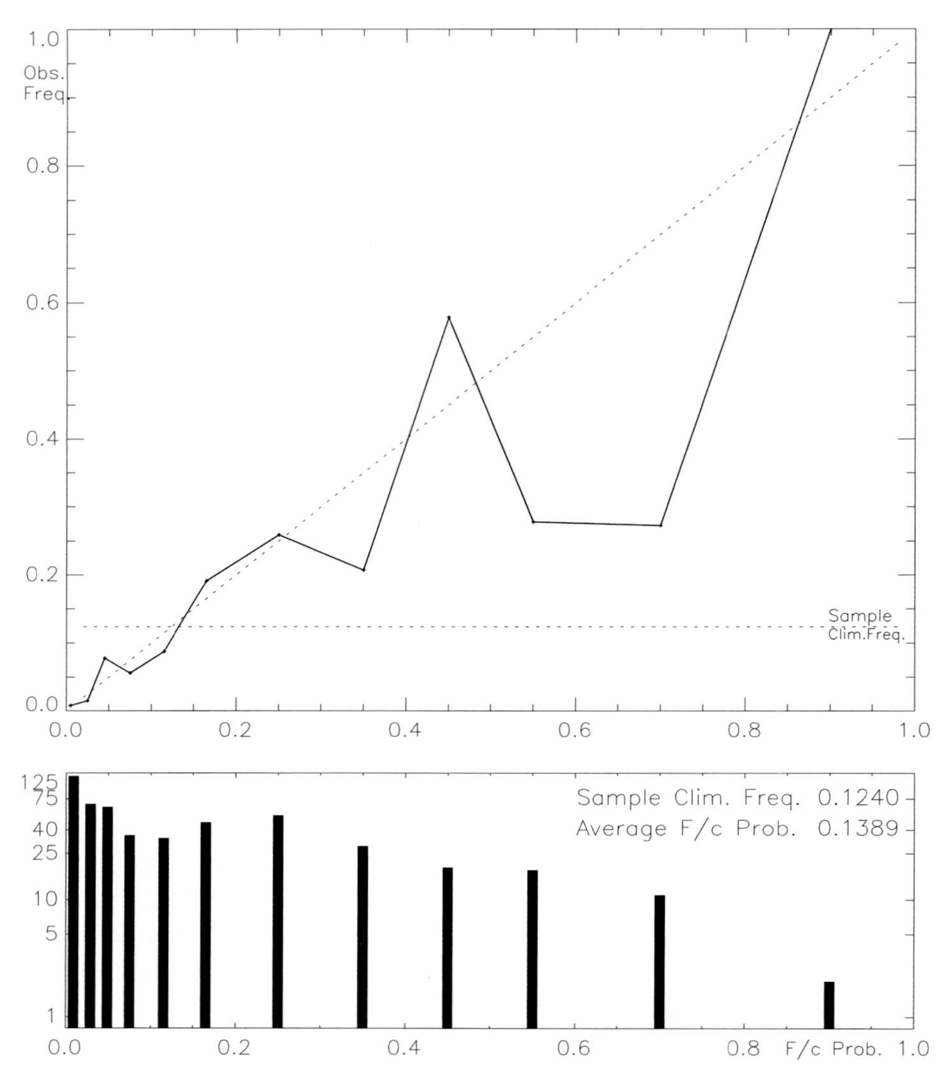

Figure 23.3 Reliability (top) and sharpness (bottom) diagrams for FGEW warnings of heavy rainfall events in the UK four days ahead. Data period: 1 Oct 2001 to 12 Feb 2003.

the severe weather event threshold to optimise the overall probability bias over an initial training period in the winter 2000–1; verification results shown here are taken from a subsequent period from October 2001 to February 2003, and confirm that the calibration was quite successful. Sample climatology for the verification period is 0.124 and mean forecast probability was 0.139, giving a forecast frequency bias of 1.10, close to the ideal value of 1.0.

The ROC (Stanski *et al.*, 1989) curve shown in Figure 23.4 for 4-day severe gale warnings shows that the system has considerable ability to discriminate occasions

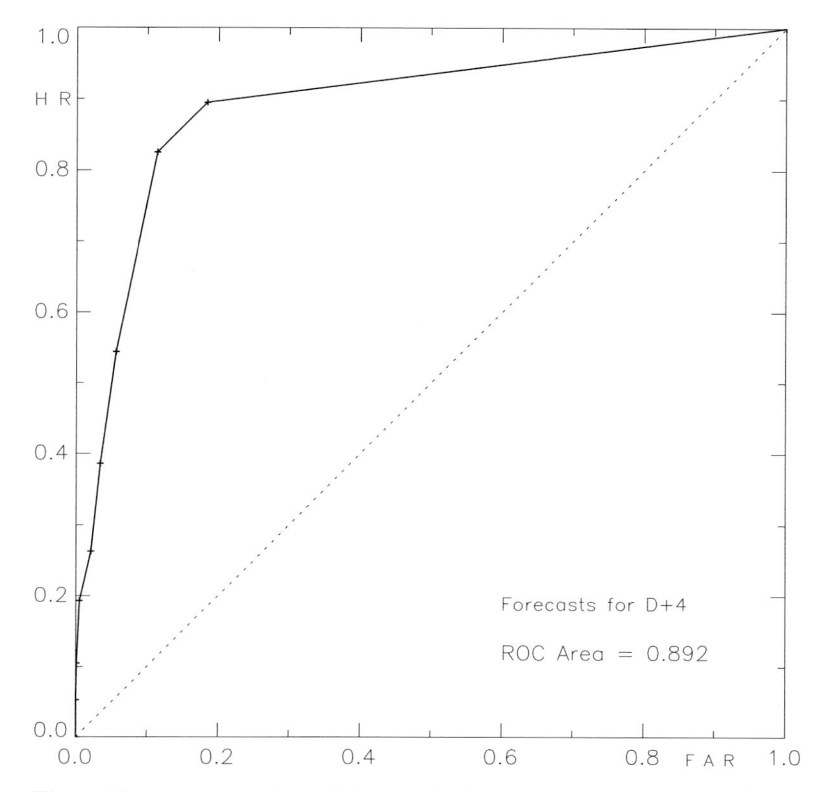

Figure 23.4 ROC curve for FGEW probabilities of severe gale events in 12 local regions four days ahead. Data period: 1 Oct 2001 to 12 Feb 2003.

when gales are more likely to occur. (ROC points lying on the main diagonal represent 'no-skill'; points above the line show discrimination ability.) However, the points in a ROC curve nearer the top right of the graph give hit rates and false alarm rates corresponding to the lowest probability thresholds, so most of the discrimination ability is due to low probability forecasts. Note that although quite high hit rates (vertical axis) can be achieved for these low probability forecasts, this is at the cost of large numbers of false alarms. Although the false alarm rates plotted (horizontal axis) look quite small, they represent large numbers of false alarms because they are expressed as a fraction of all non-events.

Together, the reliability and ROC diagrams show that the first-guess system is able to provide some reliable probabilistic information on the likelihood of severe weather, but only on rare occasions is it able to provide the high probabilities that most customers would require before taking any protective action.

Verification results shown are for 4-day FGEW forecasts. ROC scores for different lead times show useful discrimination of events up to six days ahead. As lead time reduces, the main difference is that more events are predicted with higher probabilities. A common problem for forecasters using deterministic NWP is that the

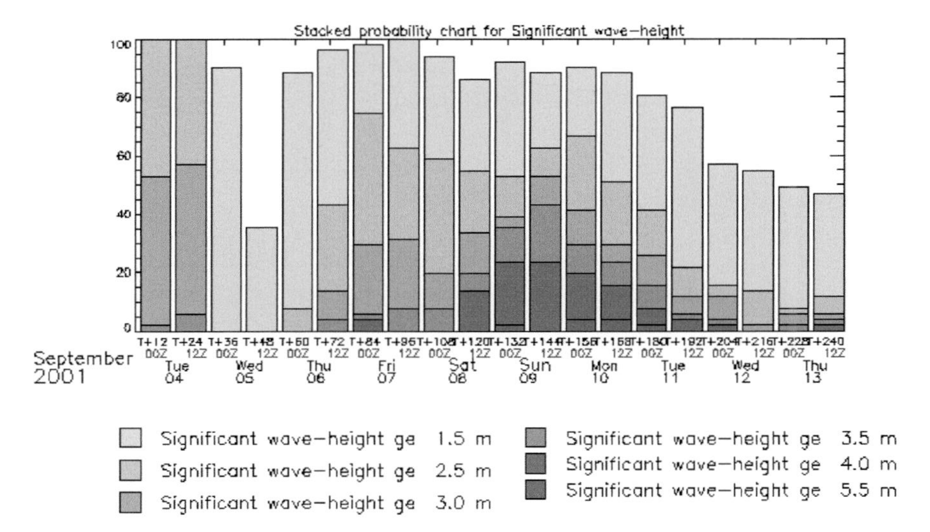

Figure 23.5 Example of a 'stacked probability chart' generated from the marine wave model within the EPS, showing the risks of exceeding various significant wave-height thresholds at a site in the North Sea for days 1 to 10 ahead.

forecast can change dramatically between runs of the model, particularly when the atmosphere is in an unstable state during development of severe weather. Before the advent of FGEW, forecasters rarely had the confidence to issue early warnings until very close to the event. By using the EPS, FGEW provides early alerts with low probabilities, with probabilities usually increasing as an event comes nearer. Combined with the verifiable skill of FGEW, this gives forecasters confidence to issue warnings earlier and more frequently.

The FGEW system also illustrates the difficulty of meeting customer requirements for high-confidence forecasts where predictability is low. While the system has some considerable skill in identifying the possibility of severe weather, and some ability to produce unbiased, reliable probabilities, the fundamental low predictability of the severe weather means that on most occasions early warning can only be given at low probability. If users were able to make use of such warnings their ability to mitigate the effects of severe weather could be much increased, and this will be discussed later in Section 23.6.

23.4.4 Site-specific forecasts

Most weather forecast customers require site-specific forecasts for their particular locations, so the Met Office has invested significant effort in extracting site-specific weather parameters from each EPS member to allow the generation of probability forecasts. Ensemble forecasts for around 300 sites are held in a database for product generation, and several graphical tools are available to display forecasts for customers. Figure 23.5 shows an example of a stacked probability chart generated from the marine wave model within the EPS, showing the risks of exceeding various significant wave-height thresholds at a site in the North Sea. This chart, generated routinely for use by offshore oil industry customers, is designed for risk assessment and is ideal for the identification of 'weather windows' in which weather-sensitive work can be carried out. This example shows a safe window between about 36 and 72 hours ahead for operations requiring a safety threshold of 4.0 m, and up to 120 h ahead if the operator is prepared to accept, for example, a 10% risk of exceeding this threshold.

Interpolating weather parameters directly from NWP model grids to specific locations can be subject to large errors as the NWP model cannot resolve the subgrid-scale features which are important in generating the microclimate of a real site. An NWP model can only attempt to represent weather parameters as an average over a grid box, and the true resolution of a model is around five grid lengths. To allow for this the Met Office applies a multivariate Kalman filter to relate values interpolated from model fields to observed weather parameters statistically (Mylne *et al.*, 2002). Parameters such as temperature are related to model temperatures, but also to wind direction, which is particularly important in coastal locations, so the Kalman filter provides more than just a simple bias correction. Use of the Kalman filter also allows the derivation of parameters which may not be available directly from the model, but which are available from site observations and which are required by customers, such as maximum and minimum temperatures. Figure 23.6 shows an example of maximum and minimum temperature forecasts derived from model fields using the Kalman filter for days 1 to 10 ahead. Temperatures are here presented using a 'box and whisker' plot to indicate the spread of uncertainty. Some end users prefer to be presented with an uncertainty range as indicated by this type of plot, with a given level of confidence, rather than probabilities for given events.

Figure 23.7 shows an example of a reliability diagram for 3-day forecasts of wind speed exceeding Beaufort Force 7. The line marked with triangles indicates wind-speed probabilities interpolated directly from model fields and shows that the wind is significantly over-forecast, as forecast probabilities are consistently too high. This bias is very largely corrected by the Kalman filter (diamonds) which produces a curve close to the ideal slope of 45 degrees. The improvement is also shown by the marked reduction in the reliability score.

23.5 Short-range predictability

So far we have been discussing medium- and long-range prediction, but most forecast customers are primarily interested in short-range forecasts (1–2 days). At this range operational NWP is still largely deterministic. Forecasts have improved steadily due to increased resolution, improved model formulation and data assimilation, and better use of observations, but there are still many uncertainties in the forecasts issued.

EPS Meteogram
HEATHROW (03772) 51.5° N .4° W
RECAL - EPS Forecasts : 2 February 2004 12 UTC

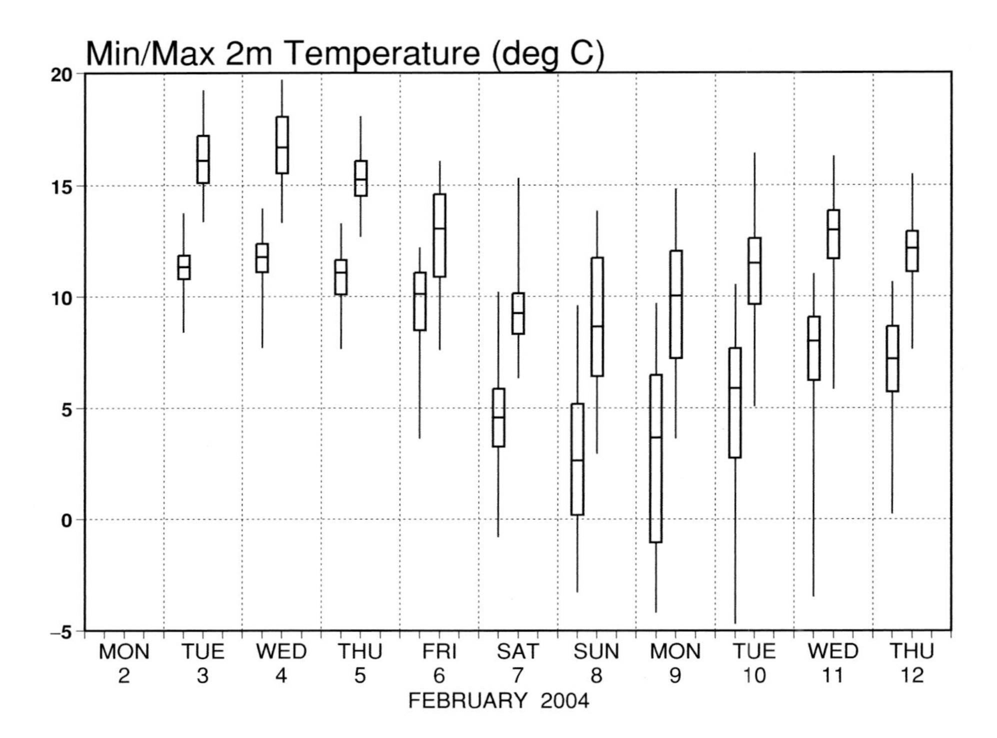

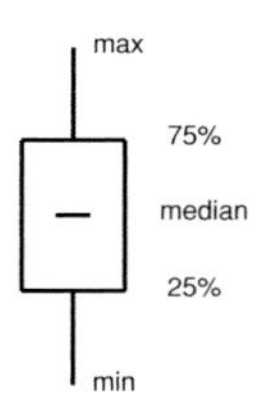

Figure 23.6 Example of maximum and minimum temperature forecasts for 10 days ahead presented as box and whisker plots. The central box represents the inter-quartile range, and the whiskers represent the 95% confidence range, while the horizontal line in the box gives the median value.

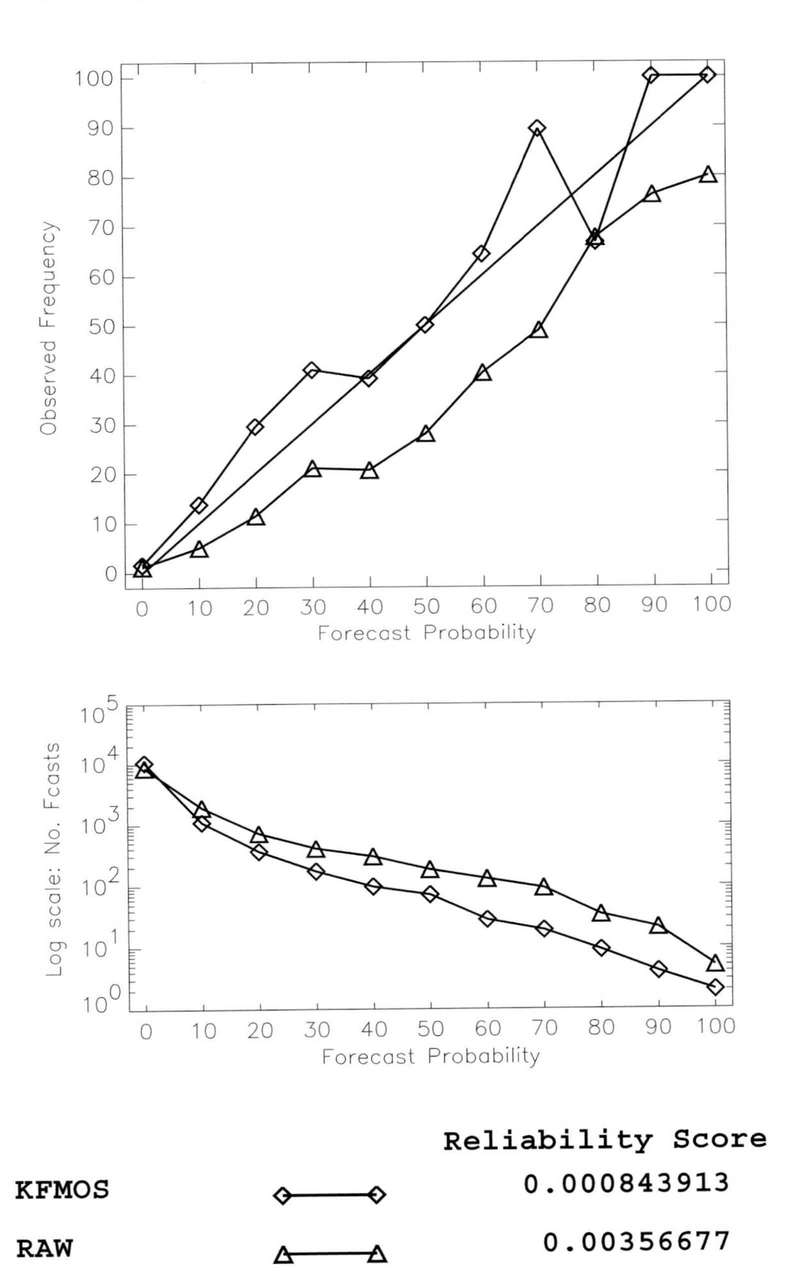

Figure 23.7 Reliability (top) and sharpness (bottom) diagrams for ensemble forecasts of wind-speed exceeding Beaufort force 7 at around 50 sites in the British Isles for the year ending 31 August 2004. Triangles indicate wind-speed probabilities from raw model fields; diamonds are after application of the Kalman filter.

Uncertainty in the short-range detail is assessed subjectively by forecasters with few objective tools to help them, but research is now starting into whether these issues can be addressed with ensembles.

Large synoptic-scale errors are rare in the short range, but typically involve rapid cyclogenesis and are therefore critically important. Much more common are errors in subsynoptic details such as frontal waves, quantitative precipitation, convection and weather parameters of importance to customers like cloud height and visibility. The latest computers now allow the possibility of NWP models with grid lengths as low as 4 km or even 1 km, which can start to resolve processes such as convection. While these models offer the possibility of forecasting great detail in mesoscale weather systems, such details are likely to be highly sensitive to small errors, and the long-term vision for the future of forecasting is to run such models in ensembles to assess the predictability of mesoscale weather systems. As a step towards this goal a number of NWP centres are now developing regional ensembles. The Met Office is developing an ensemble based on a regional model covering the Atlantic and Europe with a horizontal resolution of around 20 km. This resolution will only start to address some of the uncertainty issues, particularly rapidly developing synoptic-scale systems, but also lays the foundation for the future development of high resolution ensembles to address issues such as probabilities of localised heavy precipitation.

23.6 Real-world customers and probabilities

Customers ideally want high confidence categorical forecasts on which to base decisions. Limited predictability of many parameters means that this is often not possible, and the most informative products we can provide are probabilities such as 2%, 50% or 80%. Some of the most important forecasts for society and for business are for severe or extreme events. As demonstrated with the FGEW forecasts, it is frequently only possible to predict such events with low probabilities, especially for small localised areas. To balance customer requirements with scientific predictability we need to ask what users really need, and the answer is normally 'decisions'. So how can we help them make decisions from probabilities? In another chapter, David Richardson uses measures of economic value to assess forecasts, and the same approach can be used to guide decision-making with probability forecasts. By working with a customer to analyse their losses L associated with a weather event, and their costs C of protecting against that event, we can identify the cost–loss ratio C/L. Given this, the user's best strategy is to protect against the event whenever the probability is greater than C/L. Averaged over many occasions, and provided the forecast probabilities are reliable, this strategy will maximise savings. Even if forecasts are not perfectly reliable, analysis of past forecast performance can allow us to identify the optimal decision threshold for a particular customer. Figure 23.8 presents an example of forecast value for a particular customer with $C/L = 0.2$ plotted against a decision

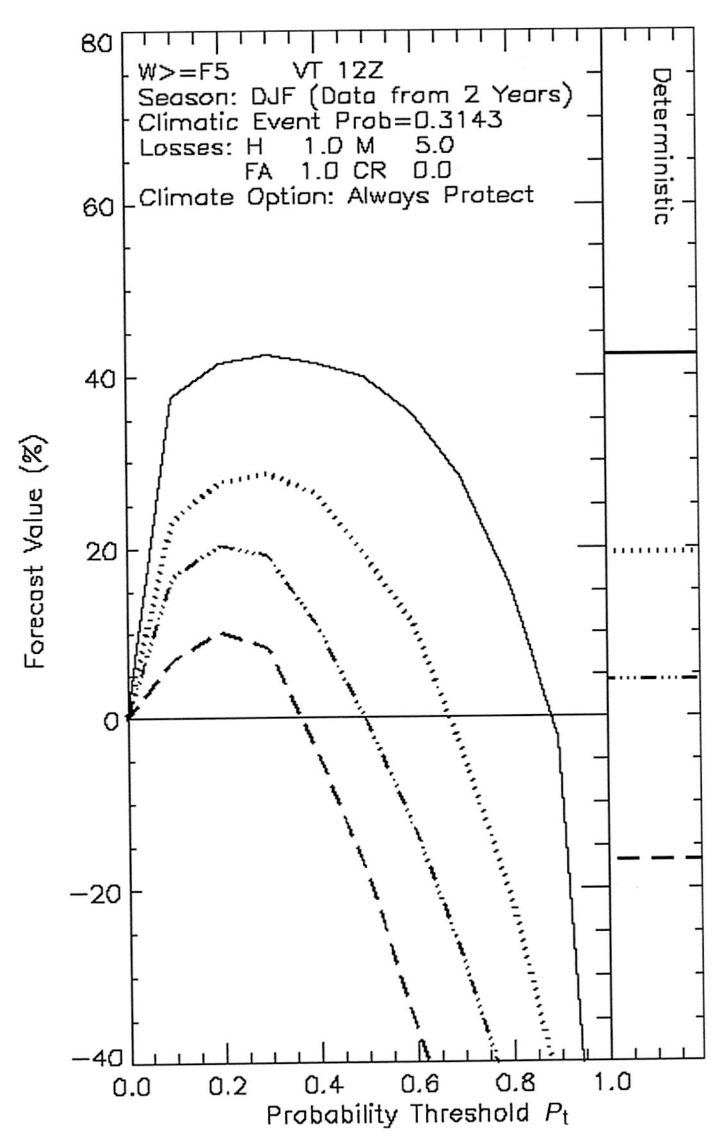

Figure 23.8 Examples of relative forecast value plotted against probability threshold P_t for probability forecasts of wind speed exceeding force 5. The values of equivalent deterministic forecasts are shown in the columns at the right-hand side of each graph. Forecast lead times are: 48 h (solid line), 96 h (dotted), 144 h (dot-dot-dot-dash) and 192 h (dashed).

threshold P_t. These curves are based on verification of uncalibrated probability forecasts interpolated directly from the EPS with no post-processing. For the user the best decision threshold is the P_t which maximises the value, so for 48 h or 96 h forecasts (solid and dotted lines) the user's best strategy is to protect against the weather when the probability exceeds 30%. Normally one would expect the customer's optimum threshold to be 20%, because $C/L = 0.2$, but because the forecasts are imperfectly calibrated they actually do better using 30%. (For further details see Mylne, 2002.)

In practice, real-world decision-making is usually much more complex, and more sophisticated decision tools are required, but methods such as this point the way to how forecast providers can work with their customers to maximise the benefit from forecasts where predictability is low. With increasing automation of forecast products, providers like the Met Office are increasingly working with customers to help them optimise decision-making, rather than simply providing a best-guess weather forecast. The use of probability products is a key part of this optimisation. After several years of introducing the ideas to customers we are now making significant progress in several sectors, notably severe weather warnings, the offshore oil industry (where potential losses are often massive) and weather derivatives traders who are very used to managing risk and basing decisions on small probabilities. Nevertheless, there are still very few customers who are prepared to take action on the basis of a probability as low as 10%, and much more work is required before society can take full advantage of the potential benefits of ensemble forecasts.

23.7 Conclusions

Predictability is an issue for forecasters and customers on all time-scales, and ensembles are now well-established tools to aid assessment of predictability at long and medium ranges. Ensembles are used to improve the quality of deterministic forecasts by identifying the most probable solutions, and to supplement them with confidence, information and alternative solutions. Forecasters are also provided with probabilistic guidance to help with risk assessment of severe weather. Many tools are now available to provide high-quality automatic probability forecasts to customers, who are starting to see the benefits in some sectors. Research is progressing to predictability issues in short-range forecasting.

Acknowledgements

I am grateful to numerous colleagues at the Met Office who have performed much of the work described, most notably Caroline Woolcock, Tim Legg and Tim Hewson. Also to Tim Palmer, François Lalaurette and colleagues at ECMWF who have provided much inspiration as well as the excellent forecasts from the EPS.

References

Carroll, E. B. (1997). A technique for consistent alteration of NWP output fields. *Meteorol. Appl.*, **4**, 171–8.

Hewson, T. (2004). *The Value of Field Modifications in Forecasting*. Forecasting Research Technical Report 437. Met Office.

Jolliffe, I. T. and D. B. Stephenson (2003). *Forecast Verification: A Practitioner's Guide in Atmospheric Science*. Wiley.

Legg, T. P. and K. R. Mylne (2004). Early warnings of severe weather from ensemble forecast information. *Weather and Forecasting*, **19**, 891–906.

Murphy, A. H. (1973). A new vector partition of the probability score. *J. Appl. Meteorol.*, **12**, 595–600.

Mylne, K. R. (2002). Decision-making from probability forecasts based on forecast value. *Meteorol. Appl.*, **9**, 307–16.

Mylne, K. R., C. Woolcock, J. C. W. Denholm-Price and R. J. Darvell (2002). Operational calibrated probability forecasts from the ECMWF Ensemble Prediction System: implementation and verification. In *Preprints of Symposium on Observations, Data Assimilation, and Probabilistic Prediction*, 13–17 January 2002, Orlando, FL, pp. 113–18. American Meteorological Society.

Stanski, H. R., L. J. Wilson and W. R. Burrows (1989). *Survey of Common Verification Methods in Meteorology*. WWW Technical Report 8, WMO TD 358, WMO.

Wilks, D. S. (1995). *Statistical Methods in the Atmospheric Sciences*. Academic Press.

24

Ensemble forecasts: can they provide useful early warnings?

François Lalaurette
Ecole Nationale de la Météorologie, Toulouse

Gerald van der Grijn
European Centre for Medium-Range Weather Forecasts, Reading

24.1 Medium-range forecasts

The design of ensemble methods for medium-range weather forecasting ten years ago was aiming mainly at addressing the problem of limited predictability of supra-synoptic weather regimes within the 6–10 days range. From this point of view, it has been shown since then that the ensembles have delivered improved forecasts compared with a purely deterministic approach, both improving single-value estimates by removing stochastic errors (ensemble mean) and providing reliable and still sharp estimates of the probability distributions of large scale flow patterns such as blocking (Chessa and Lalaurette, 2001; Pelly and Hoskins, 2003).

Severe weather forecasts seem, however, to be clearly beyond reach of such medium-range, global and therefore relatively coarse grid models. Indeed the experience in trying to use medium-range forecasting systems to be alerted of the risk of severe weather more than one day in advance is very limited. On most occasions, civil security services are alerted not earlier than the day before the event, while public warnings are only issued on the same day – tropical cyclones being a notable exception to this common rule. The reasons usually given by forecasters as to why they do not use numerical forecasts in the early medium range are mainly twofold:

1. The global numerical models are generating nothing looking remotely like severe weather.

Predictability of Weather and Climate, ed. Tim Palmer and Renate Hagedorn. Published by Cambridge University Press.

2. If one takes signatures from the global models that are associated with severe weather, there is no consistency in the forecast from one day to the next: the rate of false alarms would be far too high to be considered.

The current status of development of global models used for medium-range forecasting is such that the first item can be debated. ECMWF has been running in spectral space with truncation T511 (roughly 40 km) and 60 vertical levels since 2000 and is expected to go for T799 (25 km) and 91 levels by 2005/6, a resolution that was considered not so long ago to be suitable for mesoscale modelling. Running a global model also has many advantages compared with a limited area approach, as it is easier to keep the forecasts under control of a global observing system without having to take care of the propagation of information through lateral boundary conditions.

The following sections will aim at providing evidence that global numerical weather prediction (NWP) models nowadays generate weather patterns that can be associated with severe weather. It will then be shown that there is some predictive signal to be detected up to between three to five days ahead using the improved signal detection capability offered by running an ensemble of forecasts. Indeed, although there is certainly a degree of chaotic behaviour in the generation of extreme weather, some signal can be extracted out of this chaos by using these dynamical Monte Carlo techniques that could prove useful for some actions to be based on early warnings.

24.2 Severe weather in medium-range forecast models

Rainfall is one of the weather events that is likely, when coming in large amounts, to cause severe disruptions and loss of human life due to the floods that it can generate on different timescales. Because it is so important in the hydrological cycle to ensure a correct energy balance in global models, considerable efforts have been spent in recent years to try to improve their representation, with the result that they now compare much better to real observations. As an example of such a trend, the distribution of rain events in the model is compared for two different years with observations in Figure 24.1. As expected, events with daily rainfall in excess of 60 mm/day are poorly represented by the model. For such events a more accurate representation of small-scale processes would be needed, when small-scale convection and local orographic enhancement effects are of paramount importance in order to capture the essence of the most extreme scenarios. The trend over recent years though has clearly been towards an improvement, with a larger proportion of forecasts generating events in excess of 50 mm/day, bringing them closer to observations. At the other end of the spectrum, smaller amounts are generally generated in excessively large proportions. This is partly due to a resolution effect – such small amounts are usually resulting from an average over the model grid of more intense but localised events. Using more than single observations to validate the model precipitation fluxes usually validates

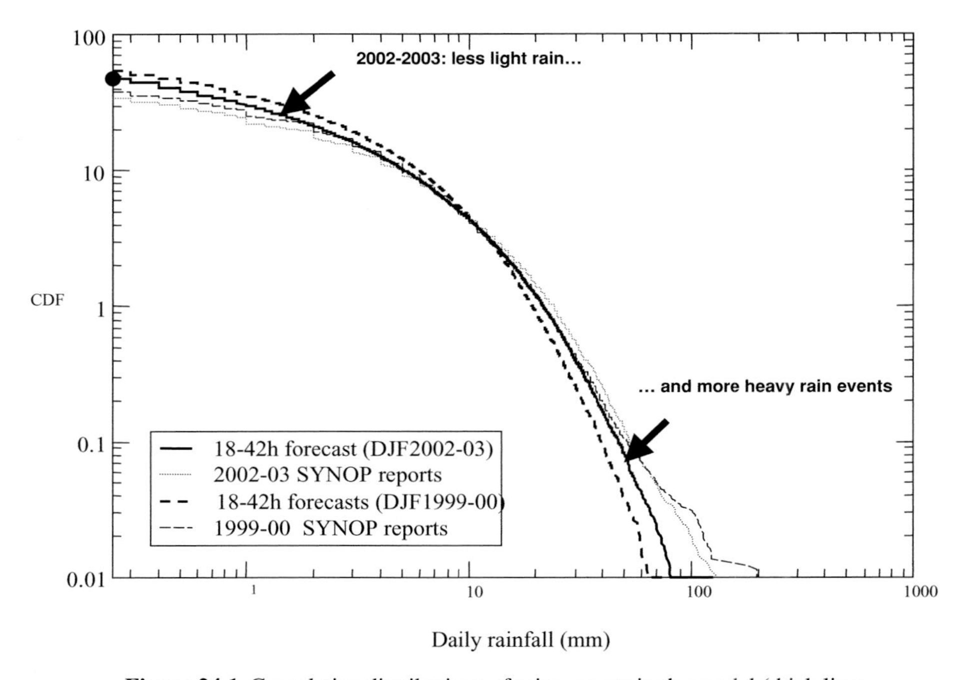

Figure 24.1 Cumulative distributions of rainy events in the model (thick lines, 1999–2000 dashed and 2002–3 full) and observations (thin lines, 1999–2000 dashed, 2002–3 dotted).

this interpretation by showing a much better agreement between model rainfall and observation aggregated within the model grid (Cherubini *et al.*, 2002).

The ability for the model to generate severe storms has improved in recent years, although the direct comparison from model wind speed over land with observations shows a relatively large, negative bias. Among the reasons why this occurs is that modellers are more concerned when designing boundary layer representations to have a good momentum budget than to optimise on-site validation of local effects. However, a step towards the post-processing of maximum wind-gust values based both on explicit model winds and the subgrid-scale representation of turbulent fluxes was taken in 2000 at ECMWF, resulting in a better adequacy between model and observations. An example of signatures now to be found for a typical, small-scale storm over western Europe is given in Figure 24.2, where a good agreement between the short-range forecast and observations can be seen.

To identify the ability for models to generate severe weather signatures is only one side of the story of course, which tells nothing of the rate of success and failure with which the model then generates three or four days in advance. Severe weather events are usually not seen by the models as the most likely scenarios at these ranges, but forecasters have expressed an interest in using even small probabilities of severe weather occurrence as a useful early warning that will help them focus their

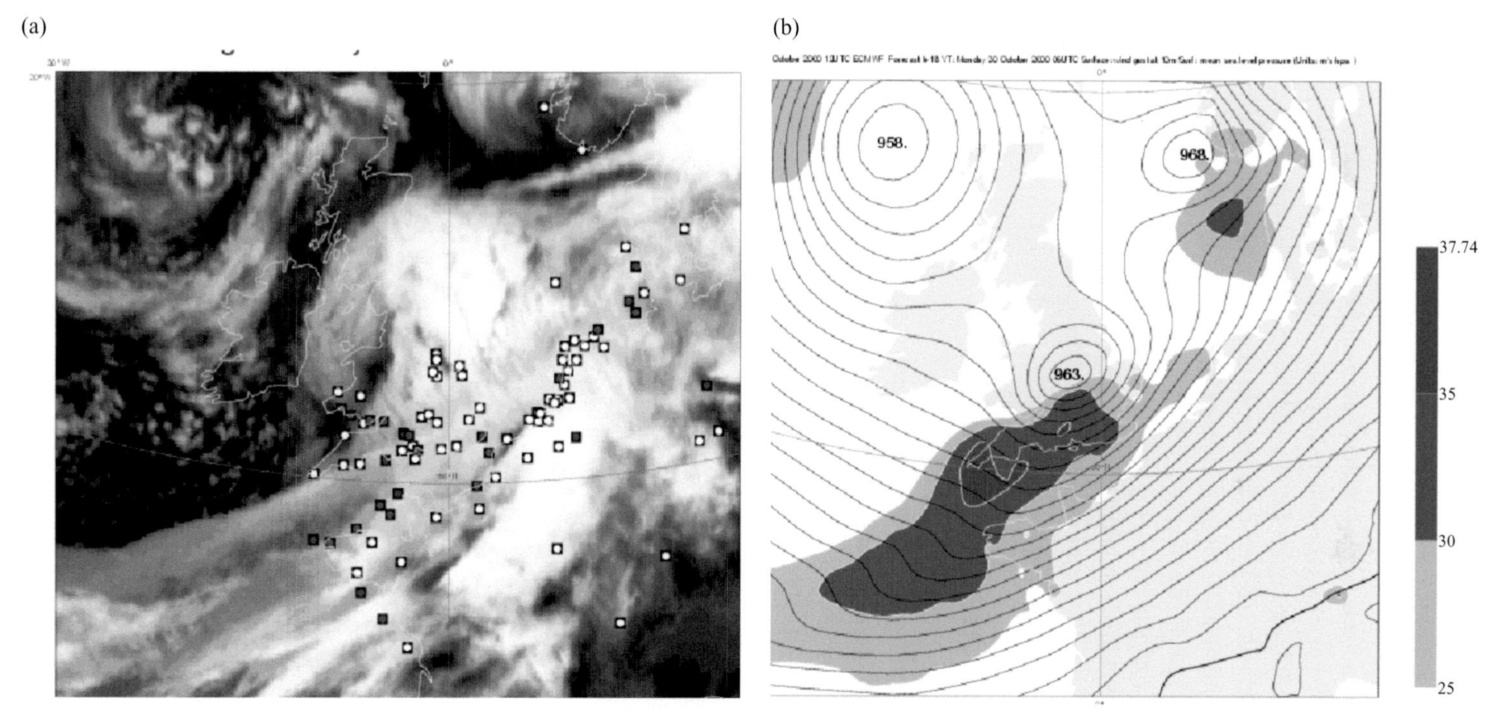

Figure 24.2 'Halloween' storm over the British Isles on 06 UTC 30 October 2000. (a) Meteosat7 Infrared picture+ SYNOP wind gust reports (white squares > 25 m/s; grey>30 m/s; dark grey>35 m/s). (b) 29 October 12 UTC forecast for mean sea level pressure and wind gust.

monitoring of the situation when the severe weather eventually comes closer. The definition of thresholds on which to base such probabilities is, however, difficult, as it is very dependent on the local climate. In fact, it could be argued that in any populated area indigenous populations have remarkably well adapted their activities to the local climate: what is considered a severe cold outbreak in Montreal or Cairo is not likely to be the same type of event, and the same could probably be said for severe storms in Reykjavik and Berlin. In order to generate maps where all locations are a priori equally likely to be hit by an unusual event, it has been proposed (Lalaurette, 2003) to map the severity of events with respect to the local climate distribution. In this way, and provided that we use the model to picture the local climate, orographic or land–sea map effects should be handled consistently in the climate and the model forecast, making the measure of departures more meaningful.

24.3 An extreme forecast index

Although the models currently generate severe weather systems consistently throughout the medium range, their forecast skill quickly declines over the first few days. In the case of the Halloween storm shown in Figure 24.2, this resulted in both an error in the location and the intensity of the storm (Figure 24.3) – although the model correctly predicted the large-scale rapid, perturbed flow that was to affect the area. Such errors in the smaller scales are to be expected for unstable systems where small errors in the analysis quickly amplify. It is therefore of interest to see whether the Ensemble Prediction System (EPS) is able to tackle these uncertainties in a probabilistic way.

The distribution of ensemble wind forecasts valid for the same time near Dover is also reported in Figure 24.3(b). Although far from the 25 m/s sustained wind speed that was observed, some of the members did indicate that the situation was deviating from 'normal'. This is easier to see when the climate distribution based on EPS forecasts valid for this time of the year and this location are also reported, as is the case here. For example, although a value of 15 m/s that is exceeded by 33 members out of 50 here would not be considered as a severe weather event in Dover, it is to be found only in slightly more than 1% of the EPS records at this time of the year and this location.

The inspection of such local, empirical EPS distributions of parameters such as wind, temperature or rainfall is, however, not something that could easily become part of the routine work of any forecaster: the amount of information that would be needed before an assessment of the situation can be made would be far too large to be achievable in time for the forecast to be of any use. As has been the case each time a practical use of ensemble forecasts has been considered, some aggregation of the available information has to be made. Early attempts have mostly aimed at *clustering* large-scale scenarios on the basis of their similarity over subcontinental areas. Such an approach, although helpful to describe the large-scale evolution of the weather, is

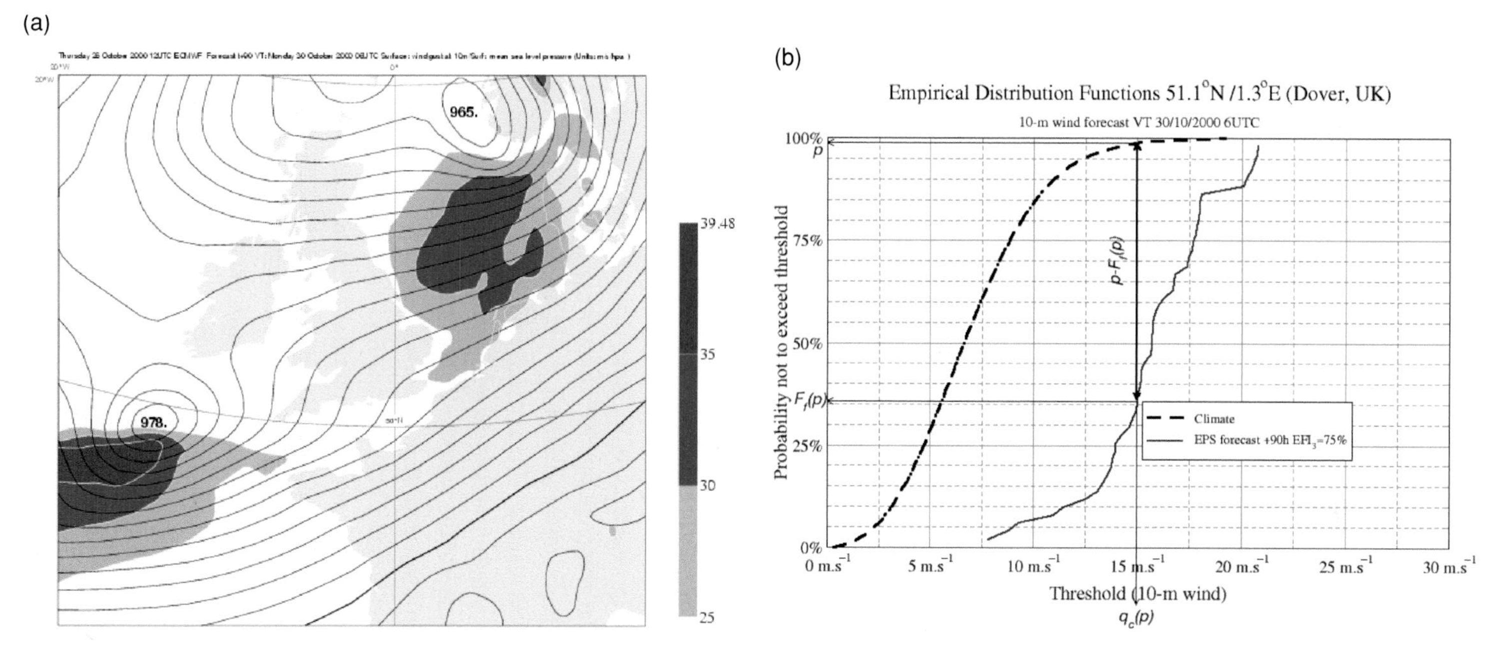

Figure 24.3 (a) Ninety-hour forecast of the 'Halloween' storm. (b) Climate (dashed) and EPS 90-h forecast (solid) distributions of wind at Dover (same valid time).

unlikely to provide a useful information for severe events, as scenarios that can look similar on the large scale can generate very different kinds of local events. It is rather proposed here to scale the EPS distributions with respect to the climate.

To each proportion p of the climate records is attached a *quantile* $q_c(p)$ (Wilks, 1995). How much a given EPS forecast deviates from the climate therefore can be measured by $p - F_f(p)$ where $F_f(p)$ is the proportion of EPS members that lies below $q_c(p)$. These differences are then integrated in probability space to provide the *Extreme Forecast Index* (EFI; Lalaurette, 2003):

$$\mathrm{EFI}_n = (n+1)\int_0^1 (p - F_f(p))^n dp \tag{24.1}$$

The result from such a procedure is that differences between any climate and forecast probability distributions functions (pdf) are scaled between -1 and $+1$ (-1 if the pdf in the forecast is entirely shifted below the minimum climate record, $+1$ if it is entirely shifted beyond the maximum). If the pdf from the forecast is the same as that for the climate, then $\mathrm{EFI} = 0$. n is an odd number that controls how selective the index is to large values. An alternative formulation offering better selection near the tails of the distribution has arbitrarily been given the order $n = 0$:

$$\mathrm{EFI}_0 = \frac{2}{\pi}\int_0^1 \frac{p - F_f(p)}{\sqrt{p(1-p)}} dp \tag{24.2}$$

An example generated from 90 h forecast EPS distributions based from 26 October 12 UTC is shown in Figure 24.4. It shows that although the deterministic forecast was misplacing the event, enough EPS members did have it at the right location for the EFI to reach high values where the event happened (Figure 24.3b). Indeed the EFI_3 value at Dover reached +75%, clearly indicating a well above normal level of risk for gale force winds four days in advance.

Geographical maps of the type shown in Figure 24.3(b) are available for ECMWF Member States on www.ecmwf.int. Several case studies have been looked at, and forecasters have expressed a keen interest for this type of product. It is not intended of course as an automated warning system – rather a 'warning light' that ensures a potentially dangerous event does not go unnoticed by the forecaster. From this point, more detailed investigations of the full probability distributions, either locally or by isolating 'worse case' synoptic scenarios should help in detecting dynamical signatures from future observations and making well-informed decisions such as issuing public warnings.

Another useful application of the EFI is to explore the predictability of severe weather in the early medium range. Case studies are indeed well known to be biased estimators of severe weather forecast performance: there is always some kind of signal

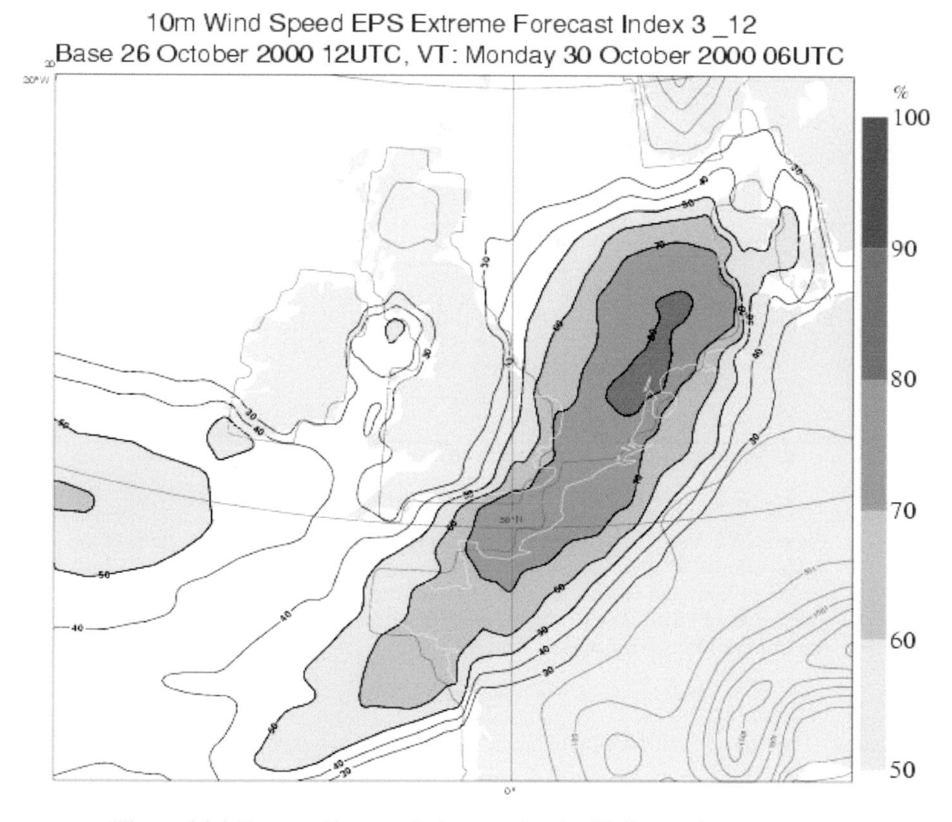

Figure 24.4 Extreme Forecast Index map for the 'Halloween' storm.

that the forecaster should in retrospect have been aware of. By contrast, conducting a verification study using forecasters' expertise in real time or delayed mode is both costly in terms of human resources and biased in its own way by the forecaster's perspective. Day 3 to day 5 forecasts are moreover hardly ever looked at in the context of severe weather. Strategies to explore the potential benefit of using such forecasts should be explored first before any serious consideration is given as to whether or not the range of warnings can be extended for some users.

As a first step in this direction, 6–30 h model forecasts have been used as a proxy for daily precipitation analyses – a reasonable choice in the absence of a comprehensive verification network (Rubel and Rudolf, 2001). The events targeted were those with daily rainfall exceeding 95% of the model climate records, and the verification period was December 2001 to April 2002. Results in terms of hit rates and false alarm rates are shown for Europe in Figure 24.5 for different values of the EFI, both for the EPS and the control, the single-value forecast. Results in terms of the ROC curves look rather positive, with a large portion of the curve lying well above the zero-skill

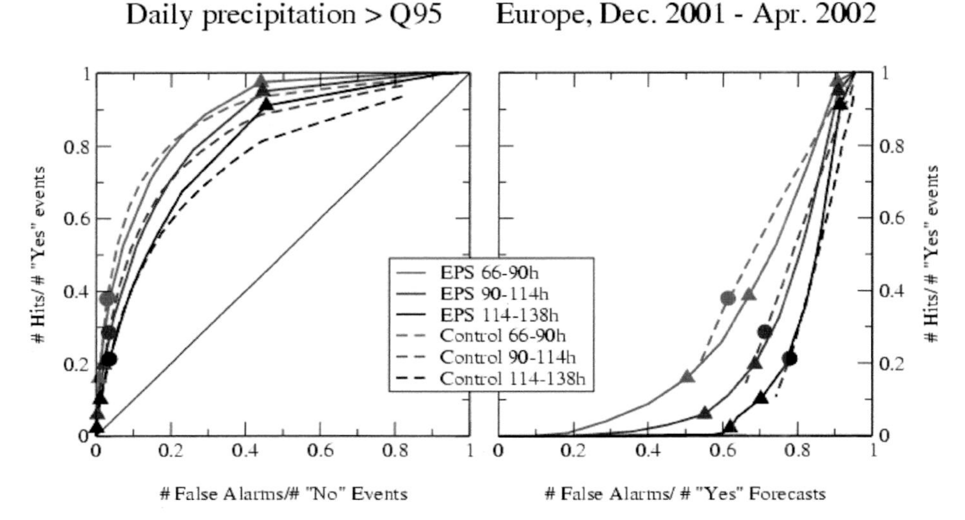

Figure 24.5 Verification of EPS forecasts that daily rainfall exceeds 95% of the model climate records (Q95). Left: ROC curves. Right: Same, but false alarms are scaled with respect to total number of forecast occurrence; each point corresponds to a different EFI threshold (triangles are 50%, 30% and 0%, while the circles are for the Control forecasting exactly the Q95 value)

diagonal. The longer the forecast range, the larger becomes the benefit from using an ensemble – indeed at the 6–30 h range, the verification used here is biased in such a way that the control forecast would be perfect, while the ensemble would still generate some false alarms and miss some events.

Although these results indicate without doubt that there is a skill in the early medium range to forecast moderately severe weather events, it should be realised that to achieve large hit rates, a significant number of false alarms will be generated as well. This fact is in a way hindered by the rarity of the event on the ROC curves. If instead of a false alarm rate per number of non-occurring events, the curves are shown with respect to the number of warnings issued, it can be seen on Figure 24.5 that to achieve a hit rate of 50%, 60 to 80% of the ‘warnings’ would be wrong. Whether or not this is an acceptable result is certainly user dependent. It stresses in any case that using early warnings for severe weather has to be carefully balanced. There is certainly more than pure randomness in the skill achieved, as a ‘no-skill’ rate of false alarms per warning issued should be 95% when one forecasts an event occurring only in 5% of the cases. The rate of hits/false alarms generated by using such an early warning system would clearly be very far from what can be obtained by waiting until a few hours before the event. But because there are protection measures that cannot wait until the last minute to be taken – for example, releasing water from reservoirs in the case of floods, evacuating of populations for tropical cyclones, interrupting air and rail traffic for a major storm event – there may be some value attached to such early warning procedures. This would mostly be the case if such warnings were used for early planning purposes such as putting extra staff in pre-alert mode or moving rescue facilities closer to the areas expected to be affected by severe weather.

24.4 Tracking tropical cyclones in the EPS

Earlier studies (Barkmeijer *et al.*, 2001; Puri *et al.*, 2001) have shown that tropical cyclone (TC) forecasts could benefit from ensemble techniques, provided that singular vector (SV) perturbations include some representation of diabatic processes and target areas for error growth are defined in the vicinity of TC locations rather than using the entire tropical region. In order to validate and disseminate TC forecast products, a tracker has been developed in line with developments in other centres (van der Grijn, 2002). No TC genesis is handled at present – only those TCs that have been reported by the WMO regional centres (RSMCs) with responsibility for TC are tracked. The algorithm currently uses model data on a 0.5° by 0.5° latitude-longitude grid. Starting from the analysis, TCs are tracked for 120 h every 12 h (EPS) and 6 h (T511) using an algorithm adapted from Sinclair (1994).

The skill of the deterministic TC forecasts has been assessed for April 2002 to June 2004. The results are shown in Figure 24.6. Scores are given for the high resolution T511 model (OPER) and lower resolution EPS control model (CTRL). The sample size decreases rapidly with increasing forecast step due to all TCs not surviving for the 5-day period considered in the verification, which is partly an observed feature (some TCs die during this period) and partly a model failure to maintain the TC activity. The forecast error in core pressure is always positive: TCs in the analysis and forecasts are on average weaker than observed. This is especially the case for the CTRL. However, this positive bias in core pressure seems to decrease, or at least to saturate, later in the forecast. Apparently the model is more capable of developing TCs with a realistic core pressure in the forecast than analysing them in the initial conditions, a feature that may be related to the limited resolution of both structure (background) functions and incremental 4D-Var (T159 inner loops).

Of particular importance for practical applications is the distance error in tracking TCs. It is on average 100 km in the analysis and increases almost linearly to around 500 km at D + 5. Model resolution does not seem to have much impact on this error. Both resolution forecasts have a slow bias and a tendency to recurve too quickly.

24.5 'Strike probability' and probabilistic verification of TC forecasts

Conventional probability maps are useful in assessing the likelihood of a certain event at a specific location. However, a drawback of such a probability map is that probability values do not always 'add up'. In theory, probability values can be very

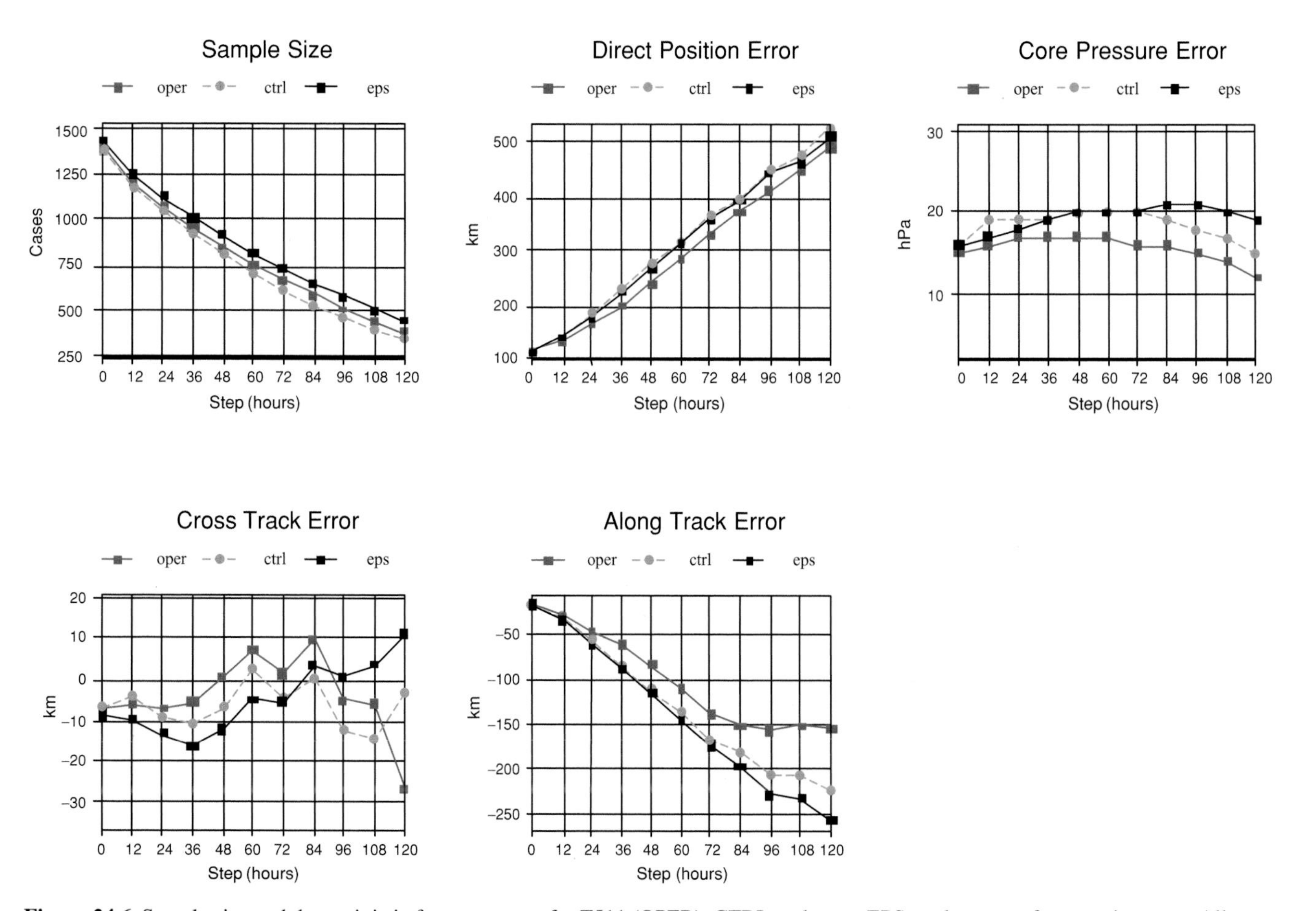

Figure 24.6 Sample size and deterministic forecast errors for T511 (OPER), CTRL and mean EPS tracks versus forecast time step. All forecasts from 6 April 2002 to 21 June 2004.

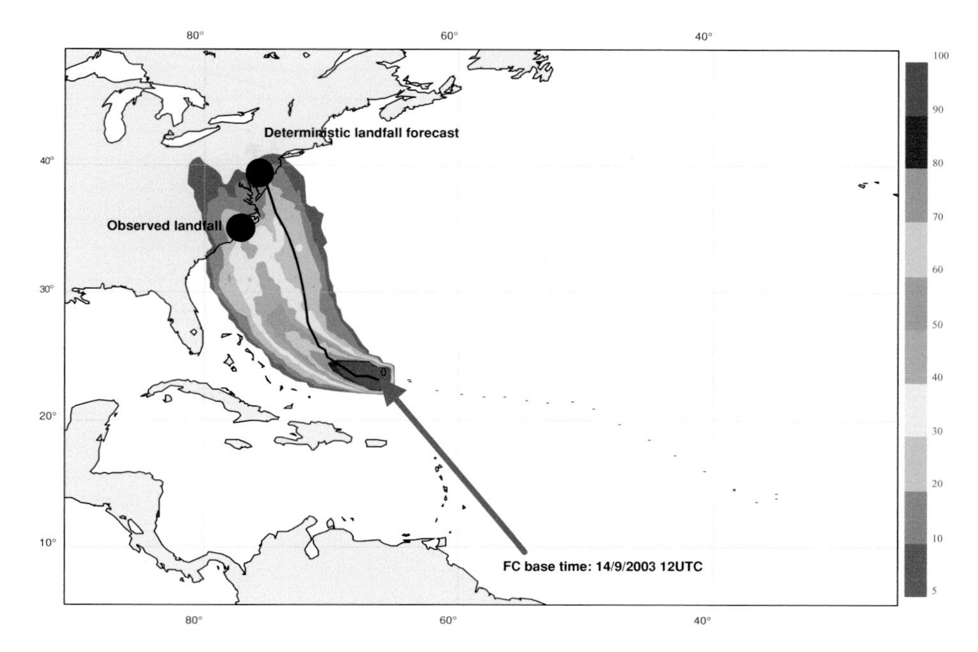

Figure 24.7 Tracking Tropical Cyclone Isabel in the EPS: probability to get closer than 120 km from the TC centre within 5 days from forecast start date. The observed location of Isabel landfall is also reported.

low even when all EPS members support a specific event (e.g. 10 m wind speed exceeding 25 m/s). This is the case when the EPS members predict the event at different locations for a given time step. To enhance the signal for severe weather one must think of a different type of probability: a forecaster is often more interested in whether a TC will affect a certain area than when that TC will hit a specific location; the exact location of the TC is of less importance – within a certain margin, the TC is likely to be equally (or even more) devastating if its centre does not exactly follow the forecasted track, or passes along it with a slight time delay.

The concept of 'strike probability' originates from this idea. The strike probability is defined as the probability that a TC will pass within 120 km from a given location at any time during the next 120 hours. The strike probability is based on the relative number of members that predict this event, each member having an equal weight.

One of the features of a strike probability map as presented in Figure 24.7 is the elimination of the time dimension and therefore its advantage is its simplicity. It allows the forecaster to make a quick assessment of the high-risk areas regardless of the exact timing of the event. Another feature of a strike probability map is that it gives the forecaster an estimate of the skill that can be expected from the CTRL. This is because on average the CTRL track error should be equal to the ensemble track spread. In other words, the width of the probability plume is a direct measure of the spread in the EPS and would ideally be a good indicator of the expected

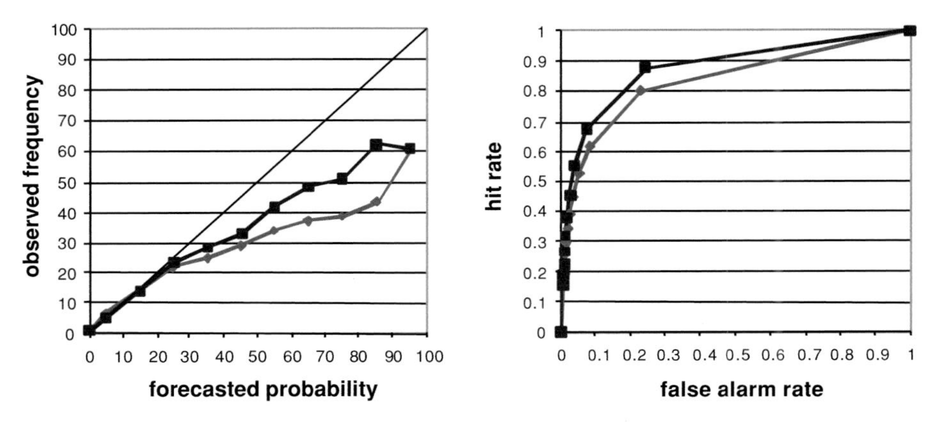

Figure 24.8 Reliability diagram (left) and ROC (right) for the forecast probability that a TC will pass within 120 km any time during the first 120 hours of the forecast. Grey lines: cycle 23r4 (no TC targeting), black lines: cycle 24r3 (with TC targeting). From van der Grijn (2002).

error in the CTRL. In the case shown in Figure 24.6 though, rather than the spread of the EPS tracks it is their bimodal mode that is more remarkable – although the T511 recurved the hurricane track too much, the EPS gave an early indication of an alternative location for landfall that might have been a useful indication for a forecaster planning early warnings.

The strike probability can be used to assess the skill of the EPS with respect to TC forecasting. Figure 24.8 illustrates this both in terms of reliability and signal detection (ROC). A clear improvement in reliability can be seen from the pre-operational testing of the model change (cy24r3) that was to introduce new TC targeting in January 2002. However, the system is overconfident in the high probability range. A 95% probability forecast only verifies in 60% of the cases. This might be an indication that the spread is still too low in the early forecast steps. Relative operating characteristics (ROC) curves also show the positive impact of targeting on the detection of TC.

24.6 Summary and perspectives

Preliminary work aiming at designing new products that could be used for early warnings of severe weather conditions has been described in this chapter. Although a probabilistic approach is likely to be desirable in order to be able to extract the signal from the numerical forecasts in a way that can be tailored to the user's needs, the quality of the product ultimately relies on the quality of the numerical model itself – from this viewpoint, the forecast quality has improved quite significantly in recent years, and this is what makes it possible to think of extending the range of warnings for severe weather into the early medium range. Development of the EPS system is also of paramount importance if one expects to sample properly the tails of the forecast distributions – those that matter in the early medium range when severe weather is only a possible, but low probability, scenario. Some results showing the impact of improving the sampling of tropical cyclone perturbations have been illustrated here in Figure 24.8.

Signal detection of severe weather events in the early medium range is likely to be difficult. Although one can find some comfort in the fact that the preliminary results shown here for tropical cyclones and large precipitation events indicate high rates of successful forecasts, it cannot be disputed that these are achieved provided that preliminary action is taken on the basis of very small probabilities – thus generating very high false alarm rates. The skill achieved by the EPS system in this context is, however, not negligible: it has been shown that the false alarms are reduced significantly compared with a random forecast system, and even compared with a single-value (EPS control) system (Figure 24.5). Further, careful evaluations of the impact of such products on early decision-making in pre-operational environments should convince some users of the benefit of such products.

References

Barkmeijer, J., R. Buizza, T. N. Palmer, K. Puri and J.-F. Mahfouf (2001). Tropical singular vectors computed with linearized diabatic physics. *Quart. J. Roy. Meteor. Soc.*, **127**, 685–708.

Cherubini, T., A. Ghelli and F. Lalaurette (2002). Verification of precipitation forecasts over the Alpine region using a high-density observing network. *Weather Forecast.*, 238–49.

Chessa, P. and F. Lalaurette (2001). Verification of the ECMWF Ensemble Prediction System forecasts: a study of large-scale patterns. *Weather Forecast.*, **16**, 611–19.

Lalaurette, F. (2003). Early detection of abnormal weather conditions using a probabilistic extreme forecast index. *Quart. J. Roy. Meteor. Soc.*, 3037–57.

Pelly, J. L. and B. J. Hoskins (2003). How well does the ECMWF ensemble prediction system predict blocking? *Quart. J. Roy. Meteor. Soc.*, **129**, 1683–702.

Puri, K., J. Barkmeijer and T. N. Palmer (2001). Ensemble prediction of tropical cyclones using targeted diabatic singular vectors. *Quart. J. Roy. Meteor. Soc.*, **127**, 709–32.

Rubel, F. and B. Rudolf (2001). Global daily precipitation estimates proved over the European Alps. *Meteorologische Zeitschrift*, vol. 10, pp. 407–18.

Sinclair, M. R. (1994). An objective cyclone climatology study for the southern hemisphere. *Mon. Weather Rev.*, **122**, 2239–56.

van der Grijn, G. (2002). *Tropical Cyclones Forecasting at ECMWF: New Products and Validation*. ECMWF Technical Memorandum, 386. Recent ECMWF Technical Memoranda can be retrieved in electronic format from http://www.ecmwf.int/publications/library/ecpublications/techmemos/tm00.html.

Wilks, D. S. (1995). *Statistical Methods in the Atmospheric Sciences*. Academic Press.

25

Predictability and economic value

David S. Richardson
European Centre for Medium-Range Weather Forecasts, Reading

25.1 Introduction

To many people, probability forecasts are still much less familiar than traditional deterministic forecasts. Two issues are often raised as practical problems for the use of probabilities. First, there is a common perception that probability forecasts have no place in the real world, where users need to make hard yes/no decisions. Secondly there is the feeling that probability forecasts are difficult to assess – 'probability forecasts are never wrong', the scores are complicated, and different scores tend to show different 'skill'. As an illustration of this last point, Figure 25.1 shows two examples of the evaluation of probabilistic skill for the ECMWF Ensemble Prediction System (EPS; Buizza, this volume; Palmer *et al.*, 1993; Molteni *et al.*, 1996; Buizza *et al.*, 2003). The ROC skill score (based on the area under the relative operating characteristic (ROC) curve; Richardson, 2000, 2003) shows substantial skill, remaining above 40% throughout the 10-day forecast range. However, the Brier skill score (BSS; Wilks, 1995) decreases quickly so that there is no skill at all beyond day 8. Clearly, the two skill measures present contrasting perceptions of the performance of the EPS. This raises the obvious question of whether the forecasts are skilful or not and, perhaps more importantly, are the forecasts useful or not? It should be noted that these questions are not restricted to probability forecasts but are equally relevant to the more traditional deterministic forecasts. It is perhaps just

Predictability of Weather and Climate, ed. Tim Palmer and Renate Hagedorn. Published by Cambridge University Press.

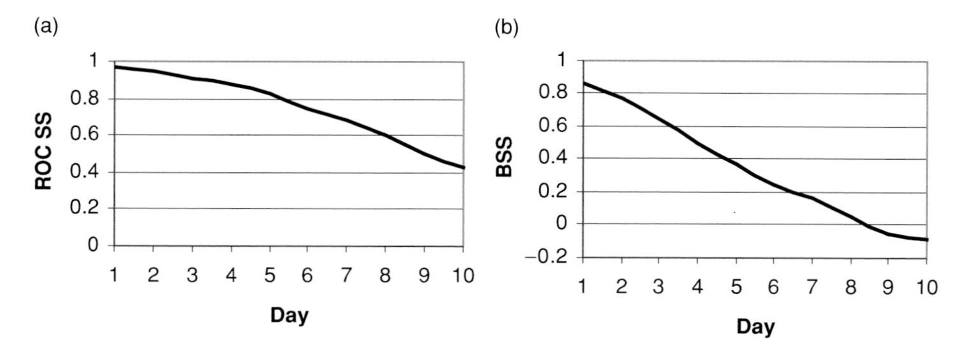

Figure 25.1 Skill of ECMWF EPS probability forecasts of 500 hPa height anomalies more than 2 standard deviations above normal over the extratropical northern hemisphere for spring 2002. (a) ROC area skill score (ROCSS) and (b) Brier (BSS) skill score.

unfamiliarity with using probability forecasts and with the scores used to evaluate them that makes the issues more apparent.

To understand how probability forecasts can be used, whether they give benefits to the user and how this relates to the various skill measures, it is necessary to consider the decision-making process of individual users. In general this is a complex process, specific to each user, which may not easily be modelled (e.g. Smith, this volume, and references therein). Nevertheless, it is useful to study a simple decision-making model in order to illustrate some of the important concepts. This provides an initial user perspective on the use and evaluation of probability (and deterministic) forecasts.

First the simple cost–loss decision model is introduced. This provides a useful introduction and framework for understanding the concept of forecast value and how probability forecasts can be used in making yes/no decisions. The model is used to explore how the benefit of forecast information can vary between users, the value of probability forecasts compared with deterministic forecasts, and also the difference between Brier and ROC skill scores. Despite the benefits of the cost–loss model, there are a number of obvious limitations. Two apparent deficiencies of the model are that it assumes that the consequences of the user's actions can be expressed numerically (for example as financial costs and losses) and that the users are assumed to be risk-neutral (they are only concerned with long-term average expense). In the final section, the concept of 'utility' is introduced to generalise the decision model and show that the results from the earlier sections can have wider application.

25.2 The cost–loss decision model

The simple cost–loss decision model has a history dating back to the early twentieth century (Ångström, 1922; Liljas and Murphy, 1994). For a recent comprehensive

Table 25.1 *Costs and losses associated with different actions and outcomes in the cost-loss model*

		Event occurs	
		Yes	No
Action taken	Yes	C	C
	No	L	0

introduction see Richardson (2003). Consider a user, or decision-maker, who is sensitive to a specific adverse weather event X. Palmer (this volume) introduced the case of the builder Charlie whose concrete-laying work is disrupted by the occurrence of frost. For other users, X might be the occurrence of ice on the road, or more than a certain amount of precipitation in a given period. If this event occurs and the user has not taken any preventative action then they suffer a financial loss L. Alternatively, the user could take action at a cost C that would protect against this potential loss. The costs and losses of the various combinations of action and outcome are shown in Table 25.1. The aim of the user is to minimise their overall expense by deciding on each occasion whether to protect or not.

Over a large number of cases, let $\bar{o}$ be the fraction of occasions when X occurs. If the user always protects then the cost will be C on every occasion and the average expense (per case) will be

$$E_{\mathrm{A}} = C \qquad (25.1)$$

Alternatively, if the user never takes action, the loss L will only be incurred when X happens, so the average expense will be

$$E_{\mathrm{N}} = \bar{o}L. \qquad (25.2)$$

Assuming the user knows $\bar{o}$ but has no additional forecast information, then the optimal strategy is either always or never protect, depending on which gives the lower overall expense. This gives a baseline against which improvements from using forecast information can be judged, which we will call the climatological expense, E_{C}, where

$$E_{\mathrm{C}} = \min(C, \bar{o}L). \qquad (25.3)$$

Another useful reference point is provided by the expense associated with perfect forecast information: the user would only protect if X was going to occur and the average expense would be

$$E_{\mathrm{P}} = \bar{o}C. \qquad (25.4)$$

Table 25.2 *Contingency table for deterministic forecast of specified event over a set of cases, showing fraction of occasions for each combination of forecast and outcome*

		Event occurs	
		Yes	No
Event forecast	Yes	a	b
	No	c	d
		$\bar{o}$	$1-\bar{o}$

A deterministic forecasting system gives a simple yes/no prediction for X to occur. The performance of the forecast system over a large set of cases can be summarised in a contingency table as shown in Table 25.2. Note that a, b, c and d are the fraction of occasions on which the various combinations of forecast and observation occurred, so they sum to 1. The average expense of using the deterministic forecast is obtained by multiplying the corresponding cells of Tables 25.1 and 25.2:

$$E_{\mathrm{F}} = aC + bC + cL \tag{25.5}$$

The difference in expense between E_{F} and E_{C} is a measure of the saving the user can make by using the forecast, compared with having only climatological information. We can define the relative value of the forecasts by comparing this saving with the maximum possible saving that could be made from perfect deterministic forecasts

$$V = \frac{E_{\mathrm{C}} - E_{\mathrm{F}}}{E_{\mathrm{C}} - E_{\mathrm{P}}}. \tag{25.6}$$

From now on we will refer to V simply as the 'value' of the forecasts.

In Table 25.2, the entries a, b, c, and d together describe the quality of the forecasts. Various measures of quality can be defined using these terms. Two such measures are the hit rate $H = a/(a+c)$ and the false alarm rate $F = b/(b+d)$. It is often convenient to write the expression for V in terms of H and F. Using the various expressions for the different expenses, this gives

$$V = \frac{\min(\alpha, \bar{o}) - F(1-\bar{o})\alpha + H\bar{o}(1-\alpha) - \bar{o}}{\min(\alpha, \bar{o}) - \bar{o}\alpha} \tag{25.7}$$

where $\alpha \equiv C/L$ is known as the cost–loss ratio of the user. A number of points should be noted from this expression for V. Value depends on two independent measures of forecast quality (H and F here, but you could choose other measures). This is key to differences noted previously (Figure 25.1) between different skill scores – a single score is not enough to give information relevant to a range of different users.

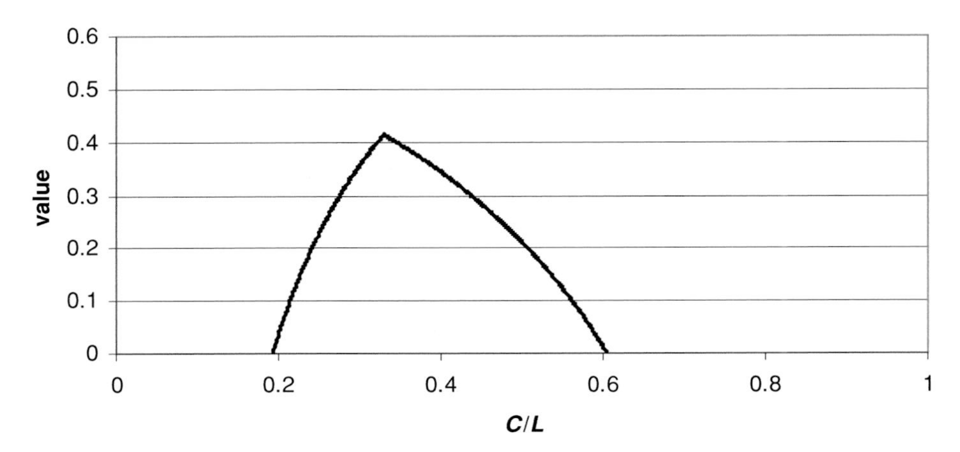

Figure 25.2 Value V of ECMWF deterministic control forecasts of more than 1 mm precipitation in 24 hours at day 5 over northern hemisphere extratropics for winter 2001–2.

Value also depends on the event through the climatological frequency $\overline{o}$, and on the individual user through the cost–loss ratio.

For a given weather event and forecast system, $\overline{o}$, H and F are fixed and V is then only a function of C/L. Figure 25.2 shows the variation of V with C/L for precipitation forecasts from the deterministic control forecast of the ECMWF EPS. It is clear from Figure 25.2 that forecast value varies considerably between users with different cost–loss ratios. While users with $C/L \sim 0.3$ may receive up to 40% of the value they would obtain from perfect deterministic forecasts, others (C/L above 0.6 or below 0.2) will get no benefit from these forecasts.

The shape of the value curve is typical. It is straightforward to show that maximum value is obtained for users with $C/L = \overline{o}$ and that the maximum value is given by

$$V_{\max} = H - F \equiv P \tag{25.8}$$

where P is the Peirce skill score (also known as the Kuipers score or true skill statistic; Mason, 2003; see Richardson, 2000, for the derivation of this equivalence). This provides a first link between skill and value. P can be interpreted as a measure of potential forecast value as well as being a skill score. But P only indicates the maximum possible value and does not show how the forecasts will benefit users with C/L different from $\overline{o}$. It can also be shown that the range of C/L for which the forecasts give positive value is related to a different skill measure, the Clayton skill score (Wandishin and Brooks, 2002). This is just one example illustrating that different skill measures can be related to different aspects of forecast value. It may therefore be important when evaluating forecast performance to bear in mind the purpose of the evaluation and to use a range of performance measures.

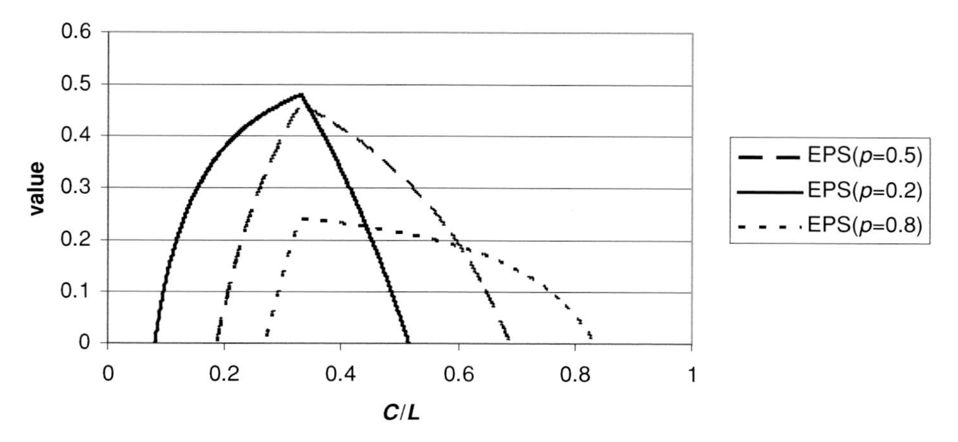

Figure 25.3 Value of the ECMWF probability forecast for more than 1 mm precipitation in 24 hours at day 5 over the northern hemisphere extratropics for winter 2001–2. The curves show the variation of V with C/L for probability thresholds $p_t = 0.2$ (solid), 0.5 (dashed) and 0.8 (dotted).

25.3 Probability forecasts

The use of deterministic forecasts in the cost–loss decision framework is straightforward: take action whenever the event is forecast; otherwise do nothing. In contrast, probability forecasts may at first sight appear inappropriate when a firm yes/no decision needs to be made. The cost–loss model gives a simple illustration of the way that probability forecasts can be used in such situations, and the benefits that may be obtained.

Given forecast information as a probability, the user is faced with the decision of whether the probability of adverse weather is high enough for protective action to be needed. The user needs to set a probability threshold p_t and take action if the forecast probability exceeds that threshold. This choice of p_t effectively converts the probability forecast to a deterministic one: forecasts with probability higher than p_t become 'yes' forecasts while the remainder become 'no' forecasts. The resulting deterministic forecast can be evaluated as in the previous section to obtain $H(p_t)$, $F(p_t)$ and $V(p_t)$. Different choices of p_t will result in different values for H, F and V.

Figure 25.3 shows $V(C/L)$ for three choices of p_t for EPS probability forecast corresponding to the control forecast of Figure 25.2. Different users will benefit more from different choices of p_t. In general users with lower C/L will benefit more from acting at lower probability thresholds while those with high C/L will gain more from acting only when there is greater certainty about the event. For example a user with $C/L = 0.2$ will have a relatively high potential loss and will benefit substantially by taking action at $p_t = 0.2$. Users with much higher relative costs would lose out

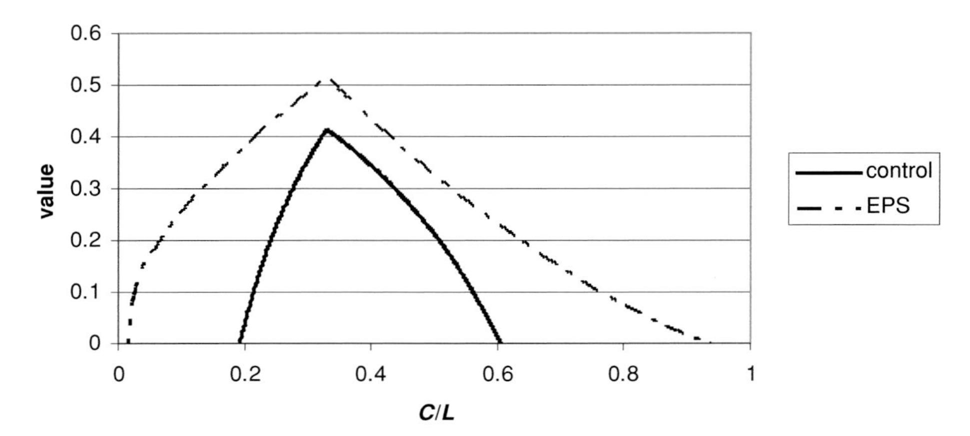

Figure 25.4 Value of the ECMWF deterministic control forecast (solid) and EPS probability forecast (dash-dotted) for more than 1 mm precipitation in 24 hours at day 5 over the northern hemisphere extratropics for winter 2001–2. The probability forecast curve shows the optimal value, obtained when each user chooses the most appropriate p_t.

by taking action when the chance of the event is so low, but they would benefit from using a higher threshold probability.

It is, then, important that probability forecasts are used with care. An inappropriate or arbitrary choice of p_t may substantially reduce the benefit of the forecasts to a user. Moreover, it is not possible to choose a suitable p_t without knowing the cost–loss ratio of the user. It is straightforward to show that for reliable probability forecasts (Wilks, 1995; Toth *et al.*, this volume) the optimal choice is $p_t = C/L$ (Richardson, 2000). In other words, each user should take action when the probability of the event exceeds their own cost–loss ratio. Consider just those occasions when the probability of the event is a specific value q. Then taking action incurs the constant cost C. The alternative is not to act and accept the consequences: if the event occurs the loss is L, while if it does not occur there is no expense. The average expense of not acting is then Lq. So to minimise the average expense the user should act if $C/L < q$, but accept the risk otherwise. In this way, $p_t = C/L$ can be seen to be the appropriate choice for threshold probability. If the probability is higher than p_t the user should act; if not it is better to risk the loss.

Probability forecasts that are not completely reliable can generally be corrected for these biases by calibrating, or relabelling, the forecast probabilities (Zhu *et al.*, 1996, 2002). In this way the users can treat the forecast probabilities at face value and gain maximum benefit. In general, users should be presented with reliable, calibrated forecasts. Each user should then take action when the probability of the event exceeds their own cost–loss ratio.

The optimal value that would be obtained when each user chooses the most appropriate p_t is shown in Figure 25.4, together with the corresponding curve for the deterministic control forecast. All users will gain more from using the probabilities than from the control forecast, and by allowing different users to take action at different probability thresholds the ensemble probabilities benefit a much wider range of users. This benefit should also be compared with the limited benefits provided by any single choice of p_t (Figure 25.3).

25.4 Skill and value

We have already noted the relationship between value and certain skill measures for deterministic forecasts. In this section we explore the link between skill and value for probability forecasts. This will for instance help us understand the striking difference, shown in Figure 25.1, between the perception of EPS skill that would be obtained when using the Brier skill score (BSS) or the ROC area skill score (ROCSS).

The skill scores are single overall summary measures of forecast performance. Yet we have seen that forecast value varies greatly between users depending on their particular costs and losses. There is no simple relationship between either of the skill scores and the value to individual users. However, we can imagine that it could be useful to have some summary measure of value that reflects the overall benefit to the range of users of the forecast data. For example, if we knew the appropriate C and L for each user we could work out the total saving that the group of users would make from the forecasts. In general little is known about the circumstances of individual users and of course the cost–loss model will not be appropriate in many cases. It turns out that the Brier skill score is equal to the overall value that would be obtained for a set of users distributed uniformly through the range of possible C/L (see Richardson, 2001 for details).

Figure 25.5 shows EPS value for a heavy precipitation event. This is a rather rare event ($\overline{o} = 0.02$) and the value is concentrated around low C/L users. The Brier skill score, being a measure of the overall value for all possible users, is low: $BSS = 0.06$. However, $ROCSS = 0.65$, similar to the maximum value of around 0.6. In fact, ROCSS is closely linked to maximum value. For deterministic forecasts, the relationship is exact ($ROCSS = V_{\max}$). For probability forecasts ROCSS is always greater than $V_{\max}$ (see Richardson 1999, 2000, for more discussion on this relationship).

Which skill measure is more appropriate depends on the distribution of users. Although little is known about real-world costs and losses, general economic considerations tend to suggest that lower values of C/L are more likely than higher values (Roebber and Bosart, 1996). The few studies that have applied the simple cost–loss model to financial decisions seem to support this. Examples include C/L of 0.03 for raisin drying (Kolb and Rapp, 1962), 0.02–0.05 for orchardists (Murphy, 1977), 0.01–0.12 for fuel loading of aircraft (Leigh, 1995; Keith 2003), 0.125 for winter road gritting (Thornes and Stephenson, 2001). With this in mind, the BSS and ROCSS can be seen as indicating lower and upper bounds respectively for the value

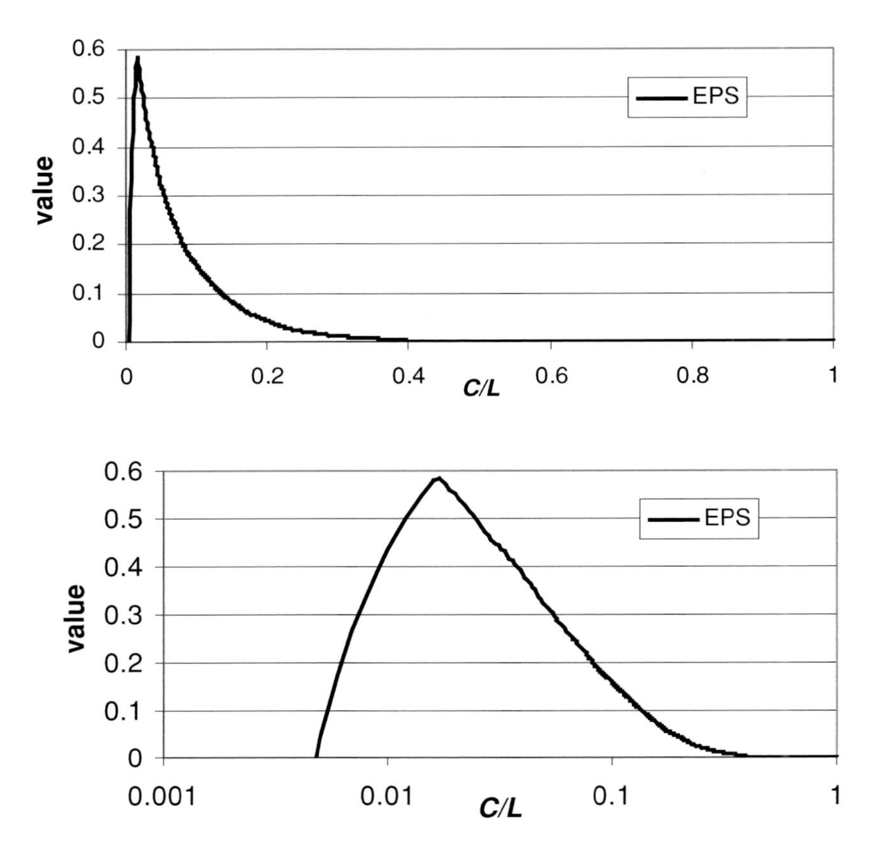

Figure 25.5 Value of the ECMWF EPS probability forecast for more than 20 mm precipitation in 24 hours at day 5 over the northern hemisphere extratropics for winter 2001–2. The probability forecast curve shows the optimal value, obtained when each user chooses the most appropriate p_t. (a) Plotted on regular axis for C/L. (b) Logarithmic axis for C/L.

of the forecasts. The important factor to remember is that value is very much user dependent. Any summary measure condenses information and will not reflect the benefits to be gained by individual users. Unless details of the decision making and associated costs of specific users or groups of users are known, caution is needed in interpreting summary skill measures. In general, a range of measures, including for example both BSS and ROCSS, should be used.

25.5 Effect of resolution and ensemble size

Just as different users will benefit differently from the EPS, so the benefits of changes to the EPS configuration will also be different. Figure 25.6 shows an example of the potential improvement for different users from the introduction of a higher-resolution

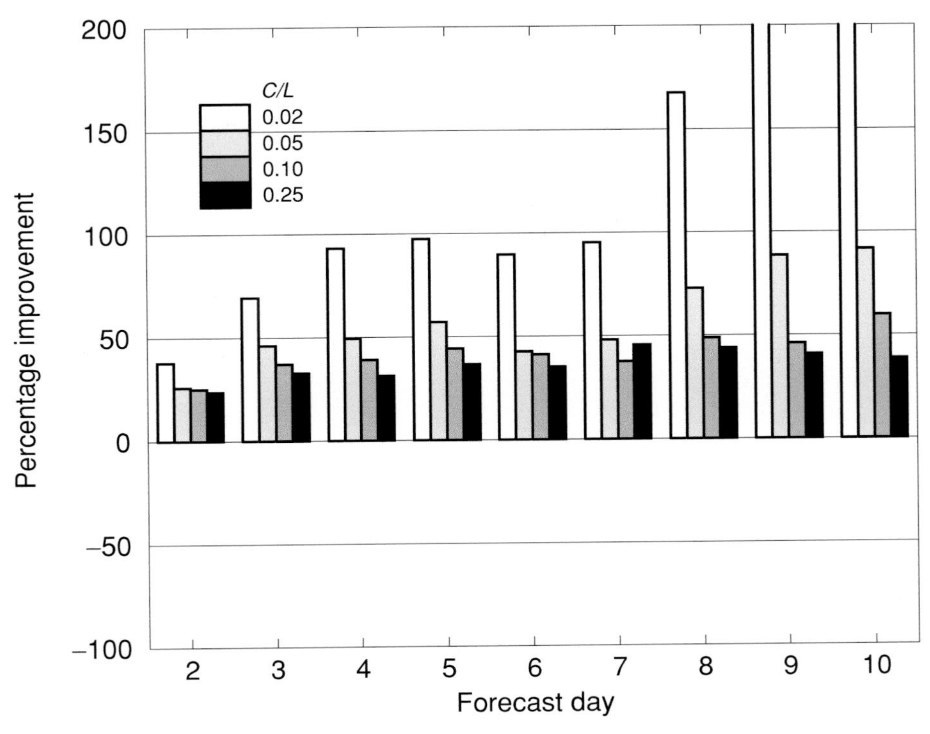

Figure 25.6 Improvement in value for high-resolution EPS compared with previous system for moderate 850 temperature anomalies over 57 winter cases. The benefit of the new system is shown using a Relative Improvement index (RI, where $RI = 100\%$ is equivalent to a 1-day gain in skill) for selected cost/loss ratios: $C/L = 0.02$ (white), $C/L = 0.05$ (light grey), $C/L = 0.10$ (dark grey) and $C/L = 0.25$ (black).

system for the ECMWF EPS (Buizza *et al.*, 2003; Buizza, this volume). This is shown as a percentage relative improvement in value $RI(V)$, defined so that $RI = 0$ indicates no change compared to the previous system and $RI = 100\%$ indicates an improvement equivalent to a one-day gain in skill (Buizza *et al.*, 2003). In this example, while all users benefit substantially the gains are substantially higher for users with low C/L.

The potential benefits of additional ensemble members will also be different in different circumstances. Figure 25.7 shows the effect of ensemble size on the value of an otherwise perfectly specified (completely reliable) ensemble system (Richardson, 2001). Figure 25.7(a) shows the value of 10-member (dashed line) and 50-member (solid line) ensembles for a common event (climatological frequency 0.5); the dash-dotted curve shows the maximum potential value that would be achieved with larger ensemble size (the large-ensemble limit). Increasing ensemble size from 10 to 50 members benefits all users. There is less to be gained from further increases in ensemble size. Figure 25.7(b) shows corresponding curves for a less common event

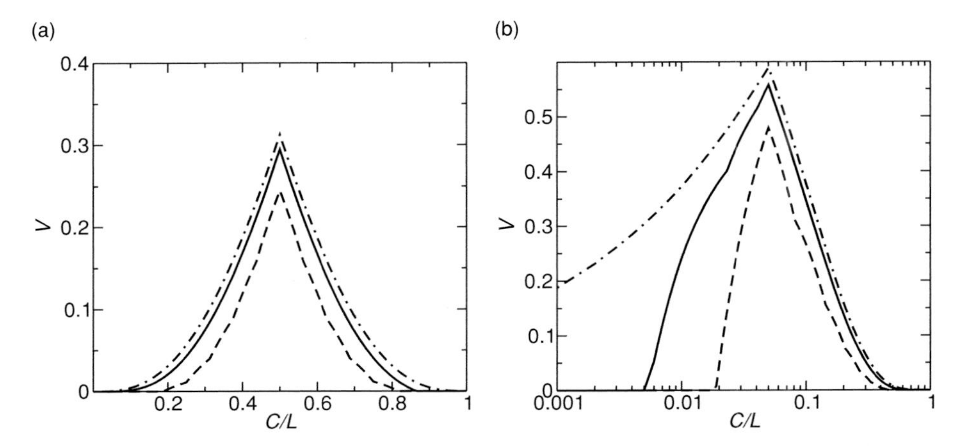

Figure 25.7 Variation of value with ensemble size. Curves show V as a function of C/L for 10-member (dashed) and 50-member (solid) ensembles compared with the potential value for the underlying distribution (large-ensemble limit; dash-dotted) for an idealised perfectly reliable probability forecast system. The two panels correspond to a common event (a; $\overline{o} = 0.5$) and a rarer event (b; $\overline{o} = 0.05$); the underlying predictability, given by the Brier skill score for the underlying probability distribution, is the same in both cases.

(climate frequency 0.05; note the logarithmic x-axis). Here the benefit of increasing ensemble size depends considerably on the cost–loss ratio of the users. While increasing from 10 to 50 members gives general benefit, the main gains for further increases in size are for low C/L. The Brier skill scores for the ensembles in both cases is the same (1/7). Less predictable events (lower Brier skill score) will show more sensitivity to ensemble size; for more predictable events, the importance of ensemble size is less. Further details are given in Richardson (2001).

As noted in the previous section, the Brier skill score is equal to the overall value for a set of users whose cost–loss ratios C/L are distributed uniformly throughout (0,1) (Murphy, 1966; Richardson, 2001). Figure 25.7 shows that for much of this interval the benefit of more than around 50 members will be slight. Both theoretical (Richardson, 2001) and empirical (Talagrand *et al.*, 1997) studies confirm that increasing ensemble size to around 50 members can improve Brier skill scores especially for relatively low predictability events, but that further increases in size have little impact on this skill measure.

However, Figure 25.7 also shows that larger ensembles may give significant gains for users with low cost–loss ratios. The overall value for groups of users whose C/L distribution is not uniform can be calculated using a generalised skill score, G (Richardson, 2001). The effect of ensemble size on G is shown in Figure 25.8 (from Richardson, 2001) for four distributions of users, weighted toward low C/L for a rare event (climatological frequency $\overline{o} = 0.01$). Each curve shows how skill changes as the ensemble size is increased; the different curves represent different

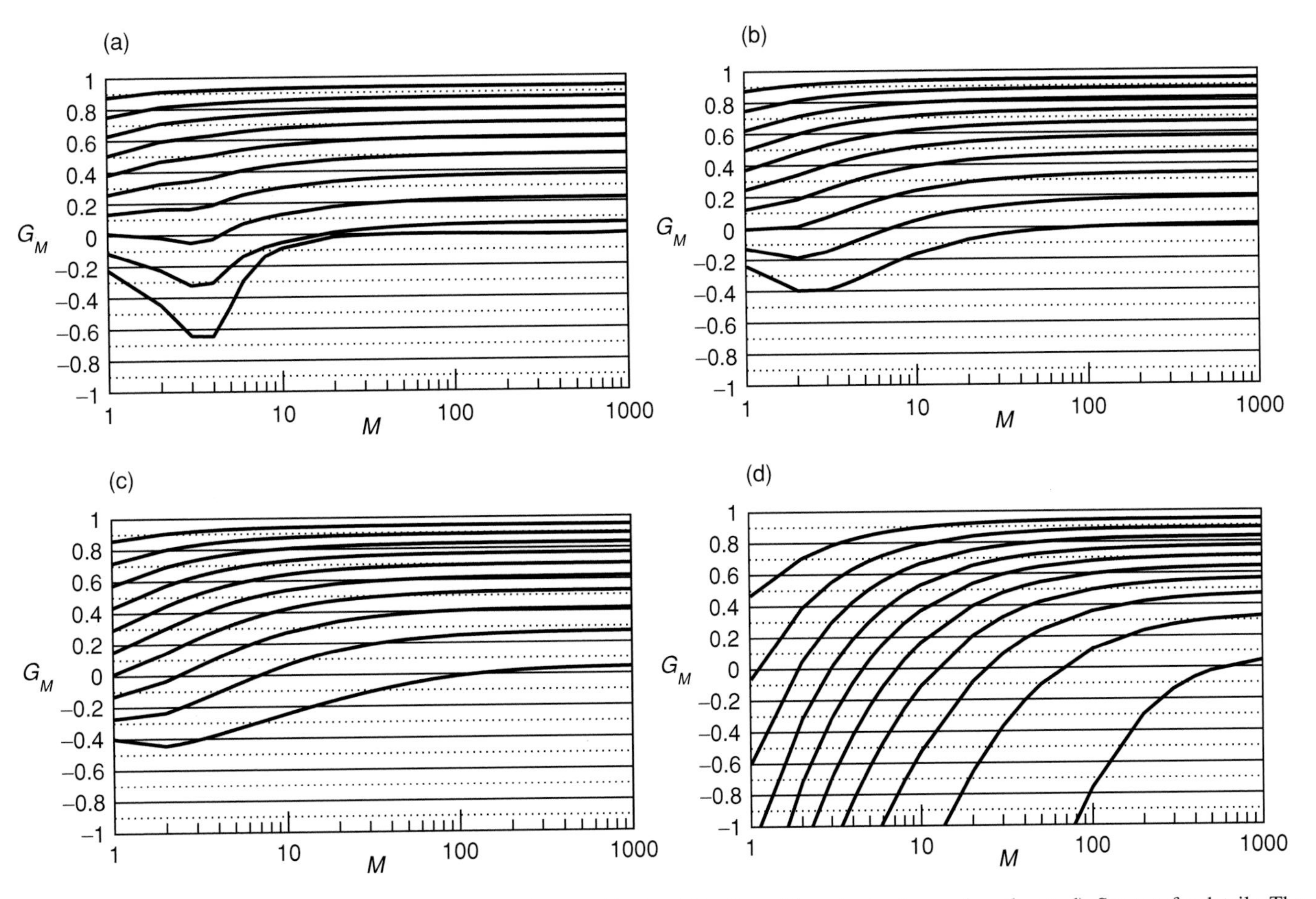

Figure 25.8 Variation of a generalised skill score G_M with ensemble size M for four different sets of users (panels a to d). See text for details. The ten curves in each panel correspond to different underlying levels of predictability; each curve shows how G_M approaches the corresponding limiting value as the number of ensemble members, M, increases. From Richardson (2001); see text for details.

Table 25.3 *Costs and losses associated with different actions and outcomes in the general cost–loss model*

		Event occurs	
		Yes	No
Action taken	Yes	C_{11}	C_{12}
	No	C_{21}	C_{22}

levels of predictability of the underlying distribution – lower curves correspond to lower intrinsic predictability. The behaviour is not always straightforward. In Figures 25.8(a) and 25.8(b), overall skill initially decreases with increasing ensemble size (see Richardson, 2001). Although not seen in these figures, it is even possible that the curves may cross, indicating that an intrinsically more predictable system may have lower value to some users (Murphy and Ehrendorfer, 1987). Figure 25.8(d) shows (as in Figure 25.7b) that for some sets of users the sensitivity to ensemble size is much greater than is apparent from the Brier skill score, with potential benefits for ensembles of several hundred members. Which measure is most appropriate depends of course on the users of the product, but the properties of the scores being used must always be borne in mind.

25.6 Utility

We have used the cost–loss model as a simple example to illustrate the decision-making process of a user. In this model we assumed that the consequences of the actions can be expressed directly as financial costs and losses, that the cost C gives complete protection against the loss L, and that the user is only concerned with minimising the long-term average expense. In this section we show that the results presented so far are also valid in the more general situation where each of these assumptions is relaxed.

First, consider the general case where each combination of action and outcome results in a different expense (Table 25.3). For a fixed forecast probability q, the mean expense of taking action is

$$E_{\mathrm{A}} = C_{11}q + C_{12}(1-q) \tag{25.9}$$

and the mean expense of not acting is

$$E_{\mathrm{N}} = C_{21}q + C_{22}(1-q). \tag{25.10}$$

As before, these expressions can be used to find the probability threshold p_{t} above which the user should take action:

$$p_{\mathrm{t}} = \frac{C_{12} - C_{22}}{C_{21} - C_{22} + C_{12} - C_{11}} = \beta. \tag{25.11}$$

β is a generalised version of the simple cost–loss ratio $\alpha = C/L$. Each term in the above expression can be written as a cost relative to the expense of the 'default' outcome where no action is taken and the adverse weather does not occur, so there is not loss of generality in setting this term C_{22} to zero as in Table 25.1. The remaining terms can be interpreted as follows. C_{12} is simply the cost of taking the protective action, and C_{21} is the total loss that would be experienced if the adverse event occurred and no action was taken. In this more general situation we allow for the possibility that the action does not fully protect against the loss, so that C_{11} is greater than C_{12}. The denominator of β is the part of the loss which can be protected by taking action, so β still has the interpretation of the ratio of the cost of taking action to the protectable loss.

The expression for forecast value can be derived in this more general situation (Richardson, 2000). The expression for V is the same as in Eq. (25.7), just with β replacing α. All the subsequent results apply equally to this general situation.

So far we have considered the different outcomes in terms of the financial consequences and have assumed the user acts to minimise the long-term mean expense. A further generalisation is to move away from the strict financial interpretation of the outcomes and introduce the concept of the utility of the outcomes. Rather than working directly with the costs C_{ij} in Table 25.3, consider a function $U(C_{ij})$ that represents in a more general sense the usefulness of this consequence to the user. All the previous derivations can again be made using the utilities $U(C_{ij})$ instead of the costs C_{ij}. In this case, the user aims to maximise the expected utility of his decisions (utility is a positive concept) rather than minimise the expected expense. Again, the same expression is found for value, with α now representing a ratio of the utilities instead of costs.

In terms of utility, the threshold probability above which the user should take action is

$$p_t = \frac{U(C_{12}) - U(C_{22})}{U(C_{21}) - U(C_{22}) + U(C_{12}) - U(C_{11})} = \gamma. \tag{25.12}$$

If U is a linear function of the costs then $\beta = \gamma$, this utility ratio is the same as the cost–loss ratio: the threshold probability for maximising utility is the same as for minimising the expense. Users with linear utility functions are described as risk-neutral.

The relationship between utility and threshold probability is easiest to understand in the original cost–loss situation. On each occasion the user has to choose between the certainty of paying the cost C for protection and the risk of a greater potential loss L; p_t is the probability at which the user is indifferent between these two alternatives. For the risk-neutral user, this threshold was measured simply in terms of the mean expense of the two options and the threshold probability was shown to be $p_t = C/L$.

Now consider the threshold probability in terms of utility

$$p_t = \frac{U(C) - U(0)}{U(L) - U(0)} = \gamma. \tag{25.13}$$

Consider the utilities of the four possible outcomes. The best outcome for the user is where there is no expense (no action is taken and the event does not occur). This then has maximum utility and we are free to choose $U(0) = 1$. The worst outcome for the user is when the expense is greatest (loss L). This outcome has minimum utility and we choose $U(L) = 0$. These choices for maximum and minimum utility allow utility to be interpreted as a probability: the threshold probability and utility ratio become

$$p_t = 1 - U(C) = \gamma. \tag{25.14}$$

So the utility of taking action is equivalent to the threshold probability at which the user is prepared to act. For a linear utility function (risk-neutral user) our choices for $U(0)$ and $U(L)$ mean that $U(C) = 1 - C/L$, so that $p_t = C/L$. However, a user may be concerned about the immediate impact of a substantial loss L and may be prepared to pay C at lower probabilities to reduce the occurrence of this loss. This concern can be described by giving the user a higher utility for taking action $U(C)$. This will lower the threshold probability, so the user will take action at lower probabilities than would be indicated by a simple cost–loss analysis. Such a user is described as risk-averse: they are unwilling to risk the loss simply on the basis of minimising long-term expense.

In this section we have introduced a number of generalisations of the cost–loss model. The expression for value from the cost–loss model is equally valid for all two-state, two-action decisions (as long as the consequences or utilities do not change with time). All the results presented in earlier sections are equally valid for the general case. Users who are averse to risk (probably the majority) will have higher utility ratios (and so act at lower probability thresholds) than would be suggested by the cost–loss ratios in the original model.

From a practical perspective the use of utility has several advantages. Users do not need to be able to express the consequences of their actions in financial terms. They do not even need to know precisely what the consequences are. Users do need to be able to express their preference for different outcomes. It is assumed that the user can choose the best and worst consequences. These are given utility 1 and 0 respectively. The utility of intermediate consequences is then deduced by preferences expressed by users when they are offered a (hypothetical) choice between the certainty of a particular outcome or the possibility of either the best or worst outcome with a given probability. This is perhaps easier to obtain from the user than a precise evaluation of the costs and losses associated with each action.

References

Ångström, A. (1922). On the effectivity of weather warnings. *Nord. Statist. Tidskr.*, **1**, 394–408.

Buizza, R., D. S. Richardson and T. N. Palmer (2003). Benefits of increased resolution in the ECMWF ensemble system and comparison with poor-man's ensembles. *Quart. J. Roy. Meteor. Soc.*, **129**, 1269–88.

Keith, R. (2003). Optimization of value of aerodrome forecasts. *Weather Forecast.*, **18**, 808–24.

Kolb, L. L. and R. R. Rapp (1962). The utility of weather forecasts to the raisin industry. *J. Appl. Meteorol.*, **1**, 8–12.

Liljas, E. and A. H. Murphy (1994). Anders Angstrom and his early papers on probability forecasting and the use/value of weather forecasts. *Bull. Am. Meteorol. Soc.*, **75**, 1227–36.

Leigh, R. J. (1995). Economic benefits of terminal aerodrome forecasts (TAFs) for Sydney Airport, Australia. *Meteorol. Appl.*, **2**, 239–47.

Mason, I. B. (2003). Binary events. In *Forecast Verification: a Practitioner's Guide in Atmospheric Science*, ed. I. T. Jolliffe and D. B. Stephenson. Wiley.

Molteni, F., R. Buizza, T. N. Palmer and T. Petroliagis (1996). The ECMWF ensemble prediction system: methodology and validation. *Quart. J. Roy. Meteor. Soc.*, **122**, 73–119.

Murphy, A. H. (1966). A note on the utility of probabilistic predictions and the probability score in the cost-loss ratio decision situation. *J. Appl. Meteorol.*, **5**, 534–7.

Murphy, A. H. (1977). The value of climatological, categorical and probabilistic forecasts in the cost-loss ratio situation. *Mon. Weather Rev.*, **105**, 803–16.

Murphy, A. H. and M. Ehrendorfer (1987). On the relationship between the accuracy and value of forecasts in the cost-loss ratio situation. *Weather Forecast.*, **2**, 243–51.

Palmer, T. N., F. Molteni, R. Mureau and R. Buizza (1993). Ensemble prediction. In *ECMWF Seminar Proceeding: Validation of models over Europe*, vol. 1. ECMWF, Shinfield Park, Reading, RG2 9AX, UK.

Richardson, D. S. (1999). The application of cost-loss models to forecast verification. In *Proceedings of the 7th Workshop on Meteorological Systems*. ECMWF, Shinfield Park, Reading, RG2 9AX, UK.

(2000). Skill and relative economic value of the ECMWF Ensemble Prediction System. *Quart. J. Roy. Meteor. Soc.*, **126**, 649–68.

(2001). Measures of skill and value of Ensemble Prediction Systems, their interrelationship and the effect of ensemble size. *Quart. J. Roy. Meteor. Soc.*, **127**, 2473–89.

(2003). Economic value and skill. In *Forecast Verification: a Practitioner's Guide in Atmospheric Science*, ed. I. T. Jolliffe and D. B. Stephenson. Wiley.

Roebber, P. J. and L. F. Bosart (1996). The complex relationship between forecast skill and forecast value: a real-world analysis. *Weather Forecast.*, **11**, 544–59.

Talagrand, O., R. Vautard and B. Strauss (1997). Evaluation of Probabilistic Prediction Systems. In *Proceedings of the ECMWF Workshop on Predictability, 20–22 October 1997*, pp. 157–66. ECMWF, Shinfield Park Reading, UK.

Thornes, J. E. and D. B. Stephenson (2001). How to judge the quality and value of weather forecast products. *Meteorol. Appl.*, **8**, 307–14.

Wandishin, M. S. and H. E. Brooks (2002). On the relationship between Clayton's skill score and expected value for forecasts of binary events. *Meteorol. Appl.*, **9**, 455–9.

Wilks, D. S. (1995). *Statistical Methods in the Atmospheric Sciences*. Academic Press.

Wilks, D. S. and T. M. Hamill (1995). Potential economic value of ensemble forecasts. *Mon. Weather Rev.*, **123**, 3565–75.

Zhu, Y., G. Iyengar, Z. Toth, M. S. Tracton and T. Marchok (1996). Objective evaluation of the NCEP global ensemble forecasting system. In *Preprints, 15th Conference on Weather Analysis and Forecasting, Norfolk, VA*, pp. J79–J82. American Meteorological Society.

Zhu, Y., Z. Toth, R. Wobus, D. Richardson and K. Mylne (2002). The economic value of ensemble-based weather forecasts. *Bull. Am. Meteorol. Soc.*, **83**, 73–83.

26

A three-tier overlapping prediction scheme: tools for strategic and tactical decisions in the developing world

Peter J. Webster
School of Earth and Atmospheric Sciences, Atlanta

T. Hopson
Program in Atmospheric and Oceanic Sciences, University of Colorado, Boulder

C. Hoyos
School of Earth and Atmospheric Sciences, Atlanta

A. Subbiah
Asian Disaster Preparedness Centre, Bangkok

H.-R. Chang
School of Earth and Atmospheric Sciences, Atlanta

R. Grossman
Colorado Research Associates, Boulder

During the last decade our understanding of processes that determine the variability of the atmosphere and the climate system have improved to the extent that predictability of some phenomena has become established. The predictability, at least in the short- and long-term timescales, has on rare occasions been translated into useful predictions. However, the value of forecasts (i.e. how the forecasts are interpreted and their value to a user group) has not improved at the same rate. Arguably, the problem lies with the psychological and physical separation of the scientist or technician who makes the forecasts, and decision-makers and user communities who utilise the forecasts. Furthermore, the separation is exaggerated by the fact that for any one forecast there are many potential applications for different user communities each with unique needs. This added complexity makes it impossible in a practical sense for a forecaster to communicate with all users. Instead, for a given probabilistic

Predictability of Weather and Climate, ed. Tim Palmer and Renate Hagedorn. Published by Cambridge University Press.

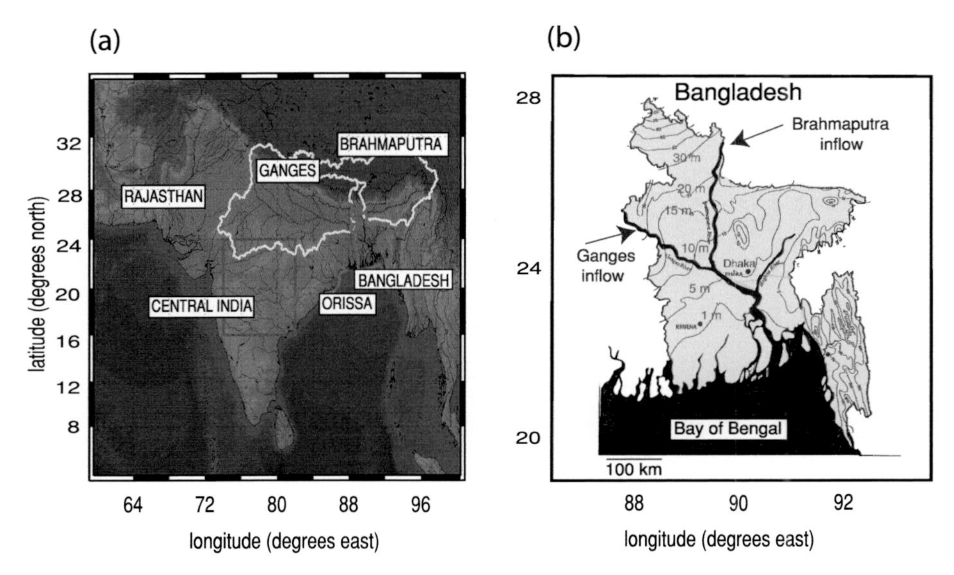

Figure 26.1 (a) The South Asian region showing the Brahmaputra and Ganges catchment areas and Bangladesh. Areas for which precipitation forecasts are routinely made (central India, the Indian states of Orissa and Rajasthan, Bangladesh and the two major river catchment areas) are indicated on the map. (b) Entry points of the Ganges and the Brahmaputra into Bangladesh. River discharge is forecast on all three timescales at these points.

forecast sets of decision tools need to be constructed for each user class. Clearly, this cannot be accomplished by a forecasting centre but requires input of intermediaries (organisational or personnel) and the user communities themselves.

We describe one attempt to provide probabilistic forecasts over an overlapping set of timescales with decision models to bridge the forecaster–user separation. Such a model allows user communities to make interrelated longer-term strategic and shorter-term tactical decisions and to hedge in an iterative sense against uncertainty. Central to this scheme is the marriage of quantitative user community information (a cost function) with probabilistic forecasts to produce risk-based decision tools. We refer to this model as a 'user metric'. To illustrate the philosophy of the scheme, we use as examples applications of rainfall forecasts over India and predictions of river discharge for the Brahmaputra and the Ganges, both of which have been utilised experimentally in Bangladesh in a quasi-operational setting since the summer of 2004 (see Figure 26.1 for locations). The scheme consists of three overlapping sets of forecasts: seasonal (1–6 months) which commences in April and is issued each month, intraseasonal (20–30 days) issued every 5 days and short-term forecasts (1–10 days) issued daily. The short-term and seasonal forecasts use ensemble information from the European Centre for Medium-Range Weather Forecasts' (ECMWF) operational and experimental models, statistical dressing of the output and, where necessary, a suite of hydrological models. For the intermediate timescale predictions (20–30 days) we use a Bayesian physically based empirical model.

26.1 Introduction

26.1.1 Environmental forecasting in South Asian monsoon climates

Of all populations, those who inhabit the monsoon regions are most vulnerable to weather and climate variability and most in need of accurate and timely environmental forecasts. In India, for example, farming areas that depend upon rainfall (non-irrigated land) map directly on to the most impoverished regions in the country. Inhabitants of these 'rain-fed' regions are most susceptible to weather and climate anomalies where unexpected periods of drought or heavy rainfall can have dire consequences. Irrigated sectors of the country are less susceptible to short-term weather variability, at least in terms of short-term droughts. But there is an equal susceptibility to periods of heavy rainfall especially near harvesting times, or to prolonged drought such as occurred in the summer of 2002. In fact, all sectors of the country are susceptible to intraseasonal variability. With some certainty, it can be argued that in all regions of South Asia timely and skilful forecasts of weather and climate rendered into a form useful to the user communities would reduce the impact of extreme meteorological and hydrological events and lead towards an improvement of agricultural practices. In irrigated lands skilful and timely forecasts would lead to better water resource management and the partition of use between irrigation and power generation.

Figure 26.2(a) shows the variability of central Indian rainfall and illustrates the multiple timescales that dominate the weather and climate of South Asia. Precipitation is displayed for four years (1999–2002) as 5-day averages from the spring through to autumn. The background curve is the long-term climatological precipitation. Each year is different, indicating distinct interannual variability. Within each season there are marked periods of prolonged precipitation ('active' periods) and rainfall minima ('break' periods). These occur randomly throughout the monsoon summer season as indicated by the smoothness of the long-term climatological precipitation. Within each of the active periods there is considerable rainfall variability (Figure 26.2b, c). These higher frequency rainfall events denote monsoon weather.

During the last decade there have been marked advances in our understanding of the variability of monsoon rainfall over a wide range of timescales (see, e.g., review by Webster *et al.*, 1998) perhaps because of the attention the tropics has received in the World Climate Research Programme's (WCRP) Tropical Ocean Global Atmosphere project (TOGA) and the more recent Climate Variability and Predictability project (CLIVAR). A major scientific objective of CLIVAR is to understand the interannual and intraseasonal dynamics of the atmosphere–ocean–land system of the monsoon with the aim of translating this scientific understanding into a predictive capability.

Despite the scientific advances, predictability has remained elusive on timescales longer than a few days, In essence, forecasting of monsoon weather embedded in the large-scale monsoon circulation is little different from forecasting weather at

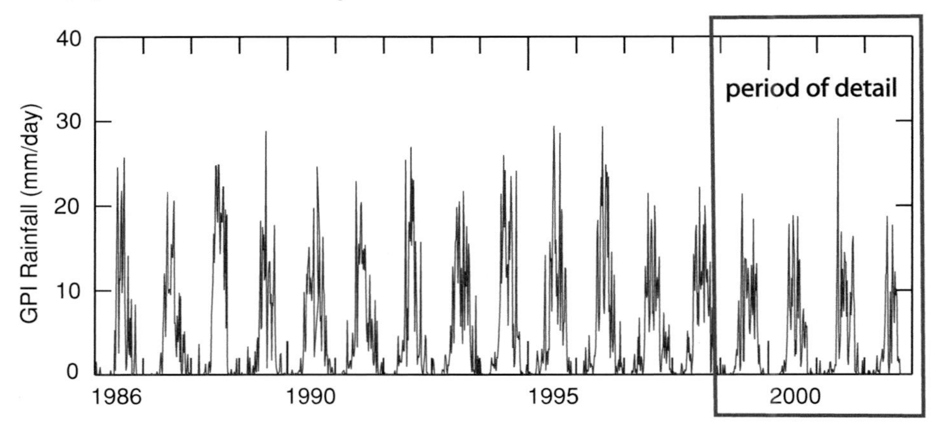

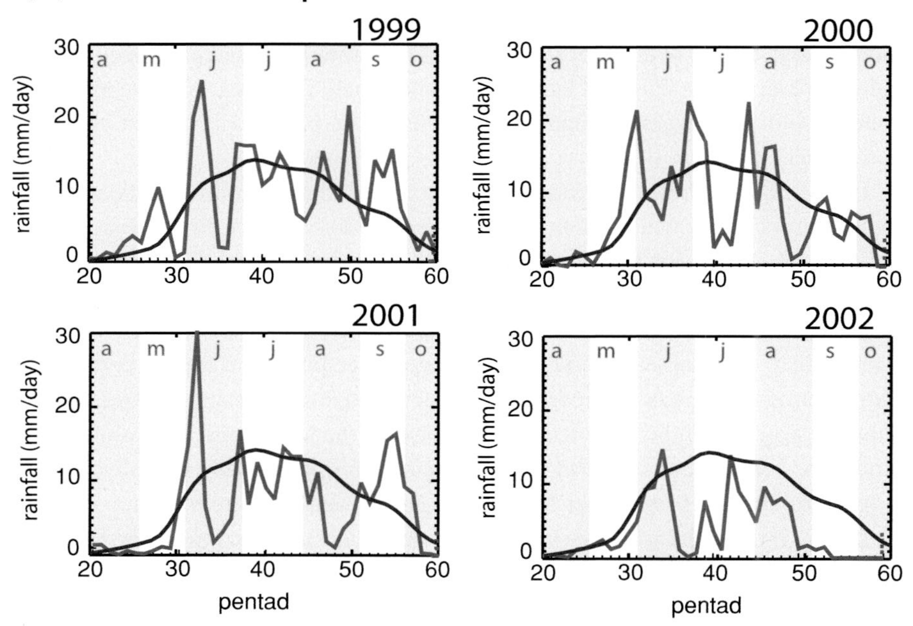

Figure 26.2 (a) Pentad rainfall for central India (see Figure 26.1) from GPI rainfall for the period 1986–2002. Units mm/day. Adapted from Webster and Hoyos (2004). (b) Details of the summer pentad rainfall for central India for the years 1999 to 2002. The background smooth curve represents the long-term climatological average. Sequences of high and low periods of rainfall dominate the seasonal rainfall. These are the active (wet) and break (dry) periods of the monsoon. Adapted from Webster and Hoyos (2004).

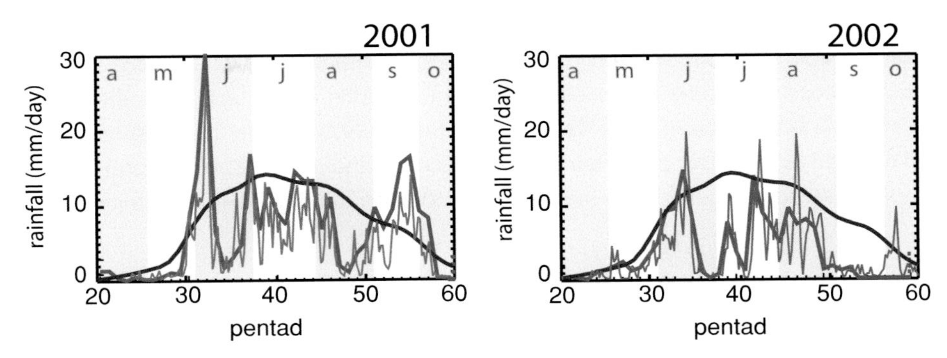

Figure 26.2 (*cont.*) (c) Depiction of the daily rainfall (grey lines) and the pentad rainfall (black lines). Within each active period there is higher frequency variability. These peaks and valleys constitute monsoon weather.

higher latitudes and depends to a large degree on the quality of the model and the initial data. Forecasting or even simulating intraseasonal 20–40 day variability by numerical techniques has proven extremely difficult despite the fact that next to the annual cycle variance in this spectral band the monsoon dominates.

Numerical attempts to foreshadow seasonal variations of the monsoon have not been particularly successful at this stage either. Empirical schemes, which have been the main method of forecasting interannual monsoon variability, attempt to relate large-scale forcing from the El Niño–Southern Oscillation (ENSO) phenomenon to the future state of the monsoon. However, since the mid 1980s, skill using empirical methods has diminished significantly. For example, when the largest El Niño of the twentieth century occurred in 1997–8, the Indian monsoon was essentially normal. During a relatively weak El Niño (2002–3) India suffered the worst drought in decades. Both of these recent years defied the canonical El Niño–Indian monsoon–drought relationship suggested by Rasmusson and Carpenter (1982). Similar anomalous responses (or lack of response) were found in Australia, which was spared the canonical drought in 1997–8 but faced the worst drought in a hundred years in 2002. It may be argued that the impact of these two El Niños depended on subtle differences in the westward extent of the anomalous warm pool. For example, recent research results suggest:

> 'The differences in ocean and large-scale atmospheric structure over the tropics and Australia during the 2002/3 and 1997/8 El Niño have been compared using the NCEP re-analysis. There were large scale differences in the tropical circulation that may explain the different impacts of the two El Niño's, which may be related to the location of the SST maxima in the tropical Pacific.'
>
> (Personal communication: Drs. Guomin Wang and Harry Hendon, Bureau of Meteorology Research Centre, Melbourne, Australia)

It is true that our ability to forecast ENSO variability is one of the great triumphs of climate dynamics but forecasting future subtle spatial differences in the location and magnitude of the Pacific sea surface temperature anomaly may have to await considerable technological advances and may lie somewhere in the future. Even if such forecasts could be produced, there is the question of the lack of predictability across the boreal spring (e.g. Webster and Yang, 1992) and the minimal lead time that would be available to implement the forecast in India. These are the realities that must be faced in dealing with interannual prediction in the monsoon regions.

26.1.2 The concept of a 'useful' forecast

Even if a forecast of El Niño, and its influence on the South Asian monsoon, were available, would that in itself constitute a useful prediction? Here we define 'useful prediction' in its broadest sense. A useful forecast, for example, requires a bridging of the gap between a 'broad-brush' seasonal forecast over a large area (e.g. a forecast of the seasonal anomaly in the All India Rainfall, an index representing rainfall over the entirety of India) to a forecast that can help a decision maker in a particular location (e.g. a farmer, water resource manager, district agricultural extension officer, governmental official, politician, etc.) to take action to hedge against uncertainty in the future state of weather and climate. For example, consider the role of the agricultural extension officer whose purpose is to transfer highly technical information to members of an agricultural community who may not be privy to environmental considerations or technical advances that would help in agricultural practices. The officer's role is to provide advice on issues such as type of crop or species to be used in a given state of the environment, what type of pesticide or fertiliser should be applied and to advise on the quantity and the timing of their application. But from where, and in what form, does the agricultural officer receive information to make these suggestions to the farming community? And how does the officer receive training in order to convey this information in a confident and credible manner? In the Climate Forecast Application in Bangladesh (CFAB) project, we have attempted to bridge the gap between probabilistic forecasts and their application in real circumstances.

Zhu *et al.* (2002) postulated a necessary condition for the provision of a useful forecast. They note that each day individuals, communities, administrations and so on have to make decisions to hedge against uncertainty. For example, should one plant a crop today or wait until tomorrow or the following week to take advantage of a proposed rainy period or avoid a rainless period following planting? Should one spray pesticide on one particular day or another in order to optimise the impact of the chemical and minimise the loss through excessive rainfall? Consider a farmer who needs to obtain the maximum yield from his crop. He is faced with many choices: he could wait and harvest a crop at full maturation (say in ten days) and possibly achieve full yield, but would be taking the chance that adverse weather or floods may seriously reduce the yield. He could harvest partially throughout the forecast period, perhaps taking advantage of current weather and the potential that part of his crop might reach maturation, or he could harvest totally and immediately knowing that there would be a reduction in total yield but that there would be no further risk of falling below that level. Without environmental information, the strategy that the farmer may decide on can be chosen at random. Without probabilistic information about the future state of the environment, it is not possible to undertake a cost–benefit or risk analysis and hedge against uncertainty. Following Zhu *et al.* (2002), only probabilistic information is useful.

But, the problem of providing 'useful forecasts' goes beyond producing a probabilistic forecast of the rainfall for the coming season. No matter how skilful the forecast may be, there is no guarantee that the forecast will be useful to a regional user group. The problem also goes beyond the downscaling of large-scale forecasts to some region. That is itself a major problem. Part (a) of Figure 26.3 (colour plate; from Webster and Hoyos, 2004) underlines the problem of inferring regional anomalies from macroscale forecasts. Even when the overall rainfall is decidedly above or below average, there are many regions which are of the opposite in sign to the average. Only in the cases of extreme anomalies (e.g. 2002) would most districts in India tend to be below average. The same problem exists in the temporal variation of rainfall. Figure 26.3(b) shows the variability of rainfall over central India binned as functions of the overall monsoon rainfall. The active and break periods possess minimal temporal clustering relative to the overall seasonal rainfall. That is, irrespective of the total rainfall for the season, there is no knowing when the first active or break period or any subsequent variability of the monsoon will occur. This means that even in a 'good' monsoon year, it is possible that the first break will be so timed as to adversely impact the crop yield. Thus a useful forecast is one that provides relevant climate variables at the local level on a timescale that allows changes in plans if necessary.

The definition of what the relevant climate variables are is made by the user community. For example, the agricultural extension officer is aware of the environmental conditions that will allow some pest to thrive. The officer is also aware of the effectiveness of a particular pesticide and the environmental conditions that will allow it to eradicate the pest. The officer may want to know if there is a window of four rainless days that can be used for application in order to compare the cost of applying the pesticide versus the profit from a successful application relative to the predicted weather. Of course, providing all of the information that a user may want may not be possible for all users in all environmental circumstances. But the message here is that the user is an essential partner in the development of useful forecasts and that the user can provide quantitative information.

It is obvious from Figures 26.2 and 26.3 that monsoon variability occurs on multiple timescales. Within the summer rainy season there are successive active (wet) and break (dry) monsoon periods on timescales of 20–40 days acting in a sense

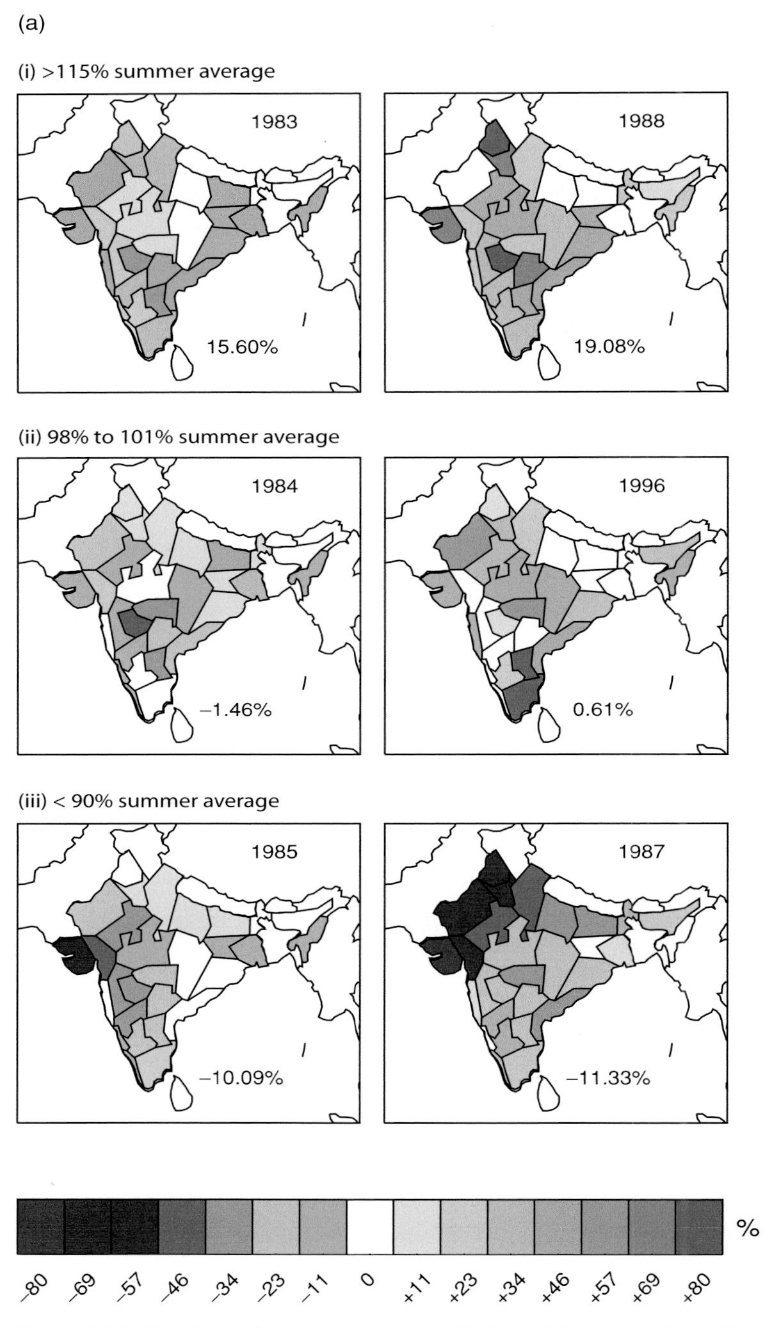

Figure 26.3 (See also colour plate section.) (a) Maps of departures from climatology of the total regional summer precipitation for three classifications of overall monsoon rainfall. (i) All India rainfall >115%, (ii) All India rainfall about average and (iii) All India rainfall <90%. Irrespective of the All India rainfall, there are regions of drought or flood in all classifications. Noting that most seasons fall within ±10% from normal, it is clear that even an excellent seasonal rainfall forecast will be difficult to downscale to the regional level.

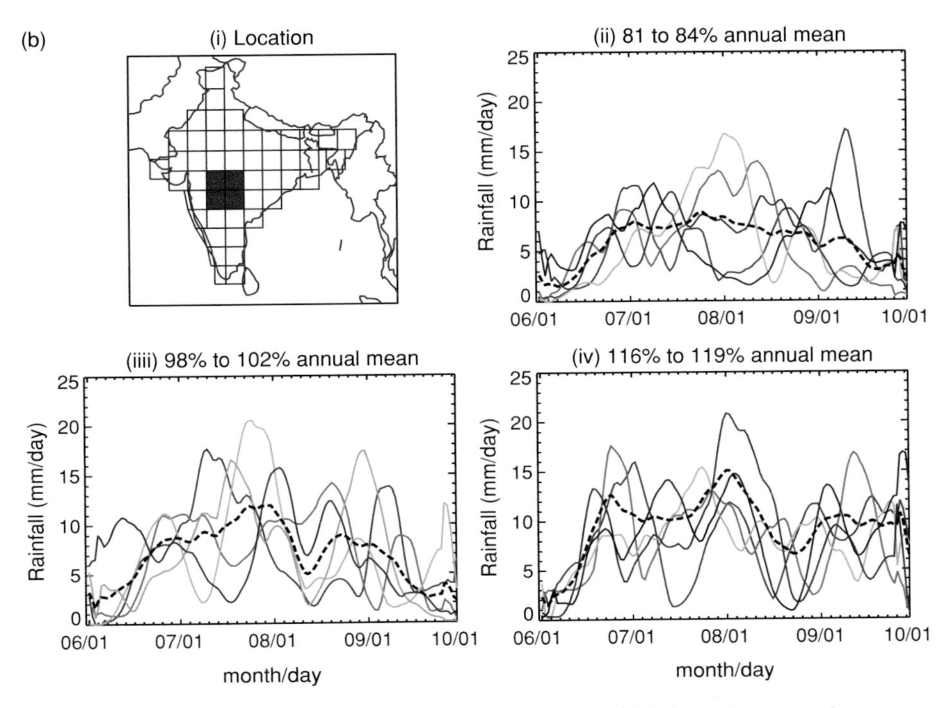

Figure 26.3 (See also colour plate section.) (*cont.*) (b) Schematic temporal variability of summer monsoon precipitation: a seasonal anomaly in precipitation can be made up from many temporal combinations of precipitation. Each panel shows a 20% deficit (matching the overall of 2002) distributed in different ways. Panel (i) uniformly reduced throughout the summer; (ii) a late monsoon and an early withdrawal; (iii) a normal early and late monsoon with a prolonged mid-season break (similar to 2002); and, (iv) a series of short-lived active and break periods. A perfect forecast of the overall seasonal anomaly provides no information about how the precipitation will be distributed throughout the summer season.

of an evolving envelope within which weather is modulated. Within an active period extreme rainfall events may be embedded, increasing the potential for flooding. By contrast, breaks in the monsoon, occurring at times that are agriculturally sensitive (particularly planting and harvesting), may be devastating, such as in 2002 over most of India. Thus, the information needed for a decision-maker (e.g. our agricultural extension officer) also has different timescales. Therefore, another necessary condition for a useful forecast is that it must cover the major variance intervals in a particular climate system. For the monsoons, this means seasonal variability, intraseasonal variability and monsoon weather.

For example, consider the problem of a decision-maker who has been told that there is a high probability to expect a slightly below average seasonal monsoon rainfall on the scale of the subcontinent. The agricultural expert may decide to choose a drought-resistant seed but he is still faced with the immediate problem of when to plant. To optimise planting it is necessary to know the probability of not only when the first rains will occur but also the probable duration of these early season rains. That is, besides an indication of the overall seasonal rainfall, the timing of the onset must be known *and* the timing of the first break in the monsoon and how long it will persist.

26.1.3 An example: the drought of 2002

The summer of 2002 provides a useful example (Figure 26.2a) where the overall seasonal precipitation for the season turned out to be 20% below average although official forecasts predicted only slightly below average seasonal rainfall. Note that the official Indian Meteorological Department forecast was deterministic: that is, there was no probability attached to the forecast. But the problem encountered by India during 2002 came not from the failure of the overall forecast (or its determinism) following the monsoon onset. With a seemingly successful onset of rain in June (albeit slightly later than average) planting commenced. However, the prolonged break period in July was not predicted. Subbiah (2004) comments on the midsummer drought and what might have occurred if forecasts on intraseasonal timescales had been available.

> The dry spell starting from mid-July to the first week of August 2002 in most parts of India caused serious dislocations in water management and agricultural operations. The revival of monsoon conditions in the second week of August [see Figure 26.2c] eased the water stress situation to some extent. Assuming that a prediction of the July drought had been available by the third week of June 2002, and of the revival of the monsoon rains by second week of July 2002, the forecasts would have made the following differences. In most parts of India agriculture operations start in second week of June and farmers make heavy investments during this period for land preparation, seedbed preparation, nursery raising and transplanting of seedlings. The water resource managers make decisions on allocation of water for various purposes (irrigation, hydroelectricity generation) on the assumption of normal rains. The prediction of likely dry spell in mid-June with a lead-time of weeks could have motivated farmers to postpone agriculture operations, saving investments worth billions of dollars. The water resource managers could have introduced water budgeting measures, such as minimizing water availability for water consuming crops and maximizing water for low water consuming crops, and by rationing water use for hydroelectric power. Similarly, the prediction of the revival of monsoon rains by the second week of July would have motivated the planners and farmers to undertake contingency crop-planning by mobilizing resources such as seed availability and credit for choosing suitable crop varieties, carrying out mid-season corrections and undertaking crop life saving measures. These actions would have helped to preserve farm income and ensured food security and reduce relief expenditure by at least 60% of the present cost (i.e., around 6 billion US $). Water resources

> could have been used to raise fodder crop in northwest India thus reducing the need for transportation of fodder from distant places at a huge cost. In summary, a 20-day forecast during monsoon 2002 in India could have mitigated the impacts of the droughts in several parts of India to a significant extent.

Subbiah's comments on the 2002 drought in India describe well the forecast requirements of the agricultural and water resource managers in South Asia. There are no simple solutions and it is necessary to think beyond normal techniques and procedures to gain useful information.

26.1.4 Summary of requirements

We can summarise the list of general requirements for functional environmental predictions for a user community.

(a) The forecasts must match the timescales of the major phenomenological time periods in the particular region if they are pertinent to the user community in question. For example, in the monsoon regions forecasts should include predictions of seasonal anomalies, intraseasonal variability, and weather.

(b) A suite of forecasts should be constructed that are temporally overlapping in order to allow strategic decisions to be made at the longest time period and tactical corrections to be made on the shorter timescales.

(c) The forecasts must be probabilistic. Only in this manner can a user of the forecast make a reasoned cost–loss analysis.

(d) The forecast should be user specific or be rendered into information that is useful and understandable to the user.

(e) User information should be included into the forecast process.

(f) The expectations of a user community should understand the rule that the longer the lead time of a forecast, the less regionally specific a forecast will be. Statistical downscaling techniques or historical data may help but it should be realised that there are basic uncertainty issues that limit the veracity of a forecast.

It is clear that the problem of creation of useful forecasts comes from an interaction between a user community and the provider of the forecast. It is clear also that quite often the desires of the user and the abilities of the forecaster may not match. However, through the interaction of the forecaster and the user, the question of what is possible can be addressed.

In the following paragraphs we will outline a forecasting system that provides the user community with a best use of available information. We will use the example of an operational system that we have implemented in Bangladesh for the forecasting of river discharge into the country in addition to regional precipitation forecasts. This system, broadly described as the 'three-tier' forecasting system, produces overlapping forecasts on seasonal, 20–30 days and 1–10 days. We will then address the problem of how to interface the products of the physical scientist with the needs of the user community through the development of a 'user metric' which allows a simple depiction of hedging strategies.

26.2 Examples of tiered overlapping forecasts

We provide two examples of forecasts utilising the three-tier system described earlier. We choose Bangladesh (where summer flooding is a major problem) and rainfall in the Ganges catchment region and in central India. The latter two regions are major agricultural regions in India. Bangladesh is a deltaic country that lies at the confluence of three major rivers – the Ganges, the Brahmaputra and the Meghna – and is thus susceptible to flooding from one or a combination of all three rivers. Flooding occurs each year in Bangladesh but in different parts of the country and with sufficient irregularity throughout summer to disrupt planting and harvesting cycles and cause local social disruptions. Occasionally the flooding is severe and prolonged as in the summer of 1998 when 90% of the country was inundated for nearly three months. Bangladesh flooding is often out-of-phase with rainfall over peninsular and central India. For example, during the summer of 2002, while most of India was under severe drought conditions, floods occurred over Bangladesh as the Brahmaputra passed its critical discharge levels. This occurred because during break periods of the monsoon, precipitation shifts north over the foothills of the Himalayas in the catchment area of the Brahmaputra. Flooding during active periods of the monsoon comes from the overflow of the Ganges.

We discuss briefly the progress that has been made during the last three years in developing a three-tier forecast system of river discharge, flood warning and precipitation for Bangladesh and a number of regions of India under the auspices of CFAB. The CFAB project was formed as a joint effort between Georgia Institute of Technology, University of Colorado, the Asian Disaster Preparedness Centre (ADPC) and ECMWF. The basic aims of CFAB lie in four main areas:

(a) The generation of a river discharge and precipitation operational forecasting system that would be available in real time with forecasts provided on a three-tier time system: seasonal outlooks (1–6 months), intermediate (20–30 days) and short term (1–10 days) using state-of-the-art models or with models developed specifically for the regional problem. These timescales were chosen as they match statistically significant spectral maxima that are found in monsoon variables. In addition, they were chosen to allow strategic decisions to be made relative to seasonal outlooks and tactical decisions or reorientations at the intermediate and short timescale.

(b) Creation of a collaborative enterprise between international (US and Europe) and Bangladeshi (and eventually Indian) partners for the forecasting of the probability of floods on timescales of days to months leading to the transfer of the techniques and technology to our appropriate partners.
(c) The development of an infrastructure that allows the application of the forecasts by regional scientists, engineers, agricultural extension workers, disaster relief organisations.
(d) The development of methods and decision tools so that the forecasts are directly applicable to the user sectors.
(e) The transfer of the forecasting technology to the Bangladeshis in a form that is immediately usable in an operational sense and modifiable for other uses and eventually to the larger monsoon community of Asia and Africa.

Considerable progress has been achieved in the implementation of (a), (b) and (c). During the summer of 2003 and 2004, operational forecasts were made available for the long-term and short-term forecasts during the entire season on an experimental basis. Seasonal outlooks (i.e. river discharge forecasts at 1, 2, 3, 4, 5, 6 months) were provided each month. Short-term forecasts (1–10 days) were issued each day for both 2003 and 2004. These latter forecasts were used extensively by various water resource groups in Bangladesh. Intermediate 20–30 day forecasts were issued every five days starting in the middle of the 2004 season.[1]

Forecasting river discharge and translating these forecasts to flood forecasts is a special challenge in Bangladesh. If floods in Bangladesh can be forecast with sufficient lead time and accuracy, actions could be taken across the country that could lessen the impact of the floods. However, until recently, the ability to forecast floods in Bangladesh has not existed for the following reasons:

(a) Floods can be forecast at a point downstream by knowing the river flow at some point upstream in conjunction with a local precipitation forecast coupled to a hydrological/land use model. Based on this information, simple regression forecasts can give fairly accurate short-term estimates of river discharge. However, Bangladesh does not receive any upstream river flow information from India and the only information that Bangladeshi flood forecasters have are the river flows they measure at staging points where the two major rivers enter Bangladesh and at other points within Bangladesh (Figure 26.1). From these data it has been possible to forecast flood levels in the interior and in the south of Bangladesh but with only two days lead. CFAB decided to extend the lead time by assuming that the Ganges and the Brahmaputra were ungauged river basins and by using a variety of model types.
(b) The physical factors that determine the rainfall over the Ganges/Brahmaputra catchments have only recently been understood. Hitherto, numerically based

deterministic (or probabilistic) forecasts of rainfall on any timescales have not been available to the Bangladeshis. In fact, to date the Bangladeshis do not have any numerical meteorological capability. India has some but this is restricted to relatively short range.

In the following sections of the chapter, we show examples of the three-tier forecasting system for Bangladesh and India. In essence, a good hydrological forecast must arise essentially from a good precipitation forecast, especially since we have to treat the Ganges and Brahmaputra as ungauged river basins. In the next section, we will outline briefly the techniques employed in each of the three tiers but leave details of the schemes to published papers and the websites listed at the end of the chapter.

26.3 Techniques for tiered forecasting

There exists within the World Climate Research Programme (WCRP) a large number of components, each dealing with different timescales and each, to some degree, with simulation and prediction. A recently adopted programme, Predictability Assessment of the Climate System (PACS), has been to set up in order to address the common needs of the diverse WCRP programs. PACS calls for a 'seamless modelling approach' in which one 'unified' model would attempt to forecast variability and climate on all timescales. With this vision, variability on the timescales of weather (1–10 days) would be predicted with the same model as intraseasonal variability (20–40 days) or interannual variability and so on. Such a proposal has many positive aspects. For example, Palmer and Webster (1994) suggested that a unified model approach would lead to better predictions on all timescales on the assumption that the best climate model would be a weather prediction model and, conversely, the best weather prediction model would be the best climate model. The rationale was that a model used for the dual purpose of weather and climate prediction would undergo significant and continual scrutiny on many timescales. In this manner, systematic errors in seasonal means, for example, which could influence the frequency and timing of weather events, could be minimised. At the other end of the scale, systematic high frequency errors could be minimised before they produce errors that erode lower frequency spectral bands.

In the future, through the concept of the unified model one may look forward to significant advances in prediction, on the one hand, and efficient utilisation of resources, on the other. Unfortunately, there are real and pressing problems that need to be addressed immediately and we do not have the luxury of waiting until the unified model concept is implemented. In addition, there are distinct model problems that preclude the use of a unified model at this time. In particular, models have great difficulty in predicting, or even simulating, intraseasonal variability. Given the importance of the intraseasonal climate mode in the monsoon regions (see Figures 26.2a, b), this is a considerable problem. Clearly, other techniques will have to be used to fill this predictability gap.

Rather than using one model to predict the three pertinent timescales, we take the more pragmatic approach and use three different techniques for each of the tiers. The shortest and longest predictions use ECMWF model output although from different models. The intermediate prediction timescale uses a Bayesian empirical technique using sets of observed data. The overall philosophy of the overlapping approach is shown schematically in Figure 26.4.

26.3.1 Seasonal trends (1–6 months)

We have the choice of developing empirical forecasts for the prediction on seasonal timescales or using output from coupled ocean–atmosphere models. We choose the latter as estimates of probability flow in a simpler manner from the numerical model. The system is summarised below:

Models

The extended forecasts are based on the European Centre for Medium-Range Weather Forecasts, 41-member seasonal ensemble precipitation forecast. The ECMWF seasonal model utilises an ocean model based on HOPE (Hamburg Ocean Primitive Equation model) version 2 (Latif *et al*., 1994).

The atmospheric component of the coupled ECMWF seasonal model is the ECMWF IFS (Integrated Forecast System) model version 23r4. Except for resolution, this is the same model as was used for numerical weather prediction (NWP), in early 2001. It is also the same cycle as is used in ERA-40, except that there are 40 levels in the vertical, compared with the 60 used in ERA-40 and the horizontal resolution used for the atmospheric component is TL95. The spectral representation is used only for the dynamical part of the model calculations. All of the model physical parametrisations (including clouds, rain and the land surface) are calculated on a Gaussian grid with about 1.875° spacing. The atmospheric model uses a two time-level semi-Lagrangian scheme for its dynamics with a 1-hour time step.

Precipitation forecasts

CFAB computes 'relative' forecasts of monthly-average precipitation for Bangladesh and regions in India. CFAB distributes the rainfall forecasts to Bangladesh. The forecasts assess the probability that precipitation in a given month will be in the lowest, second, third, fourth, or highest quintile. To produce these relative forecasts, the first

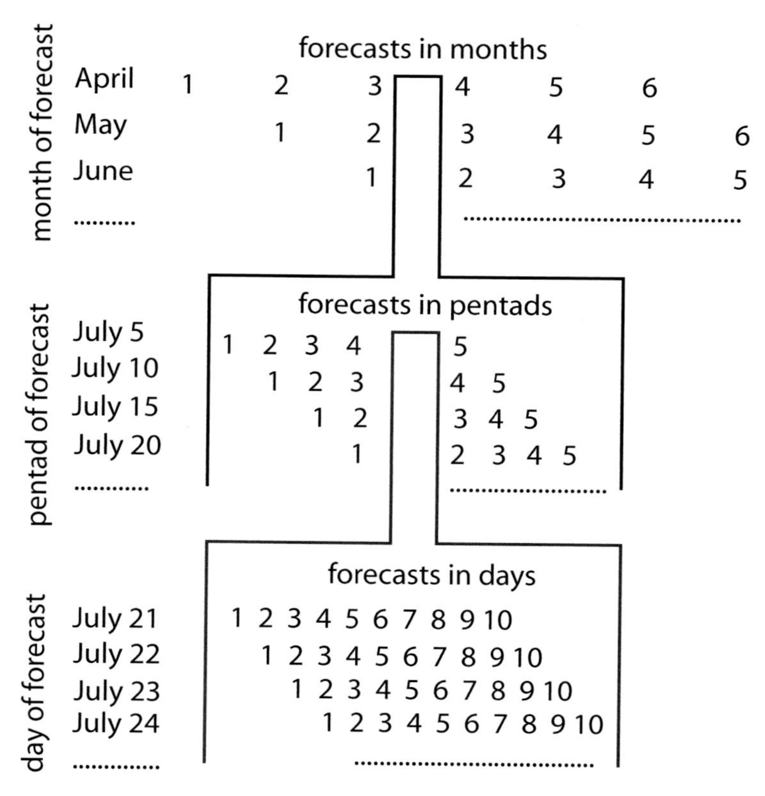

Figure 26.4 Schematic of overlapping three-tier forecast scheme. The scheme consists of distinct sets of forecasts. At the longest timescale, 6-month projections are made each month starting in April and continuing through September. The forecasts are made with a combination of the ECMWF coupled ocean–atmosphere climate model and a statistical-hydrological model mix. Within any one-month period, 5-pentad forecasts are made using the Bayesian empirical scheme of Webster and Hoyos (2004). This second tier of forecasts is enclosed in the grey box. Within any 5-day period, there are 1–10 day forecasts made each day using a combination of ECMWF operational forecast output and statistical-hydrological models. These forecasts are enclosed within the black box.

step is to develop 'model-space' climatology of catchment-averaged precipitation using the ECMWF seasonal model to compare against observations (see Hopson, 2005). 'Model-space' climatologies are developed for each month of the monsoon season and for each forecast lead time (i.e. 1-month, 2-month, etc.) individually, and are derived using the 5-member ensemble precipitation hindcasts from 1987 to 2001 and 40-member ensembles from 2002 to the present of the ECMWF seasonal model using the 'kernel' approach (see Hopson, 2005). The implicit assumption is made that taking all ensemble members together over this time span sufficiently defined the climatological 'attractor'. Using this model-space climatology, the current year's forecasts are compared with the climatology to determine the forecast's relative ranking (nearest quantile) within this model space. Once the quantiles of all the forecasts are determined, they are binned into the larger categories if desired (i.e. 0 to 20). Note that these calculations were done in 'model space' solely because large biases in the forecasts mean that the forecasts cannot directly be compared with the 'observed' climatology.

To produce 'actual-valued' precipitation forecasts of monthly averaged precipitation (as opposed to the relative forecasts discussed above), an observational-climatology was also determined using rain-gauge data from 1979 to 1996, and combined rain-gauge and satellite-derived precipitation data from 1996 to the present using data from the Global Precipitation Climatology Project (1996 to the present) and from the CMORPH project (2002 to the present). Using a 'quantile-to-quantile' mapping technique developed by Hopson (2005), the forecast quantiles discussed above were used to extract the equivalent quantiles in 'observational space'. In this way significant biases were removed and forecasts were made that corresponded statistically to the 'observed' climatology.

Discharge forecasts

The principle behind CFAB's discharge forecasts is that monthly averaged precipitation is significantly correlated with monthly averaged discharge. Therefore, if, say, above or below average monthly precipitation is forecast, this directly implies a forecast of above(below)-average monthly discharge. To explore this relationship between precipitation and discharge, monthly averaged observed precipitation was correlated with lagged monthly averaged discharges for the Ganges (at the entry point of the Ganges into Bangladesh at Hardinge Bridge), the Brahmaputra (at the border entry point at Bahadurabad), and the combined border Ganges–Brahmaputra river discharge were derived for the months of June through September for 17 years (1987–2003). The following correlations were calculated and are shown in the following tables.

In Table 26.1, it should be noticed that the optimal lag for discharge after precipitation for the Ganges catchment is at 23 days, with a Pearson correlation coefficient of 0.76; the optimal lag for the Brahmaputra is 10 days, with a correlation of 0.60; for the combined catchments the optimal lag is 17 days with a correlation of 0.65. These are all statistically significant. Armed with these correlations, forecasts of precipitation can then be used to derive discharge.

26.3.2 Intraseasonal variability (20–30 days)

Given the difficulty numerical models have in simulating and predicting intraseasonal variability in the tropics and subtropics, we use empirical schemes to forecast on

Table 26.1 *Correlations of monthly-averaged precipitation with monthly-averaged discharge at different lags for the (a) Ganges, (b) Brahmaputra and (c) combined basins*

The Pearson correlation coefficient, Spearman rank correlation, and Kendall rank correlation are given at 0-day lag, optimal-day lag, and 29-day lag.

		Correlations		
	Lag	Pearson	Spearman	Kendall
(a) Ganges	0-day	0.39	0.37	0.25
	23-day	0.76	0.81	0.61
	29-day	0.71	0.75	0.54
(b) Brahmaputra	0-day	0.41	0.37	0.26
	10-day	0.60	0.56	0.41
	29-day	0.42	0.42	0.31
(c) Ganges and Brahmaputra	0-day	0.42	0.39	0.27
	17-day	0.65	0.66	0.28
	29-day	0.54	0.52	0.37

the intermediate tier. Specifically, we use the physically based Bayesian scheme of Webster and Hoyos (2004).

Model

The model is based on the concept that separating the time series of predictands and predictors into significant physical bands determined from the predictand time series and treating each band separately will minimise the influence of high frequency errors projecting onto the lower frequency signal. This is accomplished by first selecting a predictand such as precipitation over Ganges catchment or Brahmaputra discharge into Bangladesh. Long-term time series of the data is analysed to assess the major spectral bands. In the monsoon regions there is a strong signal in the intraseasonal band. Composite analysis (e.g. Lawrence and Webster 2002) is then undertaken to assess the morphology of the intraseasonal variability relative to the intraseasonal peaks and valleys such as those observed in Figure 26.2. From the composite analysis a set of predictors can be determined. Both the predictor set (Table 26.2) and the predictand is band-passed to produce time series. These sets are regressed using an evolving regression scheme (see Webster and Hoyos, 2004) and recombined to form the prediction.

Discharge and precipitation predictions

The precipitation fields that are predicted are on the scale of Indian states (e.g. Orissa, Rajasthan) or regions (Ganges catchment, central India). Forecasts are made of 5-day (pentad) fields so that for a 20-day forecast, four lags are advanced. The forecasts are

Table 26.2 *Predictors used in the multiple input single output (MISO) statistical prediction scheme. Predictors are chosen so as to constitute a complete description of the evolution of the MISO*

Predictors			
#	Field	Region	Data source
1	Arabian Sea 10 m zonal wind	10°N–15°N, 65°E–75°E	NCEP[a]
2	Central India OLR	15°N–25°N, 70°E–85°E	NOAA[b]
3	Central India soil moisture	15°N–25°N, 70°E–85°E	NCEP[a]
4	Equatorial Indian Ocean 200 mb zonal wind	5°S–5°N, 70°E–85°E	NCEP[a]
5	Somalia Jet 925 mb meridional wind	2°S–8°N, 65°E–75°E	NCEP[a]
6	Arabian Sea 10 m meridional wind	10°N–15°N, 65°E–75°E	NCEP[a]
7	Central India surface pressure	15°N–25°N, 70°E–85°E	NCEP[a]
8	Equatorial Indian Ocean 10 m zonal wind	5°S–5°N, 70°E–85°E	NCEP[a]
9	Equatorial Indian Ocean OLR	5°S–5°N, 70°E–85°E	NOAA[b]
10	Tropical 200 mb easterly jet index	20°N–30°N, 70°E–100°E	NCEP[a]

[a] NCEP/NCAR reanalysis data has been obtained from http://www.cdc.noaa.gov/cdc/reanalysis/. The data are described by Kalnay *et al.* (1996).
[b] Estimates of outgoing longwave radiation from the National Oceanic and Atmospheric Administration (NOAA) polar-orbiting satellites (Gruber and Krueger, 1984; Liebmann and Smith, 1996).

made for the Ganges and Brahmaputra river discharge and for Indian rainfall districts every five days throughout the summer commencing in May.

26.3.3 Short-term (1–10 days)

Model

Predictions of precipitation surface energy fluxes come from the operational ECMWF model. Each day, 51 ensemble members are used to determine future states over prescribed regions of South Asia. Statistical corrections similar to those discussed in Section 26.3.1 above are employed. The manner in which the output from the ECMWF model is statistically dressed is complicated and beyond the detail required in this overview. Details may be found in Hopson (2005).

Hydrology Models

The ECMWF model provides ensembles of precipitation forecasts that are used to force hydrological models. Two distinct hydrologic modelling approaches are used in a multimodel format: the 'data-based modelling' (Beven, 2000) and 'distributed modeling' (similar to the US National Weather Service). Both models are described

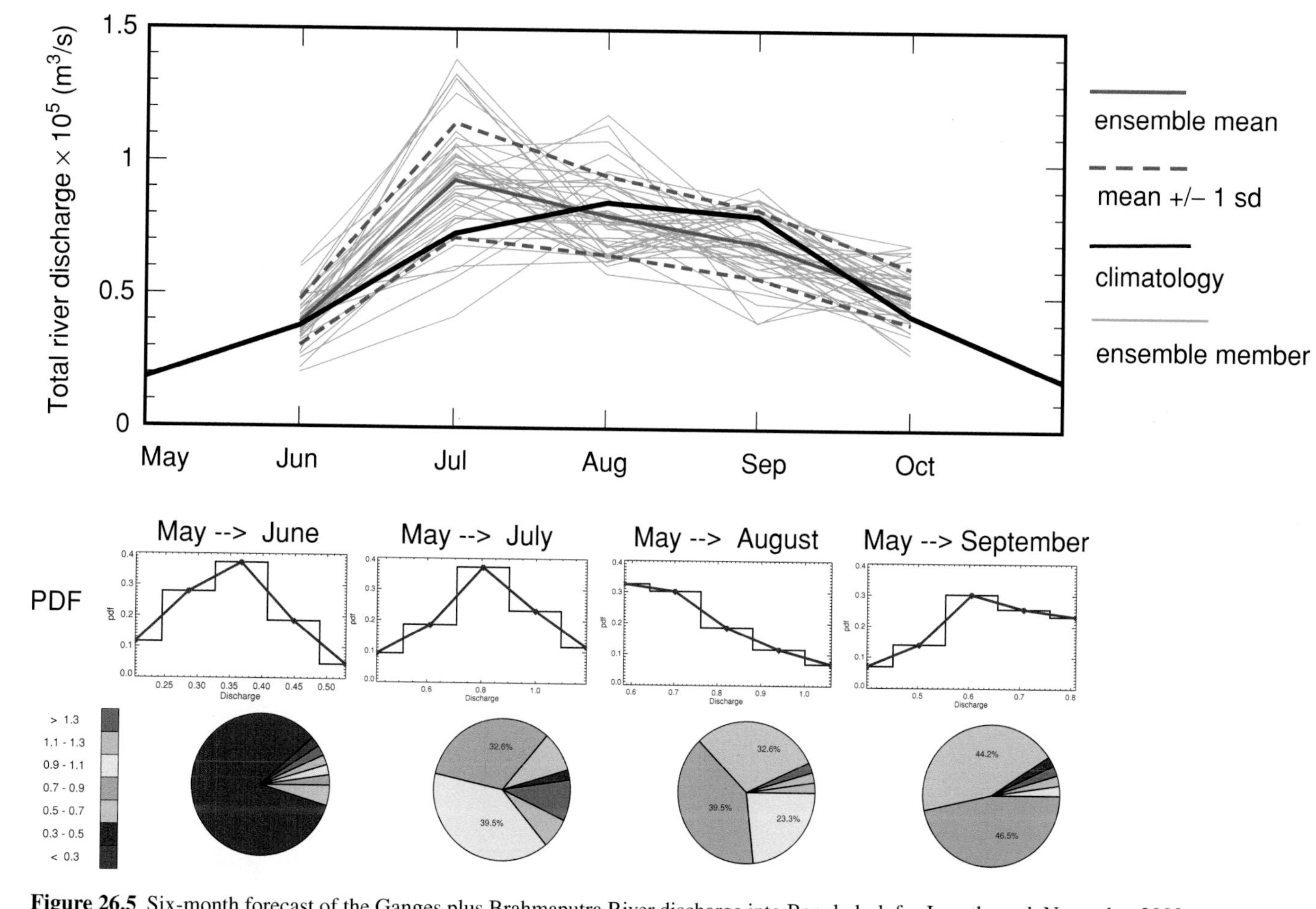

Figure 26.5 Six-month forecast of the Ganges plus Brahmaputra River discharge into Bangladesh for June through November 2003, commencing in May 2003. Forecasts are made every month for a six-month period using the 41 members of the ECMWF coupled climate model. Thick grey lines show the behaviour of each ensemble member. The ensemble mean is shown as the solid grey line and the two dashed lines indicate plus and minus one standard deviation about the ensemble mean. The long thick line is the climatological river discharge. The lower panel shows the PDFs of river discharge relative to the shade-coding.

in detail by Hopson (2005). The two-model approach, employed for both the Ganges and Brahmaputra basins, has certain attributes and drawbacks. Hopson (2005) shows that the results from the combination of models are better than either model used singly. Observed discharge of the two rivers is used to train models.

Prediction

Forecasts of regional precipitation and river discharge are made each day with lags of 1 to 10 days. Considerable thought has gone into producing output that is easily understood and useful to the user.

26.4 Examples of the three-tiered forecasts

An example of the seasonal outlook for the combined Brahmaputra and Ganges river discharge can be seen in Figure 26.5. The forecasts were initialised in May 2004. The upper panel shows the ‘plume’ of forecasts from the 41 member ensemble while the lower panel shows the probability density functions (pdf) in the form of graphs and pie charts for the different forecast lags. The general expectation for seasonal forecasts for areas as small as Bangladesh (1.4×10^5 km^2 or roughly the size of Wisconsin or half the area of the UK) would be that they would possess low skill simply because as the length of the forecast increases, uncertainty increases as the inverse of the area of the forecast. However, almost all the river inflow into Bangladesh is accumulated in a catchment area 12 times the size of Bangladesh. As river discharge is essentially a weighted spatial and temporal integral of the rainfall over the catchment, a greater skill can be expected in seasonal outlooks of river discharge. In essence, the skill of forecasts of river discharge into Bangladesh is the integrated skill of the precipitation forecast over the much larger catchment areas of the Ganges and the Brahmaputra.

Figure 26.5 shows that as early as May 2004, the model predicted excessive discharge in the July–August period. Although 2004 was a relatively normal year compared with the great flood year of 1998, extensive flooding did occur throughout the country during July and early August. The flooding can be inferred from Figure 26.6, which plots the Ganges and Brahmaputra river discharge throughout the monsoon season. The danger (or flooding) level was reached by the Brahmaputra and shows that the danger level (dashed line) was exceeded during this period. Even though the current seasonal model is configured to give forecasts of the combined discharge of the Brahmaputra and the Ganges, it is clear that there is some skill in the forecast.

Figure 26.7 shows a summary of the 2004 forecast of the central India region (defined in Figure 26.1) issued every five days for 20 days in advance. These are depicted as the black curve starting in mid June. The shaded swarth shows the

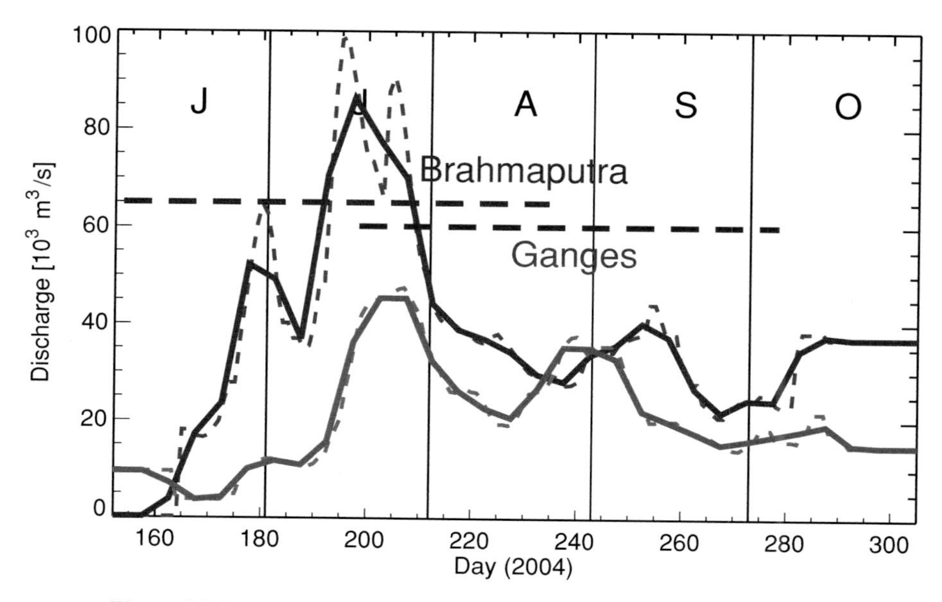

Figure 26.6 Brahmaputra and Ganges discharge into Bangladesh during 2004. Solid curves show the pentad discharge while the dashed curves show the daily values. The two horizontal lines show the Brahmaputra and Ganges danger levels. The Brahmaputra exceeded danger levels in July/August whereas the Ganges remained below flood levels throughout the summer.

probability distribution of the forecast made at the end of July. The pdfs are shown below in the same form as Figure 26.6.

Short-term forecasts (1–10 days) are shown in Figure 26.8(a) and (b) in two formats. Both panels refer to 10-day forecasts of river discharge. Figure 26.8(a) shows the results of 10-day forecasts in ensemble mode. The observed Brahmaputra discharge (solid curve) falls within the spread of the ensemble members throughout most of the summer of 2004. The scheme also predicts with considerable accuracy exceedance of the danger level (horizontal dashed line: see Figure 26.6) 10 days in advance. Using the spread of the ensemble members, it is an easy task to compute the probability of the exceedance of the danger level. This is shown in Figure 26.8(b). We have found that the threshold probability forecasting format depicted in Figure 26.8(b) is the easiest form of presentation for a user community to understand.

26.5 Communication of forecasts

Whereas the forecasting of an environmental event or the prediction of the probability of the exceedance of some limit (e.g. Figure 26.8b) is an academic achievement, there is no value unless the forecast is understood and applied by a user community.

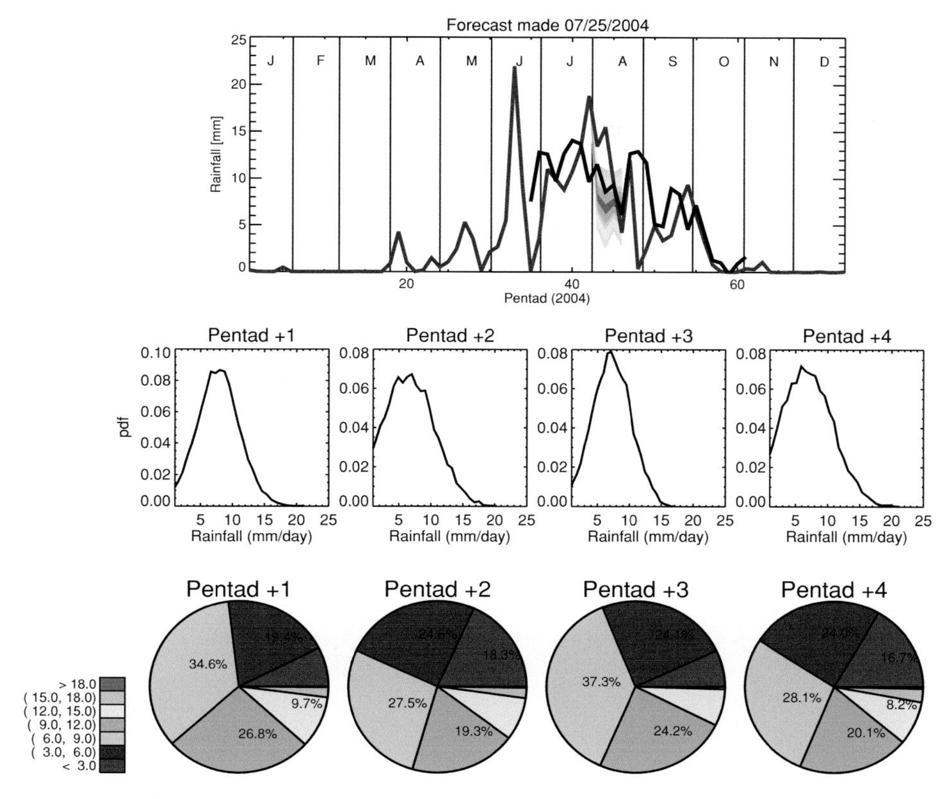

Figure 26.7 20-day forecast of Central Indian rainfall rate for the year 2004 using the Bayesian scheme developed by Webster and Hoyos (2004). Upper panel shows the observed pentad rainfall in the 20–40 day period band (grey) and the forecast (black line) made 20 days prior. The figure also shows a probability distribution of the forecast made in late July. The second and third panels show the probability density functions in two formats for 1, 2, 3, and 4 pentads following the forecast day.

Providing an understandable probability forecast is a challenge in both developed and developing societies. We have approached this problem by the development of a utility called the User Metric (Figure 26.9). The principal aim of a User Metric is to allow the transformation of probabilistic forecasts (difficult to understand and apply) to a usable assessment of aggregate risk (easy to understand) so that a deterministic decision of future action can be made (easy to apply). A User Metric should have the following properties:

(a) *Incorporation of a probabilistic forecast of some pertinent parameter* (e.g. river discharge, rainfall variability; upper left panel of Figure 26.9). This is supplied by the physical scientists/forecast offices using the forecast modules described above. We note that the probability density function will change with each forecast.

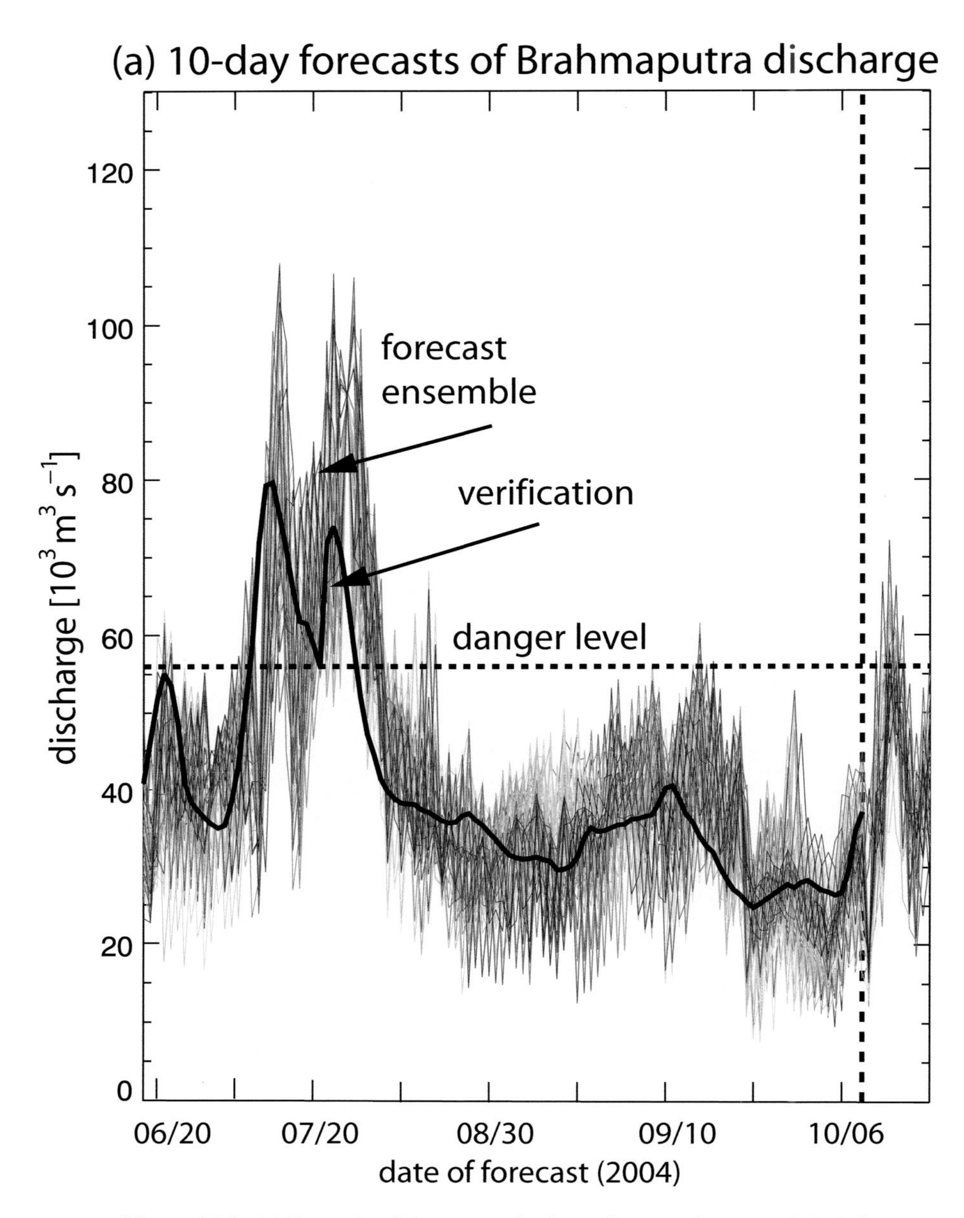

Figure 26.8 (a) Example of short-term discharge forecasts into Bangladesh for Brahmaputra. Ten-day forecasts made in the June–October time period in 2004. The forecasts use the ECMWF ensemble precipitation forecasts in combination with statistical and distributed hydrological models. The members of the ensemble relative to the observed discharge (solid curve) are shown as the shaded lines. The method successfully forecasts the exceedance of the danger or flood level 10 days in advance.

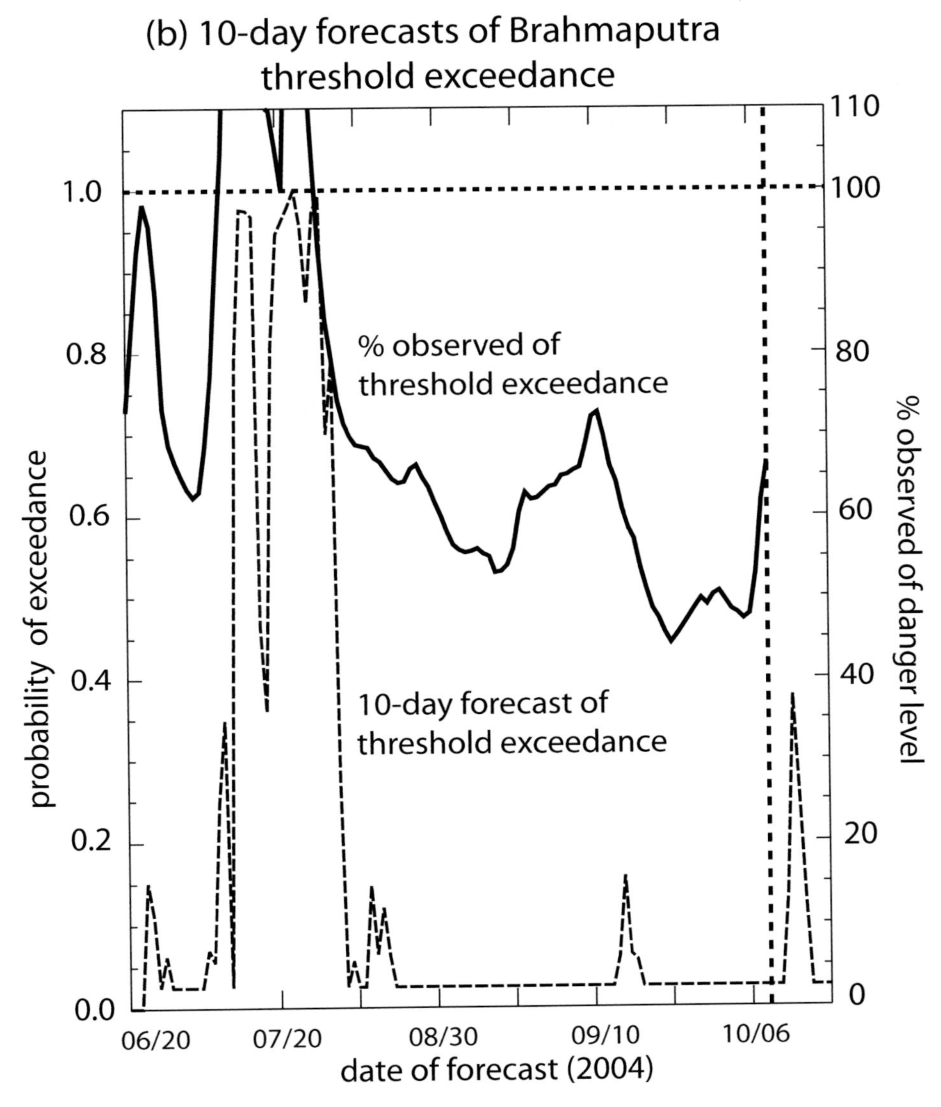

Figure 26.8 (*cont.*) (b) Alternative representation of (a) in the form of the probability of exceedance of flood level (dashed curve) made 10 days in advance. The observed fraction of discharge relative to flood level is shown as the solid curve. This form of representation of a probabilistic forecast has been shown to be easily understood and, hence, very useful to a user community.

(b) *Incorporation of local knowledge of the impact of an environmental event of a given severity.* This can be in the form of a costing function provided by the user community (top right panel of Figure 26.9). The costing function provides a quantification of the impact of a range of meteorological events (impact of no rain, moderate rain, heavy rain, etc., on yield of a particular

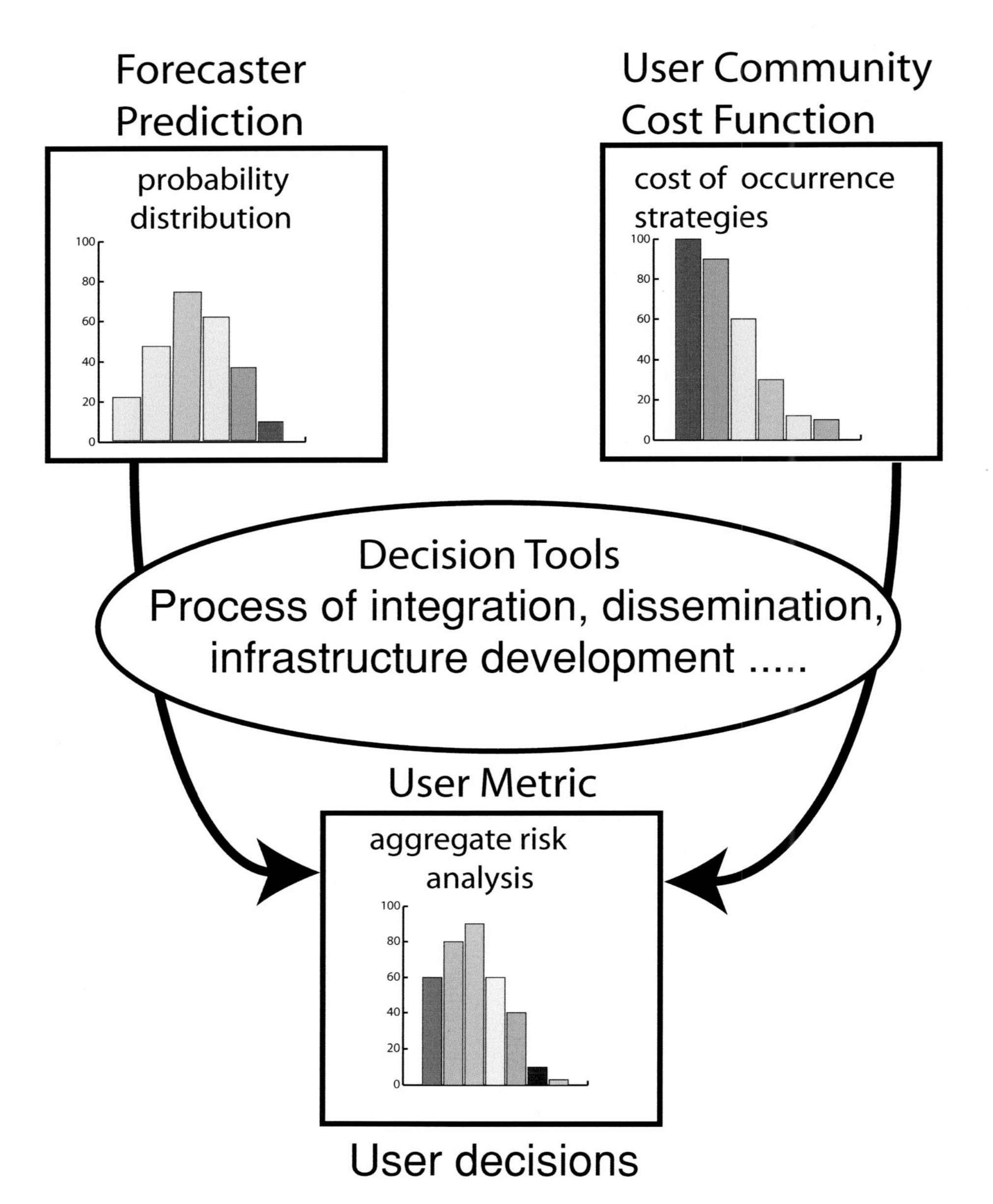

Figure 26.9 The components of the User Metric. The upper left panel shows the probability density function of some phenomenon (e.g. rain rate) produced by an environmental prediction group. Different user groups or the same user group at different times will have a cost function associated with each of the probabilities. This family of user dependent outcomes can be seen in the upper right hand panel. Using some institutional context (e.g. individual, market based) a family of aggregate risk analyses can be made which reflect the optimal decision for the particular user group. For the same forecast pdf, the aggregate risk analysis may be different. By contrast for one user group and a different forecast, there will be a different optimal strategy. The purpose of the bottom panel is to provide the user with one readily understandable diagram that takes into account the forecasts, pdf and the particular user circumstances.

crop) of differing severity on a particular application. The cost function is independent of a particular forecast and merely states the user's view of the impact of an environmental state. For example, at the time of planting of a crop, no rainfall would be disastrous, moderate rainfall is beneficial, but too much rainfall may cause floods. However, later in the season, say at harvest time, there would be a completely different costing function on the same parameters.

(c) *An easily comprehensible and visually decipherable representation of risk.* It is necessary to combine the probabilistic predictions with the costing functions to provide an aggregate risk analysis (bottom panel of Figure 26.9). Such a measure provides a user community with an optimal at a particular time for a given circumstance. This visual analysis will aid the user community in making reasoned decisions by the generation of an aggregate risk analysis.

The concept illustrated in Figure 26.9 is relatively straightforward. The question may be whether or not a farmer should harvest all of his crop ahead of maturation (thus having zero risk of environmental damage but accepting a lower yield), wait until maturation (taking a chance of reduced yield due to environmental factors but noting that there is a chance that full yield will be achieved) or harvesting partially in order to spread risk and benefits. The problem then is choosing the best strategy to hedge against uncertainty. First, the farmer knows that if heavy rains occur there will be a certain quantifiable reduction of yield. Also, if the forecast calls for a very high probability of dry weather then the harvesting strategy will be clear. Both of these are components of a user-produced costing function. But for a wider range of possible future states, the strategy is less clear and it is necessary to combine the probabilistic forecast with the costing strategy to come up with an optimal hedging strategy.

A basic tenet of our work is that we believe that there is important and valuable information in estimating risk of the occurrence of some event to which the user community is sensitive (e.g. floods), even when this risk is small but non-zero. Probabilistic forecasts offer the only way in which reasoned decisions can be made by the user community or relief organisation. There appears to us no need to make decisions without computing probabilities of occurrence and ascertaining the cost/benefit relationship of a particular event in agreement with Zhu *et al.* (2002). Finally, the User Metric offers a simple way to incorporate information from the user community, combine it with probabilistic forecasts from numerical or statistical models, and provide an easily interpretable graphic from which reasoned decisions can be made.

26.6 Concluding remarks

Perhaps the most important conclusion that can be made from this study is that the creation of a useful forecast is not an easy task. Clearly, it is not possible for an operational entity such as ECMWF to be able to anticipate the needs and incorporate the cost functions of all user communities. A successful end-to-end system requires the injection of engineering decision tools such as those introduced into the Bangladesh system by the CFAB group. That is, intermediate groups are needed between the forecaster and the user community. With the advent of ENSO forecasts in the late 1990s, the first attempts were made to produce end-to-end prediction systems but these rarely resulted in satisfactory results. The intermediary groups in these schemes were principally social scientists lacking perhaps in the engineering approach necessary to produce quantitative interpretations of probabilistic forecasts and the development of quantitative decision tools with which engineers are familiar. Whereas we can point to some success with the CFAB project it should be remembered that the bridge to other user groups would require different decision models and interpretations of probabilistic forecasts. CFAB, perhaps in the mode of linking forecasts and users through engineering systems, stands as a template upon which other systems can be patterned or improved.

Acknowledgements

Basic concepts were developed with funding from the US NOAA office of Global Programs Grant NAØ40AR 4310111. Funding for CFAB was from USAID/OFDA with additional support from the European Centre for Medium-Range Forecasts (ECMWF) who have generously provided the CFAB group and Government of Bangladesh agencies data that have allowed the experimental operational system to be constructed. We are especially grateful to Drs. T. N. Palmer and D. L. T. Anderson of ECMWF.

Note

1. The CFAB project is described in detail at: http://cfab.eas.gatech.edu/cfab/cfab.html, while real-time forecasts can be found at http://cfab2.eas.gatech.edu.

References

Beven, K. J. (2000). *Rainfall-Runoff Modelling: A Primer*. John Wiley.

Gruber A. and A. F. Krueger (1984). The status of the NOAA outgoing longwave radiation data set. *Bull. Am. Meteorol. Soc.*, **65**, 958–62.

Hopson, T. M. (2005). Operational flood forecasting for Bangladesh. PhD dissertation, University of Colorado.

Kalnay, E. and 21 co-authors (1996). The NCEP/NCAR 40-year reanalysis project. *Bull. Am. Meteorol. Soc.*, **77**, 437–71.

Latif, M., T. Stockdale and J. Wolff (1994). Climatology and variability in the echo coupled GCM. *Tellus*, **46**, 351–66.

Lawrence, D. and P. J. Webster (2002). The boreal summer intraseasonal oscillation and the South Asian monsoon. *J. Atmos. Sci.*, **59**, 1593–606.

Liebmann and C. A. Smith (1996). Description of a complete (interpolated) outgoing longwave radiation dataset. *Bull. Am. Meteorol. Soc.*, **77**, 1275–7.

Palmer, T. N. and P. J. Webster (1994). Towards a unified approach to climate and weather prediction. *Global Climate Change*, **1**, 265–80.

Rasmusson E. M. and T. H. Carpenter (1982). Variations in tropical sea-surface temperature and surface wind fields associated with the Southern Oscillation El-Niño. *Mon. Weather Rev.*, **110**(5), 354–84.

Subbiah, A. (ed.) (2004). *Initial Report on the Indian Monsoon Drought of* 2002. Available from Asian Disaster Preparedness Center, Bangkok, Thailand.

Webster, P. J. and C. Hoyos (2004). Forecasting monsoon rainfall and river discharge variability on 20–25 day timescales. *Bull. Am. Meteorol. Soc.*, **85**, 1745–65.

Webster, P. J. and S. Yang (1992). Monsoon and ENSO: selectively interactive systems. *Quart. J. Roy. Meteor. Soc.*, **118**, 877–926.

Webster, P. J., T. Palmer, M. Yanai, *et al.* (1998). Monsoons: processes, predictability and the prospects for prediction. *J. Geophys. Res.*, **103** (TOGA special issue), 14451–510.

Zhu, Y. J., Z. Toth, R. Wobus, D. Richardson and K. Mylne (2002). The economic value of ensemble-based weather forecasts. *Bull. Am. Meteorol. Soc.*, **83**, 73–83.

27

DEMETER and the application of seasonal forecasts

Renate Hagedorn, Francisco J. Doblas-Reyes, T. N. Palmer
European Centre for Medium-Range Weather Forecasts, Reading

A multimodel ensemble-based system for seasonal-to-interannual prediction has been developed in a joint European project known as DEMETER (Development of a European Multi-Model Ensemble System for Seasonal to Interannual Prediction). The DEMETER system comprises seven global coupled atmosphere–ocean models, each running from an ensemble of initial conditions. Comprehensive hindcast evaluation demonstrates the enhanced reliability and skill of the multimodel ensemble over a more conventional single-model ensemble approach. In addition, applications of seasonal ensemble forecasts have been incorporated into the DEMETER system. As an example of this innovative end-to-end system strategy, the use of DEMETER data in malaria forecasting processes is discussed. The strategy followed in DEMETER deals with important problems such as communication across disciplines, downscaling of climate simulations, and use of probabilistic forecast information. This illustrates the economic value of seasonal-to-interannual prediction for society as a whole.

27.1 Introduction

Seasonal-timescale climate predictions are now made routinely at a number of operational meteorological centres around the world, using comprehensive coupled models of the atmosphere, oceans, and land surface (Stockdale *et al.*, 1998; Mason *et al.*, 1999; Alves *et al.*, 2002; Kanamitsu *et al.*, 2002). They are clearly of value to a

Predictability of Weather and Climate, ed. Tim Palmer and Renate Hagedorn. Published by Cambridge University Press.

wide cross-section of society, for personal, commercial and humanitarian reasons (Thomson *et al.*, 2000; Hartmann *et al.*, 2002b). However, the successful transition from research activity to full operational practice has led some potential users of seasonal forecasts to have unrealistic expectations of what is practicable ('We are getting married in six months time – should we order a marquee for the wedding reception, or will it be dry that day?'). Notwithstanding predictable signals arising from atmosphere–ocean coupling (Shukla and Kinter, this volume; Timmermann and Jin, this volume), the overlying atmosphere is intrinsically chaotic, implying that predicted day-to-day evolution of weather is necessarily sensitive to initial conditions (Palmer, 1993; Shukla, 1998). In practice, the impact of such initial-condition sensitivity can be determined by integrating forward in time ensembles of forecasts of coupled ocean–atmosphere models; the individual members of the forecast differing by small perturbations to the starting conditions of the atmosphere and underlying oceans. The phase-space dispersion of the ensemble gives a quantifiable flow-dependent measure of the underlying predictability of the flow.

However, if uncertainties in initial conditions are the only perturbations represented in a seasonal-forecast ensemble, then the resulting measures of predictability will not be reliable; the reason being that the model equations are also uncertain. More specifically, although the equations for the evolution of climate are well understood at the level of partial differential equations, their representation as a finite-dimensional set of ordinary differential equations, for integrating on a digital computer, inevitably introduces inaccuracy (Palmer, this volume).

At present, there is no underlying theoretical formalism from which a probability distribution of model uncertainty can be estimated – as such a more pragmatic approach must be sought. One approach relies on the fact that global climate models have been developed somewhat independently at different climate institutes. An ensemble comprising such quasi-independent models is referred to as a multimodel ensemble.

In order to advance the concept of multimodel ensemble prediction and to explore the utility of such a forecast system for potential end users, the DEMETER project (Development of a European Multi-Model Ensemble System for Seasonal to Interannual Prediction) was conceived, and successfully funded under the European Union Vth Framework Environment Programme. A description of the DEMETER coupled models, the DEMETER hindcast integrations, the archival structure, and the common diagnostics package used to evaluate the hindcasts, is described in Section 27.2. Some meteorological and oceanographic results, comparing these single and multimodel ensemble hindcasts, are described in Section 27.3. As mentioned at the beginning of this chapter, there is considerable interest amongst a wide cross-section of society for seasonal climate forecast information. However, as also mentioned, some users will be disappointed in what can realistically be achieved, whilst others may find great economic value in the predictions. How can one distinguish viable applications from unrealistic applications? It might be easy to dismiss as unrealistic the potential customer who wants to know whether it will rain in the afternoon six months from today, but is the demand of a health official who wants to use seasonal predictions to predict malaria prevalence six months ahead, and whose malaria model requires daily weather parameters as input, also unrealistic?

A general methodology for assessing the value of ensemble forecasts for such users was introduced in Richardson (2000) and further discussed in Richardson (this volume). In particular, if these users have quantitative application models requiring forecast weather information as input (Hartmann *et al.*, 2002a), these models can be directly linked to the output of individual members of the forecast ensemble. The net result is a probability forecast, not of weather as such, but of a variable directly relevant to the user. Hence, in the case of the health official, the ensemble will produce a probability distribution of malaria prevalence. The potential usefulness of the ensemble forecasts can then be judged by asking whether the forecast probability distributions of malaria prevalence are sufficiently different from climatological probability distributions for the health official to be able to make decisions or recommendations on preventive measures, for example distribution of bed nets or anti-malaria spraying actions. In the DEMETER project, there are applications partners both in tropical disease prediction and also in agronomy. Some of the results of these end users in DEMETER are described in Section 27.4. As a result of DEMETER, real-time multimodel ensemble seasonal predictions are now routinely made at the European Centre for Medium-Range Weather Forecasts (ECMWF). This development, and other plans that derive from DEMETER, are outlined in the concluding section of this chapter.

27.2 The DEMETER system

27.2.1 Coupled models and initialisation procedures

The DEMETER system comprises seven global coupled ocean–atmosphere models. A brief summary[1] of the different coupled models used in DEMETER is given in Table 27.1

For each model, except that of the Max Planck Institute (MPI), uncertainties in the initial state are represented through an ensemble of nine different ocean initial conditions. This is achieved by creating three different ocean analyses: a control ocean analysis is forced with momentum, heat and mass flux data from the ECMWF 40-year reanalysis (Uppala, 2005; ERA-40 henceforth), and two perturbed ocean analyses are created by adding daily wind stress perturbations to the ERA-40 momentum fluxes. The wind stress perturbations are randomly taken from a set of monthly differences between two quasi-independent analyses. In addition, in order to represent the uncertainty in sea surface temperatures (SSTs), four SST perturbations are added and subtracted at the start of the hindcasts. As in the case of the wind perturbations, the SST

Table 27.1 *Horizontal and vertical resolution of the atmospheric and ocean components of the seven individual global coupled models forming the DEMETER multimodel system*

	Atmospheric component			Ocean component			
		Resolution			Resolution		
Partner	Model	Horizontal	Vertical	Model	Longitudinal	Latitudinal	Vertical
CERFACS	ARPEGE	T63	31 L	OPA 8.2	2.0°	2.0°	31 L
ECMWF	IFS	T95	40 L	HOPE-E	1.4°	0.3°–1.4°	29 L
INGV	ECHAM-4	T42	19 L	OPA 8.1	2.0°	0.5°–1.5°	31 L
LODYC	IFS	T95	40 L	OPA 8.2	2.0°	2.0°	31 L
Météo-France	Arpege	T63	31 L	OPA 8.0	182 GP	152 GP	31 L
Met Office	HadAM3	2.5° × 3.75°	19 L	HadCM3	1.25°	0.3°–1.25°	40 L
MPI	ECHAM-5	T42	19 L	MPI-OMI	2.5°	0.5°–2.5°	23 L

perturbations are based on differences between two quasi-independent SST analyses. Atmospheric and land-surface initial conditions are taken directly from the ERA-40 reanalyses. A separate ensemble initialisation procedure is used for the MPI model.

27.2.2 Definition of hindcast experiments

The performance of the DEMETER system has been evaluated from a comprehensive set of hindcasts over a substantial part of the ERA-40 period. Only hindcasts for the period 1980 to 2001 will be discussed in this chapter.

In order to assess seasonal dependence on skill, the DEMETER hindcasts have been started from 1 February, 1 May, 1 August, and 1 November initial conditions. Each hindcast has been integrated for six months and comprises an ensemble of nine members. In its simplest form, the multimodel ensemble is formed by merging the ensemble hindcasts of the seven models, thus comprising 7×9 ensemble members. To enable a fast and efficient post-processing and analysis of this complex data set, much attention was given to the definition of a common archiving strategy for all models; the ECMWF's Meteorological Archival and Retrieval System (MARS) was used for this purpose. A subset of atmosphere and ocean variables, both daily data and monthly means, have been stored into MARS. Special attention was given to the time-consuming task of ensuring that all model output complies with agreed data formats and units.

A significant part of the DEMETER data set (monthly averages of a large subset of surface and upper-air fields) is freely available for research purposes through an online data retrieval system installed at ECMWF.[2]

27.2.3 Diagnostics and evaluation tools

The need to provide a common verification methodology has been recognised by the World Meteorological Organization Commission for Basic Systems (WMO-CBS), and an internationally accepted standardised verification system (SVS) is being prepared. A comprehensive verification system to evaluate all DEMETER single models as well as the multimodel DEMETER ensemble system has been set up at ECMWF. It has been run periodically to monitor hindcast production, to check correct archiving and to calculate a common set of diagnostics.

The DEMETER verification system is designed with a modular structure so as to easily incorporate new evaluation tools provided by project partners or other sources. The basic set of diagnostics is summarised as follows:

- Global maps and zonal averages of the single-model bias are shown relative to a model climatology. Hindcast anomalies are computed by removing the model climatology for each grid point, each initial month, and each lead time from the original ensemble hindcasts. A similar process is used to produce the verification anomalies.

- Time series of specific climate indices, e.g. related to area averaged SSTs, precipitation and circulation patterns are displayed.
- Standard deterministic ensemble-mean scores, such as anomaly correlation coefficient (ACC), root mean square skill score (RMSSS), and mean square skill score (MSSS) are shown.
- Probabilistic skill measures: reliability diagrams, relative operating characteristic (ROC) score, Brier score, ranked probability skill score (RPSS), and potential economic value curves are calculated and displayed. Significance tests are applied to most of the skill measures.
- The skill of single-model ensembles is compared with that of multi-model ensembles using scatter diagrams of area-averaged skill measures and probability density functions (pdfs) of grid-point skill scores.

Both anomalies and scores have been computed using a cross-validation 'leave-one-out' method. To generate the anomaly or the score for a particular time t, only data at other times different from t have been used.

The main verification data set used in this system is ERA-40. This is consistent with the general concept of producing the DEMETER hindcasts, in which ERA-40 is used as forcing for the ocean analyses and as atmospheric and land-surface initial conditions. Effectively, it is assumed that we are 'living in the ERA-40 world'. However, because of the modularity of the validation system, it is possible to validate the model data with more than one verification data set. In fact, precipitation has been verified against the Global Precipitation Climatology Project (GPCP) dataset.[3]

27.3 Hindcast skill assessment

A sample of results from the DEMETER standard verification system is presented in this section. To view a more comprehensive set of verification diagnostics the reader is referred to the DEMETER website.[4]

The scientific basis for seasonal atmospheric prediction relies on the premise that the lower boundary forcing, in particular SST, can impart significant predictability on atmospheric development (Palmer and Anderson, 1994). Thus, one of the prerequisites for successful seasonal forecasts is the ability to represent and predict accurately the state of the ocean. A basic problem, faced when attempting to predict SST with coupled models, is the bias in the model forecasts, which may be comparable to the magnitude of the interannual anomalies to be predicted. Since SSTs in the tropical Pacific are a major source of predictability in the atmosphere on seasonal timescales, model performance in the tropical Pacific is of particular interest. To demonstrate the typical level of skill in this area, Table 27.2 shows the anomaly correlation coefficient (ACC) of the ensemble mean for the single-model ensembles and the multimodel ensemble for the SSTs averaged over the NINO-3.4 area. The correlation has been computed for the 1-month and 3-month lead seasonal hindcasts starting in February,

Table 27.2 *Ensemble-mean bias and anomaly correlation coefficient (ACC) for the 1-month and 3-month lead seasonal average of sea surface temperature over the NINO-3.4 area calculated using all start dates for the years 1980–2001*

Note that the bias for the multimodel ensemble and the persistence hindcast are not defined since the multimodel ensemble is based on single-model anomalies, which are constructed with regard to the single-model bias, and persistence uses observed anomalies.

Model	1-month lead		3-month lead	
	Bias/K	ACC	Bias/K	ACC
DEMETER multimodel	–	0.95	–	0.89
CERFACS	−0.34	0.94	0.07	0.86
ECMWF	−0.87	0.93	−1.50	0.86
INGV	−0.60	0.92	−0.76	0.82
LODYC	−0.96	0.95	−1.52	0.89
Météo-France	−0.03	0.93	0.43	0.83
Met Office	−0.53	0.92	1.45	0.81
MPI	−2.07	0.86	−3.42	0.66
Persistence	–	0.80	–	0.62

May, August, and November. Therefore, the values verify during the seasons MAM, JJA, SON, and DJF for the 1-month lead hindcasts, and MJJ, ASO, NDJ, and FMA for the 3-month lead hindcasts. Results suggest that the single-model ensembles generally perform well as El Niño–Southern Oscillation (ENSO) prediction systems. For the sake of comparison, the ACC for a persisted-SST hindcast has been included. This hindcast is made by persisting initial SST anomaly for the six months corresponding to the coupled model integration. For instance, the 6-month-long persistence hindcasts starting on 1 February are obtained from the anomaly on the previous January. Both the multimodel ensemble and the single models perform at levels comparable to dedicated ENSO prediction models and much better than persistence, especially in the 3-month lead time range. In addition, note the high correlation of the multimodel ensemble for both lead times, proving it to be the most skilful system for the 3-month lead hindcasts. The coupled model climate may differ from the observed climatology as a result of model ocean–atmosphere interactions. The bias of the single models is generally in the range of ±1 K (Table 27.2). These are typical figures for current leading coupled models. As is the case for most variables and areas, there appears to be no clear relationship between bias and anomaly forecast skill, though this is a topic that needs further investigation.

Figure 27.1 shows 1980–2001 time series of ACC of precipitation for all single models and the multimodel ensemble, for summer (JJA, May start date) over the

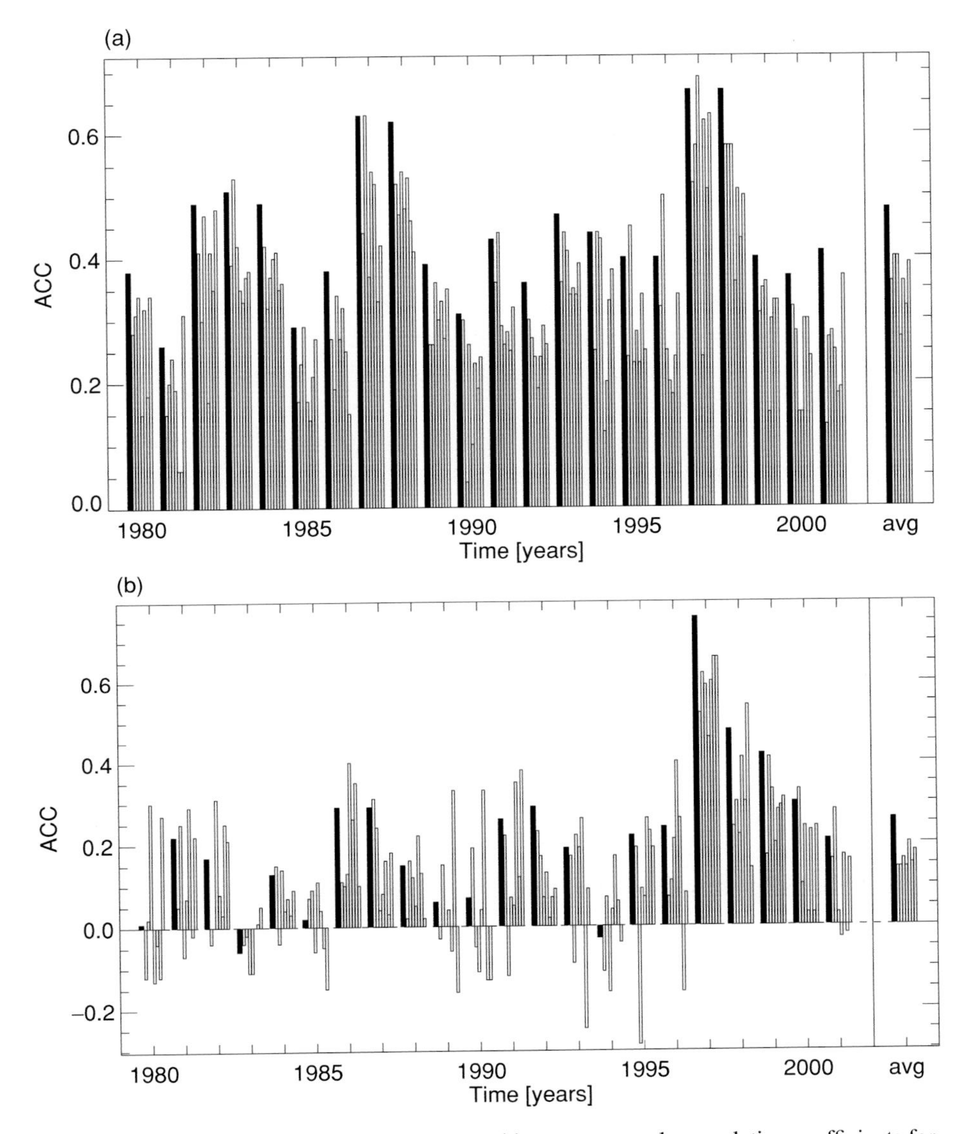

Figure 27.1 Time series of the ensemble-mean anomaly correlation coefficients for the multimodel (thick filled bars) and all individual models (thin open bars). (a) 1-month lead summer (JJA) precipitation in the tropics (latitudinal band of 30° S –30° N); (b) 1-month lead winter (DJF) precipitation in the northern extratropics (latitudinal band of 30° N–90° N). Additionally, the time average over the whole period 1980–2001 is shown at the end of each plot.

tropics (Figure 27.1a) and winter (DJF, November start date) over the northern extratropics (Figure 27.1b). The skill in the northern extratropics is considerably less than in the tropics. In both regions the variability in prediction skill, both from year to year and between different models, is clearly evident. The occurrence of higher skill during ENSO events is consistent with relatively large ACC for 1982–83, 1987–88 and 1997–98. This, in turn, is consistent with the link between ENSO activity and seasonal predictability found in many studies (e.g. Brancovic and Palmer, 2000). In general, the identity of the most skilful single model varies with region and year. Finally, this figure illustrates the relatively skilful performance of the multimodel ensemble. In most years the multimodel ensemble skill is close to the best single-model skill and is the most skilful when performance is averaged over all years. This highlights the greater reliability of the multimodel ensemble system.

To further summarise atmospheric hindcast skill, Figure 27.2 shows indices of the winter (DJF, November start date) Pacific North American (PNA) and North Atlantic Oscillation (NAO) patterns for the multimodel ensemble. The indices are computed by projecting every ensemble member anomaly onto a predefined pattern. To compute the reference patterns, an empirical orthogonal function (EOF) analysis of the 500-hPa geopotential height has been performed for the winter monthly mean anomalies using National Centers for Environmental Prediction (NCEP) reanalyses for the period 1949–2000. The EOF analysis was carried out using data over the regions 20° N–87.5° N and 110° E–90° W for the PNA and 20° N–87.5° N and 90° W–60° E for the NAO, and the leading EOF retained. The spatial covariance between the monthly anomaly patterns was then calculated for every single member of the hindcast ensemble and the reference pattern was computed. The monthly covariances were averaged to produce seasonal means. Figure 27.2 displays the index against time using a box-and-whisker representation in which the central box and each whisker contain one third of the ensemble members. The value obtained computing the spatial covariance between the reference pattern and the ERA-40 anomalies is also displayed. Comparison of the interannual variations of ERA-40 and ensemble-mean values gives a visual impression of ensemble-mean hindcast skill. The verification lies within the ensemble range in all but two cases for both indices. Table 27.3 shows the correlation between the two time series for the multimodel and the single-model ensembles. The multimodel ensemble shows one of the highest correlations among all the models for both indices. In addition, the multimodel ensemble correlation can be considered non-zero with a 95% confidence level using a two-sided *t*-test, which is not always the case for the single-model ensembles. However, it should be noted that scores based on indices are less robust than scores based on large area correlations, when calculated with relatively short time series. For example, the high PNA correlation for some single models may be explained by good predictions in 1982 and 1997. Note that, while PNA index hindcast skill tends to be quite satisfactory (Figure 27.2a), NAO index skill is lower but always positive. Figure 27.2b indicates that the multimodel ensemble can produce a useful signal in years when the observed

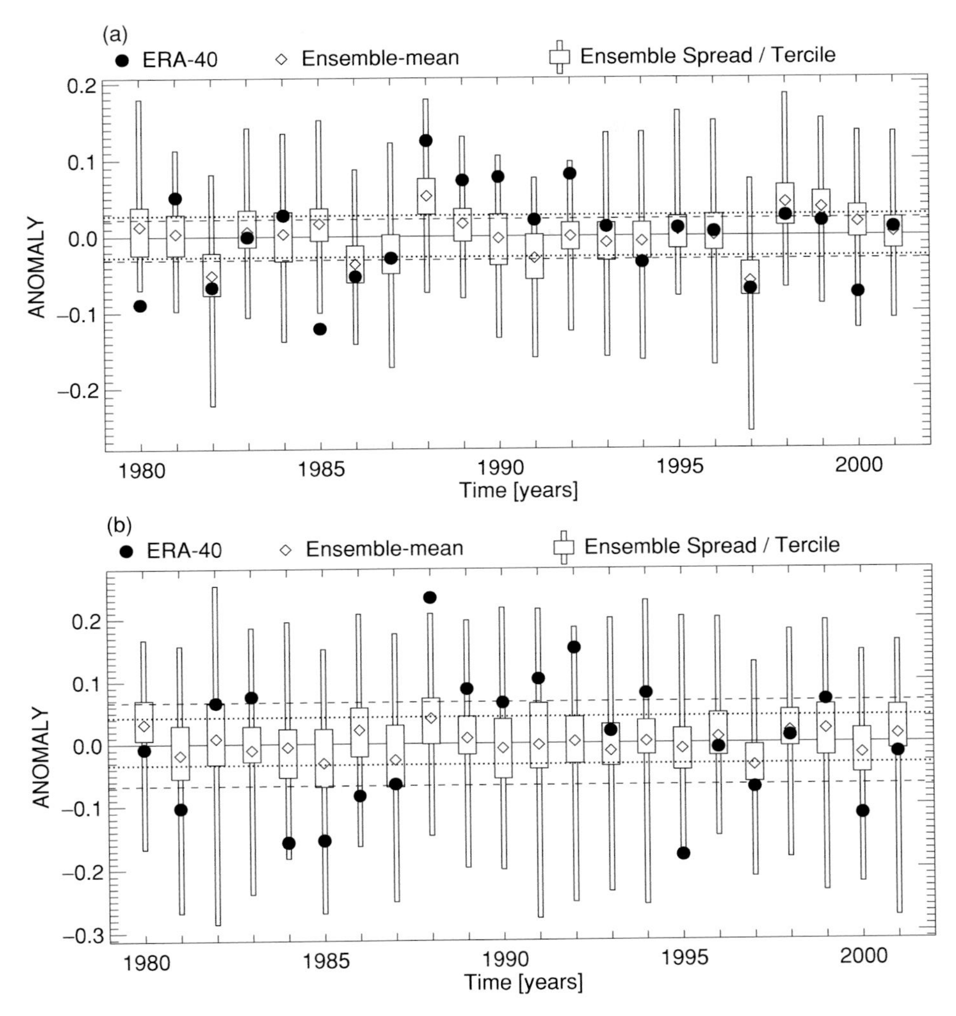

Figure 27.2 Time series of the 1-month lead winter (DJF) PNA (a) and NAO (b) index. The multimodel ensemble spread is depicted by the box-and-whisker representation with the whiskers containing the lower and upper tercile of the ensemble. The diamonds represent the ensemble mean, the ERA-40 anomalies being displayed by black bullets. The horizontal lines around the solid zero line mark the tercile boundaries of the ERA-40 (dashed) and hindcast data (dotted).

NAO index is large in magnitude, such as 1987, 1988 and 1997. These years may in themselves account for the high correlation coefficient obtained in Table 27.3. Nevertheless, the model signal in some years is weak (little shift of the predicted index away from zero) as in 1992 and 1995, when the observed index was large in magnitude.

Considerable effort has been devoted to the validation of the ensembles as probability forecasts. The dotted and dashed lines in Figure 27.2 correspond to the ensemble

Table 27.3 *Ensemble-mean correlation and ranked probability skill score for the PNA and NAO indices calculated from the 1-month lead hindcasts started in November (DJF seasonal average) for the years 1980–2001*
Statistically significant values (95% confidence level) are printed in bold letters

	Correlation		RPSS	
Model	PNA	NAO	PNA	NAO
DEMETER multimodel	**0.41**	**0.54**	**0.18**	**0.10**
CERFACS	0.16	0.30	−0.19	**0.16**
ECMWF	**0.39**	0.10	**0.24**	−0.12
INGV	0.23	0.20	−0.02	**0.04**
LODYC	**0.46**	**0.43**	**0.10**	0.00
Météo-France	0.23	**0.55**	−0.11	**0.01**
Met Office	**0.31**	0.18	**0.22**	−0.12
MPI	**0.32**	0.14	**0.10**	**0.02**

and ERA-40 tercile boundaries. The corresponding probabilistic skill measure used is the ranked probabilistic skill score (RPSS) based on these tercile categories. Hindcast performance is summarised in Table 27.3. RPSS is defined so that positive values imply higher skill than climatology forecasts and perfect forecasts have a skill score of 1. The skill of the multimodel ensemble for the PNA index is close to the skill of the best models and statistically significant at the 95% confidence level, in good agreement with the correlation results. Also, for the NAO index, RPSS values are generally high and tend to be statistically significant, which was not the case for the correlation. RPSS statistical confidence has been assessed by computing the distribution of the skill score from a random set of hindcasts obtained from scrambling the available hindcasts and verifications.

The above demonstrated superiority of the multimodel compared with single-model performance can also be found when considering the reliability of the predictions, with reliability having a precise technical meaning in this context. A forecast system is called reliable if the predicted probability of an event matches its frequency of occurrence when it was forecast. That is, when considering all cases where an event is predicted to occur with a 40% probability, this event should verify in exactly 40% of these cases, not less and not more. As one example (out of many), the reliability diagrams of the seven single models as well as the multimodel are shown in Figure 27.3 for the seasonal averages of the 2m-temperature in summer (May start date, 1-month lead time) averaged over the tropical band ($\pm 30^{\circ}$). The reliability diagram displays the accumulated proportion of forecast probabilities versus the accumulated observed frequency of the event. Every single-model ensemble proves to be overconfident, which is characterised by a too shallow slope of the line joining

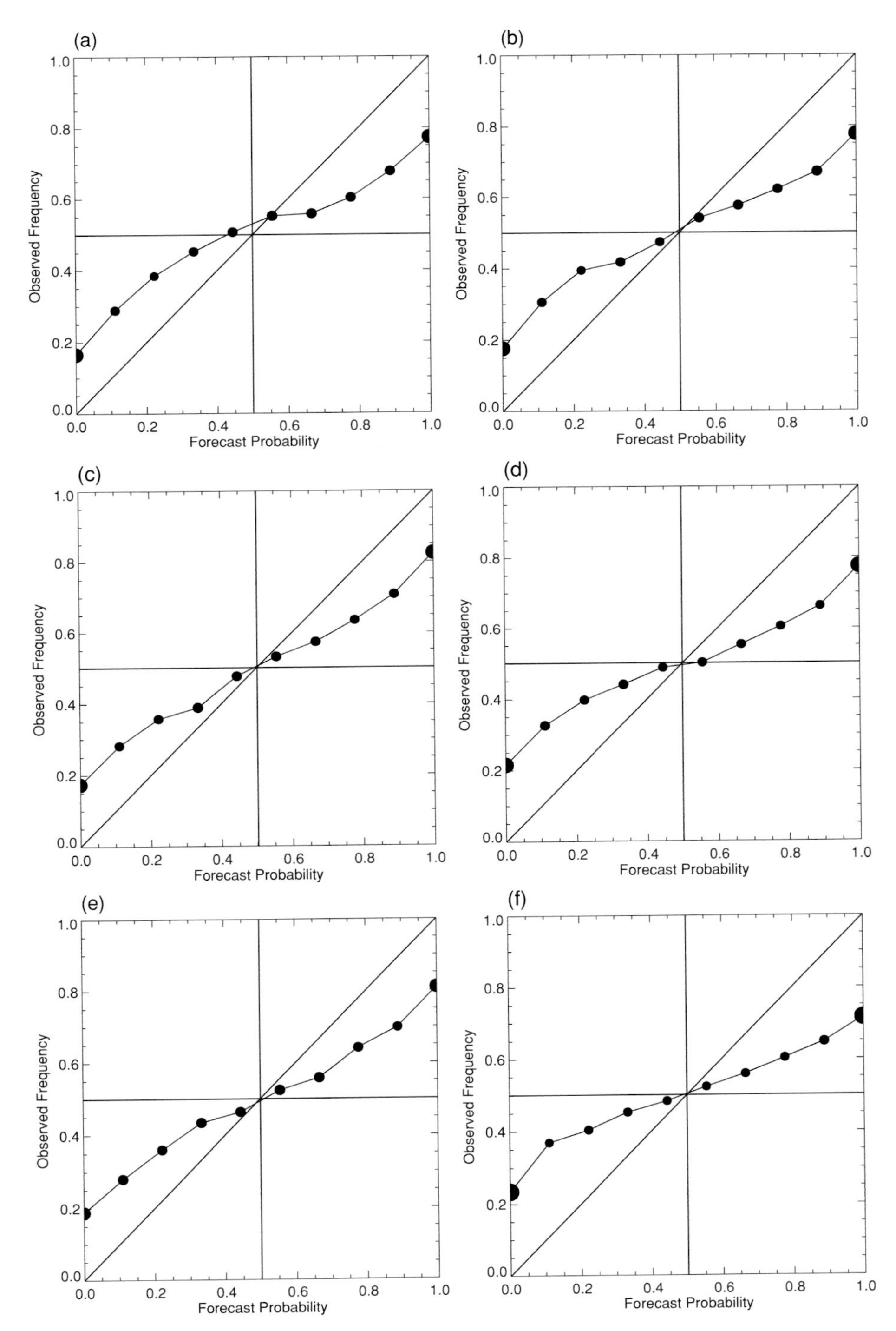

Figure 27.3 Reliability diagrams for the positive anomalies of seasonal averages of the 2m-temperature in summer (May start date, 1-month lead time) averaged over the tropical band (± 30°) for the period 1980–2001. The straight horizontal and vertical lines display the average observed frequency and forecast probability of the event. The size of the bullets represents the relative forecast frequency. The seven single-model results are given in (a) to (g), the multimodel reliability diagram is shown in (h).

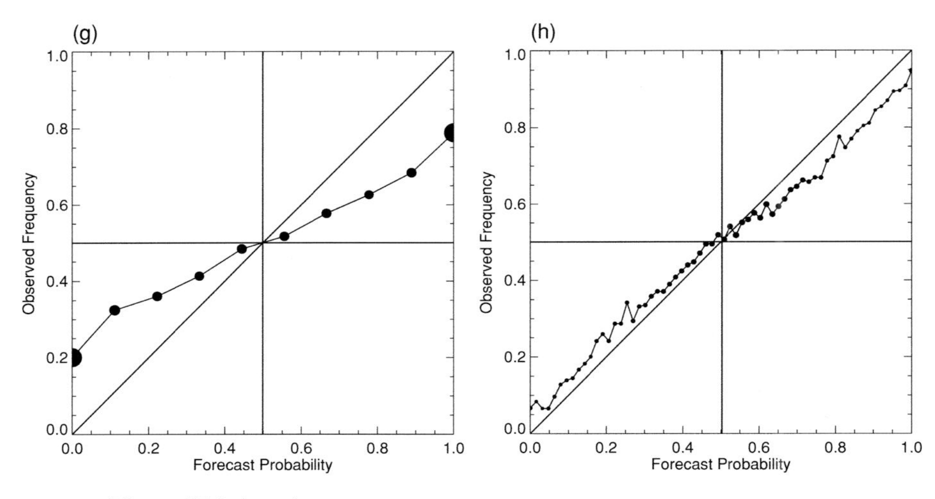

Figure 27.3 (*cont.*)

the points in the diagram (Figure 27.3a–g). By contrast, the reliability diagram for the multimodel ensemble fits much better the diagonal (Figure 27.3h). This implies that, given a prediction with a specific probability, the multimodel will verify on average the same proportion of observed events, while the single-model ensembles will assign low (high) probabilities to cases that are observed a higher (lower) proportion of times.

In spite of the clear improvement of the multimodel ensemble performance over the single-model ensembles, an important question arises. Is the improvement in the multimodel ensemble merely due to increased ensemble size or does the additional information from different models add to the performance? In order to separate the benefits that derive from combining models of different formulation to those derived simply from the accompanying increase in ensemble size, a 54-member ensemble hindcast has been generated with the ECMWF model alone for the period 1987–99 using the May start date. Figure 27.4 shows the reliability diagram for the same case as in Figure 27.3 (1-month lead positive anomalies of 2m-temperature in summer over the tropical band (±30°)), but here for the 54-member single-model ensemble and an equally sized multimodel ensemble. The multimodel ensemble for this example was constructed by randomly selecting 54 members out of the 63 available from the seven single-model ensembles. Although the increase in ensemble size in the single-model results in improved reliability compared with the 9-member ensemble predictions (Figure 27.3), it still does not outperform the multimodel with the same ensemble size. This indicates that the additional information coming from the other single models adds to the improvement seen in the multimodel results.

The different rate of increase in skill related to adding more ensemble members either from different models or the same model can be seen in Figure 27.5. As expected, for both single and multimodel hindcasts, the skill generally increases when adding more ensemble members. In the case of the 18-member/2-model ensembles,

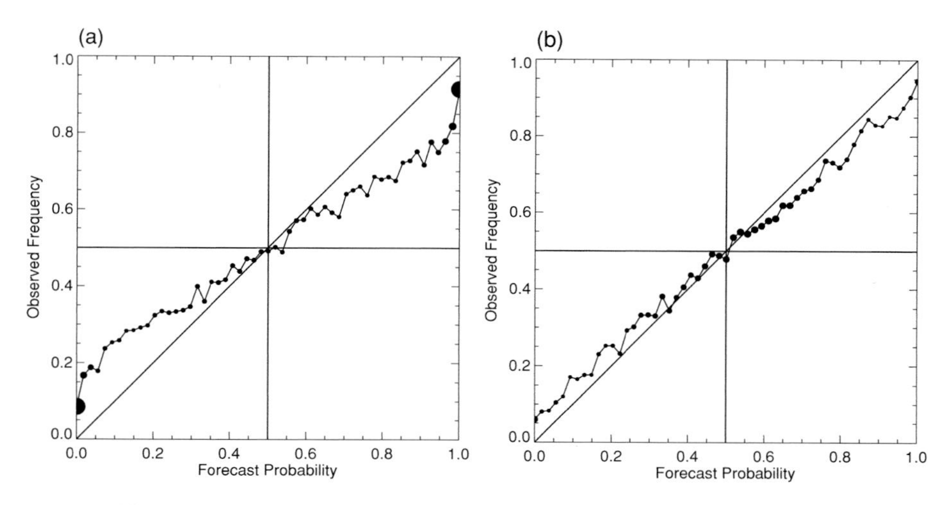

Figure 27.4 As Figure 27.3 but for the period 1987–1999. (a) Single-model, (b) multimodel, both ensembles consisting of 54 members.

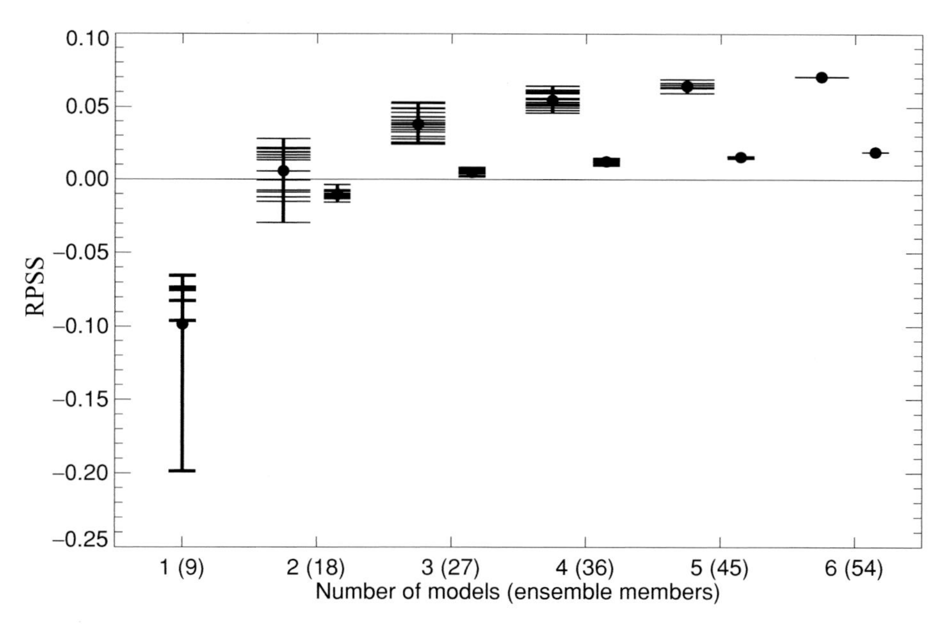

Figure 27.5 RPSS of 1-month lead time summer precipitation hindcasts for the period 1987–99 calculated over the tropical band (± 30°). The RPSS of the single models is given in the first column, each horizontal bar representing the value of one single model with 9 ensemble members. The next columns of wide horizontal bars mark the RPSS of all possible multimodel combinations composed of 2, 3, 4, 5, and 6 models. The slim horizontal bars beside the wide multimodel bars mark the RPSS of a single model with the same ensemble size as the respective multimodel (18, 27, 36, 45, and 54). For each multimodel realisation a single model was constructed by randomly choosing the same number of members as in the corresponding multimodel.

the multimodel skill varies considerably depending on the quality of the contributing single models. Combining two of the poorer models leads to a lower RPSS compared with the 18-member single-model ensembles constructed from one of the best single models. However, already in the case of the 27-member/3-model ensembles, the single and multimodel results are well separated. Every multimodel combination of three single models beats all single-model realisations with the same ensemble size. Furthermore, the gap between single and multimodel increases even more when including further models into the multimodel, though it seems to stabilise with six or more models included.

The above demonstration of the superiority of the multimodel compared with single-model results has been made under the assumption of using a simple equal weight multimodel. The key for the success of the multimodel concept is that the combination of the single models, with all its strengths and weaknesses, leads to a more reliable forecast system. A logical question arising from this argument is: Why do we have to combine strengths *and weaknesses* of the single models? Is it not possible to eliminate the weaknesses and keep only the strengths? In a more detailed analysis of the performance of the single models (not shown here) it seemed that, for example, the SST predictions of one particular single model are often worse than average. If this were a robust feature, giving a lower weight to the SST forecasts of this particular model might be a way of improving the multimodel ensemble even more. However, this concept of applying different weights to the single models when combining them to the multimodel ensemble forecast, is not as straightforward as it might seem at first glance. Various methods of finding optimal weights exist, and for a detailed analysis of all constraints and pitfalls related to these methods the reader is referred to Doblas-Reyes *et al.* (2005).

27.4 Applications

One of the main objectives of DEMETER is a demonstration of the utility of seasonal climate forecasts through the coupling of quantitative application models to the global climate prediction models. DEMETER data have been used as forcing for a variety of end-user applications such as crop yield forecasting in Europe (Cantelaube and Terres, 2005), ground-nut yield in India (Challinor *et al.*, 2005), malaria prevalence in South Africa (Morse *et al.*, 2005) or predictions of river streamflow in southern Brazil (C. A. S. Coelho, personal communication). In view of the limited space available in this chapter, we will concentrate on a brief description of results from a malaria transmission simulation model (MTSM) integrated into the DEMETER end-to-end system.

Malaria is estimated to kill between 700 000 and 2 700 000 annually with over 75% of the victims being African children. The disease is caused by a parasite that is passed between humans by mosquitoes, the vector. Malaria only occurs in areas where environmental conditions are suitable for both the parasite and vector and these conditions are sustained for a number of months. Epidemics are often the consequence of climate anomalies like higher than normal precipitation and temperatures which increase vector breeding and survivorship as well as parasite development rates. The temperature drives the development of the parasite within the vector and it also drives the developmental life cycle of the vector. For both developmental cycles there are lower temperature thresholds and, within certain upper bounds, higher temperatures lead to greater rates of development. Precipitation is important in providing breeding sites for mosquitoes and for increasing the humidity of the air, which increases the survivorship of the vectors.

Using the whole 63-member DEMETER hindcast data set, probabilistic hindcasts of simulated malaria prevalence scenarios for regions of tropical Africa have been produced. The MTSM has been run out to 180 days with bias-corrected 2m-temperature and precipitation data from each of the 63 DEMETER ensemble members as forcing. These 6-month integrations have been performed for four start dates a year (1 February, 1 May, 1 August, and 1 November) over the period 1987–2001. For a thorough assessment of the actual value of these predictions, a comparison of simulated and observed malaria prevalence should be performed. However, for most parts of Africa adequate clinical data are not available, so that our assessment is restricted to a so-called tier-2 validation (see Morse *et al.*, 2005, for a more detailed explanation of the tier-2 validation concept). That is, instead of comparing the application model output with real observations, the probabilistic MTSM output produced with DEMETER data as input is compared with MTSM output produced with ERA-40 reanalyses as forcing. In this way, the performance of the malaria model itself is of secondary importance, but the usefulness of the forcing data, in our case the DEMETER multimodel ensemble hindcasts, is evaluated.

An example for a comparison between DEMETER and ERA-40 driven malaria predictions at a grid point in South Africa (17.5° S, 25.0° E) is given in Figure 27.6. All of these 1-month lead time predictions of the March/April/May seasonal average malaria prevalence capture the ERA-40 driven MTSM reference output. The spread of the ensemble members varies between lower and upper tercile, with the largest spread generally found in the lower tercile. Most of the lower tercile values have temperatures close to 18°C during the critical stages of the parasite development, which would lead to very slow development rates.

The results presented here cannot be assumed to be universal, as different results may be found in different African regions or other parts of the world or, obviously, other application models. However, extended studies have shown that skilful forecasts of actual malaria prevalence are also possible for other parts of Africa, and that decision-makers can use this information for improved resource allocation (Thomson *et al.*, 2000).

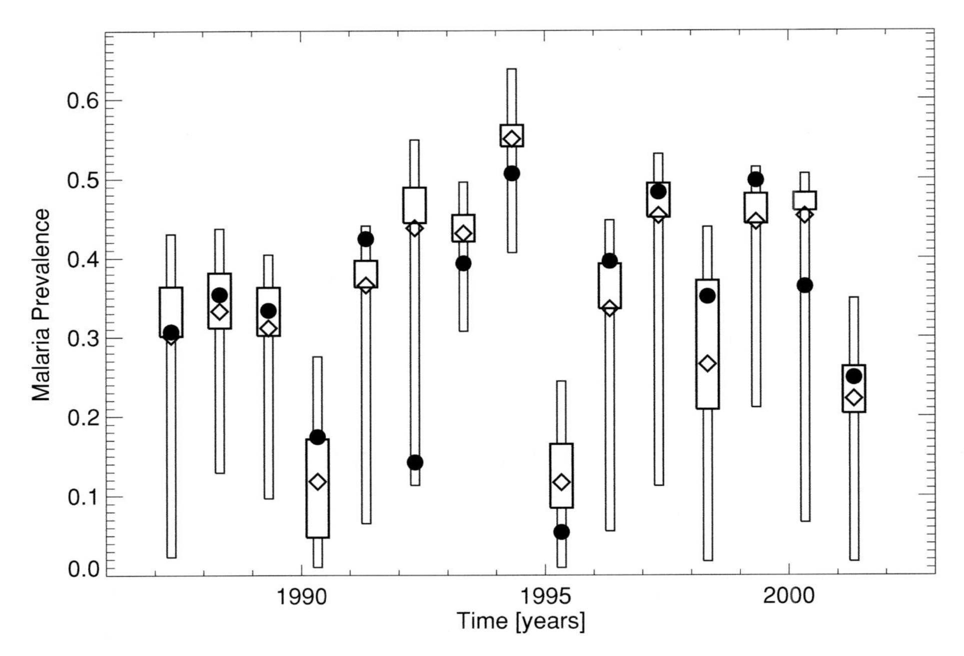

Figure 27.6 Predicted malaria prevalence from the 1-month lead February start date MTSM forecast (March/April/May prediction). The spread of the MTSM forecasts driven by DEMETER ensemble hindcasts is depicted by the box-and-whisker representation with the whiskers containing the lower and upper tercile of the ensemble. The diamonds represent the ensemble mean, the reference MTSM forecast driven by ERA-40 data being displayed by black bullets.

27.5 Summary

As part of the European-Union funded DEMETER project, a multimodel ensemble system based on seven European global coupled ocean–atmosphere models has been described and validated in hindcast mode using ECMWF ERA-40 reanalysis data. Output from the DEMETER system, suitably downscaled, has been applied to end-user models like malaria prediction or crop yield models. Results indicate that the multimodel ensemble is a viable pragmatic approach to the problem of representing model uncertainty in seasonal-to-interannual prediction, and will lead to a more reliable forecasting system than that based on any one single model.

In the limited space available in this chapter, a few illustrative examples of results from the DEMETER project have been given. However, we invite readers to visit the DEMETER website[4] where an extensive range of diagnostics and skill scores used to evaluate the DEMETER system are presented.

In addition to these specific diagnostics and skill scores, visitors to the DEMETER website can download (in GRIB or NetCDF format) gridded data from a large data set comprising monthly mean fields for a large number of variables from the

DEMETER hindcasts, including ERA-40 verification. We thus encourage scientists and potential users of seasonal forecasts to perform their own analysis of the DEMETER data (perhaps to assess skill for specific regions and variables of interest not covered in our standard analysis). More generally, we offer this DEMETER data set for education training purposes, both in the developed and developing world.

As a result of the success of DEMETER, real-time multimodel forecasting is now being established as part of the operational seasonal forecast suite at ECMWF (Anderson, this volume). At the time of writing, the coupled systems of ECMWF, Met Office and Météo-France are included in this multimodel mix. It is possible that other models may be included at a later stage.

In the European Union FP-6 project ENSEMBLES, a successor system to DEMETER will be used to explore the use of multimodel ensembles not only for seasonal-to-interannual timescales, but also for decadal timescales for which scientific evidence of predictability has emerged in recent years. For this purpose it is planned to ensure that the model components used for seasonal-to-decadal ensemble prediction, are, as far as practicable, identical to those used for century-timescale anthropogenic climate change. In this way, the reliability of century-timescale climate change projections can be assessed by running essentially the same ensemble systems on timescales for which verification data exists. We believe that a unification and rationalisation of research and development across these timescales will enhance enormously the credibility of our science.

Acknowledgements

The DEMETER project has been funded by the European Union under the contract EVK2-CT-1999–00024. The authors would like to thank M. Alonso-Balmaseda, D. Anderson, L. Ferranti, M. Fuentes, C. Gibert, D. Lucas, T. Stockdale, J. Vialard, F. Vitart, for their invaluable help and support. The project would not have been possible without the technical support of the ECMWF staff and consultants. Furthermore, we would like to thank all DEMETER partners for their contributions to the project, in particular Andy Morse and Moshe Hoshen for the MTSM results described in more detail in this contribution.

Notes

1. Detailed information on the models and the initialisation procedures can be found on the DEMETER website: www.ecmwf.int/research/demeter/general/docmodel/index.html.
2. Model hindcasts can be retrieved in GRIB and NetCDF formats from www.ecmwf.int/research/demeter/data. A tool to display the fields is also available.
3. The GPCP dataset can be found online at http://cics.umd.edu~yin/GPCP.
4. www.ecmwf.int/research/demeter/verification

References

Alves, O., G. Wang, A. Zhong, *et al.* (2002). Poama: Bureau of Meteorology operational coupled model seasonal forecast system. In *Seminar Proceedings: The Role of the Upper Ocean in Medium and Extended Range Forecasting*. ECMWF.

Brankovic, C. and T. N. Palmer (2000). Seasonal skill and predictability of ECMWF PROVOST ensembles. *Quart. J. Roy. Meteor. Soc.*, **126**, 2035–68.

Cantelaube, P. and J.-M. Terres (2005). Use of seasonal weather forecasts in crop yield modelling. *Tellus*, **57A**, 476–87.

Challinor, A. J., J. M. Slingo, T. R. Wheeler and F. J. Doblas-Reyes (2005). Probabilistic simulations of crop yield over western India using the DEMETER seasonal hindcast ensembles. *Tellus*, **57A**, 498–512.

Doblas-Reyes, F. J., R. Hagedorn and T. N. Palmer (2005). The rationale behind the success of multimodel ensembles in seasonal forecasting. II: Calibration and combination. *Tellus*, **57A**, 234–252.

Hartmann, H. C., R. Bales and S. Sorooshian (2002a). Weather, climate, and hydrologic forecasting for the US Southwest: a survey. *Climate Res.*, **21**, 239–58.

Hartmann, H. C., T. C. Pagano, S. Sorooshian and R. Bales (2002b). Confidence builders: evaluating seasonal climate forecasts for user perspectives. *Bull. Am. Meteorol. Soc.*, **83**, 683–98.

Kanamitsu, M., A. Kumar, H.-M. H. Juang, *et al.* (2002). NCEP dynamical seasonal forecast system 2000. *Bull. Am. Meteorol. Soc.*, **83**, 1019–37.

Mason, S. J., L. Goddard, N. E. Graham, E. Yulaeva, L. Sun and P. A. Arkin (1999). The IRI seasonal climate prediction system and the 1997/98 El Niño event. *Bull. Am. Meteorol. Soc.*, **80**, 1853–73.

Morse, A. P., F. Doblas-Reyes, M. B. Hoshen, R. Hagedorn and T. N. Palmer (2005). A forecast quality assessment of an end-to-end probabilistic multi-model seasonal forecast system using a malaria model. *Tellus*, **57A**, 464–75.

Palmer, T. N. (1993). Extended-range atmospheric prediction and the Lorenz model. *Bull. Am. Meteorol. Soc.*, **74**, 49–65.

Palmer, T. N. and D. L. T. Anderson (1994). The prospects for seasonal forecasting. *Quart. J. Roy. Meteor. Soc.*, **120**, 755–93.

Richardson, D. S. (2000). Skill and relative economic value of the ECMWF ensemble prediction system. *Quart. J. Roy. Meteor. Soc.*, **126**, 649–68.

Shukla, J. (1998). Predictability in the midst of chaos: a scientific basis for climate forecasting. *Science*, **282**, 728–31.

Stockdale, T. N., D. L. T. Anderson, J. O. S. Alves and M. A. Balmaseda (1998). Global seasonal rainfall forecasts using a coupled ocean–atmosphere model. *Nature*, **392**, 370–3.

Thomson, M. C., T. N. Palmer, A. P. Morse, M. Cresswell and S. J. Connor (2000). Forecasting disease risk with seasonal climate predictions. *Lancet*, **355**, 1559–60.

Thomson, M. C., F. J. Doblas-Reyes, S. J. Mason et al. (2006). Malaria early warnings based on seasonal climate forecasts from multi-model ensembles. *Nature*, **439**, 576–9.

Uppala, S. M. and coauthors (2005). The ERA-40 re-analysis. *Quart. J. Roy. Meteor. Soc.*, **131**, 2961–3012.

Index

Page numbers in *italic* denote figures. Page numbers in **bold** denote tables.

Printed in the United States
By Bookmasters